2.8 ABSOLUTE VALUE

$$|x| = \begin{cases} x & \text{when } x \geq 0 \\ -x & \text{when } x < 0 \end{cases}$$

If $a > 0$, then
$$|a| = \sqrt{a^2}$$

$|x| = a$ is equivalent to $x = a$ or $x = -a$.

$|x| < a$ is equivalent to $-a < x < a$.

$|x| > a$ is equivalent to $x > a$ or $x < -a$.

$$|ab| = |a||b| \qquad \left|\frac{a}{b}\right| = \frac{|a|}{|b|} \qquad |a + b| \leq |a| + |b|$$

3.1 THE RECTANGULAR COORDINATE SYSTEM

Distance formula: $d(PQ) = \sqrt{(x_2 - x_1)^2 + (y_2 - y_1)^2}$

Midpoint formula: $\left(\dfrac{x_1 + x_2}{2}, \dfrac{y_1 + y_2}{2}\right)$

3.2 THE SLOPE OF A NONVERTICAL LINE

Formula for slope: $m = \dfrac{y_2 - y_1}{x_2 - x_1} \qquad (x_2 \neq x_1)$

Lines with equal slopes are parallel.
Lines with slopes that are negative reciprocals are perpendicular.

3.3 EQUATIONS OF LINES

Point–slope form: $y - y_1 = m(x - x_1)$
Slope–intercept form: $y = mx + b$
General form: $Ax + By = C$

3.4 GRAPHS OF OTHER EQUATIONS

Tests for symmetry:

If $(-x, y)$ lies on a graph whenever (x, y) does, the graph is symmetric about the y-axis.

If $(x, -y)$ lies on a graph whenever (x, y) does, the graph is symmetric about the x-axis.

If $(-x, -y)$ lies on a graph whenever (x, y) does, the graph is symmetric about the origin.

Equations of a circle with radius r:

$(x - h)^2 + (y - k)^2 = r^2$; center at (h, k)

$x^2 + y^2 = r^2$; center at origin

3.6 PROPORTION AND VARIATION

$y = kx \qquad y$ varies directly with x

$y = \dfrac{k}{x} \qquad y$ varies inversely with x

$y = kxz \qquad y$ varies jointly with x and z

4.2 QUADRATIC FUNCTIONS

The graph of
$$y - k = a(x - h)^2 \quad (a \neq 0)$$
is a parabola with vertex at (h, k).

The graph of
$$y = ax^2 + bx + c \quad (a \neq 0)$$
is a parabola with vertex at
$$\left(-\frac{b}{2a}, c - \frac{b^2}{4a}\right).$$

4.4 TRANSLATING AND STRETCHING GRAPHS

If $k > 0$, the graph of $\begin{cases} y = f(x) + k \\ y = f(x) - k \end{cases}$ is identical to the graph of

$y = f(x)$, except it is translated k units $\begin{cases} \text{upward} \\ \text{downward} \end{cases}$.

If $k > 0$, the graph of $\begin{cases} y = f(x - k) \\ y = f(x + k) \end{cases}$ is identical to the graph of

$y = f(x)$, except it is translated k units to the $\begin{cases} \text{right} \\ \text{left} \end{cases}$.

4.6 OPERATIONS ON FUNCTIONS

If the ranges of functions f and g are subsets of the real numbers, then

$$(f + g)(x) = f(x) + g(x)$$
$$(f - g)(x) = f(x) - g(x)$$
$$(f \cdot g)(x) = f(x) \cdot g(x)$$
$$(f/g)(x) = \frac{f(x)}{g(x)} \quad (g(x) \neq 0)$$
$$(f \circ g)(x) = f(g(x))$$

5TH EDITION

COLLEGE ALGEBRA

Books in the Gustafson/Frisk Series

Essential Mathematics with Geometry   Second Edition
Beginning Algebra   Third Edition
Intermediate Algebra   Third Edition
Algebra for College Students   Third Edition
College Algebra   Fifth Edition
Plane Trigonometry   Fourth Edition
College Algebra and Trigonometry   Fourth Edition
Functions and Graphs   Second Edition

5TH
E D I T I O N

COLLEGE
ALGEBRA

R. David Gustafson
Peter D. Frisk

Rock Valley College

Brooks/Cole Publishing Company
Pacific Grove, California

The ITP logo is a registered trademark under license.

Brooks/Cole Publishing Company
A division of International Thomson Publishing Inc.

© 1994, 1990, 1986, 1983, 1980 by **Brooks/Cole Publishing Company.**
All rights reserved.
No part of this book may be reproduced, stored in a retrieval system, or transcribed,
in any form or by any means—electronic, mechanical, photocopying, recording,
or otherwise—without the prior written permission of the publisher,
Brooks/Cole Publishing Company, Pacific Grove, California 93950,
a division of **International Thomson Publishing Inc.**

Printed in the United States of America

10 9 8 7 6 5 4 3

Library of Congress Cataloging-in-Publication Data

Gustafson, R. David (Roy David), [date]
    College algebra/R. David Gustafson, Peter D. Frisk—5th ed.
      p.   cm.
    Includes index.
    ISBN 0-534-20880-0
    1. Algebra.   I. Frisk, Peter D., [date].   II. Title.
QA154.2.G87   1993
512′.9—dc20   93-8376
                                                   CIP

Sponsoring Editors: *Craig Barth, Gary Ostedt*
Editorial Assistant: *Carol Ann Benedict*
Production Editor: *Ellen Brownstein*
Production Service: *Hoyt Publishing Services*
Manuscript Editor: *David Hoyt*
Permissions Editor: *Carline Haga*
Interior and Cover Design: *Roy R. Neuhaus*
Cover and Chapter Opening Photos: *Ed Young*
Interior Illustration: *Lori Heckelman*
Typesetting: *Weimer Graphics, Inc.*
Cover Printing: *Phoenix Color Corp.*
Printing and Binding: *R. R. Donnelley & Sons, Crawfordsville*

About the Cover: Located at the Jamesburg Earth Station in Cachagua, California, this Philco Ford satellite antenna was built in 1968 by the U.S. government for its space program. The dish, now owned by AT&T, is 97 feet in diameter and houses an 8-foot hyperboloid in its center (see back cover). Unlike domestic communication antennae, which use linear polarization signals, this antenna creates a circular polarization signal for international use. The signal is directed to the hyperboloid, bounces back to the paraboloid, and then is beamed to a satellite. The signal has a maximum strength of 10^{10} watts of power or 100 dbw. There are currently only 10 satellite antennae of this class operating in the U.S.

Photo Credits: Cover, Courtesy of AT&T and the Jamesburg Earth Station; **page 136,** Courtesy of DeltaPoint, Monterey, California; **page 339,** Courtesy of Monterey Cypress Stained Glass Studio; **page 418,** Courtesy of The Nature Company, Carmel, California; **page 453,** Courtesy of 4S Casino Party Supplies, San Carlos, California.

To our wives, Carol and Martha,
and our children, Kristy and Steven;
Sarah, Heidi, and David

PREFACE

To THE INSTRUCTOR

College Algebra has been extensively revised in this fifth edition. This revision was motivated by the need to prepare students better for the mathematics of the next century.

Although the changes have been substantial, our fundamental philosophy as teachers remains the same. Consequently, the goal of this book remains unchanged: to hold attrition to a minimum and prepare students to succeed at the next stage, whether it is trigonometry, precalculus, statistics, liberal arts mathematics, or everyday life.

We believe that this fifth edition accomplishes this goal through a successful blending of content and pedagogy. We present comprehensive, in-depth, precise coverage of the topics of college algebra, incorporated into a framework of tested teaching strategy and combined with carefully selected pedagogical features.

Changes for the Fifth Edition

The overall effects of the changes we have made in the fifth edition have been as follows.

- To increase the emphasis on learning mathematics through graphing. Although graphing calculators are incorporated throughout the book, their use is not required. All of the topics are fully discussed in traditional ways. Of course, we recommend that instructors try the graphing calculator material.

- To increase the emphasis on problem solving through realistic applications. The variety of applications problems has been increased significantly, and all application problems are now labeled with special headings.

- To fine-tune the presentation of certain topics for better flow of ideas and for clarity.

- To increase the visual interest by the use of a four-color format. We continue to use color not just as a design feature, but in a functional way. Color is still used to highlight terms that you would point to in a classroom discussion.

Some of the specific changes made to chapters are listed below.

1. Chapter 1, which is mainly review, has been condensed so that instructors can get to college algebra topics more quickly. The work on complex numbers has been moved to Chapter 2, after a discussion of quadratic equations. Finding solutions of quadratic equations now provides the motivation for discussing complex numbers.

2. Chapter 2, dealing with equations and inequalities, has been reorganized. It now incorporates much of the work with inequalities, rational inequalities, and absolute value. The applications sections have been extensively revised to provide more authentic and varied applied problems.

3. The fourth edition's Chapter 3 has been divided into two chapters. Chapter 3 now deals with graphing lines, slope, writing equations of lines, general graphing of other relations, and ratio and proportion. Graphing calculators are introduced in Chapter 3.

4. Chapter 4 deals with the more formal aspects of functions and their inverses. A new section on translations of graphs has been added.

5. Chapter 5 now deals with the remainder and factor theorems, synthetic division, and finding rational roots of polynomial equations. After a brief discussion of the bisection method, graphing calculators are used to find approximations for irrational roots of polynomial equations.

6. Chapter 6 now deals with logarithms. Coverage of this topic has been thoroughly revised to obtain a better flow of ideas. Many more application problems have been added.

7. Chapter 7 now deals with systems of linear equations. Graphing calculators are used to enhance the discussion of solving systems by graphing. The application problems have been thoroughly revised and made more relevant. Matrix methods continue to be emphasized. The material on graphing linear inequalities in two variables has been included in the section that covers systems of linear inequalities. A separate section on linear programming is now included.

8. Chapter 8 deals with conic sections and quadratic systems. This chapter now contains more application problems.

The following are some of the specific pedagogical changes.

1. Cumulative review exercises have been added after every three chapters.

2. ⬧ Warning! Students are now warned about common errors by a special symbol.

3. All sections are now divided into subsections with headings. When a section deals with more than one topic, the headings will help students focus on each specific topic.

4. ▦ All exercises requiring scientific calculators are marked with a special logo.

   ▣ All exercises requiring graphing calculators are marked with a different logo.

At the same time, we have kept the pedagogical features that made previous editions of the book so successful, as outlined in the paragraphs that follow.

Solid Mathematics   The treatment of college algebra is direct and straightforward. Although the treatment is mathematically sound, it is not so rigorous that it will confuse students. Every effort has been made to ensure the accuracy of the mathematics and of the answers to the exercises. The book has been critiqued by dozens of reviewers. Each author and a problem checker have worked every exercise. Although the exercise sets are designed primarily to provide practice and drill, they also contain problems that will challenge the best students. The book contains over 4000 exercises.

Accessibility to Students   The book is written for students to read and understand. The numerous problems within each exercise set are carefully keyed to over 400 worked examples, in which author's notes explain many of the steps used in the problem-solving process. In the student edition, Appendix III contains the answers to the odd-numbered exercises, as well as all answers to the chapter review exercises, chapter tests, and cumulative review exercises. In the instructor's edition, Appendix III provides the answers to all exercises.

Review is incorporated into the book in many ways: There are chapter summaries, review exercises at the end of each chapter, cumulative review exercises at the end of every three chapters, and endpapers that list (in order of presentation) the important formulas developed in the book.

Emphasis on Applications   To show that mathematics is useful, we include a large number of word problems and applications throughout the book.

■ Organization and Coverage

The book can be used in a variety of ways. For optimum flexibility, many of the chapters have been designed to be sufficiently independent that you can pick and choose topics that are relevant to your students' needs. The accompanying diagram shows how the chapters are interrelated.

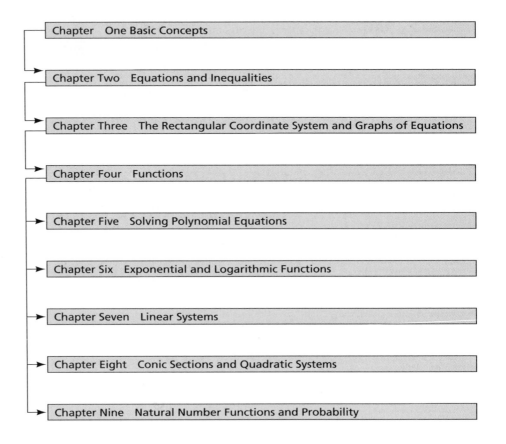

Chapter One   Basic Concepts

Chapter Two   Equations and Inequalities

Chapter Three   The Rectangular Coordinate System and Graphs of Equations

Chapter Four   Functions

Chapter Five   Solving Polynomial Equations

Chapter Six   Exponential and Logarithmic Functions

Chapter Seven   Linear Systems

Chapter Eight   Conic Sections and Quadratic Systems

Chapter Nine   Natural Number Functions and Probability

■ Calculators and Computers

The use of calculators is assumed throughout the book. We believe that students should learn calculator skills in the mathematics classroom. They will then be prepared to use calculators in science and business classes and for nonacademic purposes. The directions within each exercise set indicate which exercises require calculators.

After much deliberation, we decided not to give keystrokes for any specific model of graphing calculator. Several manufacturers told us that they will soon be introducing new models that have different keystrokes from the models currently available. We do not want to confuse students by giving obsolete keystrokes. Instead, the book covers the features that are common to all graphing calculators.

■ Features of This Edition

14 CHAPTER 1 BASIC CONCEPTS

EXAMPLE 1
a. $4^2 = 4 \cdot 4 = 16$ Read 4^2 as "four squared."
b. $(-4)^2 = (-4)(-4) = 16$
c. $-4^2 = -(4 \cdot 4) = -16$
d. $(5)^3 = 5 \cdot 5 \cdot 5 = 125$ Read 5^3 as "five cubed."
e. $(-5)^3 = (-5)(-5)(-5) = -125$
f. $3x^4 = 3 \cdot x \cdot x \cdot x \cdot x$ Read x^4 as "x to the fourth power."
g. $(3x)^4 = (3x)(3x)(3x)(3x) = 81x^4$ ■

Warning! It is important to note the distinction between ax^n and $(ax)^n$, and between $-x^n$ and $(-x)^n$:

$$ax^n = a \cdot \overbrace{x \cdot x \cdot x \cdot \cdots \cdot x}^{n \text{ factors of } x} \qquad (ax)^n = \overbrace{(ax)(ax)(ax) \cdot \cdots \cdot (ax)}^{n \text{ factors of } ax}$$

$$-x^n = -\overbrace{(x \cdot x \cdot x \cdot \cdots \cdot x)}^{n \text{ factors of } x} \qquad (-x)^n = \overbrace{(-x)(-x)(-x) \cdot \cdots \cdot (-x)}^{n \text{ factors of } -x}$$

◀ Warnings to students are highlighted.

■ Rules of Exponents

We begin to develop the rules of exponents by considering the product $x^m x^n$. Because x^m indicates that x is to be used as a factor m times and because x^n indicates that x is to be used as a factor n times, there are $m + n$ factors of x in the product $x^m x^n$.

◀ Sections are divided into subsections.

$$x^m x^n = \overbrace{\underbrace{x \cdot x \cdot x \cdot \cdots \cdot x}_{m \text{ factors of } x} \cdot \underbrace{x \cdot \ }_{n \text{ factors of } x}}^{m + n \text{ factors of } x}$$

Thus, to multiply exponential expressions add the exponents.

The Product Rule of Exponents	If m and n are natural numbers, then $$x^m x^n = x^{m+n}$$

Warning! The product rule app the same base. A product of two p cannot be simplified.

To find another property of exponents, $(x^m)^n$. The exponent n indicates that x^m is t that x is to be used as a factor mn times.

$$(x^m)^n = \overbrace{\underbrace{(x^m)(x^m)(x^m) \cdot \cdots \cdot (x^m)}_{n \text{ factors of } x^m}}^{mn \text{ factors of } x}$$

All definitions and theorems are clearly boxed. ▲

Application problems have been ▶
extensively revised and updated.

All art has been redrawn. ▶

Each application has a title. ▶

---

210 CHAPTER 4 FUNCTIONS

4.2 EXERCISES

In Exercises 1–8, graph each quadratic equation.

1. $y = x^2 - x$ **2.** $y = x^2 + 2x$ **3.** $y = -3x^2 + 2$ **4.** $y = -3x^2 + 4$

5. $y = -\frac{1}{2}x^2 + 3$ **6.** $y = \frac{1}{2}x^2 - 2$ **7.** $y = x^2 - 4x + 1$ **8.** $y = -x^2 - 4x + 1$

In Exercises 9–16, find the vertex of each parabola.

9. $y = x^2 - 1$ **10.** $y = -x^2 + 2$ **11.** $y = x^2 - 4x + 4$ **12.** $y = x^2 - 10x + 25$

13. $y = x^2 + 6x - 3$ **14.** $y = -x^2 + 9x - 2$ **15.** $y = -2x^2 + 12x - 17$ **16.** $y = 2x^2 + 16x + 33$

In Exercises 17–30, find each maximum.

17. *Architecture* A parabolic arch has an equation of $x^2 + 20y - 400 = 0$, where x is measured in feet. Find the maximum height of the arch.

18. *Ballistics* An object is thrown from the origin of a coordinate system with the x-axis along the ground and the y-axis vertical. Its path, or **trajectory**, is given by the equation $y = 400x - 16x^2$. Find the object's maximum height.

19. *Ballistics* A child throws a ball up a hill that makes an angle of 45° with the horizontal. The ball lands 100 feet up the hill. Its trajectory is a parabola with equation $y = -x^2 + ax$ for some number a. Find a. (See Illustration 1.)

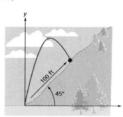

ILLUSTRATION 1

20. *Maximizing area* The rectangular garden in Illustration 2 has a width of x and a perimeter of 100 feet. Find x so that the area of the rectangle is maximum.

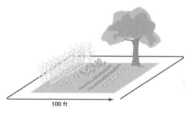

ILLUSTRATION 2

21. *Maximizing storage area* A farmer wants to partition a rectangular feed-storage area in a corner of his barn. The barn walls form two sides of the stall, and the farmer has 50 feet of partition for the remaining two sides. What dimensions will maximize the area? (See Illustration 3.)

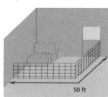

ILLUSTRATION 3

4.4   TRANSLATING AND STRETCHING GRAPHS

■ Vertical and Horizontal Translations   ■ Vertical and Horizontal Stretchings

◀ A new section on translations has been included.

■ Vertical and Horizontal Translations

The graphs of different equations may be identical except for their position in the *xy*-plane. For example, Figure 4-18(a) shows the graph of the function $y = x^2 + k$ for three different values of k. The graph of $y = x^2 + 2$ is identical to the graph of $y = x^2$ except that it is shifted 2 units upward. Similarly, the graph of $y = x^2 - 3$ is identical to the graph of $y = x^2$ except that it is shifted 3 units downward. Such shifts are called **vertical translations**.

Figure 4-18(b) shows the graph of the equation $y = (x + h)^2$ for three different values of h. The graph of $y = (x - 2)^2$ is identical to the graph of $y = x^2$ except that it is shifted 2 units to the right. The graph of $y = (x + 3)^2$ is identical to the graph of $y = x^2$ except that it is shifted 3 units to the left. Such shifts are called **horizontal translations**.

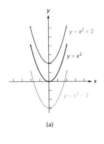

(a)

FIGUR

In general, we can make the following

Vertical Translations	If f is a function and k is a positive numb
	• The graph of $y = f(x) + k$ is identical translated k units upward.
	• The graph of $y = f(x) - k$ is identical translated k units downward.

Many summary boxes have been added. ▲

Graphing calculators are included often ▶ to illustrate mathematical concepts.

Calculator exercises are marked clearly ▶ with either a calculator logo or a graphing calculator logo.

We substitute 5 for x and calculate I.

$$I = 12(0.6)^5$$
$$I \approx 0.93312$$

At a depth of 5 meters, the intensity of the light is slightly less than 1, or about one-twelfth of the intensity at the surface. ∎

■ Watching Money Grow

EXAMPLE 8   If \$1 is deposited in an account earning 9% annual interest, compounded monthly, estimate how much will be in the account in 70 years.

Solution   We can substitute 1 for A_0, .09 for r, and 12 for k into the formula

$$A = A_0\left(1 + \frac{r}{k}\right)^{kt}$$

and simplify to get

$$A = (1.0075)^{12t}$$

We now use a graphing calculator to see how the money grows year by year. We graph the function $A = (1.0075)^{12t}$ in the viewing window $0 \le t \le 100$ and $0 \le A \le 750$ to obtain the graph shown in Figure 6-5. We can then use the TRACE and ZOOM features to estimate that \$1 grows to the surprising amount of approximately \$532 in 70 years. ∎

FIGURE 6-5

6.1   EXERCISES

▦ *In Exercises 1–4, find each value to four decimal places.*

1. $4^{\sqrt{3}}$ **2.** $5^{\sqrt{2}}$ **3.** 7^{π} **4.** $3^{-\pi}$

In Exercises 5–12, graph each exponential function.

5. $y = 3^x$ **6.** $y = 5^x$ **7.** $y = \left(\frac{1}{5}\right)^x$ **8.** $y = \left(\frac{1}{3}\right)^x$

9. $y = -2^x$ **10.** $y = -3^x$ **11.** $y = \left(\frac{3}{4}\right)^x$ **12.** $y = \left(\frac{4}{3}\right)^x$

In Exercises 13–28, graph each function.

13. $y = 3^x - 1$ **14.** $y = 2^x + 3$ **15.** $y = 2^x + 1$ **16.** $y = 4^x - 4$

17. $y = 3^{x-1}$ **18.** $y = 2^{x+3}$ **19.** $y = 3^{x+1}$ **20.** $y = 2^{x-3}$

21. $y = 2^{x+1} - 2$ **22.** $y = 3^{x-1} + 2$ **23.** $y = 3^{x-2} + 1$ **24.** $y = 3^{x+2} - 1$

25. $y = 5(2^x)$ **26.** $y = 2(5^x)$ **27.** $y = 3^{-x}$ **28.** $y = 2^{-x}$

6.2 BASE-e EXPONENTIAL FUNCTIONS **299**

6.2 BASE-e EXPONENTIAL FUNCTIONS

■ Graphing the Exponential Function ■ Applications of Exponential
Functions ■ The Malthusian Theory

Leonhard Euler
(1707–1783)
Euler first used the letter i to
represent √−1, the letter e
for the base of natural
logarithms, and the symbol
Σ for summation. Euler was
one of the most prolific
mathematicians of all time,
contributing to almost all
areas of mathematics.
Much of his work was
accomplished after he
became blind.

In mathematical models of natural events, the number $e = 2.71828182845904\ldots$
appears often as the base of an exponential function. We introduce this important
number by recalling the formula for compound interest,

$$A = A_0\left(1 + \frac{r}{k}\right)^{kt}$$

and allowing k, representing the number of compounding periods per year, to
become very large. To see what happens, we let $k = rp$, where p is a new variable.

$$A = A_0\left(1 + \frac{r}{k}\right)^{kt}$$

$$A = A_0\left(1 + \frac{r}{rp}\right)^{rpt} \qquad \text{Substitute } rp \text{ for } k.$$

$$A = A_0\left(1 + \frac{1}{p}\right)^{rpt} \qquad \text{Simplify } \frac{r}{rp}.$$

$$A = A_0\left[\left(1 + \frac{1}{p}\right)^{p}\right]^{rt} \qquad \text{Remember that } (x^m)^n = x^{mn}.$$

Because the annual r
becomes very large, the
becomes tied to the ques

$$\left(1 + \frac{1}{p}\right)^{p}$$

as p becomes very large
Some results calculat

◀ Sections are divided into subsections.

◀ Historical notes appear throughout the
text.

◀ Color is used in a functional way.

◀ Author's notes explain steps.

CUMULATIVE REVIEW EXERCISES

In Cumulative Review Exercises 1–4, decide whether each equation defines a function.

1. $y = 3x - 1$ **2.** $y = x^2 + 3$ **3.** $y = \dfrac{1}{x - 2}$ **4.** $y^2 = 4x$

In Cumulative Review Exercises 5–8, find the domain and range of each function.

5. y

In Cum

9. y

In Cum

11. y

In Cum

15. y

In Cum

17. $(f$

In Cum

21. $(f$

4 CHAPTER TEST

In Questions 1–2, find the domain and range of each function.

1. $f(x) = \dfrac{3}{x - 5}$ **2.** $f(x) = \sqrt{x + 3}$

In Questions 3–4, find $f(-1)$ and $f(2)$.

3. $f(x) = \dfrac{x}{x - 1}$ **4.** $f(x) = \sqrt{x + 7}$

In Questions 5–8, find the vertex of each parabola.

5. $y = 3(x - 7)^2 - 3$ **6.** $y = x^2 - 2x - 3$
7. $y = 3x^2 - 24x + 38$ **8.** $y = 5 - 4x - x^2$

In Questions 9–10, graph each polynomial function.

9. $y = x^4 - x^2$ **10.** $y = x^5 - x^3$

In Questions 11–12, assume that an object tossed vertically upward reaches a height of h feet after t seconds, where
$h = 100t - 16t^2$.

11. In how many seconds does the object reach its maximum height?

12. What is that maximum height?

13. *Suspension bridge* The cable of a suspension bridge
is in the shape of the parabola $x^2 - 2500y + 25,000$
$= 0$ in the coordinate system shown in Illustration 1.
Distances are in feet. How far above the roadway is the
cable's lowest point?

14. Refer to Question 13. How far above the roadway does
the cable attach to the vertical pillars?

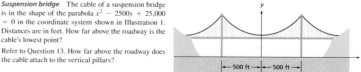

|← 500 ft →|← 500 ft →|

ILLUSTRATION 1

▲
Sample Chapter Tests and Cumulative ▶
Review Exercises have been added.

■ Topics Covered

Review   Chapters 1 and 2 review topics from basic algebra—the real number system, exponents and radicals, polynomial arithmetic, solutions of linear and quadratic equations, complex numbers, radical equations, inequalities, and absolute value.

Inequalities   Inequalities with one variable are now discussed in Chapter 2. Rational inequalities are solved by both the test-point and sign-graph methods. Inequalities with two variables, along with their graphical interpretations, are covered in Chapter 7, before discussing systems of inequalities.

Functions   The concept of the rectangular coordinate system is introduced in Chapter 3, with emphasis on graphing lines, writing equations of lines, and graphing various relations. Graphing calculators are also introduced in Chapter 3.

Translations of graphs and the more formal aspects of functions, function notation, rational functions, algebra of functions, composition of functions, and inverse functions are now covered in Chapter 4.

Roots of Polynomial Equations   Chapter 5 discusses methods for finding rational roots of polynomial equations. Several examples illustrate the interplay between the fundamental theorem of algebra, Descartes' rule of signs, the remainder and factor theorems, the rational-root theorem, and the conjugate-pairs result. The bisection method is briefly discussed before graphing calculators are used to find irrational roots.

Exponential and Logarithmic Functions   Chapter 6 covers exponential functions and logarithms and many of their applications. The use of calculators is emphasized in this chapter.

Systems of Equations and Inequalities   Chapter 7 includes techniques for solving systems of linear equations. Matrix methods are developed, and some matrix algebra is presented. Linear programming using graphical methods has been expanded into a complete section. The topic of partial fractions is introduced as an application of systems of linear equations.

Conic Sections   Chapter 8 develops the basic forms of the equations for conic sections and provides opportunities for graphing these equations. Solutions of simultaneous second-degree equations are obtained both graphically and algebraically.

Natural Number Functions and Probability   The binomial theorem, permutations, and combinations lead to a presentation of simple and compound probabilities, odds, and mathematical expectation. The chapter includes an introduction to mathematical induction. An induction proof of the binomial theorem is given in Appendix I.

■ Ancillaries for the Instructor

Instructor's Edition   The *Instructor's Edition* contains an expanded answer section, which includes the answers for *all exercises in the text*.

Test Manual
Teresa Bittner   The *Test Manual* contains three ready-to-use forms of every chapter test. Answer keys are also provided.

Computer Testing Software

Available with our text are two extensive electronic question banks, one short-answer and one multiple-choice. Each form contains approximately 1700 test items and is available for IBM-compatible and Macintosh computers. The testing program gives you all the features of state-of-the-art word processors and more, including the ability to see all technical symbols, fonts, and formatting on the screen just the way they will appear when printed. The question banks can be edited.

EXPTEST™ runs on IBM and compatible computers.

ExamBuilder™ runs on Macintosh computers.

Transparencies

Transparencies of 37 key graphics from the text are available to assist the instructor in the classroom.

Videos
Educational Video Resources

A complete set of videos, produced by Educational Video Resources, is available upon adoption of the text.

■ Ancillaries for the Student

Study Guide
George Grisham

For each chapter in the text, the *Study Guide* contains a message to the student, completely worked examples for each section, additional exercises, cautions and hints, and completely worked solutions for selected odd-numbered exercises from the text. Available for sale to students.

Student Solutions Manual
Michael Welden

The *Student Solutions Manual* provides worked-out solutions for every odd-numbered exercise in the text. Available for sale to students.

Visual Precalculus
David Schneider

This fun-to-use software for IBM and compatible computers is customized to the text. *Visual Precalculus* uses animation to explore relevant exercises and examples from the text. A mouse is optional. A *Student User's Guide* accompanies the software. Available for sale to students. Site licenses are also available for sale.

Graphing Calculator Activities
for Algebra
Miller, Perry, and Tveten

Graphing Calculator Activities for Algebra provides instruction, problems, exploratory exercises, and projects, plus instruction and keystrokes for using the TI-81 and Casio 7700 graphing calculators. Available for sale to students.

To the Student

Congratulations. You now own a state-of-the-art textbook that has been written especially for you. We, the authors, have tried to write a book that you can read and understand. The book is carefully written and includes an extensive number of worked examples. However, if you intend to get the most out of your algebra course, you must read and study the textbook properly. We recommend that you work the examples on paper and be sure that you understand them before attempting to work the exercises.

A *Student Solutions Manual* is available for sale, containing worked-out solutions to all of the odd-numbered exercises. Also available for sale is a *Student Study Guide*, designed to enhance your study techniques.

The material presented in *College Algebra*, 5th edition, will be of value to you in later years. Therefore, we suggest that you keep this book after completing the course. It will be a good source of reference and will keep at your fingertips the material that you have learned here.

We wish you well.

ACKNOWLEDGMENTS

We are grateful to the following people who reviewed the manuscript in its various stages. They all had valuable suggestions that have been incorporated into the text.

The reviewers include Richard Andrews, University of Wisconsin; James Arnold, University of Wisconsin; Wilson Banks, Illinois State University; Jerry Bloomberg, Essex Community College; Elaine Bouldin, Middle Tennessee State University; Dale Boye, Schoolcraft College; Lee R. Clancy, Golden West College; Jan Collins, Embry Riddle College; Cecilia Cooper, William & Harper College; Romae J. Cormier, Northern Illinois University; John S. Cross, University of Northern Iowa; M. Hilary Davies, University of Alaska at Anchorage; Grace DeVelbiss, Sinclair Community College; Lena Dexter, Faulkner State Junior College; Emily Dickinson, University of Arkansas; Robert E. Eicken, Illinois Central College; Eric Ellis, Essex Community College; Eunice F. Everett, Seminole Community College; Dale Ewen, Parkland College; Ronald J. Fischer, Evergeen Valley College; Mary Jane Gates, University of Arkansas at Little Rock; Marvin Goodman, Monmouth College; Jerry Gustafson, Beloit College; Jerome Hahn, Bradley University; Douglas Hall, Michigan State University; Robert Hall, University of Wisconsin; David Hansen, Monterey Peninsula College; Kevin Hastings, University of Delaware; William Hinrichs, Rock Valley College; Arthur M. Hobbs, Texas A & M University; Jack E. Hofer, California Polytechnic State University; Ingrid Holzner, University of Wisconsin; Warren Jaech, Tacoma Community College; Nancy Johnson, Broward Community College; William B. Jones, University of Colorado; Barbara Juister, Elgin Community College; David Kinsey, University of Southern Indiana; Helen Kriegsman, Pittsburg State University; Marjorie O. Labhart, University of Southern Indiana; Jaclyn LeFebvre, Illinois Central College; Judy McKinney, California Polytechnic Institute at Pomona; Sandra McLaurin, University of North Carolina; Marcus McWaters, University of Southern Florida; Donna Menard, University of Massachusetts, Darmouth; James W. Mettler, Pennsylvania State University; Eldon L. Miller, University of Mississippi; Stuart E. Mills, Louisiana State University, Shreveport; Gilbert W. Nelson, North Dakota State; Marie Neuberth, Catonsville City College; Anthony

Peressini, University of Illinois; David L. Phillips, University of Southern Colorado; William H. Price, Middle Tennessee State University; Janet P. Ray, Seattle Central Community College; Barbara Riggs, Tennessee Technological University; Paul Schaefer, SUNY, Geneseo; Vincent P. Schielack, Jr., Texas A & M University; Robert Sharpton, Miami Dade Community College; L. Thomas Shiflett, Southwest Missouri State University; Richard Slinkman, Bemidji State University; Merreline Smith, California Polytechnic Institute at Pomona; John Snyder, Sinclair Community College; Warren Strickland, Del Mar College; Ray Tebbetts, San Antonio College; Faye Thames, Lamar State University; Douglas Tharp, University of Houston-Downtown; Carol M. Walker, Hinds Community College; Carroll G. Wells, Western Kentucky University; William H. White, University of South Carolina at Spartanburg; Charles R. Williams, Midwestern State University; Harry Wolff, University of Wisconsin; Clifton Whyburn, University of Houston; Albert Zechmann, University of Nebraska.

We wish to thank the staff at Brooks/Cole, especially Craig Barth, Gary Ostedt, Sue Ewing, Ellen Brownstein, and Roy Neuhaus for their competent work and support. We give special thanks to David Hoyt for his excellent editing of the manuscript.

We also thank the following people for their skillful preparation of the ancillaries for the text: Teresa Bittner, George Grisham, David Schneider, and Michael Weldon.

We are especially grateful to Robert Hessel, Diane Koenig, and Michael Weldon (all of Rock Valley College) for checking the answers to the problems.

R. David Gustafson
Peter D. Frisk

CONTENTS

CHAPTER

1

BASIC CONCEPTS

The concept of number is fundamental to mathematics. For this reason, we begin by discussing various sets of numbers and their properties.

SETS OF NUMBERS AND THEIR PROPERTIES

■ Sets of Numbers   ■ Graphs of Sets of Real Numbers   ■ Intervals
■ Properties of Negatives   ■ Absolute Value   ■ Properties of
Real Numbers

A **set** is a collection of objects. The notation $\{a, b, c\}$ represents the set whose objects, or **elements**, are a, b, and c. The order of the letters a, b, and c is not important. To indicate that b is an element of this set, we write

$b \in \{a, b, c\}$ Read as "b is an element of the set containing a, b, and c."

To indicate that d is not an element of this set, we write $d \notin \{a, b, c\}$.

The expression

$\mathbf{A} = \{a, e, i, o, u\}$ Capital letters are often used to name sets.

indicates that $\mathbf{A}$ is the set containing the vowels a, e, i, o, and u.

In **set-builder notation**, the set of vowels in the English alphabet can be denoted as

$\mathbf{V} = \{x \mid x \text{ is a vowel of the English alphabet.}\}$ Read as "$\mathbf{V}$ is the set of all letters x such that x represents a vowel of the English alphabet."

In this notation, the letter x, called a **variable**, represents any one of the vowels in the English alphabet.

Because sets $\mathbf{A}$ and $\mathbf{V}$ above have the same elements, they are equal, and we write $\mathbf{A} = \mathbf{V}$. If two sets $\mathbf{A}$ and $\mathbf{B}$ have different elements, then $\mathbf{A} \neq \mathbf{B}$.

If $\mathbf{B} = \{a, c, e\}$ and $\mathbf{A} = \{a, b, c, d, e\}$, each element of set $\mathbf{B}$ is also an element of set $\mathbf{A}$. When every element of set $\mathbf{B}$ is also an element of set $\mathbf{A}$, we say that $\mathbf{B}$ is a **subset** of $\mathbf{A}$. In symbols, we write

$\mathbf{B} \subseteq \mathbf{A}$ Read as "$\mathbf{B}$ is a subset of $\mathbf{A}$."

Because every element in set $\mathbf{A}$ is an element in set $\mathbf{A}$, we have $\mathbf{A} \subseteq \mathbf{A}$. In general, any set is a subset of itself.

A set with no elements is called the **empty set**, denoted by $\emptyset$. Thus, $\emptyset = \{\ \}$. The empty set is a subset of every set.

If the elements of some set $\mathbf{A}$ are united with the elements of some set $\mathbf{B}$, the **union** of set $\mathbf{A}$ and set $\mathbf{B}$ is formed. The union of set $\mathbf{A}$ and set $\mathbf{B}$ is denoted as

$\mathbf{A} \cup \mathbf{B}$ Read as "the union of set $\mathbf{A}$ and set $\mathbf{B}$."

The elements in $\mathbf{A} \cup \mathbf{B}$ are *either* elements of set $\mathbf{A}$, *or* elements of set $\mathbf{B}$, *or* elements of *both* set $\mathbf{A}$ and set $\mathbf{B}$.

The set of elements that are common to set $\mathbf{A}$ and set $\mathbf{B}$ is called the **intersection** of $\mathbf{A}$ and $\mathbf{B}$. The intersection of set $\mathbf{A}$ and set $\mathbf{B}$ is denoted as

$$\mathbf{A} \cap \mathbf{B} \qquad \text{Read as "the intersection of set } \mathbf{A} \text{ and set } \mathbf{B}.\text{"}$$

The elements of $\mathbf{A} \cap \mathbf{B}$ are those elements that are in *both* set $\mathbf{A}$ and set $\mathbf{B}$. If $\mathbf{A}$ and $\mathbf{B}$ have no elements in common, then $\mathbf{A} \cap \mathbf{B} = \emptyset$. When the intersection of two sets is the empty set, we say that the two sets are **disjoint**.

EXAMPLE 1   If $\mathbf{A} = \{a, b, c, d, e\}$ and $\mathbf{B} = \{a, d, g\}$, find   **a.** $\mathbf{A} \cup \mathbf{B}$,   **b.** $\mathbf{A} \cap \mathbf{B}$,   and **c.** $\mathbf{A} \cup (\mathbf{B} \cap \emptyset)$.

Solution   **a.** $\mathbf{A} \cup \mathbf{B} = \{a, b, c, d, e, g\}$ **b.** $\mathbf{A} \cap \mathbf{B} = \{a, d\}$

c. $\mathbf{A} \cup (\mathbf{B} \cap \emptyset) = \mathbf{A} \cup \emptyset$ Do the work in parentheses first.
$\qquad\qquad\qquad\qquad = \mathbf{A}$ ∎

Sets of Numbers

A basic set in mathematics is the set of **natural numbers**, the numbers that we use for counting.

$$\mathbf{N} = \{1, 2, 3, 4, 5, 6, 7, 8, 9, 10, 11, \ldots\} \qquad \begin{array}{l}\text{The three dots, called the } \textbf{ellipsis,}\\ \text{indicate that the list continues}\\ \text{on forever.}\end{array}$$

Since we can add 1 to any natural number to obtain a larger one, there is no largest natural number. Because the set of natural numbers has an infinite number of elements, it is called an **infinite set**. A set with a limited number of elements is called a **finite set**.

A **prime number** is any natural number that is greater than 1 and is exactly divisible only by 1 and itself. The prime numbers form the set

$$\mathbf{P} = \{2, 3, 5, 7, 11, 13, 17, 19, \ldots\}$$

Because every prime number is a natural number, $\mathbf{P} \subseteq \mathbf{N}$.

A **composite number** is any natural number greater than 1 that is not prime. The composite numbers form the set

$$\mathbf{C} = \{4, 6, 8, 9, 10, 12, 14, 15, 16, \ldots\}$$

Because every composite number is a natural number, $\mathbf{C} \subseteq \mathbf{N}$. Furthermore, since no composite numbers are prime, $\mathbf{P} \cap \mathbf{C} = \emptyset$.

If we include 0 with the natural numbers, we have the set, $\mathbf{W}$, of **whole numbers**:

$$\mathbf{W} = \{0, 1, 2, 3, 4, 5, 6, 7, 8, 9, \ldots\}$$

The union of the set $\{-1, -2, -3, \ldots\}$ and $\mathbf{W}$ gives the set, $\mathbf{Z}$, of **integers**.

$$\mathbf{Z} = \{\ldots, -6, -5, -4, -3, -2, -1, 0, 1, 2, 3, 4, 5, 6, \ldots\}$$

Integers that are exactly divisible by 2 are called **even integers**, and those that are not are called **odd integers**. If **E** is the set of even integers and **O** is the set of odd integers, then

$$\mathbf{E} = \{\ldots, -8, -6, -4, -2, 0, 2, 4, 6, 8, \ldots\}$$
$$\mathbf{O} = \{\ldots, -7, -5, -3, -1, 1, 3, 5, 7, \ldots\}$$

A **rational number** is any number that can be written as a fraction with an integer for its numerator and a nonzero integer for its denominator. To denote the set **Q** of rational numbers, we use set-builder notation.

$$\mathbf{Q} = \left\{ x \mid x \text{ is a number that can be written in the form } \frac{a}{b}, \right.$$
$$\left. \text{where } a \text{ and } b \text{ are integers and } b \neq 0. \right\}$$

Some examples of rational numbers are

$$\frac{3}{4}, \quad \frac{-1}{3}, \quad \frac{5}{1}, \quad -\frac{8}{4}, \quad \frac{0}{5}, \quad \text{and} \quad \frac{99}{113}$$

The numbers 3, 0, -0.25, and $0.333\ldots$ are also rational numbers, because each one can be written as a fraction with an integer numerator and a nonzero integer denominator.

$$3 = \frac{3}{1}, \quad 0 = \frac{0}{7}, \quad -0.25 = -\frac{1}{4}, \quad \text{and} \quad 0.333\ldots = \frac{1}{3}$$

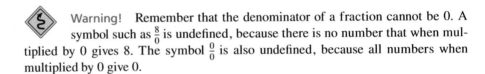 **Warning!** Remember that the denominator of a fraction cannot be 0. A symbol such as $\frac{8}{0}$ is undefined, because there is no number that when multiplied by 0 gives 8. The symbol $\frac{0}{0}$ is also undefined, because all numbers when multiplied by 0 give 0.

Rational numbers have decimal forms that are either **terminating** or **repeating decimals**. The overbars in the following examples indicate that the block of digits under the bar repeats forever.

$$\frac{3}{4} = 0.75 \qquad\qquad \frac{13}{25} = 0.52$$

$$\frac{4}{15} = 0.2666\ldots = 0.2\overline{6} \qquad \frac{5}{11} = 0.4545454\ldots = 0.\overline{45}$$

The set of rational numbers can be described as the set of all decimals that either terminate or repeat.

$$\mathbf{Q} = \{x \mid x \text{ can be written as a terminating or a repeating decimal.}\}$$

Numbers whose decimal forms neither terminate nor repeat are called **irrational numbers**. Numbers such as

$$\sqrt{2} = 1.414213562\ldots \qquad \text{and} \qquad \pi = 3.141592653\ldots$$

are examples of irrational numbers. Their decimal forms are nonterminating, non-repeating decimals. To express the set of irrational numbers, we write

$$\mathbf{H} = \{x \mid x \text{ is a nonterminating, nonrepeating decimal.}\}$$

The union of the set of rational numbers and the set of irrational numbers is the set of all decimals, called the set of **real numbers**.

$$\Re = \{x \mid x \text{ is a decimal number.}\}$$ The symbol $\Re$ denotes the set of real numbers.

The set of rational numbers and the set of irrational numbers are both subsets of the real numbers, and $\Re = \mathbf{Q} \cup \mathbf{H}$.

Graphs of Sets of Real Numbers

We can graph sets of numbers on a **number line**. The number line shown in Figure 1-1 and the number labels on it continue forever in both directions. The point labeled 0 is called the **origin**.

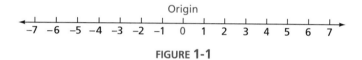

FIGURE 1-1

Figure 1-1 shows that if a and b are real numbers and $a < b$ (a is less than b), the graph of a is to the left of the graph of b. If $a > b$ (a is greater than b), the graph of a is to the right of the graph of b. The **positive numbers** represent points to the right of 0, and the **negative numbers** represent points to the left of 0.

 Warning!   Zero is neither positive nor negative.

Figure 1-2(a) shows the graph of the natural numbers from 1 to 10. The point associated with each number is called the **graph** of that number. The number is called the **coordinate** of its corresponding point. Figure 1-2(b) shows the graph of the prime numbers less than 12, and Figure 1-2(c) shows the graph of the whole numbers less than 10. Figure 1-2(d) shows the graph of the integers from -7 to 7.

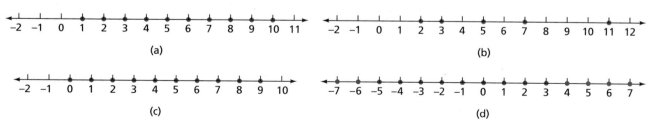

FIGURE 1-2

We can also graph fractions on a number line. For example, the point with coordinate of $\frac{3}{2}$ is at a distance halfway between the points with coordinates of 1 and 2.

The point with a coordinate of $\frac{1}{3}$ is at a distance one-third of the way from 0 to 1. These points and others are shown in Figure 1-3.

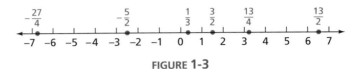

FIGURE 1-3

■ Intervals

Many graphs of sets of real numbers are portions of the number line called **intervals**.

Figure 1-4(a) shows the graph of the real numbers x that are between -2 and 4. We describe this set as

$$\{x \mid -2 < x < 4\} \qquad \text{or} \qquad \{x \mid -2 < x \text{ and } x < 4\}$$

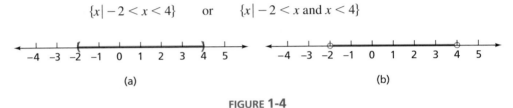

(a)   (b)

FIGURE 1-4

The inequalities $-2 < x$ and $x < 4$ indicate that neither -2 nor 4 is in the set. The parentheses at the points with coordinates of -2 and 4 show that the endpoints are not included in the graph. When neither endpoint is included, the interval is called an **open interval**. The open interval between -2 and 4 can be denoted as $(-2, 4)$. The statement $-2 < x$ and $x < 4$ means that both conditions on x are true at the same time.

Figure 1-4(b) uses a second notation for showing open intervals. The open circles show that -2 and 4 are not included in the graph.

The graph of the set of real numbers from 1 to 6 is shown in Figure 1-5(a). This set includes both endpoints. We describe the set as

$$\{x \mid 1 \le x \le 6\} \qquad \text{or} \qquad \{x \mid 1 \le x \text{ and } x \le 6\} \qquad \text{Read } \le \text{ as ``is less than or equal to.''}$$

The brackets at the points with coordinates of 1 and 6 indicate that these points are included in the graph. When both endpoints are included, the interval is called a **closed interval**. The closed interval from 1 to 6 can be denoted as $[1, 6]$.

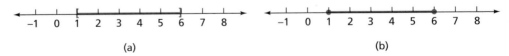

(a)   (b)

FIGURE 1-5

This closed interval is also shown in Figure 1-5(b). The closed circles show that 1 and 6 are included in the graph.

Figure 1-6(a) shows the graph of the **half-open interval** $(-6, -1]$. This interval can be described with the expressions

$$\{x \mid -6 < x \le -1\} \quad \text{or} \quad \{x \mid -6 < x \text{ and } x \le -1\}$$

Figure 1-6(b) shows the graph of all real numbers x such that $x < -5$ or $x > 2$. The statement $x < -5$ or $x > 2$ means that only one of the conditions on x needs to be true.

(a)   (b)

FIGURE **1-6**

To show these intervals with open and closed circles, we would replace the parentheses with an open circle and the brackets with a closed circle.

EXAMPLE 2   If $\mathbf{A} = [-1, 3]$ and $\mathbf{B} = [1, 4)$, find the graph of   **a.** $\mathbf{A} \cup \mathbf{B}$ and   **b.** $\mathbf{A} \cap \mathbf{B}$.

Solution   **a.** The union of intervals **A** and **B** is the set of all real numbers that are elements of either set **A** or set **B** or both. Numbers from -1 to 3 are in set **A**, and numbers from 1 to 4, not including 4, are in set **B**. Numbers from -1 to 4, not including 4, are in at least one of these sets. See Figure 1-7(a). Thus,

$$\mathbf{A} \cup \mathbf{B} = [-1, 3] \cup [1, 4) = [-1, 4)$$

b. The intersection of intervals **A** and **B** is the set of all real numbers that are elements of both set **A** and set **B**. The numbers that are in both of these sets are those numbers from 1 to 3. See Figure 1-7(b). Thus,

$$\mathbf{A} \cap \mathbf{B} = [-1, 3] \cap [1, 4) = [1, 3]$$

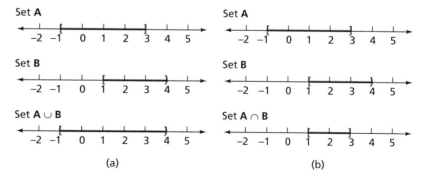

(a)   (b)

FIGURE **1-7**

Properties of Negatives

Two real numbers are called **negatives** of each other if their graphs are on opposite sides of the origin and are the same distance from the origin. For example, the negative of 2 is negative 2, denoted as -2, and the negative of -2, denoted as $-(-2)$, is 2. (See Figure 1-8.) In general, for any real number x,

$$-(-x) = x$$

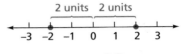

FIGURE **1-8**

This property of negatives and others are summarized as follows:

Properties of Negatives	**1.** $-(-x) = x$ **2.** $(-x) + (-y) = -(x + y)$ **3.** $x - y = x + (-y)$ **4.** $(-1)x = -x$ $(-x)y = x(-y) = -xy$ $(-x)(-y) = xy$ **5.** $\dfrac{-x}{y} = \dfrac{x}{-y} = -\dfrac{x}{y}$ and $\dfrac{-x}{-y} = \dfrac{x}{y}$ $\Big\}$ provided that $y \neq 0$

■ Absolute Value

The **absolute value** of a real number a, denoted as $|a|$, is the distance on a number line between 0 and the point with a coordinate of a. Because points with coordinates of 4 and -4 both lie 4 units from 0, it follows that $|4| = |-4| = 4$. In general, for any real number a,

$$|-a| = |a|$$

We can define absolute value more formally as follows.

Absolute Value	If x is a real number, $\quad |x| = x \quad$ if $x \geq 0$. $\quad |x| = -x$ if $x < 0$.

This definition indicates that when x is a positive number or 0, x is its own absolute value. However, when x is a negative number, $-x$ (which is positive) is its absolute value. Thus, $|x|$ always represents a nonnegative number.

$$|x| \geq 0 \qquad \text{for all real numbers } x$$

 Warning!   Remember that x is not always positive and $-x$ is not always negative.

EXAMPLE 3 Find **a.** $|3|$, **b.** $|-4|$, **c.** $|0|$, and **d.** $|1 - \pi|$.

Solution **a.** $|3| = 3$ **b.** $|-4| = -(-4)$

c. $|0| = 0$ $= 4$

d. $|1 - \pi| = -(1 - \pi)$ Because $1 < \pi$, the number $1 - \pi$ is negative.

              $= \pi - 1$ ∎

On the number line shown in Figure 1-9, the distance between the points with coordinates of 1 and 4 is $4 - 1$, or 3, units. However, if the subtraction were done in the other order, the result would be $1 - 4$, or -3, units. To guarantee that the distance between two points is always positive, we use absolute value symbols. Thus, the distance, d, between points with coordinates 1 and 4 is

$$d = |4 - 1| = |1 - 4| = 3$$

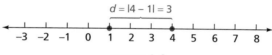

FIGURE 1-9

This suggests that we define the distance d between two points with coordinates a and b on the number line by using the concept of absolute value.

Distance between Two Points	If a and b are the coordinates of two points on a number line, then the distance between those points is given by the formulas $$d = |a - b| \quad \text{or} \quad d = |b - a|$$

EXAMPLE 4 Find the distance d on a number line between points with coordinates of **a.** 3 and 5, **b.** -2 and 3, and **c.** -5 and -1.

Solution **a.** $d = |5 - 3| = |2| = 2$ or $d = |3 - 5| = |-2| = 2$

b. $d = |3 - (-2)| = |5| = 5$ or $d = |-2 - 3| = |-5| = 5$

c. $d = |-5 - (-1)| = |-4| = 4$ or $d = |-1 - (-5)| = |4| = 4$ ∎

▪ Properties of Real Numbers

Any two real numbers a and b can be compared. Either a is equal to b, a is greater than b, or a is less than b. This fact is called the **trichotomy property**.

The Trichotomy Property	For any two real numbers a and b, exactly one of the following relationships must hold: $$a = b \quad \text{or} \quad a < b \quad \text{or} \quad a > b$$

To indicate that x is not equal to y, we write $x \neq y$. If $x \not< y$ (x is not less than y), then $x \geq y$. Similarly, if $x \not> y$, then $x \leq y$.

Properties of Equality

If a, b, and c are real numbers, then

$a = a$	(reflexive property)
If $a = b$, then $b = a$.	(symmetric property)
If $a = b$ and $b = c$, then $a = c$	(transitive property)

EXAMPLE 5   Determine whether the relation $<$ is **a.** reflexive, **b.** symmetric, and **c.** transitive.

Solution   **a.** To determine whether the relation $<$ is reflexive, we ask the question, "Is a real number less than itself? (Is $a < a$?)" Because the answer is no, the relation is not reflexive.

b. To determine whether the relation $<$ is symmetric, we ask the question, "If one real number is less than a second, is it also true that the second number is less than the first? (If $a < b$, is $b < a$?)" Because the answer is no, the relation is not symmetric.

c. To determine whether the relation $<$ is transitive, we ask the question, "If one real number is less than a second, and if the second real number is less than a third, does this mean that the first real number is less than the third? (If $a < b$ and $b < c$, is $a < c$?)" Because the answer is yes, the relation is transitive.   ■

When we work with real numbers, we will use the following properties.

The Associative, Commutative, and Distributive Properties

If a, b, and c are real numbers, then

The Associative Properties for Addition and Multiplication:

$$(a + b) + c = a + (b + c) \qquad (ab)c = a(bc)$$

The Commutative Properties for Addition and Multiplication:

$$a + b = b + a \qquad ab = ba$$

The Distributive Property of Multiplication over Addition:

$$a(b + c) = ab + ac$$

EXAMPLE 6   The statements in the left-hand column are true because of the property of equality or the property of real numbers listed in the right-hand column.

$32 = 32$	Reflexive property of equality
If $a = b + c$, then $b + c = a$.	Symmetric property of equality
If $a = 37$ and $37 = b$, then $a = b$.	Transitive property of equality
$a < b$ or $a = b$ or $a > b$	Trichotomy property
$9 + (2 + 3) = (9 + 2) + 3$	Associative property for addition

$2(xy) = (2x)y$	Associative property for multiplication
$9 + 3 = 3 + 9$	Commutative property for addition
$8 \cdot 3 = 3 \cdot 8$	Commutative property for multiplication
$(a + b) + c = c + (a + b)$	Commutative property for addition
$a(bc) = a(cb)$	Commutative property for multiplication
$2(x + 3) = 2x + 2 \cdot 3$	Distributive property ■

1.1 EXERCISES

In Exercises 1–8, **A** $= \{2, 4, 6, 8\}$, **B** $= \{2, 4, 6, 7\}$, *and* **C** $= \{x \,|\, x$ *is an even integer.*$\}$. *Insert either an* $\in$ *or a* $\subseteq$ *symbol to make a true statement.*

1. **A** __ **C**   **2.** 4 __ **A**   **3.** 7 __ **B**   **4.** $\{4, 6\}$ __ **C**

5. $\emptyset$ __ **B**   **6.** $\emptyset$ __ **A**   **7.** **B** __ **B**   **8.** **A** __ **A**

In Exercises 9–16, **A** $= \{2, 3, 5, 7\}$, **B** $= \{1, 3, 5, 7, 9\}$, *and* **C** $= \{2, 4, 6, 8, 10\}$. *Find each set.*

9. **A** $\cap$ **B**   **10.** **A** $\cup$ **C**   **11.** **B** $\cup$ **C**   **12.** **A** $\cap$ **C**

13. $(\mathbf{A} \cap \emptyset) \cup \mathbf{B}$ **14.** $\mathbf{B} \cap (\emptyset \cup \mathbf{C})$ **15.** $(\mathbf{A} \cap \mathbf{B}) \cup \mathbf{C}$ **16.** $(\mathbf{A} \cup \mathbf{C}) \cap \mathbf{B}$

In Exercises 17–20, list the elements in each set, if possible.

17. $\{x \,|\, x$ is a composite number between 9 and 19.$\}$

18. $\{x \,|\, x$ is a prime number between 10 and 20.$\}$

19. $\{x \,|\, x$ is a number that is both even and prime.$\}$

20. $\{x \,|\, x$ is a natural number that is neither prime nor composite.$\}$

In Exercises 21–26, tell whether each set is a finite or an infinite set.

21. $\{x \,|\, x$ is a composite number.$\}$

22. $\{x \,|\, x$ is an even integer.$\}$

23. $\{1, 2, 3, 4, 5, 6, 7, 8, 9, 10\}$

24. $\{x \,|\, x$ is a rational number.$\}$

25. $\{x \,|\, x$ is a country in the world.$\}$

26. $\{x \,|\, x$ is an integer that is both prime and composite.$\}$

27. Find the first prime larger than 50.

28. Find the largest prime less than 100.

In Exercises 29–38, indicate whether each statement is true. If a statement is false, explain why.

29. The number 27 is a prime number.

30. The number 2 is a prime number.

31. The set of composite numbers is a subset of the set of integers.

32. The set of prime numbers is a subset of the set of odd integers.

33. The number 0 is not a rational number.

34. The symbol $\dfrac{16}{0}$ represents a rational number.

35. All integers are rational numbers.

36. Some integers are greater than 1.

37. The number 1 is a prime number.

38. The number 0 is neither even nor odd.

In Exercises 39–52, indicate whether each statement is true. If a statement is false, give an example to show that it is false.

39. The product of two prime numbers is a prime number.

40. The sum of two prime numbers is a prime number.

41. The square of a prime number is a prime number.

42. The sum of two prime numbers can be a composite number.

43. The sum of two composite numbers is a prime number.

44. All natural numbers are integers.

45. No even integers are prime numbers.

46. All odd integers are prime numbers.

47. The sum of an even integer and an odd integer is an odd integer.

48. The sum of three odd integers must be another odd integer.

49. If the product of several integers is an even integer, then at least one of the integers is an even integer.

50. If the product of several integers is an odd integer, then each of the integers must be an odd integer.

51. If the sum of three integers is an even integer, then each of the integers must be an even integer.

52. The set of even integers combined with the set of odd integers forms the set of integers.

In Exercises 53–60, indicate whether each statement is true. If a statement is false, explain why.

53. The rational numbers are a subset of the real numbers.

54. Every real number is an irrational number.

55. Some irrational numbers are integers.

56. Some real numbers are integers.

57. There are real numbers that are neither rational nor irrational.

58. The smallest irrational number is $\sqrt{2}$.

59. The intersection of the set of real numbers and the set of rational numbers is the set of decimals that either terminate or repeat.

60. The union of the set of rational numbers and the set of irrational numbers is the set of real numbers.

In Exercises 61–68, graph each set on a number line, if possible.

61. The set of even integers between 19 and 31.

62. The set of integers from -5 to -1.

63. The set of prime numbers between 10 and 20.

64. The set of even composite numbers between 20 and 30.

65. The set of natural numbers that are neither prime nor composite.

66. The set of integers that are both even and prime.

67. The set of numbers that are both even and odd.

68. The set of numbers that are odd and composite and that are less than 10.

In Exercises 69–90, graph each interval on the number line, if possible.

69. $\{x \mid 3 < x < 10\}$

70. $\{x \mid -2 \le x \le 3\}$

71. $\{x \mid -4 \le x < 0\}$

72. $\{x \mid 5 < x \le 15\}$

73. $\{x \mid 3 < x \text{ and } x \le 7\}$

74. $\{x \mid x \le -4 \text{ or } x > 8\}$

75. $\{x \mid x < -2 \text{ or } x > 10\}$

76. $\{x \mid x \le 0 \text{ and } x \ge 5\}$

77. $\{x \mid x \ge 0 \text{ and } x \le 0\}$

78. $\{x \mid x > 0 \text{ or } x < 0\}$

79. $(-2, 3)$

80. $[-5, 5]$

81. $[4, 8)$

82. $(-10, -2]$

83. $(2, 6) \cup (1, 5)$

84. $(3, 5] \cup [4, 6]$

85. $[-2, 2) \cap [1, 2]$

86. $(-5, -1) \cap (-3, 0)$

87. $(-5, -3] \cup [-3, -2)$

88. $[-5, 2) \cap (-2, 0]$

89. $\{x \mid x \ne 0\} \cap (-3, 2]$

90. $\{x \mid x > 4\} \cup [0, 4)$

In Exercises 91–98, simplify each expression.

91. $-(-12)$

92. $(-2) + (-3)$

93. $4 - (-2)$

94. $(-4)(3)$

95. $(-12)(-3)$

96. $(-4)(-2)$

97. $\dfrac{-36}{12}$

98. $\dfrac{-100}{-50}$

In Exercises 99–106, find the value of each expression.

99. $|-7|$

100. $-|-6|$

101. $|7| - |-6|$

102. $|8 - 5| + |5 - 8|$

103. $|\sqrt{2} - 1|$

104. $|1 - \sqrt{3}|$

105. $|1 - \pi|$

106. $|\pi - 2|$

In Exercises 107–108, assume that $x = -5$, $y = -2$, and $z = 3$.

107. Find the distance on the number line between two points with coordinates of x and y.

108. Find the distance on the number line between two points with coordinates of x and z.

In Exercises 109–116, tell whether the given relation is reflexive, symmetric, or transitive.

109. $>$ **110.** $\geq$ **111.** $\leq$ **112.** $\neq$

113. $\cong$ ("is congruent to" from geometry)

114. "is divisible by" on the set of natural numbers

115. "is the same color as"

116. "is taller than"

In Exercises 117–128, tell which property justifies each expression.

117. $(8 + a) + b = b + (8 + a)$

118. $3(x + y) = 3x + 3y$

119. If $3 = a$ and $a = c$, then $3 = c$.

120. $(a + b)c = c(a + b)$

121. $-5 = -5$

122. $3(4 + t) = 3 \cdot 4 + 3t$

123. $a + (b + c) = a + (c + b)$

124. $a(bc) = (bc)a$

125. If $a = b$, then $b = a$.

126. $a + (b + c) = (a + b) + c$

127. $[a(b + c)]d = a[(b + c)d]$

128. $5 + 4 + (2 + 1) = 5 + (4 + 2) + 1$

1.2 INTEGER EXPONENTS AND SCIENTIFIC NOTATION

■ Rules of Exponents ■ Order of Operations ■ Scientific Notation

When two or more quantities are multiplied together, each quantity is called a **factor** of the product. The expression x^4 indicates that x is to be used as a factor four times. Thus,

$$x^4 = x \cdot x \cdot x \cdot x$$

In general, the following is true.

Natural-Number Exponents

For any natural number n,

$$x^n = \overbrace{x \cdot x \cdot x \cdot \cdots \cdot x}^{n \text{ factors of } x}$$

The number x in the **exponential expression** x^n is called the **base**, and the number n is called the **exponent** or the **power** to which the base is raised. The expression x^n is called a **power of x**. A natural number exponent tells how many times the base of an exponential expression is to be used as a factor in a product. If an exponent is 1, it is usually not written. We understand that

$$x^1 = x$$

EXAMPLE 1
  a. $4^2 = 4 \cdot 4 = 16$ Read 4^2 as "four squared."

  b. $(-4)^2 = (-4)(-4) = 16$

  c. $-4^2 = -(4 \cdot 4) = -16$

  d. $(5)^3 = 5 \cdot 5 \cdot 5 = 125$ Read 5^3 as "five cubed."

  e. $(-5)^3 = (-5)(-5)(-5) = -125$

  f. $3x^4 = 3 \cdot x \cdot x \cdot x \cdot x$ Read x^4 as "x to the fourth power."

  g. $(3x)^4 = (3x)(3x)(3x)(3x) = 81x^4$ ■

 Warning! It is important to note the distinction between ax^n and $(ax)^n$, and between $-x^n$ and $(-x)^n$:

$$ax^n = a \cdot \overbrace{x \cdot x \cdot x \cdot \cdots \cdot x}^{n \text{ factors of } x} \qquad\qquad (ax)^n = \overbrace{(ax)(ax)(ax) \cdot \cdots \cdot (ax)}^{n \text{ factors of } ax}$$

$$-x^n = -\overbrace{(x \cdot x \cdot x \cdot \cdots \cdot x)}^{n \text{ factors of } x} \qquad\qquad (-x)^n = \overbrace{(-x)(-x)(-x) \cdot \cdots \cdot (-x)}^{n \text{ factors of } -x}$$

■ Rules of Exponents

We begin to develop the rules of exponents by considering the product $x^m x^n$. Because x^m indicates that x is to be used as a factor m times and because x^n indicates that x is to be used as a factor n times, there are $m + n$ factors of x in the product $x^m x^n$.

$$x^m x^n = \overbrace{\underbrace{x \cdot x \cdot x \cdot \cdots \cdot x}_{m \text{ factors of } x} \cdot \underbrace{x \cdot x \cdot x \cdot \cdots \cdot x}_{n \text{ factors of } x}}^{m + n \text{ factors of } x}$$

Thus, to multiply exponential expressions with the same base, we keep the base and add the exponents.

The Product Rule of Exponents	If m and n are natural numbers, then $$x^m x^n = x^{m+n}$$

 Warning! The product rule applies only to exponential expressions with the same base. A product of two powers with different bases, such as $x^4 y^3$, cannot be simplified.

To find another property of exponents, we consider the exponential expression $(x^m)^n$. The exponent n indicates that x^m is to be used as a factor n times. This implies that x is to be used as a factor mn times.

$$(x^m)^n = \overbrace{\underbrace{(x^m)(x^m)(x^m) \cdot \cdots \cdot (x^m)}_{n \text{ factors of } x^m}}^{mn \text{ factors of } x} = x^{mn}$$

Thus, to raise an exponential expression to a power, we keep the base and multiply the exponents.

To raise a product to a power, we raise each factor to that power.

$$
(xy)^n = \overbrace{(xy)(xy)(xy) \cdot \cdots \cdot (xy)}^{n \text{ factors of } xy} = \overbrace{(x \cdot x \cdot x \cdots \cdots x)}^{n \text{ factors of } x}\overbrace{(y \cdot y \cdot y \cdots \cdots y)}^{n \text{ factors of } y} = x^n y^n
$$

To raise a fraction to a power, we raise both the numerator and the denominator to that power. If $y \neq 0$, then

$$
\left(\frac{x}{y}\right)^n = \overbrace{\left(\frac{x}{y}\right)\left(\frac{x}{y}\right)\left(\frac{x}{y}\right) \cdots \cdots \left(\frac{x}{y}\right)}^{n \text{ factors of } \frac{x}{y}}
$$

$$
= \frac{\overbrace{xxx \cdots \cdots x}^{n \text{ factors of } x}}{\underbrace{yyy \cdots \cdots y}_{n \text{ factors of } y}}
$$

$$
= \frac{x^n}{y^n}
$$

The previous three results are often called **power rules of exponents**.

The Power Rules of Exponents

If m and n are natural numbers, then

$$
(x^m)^n = x^{mn} \qquad (xy)^n = x^n y^n \qquad \left(\frac{x}{y}\right)^n = \frac{x^n}{y^n} \qquad (y \neq 0)
$$

EXAMPLE 2

a. $x^5 x^7 = x^{5+7} = x^{12}$

b. $x^2 y^3 x^5 y = x^{2+5} y^{3+1} = x^7 y^4$

c. $(x^4)^9 = x^{4 \cdot 9} = x^{36}$

d. $(x^2 x^5)^3 = (x^7)^3 = x^{21}$

e. $\left(\dfrac{x}{y^2}\right)^5 = \dfrac{x^5}{(y^2)^5} = \dfrac{x^5}{y^{10}} \quad (y \neq 0)$

f. $\left(\dfrac{5x^2 y}{z^3}\right)^2 = \dfrac{5^2 (x^2)^2 y^2}{(z^3)^2} = \dfrac{25x^4 y^2}{z^6} \quad (z \neq 0)$

■

We can extend the definition of natural-number exponents to include other exponents. For example, if we assume that the rules for natural-number exponents hold for exponents of 0, we can write

$$
x^0 x^n = x^{0+n} = x^n = 1x^n
$$

Because $x^0 x^n = 1x^n$, it follows that if $x \neq 0$, then $x^0 = 1$. Thus, we make the following definition.

Zero Exponent	If $x \neq 0$, then
	$$x^0 = 1$$

Furthermore, if we assume that the rules for natural-number exponents hold for exponents that are negative integers, we can write

$$x^{-n}x^n = x^{-n+n} = x^0 = 1$$

provided that $x \neq 0$. However, because $\dfrac{1}{x^n} \cdot x^n = \dfrac{x^n}{x^n}$ and any nonzero number divided by itself is 1, we have

$$\frac{1}{x^n} \cdot x^n = 1 \qquad (x \neq 0)$$

Because $x^{-n}x^n = \dfrac{1}{x^n} \cdot x^n$, it follows that if $x \neq 0$, then $x^{-n} = \dfrac{1}{x^n}$. Thus, we make the following definition.

Negative Exponents	If n is an integer and $x \neq 0$, then
	$$x^{-n} = \frac{1}{x^n} \qquad \text{and} \qquad \frac{1}{x^{-n}} = x^n$$

Because of the two previous definitions, all of the rules for natural-number exponents hold for integer exponents.

EXAMPLE 3

a. $(3x)^0 = 1$

b. $3(x^0) = 3(1) = 3$

c. $x^{-4} = \dfrac{1}{x^4}$

d. $\dfrac{1}{x^{-6}} = x^6$

e. $x^{-3}x = x^{-3+1}$
$= x^{-2}$
$= \dfrac{1}{x^2}$

f. $(x^{-4}x^8)^{-5} = (x^4)^{-5}$
$= x^{-20}$
$= \dfrac{1}{x^{20}}$ ∎

To develop the quotient rule for exponents, we proceed as follows:

$$\frac{x^m}{x^n} = x^m\left(\frac{1}{x^n}\right) = x^m x^{-n} = x^{m+(-n)} = x^{m-n} \qquad (x \neq 0)$$

Thus, to divide two exponential expressions with the same nonzero base, we keep the base and subtract the exponent in the denominator from the exponent in the numerator.

The Quotient Rule of Exponents	If m and n are integers and $x \neq 0$, then $$\frac{x^m}{x^n} = x^{m-n}$$

EXAMPLE 4

a. $\dfrac{x^8}{x^5} = x^{8-5}$

      $= x^3$

b. $\dfrac{x^{-6}}{x^2} = x^{-6-2}$

      $= x^{-8}$

      $= \dfrac{1}{x^8}$

c. $\dfrac{x^2 x^4}{x^{-5}} = \dfrac{x^6}{x^{-5}}$

      $= x^{6-(-5)}$

      $= x^{11}$

d. $\dfrac{x^m x^2}{x^3} = \dfrac{x^{m+2}}{x^3}$

      $= x^{m+2-3}$

      $= x^{m-1}$

e. $\left(\dfrac{x^3 y^{-2}}{x^{-2} y^3}\right)^{-2} = (x^{3-(-2)} y^{-2-3})^{-2}$

      $= (x^5 y^{-5})^{-2}$

      $= x^{-10} y^{10}$

      $= \dfrac{y^{10}}{x^{10}}$

f. $\left(\dfrac{x}{y}\right)^{-n} = \dfrac{x^{-n}}{y^{-n}}$

      $= \dfrac{x^{-n} x^n y^n}{y^{-n} x^n y^n}$

      $= \dfrac{x^0 y^n}{y^0 x^n}$

      $= \dfrac{y^n}{x^n}$

      $= \left(\dfrac{y}{x}\right)^n$ ∎

Part **f** of Example 4 establishes this useful theorem.

Theorem	If n is a natural number and neither x nor y is 0, then $$\left(\frac{x}{y}\right)^{-n} = \left(\frac{y}{x}\right)^n$$

■ Order of Operations

If several mathematical operations occur in one expression, the order in which those operations are performed might affect the result. To avoid conflicting answers, arithmetic operations should be performed in the following order.

Order of Operations	If an expression does not contain grouping symbols such as parentheses or brackets, follow these steps:

1. Find the values of any exponential expressions.

2. Do all multiplications and/or divisions as they are encountered while working from left to right.

3. Do all additions and/or subtractions as they are encountered while working from left to right.

If an expression contains grouping symbols, use the rules above to perform the calculations within each pair of grouping symbols, working from the innermost pair to the outermost pair.

In a fraction, simplify the numerator and the denominator separately. Then simplify the fraction, whenever possible.

EXAMPLE 5   If $x = -2$, $y = 3$, and $z = -4$, find the value of **a.** $-x^2 + y^2z$ and **b.** $\dfrac{2z^3 - 3y^2}{5x^2}$.

Solution   **a.**
$$-x^2 + y^2z = -(-2)^2 + 3^2(-4)$$
$$= -(4) + 9(-4) \qquad \text{Evaluate the powers.}$$
$$= -4 + (-36) \qquad \text{Do the multiplication.}$$
$$= -40 \qquad \text{Do the addition.}$$

b.
$$\frac{2z^3 - 3y^2}{5x^2} = \frac{2(-4)^3 - 3(3)^2}{5(-2)^2}$$
$$= \frac{2(-64) - 3(9)}{5(4)} \qquad \text{Evaluate the powers.}$$
$$= \frac{-128 - 27}{20} \qquad \text{Do the multiplications.}$$
$$= \frac{-155}{20} \qquad \text{Do the subtraction.}$$
$$= -\frac{31}{4} \qquad \text{Simplify the fraction.} \qquad \blacksquare$$

Scientific Notation

Scientists and engineers often work with numbers that are either very large or very small. These numbers can be written compactly by expressing them in **scientific notation**.

Scientific Notation	A number is written in **scientific notation** when it is written as the product of a number whose absolute value is between 1 and 10 (including 1 but not 10) and an integer power of 10.

EXAMPLE 6 Light travels 29,980,000,000 centimeters per second. Express this number in scientific notation.

Solution To write 29,980,000,000 in scientific notation, we must express the number as the product of a number between 1 and 10 and some integer power of 10. The number 2.998 lies between 1 and 10. To get 29,980,000,000, the decimal point in 2.998 must be moved ten places to the right. This is accomplished by multiplying 2.998 by 10^{10}. Thus,

$$29,980,000,000 = 2.998 \times 10^{10}$$

∎

EXAMPLE 7 Write the numbers **a.** 0.00062 and **b.** −0.0027 in scientific notation.

Solution **a.** To write 0.00062 in scientific notation, we must express the number as a product of a number between 1 and 10 and some integer power of 10. The number 0.00062 can be obtained by moving the decimal point in 6.2 four places to the left. This is accomplished by multiplying 6.2 by 10^{-4}, because

$$6.2 \times 10^{-4} = 6.2 \times \frac{1}{10^4} = \frac{6.2}{10,000} = 0.00062$$

Thus,

$$0.00062 = 6.2 \times 10^{-4}$$

b. To write −0.0027 in scientific notation, we must express the number as a product of a number whose absolute value is between 1 and 10 and some integer power of 10. The number −2.7 has an absolute value between 1 and 10. To get the number −0.0027, the decimal point in −2.7 must be moved three places to the left. Thus,

$$-0.0027 = -2.7 \times 10^{-3}$$

∎

EXAMPLE 8 Express 3.27×10^{-5} in standard notation.

Solution The factor of 10^{-5} indicates that 3.27 is to be divided by 5 factors of 10. Because each division by 10 moves the decimal point one place to the left, we must move the decimal point in 3.27 five places to the left. In standard notation,

$$3.27 \times 10^{-5} = 0.0000327$$

∎

Study each of the following numbers written in both scientific and standard notation. In each case, the exponent gives the number of places the decimal point moves, and the sign of the exponent indicates the direction in which it moves.

a. $3.72 \times 10^5 = 372,000$ 5 places to the right

b. $9.93 \times 10^9 = 9,930,000,000$ 9 places to the right

c. $5.37 \times 10^{-4} = 0.000537$ 4 places to the left

d. $-1.529 \times 10^{-1} = -0.1529$ 1 place to the left

e. $7.36 \times 10^0 = 7.36$ No movement of the decimal point

EXAMPLE 9   Use scientific notation to calculate $\dfrac{(3,400,000)(0.00002)}{170,000,000}$.

Solution $\dfrac{(3,400,000)(0.00002)}{170,000,000} = \dfrac{(3.4 \times 10^6)(2.0 \times 10^{-5})}{1.7 \times 10^8}$

$= \dfrac{6.8}{1.7} \times 10^{6+(-5)-8}$

$= 4.0 \times 10^{-7}$

$= 0.0000004$ ■

1.2   EXERCISES

In Exercises 1–8, write each number without using exponents and simplify, when possible.

1. 13^2 **2.** 10^3 **3.** -5^2 **4.** $(-5)^2$

5. $4x^3$ **6.** $(4x)^2$ **7.** $(-5x)^4$ **8.** $-6x^2$

In Exercises 9–16, write each expression using exponents and simplify, when possible.

9. $7xxx$ **10.** $-8yyyy$ **11.** $(-x)(-x)$ **12.** $(2a)(2a)(2a)$

13. $(3t)(3t)(-3t)$ **14.** $-(2b)(2b)(2b)(2b)$ **15.** $xxxyy$ **16.** $aaabbbb$

In Exercises 17–76, simplify each expression. If the base is a variable, express answers with positive exponents only. Assume that all variables are restricted to those numbers for which the expression is defined.

17. x^2x^3 **18.** y^3y^4 **19.** $(z^2)^3$ **20.** $(t^6)^7$

21. $(y^5y^2)^3$ **22.** $(a^3a^6)a^4$ **23.** $(z^2)^3(z^4)^5$ **24.** $(t^3)^4(t^5)^2$

25. $(3x)^3$ **26.** $(-2y)^4$ **27.** $(x^2y)^3$ **28.** $(x^3z^4)^6$

29. $\left(\dfrac{a^2}{b}\right)^3$ **30.** $\left(\dfrac{x}{y^3}\right)^4$ **31.** $(-x)^0$ **32.** $4x^0$

33. $(4x)^0$ **34.** $-2x^0$ **35.** z^{-4} **36.** $\dfrac{1}{t^{-2}}$

37. $y^{-2}y^{-3}$ **38.** $-m^{-2}m^3$ **39.** $(x^3x^{-4})^{-2}$ **40.** $(y^{-2}y^3)^{-4}$

41. $\dfrac{x^7}{x^3}$ **42.** $\dfrac{r^5}{r^2}$ **43.** $\dfrac{a^{21}}{a^{17}}$ **44.** $\dfrac{t^{13}}{t^4}$

45. $\dfrac{(x^2)^2}{x^2x}$ **46.** $\dfrac{s^9s^3}{(s^2)^2}$ **47.** $\left(\dfrac{m^3}{n^2}\right)^3$ **48.** $\left(\dfrac{t^4}{t^3}\right)^2$

49. $\dfrac{(a^3)^{-2}}{aa^2}$

50. $\dfrac{r^9 r^{-3}}{(r^{-2})^3}$

51. $\left(\dfrac{a^{-3}}{b^{-1}}\right)^{-4}$

52. $\left(\dfrac{t^{-4}}{t^{-3}}\right)^{-2}$

53. $\left(\dfrac{r^4 r^{-6}}{r^3 r^{-3}}\right)^2$

54. $\dfrac{(x^{-3}x^2)^2}{(x^2 x^{-5})^{-3}}$

55. $\left(\dfrac{x^5 y^{-2}}{x^{-3}y^2}\right)^4$

56. $\left(\dfrac{2x^{-7}y^5}{x^7 y^{-4}}\right)^3$

57. $\left(\dfrac{5x^{-3}y^{-2}}{3x^2 y^{-2}}\right)^{-2}$

58. $\left(\dfrac{3x^{-2}y^{-5}}{2x^{-2}y^{-6}}\right)^{-3}$

59. $\left(\dfrac{x^2 + 3x + y}{2x^5 - y}\right)^0$

60. $\left(\dfrac{x^0 + y^0 - z^0}{x^0 - y^0 + z^0}\right)^6$

61. $\left(\dfrac{3x^5 y^{-3}}{6x^{-5}y^3}\right)^{-2}$

62. $\left(\dfrac{12x^{-4}y^3 z^{-5}}{4x^4 y^{-3}z^5}\right)^3$

63. $(-x^2 y^x)^5$

64. $(-x^2 y^y)^6$

65. $-(x^2 y^3)^{4xy}$

66. $-(x^2 y)^{3xy}$

67. $x^{m+1}x^{3-m}$

68. $y^{n-3}y^{3-n}$

69. $(x^{2n}x^{3n})^n$

70. $(a^{-3m}a^m)^{-m}$

71. $\dfrac{x^{3m+5}}{x^{3m}}$

72. $\dfrac{12x^{m+3}}{3x^m}$

73. $\dfrac{(8^{-2}z^{-3}y)^{-1}}{(5y^2 z^{-2})^3 (5yz^{-2})^{-1}}$

74. $\dfrac{(m^{-2}n^3 p^4)^{-2}(mn^{-2}p^3)^4}{(mn^{-2}p^3)^{-4}(mn^2 p)^{-1}}$

75. $\left[\dfrac{(m^{-2}n^{-1}p^3)^{-2}}{(mn^2)^{-3}(p^{-3})^4}\right]^{-2}$

76. $\left[\dfrac{(3x^2)^{-4}(3p)^2}{(x^{-2}p^4)^{-3}(3x^{-3})^{-2}}\right]^{-3}$

In Exercises 77–92, let $x = -2$, $y = 0$, and $z = 3$, and evaluate each expression.

77. x^2

78. $-x^2$

79. x^3

80. $-x^3$

81. $(-xz)^3$

82. $-xz^3$

83. $\dfrac{-(x^2 z^3)}{z^2 - y^2}$

84. $\dfrac{z^2(x^2 - y^2)}{x^3 z}$

85. $5x^2 - 3y^3 z$

86. $3(x - z)^2 + 2(y - z)^3$

87. $3x^3 y^7 z^{15}$

88. $5x^{-2}z^3 + y^3$

89. $\dfrac{-3x^{-3}z^{-2}}{6x^2 z^{-3}}$

90. $\dfrac{(-5x^2 z^{-3})^2}{5xz^{-2}}$

91. $x^z z^x y^z$

92. $(x^y z^y)^{-z}$

In Exercises 93–106, express each number in scientific notation.

93. 372,000

94. 89,500

95. $-177{,}000{,}000$

96. $-23{,}470{,}000{,}000$

97. 0.007

98. 0.00052

99. -0.000000693

100. -0.000000089

101. one trillion

102. sixty-three billion

103. one trillionth

104. forty-three billionths

105. 99.7×10^{-4}

106. 0.0085×10^5

In Exercises 107–114, express each number in standard notation.

107. 9.37×10^5

108. 4.26×10^9

109. 2.21×10^{-5}

110. 2.774×10^{-2}

111. 0.00032×10^4

112. 9300×10^{-4}

113. -3.2×10^{-3}

114. -7.25×10^3

In Exercises 115–118, use the method of Example 9 to do each calculation. Write all answers in scientific notation.

115. $\dfrac{(65{,}000)(45{,}000)}{250{,}000}$

116. $\dfrac{(0.000000045)(0.00000012)}{45{,}000{,}000}$

117. $\dfrac{(0.00000035)(170{,}000)}{0.00000085}$

118. $\dfrac{(0.0000000144)(12{,}000)}{600{,}000}$

In Exercises 119–122, use scientific notation to compute each answer. Write all answers in scientific notation.

119. The speed of sound in air is 3.31×10^4 centimeters per second. Compute the speed of sound in meters per minute.

120. Calculate the volume of a box that has dimensions of 6000 by 9700 by 4700 millimeters.

121. The mass of one proton is 0.00000000000000000000000167248 gram. Find the mass of one billion protons.

122. The speed of light in a vacuum is approximately 30,000,000,000 centimeters per second. Find the speed of light in miles per hour. (160,934.4 cm = 1 mile.)

In Exercises 123–128, use a scientific calculator to work each problem.

123. Show that $4.57^0 = 1$.

124. Show that $(1.2)^3(3.2)^3 = [(1.2)(3.2)]^3$.

125. Show that $(4.1)^2 + (5.2)^2 \neq (4.1 + 5.2)^2$.

126. Show that $(3.7^2)^3 = 3.7^6$.

127. Show that $(3.2)^{-3}(3.2)^7 = (3.2)^4$.

128. Show that $4.75^{-4} = \dfrac{1}{4.75^4}$.

1.3 FRACTIONAL EXPONENTS AND RADICALS

- Radical Expressions
- Simplifying Radicals
- Combining Radicals
- Rationalizing Denominators and Numerators
- Multiplying and Dividing Radicals with Different Indexes

With the following definitions, we extend the rules for integral exponents to include fractional exponents.

Rational Exponents

If $a \geq 0$ and n is a natural number, then $a^{1/n}$ is the nonnegative real number whose nth power is a. In symbols,

$$(a^{1/n})^n = a$$

The nonnegative real number $a^{1/n}$ is called the **principal nth root of a**.

EXAMPLE 1

a. $16^{1/2} = 4$ because $4^2 = 16$.   Read $16^{1/2}$ as "the square root of 16."

b. $27^{1/3} = 3$ because $3^3 = 27$.   Read $27^{1/3}$ as "the cube root of 27."

c. $\left(\dfrac{1}{81}\right)^{1/4} = \dfrac{1}{3}$ because $\left(\dfrac{1}{3}\right)^4 = \dfrac{1}{81}$.   Read $\left(\dfrac{1}{81}\right)^{1/4}$ as "the fourth root of $\dfrac{1}{81}$."

d. $(64a^6)^{1/3} = 4a^2$ because $(4a^2)^3 = 64a^6$.

e. $-32^{1/5} = -(32^{1/5}) = -(2) = -2$ Read $32^{1/5}$ as "the fifth root of 32."

f. $(-25)^{1/2}$ is not a real number, because no real number squared is -25. In the exponential expression $a^{1/n}$, there is no real number nth root of a when a is negative and n is even. ∎

In the exponential expression $a^{1/n}$, we can remove the restriction that $a \geq 0$ when n is an odd natural number.

Rational Exponents with Odd-Number Denominators	If n is an odd natural number, then $a^{1/n}$ is the real number whose nth power is a. In symbols, $$(a^{1/n})^n = a$$

EXAMPLE 2
    a. $(-8)^{1/3} = -2$ because $(-2)^3 = -8$.
    b. $15,625^{1/6} = 5$ because $5^6 = 15,625$.
    c. $(-32x^5)^{1/5} = -2x$ because $(-2x)^5 = -32x^5$.
    d. $(256x^{16})^{1/4} = 4x^4$ because $(4x^4)^4 = 256x^{16}$. ∎

We summarize the definitions concerning $a^{1/n}$ as follows.

Summary of Definitions of $a^{1/n}$	If n is a natural number and a is a real number, then If $a > 0$, then $a^{1/n}$ is the positive number such that $(a^{1/n})^n = a$. If $a = 0$, then $a^{1/n} = 0$. If $a < 0$ $\begin{cases} \text{and } n \text{ is odd, then } a^{1/n} \text{ is the real number such that } (a^{1/n})^n = a. \\ \text{and } n \text{ is even, then } a^{1/n} \text{ is not a real number.} \end{cases}$

All positive numbers have two square roots. For example, the two square roots of 121 are 11 and -11, because 11^2 and $(-11)^2$ both equal 121. The symbol $121^{1/2}$ represents the positive square root of 121:

$$121^{1/2} = 11$$

and the symbol $-121^{1/2}$ represents the negative square root of 121:

$$-121^{1/2} = -(\mathbf{121^{1/2}}) = -\mathbf{11}$$

In general, if $a \neq 0$, the number a^2 has a and $-a$ as its two square roots. One of these square roots is positive and the other is negative. Because the expression $a^{1/2}$ represents the positive square root, we cannot write $(a^2)^{1/2} = a$ if a could be negative. Instead, we must write

$$(a^2)^{1/2} = |a|$$ The absolute value symbols are needed to guarantee that the square root is positive.

For example, if $a = -5$, we have

$$[(-5)^2]^{1/2} = |-5| = 5$$

EXAMPLE 3   **a.** If $x \geq 0$, find $(25x^2)^{1/2}$.   **b.** If $x < 0$, find $(36x^2)^{1/2}$.   If x can be any real number find   **c.** $(49x^2)^{1/2}$ and   **d.** $(x^4)^{1/2}$.

Solution   **a.** If $x \geq 0$, then

$$(25x^2)^{1/2} = 5x$$

Because $(5x)^2 = 25x^2$ and since $5x$ cannot be negative, absolute value symbols are not needed.

b. If $x < 0$, then

$$(36x^2)^{1/2} = -6x$$

Because $(-6x)^2 = 36x^2$ and because x is negative, $-6x$ is positive.

c. $(49x^2)^{1/2} = 7|x|$

Because x is unrestricted, $7x$ could be negative. In this case, we must use absolute value symbols to guarantee that the square root is nonnegative.

d. $(x^4)^{1/2} = x^2$

Because x^2 is always nonnegative, absolute value symbols are not needed.   ∎

The definition of $a^{1/n}$ can be extended to include fractional exponents whose numerators are not 1. For example, because of a power rule of exponents, $4^{3/2}$ can be written as either

$$(4^{1/2})^3 \qquad \text{or} \qquad (4^3)^{1/2}$$

In general, we have the following rule.

Theorem	If m and n are positive integers, $a^{1/n}$ is a real number, and the fraction $\frac{m}{n}$ is in lowest terms, then $$a^{m/n} = (a^{1/n})^m = (a^m)^{1/n}$$

Because of the previous rule, we can view the expression $a^{m/n}$ in two ways:

1. $a^{m/n}$ represents the mth power of the nth root of a: $(a^{1/n})^m$.
2. $a^{m/n}$ represents the nth root of the mth power of a: $(a^m)^{1/n}$.

For example, $16^{3/4}$ and $(-27)^{2/3}$ can be simplified in the following two ways:

$$16^{3/4} = (16^{1/4})^3 = 2^3 = 8 \quad \text{or} \quad 16^{3/4} = (16^3)^{1/4} = (4096)^{1/4} = 8$$
$$(-27)^{2/3} = [(-27)^{1/3}]^2 = (-3)^2 = 9 \quad \text{or} \quad (-27)^{2/3} = [(-27)^2]^{1/3} = (729)^{1/3} = 9$$

As these examples suggest, it is usually easier to take the root of the base first in order to avoid large numbers.

Negative Rational Exponents

If m and n are positive integers and $a \neq 0$, then

$$a^{-m/n} = \frac{1}{a^{m/n}} \quad \text{and} \quad \frac{1}{a^{-m/n}} = a^{m/n}$$

EXAMPLE 4

a. $25^{3/2} = (25^{1/2})^3 = 5^3 = 125$

b. $\left(-\dfrac{x^6}{1000}\right)^{2/3} = \left[\left(-\dfrac{x^6}{1000}\right)^{1/3}\right]^2 = \left(-\dfrac{x^2}{10}\right)^2 = \dfrac{x^4}{100}$

c. $32^{-2/5} = \dfrac{1}{32^{2/5}} = \dfrac{1}{(32^{1/5})^2} = \dfrac{1}{2^2} = \dfrac{1}{4}$

d. $\dfrac{1}{81^{-3/4}} = 81^{3/4} = (81^{1/4})^3 = 3^3 = 27$　■

EXAMPLE 5 Assume that all variables represent positive numbers. Write all answers without using negative exponents.

a. $(36x)^{1/2} = 36^{1/2}x^{1/2}$
$= 6x^{1/2}$

b. $\dfrac{(a^{1/3}b^{2/3})^6}{(y^3)^2} = \dfrac{a^{6/3}b^{12/3}}{y^6}$
$= \dfrac{a^2 b^4}{y^6}$

c. $\dfrac{b^{3/7}b^{2/7}}{b^{4/7}} = b^{3/7 + 2/7 - 4/7}$
$= b^{1/7}$

d. $\dfrac{a^{x/2}a^{x/4}}{a^{x/6}} = a^{6x/12 + 3x/12 - 2x/12}$
$= a^{7x/12}$

e. $\left[\dfrac{-c^{-2/5}}{c^{4/5}}\right]^{5/3} = (-c^{-2/5 - 4/5})^{5/3}$
$= [(-1)(c^{-6/5})]^{5/3}$
$= (-1)^{5/3}(c^{-6/5})^{5/3}$
$= (-1)c^{-30/15}$
$= -c^{-2}$
$= -\dfrac{1}{c^2}$

f. $\dfrac{(9r^2s)^{1/2}}{rs^{-3/2}} = \dfrac{(9r^2s)^{1/2}s^{3/2}}{r}$
$= \dfrac{9^{1/2}(r^2)^{1/2}s^{1/2}s^{3/2}}{r}$
$= \dfrac{3rs^{1/2 + 3/2}}{r}$
$= 3r^{1-1}s^{4/2}$
$= 3r^0 s^2$
$= 3s^2$　■

Radical Expressions

Another notation for expressing roots of numbers uses the **radical sign**.

Radicals

If n is a natural number greater than 1 and if $a^{1/n}$ is a real number, then

$$\sqrt[n]{a} = a^{1/n}$$

The symbol $\sqrt[n]{a}$ is called a **radical expression**. In this radical, the symbol $\sqrt{}$ is called the **radical sign**, a is called the **radicand**, and n is called the **index** (or the **order**) of the radical. If the order of a radical is 2, the expression is a **square root**, and we do not write the index. Thus,

$$\sqrt{a} = \sqrt[2]{a}$$

If the index of a radical is 3, the radical is a **cube root**.

We can restate the definition for the nth root of a number using radical notation.

Principal nth Root of a

If n is a natural number greater than 1 and $a \geq 0$, then $\sqrt[n]{a}$ is the nonnegative number whose nth power is a. In symbols,

$$(\sqrt[n]{a})^n = a$$

The nonnegative number $\sqrt[n]{a}$ is called the **principal nth root** of a.

If 2 is substituted for n in the equation $(\sqrt[n]{a})^n = a$, we have

$$(\sqrt[2]{a})^2 = (\sqrt{a})^2 = \sqrt{a}\sqrt{a} = a$$

Thus, if a number a can be factored into two equal factors, either of those factors is a square root of a. Likewise, if a can be factored into n equal factors, any of those factors is an nth root of a.

In the expression $\sqrt[n]{a}$, if n is an odd natural number greater than 1, we can remove the restriction that $a \geq 0$.

$\sqrt[n]{a}$ When n Is an Odd Natural Number

If n is an odd natural number greater than 1 and if a is any real number, then $\sqrt[n]{a}$ is the real number whose nth power is a. In symbols,

$$(\sqrt[n]{a})^n = a$$

EXAMPLE 6   **a.** $\sqrt{81} = 9$ because $9^2 = 81$.

b. $\sqrt[3]{-125} = -5$ because $(-5)^3 = -125$.

c. $-\sqrt[4]{256} = -(\sqrt[4]{256}) = -(4) = -4$

d. $-\sqrt[5]{-243} = -(\sqrt[5]{-243}) = -(-3) = 3$ ∎

We summarize the definitions concerning $\sqrt[n]{a}$ as follows.

Summary of Definitions of $\sqrt[n]{a}$	If n is a natural number greater than 1 and a is a real number, then If $a > 0$, then $\sqrt[n]{a}$ is the positive number such that $(\sqrt[n]{a})^n = a$. If $a = 0$, then $\sqrt[n]{a} = 0$. If $a < 0$ $\begin{cases}\text{and } n \text{ is odd, then } \sqrt[n]{a} \text{ is the real number such that } (\sqrt[n]{a})^n = a. \\ \text{and } n \text{ is even, then } \sqrt[n]{a} \text{ is not a real number.}\end{cases}$

We have seen that $a^{m/n} = (a^{1/n})^m = (a^m)^{1/n}$. This same fact, stated in radical notation, is

$$a^{m/n} = (\sqrt[n]{a})^m = \sqrt[n]{a^m}$$

Thus, the mth power of the nth root of a is the same as the nth root of mth power of a. For example, to find $\sqrt[3]{27^2}$, we can proceed in one of two ways:

$$\sqrt[3]{27^2} = (\sqrt[3]{27})^2 = 3^2 = 9 \qquad \text{or} \qquad \sqrt[3]{27^2} = \sqrt[3]{729} = 9$$

Either way, the result is the same.

The symbol $\sqrt{a^2}$ represents a nonnegative number. If a could be any real number, we must use absolute value symbols to guarantee that $\sqrt{a^2}$ is nonnegative. Thus, if a is not restricted,

$$\sqrt{a^2} = |a|$$

A similar argument holds when the index is any even natural number. The symbol $\sqrt[4]{a^4}$, for example, means the *positive* fourth root of a^4. Thus, if a is not restricted,

$$\sqrt[4]{a^4} = |a|$$

EXAMPLE 7   If x could be any real number, simplify   **a.** $\sqrt[6]{x^6}$,   **b.** $\sqrt[3]{x^3}$,   and   **c.** $\sqrt{x^8}$.

Solution   **a.** Since the symbol $\sqrt[6]{x^6}$ represents the positive sixth root of x^6, we must use absolute value symbols to guarantee that the result is nonnegative.

$$\sqrt[6]{x^6} = |x|$$

b. Because the index is odd, no absolute value symbols are needed.

$$\sqrt[3]{x^3} = x$$

c. Because x^4 is always positive, no absolute value symbols are needed.

$$\sqrt{x^8} = x^4$$

■

■ Simplifying Radicals

Many properties of exponents have counterparts in radical notation. For example, since $a^{1/n}b^{1/n} = (ab)^{1/n}$, we have

$$\sqrt[n]{a}\sqrt[n]{b} = \sqrt[n]{ab}$$

As long as all expressions represent real numbers, the product of two nth roots is equal to the nth root of their product.

Furthermore, since

$$\frac{a^{1/n}}{b^{1/n}} = \left(\frac{a}{b}\right)^{1/n}$$

it follows that

$$\frac{\sqrt[n]{a}}{\sqrt[n]{b}} = \sqrt[n]{\frac{a}{b}} \qquad (b \neq 0)$$

As long as all expressions represent real numbers, the quotient of two nth roots is equal to the nth root of their quotient.

Properties of Radicals	If $\sqrt[n]{a}$ and $\sqrt[n]{b}$ are real numbers, then
	1. $\sqrt[n]{ab} = \sqrt[n]{a}\,\sqrt[n]{b}$
	2. $\sqrt[n]{\dfrac{a}{b}} = \dfrac{\sqrt[n]{a}}{\sqrt[n]{b}} \qquad (b \neq 0)$

Warning! Properties 1 and 2 of radicals involve either the nth root of the product of two numbers or the nth root of the quotient of two numbers. There is no such property for sums or differences. For example, $\sqrt{9 + 4} \neq \sqrt{9} + \sqrt{4}$, because

$$\sqrt{9 + 4} = \sqrt{13} \qquad \text{but} \qquad \sqrt{9} + \sqrt{4} = 3 + 2 = 5$$

and $\sqrt{13} \neq 5$. In general,

$$\sqrt{a + b} \neq \sqrt{a} + \sqrt{b} \qquad \text{and} \qquad \sqrt{a - b} \neq \sqrt{a} - \sqrt{b}$$

Numbers such as 1, 4, 9, 16, 25, and 36 that are squares of positive integers are called **perfect squares**. Expressions such as x^2 and x^6 are also perfect squares, because each one is the square of another expression.

Numbers such as 1, 8, 27, 64, 125, and 216 that are cubes of positive integers are called **perfect cubes**. Likewise, x^3 and x^9 are perfect cubes. There are also perfect fourth powers, perfect fifth powers, and so on.

We can use these perfect powers and Property 1 of radicals to simplify many radical expressions. For example, to simplify the radical $\sqrt{12x^5}$, we factor $12x^5$ so that one factor is the largest perfect square that divides $12x^5$. In this case, the largest perfect square is $4x^4$. We then rewrite $12x^5$ as $4x^4 \cdot 3x$, use Property 1 of radicals, and simplify.

$$\sqrt{12x^5} = \sqrt{4x^4 \cdot 3x} = \sqrt{4x^4}\sqrt{3x} = 2x^2\sqrt{3x}$$

To simplify the radical $\sqrt[3]{432x^9y}$, we find the largest perfect cube factor of $432x^9y$, which is $216x^9$, and proceed as follows:

$$\sqrt[3]{432x^9y} = \sqrt[3]{216x^9 \cdot 2y} = \sqrt[3]{216x^9}\sqrt[3]{2y} = 6x^3\sqrt[3]{2y}$$

◼ Combining Radicals

Two radical expressions with the same index and the same radicand are called **like** or **similar radicals**. We can combine the like radicals in the expression $3\sqrt{2} + 2\sqrt{2}$ by using the distributive property.

$$3\sqrt{2} + 2\sqrt{2} = (3 + 2)\sqrt{2} = 5\sqrt{2}$$

Thus, to combine like radicals, we simply add their coefficients and keep the same radical.

When two radicals have the same index but different radicands, we can often change them to equivalent forms having the same radicand. We can then combine them. For example, to simplify the expression $\sqrt{27} - \sqrt{12}$, we simplify both radicals and combine like radicals.

$$\begin{aligned}
\sqrt{27} - \sqrt{12} &= \sqrt{9 \cdot 3} - \sqrt{4 \cdot 3} \\
&= \sqrt{9}\sqrt{3} - \sqrt{4}\sqrt{3} \\
&= 3\sqrt{3} - 2\sqrt{3} \\
&= \sqrt{3}
\end{aligned}$$

EXAMPLE 8 Simplify **a.** $\sqrt{50} + \sqrt{200}$, **b.** $3z\sqrt[5]{64z} - 2\sqrt[5]{2z^6}$, and **c.** $\dfrac{a\sqrt{16a} - \sqrt{a^3}}{\sqrt{9a} + \sqrt{36a}}$, for $a > 0$.

Solution **a.** $\begin{aligned}[t]
\sqrt{50} + \sqrt{200} &= \sqrt{25 \cdot 2} + \sqrt{100 \cdot 2} \\
&= \sqrt{25}\sqrt{2} + \sqrt{100}\sqrt{2} \\
&= 5\sqrt{2} + 10\sqrt{2} \\
&= 15\sqrt{2}
\end{aligned}$

b. $3z\sqrt[5]{64z} - 2\sqrt[5]{2z^6} = 3z\sqrt[5]{32 \cdot 2z} - 2\sqrt[5]{z^5 \cdot 2z}$

$$= 3z\sqrt[5]{32}\sqrt[5]{2z} - 2\sqrt[5]{z^5}\sqrt[5]{2z}$$

$$= 3z(2)\sqrt[5]{2z} - 2z\sqrt[5]{2z}$$

$$= 6z\sqrt[5]{2z} - 2z\sqrt[5]{2z}$$

$$= 4z\sqrt[5]{2z}$$

c. $\dfrac{a\sqrt{16a} - \sqrt{a^3}}{\sqrt{9a} + \sqrt{36a}} = \dfrac{a\sqrt{16}\sqrt{a} - \sqrt{a^2}\sqrt{a}}{\sqrt{9}\sqrt{a} + \sqrt{36}\sqrt{a}}$

$$= \dfrac{4a\sqrt{a} - a\sqrt{a}}{3\sqrt{a} + 6\sqrt{a}}$$

$$= \dfrac{3a\sqrt{a}}{9\sqrt{a}}$$

$$= \dfrac{a \cdot 3\sqrt{a}}{3 \cdot 3\sqrt{a}}$$

$$= \dfrac{a}{3} \qquad\qquad \dfrac{3\sqrt{a}}{3\sqrt{a}} = 1 \quad \text{because any nonzero number}$$
$$\text{divided by itself is 1.} \quad \blacksquare$$

■ Rationalizing Denominators and Numerators

We now consider radicals such as $\sqrt{\tfrac{5}{3}}$, whose radicands are fractions. Because of Property 2 of radicals, this radical can be written in the form $\dfrac{\sqrt{5}}{\sqrt{3}}$. There is a process, called **rationalizing the denominator**, that enables us to write this fraction as a fraction with a rational number in the denominator. All that we must do is multiply both the numerator and the denominator of the fraction by $\sqrt{3}$.

$$\frac{\sqrt{5}}{\sqrt{3}} = \frac{\sqrt{5}\sqrt{3}}{\sqrt{3}\sqrt{3}} = \frac{\sqrt{15}}{3}$$

To rationalize the numerator, we multiply the numerator and denominator by $\sqrt{5}$.

$$\frac{\sqrt{5}}{\sqrt{3}} = \frac{\sqrt{5}\sqrt{5}}{\sqrt{3}\sqrt{5}} = \frac{5}{\sqrt{15}}$$

EXAMPLE 9   Rationalize each denominator and simplify, whenever possible. Assume that all variables represent positive numbers.

   a. $\dfrac{1}{\sqrt{7}}$,   **b.** $\sqrt{\dfrac{3}{x}}$,   **c.** $\sqrt{\dfrac{3a^3}{5x^5}}$,   **d.** $\sqrt{\dfrac{1}{2}} + \sqrt{\dfrac{1}{8}}$,   **e.** $\sqrt[3]{\dfrac{1}{y^2}} - \sqrt[3]{\dfrac{16}{x^4}}$

Solution   **a.** $\dfrac{1}{\sqrt{7}} = \dfrac{1\sqrt{7}}{\sqrt{7}\sqrt{7}} = \dfrac{\sqrt{7}}{7}$

b. $\sqrt{\dfrac{3}{x}} = \dfrac{\sqrt{3}}{\sqrt{x}} = \dfrac{\sqrt{3}\sqrt{x}}{\sqrt{x}\sqrt{x}} = \dfrac{\sqrt{3x}}{x}$

c. $\sqrt{\dfrac{3a^3}{5x^5}} = \dfrac{\sqrt{3a^3}}{\sqrt{5x^5}}$

$\quad = \dfrac{\sqrt{3a^3}\sqrt{5x}}{\sqrt{5x^5}\sqrt{5x}}$

$\quad = \dfrac{\sqrt{15a^3x}}{\sqrt{25x^6}}$

$\quad = \dfrac{\sqrt{a^2}\sqrt{15ax}}{5x^3}$

$\quad = \dfrac{a\sqrt{15ax}}{5x^3}$

d. $\sqrt{\dfrac{1}{2}} + \sqrt{\dfrac{1}{8}} = \dfrac{1}{\sqrt{2}} + \dfrac{1}{\sqrt{8}}$

$\quad = \dfrac{1\sqrt{2}}{\sqrt{2}\sqrt{2}} + \dfrac{1\sqrt{2}}{\sqrt{8}\sqrt{2}}$

$\quad = \dfrac{\sqrt{2}}{\sqrt{4}} + \dfrac{\sqrt{2}}{\sqrt{16}}$

$\quad = \dfrac{\sqrt{2}}{2} + \dfrac{\sqrt{2}}{4}$

$\quad = \dfrac{3\sqrt{2}}{4}$

e. $\sqrt[3]{\dfrac{1}{y^2}} - \sqrt[3]{\dfrac{16}{x^4}} = \dfrac{\sqrt[3]{1}}{\sqrt[3]{y^2}} - \dfrac{\sqrt[3]{16}}{\sqrt[3]{x^4}}$

$\quad = \dfrac{1}{\sqrt[3]{y^2}} - \dfrac{2\sqrt[3]{2}}{\sqrt[3]{x^4}}$ $\sqrt[3]{16} = \sqrt[3]{8}\sqrt[3]{2} = 2\sqrt[3]{2}.$

$\quad = \dfrac{1\sqrt[3]{y}}{\sqrt[3]{y^2}\sqrt[3]{y}} - \dfrac{2\sqrt[3]{2}\sqrt[3]{x^2}}{\sqrt[3]{x^4}\sqrt[3]{x^2}}$

$\quad = \dfrac{\sqrt[3]{y}}{\sqrt[3]{y^3}} - \dfrac{2\sqrt[3]{2x^2}}{\sqrt[3]{x^6}}$

$\quad = \dfrac{\sqrt[3]{y}}{y} - \dfrac{2\sqrt[3]{2x^2}}{x^2}$

∎

Another property of radicals can be derived from the properties of exponents. If all of the expressions represent real numbers, then

$$\sqrt[n]{\sqrt[m]{x}} = \sqrt[n]{x^{1/m}} = (x^{1/m})^{1/n} = x^{1/(mn)} = \sqrt[mn]{x}$$

Similarly,

$$\sqrt[m]{\sqrt[n]{x}} = \sqrt[mn]{x}$$

These results are summarized in the following theorem.

Theorem   If all of the expressions involved represent real numbers, then

$$\sqrt[m]{\sqrt[n]{x}} = \sqrt[n]{\sqrt[m]{x}} = \sqrt[mn]{x}$$

We can use the previous theorem to simplify many radicals. For example,

$$\sqrt[3]{\sqrt{8}} = \sqrt{\sqrt[3]{8}} = \sqrt{2}$$

Fractional exponents can be used to simplify many radical expressions, as shown in the following example.

EXAMPLE 10   Assume that x and y are positive numbers and simplify: **a.** $\sqrt[6]{4}$, **b.** $\sqrt[12]{x^3}$, and **c.** $\sqrt[9]{8y^3}$.

Solution   **a.** $\sqrt[6]{4} = 4^{1/6} = (2^2)^{1/6} = 2^{1/3} = \sqrt[3]{2}$ **b.** $\sqrt[12]{x^3} = x^{3/12} = x^{1/4} = \sqrt[4]{x}$

c. $\sqrt[9]{8y^3} = (2^3y^3)^{1/9} = (2y)^{3/9} = (2y)^{1/3} = \sqrt[3]{2y}$ ■

■ Multiplying and Dividing Radicals with Different Indexes

We can often multiply radicals with different indexes. For example, to multiply $\sqrt{3}$ by $\sqrt[3]{5}$, we first write each radical as a sixth root, because 6 is the smallest number divisible by 2 and 3.

$$\sqrt{3} = 3^{1/2} = 3^{3/6} = \sqrt[6]{3^3} = \sqrt[6]{27}$$
$$\sqrt[3]{5} = 5^{1/3} = 5^{2/6} = \sqrt[6]{5^2} = \sqrt[6]{25}$$

and then multiply the sixth roots.

$$\sqrt{3}\,\sqrt[3]{5} = \sqrt[6]{27}\,\sqrt[6]{25} = \sqrt[6]{(27)(25)} = \sqrt[6]{675}$$

EXAMPLE 11   Simplify $\dfrac{\sqrt[3]{3}}{\sqrt{5}}$.

Solution $\dfrac{\sqrt[3]{3}}{\sqrt{5}} = \dfrac{\sqrt[3]{3}\,\sqrt{5}}{\sqrt{5}\,\sqrt{5}} = \dfrac{3^{1/3}5^{1/2}}{5} = \dfrac{3^{2/6}5^{3/6}}{5} = \dfrac{\sqrt[6]{(3^2)(5^3)}}{5} = \dfrac{\sqrt[6]{1125}}{5}$ ■

1.3   EXERCISES

In Exercises 1–12, simplify each expression. Assume that all variables represent positive numbers.

1. $9^{1/2}$

2. $8^{1/3}$

3. $-16^{1/4}$

4. $-625^{1/4}$

5. $\left(\dfrac{1}{25}\right)^{1/2}$

6. $\left(-\dfrac{8}{27}\right)^{1/3}$

7. $(81r^8)^{1/4}$

8. $(243a^5)^{1/5}$

9. $(-64p^6)^{1/3}$

10. $(-125q^9)^{1/3}$

11. $\left(\dfrac{16a^4}{25b^2}\right)^{1/2}$

12. $\left(\dfrac{49t^8}{100z^4}\right)^{1/2}$

In Exercises 13–20, simplify each expression. Assume that all variables are unrestricted.

13. $(16a^2)^{1/2}$ **14.** $(25a^4)^{1/2}$ **15.** $(16a^{12})^{1/4}$ **16.** $(-64a^3)^{1/3}$

17. $(-32a^5)^{1/5}$ **18.** $(64a^6)^{1/6}$ **19.** $(-216b^6)^{1/3}$ **20.** $(256t^8)^{1/4}$

In Exercises 21–36, simplify each expression. Write all answers without using negative exponents.

21. $4^{3/2}$ **22.** $8^{2/3}$ **23.** $-16^{3/4}$ **24.** $(-8)^{2/3}$

25. $-1000^{2/3}$ **26.** $100^{3/2}$ **27.** $64^{-1/2}$ **28.** $25^{-1/2}$

29. $64^{-3/2}$ **30.** $49^{-3/2}$ **31.** $-9^{-3/2}$ **32.** $(-27)^{-2/3}$

33. $\left(\dfrac{4}{9}\right)^{5/2}$ **34.** $\left(\dfrac{25}{81}\right)^{3/2}$ **35.** $\left(-\dfrac{27}{64}\right)^{-2/3}$ **36.** $\left(\dfrac{125}{8}\right)^{-4/3}$

In Exercises 37–56, simplify each expression. Write all answers without using negative exponents. Assume that all variables represent positive numbers.

37. $(100s^4)^{1/2}$ **38.** $(64u^6v^3)^{1/3}$ **39.** $(32y^{10}z^5)^{-1/5}$ **40.** $(625a^4b^8)^{-1/4}$

41. $(x^{10}y^5)^{3/5}$ **42.** $(64a^6b^{12})^{5/6}$ **43.** $(r^8s^{16})^{-3/4}$ **44.** $(-8x^9y^{12})^{-2/3}$

45. $\left(-\dfrac{8a^6}{125b^9}\right)^{2/3}$ **46.** $\left(\dfrac{16x^4}{625y^8}\right)^{3/4}$ **47.** $\left(\dfrac{27r^6}{1000s^{12}}\right)^{-2/3}$ **48.** $\left(-\dfrac{32m^{10}}{243n^{15}}\right)^{-2/5}$

49. $\left(\dfrac{a^8c^{24}}{b^{20}d^0}\right)^{0.25}$ **50.** $\left(\dfrac{u^{10}v^0}{z^5}\right)^{0.2}$ **51.** $\dfrac{a^{2/5}a^{4/5}}{a^{1/5}}$ **52.** $\dfrac{x^{6/7}x^{3/7}}{x^{2/7}x^{5/7}}$

53. $\dfrac{(16a^{-6}s^2)^{1/4}}{a^{3/2}s^{-1/2}}$ **54.** $\dfrac{(64x^{11/3}y^{-10/3})^{1/3}}{(27x^{-2/3}y^{1/3})^{-1/3}}$ **55.** $\dfrac{b^{x/3}b^{x/4}}{b^{x/8}}$ **56.** $\dfrac{c^{x/3}c^{x/6}}{c^{x/12}}$

In Exercises 57–72, simplify each radical. Assume that all variables represent positive numbers.

57. $\sqrt{49}$ **58.** $\sqrt{81}$ **59.** $\sqrt[3]{125}$ **60.** $\sqrt[3]{-64}$

61. $-\sqrt[4]{81}$ **62.** $\sqrt[5]{-243}$ **63.** $\sqrt[5]{-\dfrac{32}{100,000}}$ **64.** $\sqrt[4]{\dfrac{256}{625}}$

65. $\sqrt{36x^2}$ **66.** $-\sqrt{25y^6}$ **67.** $\sqrt{x^2y^4}$ **68.** $\sqrt{a^4b^8}$

69. $\sqrt[3]{8y^3}$ **70.** $\sqrt[3]{-27z^9}$ **71.** $\sqrt[4]{\dfrac{x^4y^8}{z^{12}}}$ **72.** $\sqrt[5]{\dfrac{a^{10}b^5}{c^{15}}}$

In Exercises 73–80, assume that all variables are unrestricted. Simplify each expression and use absolute value symbols when appropriate.

73. $\sqrt{4x^2}$ **74.** $\sqrt{9a^4}$ **75.** $-\sqrt{16b^4}$ **76.** $-\sqrt{4a^2}$

77. $\sqrt[3]{27x^3}$ **78.** $\sqrt[4]{16x^4}$ **79.** $\sqrt[6]{x^6y^{12}}$ **80.** $\sqrt[8]{a^{16}b^{32}c^{64}}$

In Exercises 81–96, simplify each expression. Assume that all variables represent positive numbers.

81. $\sqrt{8}-\sqrt{2}$ **82.** $\sqrt{75}-2\sqrt{27}$ **83.** $\sqrt{200x^2}+\sqrt{98x^2}$ **84.** $\sqrt{128a^3}-a\sqrt{162a}$

85. $2\sqrt{48y^5}-3y\sqrt{12y^3}$ **86.** $y\sqrt{112y}+4\sqrt{175y^3}$ **87.** $2\sqrt[3]{81}+3\sqrt[3]{24}$

88. $3\sqrt[4]{32} - 2\sqrt[4]{162}$

89. $\sqrt[4]{768z^5} + \sqrt[4]{48z^5}$

90. $-2\sqrt[5]{64y^2} + 3\sqrt[5]{486y^2}$

91. $\sqrt{8x^2y} - x\sqrt{2y} + \sqrt{50x^2y}$

92. $3x\sqrt{18x} + 2\sqrt{2x^3} - \sqrt{72x^3}$

93. $\sqrt[3]{16xy^4} + y\sqrt[3]{2xy} - \sqrt[3]{54xy^4}$

94. $\sqrt[4]{512x^5} - \sqrt[4]{32x^5} + \sqrt[4]{1250x^5}$

95. $\dfrac{3x\sqrt{2x} - 2\sqrt{2x^3}}{\sqrt{18x} - \sqrt{2x}}$

96. $\dfrac{4x\sqrt{3x} + \sqrt{12x^3}}{\sqrt{27x} + \sqrt{12x}}$

In Exercises 97–114, assume that all variables are positive numbers. Rationalize each denominator and simplify.

97. $\dfrac{3}{\sqrt{3}}$

98. $\dfrac{10}{\sqrt{5}}$

99. $\dfrac{2}{\sqrt{x}}$

100. $\dfrac{3}{\sqrt{y}}$

101. $\dfrac{2}{\sqrt[3]{2}}$

102. $\dfrac{3}{\sqrt[3]{9}}$

103. $\dfrac{5a}{\sqrt[3]{25a}}$

104. $\dfrac{16}{\sqrt[3]{16b^2}}$

105. $\dfrac{2b}{\sqrt[4]{3a^2}}$

106. $\dfrac{64a}{\sqrt[5]{16b^3}}$

107. $\sqrt{\dfrac{x}{2y}}$

108. $\sqrt{\dfrac{57}{3x}}$

109. $\sqrt[3]{\dfrac{2u^4}{9v}}$

110. $\sqrt[3]{-\dfrac{3s^5}{4r^2}}$

111. $\sqrt{\dfrac{1}{3}} - \sqrt{\dfrac{1}{27}}$

112. $\sqrt{\dfrac{1}{5}} + \sqrt{\dfrac{2}{5}}$

113. $\sqrt{\dfrac{x}{8}} - \sqrt{\dfrac{x}{2}} + \sqrt{\dfrac{x}{32}}$

114. $\sqrt[3]{\dfrac{y}{4}} + \sqrt[3]{\dfrac{y}{32}} - \sqrt[3]{\dfrac{y}{500}}$

In Exercises 115–120, simplify each expression. Assume that all variables represent positive numbers.

115. $\dfrac{\sqrt[3]{x^6y^3}}{\sqrt{9x^2y}}$

116. $\dfrac{\sqrt[6]{128x^{-9}y^8z^7}}{\sqrt[6]{x^6yz^4}}$

117. $\dfrac{\sqrt[4]{81x^{-6}y^8}}{\sqrt[4]{x^{-2}y^4}}$

118. $\dfrac{\sqrt[4]{625x^8y^4}}{\sqrt{16x^3y^2}}$

119. $\dfrac{\sqrt{x^{12}y^{12}}}{\sqrt[5]{32x^6y^{11}}}$

120. $\dfrac{\sqrt[3]{-27x^{10}y^{14}}}{\sqrt[3]{64x^7y^2}}$

In Exercises 121–124, simplify each radical.

121. $\sqrt[4]{9}$

122. $\sqrt[6]{27}$

123. $\sqrt[10]{16x^6}$

124. $\sqrt[6]{27x^9}$

In Exercises 125–128, write each expression as a single radical.

125. $\sqrt{2}\sqrt[3]{2}$

126. $\sqrt{3}\sqrt[3]{5}$

127. $\dfrac{\sqrt[4]{3}}{\sqrt{2}}$

128. $\dfrac{\sqrt[3]{2}}{\sqrt{5}}$

129. For what values of x does $\sqrt{x^2} = x$?

130. For what values of x does $\sqrt[3]{x^3} = x$?

131. For what values of x does $\sqrt[4]{x^4} = -x$?

132. If all of the radicals involved represent real numbers and $y \neq 0$, prove that

$$\sqrt[n]{\dfrac{x}{y}} = \dfrac{\sqrt[n]{x}}{\sqrt[n]{y}}$$

133. If all of the radicals involved represent real numbers and there is no division by 0, prove that

$$\left(\frac{x}{y}\right)^{-m/n} = \sqrt[n]{\frac{y^m}{x^m}}$$

134. The definition of $x^{m/n}$ requires that $x^{1/n}$ be a real number. Why is this important? (*Hint:* Consider what happens when n is even, m is odd, and x is negative.)

 In Exercises 135–140, let $x = 3.5$, $y = 1.2$, and $z = 1.4$. Use a calculator to verify each statement.

135. $(xy)^z = x^z y^z$

136. $\left(\dfrac{x}{y}\right)^z = \dfrac{x^z}{y^z}$

137. $(x^y)^z = (x^z)^y$

138. $(x + y)^z \neq x^z + y^z$

139. $(z^{1/x})^{1/y} = z^{1/(xy)}$

140. $(x^y)^z = x^{yz}$

1.4 ARITHMETIC OF POLYNOMIALS

- Adding and Subtracting Polynomials ■ Multiplying Polynomials
- Conjugate Binomials ■ Dividing Polynomials

A **monomial** is a number or the product of a number and one or more variables with whole-number exponents. The number is called the **numerical coefficient** or just the **coefficient**. Some examples of monomials are

$$3x, \qquad 7ab^2, \qquad -5ab^2c^4, \qquad x^3, \quad \text{and} \quad -12$$

with coefficients of 3, 7, -5, 1, and -12, respectively.

The **degree** of a monomial is the sum of the exponents of its variables. The degree of $3x$ is 1, the degree of $7ab^2$ is 3, the degree of $-5ab^2c^4$ is 7, the degree of x^3 is 3, and the degree of -12 is 0 (because $-12 = -12x^0$). All nonzero constants have a degree of 0. However, the constant 0 does not have a defined degree.

A single monomial or a finite sum of monomials is called a **polynomial**. Each of the monomials in that sum is called a **term** of the polynomial. A polynomial with two terms is called a **binomial**, and a polynomial with three terms is called a **trinomial**.

The **degree of a polynomial** is the degree of the term in the polynomial with highest degree. The only polynomial with no defined degree is 0, which is called the **zero polynomial**. Some examples of polynomials are

- $3x^2y^3 + 5xy^2 + 7$ is a trinomial of 5th degree, because the degree of its term with highest degree (the first term) is 5.
- $3ab + 5a^2b$ is a binomial of degree 3.
- $5x + 3y^2 + \sqrt[3]{3} \, z^4 - \sqrt{7}$ is a polynomial, because its variables have whole-number exponents. It is of degree 4.

- $-7y^{1/2} + 3y^2 + \sqrt[5]{3}\,z$ is not a polynomial, because some of its variables do not have whole-number exponents.

If two terms of a polynomial have the same variables with the same exponents, they are called **like** or **similar terms**. For example, to combine the like terms in the binomial $3x^2y + 5x^2y$, we use the distributive property:

$$3x^2y + 5x^2y = (3 + 5)x^2y = 8x^2y$$

Thus, to combine like terms, we add their numerical coefficients and keep the same variables with the same exponents.

■ Adding and Subtracting Polynomials

To add and subtract polynomials, we often use the **extended distributive property**:

$$a(b + c + d + e + \cdots) = ab + ac + ad + ae + \cdots$$

With this property, we can remove parentheses enclosing several terms of a polynomial. When the sign preceding the parentheses is $+$, we simply drop the parentheses:

$$+(a + b - c) = +1(a + b - c) = 1a + 1b - 1c = a + b - c$$

Polynomials are added by removing parentheses, if necessary, and then combining any like terms that are contained within the polynomials.

EXAMPLE 1 Perform the addition $(3x^3y + 5x^2 - 2y) + (2x^3y - 5x^2 + 3x)$.

Solution To add the polynomials, we remove parentheses and combine like terms. Because the terms in each polynomial represent real numbers, we can use the commutative and associative properties of real numbers to rearrange and regroup terms.

$$\begin{aligned}
(3x^3y &+ 5x^2 - 2y) + (2x^3y - 5x^2 + 3x) \\
&= 3x^3y + 5x^2 - 2y + 2x^3y - 5x^2 + 3x \\
&= 3x^3y + 2x^3y + 5x^2 - 5x^2 - 2y + 3x \\
&= 5x^3y - 2y + 3x
\end{aligned}$$ ■

To remove parentheses enclosing several terms of a polynomial when the sign preceding the parentheses is $-$, we simply drop the parentheses and the $-$ sign and change the sign of each term within the parentheses.

$$-(a + b - c) = -1(a + b - c) = -1a + (-1)b - (-1)c = -a - b + c$$

This suggests that the way to subtract polynomials is to remove parentheses and combine like terms.

EXAMPLE 2 Perform the subtraction $(2x^2 + 3y^2) - (x^2 - 2y^2 + 7)$.

Solution To subtract the polynomials, we remove parentheses and combine like terms.

$$\begin{aligned}
(2x^2 &+ 3y^2) - (x^2 - 2y^2 + 7) \\
&= 2x^2 + 3y^2 - x^2 + 2y^2 - 7
\end{aligned}$$

$$= 2x^2 - x^2 + 3y^2 + 2y^2 - 7$$
$$= x^2 + 5y^2 - 7 \qquad\qquad\blacksquare$$

To remove parentheses enclosing several terms that are multiplied by a constant, we multiply each term within the parentheses by the constant and drop the parentheses. For example,

$$4(3x^2 - 2x + 6) = 4(3x^2) - 4(2x) + 4(6)$$
$$= 12x^2 - 8x + 24$$

Thus, to add multiples of one polynomial to another, or to subtract multiples of one polynomial from another, we remove parentheses and combine like terms.

EXAMPLE 3 Simplify the expression $7x(2y^2 + 13x^2) - 5(xy^2 - 13x^3)$.

Solution
$$7x(2y^2 + 13x^2) - 5(xy^2 - 13x^3)$$
$$= 14xy^2 + 91x^3 - 5xy^2 + 65x^3$$
$$= 14xy^2 - 5xy^2 + 91x^3 + 65x^3$$
$$= 9xy^2 + 156x^3 \qquad\qquad\blacksquare$$

■ Multiplying Polynomials

To find the product of two monomials such as $3x^2y^3z$ and $5xyz^2$, we can proceed as follows:

$$(3x^2y^3z)(5xyz^2) = 3 \cdot x^2 \cdot y^3 \cdot z \cdot 5 \cdot x \cdot y \cdot z^2$$
$$= 3 \cdot 5 \cdot x^2 \cdot x \cdot y^3 \cdot y \cdot z \cdot z^2$$
$$= 15x^3y^4z^3$$

Thus, to multiply two monomials, we first multiply the numerical coefficients and then multiply the variables.

To find the product of a monomial and a polynomial, we use the distributive property or the extended distributive property. For example,

$$3xy^2(2xy + x^2 - 7yz) = (3xy^2)(2xy) + (3xy^2)(x^2) - (3xy^2)(7yz)$$
$$= 6x^2y^3 + 3x^3y^2 - 21xy^3z$$

Thus, to multiply a polynomial by a monomial, we multiply each term of the polynomial by the monomial.

To multiply one binomial by another, we use the distributive property twice.

EXAMPLE 4 Find the products **a.** $(x + y)(x + y)$, **b.** $(x - y)(x - y)$, and **c.** $(x + y)(x - y)$.

Solution **a.** $(x + y)(x + y) = (x + y)x + (x + y)y$
$$= x^2 + xy + xy + y^2$$
$$= x^2 + 2xy + y^2$$
b. $(x - y)(x - y) = (x - y)x - (x - y)y$
$$= x^2 - xy - xy + y^2$$
$$= x^2 - 2xy + y^2$$

c. $(x + y)(x - y) = (x + y)x - (x + y)y$
$$= x^2 + xy - xy - y^2$$
$$= x^2 - y^2 \qquad \blacksquare$$

The products in Example 4 are **special products**. Because they occur so often, it is worthwhile to learn their forms.

Special Product Formulas	$(x + y)^2 = (x + y)(x + y) = x^2 + 2xy + y^2$ $(x - y)^2 = (x - y)(x - y) = x^2 - 2xy + y^2$ $(x + y)(x - y) = x^2 - y^2$

 Warning! Remember that $(x + y)^2$ and $(x - y)^2$ have trinomials for their products and that

$$(x + y)^2 \neq x^2 + y^2 \qquad \text{and} \qquad (x - y)^2 \neq x^2 - y^2$$

The results of Example 4 suggest that to multiply one binomial by another, we multiply each term of one binomial by each term of the other binomial and combine like terms, when possible.

We can use a shortcut method, called the **FOIL method**, to multiply one binomial by another. The word **FOIL** is an acronym for **F**irst terms, **O**uter terms, **I**nner terms, and **L**ast terms. To use this method to multiply $(3x - 4)$ by $(2x + 5)$, we write

First terms Last terms

$$(3x - 4)(2x + 5) = 3x(2x) + 3x(5) - 4(2x) - 4(5)$$

Inner terms

Outer terms

$$= 6x^2 + 15x - 8x - 20$$
$$= 6x^2 + 7x - 20$$

In this example, the product of the first terms is $6x^2$, the product of the outer terms is $15x$, the product of the inner terms is $-8x$, and the product of the last terms is -20. The resulting terms of the product are then combined.

EXAMPLE 5   Use the FOIL method to find the product $(\sqrt{3} + x)(2 - \sqrt{3}\,x)$.

Solution
$$(\sqrt{3} + x)(2 - \sqrt{3}\,x) = 2\sqrt{3} - \sqrt{3}\sqrt{3}\,x + 2x - x\sqrt{3}\,x$$
$$= 2\sqrt{3} - 3x + 2x - \sqrt{3}\,x^2$$
$$= 2\sqrt{3} - x - \sqrt{3}\,x^2 \qquad \blacksquare$$

To multiply a polynomial with more than two terms by another polynomial, we multiply each term of one polynomial by each term of the other polynomial and combine like terms whenever possible.

EXAMPLE 6   Find the products   **a.** $(x + y)(x^2 - xy + y^2)$ and   **b.** $(x + y)^3$.

Solution   **a.** $(x + y)(x^2 - xy + y^2) = x^3 - x^2y + xy^2 + yx^2 - xy^2 + y^3$
$$= x^3 + y^3$$

b. $(x + 3)^3 = (x + 3)(x + 3)^2$
$$= (x + 3)(x^2 + 6x + 9)$$
$$= x^3 + 6x^2 + 9x + 3x^2 + 18x + 27$$
$$= x^3 + 9x^2 + 27x + 27$$ ∎

If n is a whole number, the expressions $a^n + 1$ and $2a^n - 3$ are polynomials, and we can use the FOIL method to multiply them together.

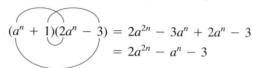

$$(a^n + 1)(2a^n - 3) = 2a^{2n} - 3a^n + 2a^n - 3$$
$$= 2a^{2n} - a^n - 3$$

We can also use the methods previously discussed to multiply expressions such as $x^{-2} + y$ and $x^2 - y^{-1}$ that are not polynomials.

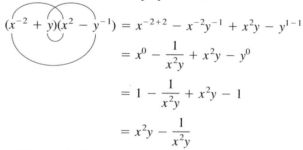

$$(x^{-2} + y)(x^2 - y^{-1}) = x^{-2+2} - x^{-2}y^{-1} + x^2y - y^{1-1}$$
$$= x^0 - \frac{1}{x^2y} + x^2y - y^0$$
$$= 1 - \frac{1}{x^2y} + x^2y - 1$$
$$= x^2y - \frac{1}{x^2y}$$

■ **Conjugate Binomials**

If the denominator of a fraction is a binomial containing radicals, we can use the product formula $(x + y)(x - y)$ to rationalize its denominator. For example, to rationalize the denominator of the fraction $\dfrac{6}{\sqrt{7} + 2}$ we multiply both the numerator and the denominator of the fraction by $\sqrt{7} - 2$ and simplify.

$$\frac{6}{\sqrt{7} + 2} = \frac{6(\sqrt{7} - 2)}{(\sqrt{7} + 2)(\sqrt{7} - 2)} = \frac{6(\sqrt{7} - 2)}{7 - 4} = \frac{6(\sqrt{7} - 2)}{3} = 2(\sqrt{7} - 2)$$

In this example, we multiplied both the numerator and the denominator of the given fraction by $\sqrt{7} - 2$. This binomial is the same as the denominator of the given fraction $\sqrt{7} + 2$, except for the sign. Such binomials are called **conjugate binomials**.

Conjugate Binomials	**Conjugate binomials** are binomials that are the same except for the sign between their terms. The conjugate of $a + b$ is $a - b$, and the conjugate of $a - b$ is $a + b$.

Amalie Noether (1882–1935)
Albert Einstein described Noether as the most creative female mathematical genius since the beginning of higher education for women. Her work was in the area of abstract algebra. Although she received a doctoral degree in mathematics, she was denied a mathematics position in Germany because she was a woman.

EXAMPLE 7 Simplify $\dfrac{\sqrt{3x} - \sqrt{2}}{\sqrt{3x} + \sqrt{2}}$ $(x > 0)$ by rationalizing the denominator.

Solution We multiply both the numerator and the denominator by $\sqrt{3x} - \sqrt{2}$ (which is the conjugate of $\sqrt{3x} + \sqrt{2}$) and simplify.

$$\frac{\sqrt{3x} - \sqrt{2}}{\sqrt{3x} + \sqrt{2}} = \frac{(\sqrt{3x} - \sqrt{2})(\sqrt{3x} - \sqrt{2})}{(\sqrt{3x} + \sqrt{2})(\sqrt{3x} - \sqrt{2})}$$

$$= \frac{\sqrt{3x}\sqrt{3x} - \sqrt{3x}\sqrt{2} - \sqrt{2}\sqrt{3x} + \sqrt{2}\sqrt{2}}{(\sqrt{3x})^2 - (\sqrt{2})^2}$$

$$= \frac{3x - \sqrt{6x} - \sqrt{6x} + 2}{3x - 2}$$

$$= \frac{3x - 2\sqrt{6x} + 2}{3x - 2}$$

∎

In calculus, we often rationalize a numerator.

EXAMPLE 8 Rationalize the numerator of the fraction $\dfrac{\sqrt{x + h} - \sqrt{x}}{h}$.

Solution We can multiply both the numerator and the denominator by the conjugate of the numerator and simplify.

$$\frac{\sqrt{x + h} - \sqrt{x}}{h} = \frac{(\sqrt{x + h} - \sqrt{x})(\sqrt{x + h} + \sqrt{x})}{h(\sqrt{x + h} + \sqrt{x})}$$

$$= \frac{x + h - x}{h(\sqrt{x + h} + \sqrt{x})}$$

$$= \frac{h}{h(\sqrt{x + h} + \sqrt{x})}$$

$$= \frac{1}{\sqrt{x + h} + \sqrt{x}} \qquad \text{Divide out the common factor of } h.$$ ∎

■ Dividing Polynomials

To divide a monomial by a monomial, we write the quotient of the monomials as a fraction and simplify the fraction by using the rules of exponents. For example,

$$\frac{6x^2y^3}{-2x^3y} = -3x^{2-3}y^{3-1}$$

$$= -3x^{-1}y^2$$

$$= -\frac{3y^2}{x}$$

To divide a polynomial by a monomial, we write the quotient of the polynomial and the monomial as a fraction, write the fraction as a sum of separate fractions,

Gottfried Wilhelm Leibniz
(1646–1716)
Leibniz, a German philosopher and logician, is principally known as one of the inventors of calculus, along with Newton. He also developed the binary numeration system, which is basic to modern computers.

and simplify each separate fraction. For example, to divide $8x^5y^4 + 12x^2y^5 - 16x^2y^3$ by $4x^3y^4$, we proceed as follows:

$$\frac{8x^5y^4 + 12x^2y^5 - 16x^2y^3}{4x^3y^4} = \frac{8x^5y^4}{4x^3y^4} + \frac{12x^2y^5}{4x^3y^4} + \frac{-16x^2y^3}{4x^3y^4}$$

$$= 2x^2 + \frac{3y}{x} - \frac{4}{xy}$$

To divide two polynomials, we can use a process similar to long division of whole numbers. To illustrate, we consider the division

$$\frac{2x^2 + 11x - 30}{x + 7}$$

which can be written in long division form as

$$x + 7)\overline{2x^2 + 11x - 30}$$

The binomial $x + 7$ is called the **divisor**, and the trinomial $2x^2 + 11x - 30$ is called the **dividend**. The final answer, called the **quotient**, will appear above the long division symbol.

We begin the division by asking, "What monomial when multiplied by x gives $2x^2$?" Because $x \cdot 2x = 2x^2$, the answer is $2x$. We place $2x$ in the quotient, multiply each term of the divisor by $2x$, subtract, and bring down the -30.

$$\begin{array}{r} 2x \\ x + 7)\overline{2x^2 + 11x - 30} \\ \underline{2x^2 + 14x} \\ -3x - 30 \end{array}$$

We continue the division by asking "What monomial when multiplied by x gives $-3x$?" We place the answer, -3, in the quotient, multiply each term of the divisor by -3, and subtract. This time, there is no number to bring down.

$$\begin{array}{r} 2x -3 \\ x + 7)\overline{2x^2 + 11x - 30} \\ \underline{2x^2 + 14x} \\ -3x - 30 \\ \underline{-3x - 21} \\ -9 \end{array}$$

Because the degree of the remainder, -9, is less than the degree of the divisor, the division process stops, and we can express the result in *quotient* $+ \frac{remainder}{divisor}$ form. Thus,

1. $$\frac{2x^2 + 11x - 30}{x + 7} = 2x - 3 + \frac{-9}{x + 7}$$

If we denote $2x^2 + 11x - 30$ by $R(x)$, $x + 7$ by $S(x)$, $2x - 3$ by $q(x)$, and -9 by $r(x)$ in Equation 1, we can write the equation in the form

$$\frac{R(x)}{S(x)} = q(x) + \frac{r(x)}{S(x)} \qquad (S(x) \neq 0)$$

After multiplying both sides by $S(x)$, we have

2. $\qquad R(x) = S(x)q(x) + r(x) \qquad (S(x) \neq 0)$

Equation 2 illustrates a well-known theorem called the **division algorithm**.

The Division Algorithm	If $R(x)$ and $S(x)$ are polynomials, there are unique polynomials $q(x)$ and $r(x)$ such that

$$R(x) = S(x)q(x) + r(x) \qquad (S(x) \neq 0)$$

where $r(x) = 0$ or is of lower degree than $S(x)$.

The polynomial $q(x)$ is the **quotient**, and the polynomial $r(x)$ is the **remainder** in the division of $R(x)$ by $S(x)$.

When the remainder $r(x) = 0$, we say that the polynomial $S(x)$ divides the polynomial $R(x)$ evenly.

EXAMPLE 9 Divide $6x^3 - 11$ by $2x + 2$.

Solution We set up the division, leaving spaces for the missing powers of x in the dividend.

$$2x + 2 \overline{)6x^3 + 0x^2 + 0x - 11}$$

The division process continues as usual, with the following results:

$$
\begin{array}{r}
3x^2 - 3x + 3 \\
2x + 2 \overline{)6x^3 \qquad\qquad - 11} \\
\underline{6x^3 + 6x^2} \\
-6x^2 \\
\underline{-6x^2 - 6x} \\
+6x - 11 \\
\underline{+6x + 6} \\
-17
\end{array}
$$

Thus,

$$\frac{6x^3 - 11}{2x + 2} = 3x^2 - 3x + 3 + \frac{-17}{2x + 2}$$

■

EXAMPLE 10 Divide $-3x^3 - 3 + x^5 + 4x^2 - x^4$ by $x^2 - 3$.

Solution The division process works most efficiently when the terms in both the divisor and dividend are written with their exponents in descending order. The long division is set up and accomplished as follows:

$$x^2 - 3 \overline{\smash{\big)}\, x^5 - x^4 - 3x^3 + 4x^2 + 0x - 3}$$

with quotient $x^3 - x^2 + 1$

Thus,

$$\frac{-3x^3 - 3 + x^5 + 4x^2 - x^4}{x^2 - 3} = x^3 - x^2 + 1$$

∎

1.4 EXERCISES

In Exercises 1–12, tell whether the given expression is a polynomial. If so, give its degree and tell whether it is a monomial, a binomial, a trinomial, or none of these.

1. $x^2 + 3x + 4$

2. $x^3 - 5xy$

3. $x^3 + y^{1/2}$

4. $x^{3/2} - 5y$

5. $\sqrt{5}\, x^3 + 4x^2$

6. $x^2 y^3$

7. $\sqrt{15}$

8. $\dfrac{5}{x} + \dfrac{x}{5} + 5$

9. 0

10. $\sqrt{y}$

11. $4x^{-3} - 3x^{-2} + 2x$

12. $3y^3 - 4y^2 + 2y + 2$

In Exercises 13–70, perform the indicated operations and simplify.

13. $(x^3 - 3x^2) + (5x^3 - 8x)$

14. $(2x^4 - 5x^3) + (7x^3 - x^4 + 2x)$

15. $(y^5 + 2y^3 + 7) - (y^5 - 2y^3 - 7)$

16. $(3t^7 - 7t^3 + 3) - (7t^7 - 3t^3 + 7)$

17. $2(x^2 + 3x - 1) - 3(x^2 + 2x - 4) + 4$

18. $5(x^3 - 8x + 3) + 2(3x^2 + 5x) - 7$

19. $8(t^2 - 2t + 5) + 4(t^2 - 3t + 2) - 6(2t^2 - 8)$

20. $-3(x^3 - x) + 2(x^2 + x) + 3(x^3 - 2x)$

21. $y(y^2 - 1) - y^2(y + 2) - y(2y - 2)$

22. $-4a^2(a + 1) + 3a(a^2 - 4) - a^2(a + 2)$

23. $x(x^2 - 1) - x^2(x + 2) - x(2x - 2)$

24. $x(x - 4) - (x^2 + 3) + x(2x + 3)$

25. $xy(x - 4y) - y(x^2 + 3xy) + xy(2x + 3y)$

26. $3mn(m + 2n) - 6m(3mn + 1) - 2n(4mn - 1)$

27. $2x^2 y^3(4xy^4)$

28. $-15a^3 b(-2a^2 b^3)$

29. $-3m^2 n(2mn^2)\left(-\dfrac{mn}{12}\right)$

30. $-\dfrac{3r^2 s^3}{5}\left(\dfrac{2r^2 s}{3}\right)\left(\dfrac{15rs^2}{2}\right)$

31. $-4rs(r^2 + s^2)$

32. $6u^2 v(2uv^3 - y)$

33. $6ab^2 c(2ac + 3bc^2 - 4ab^2 c)$

34. $-\dfrac{mn^2}{2}(4mn - 6m^2 - 8)$

35. $(a + 2)(a + 2)$

36. $(y - 5)(y - 5)$

37. $(a - 6)^2$

38. $(t + 9)^2$

39. $(x + 4)(x - 4)$

40. $(z + 7)(z - 7)$

41. $(a - 2b)^2$

42. $(4a + 5b)(4a - 5b)$

43. $(3m + 4n)(3m - 4n)$ **44.** $(4r + 3s)^2$ **45.** $(x - 3)(x + 5)$ **46.** $(z + 4)(z - 6)$

47. $(u + 2)(3u - 2)$ **48.** $(4x + 1)(2x - 3)$ **49.** $(2x + 3)(3x - 5)$ **50.** $(2a - 3)(2 + 3a)$

51. $(5x - 1)(2x + 3)$ **52.** $(4x - 1)(2x - 7)$ **53.** $(3v - 4)(4v + 3)$ **54.** $(4c + 5)(3c - 4)$

55. $(2y - 4x)(3y - 2x)$ **56.** $(-2x + 3y)(3x + y)$ **57.** $(9x - y)(x^2 - 3y)$ **58.** $(8a^2 + b)(a + 2b)$

59. $(5z + 2t)(z^2 - t)$ **60.** $(y - 2x^2)(x^2 + 3y)$ **61.** $(3x - 1)^3$ **62.** $(2x - 3)^3$

63. $(xy^2 - 1)(x^2y + 2)$ **64.** $(xy - z^2)(2xy + z^2)$

65. $(3x + 1)(2x^2 + 4x - 3)$ **66.** $(2x - 5)(x^2 - 3x + 2)$

67. $(3x + 2y)(2x^2 - 3xy + 4y^2)$ **68.** $(4r - 3s)(2r^2 + 4rs - 2s^2)$

69. $(x^2 + x + 1)(x^2 + x - 1)$ **70.** $(x^2 - x + 1)(x^2 + 2x + 3)$

In Exercises 71–82, multiply the expressions as you would multiply polynomials.

71. $2y^n(3y^n - 2y^2 + y^{-n})$ **72.** $3a^{-n}(2a^n + 3a^{n-1} - 7a^{n-2})$ **73.** $-5x^{2n}y^n(2x^{2n}y^{-n} + 3x^{-2n}y^n)$

74. $-2a^{3n}b^{2n}(5a^{-3n}b - ab^{-2n})$ **75.** $(x^n + 3)(x^n - 4)$ **76.** $(a^n - 5)(a^n - 3)$

77. $(2r^n - 7)(3r^n - 2)$ **78.** $(4z^n + 3)(3z^n + 1)$ **79.** $x^{1/2}(x^{1/2}y + xy^{1/2})$

80. $ab^{1/2}(a^{1/2}b^{1/2} + b^{1/2})$ **81.** $(a^{1/2} + b^{1/2})(a^{1/2} - b^{1/2})$ **82.** $(x^{3/2} + y^{1/2})^2$

In Exercises 83–98, rationalize each denominator.

83. $\dfrac{2}{\sqrt{3} - 1}$ **84.** $\dfrac{1}{\sqrt{5} + 2}$ **85.** $\dfrac{3x}{\sqrt{7} + 2}$ **86.** $\dfrac{14y}{\sqrt{2} - 3}$

87. $\dfrac{6}{\sqrt{5} - \sqrt{2}}$ **88.** $\dfrac{15}{\sqrt{7} + \sqrt{2}}$ **89.** $\dfrac{x}{x - \sqrt{3}}$ **90.** $\dfrac{y}{2y + \sqrt{7}}$

91. $\dfrac{y + \sqrt{2}}{y - \sqrt{2}}$ **92.** $\dfrac{x - \sqrt{3}}{x + \sqrt{3}}$ **93.** $\dfrac{\sqrt{5} - 2}{\sqrt{5} + 2}$ **94.** $\dfrac{\sqrt{7} + 4}{\sqrt{7} - 4}$

95. $\dfrac{\sqrt{2} - \sqrt{3}}{1 - \sqrt{3}}$ **96.** $\dfrac{\sqrt{3} - \sqrt{2}}{1 + \sqrt{2}}$ **97.** $\dfrac{\sqrt{x} - \sqrt{y}}{\sqrt{x} + \sqrt{y}}$ **98.** $\dfrac{\sqrt{2x} + y}{\sqrt{2x} - y}$

In Exercises 99–100, rationalize each numerator.

99. $\dfrac{\sqrt{x + 3} - \sqrt{x}}{3}$ **100.** $\dfrac{\sqrt{2 + h} - \sqrt{2}}{h}$

In Exercises 101–108, perform each division. Express all answers without using negative exponents.

101. $\dfrac{36a^2b^3}{18ab^6}$ **102.** $\dfrac{-45r^2s^5t^3}{27r^6s^2t^8}$ **103.** $\dfrac{16x^6y^4z^9}{-24x^9y^6z^0}$ **104.** $\dfrac{32m^6n^4p^2}{26m^6n^7p^2}$

105. $\dfrac{5x^3y^2 + 15x^3y^4}{10x^2y^3}$ **106.** $\dfrac{9m^4n^9 - 6m^3n^4}{12m^3n^3}$

107. $\dfrac{24x^5y^7 - 36x^2y^5 + 12xy}{60x^5y^4}$ **108.** $\dfrac{9a^3b^4 + 27a^2b^4 - 18a^2b^3}{18a^2b^7}$

In Exercises 109–124, simplify each fraction by performing a long division. If there is a nonzero remainder, write the answer in quotient $+ \frac{remainder}{divisor}$ form.

109. $\dfrac{3x^2 + 11x + 6}{x + 3}$ **110.** $\dfrac{3x^2 + 11x + 6}{3x + 2}$ **111.** $\dfrac{2x^2 - 19x + 37}{2x - 5}$ **112.** $\dfrac{2x^2 - 19x + 35}{x - 7}$

113. $\dfrac{x^3 - 2x^2 - 4x + 3}{x^2 + x - 1}$ **114.** $\dfrac{x^3 - 2x^2 - 4x + 5}{x^2 - 3}$ **115.** $\dfrac{x^5 - 2x^3 - 3x^2 + 9}{x^3 - 2}$ **116.** $\dfrac{x^5 - 2x^3 - 3x^2 + 9}{x^3 - 3}$

117. $\dfrac{x^5 - 32}{x - 2}$ **118.** $\dfrac{x^4 - 1}{x + 1}$ **119.** $\dfrac{x^4 + 16}{x^2 - 2}$ **120.** $\dfrac{x^5 + 32}{x^3 - 2}$

121. $\dfrac{36x^4 - 121x^2 + 120 + 72x^3 - 142x}{11x - 10 + 6x^2}$ **122.** $\dfrac{-121x^2 + 72x^3 - 142x + 120 + 36x^4}{x + 6x^2 - 12}$

123. $\dfrac{11x^5 - 9x^2 + 12x^6 + 3x^4 + 3x + 10x^3 - 6}{4x^4 - 3 + 5x^3}$ **124.** $\dfrac{3x^4 + 10x^3 + 3x - 6 - 9x^2 + 12x^6 + 11x^5}{2 + 3x^2 - x}$

125. Show that any trinomial can be squared using the formula $(a + b + c)^2 = a^2 + b^2 + c^2 + 2ab + 2ac + 2bc$.

126. Show that $(a + b + c + d)^2 = a^2 + b^2 + c^2 + d^2 + 2ab + 2ac + 2ad + 2bc + 2bd + 2cd$.

1.5 FACTORING POLYNOMIALS

- ■ Factoring Out a Common Monomial
- ■ Factoring the Difference of Two Squares
- ■ Factoring the Sum and Difference of Two Cubes
- ■ Miscellaneous Factoring
- ■ Factoring by Grouping
- ■ Factoring Trinomials

When two or more polynomials are multiplied together, each polynomial is called a **factor** of the resulting product. For example, the product $7(x + 2)(x + 3)$ has polynomial factors of 7, $x + 2$, and $x + 3$. The process of writing a polynomial as the product of several factors is called **factoring**.

In this section we will discuss factoring over the set of integers. That is, the coefficients of the polynomial and its polynomial factors are integers. If a polynomial cannot be factored by using integers only, we will call the polynomial a **prime polynomial**.

■ **Factoring Out a Common Monomial**

EXAMPLE 1   Factor the binomial $3xy^2 + 6x$.

Solution   We note that each term contains a factor of $3x$:

$$3xy^2 + 6x = 3x(y^2) + 3x(2)$$

We can then use the distributive property to factor out the common factor of $3x$:

$$3xy^2 + 6x = 3x(y^2 + 2)$$

The factored form of $3xy^2 + 6x$ is $3x(y^2 + 2)$. ■

EXAMPLE 2   Factor the binomial $x^2y^2z^2 - xyz$.

Solution   We factor out the common factor of xyz:

$$x^2y^2z^2 - xyz = xyz(xyz) - xyz(1)$$
$$= xyz(xyz - 1)$$

It is important to understand where the 1 comes from. The last term in the expression $x^2y^2z^2 - xyz$ has an understood coefficient of 1. When the xyz is factored out, the understood 1 must be written.   ■

■ Factoring by Grouping

EXAMPLE 3   Factor the expression $ax + bx + a + b$.

Solution   There is no factor that is common to all four terms. However, we can factor an x out of the first two terms and write the expression as

$$ax + bx + a + b = x(a + b) + (a + b)$$

In each term on the right-hand side of the equation, there is now a common factor of $a + b$, which can be factored out.

$$ax + bx + a + b = x(a + b) + (a + b)$$
$$= x(a + b) + 1(a + b)$$
$$= (a + b)(x + 1)$$ ■

■ Factoring the Difference of Two Squares

EXAMPLE 4   Factor the binomial $49x^2 - 4$.

Solution   We observe that each term is a perfect square:

$$49x^2 - 4 = (7x)^2 - 2^2$$

The difference of the squares of two quantities is the product of two factors. One is the sum of the quantities, and the other is the difference of two quantities. Thus, $49x^2 - 4$ factors as

$$49x^2 - 4 = (7x)^2 - 2^2$$
$$= (7x + 2)(7x - 2)$$ ■

Example 4 suggests the following formula for factoring the difference of two squares.

Factoring the Difference of Two Squares	$x^2 - y^2 = (x + y)(x - y)$

> **Warning!** With only integers to work with, the sum of two squares cannot be factored. For example, $x^2 + y^2$ is a prime polynomial.

EXAMPLE 5 Factor the binomial $16m^4 - n^4$.

Solution The binomial $16m^4 - n^4$ can be factored as the difference of two squares:

$$16m^4 - n^4 = (4m^2)^2 - (n^2)^2$$
$$= (4m^2 + n^2)(4m^2 - n^2)$$

The second factor is a difference of two squares and can also be factored:

$$16m^4 - n^4 = (4m^2 + n^2)(4m^2 - n^2)$$
$$= (4m^2 + n^2)(2m + n)(2m - n)$$

With only integers to work with, the factor $4m^2 + n^2$ is prime. ∎

EXAMPLE 6 Factor the binomial $18t^2 - 32$.

Solution We begin by factoring out the common monomial factor of 2.

$$18t^2 - 32 = 2(9t^2 - 16)$$

Since $9t^2 - 16$ is the difference of two squares, it can be factored.

$$18t^2 - 32 = 2(9t^2 - 16)$$
$$= 2(3t + 4)(3t - 4)$$ ∎

■ Factoring Trinomials

Trinomials that are squares of binomials can be factored by using the following formulas:

Factoring Trinomial Squares		
	1.	$x^2 + 2xy + y^2 = (x + y)(x + y) = (x + y)^2$
	2.	$x^2 - 2xy + y^2 = (x - y)(x - y) = (x - y)^2$

For example, to factor $a^2 - 6a + 9$, we note that $a^2 - 6a + 9$ can be written in the form

$$a^2 - 2(3a) + 3^2$$

and this form matches the left-hand side of Equation 2 above. Thus,

$$a^2 - 6a + 9 = a^2 - 2(3a) + 3^2 = (a - 3)(a - 3) = (a - 3)^2$$

Factoring trinomials that are not squares of binomials is more difficult. If a trinomial with no common factors is to be factorable, it must factor into the product of two binomials.

EXAMPLE 7   Factor the trinomial $x^2 + 3x - 10$.

Solution   This trinomial will factor into the product of two binomials. To factor $x^2 + 3x - 10$, we must find two binomials $x + a$ and $x + b$ such that

$$x^2 + 3x - 10 = (x + a)(x + b)$$

where the product of a and b is -10, and the sum of a and b is 3.

$$ab = -10 \quad \text{and} \quad a + b = 3$$

To find such numbers, we list the possible factorizations of -10:

$$10(-1) \qquad 5(-2) \qquad -10(1) \qquad -5(2)$$

Only in the factorization $5(-2)$ do the factors have a sum of 3. Thus, $a = 5$ and $b = -2$, and

$$x^2 + 3x - 10 = (x + a)(x + b)$$

3. $x^2 + 3x - 10 = (x + 5)(x - 2)$

Because of the commutative property of multiplication, the order of the factors in Equation 3 is not important. Equation 3 can also be written as

$$x^2 + 3x - 10 = (x - 2)(x + 5) \qquad \blacksquare$$

EXAMPLE 8   Factor the trinomial $2x^2 - x - 6$.

Solution   Since the first term is $2x^2$, the first terms of the binomial factors must be $2x$ and x:

$$2x^2 - x - 6 = (2x + ?)(x + ?)$$

The product of the last terms must be -6, and the sum of the products of the outer terms and the inner terms must be $-x$.

$$2x^2 - x - 6 = (2x + ?)(x + ?)$$

The only factorization of -6 that will cause this to happen is $3(-2)$. Thus,

$$2x^2 - x - 6 = (2x + 3)(x - 2) \qquad \blacksquare$$

It is not easy to give specific rules for factoring trinomials, because some guesswork is often necessary. However, the following hints are helpful.

Steps for Factoring General Trinomials

To factor a general trinomial, follow these steps:

1. Write the trinomial in descending powers of one variable.
2. Factor out any greatest common factor (including -1) if that is necessary to make the coefficient of the first term positive.
3. When the sign of the first term of a polynomial is $+$ and the sign of the third term is $+$, the sign between the terms of each binomial factor is the same as the sign of the middle term of the trinomial.

continued

| **Steps for Factoring General Trinomials (continued)** | When the sign of the first term is + and the sign of the third term is −, one of the signs between the terms of the binomial factors is + and the other is −. |

4. Mentally, try various combinations of the first terms and last terms until you find one that works. If you exhaust all the possibilities, the trinomial does not factor with integer coefficients.

5. Check the factorization by multiplication.

EXAMPLE 9   Factor the trinomial $10xy + 24y^2 - 6x^2$.

Solution   We begin by writing the trinomial in descending powers of x and then factor out the common factor of -2.

$$10xy + 24y^2 - 6x^2 = -6x^2 + 10xy + 24y^2$$
$$= -2(3x^2 - 5xy - 12y^2)$$

Since the sign of the third term of $3x^2 - 5xy - 12y^2$ is −, the signs between the binomial factors will be the opposite. Since the first term is $3x^2$, the first terms of the binomial factors must be $3x$ and x:

$$-2(3x^2 - 5xy - 12y^2) = -2(3x \qquad)(x \qquad)$$

The product of the last terms must be $-12y^2$, and the sum of the outer terms and the inner terms must be $-5xy$:

$$-2(3x^2 - 5xy - 12y^2) = -2(3x \qquad ?)(x \qquad ?)$$

Of the many factorizations of $-12y^2$, only $4y(-3y)$ leads a middle term of $-5xy$. Thus,

$$10xy + 24y^2 - 6x^2 = -6x^2 + 10xy + 24y^2$$
$$= -2(3x^2 - 5xy - 12y^2)$$
$$= -2(3x + 4y)(x - 3y) \qquad ■$$

We can often factor polynomials with variable exponents. For example, if n is a natural number,

$$a^{2n} - 5a^n - 6 = (a^n + 1)(a^n - 6)$$

because

$$(a^n + 1)(a^n - 6) = a^{2n} - 6a^n + a^n - 6$$
$$= a^{2n} - 5a^n - 6$$

■ **Factoring the Sum and Difference of Two Cubes**

Two other types of factoring involve binomials that are either the sum or difference of two cubes. Like the difference of two squares, they can be factored by using a formula.

Factoring the Sum and Difference of Two Cubes	$x^3 + y^3 = (x + y)(x^2 - xy + y^2)$ $x^3 - y^3 = (x - y)(x^2 + xy + y^2)$

EXAMPLE 10   Factor the binomial $x^3 - 8$.

Solution   This binomial can be written in the form $x^3 - 2^3$, which is the difference of two cubes. Substituting into the formula for the difference of two cubes gives

$$x^3 - 2^2 = (x - 2)(x^2 + 2x + 2^2)$$
$$= (x - 2)(x^2 + 2x + 4)$$

Thus, $x^3 - 8 = (x - 2)(x^2 + 2x + 4)$ ∎

EXAMPLE 11   Factor the binomial $27x^6 + 64y^3$.

Solution   We can write this expression in the form of the sum of two cubes and factor it as follows:

$$27x^6 + 64y^3 = (3x^2)^3 + (4y)^3$$
$$= (3x^2 + 4y)[(3x^2)^2 - (3x^2)(4y) + (4y)^2]$$
$$= (3x^2 + 4y)(9x^4 - 12x^2y + 16y^2)$$ ∎

■ **Miscellaneous Factoring**

EXAMPLE 12   Factor $x^2 - y^2 + 6x + 9$.

Solution   In this example we will factor a trinomial and a difference of two squares.

$$x^2 - y^2 + 6x + 9 = x^2 + 6x + 9 - y^2 \qquad \text{Rearrange terms.}$$
$$= (x + 3)^2 - y^2 \qquad \text{Factor } x^2 + 6x + 9.$$
$$= (x + 3 + y)(x + 3 - y) \qquad \text{Factor out the difference of two squares.}$$

We could try to factor this expression in another way.

$$x^2 - y^2 + 6x + 9 = (x + y)(x - y) + 3(2x + 3) \quad \text{Factor } x^2 - y^2 \text{ and } 6x + 9.$$

However, we are unable to finish the factorization. If grouping in one way doesn't work, try various other ways.   ∎

EXAMPLE 13   Factor $z^4 - 3z^2 + 1$.

Solution   This trinomial cannot be factored as the product of two binomials, because there is no combination that will give a middle term of $-3z^2$. However, if the middle term were $-2z^2$, the trinomial would be a perfect square trinomial, and the factorization would be easy:

$$z^4 - 2z^2 + 1 = (z^2 - 1)(z^2 - 1) = (z^2 - 1)^2$$

We can change the middle term in the trinomial to $-2z^2$ by adding z^2 to it. To ensure that adding z^2 does not change the value of the trinomial, however, we must also subtract z^2. We can then proceed as follows.

$$
\begin{aligned}
z^4 - 3z^2 + 1 &= z^4 - 3z^2 + z^2 + 1 - z^2 && \text{Add and subtract } z^2. \\
&= z^4 - 2z^2 + 1 - z^2 && \text{Combine } -3z^2 \text{ and } z^2. \\
&= (z^2 - 1)^2 - z^2 && \text{Factor } z^4 - 2z^2 + 1. \\
&= (z^2 - 1 + z)(z^2 - 1 - z) && \text{Factor the difference of} \\
& && \text{two squares.}
\end{aligned}
$$

In this type of problem, we will always try to add and subtract a perfect square, in hopes of making a perfect square trinomial that will lead to factoring a difference of two squares. ∎

1.5 EXERCISES

In Exercises 1–114, completely factor each expression over the set of integers. If an expression is prime, so indicate.

1. $3x - 6$

2. $5y - 15$

3. $8x^2 + 4x^3$

4. $9y^3 + 6y^2$

5. $7x^2y^2 + 14x^3y^2$

6. $25y^2z - 15yz^2$

7. $3a^2bc + 6ab^2c + 9abc^2$

8. $5x^3y^3z^3 + 25x^2y^2z^2 - 125xyz$

9. $b(x + y) - a(x + y)$

10. $b(x - y) + a(x - y)$

11. $4a + b - 12a^2 - 3ab$

12. $x^2 + 4x + xy + 4y$

13. $3x^3 + 3x^2 - x - 1$

14. $4x + 6xy - 9y - 6$

15. $2txy + 2ctx - 3ty - 3ct$

16. $2ax + 4ay - bx - 2by$

17. $ax + bx + ay + by + az + bz$

18. $6x^2y^3 + 18xy + 3x^2y^2 + 9x$

19. $6xc + yd + 2dx + 3cy$

20. $ax + ay + az + bx + by + bz$

21. $4x^2 - 9$

22. $36z^2 - 49$

23. $4 - 9r^2$

24. $16 - 49x^2$

25. $(x + z)^2 - 25$

26. $(x - y)^2 - 9$

27. $25x^4 + 1$

28. $36t^4 + 121$

29. $x^2 - (y - z)^2$

30. $z^2 - (y + 3)^2$

31. $(x - y)^2 - (x + y)^2$

32. $(2a + 3)^2 - (2a - 3)^2$

33. $x^4 - y^4$

34. $z^4 - 81$

35. $x^8 - 64z^4$

36. $1 - y^8$

37. $3x^2 - 12$

38. $3x^3y - 3xy$

39. $18xy^2 - 8x$

40. $27x^2 - 12$

41. $x^2 + 8x + 16$

42. $a^2 - 12a + 36$

43. $b^2 - 10b + 25$

44. $y^2 + 14y + 49$

45. $m^2 + 4mn + 4n^2$

46. $r^2 - 8rs + 16s^2$

47. $x^2 + 10x + 21$

48. $x^2 + 7x + 10$

49. $x^2 - 4x - 12$

50. $x^2 - 2x - 63$

51. $x^2 - 2x + 15$

52. $x^2 + x + 2$

53. $12x^2 - xy - 6y^2$

54. $8x^2 - 10xy - 3y^2$

55. $24y^2 + 15 - 38y$

56. $10x^2 - 18 + 3x$

57. $-15 + 2a + 24a^2$

58. $-32 - 68x + 9x^2$

59. $6x^2 + 29xy + 35y^2$

60. $10x^2 - 17xy + 6y^2$

61. $-35 - x + 6x^2$

62. $-5x - 6 + 6x^2$

63. $12y^2 - 58y - 70$

64. $3x^2 - 6x - 9$

65. $-6x^3 + 23x^2 + 35x$

66. $-y^3 - y^2 + 90y$

67. $6x^4 - 11x^3 - 35x^2$

68. $12x + 17x^2 - 7x^3$

69. $-35 + 47x - 6x^2$

70. $-14x^2 - 11x + 15$

71. $x^4 + 2x^2 - 15$

72. $x^4 - x^2 - 6$

73. $2y^2z^2 - 16yz + 80$

74. $3x^2 + 30x + 75$

75. $a^{2n} - 2a^n - 3$

76. $a^{2n} + 6a^n + 8$

77. $6x^{2n} - 7x^n + 2$ **78.** $9x^{2n} + 9x^n + 2$ **79.** $4x^{2n} - 9y^{2n}$ **80.** $8x^{2n} - 2x^n - 3$

81. $10y^{2n} - 11y^n - 6$ **82.** $16y^{4n} - 25y^{2n}$ **83.** $8z^3 - 27$ **84.** $125a^3 - 64$

85. $2x^3 + 2000$ **86.** $3y^3 + 648$ **87.** $(x + y)^3 - 64$ **88.** $(x - y)^3 + 27$

89. $1 - (x + 1)^3$ **90.** $1 + (x - 1)^3$ **91.** $64a^6 - y^6$ **92.** $a^6 + b^6$

93. $a^3 - b^3 + a - b$ **94.** $(a^2 - y^2) - 5(a + y)$ **95.** $64x^6 + y^6$ **96.** $z^2 + 6z + 9 - 225y^2$

97. $x^2 - 6x + 9 - 144y^2$ **98.** $x^2 + 2x - 9y^2 + 1$ **99.** $(a + b)^2 - 3(a + b) - 10$

100. $2(a + b)^2 - 5(a + b) - 3$ **101.** $6(u + v)^2 + 11u + 4 + 11v$ **102.** $8(r + s)^2 - 10(r + s) - 3$

103. $x^6 + 7x^3 - 8$ **104.** $x^6 - 13x^4 + 36x^2$ **105.** $a(c + d) + a + b(c + d) + b$

106. $c(a + b) + 3c + 3d + d(a + b)$ **107.** $x^4 + 3x^2 + 4$ **108.** $x^4 + x^2 + 1$

109. $x^4 + 7x^2 + 16$ **110.** $y^4 + 2y^2 + 9$ **111.** $4a^4 + 1 + 3a^2$

112. $x^4 + 25 + 6x^2$ **113.** $2x^4 + 8$ **114.** $3x^4 - 21x^2 + 27$

In Exercises 115–122, factor the indicated monomial from the given expression.

115. $3x + 2; 2$ **116.** $a + b; a$ **117.** $x + x^{1/2}; x^{1/2}$ **118.** $2x + \sqrt{2}y; \sqrt{2}$

119. $ab^{3/2} - a^{3/2}b; ab$ **120.** $ab^2 + b; b^{-1}$ **121.** $a^{n+2} + a^{n+3}; a^2$ **122.** $x^4 - 5x^6; x^{-2}$

1.6   ALGEBRAIC FRACTIONS

■ Simplifying Fractions   ■ Multiplying and Dividing Fractions   ■ Adding
and Subtracting Fractions   ■ Complex Fractions

If $y \neq 0$, the number represented by the quotient $\frac{x}{y}$ is called a **fraction**. The number x is called the **numerator** and y is called the **denominator**. Because division by zero is not defined, the denominator of a fraction cannot be 0.

Quotients of algebraic expressions are called **algebraic fractions**. If the algebraic expressions are polynomials, the fraction is called a **rational expression**. Each of the following fractions is an algebraic fraction. The first two are rational expressions.

$$\frac{5y^2 + 2y}{y^2 - 3y - 7} \qquad \frac{8ab^2 - 16c^3}{2x + 3} \qquad \frac{x^{1/2} + 4x}{x^{3/2} - x^{1/2}}$$

 Warning!   As with all fractions, the denominators cannot be 0.

Many properties of fractions are summarized as follows.

Properties of Fractions

If a, b, and c are real numbers and no denominators are 0, then

Equality of Fractions: $\dfrac{a}{b} = \dfrac{c}{d}$ if and only if $ad = bc$

The Fundamental Property of Fractions: $\dfrac{ax}{bx} = \dfrac{a}{b}$

Multiplication of Fractions: $\dfrac{a}{b} \cdot \dfrac{c}{d} = \dfrac{ac}{bd}$

Division of Fractions: $\dfrac{a}{b} \div \dfrac{c}{d} = \dfrac{a}{b} \cdot \dfrac{d}{c} = \dfrac{ad}{bc}$

Addition and Subtraction of Fractions with Like Denominators:

$$\frac{a}{b} + \frac{c}{b} = \frac{a + c}{b}$$

and

$$\frac{a}{b} - \frac{c}{b} = \frac{a - c}{b}$$

Each of these properties of fractions is illustrated in the following example.

EXAMPLE 1   **a.** $\dfrac{2a}{3} = \dfrac{4a}{6}$ because $2a(6) = 3(4a)$ **b.** $\dfrac{6xy}{10xy} = \dfrac{3(2xy)}{5(2xy)} = \dfrac{3}{5}$

$\dfrac{3y}{5} \neq \dfrac{15z}{25}$ because $3y(25) \neq 5(15z)$

c. $\dfrac{2r}{7s} \cdot \dfrac{3r}{5s} = \dfrac{2r \cdot 3r}{7s \cdot 5s} = \dfrac{6r^2}{35s^2}$ **d.** $\dfrac{3mn}{4pq} \div \dfrac{2pq}{7mn} = \dfrac{3mn}{4pq} \cdot \dfrac{7mn}{2pq} = \dfrac{21m^2n^2}{8p^2q^2}$

e. $\dfrac{2ab}{5xy} + \dfrac{ab}{5xy} = \dfrac{2ab + ab}{5xy} = \dfrac{3ab}{5xy}$ **f.** $\dfrac{6uv^2}{7w^2} - \dfrac{3uv^2}{7w^2} = \dfrac{6uv^2 - 3uv^2}{7w^2} = \dfrac{3uv^2}{7w^2}$ ■

To add or subtract fractions with unlike denominators, we can write each fraction as an equivalent fraction with a common denominator. We can then add or subtract the fractions. For example,

$$\frac{3x}{5} + \frac{2x}{7} = \frac{3x(7)}{5(7)} + \frac{2x(5)}{7(5)} = \frac{21x}{35} + \frac{10x}{35} = \frac{21x + 10x}{35} = \frac{31x}{35}$$

$$\frac{4a^2}{15} - \frac{3a^2}{10} = \frac{4a^2(2)}{15(2)} - \frac{3a^2(3)}{10(3)} = \frac{8a^2}{30} - \frac{9a^2}{30} = \frac{8a^2 - 9a^2}{30} = \frac{-a^2}{30} = -\frac{a^2}{30}$$

A fraction is said to be **reduced to lowest terms** if all factors common to the numerator and the denominator have been removed. To **simplify a fraction** means to reduce it to lowest terms.

■ Simplifying Fractions

To simplify a fraction, we use the fundamental property of fractions. This property enables us to divide out all factors that are common to both the numerator and the denominator.

EXAMPLE 2   Simplify the fraction $\dfrac{x^2 - 9}{x^2 - 3x}$.

Solution   We factor the difference of two squares in the numerator, factor out the common monomial factor of x in the denominator, and divide out the common factor of $x - 3$. Since the denominator of the fraction cannot be 0, x cannot be 0 or 3. In parentheses, we note these restrictions on the variable x.

$$\frac{x^2 - 9}{x^2 - 3x} = \frac{(x + 3)(x - 3)}{x(x - 3)} = \frac{x + 3}{x} \qquad (x \neq 0, 3)$$ ■

We will encounter the following properties of fractions in the next examples.

Properties of Fractions	If a and b represent real numbers and there are no divisions by 0, then

$$\frac{a}{1} = a \qquad\qquad\qquad\qquad \frac{a}{a} = 1$$

$$\frac{a}{b} = \frac{-a}{-b} = -\frac{a}{-b} = -\frac{-a}{b} \qquad\qquad -\frac{a}{b} = \frac{a}{-b} = \frac{-a}{b} = -\frac{-a}{-b}$$

EXAMPLE 3   Simplify the fraction $\dfrac{x^2 - 2xy + y^2}{y - x}$.

Solution   We factor the trinomial in the numerator, -1 from the denominator, and divide out the common factor of $x - y$.

$$\frac{x^2 - 2xy + y^2}{y - x} = \frac{(x - y)(x - y)}{-(x - y)} \qquad (x \neq y)$$

$$= \frac{x - y}{-1}$$

$$= -\frac{x - y}{1}$$

$$= -(x - y)$$ ■

EXAMPLE 4   Simplify the fraction $\dfrac{x^2 - 3x + 2}{x^2 - x - 2}$.

Solution   We factor the numerator and the denominator and simplify:

$$\frac{x^2 - 3x + 2}{x^2 - x - 2} = \frac{(x - 1)(x - 2)}{(x + 1)(x - 2)} = \frac{x - 1}{x + 1} \qquad (x \neq -1, 2)$$ ■

Multiplying and Dividing Fractions

EXAMPLE 5 Multiply: $\dfrac{x^2 - x - 2}{x^2 - 1} \cdot \dfrac{x^2 + 2x - 3}{x - 2}$.

Solution To multiply the fractions, we multiply the numerators and multiply the denominators and simplify the result.

$$\frac{x^2 - x - 2}{x^2 - 1} \cdot \frac{x^2 + 2x - 3}{x - 2} = \frac{(x^2 - x - 2)(x^2 + 2x - 3)}{(x^2 - 1)(x - 2)}$$

$$= \frac{(x - 2)(x + 1)(x - 1)(x + 3)}{(x + 1)(x - 1)(x - 2)} \qquad (x \neq 1, -1, 2)$$

$$= x + 3 \qquad \blacksquare$$

EXAMPLE 6 Divide: $\dfrac{x^2 - 2x - 3}{x^2 - 4} \div \dfrac{x^2 + 2x - 15}{x^2 + 3x - 10}$.

Solution To divide the fractions, we multiply by the reciprocal of the divisor.

$$\frac{x^2 - 2x - 3}{x^2 - 4} \div \frac{x^2 + 2x - 15}{x^2 + 3x - 10} = \frac{x^2 - 2x - 3}{x^2 - 4} \cdot \frac{x^2 + 3x - 10}{x^2 + 2x - 15}$$

$$= \frac{(x^2 - 2x - 3)(x^2 + 3x - 10)}{(x^2 - 4)(x^2 + 2x - 15)}$$

$$= \frac{(x - 3)(x + 1)(x - 2)(x + 5)}{(x + 2)(x - 2)(x + 5)(x - 3)} \qquad (x \neq 2, -2, 3, -5)$$

$$= \frac{x + 1}{x + 2} \qquad \blacksquare$$

EXAMPLE 7 Simplify: $\dfrac{2x^2 - 5x - 3}{3x - 1} \cdot \dfrac{3x^2 + 2x - 1}{x^2 - 2x - 3} \div \dfrac{2x^2 + x}{3x}$.

Solution We can change the division to a multiplication, factor, and simplify.

$$\frac{2x^2 - 5x - 3}{3x - 1} \cdot \frac{3x^2 + 2x - 1}{x^2 - 2x - 3} \div \frac{2x^2 + x}{3x} = \frac{2x^2 - 5x - 3}{3x - 1} \cdot \frac{3x^2 + 2x - 1}{x^2 - 2x - 3} \cdot \frac{3x}{2x^2 + x}$$

$$= \frac{(2x^2 - 5x - 3)(3x^2 + 2x - 1)(3x)}{(3x - 1)(x^2 - 2x - 3)(2x^2 + x)}$$

$$= \frac{(x - 3)(2x + 1)(3x - 1)(x + 1)3x}{(3x - 1)(x + 1)(x - 3)x(2x + 1)} \qquad \left(x \neq \frac{1}{3}, -1, 3, 0, -\frac{1}{2} \right)$$

$$= 3 \qquad \blacksquare$$

Adding and Subtracting Fractions

To add fractions with like denominators, we add the numerators, keep the denominator, and simplify the result.

EXAMPLE 8 Add: $\dfrac{2x + 5}{x + 5} + \dfrac{3x + 20}{x + 5}$.

Solution

$$\dfrac{2x + 5}{x + 5} + \dfrac{3x + 20}{x + 5} = \dfrac{5x + 25}{x + 5} \qquad (x \neq -5)$$

$$= \dfrac{5(x + 5)}{1(x + 5)} \qquad \text{Factor out 5.}$$

$$= 5 \qquad \blacksquare$$

To add or subtract fractions with unlike denominators, we must find a common denominator, called the **least** (or lowest) **common denominator (LCD)**. We now consider how it can be found.

Suppose the unlike denominators of three fractions are 12, 20, and 35. First, we find the unique prime factorization of each number.

$$12 = 4 \cdot 3 = 2^2 \cdot 3$$
$$20 = 4 \cdot 5 = 2^2 \cdot 5$$
$$35 = 5 \cdot 7$$

Because the LCD is the smallest number that can be divided by 12, 20, and 35, it must contain factors of 2^2, 3, 5, and 7. Thus,

$$\text{LCD} = 2^2 \cdot 3 \cdot 5 \cdot 7 = 420$$

That is, 420 is the smallest number that can be divided without remainder by 12, 20, and 35.

When finding a LCD, we always factor each denominator and then create the LCD by using each factor the greatest number of times that it appears in any one denominator. The product of these factors is the LCD.

EXAMPLE 9 Add: $\dfrac{1}{x^2 - 4} + \dfrac{2}{x^2 - 4x + 4}$.

Solution We factor each denominator and find the LCD.

$$x^2 - 4 = (x + 2)(x - 2)$$
$$x^2 - 4x + 4 = (x - 2)(x - 2) = (x - 2)^2$$

The LCD is $(x + 2)(x - 2)^2$. We then write each fraction with its denominator in factored form, convert each fraction into an equivalent fraction with a denominator of $(x + 2)(x - 2)^2$, add the fractions, and simplify.

$$\dfrac{1}{x^2 - 4} + \dfrac{2}{x^2 - 4x + 4} = \dfrac{1}{(x + 2)(x - 2)} + \dfrac{2}{(x - 2)(x - 2)} \qquad (x \neq 2, -2)$$

$$= \dfrac{1(x - 2)}{(x + 2)(x - 2)(x - 2)} + \dfrac{2(x + 2)}{(x - 2)(x - 2)(x + 2)}$$

$$= \dfrac{1(x - 2) + 2(x + 2)}{(x + 2)(x - 2)(x - 2)}$$

$$= \frac{x - 2 + 2x + 4}{(x + 2)(x - 2)(x - 2)}$$

$$= \frac{3x + 2}{(x + 2)(x - 2)(x - 2)}$$

 Warning!   Always attempt to simplify the final result. In this case, the final fraction is already in lowest terms. ∎

EXAMPLE 10   Combine and simplify $\dfrac{x - 2}{x^2 - 1} - \dfrac{x + 3}{x^2 + 3x + 2} + \dfrac{3}{x^2 + x - 2}$.

Solution   We factor the denominators to find the LCD.

$$x^2 - 1 = (x + 1)(x - 1)$$
$$x^2 + 3x + 2 = (x + 2)(x + 1)$$
$$x^2 + x - 2 = (x + 2)(x - 1)$$

The LCD is $(x + 1)(x + 2)(x - 1)$. Now we write each fraction as an equivalent fraction with this common denominator and proceed as follows:

$$\frac{x - 2}{x^2 - 1} - \frac{x + 3}{x^2 + 3x + 2} + \frac{3}{x^2 + x - 2}$$

$$= \frac{x - 2}{(x - 1)(x + 1)} - \frac{x + 3}{(x + 1)(x + 2)} + \frac{3}{(x - 1)(x + 2)} \qquad (x \neq 1, -1, -2)$$

$$= \frac{(x - 2)(x + 2)}{(x - 1)(x + 1)(x + 2)} - \frac{(x + 3)(x - 1)}{(x + 1)(x + 2)(x - 1)} + \frac{3(x + 1)}{(x - 1)(x + 2)(x + 1)}$$

$$= \frac{(x^2 - 4) - (x^2 + 2x - 3) + (3x + 3)}{(x - 1)(x + 2)(x + 1)}$$

$$= \frac{x^2 - 4 - x^2 - 2x + 3 + 3x + 3}{(x - 1)(x + 2)(x + 1)}$$

$$= \frac{x + 2}{(x - 1)(x + 2)(x + 1)}$$

$$= \frac{1}{(x - 1)(x + 1)} \qquad\qquad ∎$$

Complex Fractions

A **complex fraction** is a fraction with a fractional numerator or a fractional denominator.

EXAMPLE 11   Simplify the complex fraction $\dfrac{\dfrac{1}{x} + \dfrac{1}{y}}{\dfrac{x}{y}}$.

Solution 1 We determine that the LCD of the three fractions in the given complex fraction is xy. We multiply the numerator and denominator of the complex fraction by xy and simplify:

$$\frac{\dfrac{1}{x} + \dfrac{1}{y}}{\dfrac{x}{y}} = \frac{xy\left(\dfrac{1}{x} + \dfrac{1}{y}\right)}{xy\left(\dfrac{x}{y}\right)} = \frac{\dfrac{xy}{x} + \dfrac{xy}{y}}{\dfrac{xxy}{y}} = \frac{y + x}{x^2}$$

Solution 2 We combine the fractions in the numerator of the complex fraction to obtain a single fraction over a single fraction.

$$\frac{\dfrac{1}{x} + \dfrac{1}{y}}{\dfrac{x}{y}} = \frac{\dfrac{1(y)}{x(y)} + \dfrac{1(x)}{y(x)}}{\dfrac{x}{y}} = \frac{\dfrac{y + x}{xy}}{\dfrac{x}{y}}$$

Then we use the fact that any fraction indicates a division:

$$\frac{\dfrac{y + x}{xy}}{\dfrac{x}{y}} = \frac{y + x}{xy} \div \frac{x}{y} = \frac{y + x}{xy} \cdot \frac{y}{x} = \frac{(y + x)y}{xyx} = \frac{y + x}{x^2}$$

$\blacksquare$

1.6 EXERCISES

In Exercises 1–4, tell whether the given fractions are equal.

1. $\dfrac{8x}{3y}, \dfrac{16x}{6y}$

2. $\dfrac{3x^2}{4y^2}, \dfrac{12y^2}{16x^2}$

3. $\dfrac{25xyz}{12ab^2c}, \dfrac{50a^2bc}{24xyz}$

4. $\dfrac{15r^2s^2}{4uv^2}, \dfrac{45uv^2}{10r^2s^2}$

In Exercises 5–12, perform the indicated operations and simplify, whenever possible.

5. $\dfrac{4x}{7} \cdot \dfrac{2}{5a}$

6. $\dfrac{-5y}{2x} \cdot \dfrac{4}{y^2}$

7. $\dfrac{8m}{5n} \div \dfrac{3m}{10n}$

8. $\dfrac{15p}{8q} \div \dfrac{-5p}{16q^2}$

9. $\dfrac{3z}{5c} + \dfrac{2z}{5c}$

10. $\dfrac{7a}{4b} - \dfrac{3a}{4b}$

11. $\dfrac{15x^2y}{7a^2b^3} - \dfrac{x^2y}{7a^2b^3}$

12. $\dfrac{8rst^2}{15m^4t^2} + \dfrac{7rst^2}{15m^4t^2}$

In Exercises 13–20, simplify each fraction.

13. $\dfrac{2x - 4}{x^2 - 4}$

14. $\dfrac{x^2 - 16}{x^2 - 8x + 16}$

15. $\dfrac{25 - x^2}{x^2 + 10x + 25}$

16. $\dfrac{4 - x^2}{x^2 - 5x + 6}$

17. $\dfrac{6x^3 + x^2 - 12x}{4x^3 + 4x^2 - 3x}$

18. $\dfrac{6x^4 - 5x^3 - 6x^2}{2x^3 - 7x^2 - 15x}$

19. $\dfrac{x^3 - 8}{x^2 + ax - 2x - 2a}$

20. $\dfrac{xy + 2x + 3y + 6}{x^3 + 27}$

In Exercises 21–62, perform the indicated operations and simplify, whenever possible.

21. $\dfrac{x^2 - 1}{x} \cdot \dfrac{x^2}{x^2 + 2x + 1}$

22. $\dfrac{y^2 - 2y + 1}{y} \cdot \dfrac{y + 2}{y^2 + y - 2}$

23. $\dfrac{3x^2 + 7x + 2}{x^2 + 2x} \cdot \dfrac{x^2 - x}{3x^2 + x}$

24. $\dfrac{x^2 + x}{2x^2 + 3x} \cdot \dfrac{2x^2 + x - 3}{x^2 - 1}$

25. $\dfrac{x^2 + x}{x - 1} \cdot \dfrac{x^2 - 1}{x + 2}$

26. $\dfrac{x^2 + 5x + 6}{x^2 + 6x + 9} \cdot \dfrac{x + 2}{x^2 - 4}$

27. $\dfrac{2x^2 + 32}{8} \div \dfrac{x^2 + 16}{2}$

28. $\dfrac{x^2 + x - 6}{x^2 - 6x + 9} \div \dfrac{x^2 - 4}{x^2 - 9}$

29. $\dfrac{z^2 + z - 20}{z^2 - 4} \div \dfrac{z^2 - 25}{z - 5}$

30. $\dfrac{ax + bx + a + b}{a^2 + 2ab + b^2} \div \dfrac{x^2 - 1}{x^2 - 2x + 1}$

31. $\dfrac{3x^2 + 5x - 2}{x^3 + 2x^2} \div \dfrac{6x^2 + 13x - 5}{2x^3 + 5x^2}$

32. $\dfrac{x^2 + 13x + 12}{8x^2 - 6x - 5} \div \dfrac{2x^2 - x - 3}{8x^2 - 14x + 5}$

33. $\dfrac{x^2 + 7x + 12}{x^3 - x^2 - 6x} \cdot \dfrac{x^2 - 3x - 10}{x^2 + 2x - 3} \cdot \dfrac{x^3 - 4x^2 + 3x}{x^2 - x - 20}$

34. $\dfrac{x^2 - 2x - 3}{21x^2 - 50x - 16} \cdot \dfrac{3x - 8}{x - 3} \div \dfrac{x^2 + 6x + 5}{7x^2 - 33x - 10}$

35. $\dfrac{x^3 + 27}{x^2 - 4} \div \left(\dfrac{x^2 + 4x + 3}{x^2 + 2x} \div \dfrac{x^2 + x - 6}{x^2 - 3x + 9} \right)$

36. $\dfrac{x(x - 2) - 3}{x(x + 7) - 3(x - 1)} \cdot \dfrac{x(x + 1) - 2}{x(x - 7) + 3(x + 1)}$

37. $\dfrac{3}{x + 3} + \dfrac{x + 2}{x + 3}$

38. $\dfrac{3}{x + 1} + \dfrac{x + 2}{x + 1}$

39. $\dfrac{4x}{x - 1} - \dfrac{4}{x - 1}$

40. $\dfrac{6x}{x - 2} - \dfrac{3}{x - 2}$

41. $\dfrac{2}{5 - x} + \dfrac{1}{x - 5}$

42. $\dfrac{3}{x - 6} - \dfrac{2}{6 - x}$

43. $\dfrac{3}{x + 1} + \dfrac{2}{x - 1}$

44. $\dfrac{3}{x + 4} + \dfrac{x}{x - 4}$

45. $\dfrac{a + 3}{a^2 + 7a + 12} + \dfrac{a}{a^2 - 16}$

46. $\dfrac{a}{a^2 + a - 2} + \dfrac{2}{a^2 - 5a + 4}$

47. $\dfrac{x}{x^2 - 4} - \dfrac{1}{x + 2}$

48. $\dfrac{b^2}{b^2 - 4} - \dfrac{4}{b^2 + 2b}$

49. $\dfrac{3x - 2}{x^2 + 2x + 1} - \dfrac{x}{x^2 - 1}$

50. $\dfrac{2t}{t^2 - 25} - \dfrac{t + 1}{t^2 + 5t}$

51. $\dfrac{2}{y^2 - 1} + 3 + \dfrac{1}{y + 1}$

52. $2 + \dfrac{4}{t^2 - 4} - \dfrac{1}{t - 2}$

53. $\dfrac{1}{x - 2} + \dfrac{3}{x + 2} - \dfrac{3x - 2}{x^2 - 4}$

54. $\dfrac{x}{x - 3} - \dfrac{5}{x + 3} + \dfrac{3(3x - 1)}{x^2 - 9}$

55. $\left(\dfrac{1}{x - 2} + \dfrac{1}{x - 3} \right) \cdot \dfrac{x - 3}{2x}$

56. $\left(\dfrac{1}{x + 1} - \dfrac{1}{x - 2} \right) \div \dfrac{1}{x - 2}$

57. $\dfrac{3x}{x - 4} - \dfrac{x}{x + 4} - \dfrac{3x + 1}{16 - x^2}$

58. $\dfrac{7x}{x - 5} + \dfrac{3x}{5 - x} + \dfrac{3x - 1}{x^2 - 25}$

59. $\dfrac{1}{x^2 + 3x + 2} - \dfrac{2}{x^2 + 4x + 3} + \dfrac{1}{x^2 + 5x + 6}$

60. $\dfrac{-2}{x - y} + \dfrac{2}{x - z} - \dfrac{2z - 2y}{(y - x)(z - x)}$

61. $\dfrac{3x - 2}{x^2 + x - 20} - \dfrac{4x^2 + 2}{x^2 - 25} + \dfrac{3x^2 - 25}{x^2 - 16}$

62. $\dfrac{3x + 2}{8x^2 - 10x - 3} + \dfrac{x + 4}{6x^2 - 11x + 3} - \dfrac{1}{4x + 1}$

In Exercises 63–90, simplify each complex fraction.

63. $\dfrac{\dfrac{3a}{b}}{\dfrac{6ac}{b^2}}$

64. $\dfrac{\dfrac{3t^2}{9x}}{\dfrac{t}{18x}}$

65. $\dfrac{\dfrac{3a^2b}{ab}}{27}$

66. $\dfrac{\dfrac{3u^2v}{4t}}{3uv}$

67. $\dfrac{\dfrac{x-y}{ab}}{\dfrac{y-x}{ab}}$

68. $\dfrac{\dfrac{x^2-5x+6}{2x^2y}}{\dfrac{x^2-9}{2x^2y}}$

69. $\dfrac{\dfrac{1}{x}+\dfrac{1}{y}}{xy}$

70. $\dfrac{xy}{\dfrac{11}{x}-\dfrac{11}{y}}$

71. $\dfrac{\dfrac{1}{x}+\dfrac{1}{y}}{\dfrac{1}{x}-\dfrac{1}{y}}$

72. $\dfrac{\dfrac{1}{x}-\dfrac{1}{y}}{\dfrac{1}{x}+\dfrac{1}{y}}$

73. $\dfrac{\dfrac{3a}{b}-\dfrac{4a^2}{x}}{\dfrac{1}{b}+\dfrac{1}{ax}}$

74. $\dfrac{1-\dfrac{x}{y}}{\dfrac{x^2}{y^2}-1}$

75. $\dfrac{x+1-\dfrac{6}{x}}{x+5+\dfrac{6}{x}}$

76. $\dfrac{2z}{1-\dfrac{3}{z}}$

77. $\dfrac{3xy}{1-\dfrac{1}{xy}}$

78. $\dfrac{x-3+\dfrac{1}{x}}{-\dfrac{1}{x}-x+3}$

79. $\dfrac{3x}{x+\dfrac{1}{x}}$

80. $\dfrac{2x^2+4}{2+\dfrac{4x}{5}}$

81. $\dfrac{\dfrac{x}{x+2}-\dfrac{2}{x-1}}{\dfrac{3}{x+2}+\dfrac{x}{x-1}}$

82. $\dfrac{\dfrac{2x}{x-3}+\dfrac{1}{x-2}}{\dfrac{3}{x-3}-\dfrac{x}{x-2}}$

83. $\dfrac{1}{1+x^{-1}}$

84. $\dfrac{y^{-1}}{x^{-1}+y^{-1}}$

85. $\dfrac{3(x+2)^{-1}+2(x-1)^{-1}}{(x+2)^{-1}}$

86. $\dfrac{2x(x-3)^{-1}-3(x+2)^{-1}}{(x-3)^{-1}(x+2)^{-1}}$

87. $\dfrac{x}{1+\dfrac{1}{3x^{-1}}}$

88. $\dfrac{ab}{2+\dfrac{3}{2a^{-1}}}$

89. $\dfrac{1}{1+\dfrac{1}{1+\dfrac{1}{x}}}$

90. $\dfrac{y}{2+\dfrac{2}{2+\dfrac{2}{y}}}$

In Exercises 91–94, simplify each expression.

91. $\dfrac{\sqrt{x^2+4}-\dfrac{5}{\sqrt{x^2+4}}}{x+1}$

92. $\dfrac{\sqrt{y^2-5}+\dfrac{4}{\sqrt{y^2-5}}}{y-1}$

93. $\dfrac{\dfrac{3}{\sqrt{z^2-7}}+\sqrt{z^2-7}}{z+2}$

94. $\dfrac{\dfrac{2}{\sqrt{t^2+1}}-\sqrt{t^2+1}}{t-1}$

95. Prove the formula: $\dfrac{a}{b}+\dfrac{c}{d}=\dfrac{ad+bc}{bd}$.

96. Prove that $\dfrac{a}{b}\div\dfrac{c}{d}=\dfrac{a}{b}\cdot\dfrac{d}{c}$.

1 CHAPTER SUMMARY

Key Words

absolute value (1.1)
algebraic fraction (1.6)
associative properties (1.1)
base (1.2)
binomial (1.4)
closed interval (1.1)
coefficient (1.4)
common denominator (1.6)
complex fraction (1.6)
composite numbers (1.1)
commutative properties (1.1)
conjugate binomials (1.4)
coordinate (1.1)
cube root (1.3)
degree (1.4)
denominator (1.6)
disjoint (1.1)
distributive property (1.1)
dividend (1.4)
divisor (1.4)
elements (1.1)
empty set (1.1)
even integers (1.1)
exponent (1.2)
exponential expression (1.2)
factor (1.2)

factoring (1.5)
finite set (1.1)
fraction (1.6)
graph (1.1)
half-open interval (1.1)
infinite set (1.1)
integers (1.1)
intersection (1.1)
intervals (1.1)
irrational numbers (1.1)
like radicals (1.3)
like terms (1.4)
monomial (1.4)
natural numbers (1.1)
negative numbers (1.1)
negatives (1.1)
number line (1.1)
numerator (1.6)
odd integers (1.1)
open interval (1.1)
origin (1.1)
perfect cubes (1.3)
perfect squares (1.3)
polynomial (1.4)
positive numbers (1.1)
power (1.2)

prime number (1.1)
prime polynomial (1.5)
quotient (1.4)
radical expression (1.3)
radical (1.3)
radical sign (1.3)
radicand (1.3)
rational number (1.1)
rationalizing the denominator (1.3)
real numbers (1.1)
remainder (1.4)
repeating decimals (1.1)
scientific notation (1.2)
set (1.1)
set-builder notation (1.1)
square root (1.3)
subset (1.1)
term (1.4)
terminating decimal (1.1)
trichotomy property (1.1)
trinomial (1.4)
union (1.1)
variable (1.1)
whole numbers (1.1)
zero polynomial (1.4)

Key Ideas

(1.1) Rational numbers are either terminating or repeating decimals.

Sets of real numbers can be graphed on the number line.

Irrational numbers are nonterminating, nonrepeating decimals.

The set $\Re$ of real numbers is the set of all decimals.

If $x \geq 0$, then $|x| = x$. If $x < 0$, then $|x| = -x$. $|x| \geq 0$.

The distance between two points a and b on a number line is $d = |a - b|$.

(1.2) For any natural number n,

$$x^n = \overbrace{x \cdot x \cdot x \cdot \cdots \cdot x}^{n \text{ factors of } x}$$

Rules of Exponents: If no denominators are 0, then

$$x^m x^n = x^{m+n} \qquad (x^m)^n = x^{mn}$$

$$(xy)^n = x^n y^n \qquad \left(\frac{x}{y}\right)^n = \frac{x^n}{y^n}$$

$$x^0 = 1 \qquad x^{-n} = \frac{1}{x^n}$$

$$\frac{x^m}{x^n} = x^{m-n} \qquad \left(\frac{y}{x}\right)^{-n} = \left(\frac{x}{y}\right)^n$$

(1.3)
$$a^{1/n} = b \text{ if and only if } b^n = a. \qquad \sqrt[n]{a} = a^{1/n}$$

$$\sqrt[n]{a} = b \text{ if and only if } b^n = a. \qquad \sqrt{a}\sqrt{a} = a$$

If m and n are positive integers and a and $a^{1/n}$ are real numbers, then

$$a^{m/n} = (\sqrt[n]{a})^m = \sqrt[n]{a^m}$$

If all radicals are real numbers and there are no divisions by 0, then

$$\sqrt[n]{ab} = \sqrt[n]{a}\sqrt[n]{b} \qquad \sqrt[n]{\frac{a}{b}} = \frac{\sqrt[n]{a}}{\sqrt[n]{b}}$$

$$\sqrt[m]{\sqrt[n]{a}} = \sqrt[n]{\sqrt[m]{a}} = \sqrt[mn]{a}$$

(1.4) Special Products: $(x + y)^2 = x^2 + 2xy + y^2$
$(x - y)^2 = x^2 - 2xy + y^2 \qquad (x + y)(x - y) = x^2 - y^2$
(1.5) Factoring Formulas: $x^2 - y^2 = (x + y)(x - y)$
$x^2 + 2xy + y^2 = (x + y)^2$
$x^2 - 2xy + y^2 = (x - y)^2$
$x^3 + y^3 = (x + y)(x^2 - xy + y^2)$
$x^3 - y^3 = (x - y)(x^2 + xy + y^2)$

(1.6) Properties of Fractions: If there are no divisions by 0, then

$$\frac{a}{b} = \frac{c}{d} \text{ if and only if } ad = bc.$$

$$\frac{a}{b} = \frac{-a}{-b} = -\frac{a}{-b} = -\frac{-a}{b}$$

$$a \cdot 1 = a \qquad \frac{a}{1} = a \qquad \frac{a}{a} = 1$$

$$-\frac{a}{b} = \frac{a}{-b} = \frac{-a}{b} = -\frac{-a}{-b} \qquad \frac{a}{b} \cdot \frac{c}{d} = \frac{ac}{bd}$$

$$\frac{a}{b} \div \frac{c}{d} = \frac{ad}{bc} \qquad \frac{a}{b} + \frac{c}{b} = \frac{a + c}{b}$$

$$\frac{a}{b} - \frac{c}{b} = \frac{a - c}{b}$$

$$\frac{a}{b} = \frac{ax}{bx}$$

---

1   CHAPTER REVIEW EXERCISES

In Review Exercises 1–4, assume that $\mathbf{A} = \{2, 3, 5, 7, 11\}$, $\mathbf{B} = \{2, 4, 6, 8, 10\}$, *and* $\mathbf{C} = \{1, 2, 3, 4, 5\}$. *Find each set.*

1. $\mathbf{A} \cap \mathbf{B}$
2. $\mathbf{A} \cup \mathbf{C}$
3. $\mathbf{A} \cap \mathbf{B} \cap \mathbf{C}$
4. $\mathbf{A} \cup (\mathbf{B} \cap \emptyset)$
5. Graph the prime numbers between 10 and 20.
6. Graph the even integers from 6 to 14.

In Review Exercises 7–10, graph each interval on the number line.

7. $\{x \mid -3 < x \le 5\}$
8. $\{x \mid x \ge 0 \text{ or } x < -1\}$
9. $(-2, 4]$
10. $[-2, 4) \cup (0, 5]$

In Review Exercises 11–14, find each absolute value.

11. $|6|$
12. $|-25|$
13. $|1 - \sqrt{2}|$
14. $|\sqrt{3} - 1|$

In Review Exercises 15–20, tell which property justifies each expression.

15. $(a + b) + 2 = a + (b + 2)$
16. $a + 7 = 7 + a$
17. $4(2x) = (4 \cdot 2)x$
18. $3(a + b) = 3a + 3b$
19. If $x = b$ and $b = 3$, then $x = 3$.
20. $5x = 5x$

In Review Exercises 21–28, simplify each expression. Assume that all variables represent positive numbers. Give all answers with positive exponents.

21. $(x^3y^2)^4$
22. $\left(\dfrac{a^4}{b^2}\right)^3$
23. $(m^{-3}n^4)^2$
24. $\left(\dfrac{p^{-2}q^2}{2}\right)^3$

25. $\left(\dfrac{3x^2y^{-2}}{x^2y^2}\right)^{-2}$
26. $\left(\dfrac{a^{-3}b^2}{ab^{-3}}\right)^{-2}$
27. $\left(\dfrac{-3x^3y}{xy^3}\right)^{-2}$
28. $\left(-\dfrac{2m^{-2}n^0}{4m^2n^{-1}}\right)^{-3}$

In Review Exercises 29–40, simplify each expression.

29. $4^{1/2}$ **30.** $8^{2/3}$ **31.** $-8^{2/3}$ **32.** $(-8)^{5/3}$

33. $\left(\dfrac{16}{81}\right)^{3/4}$ **34.** $\left(\dfrac{32}{243}\right)^{2/5}$ **35.** $\left(\dfrac{8}{27}\right)^{-2/3}$ **36.** $\left(\dfrac{16}{625}\right)^{-3/4}$

37. $\sqrt{36}$ **38.** $-\sqrt{49}$ **39.** $\sqrt{\dfrac{9}{25}}$ **40.** $\sqrt[3]{\dfrac{27}{125}}$

In Review Exercises 41–48, simplify each expression. Assume that all variables represent positive numbers.

41. $(x^{12}y^2)^{1/2}$ **42.** $\left(\dfrac{x^{14}}{y^4}\right)^{-1/2}$ **43.** $\left(\dfrac{-c^{2/3}c^{5/3}}{c^{-2/3}}\right)^{1/3}$ **44.** $\left(\dfrac{a^{-1/4}a^{3/4}}{a^{9/2}}\right)^{-1/2}$

45. $\sqrt{x^2y^4}$ **46.** $\sqrt[3]{x^3}$ **47.** $\sqrt[4]{\dfrac{m^8n^4}{p^{12}}}$ **48.** $\sqrt[5]{\dfrac{a^{15}b^{10}}{c^5}}$

In Review Exercises 49–52, assume that all variables are unrestricted. Simplify each expression.

49. $(x^{16}y^4c^2)^{1/2}$ **50.** $\left(\dfrac{a^{14}}{b^6}\right)^{-1/2}$ **51.** $\sqrt{x^4y^8}$ **52.** $\sqrt{a^2b^4}$

In Review Exercises 53–60, rationalize each denominator and simplify.

53. $\dfrac{2}{\sqrt{5}}$ **54.** $\dfrac{8}{\sqrt{8}}$ **55.** $\dfrac{1}{\sqrt[3]{2}}$ **56.** $\dfrac{2}{\sqrt[3]{25}}$

57. $\dfrac{2}{\sqrt{3}-1}$ **58.** $\dfrac{-2}{\sqrt{3}-\sqrt{2}}$ **59.** $\dfrac{2x}{\sqrt{x}-2}$ **60.** $\dfrac{\sqrt{x}-\sqrt{y}}{\sqrt{x}+\sqrt{y}}$

In Review Exercises 61–64, rationalize each numerator and simplify.

61. $\dfrac{\sqrt{2}}{5}$ **62.** $\dfrac{\sqrt{5}}{5}$ **63.** $\dfrac{\sqrt{x}+2}{5}$ **64.** $\dfrac{1-\sqrt{a}}{a}$

In Review Exercises 65–70, simplify each expression and combine terms.

65. $\sqrt{50}+\sqrt{8}$ **66.** $\sqrt{12}+\sqrt{3}-\sqrt{27}$ **67.** $(\sqrt{2}+\sqrt{3})^2$ **68.** $(2+\sqrt{3})(\sqrt{3}-2)$

69. $(\sqrt{2}+1)(\sqrt{3}+1)$ **70.** $(\sqrt[3]{3}-2)(\sqrt[3]{9}+2\sqrt[3]{3}+4)$

In Review Exercises 71–74, indicate whether the given expression is a polynomial. If it is, give its degree and tell whether it is a monomial, binomial, or trinomial.

71. x^3-8 **72.** $8x^2-8x-8$ **73.** $\sqrt{3}x^2$ **74.** $3x^2-\sqrt{x}$

In Review Exercises 75–78, perform the indicated operations and simplify.

75. $3x^2(x-1)-2x(x+3)-x^2(x+2)$ **76.** $(3x+y)(2x-3y)$

77. $(4a+2b)(2a-3b)$ **78.** $(z+3)(3z^2+z-1)$

In Review Exercises 79–80, perform each division.

79. $2x+3\overline{)2x^3+7x^2+8x+3}$ **80.** $x^2-1\overline{)x^5+x^3-2x-3x^2-3}$

In Review Exercises 81–94, factor each polynomial completely, if possible.

81. $3t^3 - 3t$

82. $5r^3 - 5$

83. $6x^2 + 7x - 24$

84. $3a^2 + ax - 3a - x$

85. $8x^3 - 125$

86. $6x^2 - 20x - 16$

87. $x^2 + 6x + 9 - t^2$

88. $3x^2 - 1 + 5x$

89. $8z^3 + 343$

90. $1 + 14b + 49b^2$

91. $121z^2 + 4 - 44z$

92. $64y^3 - 1000$

93. $2xy - 4zx - wy + 2zw$

94. $x^8 + x^4 + 1$

In Review Exercises 95–106, perform the indicated operation and simplify.

95. $\dfrac{x^2 - 4x + 4}{x + 2} \cdot \dfrac{x^2 + 5x + 6}{x - 2}$

96. $\dfrac{2y^2 - 11y + 15}{y^2 - 6y + 8} \cdot \dfrac{y^2 - 2y - 8}{y^2 - y - 6}$

97. $\dfrac{2t^2 + t - 3}{3t^2 - 7t + 4} \div \dfrac{10t + 15}{3t^2 - t - 4}$

98. $\dfrac{p^2 + 7p + 12}{p^3 + 8p^2 + 4p} \div \dfrac{p^2 - 9}{p^2}$

99. $\dfrac{x^2 + x - 6}{x^2 - x - 6} \cdot \dfrac{x^2 - x - 6}{x^2 + x - 2} \div \dfrac{x^2 - 4}{x^2 - 5x + 6}$

100. $\left(\dfrac{2x + 6}{x + 5} \div \dfrac{2x^2 - 2x - 4}{x^2 - 25}\right)\dfrac{x^2 - x - 2}{x^2 - 2x - 15}$

101. $\dfrac{2}{x - 4} + \dfrac{3x}{x + 5}$

102. $\dfrac{5x}{x - 2} - \dfrac{3x + 7}{x + 2} + \dfrac{2x + 1}{x + 2}$

103. $\dfrac{x}{x - 1} + \dfrac{x}{x - 2} + \dfrac{x}{x - 3}$

104. $\dfrac{x}{x + 1} - \dfrac{3x + 7}{x + 2} + \dfrac{2x + 1}{x + 2}$

105. $\dfrac{3(x + 1)}{x} - \dfrac{5(x^2 + 3)}{x^2} + \dfrac{x}{x + 1}$

106. $\dfrac{3x}{x + 1} + \dfrac{x^2 + 4x + 3}{x^2 + 3x + 2} - \dfrac{x^2 + x - 6}{x^2 - 4}$

In Review Exercises 107–110, simplify each complex fraction.

107. $\dfrac{\dfrac{5x}{2}}{\dfrac{3x^2}{8}}$

108. $\dfrac{\dfrac{3x}{y}}{\dfrac{6x}{y^2}}$

109. $\dfrac{\dfrac{1}{x} + \dfrac{1}{y}}{x - y}$

110. $\dfrac{\dfrac{1}{x} + \dfrac{1}{y}}{\dfrac{1}{y} - \dfrac{1}{x}}$

CHAPTER TEST

In Questions 1–2, answer either true or false.

1. $a \in \{a, b, c\}$

2. $a \subseteq \{a, b, c\}$

In Questions 3–4, $\mathbf{A} = \{1, 2, 3\}$, $\mathbf{B} = \{2, 3, 4\}$*, and* $\mathbf{C} = \{4, 5, 6\}$*. Find each set.*

3. $(\mathbf{A} \cap \mathbf{B}) \cup \mathbf{C}$

4. $(\mathbf{A} \cup \mathbf{B}) \cap \mathbf{C}$

In Questions 5–6, graph each interval on a number line.

5. $\{x \mid -4 < x \le 2\}$

6. $[-2, 3) \cup (0, 4]$

In Questions 7–8, tell which property justifies each statement.

7. $(a + b) + c = (b + a) + c$

8. $a(b + c) = ab + ac$

In Questions 9–12, simplify each expression. Assume that all variables represent positive numbers and write all answers without using negative exponents.

9. $x^4 x^5 x^2$

10. $\dfrac{r^2 r^3 s}{r^4 s^2}$

11. $\dfrac{(a^{-1} a^2)^{-2}}{a^{-3}}$

12. $\left(\dfrac{x^0 x^2}{x^{-2}}\right)^6$

In Questions 13–14, write each number in scientific notation.

13. 450,000

14. 0.000345

In Questions 15–16, write each number in standard notation.

15. 3.7×10^3

16. 1.2×10^{-3}

In Questions 17–22, simplify each expression. Assume that all variables represent positive numbers and write all answers without using negative exponents.

17. $(25a^4)^{1/2}$

18. $\left(\dfrac{36}{81}\right)^{3/2}$

19. $\left(\dfrac{8t^6}{27s^9}\right)^{-2/3}$

20. $\sqrt[3]{27a^6}$

21. $\sqrt{12} + \sqrt{27}$

22. $2\sqrt[3]{80} - 3\sqrt[3]{24}$

In Questions 23–24, rationalize each denominator.

23. $\dfrac{x}{\sqrt{x} - 2}$

24. $\dfrac{\sqrt{x} + \sqrt{y}}{\sqrt{x} - \sqrt{y}}$

In Questions 25–30, perform each operation.

25. $(a^2 + 3) - (2a^2 - 4)$

26. $(3a^3 b^2)(-2a^3 b^4)$

27. $(3x - 4)(2x + 7)$

28. $(a^n + 2)(a^n - 1)$

29. $(x^2 + 4)(x^2 - 4)$

30. $(x^2 - x + 2)(2x - 3)$

In Questions 31–32, perform each division.

31. $x - 3\overline{)6x^2 + x - 23}$

32. $2x - 1\overline{)2x^3 + 3x^2 - 1}$

In Questions 33–38, factor each polynomial.

33. $3x + 6y$

34. $x^2 - 100$

35. $10t^2 - 19tw + 6w^2$

36. $3a^3 - 648$

37. $x^4 - x^2 - 12$

38. $6x^4 + 11x^2 - 10$

In Questions 39–42, perform each operation and simplify, if possible.

39. $\dfrac{x}{x+2} + \dfrac{2}{x+2}$

40. $\dfrac{x}{x+1} - \dfrac{x}{x-1}$

41. $\dfrac{x^2+x-20}{x^2-16} \cdot \dfrac{x^2-25}{x-5}$

42. $\dfrac{x+2}{x^2+2x+1} \div \dfrac{x^2-4}{x+1}$

In Questions 43–44, simplify each complex fraction.

43. $\dfrac{\dfrac{1}{a}+\dfrac{1}{b}}{\dfrac{1}{b}}$

44. $\dfrac{x^{-1}}{x^{-1}+y^{-1}}$

CHAPTER 2

EQUATIONS AND INEQUALITIES

The topic of this chapter is equations, one of the most important concepts in algebra. Equations are used in almost every academic discipline and vocational area, especially in chemistry, physics, medicine, economics, electronics, and business.

2.1   LINEAR EQUATIONS

■ Linear Equations   ■ Rational Equations   ■ Formulas

*François Vieta (Viête)
(1540–1603)
By using letters in place of
unknown numbers, Vieta
simplified the subject of
algebra and brought its
notation closer to the
notation that we use today.
One symbol he didn't use
was the equal sign.*

An **equation** is a statement indicating that two quantities are equal. An equation can be either true or false. For example, the equation $2 + 2 = 4$ is true, and the equation $2 + 3 = 6$ is false. An equation such as $3x - 2 = 10$ is true or false depending on the value of x, called a **variable**. If $x = 4$, the equation is a true statement:

$$3x - 2 = 10$$
$$3(4) - 2 \stackrel{?}{=} 10 \qquad \text{Substitute 4 for } x.$$
$$12 - 2 \stackrel{?}{=} 10$$
$$10 = 10$$

Because 4 makes the equation true, we say that it *satisfies* the equation. The equation is false for all other replacements of x.

Any number that satisfies an equation is called a **solution** or **root** of the equation. The set of all solutions of an equation is called the **solution set** of the equation. The solution set of the equation $3x - 2 = 10$ is $\{4\}$. To solve an equation means to find all of its solutions.

We have seen that there can be restrictions on the values of a variable. For example, in the fraction

$$\frac{x^2 + 4}{x - 2}$$

x cannot be replaced by 2, because the denominator of a fraction cannot be 0.

EXAMPLE 1   Find the restrictions on the values of x in the equation $\sqrt{x} = \dfrac{2}{x - 1}$.

Solution   For $\sqrt{x}$ to be a real number, x must be nonnegative, and for $\frac{2}{x-1}$ to be a real number, x cannot be 1. Thus, the replacement set for x is restricted to the set of all nonnegative numbers, except 1. ■

For some equations, called **identities**, every meaningful replacement for the variable is a solution of the equation. The equation

$$x^2 - 9 = (x + 3)(x - 3)$$

is an identity, because every real number x is a solution. For other equations, called **impossible equations** or **contradictions**, no number is a solution. The equation

$$x = x + 1$$

has no solutions, because no number can be one greater than itself.

An equation whose solution set contains some but not all numbers is called a **conditional equation**. In this section, we will discuss how to solve many of these equations.

Equivalent Equations	If two equations, each with one variable, have the same solution set, they are called **equivalent equations**.

There are several properties used to transform equations into equivalent but less complicated equations. If we use these properties, the resulting equations will be equivalent and will have the same solution set.

The Addition and Subtraction Properties of Equality	If a, b, and c are real numbers and $a = b$, then $$a + c = b + c \quad \text{and} \quad a - c = b - c$$
The Multiplication and Division Properties of Equality	$$ac = bc \quad \text{and} \quad \frac{a}{c} = \frac{b}{c} \quad (c \neq 0)$$
The Substitution Property of Equality	In any equation, a quantity may be substituted for its equal without changing the truth of the equation.

■ Linear Equations

The easiest equations to solve are the **first-degree** or **linear equations**.

Linear Equations	A **linear equation in one variable** (say, x) is any equation that can be written in the form $$ax + c = 0 \quad (a \text{ and } c \text{ are real numbers and } a \neq 0)$$

To solve the linear equation $2x + 3 = 0$, we subtract 3 from both sides of the equation and then divide both sides by 2 to obtain

$$2x + 3 = 0$$
$$2x + 3 - \mathbf{3} = 0 - \mathbf{3}$$
$$2x = -3$$
$$\frac{2x}{2} = -\frac{3}{2}$$
$$x = -\frac{3}{2}$$

The solution of $2x + 3 = 0$ is $-\frac{3}{2}$. To show that this solution satisfies the equation, we substitute $-\frac{3}{2}$ for x and simplify:

$$2x + 3 = 0$$
$$2\left(-\frac{3}{2}\right) + 3 \stackrel{?}{=} 0$$
$$-3 + 3 \stackrel{?}{=} 0$$
$$0 = 0$$

Because both sides of the equation are equal, the solution checks.

As this example suggests, every conditional linear equation has exactly one solution.

EXAMPLE 2   Find the solution set for $3(x + 2) = 5x + 2$.

Solution   We proceed as follows:

$$3(x + 2) = 5x + 2$$
$$3x + 6 = 5x + 2 \qquad \text{Use the distributive property and remove parentheses.}$$
$$3x + 6 - \mathbf{3x} = 5x - \mathbf{3x} + 2 \qquad \text{Subtract } 3x \text{ from both sides.}$$
$$6 = 2x + 2 \qquad \text{Simplify.}$$
$$6 - \mathbf{2} = 2x + 2 - \mathbf{2} \qquad \text{Subtract 2 from both sides.}$$
$$4 = 2x \qquad \text{Simplify.}$$
$$\frac{4}{2} = \frac{2x}{2} \qquad \text{Divide both sides by 2.}$$
$$2 = x \qquad \text{Simplify.}$$

Because all of the above equations are equivalent, the solution set of the final equation is the same as the solution set of the original equation. Thus, the solution set of the original equation is $\{2\}$. Verify that 2 satisfies the equation.   ■

EXAMPLE 3   Solve the equation $\frac{3}{2}x - \frac{2}{3} = \frac{1}{5}x.$

Solution   To clear the equation of fractions, we can multiply both sides by the LCD of the three fractions and proceed as follows:

$$\frac{3}{2}x - \frac{2}{3} = \frac{1}{5}x$$

$$30\left(\frac{3}{2}x - \frac{2}{3}\right) = 30\left(\frac{1}{5}x\right) \qquad \text{Multiply both sides by 30.}$$

$$45x - 20 = 6x \qquad \text{Remove parentheses and simplify.}$$

$$45x - 20 + 20 = 6x + 20 \qquad \text{Add 20 to both sides.}$$

$$45x = 6x + 20 \qquad \text{Simplify.}$$

$$45x - 6x = 6x - 6x + 20 \qquad \text{Subtract } 6x \text{ from both sides.}$$

$$39x = 20 \qquad \text{Simplify.}$$

$$\frac{39x}{39} = \frac{20}{39} \qquad \text{Divide both sides by 39.}$$

$$x = \frac{20}{39} \qquad \text{Simplify.}$$

The root of the given equation is $\frac{20}{39}$. Verify that it satisfies the equation. ■

EXAMPLE 4   Solve the equations **a.** $3(x + 5) = 3(1 + x)$ and **b.** $5 + 5(x + 2) - 2x = 3x + 15$.

Solution   We proceed as follows:

a.
$$3(x + 5) = 3(1 + x)$$
$$3x + 15 = 3 + 3x \qquad \text{Remove parentheses.}$$
$$3x - 3x + 15 = 3 + 3x - 3x \qquad \text{Subtract } 3x \text{ from both sides.}$$
$$15 = 3 \qquad \text{Simplify.}$$

Because the final equation is false, it has no roots. Thus, the original equation has no roots either. Its solution set is Ø.

b.
$$5 + 5(x + 2) - 2x = 3x + 15$$
$$5 + 5x + 10 - 2x = 3x + 15 \qquad \text{Remove parentheses.}$$
$$3x + 15 = 3x + 15 \qquad \text{Simplify.}$$

Because both sides of the final equation are the same, they are always equal, and every value of x will make the equation true. The solution set of the original equation is the set of all real numbers. This equation is an identity. ■

■ Rational Equations

Equations containing rational expressions are called **rational equations**. When solving these equations, we will sometimes multiply both sides by a quantity containing a variable. When we do this, it is possible to inadvertently multiply both sides of an equation by 0 and obtain a solution that makes the denominator of a fraction 0. In this case, we have found a false solution, called an **extraneous solution**.

Warning!   You must be careful to exclude from the solution set of an equation any value that makes the denominator of a fraction 0.

The following equation has an extraneous solution.

$$\frac{x+1}{x-2} = \frac{3}{x-2}$$

$$(x-2)\left(\frac{x+1}{x-2}\right) = (x-2)\left(\frac{3}{x-2}\right) \qquad \text{Multiply both sides by } x-2.$$

$$x + 1 = 3 \qquad\qquad \text{Simplify.}$$

$$x = 2 \qquad\qquad \text{Subtract 1 from both sides and simplify.}$$

If we check this solution by substituting 2 for x, we obtain 0's in the denominator. Thus, 2 is not a root. The solution set of this equation is ∅.

EXAMPLE 5   Solve the equation $\dfrac{x+2}{x+3} + \dfrac{1}{x^2 + 2x - 3} = 1$.

Solution   We note that x cannot be -3, because that would cause the denominator of the first fraction to be 0. To find other restrictions, we factor the trinomial in the denominator of the second fraction.

$$x^2 + 2x - 3 = (x+3)(x-1)$$

This denominator will be 0 when $x = -3$ or $x = 1$. Thus, x cannot be either -3 or 1. We solve the equation as follows:

$$\frac{x+2}{x+3} + \frac{1}{x^2 + 2x - 3} = 1$$

$$\frac{x+2}{x+3} + \frac{1}{(x+3)(x-1)} = 1 \qquad\qquad \text{Factor } x^2 + 2x - 3.$$

$$(x+3)(x-1)\left[\frac{x+2}{x+3} + \frac{1}{(x+3)(x-1)}\right] = (x+3)(x-1)1 \qquad \begin{array}{l}\text{Multiply both sides}\\ \text{by } (x+3)(x-1).\end{array}$$

$$(x+3)(x-1)\left(\frac{x+2}{x+3}\right) + (x+3)(x-1)\frac{1}{(x+3)(x-1)} = (x+3)(x-1)1 \qquad \text{Remove brackets.}$$

$$(x-1)(x+2) + 1 = (x+3)(x-1) \qquad \text{Simplify.}$$

$$x^2 + x - 2 + 1 = x^2 + 2x - 3 \qquad \text{Multiply the binomials.}$$

$$x - 1 = 2x - 3 \qquad \begin{array}{l}\text{Subtract } x^2 \text{ from both}\\ \text{sides and combine terms.}\end{array}$$

$$2 = x \qquad \begin{array}{l}\text{Add 3 and subtract}\\ x \text{ from both sides.}\end{array}$$

Because 2 is a meaningful replacement for x, it is a root. However, it is a good idea to check it.

$$\frac{x+2}{x+3} + \frac{1}{x^2 + 2x - 3} = 1$$

$$\frac{2 + 2}{2 + 3} + \frac{1}{2^2 + 2(2) - 3} \overset{?}{=} 1 \qquad \text{Substitute 2 for } x.$$

$$\frac{4}{5} + \frac{1}{5} \overset{?}{=} 1$$

$$1 = 1$$

Since 2 satisfies the equation, it is a root. ∎

Formulas

Many equations, called **formulas**, contain several variables. For example, the formula that converts degrees Celsius to degrees Fahrenheit is $F = \frac{9}{5}C + 32$. If we need to change a large number of Fahrenheit readings to degrees Celsius, it is tedious to substitute each value of F into the formula and then repeatedly solve the formula for C. It is more efficient to solve the formula for C, substitute the values for F, and evaluate C directly.

EXAMPLE 6   Solve the formula $F = \frac{9}{5}C + 32$ for C.

Solution   We solve for C by using the techniques for solving linear equations.

$$F = \frac{9}{5}C + 32$$

$$F - 32 = \frac{9}{5}C \qquad \text{Subtract 32 from both sides.}$$

$$\frac{5}{9}(F - 32) = \frac{5}{9}\left(\frac{9}{5}C\right) \qquad \text{Multiply both sides by } \frac{5}{9}.$$

$$\frac{5}{9}(F - 32) = C \qquad \text{Simplify.}$$

This result can be written in the alternate form $C = \dfrac{5F - 160}{9}$. ∎

EXAMPLE 7   The formula $A = p + prt$ is used in business to find the amount of money in a savings account at the end of a specified time. A represents the amount, p represents the principal (the original deposit), r represents the rate of simple interest per unit time, and t represents the number of units of time. Solve this formula for p.

Solution   We factor p from both terms on the right-hand side of the equation and proceed as follows:

$$A = p + prt$$

$$A = p(1 + rt) \qquad \text{Factor out } p.$$

$$\frac{A}{(1 + rt)} = p \qquad \text{Divide both sides by } 1 + rt.$$

$$p = \frac{A}{1 + rt} \qquad \text{Use the symmetric property of equality.}$$

∎

2.1   EXERCISES

In Exercises 1–8, each quantity represents a real number. Find all restrictions on x.

1. $x + 3 = 1$

2. $\frac{1}{2}x - 7 = 14$

3. $\frac{1}{x} = 12$

4. $\frac{3}{x - 2} = 3$

5. $\sqrt{x} = 4$

6. $\sqrt[3]{x} = 64$

7. $\frac{1}{x - 3} = \frac{5}{x + 2}$

8. $\frac{24}{\sqrt{x - 3}} = 4$

In Exercises 9–22, solve each equation, if possible. Classify each one as an identity, a conditional equation, or an equation with no solutions.

9. $2x + 5 = 15$

10. $3x + 2 = x + 8$

11. $2(x + 2) = 2x + 5$

12. $3(x + 2) - x = 2(x + 3)$

13. $\frac{x + 7}{2} = 7$

14. $\frac{x}{2} - 7 \neq 14$

15. $2(a + 1) = 3(a - 2) - a$

16. $x^2 = (x + 4)(x - 4) + 16$

17. $3(x - 3) = \frac{6x - 18}{2}$

18. $x(x + 2) = (x + 1)^2$

19. $\frac{3}{b - 3} = 1$

20. $x^2 - 8x + 15 = (x - 3)(x + 5)$

21. $2x^2 + 5x - 3 = (2x - 1)(x + 3)$

22. $2x^2 + 5x - 3 = 2x\left(x + \frac{19}{2}\right)$

In Exercises 23–64, solve each equation. If an equation has no solution, so indicate.

23. $2x + 7 = 10 - x$

24. $9a - 3 = 15 + 3a$

25. $\frac{5}{3}z - 8 = 7$

26. $\frac{4}{3}y + 12 = -4$

27. $\frac{z}{5} + 2 = 4$

28. $\frac{3p}{7} - p = -4$

29. $\frac{3x - 2}{3} = 2x + \frac{7}{3}$

30. $\frac{7}{2}x + 5 = x + \frac{15}{2}$

31. $5(x - 2) = 2x + 8$

32. $5(r - 4) = -5(r - 4)$

33. $2(2x + 1) - \frac{3x}{2} = \frac{-3(4 + x)}{2}$

34. $(x - 2)(x - 3) = (x + 3)(x + 4)$

35. $7(2x + 5) - 6(x + 8) = 7$

36. $(t + 1)(t - 1) = (t + 2)(t - 3) + 4$

37. $(x - 2)(x + 5) = (x - 3)(x + 2)$

38. $\frac{3x + 1}{20} = \frac{1}{2}$

39. $\frac{3}{2}(3x - 2) - 10x - 4 = 0$

40. $a(a - 3) + 5 = (a - 1)^2$

41. $x(x + 2) = (x + 1)^2 - 1$

42. $\frac{3 + x}{3} + \frac{x + 7}{2} = 4x + 1$

43. $\frac{(y + 2)^2}{3} = y + 2 + \frac{y^2}{3}$

44. $2x - \frac{7}{6} + \frac{x}{6} = \frac{4x + 3}{6}$

45. $2(s + 2) + (s + 3)^2 = s(s + 5) + 2\left(\frac{17}{2} + s\right)$

46. $\frac{3}{x} + \frac{1}{2} = \frac{4}{x}$

47. $\dfrac{2}{x + 1} + \dfrac{1}{3} = \dfrac{1}{x + 1}$

48. $\dfrac{3}{x - 2} + \dfrac{1}{x} = \dfrac{3}{x - 2}$

49. $\dfrac{9t + 6}{t(t + 3)} = \dfrac{7}{t + 3}$

50. $x + \dfrac{2(-2x + 1)}{3x + 5} = \dfrac{3x^2}{3x + 5}$

51. $\dfrac{2}{(a - 7)(a + 2)} = \dfrac{4}{(a + 3)(a + 2)}$

52. $\dfrac{2}{n - 2} + \dfrac{1}{n + 1} = \dfrac{1}{n^2 - n - 2}$

53. $\dfrac{2x + 3}{x^2 + 5x + 6} + \dfrac{3x - 2}{x^2 + x - 6} = \dfrac{5x - 2}{x^2 - 4}$

54. $\dfrac{3x}{x^2 + x} - \dfrac{2x}{x^2 + 5x} = \dfrac{x + 2}{x^2 + 6x + 5}$

55. $\dfrac{3x + 5}{x^3 + 8} + \dfrac{3}{x^2 - 4} = \dfrac{2(3x - 2)}{(x - 2)(x^2 - 2x + 4)}$

56. $\dfrac{1}{n + 8} - \dfrac{3n - 4}{5n^2 + 42n + 16} = \dfrac{1}{5n + 2}$

57. $\dfrac{1}{11 - n} - \dfrac{2(3n - 1)}{-7n^2 + 74n + 33} = \dfrac{1}{7n + 3}$

58. $\dfrac{4}{a^2 - 13a - 48} - \dfrac{2}{a^2 - 18a + 32} = \dfrac{1}{a^2 + a - 6}$

59. $\dfrac{5}{y + 4} + \dfrac{2}{y + 2} = \dfrac{6}{y + 2} - \dfrac{1}{y^2 + 6y + 8}$

60. $\dfrac{6}{2a - 6} - \dfrac{3}{3 - 3a} = \dfrac{1}{a^2 - 4a + 3}$

61. $\dfrac{3y}{6 - 3y} + \dfrac{2y}{2y + 4} = \dfrac{8}{4 - y^2}$

62. $\dfrac{3 + 2a}{a^2 + 6 + 5a} - \dfrac{2 - 3a}{a^2 - 6 + a} = \dfrac{5a - 2}{a^2 - 4}$

63. $\dfrac{a}{a + 2} - 1 = -\dfrac{3a + 2}{a^2 + 4a + 4}$

64. $\dfrac{x - 1}{x + 3} + \dfrac{x - 2}{x - 3} = \dfrac{1 - 2x}{3 - x}$

In Exercises 65–90, solve each formula for the specified variable.

65. $k = 2.2p$; p

66. $p = 2l + 2w$; w

67. $A = \dfrac{1}{2}h(b_1 + b_2)$; b_2

68. $V = \dfrac{1}{3}\pi r^2 h$; h

69. $V = \dfrac{1}{3}\pi r^2 h$; r^2

70. $z = \dfrac{x - \mu}{\sigma}$; μ

71. $P_n = L + \dfrac{si}{f}$; s

72. $P_n = L + \dfrac{si}{f}$; f

73. $A = 2(lw + hl + hw)$; w

74. $S = \dfrac{1}{2}gt^2 + v_0 t + s_0$; g

75. $F = \dfrac{mMg}{r^2}$; m

76. $\dfrac{1}{f} = \dfrac{1}{p} + \dfrac{1}{q}$; f

77. $\dfrac{x}{a} + \dfrac{y}{b} = 1$; y

78. $\dfrac{x}{a} - \dfrac{y}{b} = 1$; a

79. $\dfrac{x^2}{a^2} - \dfrac{y^2}{b^2} = 1$; x^2

80. $\dfrac{x^2}{a^2} + \dfrac{y^2}{b^2} = 1$; y^2

81. $\dfrac{1}{r} = \dfrac{1}{r_1} + \dfrac{1}{r_2}$; r

82. $\dfrac{1}{r} = \dfrac{1}{r_1} + \dfrac{1}{r_2}$; r_1

83. $l = a + (n - 1)d$; n

84. $l = a + (n - 1)d$; d

85. $a = (n - 2)\dfrac{180}{n}$; n

86. $S = \pi h(r + h)$; r

87. $S = \dfrac{a - lr}{1 - r}$; a

88. $S = \dfrac{a - lr}{1 - r}$; r

89. $R = \dfrac{1}{\dfrac{1}{r_1} + \dfrac{1}{r_2} + \dfrac{1}{r_3}}$; r_1

90. $R = \dfrac{1}{\dfrac{1}{r_1} + \dfrac{1}{r_2} + \dfrac{1}{r_3}}$; r_3

2.2 APPLICATIONS OF LINEAR EQUATIONS

- ■ Number Problems ■ Geometric Problems ■ Investment Problems
- ■ Break-Point Analysis ■ Shared Work Problems ■ Mixture Problems
- ■ Uniform-Motion Problems

In this section, we will apply equation-solving techniques to word problems. The list of steps below provides a strategy to follow.

Strategy for Solving Applied Problems	1. Read the problem several times until you understand the given facts. What information is given? What are you asked to find? Draw a sketch, diagram, or table if possible, to help you visualize the facts of the problem.
	2. Pick a variable to represent the quantity that is to be found, and write a sentence telling what that variable represents. Express all of the other quantities mentioned in the problem as expressions involving this single variable.
	3. Organize the data and find a way to express the same quantity in two different ways. This might involve a formula from geometry, finance, or physics.
	4. Set up an equation indicating that the two quantities found in Step 3 are equal.
	5. Solve the equation.
	6. Answer the questions posed in the problem. Have you found all of the required information?
	7. Check the answers in the words of the problem.

This list does not apply to all situations, but it can be applied to a wide range of problems with only slight modifications. The following examples use these steps to solve a variety of word problems.

■ Number Problems

EXAMPLE 1 A mathematics student has scores of 74%, 78%, and 70% on three examinations. What score is needed on a fourth exam for the student to earn an average grade of 80%?

Solution We can let x represent the required grade on the fourth exam. The average grade will be one-fourth of the sum of the four grades, and we know this average is to be 80.

$$\boxed{\text{The average of the four grades}} = \boxed{\text{the required average grade.}}$$

$$\frac{74 + 78 + 70 + x}{4} = 80$$

We can now solve the equation for x.

$$\frac{222 + x}{4} = 80$$

$$222 + x = 320$$

$$x = 98$$

To score an average of 80% on the four exams, the student must score 98% on the fourth exam. ■

Geometric Problems

EXAMPLE 2 A city ordinance requires a man to install a fence around the swimming pool shown in Figure 2-1. If he wants the border around the pool to be of uniform width and has 154 feet of fencing, find the width of the border.

Solution We can let x represent the width of the border. The distance around the large rectangle, called its **perimeter**, is given by the formula $P = 2l + 2w$, where l is the length, $20 + 2x$, and w is the width, $16 + 2x$. Since the man has 154 feet of fencing, the perimeter will be 154 feet. We substitute these values into the formula for perimeter to get

$$P = 2l + 2w$$

$$\mathbf{154} = 2(\mathbf{20} + 2x) + 2(\mathbf{16} + 2x)$$

$$154 = 40 + 4x + 32 + 4x$$

$$154 = 72 + 8x$$

$$82 = 8x$$

$$10\frac{1}{4} = x$$

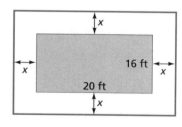

FIGURE **2-1**

The uniform border will be $10\frac{1}{4}$ feet wide. ■

Investment Problems

EXAMPLE 3 A woman invests $10,000, part at 9% annual interest and the rest at 14%, compounded annually. The total annual income from these two investments is $1275. How much is invested at each rate?

Solution We can let x represent the amount invested at 9% interest. Then $10,000 - x$ represents the amount invested at 14% interest. The annual income from any investment is the product of the interest rate and the amount invested. The total income from these two investments can be expressed in two ways: as $1275 and as the sum of the incomes of the two investments.

*George Polya
(1888–1985)*

Polya, a Hungarian, became a Professor of Mathematics at Stanford University. His approach to problem solving made him very popular with faculty and students. His book How to Solve It *became a best-seller. His problem-solving approach involves four steps:*

1. Understand the problem.
2. Devise a plan.
3. Carry out the plan.
4. Check back.

The income from the 9% investment	+	the income from the 14% investment	=	the total income.

$$9\% \text{ of } \binom{\text{amount invested}}{\text{at } 9\%} + 14\% \text{ of } \binom{\text{amount invested}}{\text{at } 14\%} = \binom{\text{total}}{\text{income}}$$

$$0.09x \qquad + \qquad 0.14(10{,}000 - x) \qquad = \qquad 1275$$

Multiply both sides of the equation by 100 to clear it of decimals and solve for x:

$$9x + 14(10{,}000 - x) = 127{,}500$$
$$9x + 140{,}000 - 14x = 127{,}500$$
$$-5x = -12{,}500$$
$$x = 2500$$

The investment at 9% was $2500, while $10,000 − $2500, or $7500, was invested at 14%. These amounts are correct, because 9% of $2500 is $225, and 14% of $7500 is $1050, and the sum of these amounts is $1275. ∎

Break-Point Analysis

Running a machine involves two types of costs—**setup costs** and **unit costs**. Setup costs include the cost of installing a machine and preparing it to do a specific job. Unit cost is the cost to manufacture one item. It includes one item's share of the costs of raw material and labor.

EXAMPLE 4 Suppose that one machine has a setup cost of $400 and a unit cost of $1.50, while a second machine has a setup cost of $500 and a unit cost of $1.25. Find the number of units to be manufactured if the cost on each machine is the same. This number is called the **break point**.

Solution We can let x represent the number of items to be manufactured. The cost c_1 of using machine 1 is

$$c_1 = 400 + 1.5x$$

and the cost c_2 of using machine 2 is

$$c_2 = 500 + 1.25x$$

The break point occurs when these two costs are equal.

The cost of using machine 1	=	the cost of using machine 2

$$400 + 1.5x = 500 + 1.25x$$
$$1.5x = 100 + 1.25x \qquad \text{Subtract 400 from both sides.}$$
$$0.25x = 100 \qquad \text{Subtract } 1.25x \text{ from both sides.}$$
$$x = 400 \qquad \text{Divide both sides by 0.25.}$$

The break point is 400 units. This result is correct because it will cost the same amount to manufacture 400 units with either machine.

$$c_1 = \$400 + \$1.5(400) = \$1000 \quad \text{and} \quad c_2 = \$500 + \$1.25(400) = \$1000$$

■

■ Shared Work Problems

EXAMPLE 5   The Illinois Tollway Authority needs to pave 100 miles of interstate tollway before freezing temperatures occur in about 60 days. Sjostrom and Sons Cement Contractors has estimated that they can do the job in 110 days. Scandroli and Sons has estimated that they can do the job in 140 days. Will the job get finished in time if the Tollway Authority hires both contractors?

Solution   We can let n represent the number of days it will take to pave the highway if both contractors are hired. In one day, the contractors working together can do $\frac{1}{n}$ of the job. In one day, Sjostrom can do $\frac{1}{110}$ of the job. In one day, Scandroli can do $\frac{1}{140}$ of the job. The work that they can do together in one day is the sum of what each can do separately in one day. This gives the equation

The part of the tollway Sjostrom can pave in one day		the part of the tollway Scandroli can pave in one day		the part of the tollway they can pave together in one day.
$\frac{1}{110}$	$+$	$\frac{1}{140}$	$=$	$\frac{1}{n}$

We now multiply both sides of this equation by $(110)(140)n$ to clear the fractions and solve for n.

$$(110)(140)n\left(\frac{1}{110} + \frac{1}{140}\right) = (110)(140)n\left(\frac{1}{n}\right)$$

$$140n + 110n = 15{,}400 \qquad \text{Simplify.}$$

$$250n = 15{,}400 \qquad \text{Combine terms.}$$

$$n = 61.6 \qquad \text{Divide both sides by 250.}$$

It will take the contractors about 62 days to pave the highway. If the Tollway Authority is lucky, the job will be done in time.   ■

■ Mixture Problems

EXAMPLE 6   A container is partially filled with 20 liters of whole milk containing 4% butterfat. How much 1% milk must be added to obtain a mixture that is 2% butterfat?

Solution   See Figure 2-2. We can let l represent the number of liters of 1% milk that must be added to the whole milk. Then $20 + l$ represents the number of liters in the final mixture. The butterfat in the final mixture is the sum of the butterfat in the whole milk and the butterfat in the 1% milk to be added. This gives the equation

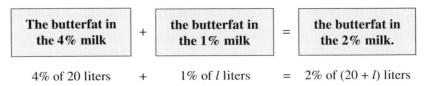

The butterfat in the 4% milk	+	the butterfat in the 1% milk	=	the butterfat in the 2% milk.
4% of 20 liters	+	1% of l liters	=	2% of $(20 + l)$ liters

We can solve this equation for l.

$$0.04(20) + 0.01(l) = 0.02(20 + l)$$
$$4(20) + l = 2(20 + l) \qquad \text{Multiply both sides by 100.}$$
$$80 + l = 40 + 2l \qquad \text{Remove parentheses.}$$
$$40 = l \qquad \text{Subtract 40 and } l \text{ from both sides.}$$

To dilute the 20 liters of 4% milk to a 2% mixture, 40 liters of 1% milk must be added. To check this answer, we note that the final mixture contains $0.02(60) = 1.2$ liters of pure butterfat, and that this is equal to the amount of pure butterfat in the 4% milk and the 1% milk: $0.04(20) + 0.01(40) = 1.2$ liters.

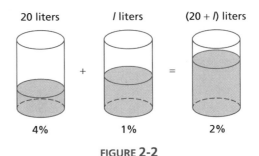

FIGURE 2-2

Uniform-Motion Problems

EXAMPLE 7   A man left home driving at the rate of 50 miles per hour. When his daughter discovered that he had forgotten his wallet, she drove after him at the rate of 65 miles per hour. How long would it take her to catch her dad if he had a 15-minute head start?

Solution   Uniform-motion problems are based on the formula $d = rt$, where d is the distance, r is the rate, and t is the time. We can organize the information given in the problem in the chart shown in Figure 2-3. In the chart, t represents the number of hours the

	d	=	r	$\cdot$	t
Man	$50\left(t + \frac{1}{4}\right)$		50		$t + \frac{1}{4}$
Daughter	$65t$		65		t

FIGURE 2-3

daughter had to drive to overtake her father. Because the father had a fifteen-minute, or $\frac{1}{4}$ hour, head start, he was on the road for $\left(t + \frac{1}{4}\right)$ hours. We can set up the following equation and solve it for r.

The distance the man drove	=	the distance the daughter drove.

$$50\left(t + \frac{1}{4}\right) = 65t$$

$$50t + \frac{25}{2} = 65t$$

$$\frac{25}{2} = 15t$$

$$\frac{5}{6} = t$$

It would take the daughter $\frac{5}{6}$ hour, or 50 minutes, to overtake her father. ■

2.2 EXERCISES

In Exercises 1–48, solve each word problem.

1. Test scores Juan scored 5 points higher on his midterm exam and 13 points higher on his final exam than he did on his first exam. If his mean (average) score was 90, what did he score on the first exam?

2. Test scores Sally took four tests in science class. On each successive test, her score improved by 3 points. If her mean score was 69.5%, what did she score on the first test?

3. Number problem One number is 3 more than twice another, and their sum is 54. Find the numbers.

4. Number problem The sum of three consecutive odd integers is 69. Find the integers. (*Hint:* Consecutive odd integers differ by 2.)

5. Replacing locks A locksmith charges $40 plus $28 for each lock installed. How many locks can be replaced for $236?

6. Delivering ads A college student earns $20 per day delivering advertising brochures door-to-door, plus 75¢ for each person he interviews. How many people did he interview on a day when he earned $56?

7. Width of a picture frame The picture frame with dimensions shown in Illustration 1 was built with 14 feet of framing material. Find its width.

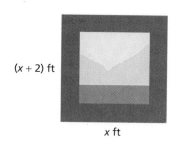

$(x + 2)$ ft

x ft

ILLUSTRATION 1

8. Tilling a garden If a gardener tills the largest rectangular area shown in Illustration 2 instead of just the square area, he will need twice as much fencing to enclose the garden. How much fencing will he need?

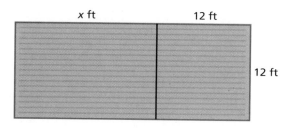

x ft   12 ft

12 ft

ILLUSTRATION 2

9. *Wading pool dimensions* The area of the triangular swimming pool shown in Illustration 3 is doubled by adding a rectangular wading pool. Find the dimensions of the wading pool. (*Hint:* Area of a triangle $= \frac{1}{2}bh$, and area of a rectangle $= lw$.)

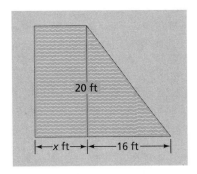

ILLUSTRATION **3**

10. *House construction* A house builder wants to install two triangular windows with the dimensions shown in Illustration 4. What angles will he have to cut to make the windows fit? (*Hint:* The sum of the angles in a triangle equals 180°.)

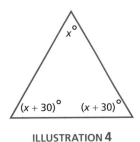

ILLUSTRATION **4**

11. *Length of a living room* If a carpenter adds the porch with dimensions shown in Illustration 5 to the living room, the living area will be increased by 50%. Find the length of the living room.

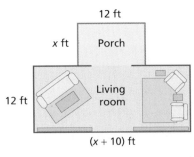

ILLUSTRATION **5**

12. *Depth of water in a trough* The trough shown in Illustration 6 has a cross-sectional area of 54 square inches. Find the depth, d, of the trough. (*Hint:* Area of a trapezoid $= \frac{1}{2}h(b + b')$.))

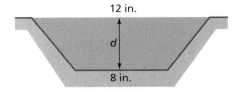

ILLUSTRATION **6**

13. *Investment problem* An executive invests $22,000, some at 7% and some at 6% annual interest. If he receives an annual return of $1420, how much is invested at each rate?

14. *Financial planning* After withdrawing the money from her IRA (individual retirement account), a woman wants to invest enough money to have an annual income of $5100. If she can invest $20,000 at 9% annual interest, how much more will she have to invest at 7.5% to achieve her goal?

15. *Ticket sales* An adult ticket for a college basketball game costs $2.50, and a student ticket costs $1.75. If 585 tickets were sold, with total receipts of $1217.25, how many tickets were student tickets?

16. *Ticket sales* Of the 800 tickets sold for a movie, 480 were adult tickets costing $3 each. If the gate receipts were $2080, what did a student ticket cost?

17. *Investment problem* A woman invests $37,000, part at 8% and the rest at $9\frac{1}{2}$% annual interest. The $9\frac{1}{2}$% investment provides $452.50 more income than the 8% investment. How much is invested at each rate?

18. *Investment problem* Equal amounts are invested at 6%, 7%, and 8% annual interest. The three investments yield a total of $2037 annual interest. Find the total investment.

19. *Discount* After being discounted 20%, a radio sells for $63.96. Find the original price.

20. *Markup* A merchant increases the wholesale cost of a washing machine by 30% to determine the selling price. If the washer sells for $588.90, find the wholesale cost.

21. *Break-point analysis* A machine to mill a brass plate has a setup cost of $600 and a unit cost of $3 for each plate manufactured. A bigger machine has a setup cost of $800 but a unit cost of only $2 for each plate manufactured. Find the break point.

22. **Break-point analysis** A machine to manufacture fasteners has a setup cost of $1200 and a unit cost of $0.005 for each fastener manufactured. A newer machine has a setup cost of $1500 but a unit cost of only $0.0015 for each fastener manufactured. Find the break point.

23. **Computer sales** A computer store has fixed costs of $8925 per month and a unit cost of $850 for every computer it sells. If the store can sell all the computers it can get for $1275 each, how many must be sold for the store to break even? (*Hint:* The break-even point occurs when costs equal income.)

24. **Restaurant management** A restaurant has fixed costs of $137.50 per day and an average unit cost of $4.75 for each meal served. If a typical check is $6, how many customers must eat at the restaurant each day for the owner to make a profit?

25. **Mowing lawns** If a woman can mow a yard with a lawn tractor in 2 hours, and her husband can mow the same lawn with a push mower in 4 hours, how long will it take to mow the lawn if they work together?

26. **Filling a swimming pool** A garden hose can fill a swimming pool in 3 days, and a larger hose can fill the pool in 2 days. How long will it take to fill the pool if both hoses are used?

27. **Filling a swimming pool** An empty swimming pool can be filled in 10 hours. When full, the pool can be drained in 19 hours. How long will it take to fill the empty pool if the drain is left open?

28. **Preparing seafood** In his job as a seafood chef, Sam stuffs shrimp. He can stuff 1000 shrimp in 6 hours. When his sister helps him, they can stuff 1000 shrimp in 4 hours. If Sam gets sick, how long will it take his sister to stuff 500 shrimp?

29. **Winterizing a car** A car radiator has a 6-liter capacity. If the liquid in the radiator is 40% antifreeze, how much liquid must be replaced with pure antifreeze to bring the mixture up to a 50% solution?

30. **Mixing milk** If a bottle holding 3 liters of milk contains $3\frac{1}{2}$% butterfat, how much skimmed milk must be added to dilute the milk to 2% butterfat?

31. **Preparing solutions** A nurse has 1 liter of a solution that is 20% alcohol. How much pure alcohol must she add to bring the solution up to a 25% concentration?

32. **Diluting solutions** If there are 400 cubic centimeters of a chemical in 1 liter of solution, how many cubic centimeters of water must be added to dilute it to a 25% solution? (*Hint:* 1000 cc = 1 liter.)

33. **Cleaning a swimming pool** A swimming pool contains 15,000 gallons of water. How many gallons of chlorine must added to "shock the pool" and bring the water to a $\frac{3}{100}$% solution?

34. **Mixing gasolines** A new automobile engine can run on a mixture of gasoline and a substitute fuel. If gas costs $1.50 per gallon and the substitute fuel costs 40¢ per gallon, what percent of a mixture must be substitute fuel to bring the cost down to $1 per gallon?

35. **Evaporation** How many liters of water must evaporate to turn 12 liters of a 24% salt solution into a 36% solution?

36. **Preparing medicine** A doctor prescribes an ointment that is 2% hydrocortisone. A pharmacist has 1% and 5% concentrations in stock. How much of each should the pharmacist use to make a 1-ounce tube?

37. **Driving rates** John drove to a distant city in 5 hours. When he returned, there was less traffic, and the trip took only 3 hours. If John drove 26 miles per hour faster on the return trip, how fast did he drive each way?

38. **Distance problem** Suzi drove home at 60 miles per hour, but her brother Jim, who left at the same time, could drive at only 48 miles per hour. When Suzi arrived, Jim still had 60 miles to go. How far did Suzi drive?

39. **Distance problem** Two cars leave Pima Community College, traveling in opposite directions. One car travels at 60 miles per hour and the other at 64 miles per hour. In how many hours will they be 310 miles apart?

40. **Bank robbery** Some bank robbers leave town, speeding at 70 miles per hour. Ten minutes later, the police give chase, traveling at 78 miles per hour. How long will it take the police to overtake the robbers?

41. **Jogging problem** Two cross-country runners are 440 yards apart and are running toward each other, one at 8 miles per hour and the other at 10 miles per hour. In how many seconds will they meet?

42. **Driving rates** One morning John drove 5 hours before stopping to eat. After lunch, he increased his speed by 10 miles per hour. He completed a 430-mile trip in 8 hours of driving time. How fast did he drive in the morning?

43. Boating problem A motorboat goes 5 miles upstream in the same time it requires to go 7 miles downstream. If the river flows at 2 miles per hour, find the speed of the boat in still water.

44. Wind velocity A plane can fly 340 miles per hour in still air. If it can fly 200 miles downwind in the same amount of time it can fly 140 miles upwind, find the velocity of the wind.

45. Coin problem A child has equal numbers of nickels, dimes, and quarters. If the coins are worth $3.20, how many of each type are there?

46. Coin problem Maria has twice as many quarters as dimes. If all the dimes were quarters and all the quarters

were dimes, she would have 60¢ less. How much money does she have?

47. Feeding cattle A farmer wants to mix 2400 pounds of cattle feed that is to be 14% protein. Barley (11.7% protein) will make up 25% of the mixture. The remaining 75% will be made up of oats (11.8% protein) and soybean oil meal (44.5% protein). How many pounds of each will the farmer use?

48. Feeding cattle If the farmer in Exercise 47 wants only 20% of the mixture to be barley, how many pounds of each will he use?

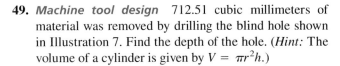 *In Exercises 49–50, use a calculator to solve each problem.*

49. Machine tool design 712.51 cubic millimeters of material was removed by drilling the blind hole shown in Illustration 7. Find the depth of the hole. (*Hint:* The volume of a cylinder is given by $V = \pi r^2 h$.)

50. Architecture The Norman window with dimensions as shown in Illustration 8 is a rectangle topped by a semicircle. If the area of the window is 68.2 square feet, find its height h.

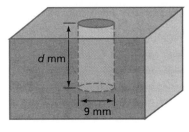

ILLUSTRATION 7

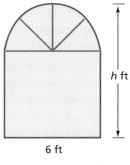

ILLUSTRATION 8

2.3 QUADRATIC EQUATIONS

■ **Completing the Square** ■ **The Quadratic Formula**. ■ **The Discriminant**
■ **Formulas**

Equations such as $2x^2 - 11x - 21 = 0$ and $3x^2 - x - 2 = 0$ are called **quadratic** or **second-degree** equations.

| **Quadratic Equation** | A **quadratic equation** is an equation that can be written in the form $ax^2 + bx + c = 0$, where a, b, and c are real numbers and $a \neq 0$. |

To solve quadratic equations by factoring, we use the following theorem.

| **The Zero-Factor Theorem** | Let a and b be real numbers.

$\quad$ If $ab = 0$, then $a = 0$ or $b = 0$. |

Proof   To prove the zero-factor theorem, we suppose that $ab = 0$. If $a = 0$, we are finished, because at least one of a or b is 0.

$\quad$ If $a \neq 0$, then a has a reciprocal $\frac{1}{a}$, and we can multiply both sides of the equation $ab = 0$ by $\frac{1}{a}$ to obtain

$$ab = 0$$

$$\frac{1}{a}(ab) = \frac{1}{a}(0)$$

$$\left(\frac{1}{a} \cdot a\right)b = 0$$

$$1b = 0$$

$$b = 0$$

Thus, if $a \neq 0$, then b must be 0, and the theorem is proved. $\qquad\square$

EXAMPLE 1   Solve the quadratic equation $2x^2 - 9x - 35 = 0$.

Solution   The left-hand side of the equation can be factored, and the equation can be written as

$$(2x + 5)(x - 7) = 0$$

If either factor is 0, the product will be 0. So we can use the zero-factor theorem and set each factor equal to 0. Then we can solve for x.

$$
\begin{array}{lll}
2x + 5 = 0 & \text{or} & x - 7 = 0 \\
2x = -5 & & x = 7 \\
x = -\dfrac{5}{2} & &
\end{array}
$$

Because the product $(2x + 5)(x - 7)$ can be 0 only if one of its factors is 0, the numbers $-\frac{5}{2}$ and 7 are the only solutions of the given equation. Verify that each of these solutions satisfies the equation $2x^2 - 9x - 35 = 0$. $\qquad\blacksquare$

$\quad$ When the quadratic expression in a quadratic equation factors, the factoring method is easy. However, many quadratic expressions do not factor over the set of integers. For example, the left-hand side of $3x^2 + 25x + 1 = 0$ is a prime polynomial and cannot be factored over the set of integers. To solve such equations, we need another method.

To develop this method, we consider the equation $x^2 = c$. If c is positive, its two real roots can be found by adding $-c$ to both sides, factoring the binomial $x^2 - c$, setting each factor to 0, and solving for x.

$$x^2 = c$$
$$x^2 - c = 0$$
$$x^2 - (\sqrt{c})^2 = 0$$
$$(x - \sqrt{c})(x + \sqrt{c}) = 0$$
$$x - \sqrt{c} = 0 \qquad \text{or} \qquad x + \sqrt{c} = 0$$
$$x = \sqrt{c} \qquad \qquad x = -\sqrt{c}$$

Thus, the roots of the equation $x^2 = c$ are $x = \sqrt{c}$ and $x = -\sqrt{c}$. This fact is called the **square root property**.

The Square Root Property	If $c > 0$, the equation $x^2 = c$ has two real roots: $$x = \sqrt{c} \qquad \text{or} \qquad x = -\sqrt{c}$$

EXAMPLE 2   Solve the equation $x^2 - 8 = 0$.

Solution   We solve the equation for x^2 and apply the square root property.

$$x^2 - 8 = 0$$
$$x^2 = 8 \qquad \qquad \text{Add 8 to both sides.}$$
$$x = \sqrt{8} \qquad \text{or} \qquad x = -\sqrt{8}$$
$$x = 2\sqrt{2} \qquad \qquad x = -2\sqrt{2} \qquad \sqrt{8} = \sqrt{4}\sqrt{2} = 2\sqrt{2}$$

Verify that each root satisfies the equation.   ∎

EXAMPLE 3   Solve the equation $(x + 4)^2 = 1$.

Solution
$$(x + 4)^2 = 1$$
$$x + 4 = \sqrt{1} \qquad \text{or} \qquad x + 4 = -\sqrt{1}$$
$$x + 4 = 1 \qquad \qquad x + 4 = -1$$
$$x = -3 \qquad \qquad x = -5$$

Verify that each root satisfies the equation.   ∎

■ Completing the Square

We now discuss another method, called **completing the square**, that can be used to solve quadratic equations. To use this method to solve the equation $x^2 - 10x + 24 = 0$, for example, we begin by subtracting 24 from both sides of the equation to get the constant term on the right-hand side of the equal sign.

$$x^2 - 10x = -24$$

If we add 25 to both sides of the equation, the left-hand side becomes a perfect-square trinomial.

$$x^2 - 10x + 25 = -24 + 25$$
$$x^2 - 10x + 25 = 1$$

When the trinomial on the left-hand side is factored, we obtain

$$(x - 5)^2 = 1$$

which we can solve by using the square root property.

$$x - 5 = 1 \quad \text{or} \quad x - 5 = -1$$
$$x = 6 \quad \quad \quad x = 4$$

EXAMPLE 4   Use the method of completing the square to solve the equation $x^2 + 4x - 5 = 0$.

Solution   We can rewrite the equation in the form

$$x^2 + 4x = 5$$

and if we add 4 to both sides, the left-hand side becomes a perfect-square trinomial that can be factored.

$$x^2 + 4x + 4 = 5 + 4$$
$$x^2 + 4x + 4 = 9$$
$$(x + 2)^2 = 9$$
$$x + 2 = 3 \quad \text{or} \quad x + 2 = -3$$
$$x = 1 \quad \quad \quad x = -5$$

Verify that each root satisfies the original equation. ∎

A question is raised by the previous two examples. What number must be added to $x^2 + bx$ to make a perfect square trinomial? There is a pattern in the following list of perfect square trinomials. We look at the right-hand side of each equation for a pattern relating the constant term and the coefficient of x.

$$(x + 1)^2 = x^2 + 2x + 1 \qquad (x + 3)^2 = x^2 + 6x + 9$$
$$(x + 5)^2 = x^2 + 10x + 25 \qquad (x - 7)^2 = x^2 - 14x + 49$$
$$(x - 8)^2 = x^2 - 16x + 64 \qquad (x - a)^2 = x^2 - 2ax + a^2$$

In each case, the constant term that completes the square is the square of one-half of the coefficient of x.

EXAMPLE 5   Solve the equation $x(x + 3) = 2$.

Solution   Remove parentheses to get

$$x^2 + 3x = 2$$

To solve this equation, we calculate the number to be added to both sides of the equation to complete the square. One-half of 3 (the coefficient of x) is $\frac{3}{2}$, and the square of $\frac{3}{2}$ is $\frac{9}{4}$. This is the number to add to both sides to complete the square.

$$x^2 + 3x + \frac{9}{4} = 2 + \frac{9}{4}$$

$$\left(x + \frac{3}{2}\right)^2 = \frac{17}{4} \qquad \text{Factor } x^2 + 3x + \frac{9}{4}.$$

$$x + \frac{3}{2} = \frac{\sqrt{17}}{2} \qquad \text{or} \qquad x + \frac{3}{2} = -\frac{\sqrt{17}}{2}$$

$$x = \frac{-3 + \sqrt{17}}{2} \qquad\qquad\qquad x = \frac{-3 - \sqrt{17}}{2}$$

Verify that each root satisfies the original equation. ∎

To solve a quadratic equation involving the variable x by completing the square, we follow these steps.

Method of Completing the Square	**1.** If the coefficient of x^2 is not 1, change it to a 1 by dividing both sides of the equation by the coefficient of x^2.
	2. If necessary, add a number to both sides of the equation to get the constant on the right-hand side of the equation.
	3. Complete the square on x: **a.** Identify the coefficient of x. **b.** Divide the coefficient of x by 2. **c.** Square the result obtained in part **b**. **d.** Add the number found in part **c**. to both sides of the equation.
	4. Factor the perfect square trinomial and combine terms.
	5. Solve the resulting quadratic equation by using the square root property.

EXAMPLE 6 Solve the equation $6x^2 + 5x - 6 = 0$.

Solution We begin by dividing both sides of the equation by 6 to make the coefficient of x^2 equal to 1. Then we proceed as follows:

$$6x^2 + 5x - 6 = 0$$

$$x^2 + \frac{5}{6}x - 1 = 0 \qquad \text{Divide both sides by 6.}$$

$$x^2 + \frac{5}{6}x = 1 \qquad \text{Add 1 to both sides.}$$

$$x^2 + \frac{5}{6}x + \frac{25}{144} = 1 + \frac{25}{144} \qquad \text{Add } \left(\frac{1}{2} \cdot \frac{5}{6}\right)^2, \text{ or } \frac{25}{144}, \text{ to both sides.}$$

$$\left(x + \frac{5}{12}\right)^2 = \frac{169}{144} \qquad \text{Factor } x^2 + \frac{5}{6}x + \frac{25}{144}.$$

Now we apply the square root property.

$$x + \frac{5}{12} = \sqrt{\frac{169}{144}} \quad \text{or} \quad x + \frac{5}{12} = -\sqrt{\frac{169}{144}}$$

$$x + \frac{5}{12} = \frac{13}{12} \qquad\qquad x + \frac{5}{12} = -\frac{13}{12}$$

$$x = \frac{8}{12} \qquad\qquad x = -\frac{18}{12}$$

$$x = \frac{2}{3} \qquad\qquad x = -\frac{3}{2}$$

Verify that each root satisfies the original equation. ■

The Quadratic Formula

We can solve the equation $ax^2 + bx + c = 0$, where $a \neq 0$, by completing the square. The result will be a formula that can be used to solve all quadratic equations.

$$ax^2 + bx + c = 0$$

$$x^2 + \frac{b}{a}x + \frac{c}{a} = \frac{0}{a} \qquad \text{Divide both sides by } a.$$

$$x^2 + \frac{b}{a}x = -\frac{c}{a} \qquad \text{Subtract } \frac{c}{a} \text{ from both sides.}$$

$$x^2 + \frac{b}{a}x + \frac{b^2}{4a^2} = \frac{b^2}{4a^2} - \frac{c}{a} \qquad \text{Add } \frac{b^2}{4a^2} \text{ to both sides.}$$

$$x^2 + \frac{b}{a}x + \frac{b^2}{4a^2} = \frac{b^2}{4a^2} - \frac{4ac}{4aa} \qquad \text{Multiply the numerator and denominator of } \frac{c}{a} \text{ by } 4a.$$

$$\left(x + \frac{b}{2a}\right)^2 = \frac{b^2 - 4ac}{4a^2} \qquad \text{Factor the left-hand side and add the fractions on the right-hand side.}$$

We can now apply the square root property.

$$x + \frac{b}{2a} = \sqrt{\frac{b^2 - 4ac}{4a^2}} \quad \text{or} \quad x + \frac{b}{2a} = -\sqrt{\frac{b^2 - 4ac}{4a^2}}$$

$$x = -\frac{b}{2a} + \frac{\sqrt{b^2 - 4ac}}{2a} \qquad\qquad x = -\frac{b}{2a} - \frac{\sqrt{b^2 - 4ac}}{2a}$$

$$x = \frac{-b + \sqrt{b^2 - 4ac}}{2a} \qquad\qquad x = \frac{-b - \sqrt{b^2 - 4ac}}{2a}$$

These two values of x are the two roots of the equation $ax^2 + bx + c = 0$. They are usually combined into a single expression, called the **quadratic formula**.

The Quadratic Formula	The solutions of the general quadratic equation $ax^2 + bx + c = 0$, where $a \neq 0$, are $$x = \frac{-b \pm \sqrt{b^2 - 4ac}}{2a}$$

 Warning! The quadratic formula should be read twice, once using the $+$ sign and once using the $-$ sign. The quadratic formula implies that

$$x = \frac{-b + \sqrt{b^2 - 4ac}}{2a} \quad \text{and} \quad x = \frac{-b - \sqrt{b^2 - 4ac}}{2a}$$

When writing the quadratic formula, be sure that the fraction bar is drawn under the entire numerator.

EXAMPLE 7 Use the quadratic formula to solve the equation $2x^2 + 8x + 7 = 0$.

Solution In this equation, $a = 2$, $b = 8$, and $c = 7$. We substitute these numbers in the quadratic formula and simplify.

$$x = \frac{-b \pm \sqrt{b^2 - 4ac}}{2a}$$

$$x = \frac{-8 \pm \sqrt{8^2 - 4(2)(7)}}{2(2)}$$

$$x = \frac{-8 \pm \sqrt{8}}{4} \qquad\qquad 8^2 - 4(2)(7) = 64 - 56 = 8$$

$$x = \frac{-8 \pm 2\sqrt{2}}{4}$$

$$x = -2 + \frac{\sqrt{2}}{2} \qquad \text{or} \qquad x = -2 - \frac{\sqrt{2}}{2}$$

Both values satisfy the original equation. ∎

EXAMPLE 8 Use the quadratic formula to solve the equation $x^2 - 5x = -3$.

Solution We first write the equation in quadratic form: $x^2 - 5x + 3 = 0$. In this equation, $a = 1$, $b = -5$, and $c = 3$.

$$x = \frac{-b \pm \sqrt{b^2 - 4ac}}{2a}$$

$$x = \frac{-(-5) \pm \sqrt{(-5)^2 - 4(1)(3)}}{2(1)}$$

$$x = \frac{5 \pm \sqrt{13}}{2} \qquad\qquad (-5)^2 - 4(1)(3) = 25 - 12 = 13$$

Both values satisfy the original equation. ∎

■ The Discriminant

It is possible to predict the type of numbers that will be roots of a given quadratic equation before solving the equation. Suppose that the coefficients a, b, and c in the quadratic equation $ax^2 + bx + c = 0$ are real numbers. Then the two roots of the equation are given by the quadratic formula

$$x = \frac{-b \pm \sqrt{b^2 - 4ac}}{2a} \qquad (a \neq 0)$$

The value of the expression $b^2 - 4ac$, called the **discriminant**, determines the nature of the roots of a given quadratic equation. When $b^2 - 4ac$ is a positive number or 0, the roots are real numbers. When $b^2 - 4ac$ is a negative number, the roots are not real numbers. The possibilities are summarized in the following table.

Nature of the Roots of a Quadratic Equation	If a, b, and c are real numbers and	
	$b^2 - 4ac$ is . . .	then the solutions are
	positive	unequal real numbers
	0	equal real numbers
	negative	not real numbers
	If a, b, and c are rational numbers and	
	$b^2 - 4ac$ is . . .	then the solutions are
	0	equal rational numbers
	a positive perfect square	unequal rational numbers
	a positive nonperfect square	unequal irrational numbers

EXAMPLE 9   Determine the nature of the roots of the quadratic equation $3x^2 + 4x + 1 = 0$.

Solution   We calculate the discriminant $b^2 - 4ac$ as follows:

$$b^2 - 4ac = 4^2 - 4(\mathbf{3})(\mathbf{1})$$
$$= 16 - 12$$
$$= 4$$

Since a, b, and c are rational numbers and the discriminant is a perfect square, the two roots of the given quadratic equation are unequal rational numbers.   ■

EXAMPLE 10   If k is a constant, many quadratic equations are represented by the equation

$$(k - 2)x^2 + (k + 1)x + 4 = 0$$

Find the values of k that will give an equation with roots that are equal real numbers.

Solution   We calculate the discriminant $b^2 - 4ac$ and demand that it be 0.

$$b^2 - 4ac = (k + 1)^2 - 4(k - 2)(\mathbf{4})$$
$$0 = k^2 + 2k + 1 - 16k + 32$$

$$0 = k^2 - 14k + 33$$
$$0 = (k - 3)(k - 11)$$

$$k - 3 = 0 \quad \text{or} \quad k - 11 = 0$$
$$k = 3 \quad \quad k = 11$$

When $k = 3$ or when $k = 11$, the given equation will have equal roots.

As a check, we let $k = 3$ and note that the equation $(k - 2)x^2 + (k + 1)x + 4 = 0$ becomes

$$(3 - 2)x^2 + (3 + 1)x + 4 = 0$$

or

$$x^2 + 4x + 4 = 0$$

When we solve this equation, we find that its roots are equal real numbers, as expected:

$$x^2 + 4x + 4 = 0$$
$$(x + 2)(x + 2) = 0$$
$$x + 2 = 0 \quad \text{or} \quad x + 2 = 0$$
$$x = -2 \quad \quad x = -2$$

Similarly, $k = 11$ will give an equation with equal real roots. ∎

There are many equations that can be put into quadratic form and then solved with the techniques for solving quadratic equations.

EXAMPLE 11 Solve the equation $\dfrac{1}{x - 1} + \dfrac{3}{x + 1} = 2$.

Solution Since the denominator of a fraction cannot be 0, x cannot be 1 or -1. If either of these numbers appears as an expected root, it is an extraneous root and must be discarded.

$$\frac{1}{x - 1} + \frac{3}{x + 1} = 2$$

$$(x - 1)(x + 1)\left[\frac{1}{x - 1} + \frac{3}{x + 1}\right] = (x - 1)(x + 1)2 \qquad \text{Multiply both sides by } (x - 1)(x + 1).$$

$$(x + 1) + 3(x - 1) = 2(x^2 - 1) \qquad \text{Remove brackets and simplify.}$$

$$4x - 2 = 2x^2 - 2 \qquad \text{Remove parentheses and simplify.}$$

$$0 = 2x^2 - 4x \qquad \text{Add } 2 - 4x \text{ to both sides.}$$

The resulting equation is a quadratic equation that we can solve as follows:

$$2x^2 - 4x = 0$$
$$2x(x - 2) = 0 \qquad \qquad \text{Factor } 2x^2 - 4x.$$
$$2x = 0 \quad \text{or} \quad x - 2 = 0$$
$$x = 0 \quad \quad x = 2$$

Because 0 and 2 are acceptable values of x, each number is a root of the original equation. Verify this by checking each root. ∎

Formulas

Many formulas involve second-degree equations. For example, if an object is fired straight up into the air with an initial velocity of 88 feet per second, its height is given by the formula $h = 88t - 16t^2$, where h represents height (in feet) and t represents elapsed time (in seconds) since it was fired.

To solve this formula for t, we use the quadratic formula.

$$h = 88t - 16t^2$$

$$16t^2 - 88t + h = 0 \qquad \text{Add } 16t^2 \text{ and } -400t \text{ to both sides.}$$

$$t = \frac{-(-88) \pm \sqrt{(-88)^2 - 4(16)(h)}}{2(16)} \qquad \text{Substitute into the quadratic formula.}$$

$$t = \frac{88 \pm \sqrt{7744 - 64h}}{32} \qquad \text{Simplify.}$$

2.3 EXERCISES

In Exercises 1–12, solve each equation by factoring. Check all answers.

1. $x^2 - x - 6 = 0$ **2.** $x^2 + 8x + 15 = 0$ **3.** $x^2 - 144 = 0$ **4.** $x^2 + 4x = 0$

5. $2x^2 + x - 10 = 0$ **6.** $3x^2 + 4x - 4 = 0$ **7.** $5x^2 - 13x + 6 = 0$ **8.** $2x^2 + 5x - 12 = 0$

9. $15x^2 + 16x = 15$ **10.** $6x^2 - 25x = -25$ **11.** $12x^2 + 9 = 24x$ **12.** $24x^2 + 6 = 24x$

In Exercises 13–20, use the square root property to solve each equation. You may need to factor an expression.

13. $x^2 = 9$ **14.** $x^2 = 20$ **15.** $y^2 - 50 = 0$ **16.** $x^2 - 75 = 0$

17. $(x - 1)^2 = 4$ **18.** $(y + 2)^2 - 49 = 0$ **19.** $a^2 + 2a + 1 = 9$ **20.** $x^2 - 6x + 9 = 25$

In Exercises 21–32, complete the square to make each binomial a perfect trinomial square.

21. $x^2 + 6x$ **22.** $x^2 + 8x$ **23.** $x^2 - 4x$ **24.** $x^2 - 12x$

25. $a^2 + 5a$ **26.** $t^2 + 9t$ **27.** $r^2 - 11r$ **28.** $s^2 - 7s$

29. $y^2 + \frac{3}{4}y$ **30.** $p^2 + \frac{3}{2}p$ **31.** $q^2 - \frac{1}{5}q$ **32.** $m^2 - \frac{2}{3}m$

In Exercises 33–44, solve each equation by completing the square.

33. $x^2 - 8x + 15 = 0$ **34.** $x^2 + 10x + 21 = 0$ **35.** $x^2 + x - 6 = 0$ **36.** $x^2 - 9x + 20 = 0$

37. $x^2 - 25x = 0$ **38.** $x^2 + x = 0$ **39.** $3x^2 + 4x = 4$ **40.** $2x^2 + 5x = 12$

41. $x^2 + 5 = -5x$ **42.** $x^2 + 1 = -4x$ **43.** $3x^2 = 1 - 4x$ **44.** $2x^2 = 3x + 1$

In Exercises 45–56, use the quadratic formula to solve each equation.

45. $x^2 - 12 = 0$
46. $x^2 - 20 = 0$
47. $2x^2 - x - 15 = 0$
48. $6x^2 + x - 2 = 0$
49. $5x^2 - 9x - 2 = 0$
50. $4x^2 - 4x - 3 = 0$
51. $2x^2 + 2x - 4 = 0$
52. $3x^2 + 18x + 15 = 0$
53. $-3x^2 = 5x + 1$
54. $2x(x + 3) = -1$
55. $5x\left(x + \dfrac{1}{5}\right) = 3$
56. $7x^2 = 2x + 2$

In Exercises 57–64, use the discriminant to determine the nature of the roots of each equation. ***Do not solve the equations.***

57. $x^2 + 6x + 9 = 0$
58. $x^2 - 5x + 2 = 0$
59. $3x^2 - 2x + 5 = 0$
60. $9x^2 + 42x + 49 = 0$
61. $10x^2 + 29x = 21$
62. $10x^2 + x = 21$
63. $-3x^2 + 2x = 21$
64. $-8x^2 - 2x = 13$

65. Find two values of k so that $x^2 + kx + 3k - 5 = 0$ will have two roots that are equal.

66. For what value(s) of b will the solutions of $x^2 - 2bx + b^2 = 0$ be equal?

67. Does $1492x^2 + 1984x - 1776 = 0$ have any roots that are real numbers?

68. Does $2004x^2 + 10x + 1994 = 0$ have any roots that are real numbers?

In Exercises 69–86, change each equation to quadratic form and solve by any method.

69. $x + 1 = \dfrac{12}{x}$
70. $x - 2 = \dfrac{15}{x}$
71. $8x - \dfrac{3}{x} = 10$
72. $15x - \dfrac{4}{x} = 4$

73. $\dfrac{5}{x} = \dfrac{4}{x^2} - 6$
74. $\dfrac{6}{x^2} + \dfrac{1}{x} = 12$
75. $x\left(30 - \dfrac{13}{x}\right) = \dfrac{10}{x}$
76. $x\left(20 - \dfrac{17}{x}\right) = \dfrac{10}{x}$

77. $(a - 2)(a + 4) = 2a(a - 3)$
78. $\dfrac{4 + a}{2a} = \dfrac{a - 2}{3}$

79. $\dfrac{1}{x} + \dfrac{3}{x + 2} = 2$
80. $\dfrac{1}{x - 1} + \dfrac{1}{x - 4} = \dfrac{5}{4}$
81. $\dfrac{1}{x + 1} + \dfrac{5}{2x - 4} = 1$
82. $\dfrac{x(2x + 1)}{x - 2} = \dfrac{10}{x - 2}$

83. $x + 1 + \dfrac{x + 2}{x - 1} = \dfrac{3}{x - 1}$
84. $\dfrac{1}{4 - y} = \dfrac{1}{4} + \dfrac{1}{y + 2}$
85. $\dfrac{24}{a} - 11 = \dfrac{-12}{a + 1}$
86. $\dfrac{36}{b} - 17 = \dfrac{-24}{b + 1}$

In Exercises 87–96, solve each formula for the indicated variable.

87. $h = \dfrac{1}{2}gt^2$; t
88. $x^2 + y^2 = r^2$; x
89. $h = 64t - 16t^2$; t
90. $y = 16x^2 - 4$; x

91. $\dfrac{x^2}{a^2} + \dfrac{y^2}{b^2} = 1$; y
92. $\dfrac{x^2}{a^2} - \dfrac{y^2}{b^2} = 1$; x
93. $\dfrac{x^2}{a^2} - \dfrac{y^2}{b^2} = 1$; a
94. $\dfrac{x^2}{a^2} - \dfrac{y^2}{b^2} = 1$; b

95. $x^2 + xy - y^2 = 0$; x
96. $x^2 - 3xy + y^2 = 0$; y

97. If r_1 and r_2 are the roots of $ax^2 + bx + c = 0$, show that $r_1 + r_2 = -\dfrac{b}{a}$.

98. If r_1 and r_2 are the roots of $ax^2 + bx + c = 0$, show that $r_1 r_2 = \dfrac{c}{a}$.

In Exercises 99–100, a stone is thrown upward, higher than the top of a tree. The stone is even with the top of the tree at times t_1 on the way up and t_2 on the way down. If the height of the tree is h, both t_1 and t_2 are solutions of $h = v_0 t - 16t^2$.

99. Show that the tree is $16t_1 t_2$ feet tall.

100. Show that v_0 is $16(t_1 + t_2)$ feet per second.

2.4 APPLICATIONS OF QUADRATIC EQUATIONS

■ Geometric Problems   ■ Uniform-Motion Problems   ■ Flying Object Problems   ■ Business Problems

The solutions of many word problems involve quadratic equations.

■ Geometric Problems

EXAMPLE 1   The length of a rectangle exceeds its width by 3 feet. If the area of the rectangle is 40 square feet, find its dimensions.

Solution   We can let w represent the width of the rectangle. Then $w + 3$ will represent its length. (See Figure 2-4.) Since the formula for the area of a rectangle is $A = lw$ (area = length × width), the area of the rectangle is $(w + 3)w$, which is equal to 40. This gives the equation

The length of the rectangle	·	**the width of the rectangle**	=	**the area of the rectangle.**
$(w + 3)$	·	w	=	40

We can solve this equation as follows:

$$w^2 + 3w = 40$$
$$w^2 + 3w - 40 = 0 \qquad \text{Subtract 40 from both sides.}$$
$$(w - 5)(w + 8) = 0 \qquad \text{Factor } w^2 + 3w - 40.$$
$$w - 5 = 0 \quad \text{or} \quad w + 8 = 0$$
$$w = 5 \quad \mid \quad w = -8$$

When the width, w, is 5, the length, $w + 3$, is 8. The solution $w = -8$ must be discarded, because a rectangle cannot have a negative width.

We can verify that this solution is correct by observing that a rectangle with dimensions of 5 feet by 8 feet does have an area of 40 square feet.   ■

w ft

$(w + 3)$ ft

FIGURE 2-4

■ Uniform-Motion Problems

EXAMPLE 2   A man drove 600 miles to a business convention. On the return trip, he increased his speed by 10 miles per hour and saved 2 hours of driving time. How fast did he drive in each direction?

Solution   We can let s represent the car's speed (in miles per hour) driving to the convention. On the return trip, its speed was $s + 10$ miles per hour. Recall that the distance traveled by an object moving at a constant rate for a certain time is given by the formula $d = rt$. Thus,

$$t = \frac{d}{r}$$

We can organize the information given in this problem as in Figure 2-5.

	d	$=$	r	$\cdot$	t
outbound trip	600		s		$\dfrac{600}{s}$
return trip	600		$s+10$		$\dfrac{600}{s+10}$

FIGURE **2-5**

Although neither the outbound nor the return travel time is given, we know the difference of those times. This fact can be used to form the equation

The longer time of the outbound trip	$-$	the shorter time of the return trip	$=$	the difference in travel times.
$\dfrac{600}{s}$	$-$	$\dfrac{600}{s + 10}$	$=$	2

We can multiply both sides of this equation by $s(s + 10)$ to clear the equation of fractions and solve for s.

$$s(s + 10)\left(\frac{600}{s} - \frac{600}{s + 10}\right) = s(s + 10)2$$

$$600(s + 10) - 600s = 2s(s + 10) \qquad \text{Simplify.}$$

$$600s + 6000 - 600s = 2s^2 + 20s \qquad \text{Remove parentheses.}$$

$$6000 = 2s^2 + 20s \qquad \text{Combine terms.}$$

$$0 = 2s^2 + 20s - 6000 \qquad \text{Subtract 6000 from both sides.}$$

$$0 = s^2 + 10s - 3000 \qquad \text{Divide both sides by 2.}$$

$$0 = (s - 50)(s + 60) \qquad \text{Factor.}$$

$$s - 50 = 0 \quad \text{ or } \quad s + 60 = 0 \qquad \text{Set each factor equal to 0.}$$

$$s = 50 \qquad\qquad s = -60$$

The solution $s = -60$ must be discarded. The man drove 50 miles per hour to the convention and $50 + 10$, or 60 miles per hour on the return trip.

These answers are correct, because a 600-mile trip at 50 miles per hour would take $\frac{600}{50}$, or 12 hours. At 60 miles per hour, the same trip would take only 10 hours, which is 2 hours less.   ∎

■ Flying Object Problems

EXAMPLE 3   If an object is thrown straight up into the air with an initial velocity of 144 feet per second, its height is given by the formula $h = 144t - 16t^2$, where h represents its height (in feet) and t represents the elapsed time (in seconds) since it was thrown. How long will it take for the object to return to the point from which it was thrown?

Solution   When the object returns to its starting point, its height is again 0. Thus, we can set h equal to 0 and solve for t.

$$h = 144t^2 - 16t^2$$
$$0 = 144t^2 - 16t^2 \qquad \text{Let } h = 0.$$
$$0 = 16t(9 - t) \qquad \text{Factor.}$$
$$16t = 0 \quad \text{or} \quad 9 - t = 0 \qquad \text{Set each factor equal to 0.}$$
$$t = 0 \qquad\qquad t = 9$$

At $t = 0$ sec, the object's height is 0 ft, because it was just released. When $t = 9$ sec, the height is again 0 ft, and the object has returned to its starting point.   ∎

■ Business Problems

EXAMPLE 4   A bus company shuttles an average of 1120 passengers daily between Rockford, Illinois and Chicago's O'Hare airport. The current one-way fare is $10. For each 25¢ increase in the fare, the company predicts that it will lose 48 passengers. What increase in fare will produce daily revenue of $10,208?

Solution   Let q represent the number of quarters the fare will be increased. Then the new fare will be $\$(10 + 0.25q)$. Since the company will lose 48 passengers for each quarter increase, $48q$ passengers will be lost when the rate increases by q quarters. The passenger load will then be $(1120 - 48q)$ passengers.

Since the daily revenue of $10,208 will be the product of the rate and the number of passengers, we have

$$(10 + 0.25q)(1120 - 48q) = 10{,}208$$
$$11{,}200 - 480q + 280q - 12q^2 = 10{,}208 \qquad \text{Remove parentheses.}$$
$$-12q^2 - 200q + 992 = 0 \qquad \text{Combine terms and subtract } 10{,}208 \text{ from both sides.}$$
$$3q^2 + 50q - 248 = 0 \qquad \text{Divide both sides by } -4.$$

We can solve this quadratic equation with the quadratic formula.

$$q = \frac{-50 \pm \sqrt{50^2 - 4(3)(-248)}}{2(3)}$$

$$q = \frac{-50 \pm \sqrt{2500 + 2976}}{6}$$

$$q = \frac{-50 \pm \sqrt{5476}}{6}$$

$$q = \frac{-50 \pm 74}{6}$$

$$q = \frac{-50 + 74}{6} = \frac{24}{6} = 4 \quad \text{or} \quad q = \frac{-50 - 74}{6} = \frac{-124}{6} = -\frac{62}{3}$$

Because the number of riders cannot be negative, the result of $-\frac{62}{3}$ must be discarded. To generate \$10,208 in daily revenues, the company should raise the fare by 4 quarters, or \$1, to \$11. ∎

2.4 EXERCISES

*In Exercises 1–34, solve each word problem. **You may use a calculator if it is helpful.***

1. Number problem   The product of two consecutive even natural numbers is 48. Find the numbers.

2. Number problem   The product of the first and last of three consecutive odd integers is 45. Find the sum of the three integers.

3. Geometric problem   A rectangle is 4 feet longer than it is wide. If its area is 32 square feet, find its dimensions.

4. Geometric problem   A rectangle is 5 times as long as it is wide. If the area is 125 square feet, find its perimeter.

5. Geometric problem   The side of a square is 4 centimeters shorter than the side of a second square. If the sum of their areas is 106 square centimeters, find the length of one side of the larger square.

6. Geometric problem   The base of a triangle is one-third as long as its height. If the area of the triangle is 24 square meters, how long is its base?

7. Metal fabrication   A piece of tin, 12 inches on a side, is to have four equal squares cut from its corners, as in Illustration 1. If the edges are then to be folded up to

make a box with a floor area of 64 square inches, find the depth of the box.

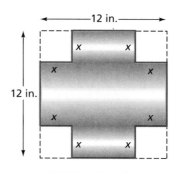

ILLUSTRATION 1

8. Making gutters   A piece of sheet metal, 18 inches wide, is bent to form the gutter shown in Illustration 2. If the cross-sectional area is 36 square inches, find the depth of the gutter.

9. Cycling rates   A cyclist rides from DeKalb to Rockford, a distance of 40 miles. His return trip takes 2 hours longer because his speed decreases by 10 miles per hour. How fast does he ride each way?

ILLUSTRATION 2

10. *Travel time*   A farmer drives a tractor from one town to another, a distance of 120 kilometers. He drives 10 kilometers per hour faster on the return trip, cutting 1 hour off the time. How fast does he drive each way?

11. *Uniform-motion problem*   If the speed were increased by 10 miles per hour, a 420-mile trip would take 1 hour less time. How long does the trip take at the slower speed?

12. *Uniform-motion problem*   By increasing her usual speed by 25 kilometers per hour, a bus driver decreases the time on a 25-kilometer trip by 10 minutes. Find the usual speed.

13. *Ballistics*   The height of a projectile fired upward with an initial velocity of 400 feet per second is given by the formula $h = -16t^2 + 400t$, where h is the height in feet and t is the time in seconds. Find the time required for the projectile to return to earth.

14. *Ballistics*   The height of an object tossed upward with an initial velocity of 104 feet per second is given by the formula $h = -16t^2 + 104t$, where h is the height in feet and t is the time in seconds. Find the time required for the object to return to its point of departure.

15. *Falling coins*   An object will fall s feet in t seconds, where $s = 16t^2$. How long will it take for a penny to hit the ground if it is dropped from the top of the Sears Tower in Chicago? (*Hint:* The tower is 1454 feet tall.)

16. *Ballistics*   The height of an object thrown upward with an initial velocity of 32 feet per second is given by the formula $h = -16t^2 + 32t$, where t is the time in seconds. How long will it take the object to reach a height of 16 feet?

17. *Setting fares*   A bus company has 3000 passengers daily, paying a 25¢ fare. For each nickel increase in fare, the company projects that it will lose 80 passengers. What fare increase will produce $994 in daily revenue?

18. *Jazz concerts*   A jazz group on tour has been drawing average crowds of 500 people. It is projected that for every $1 increase in the $12 ticket price, the average attendance will decrease by 50. At what ticket price will nightly receipts be $5600?

19. *Concert receipts*   Tickets for the annual symphony orchestra pops concert cost $15, and the average attendance at the concerts has been 1200 persons. Management projects that for each 50¢ decrease in ticket price, 40 more patrons will attend. What decrease in ticket price will result in receipts of $17,280?

20. *Projecting demand*   The *Vilas County News* earns a profit of $20 per year for each of its 3000 subscribers. Management projects that the profit per subscriber would increase by 1¢ for each additional subscriber over the current 3000. How many subscribers are needed to bring a total profit of $120,000?

21. *Filling a storage tank*   Two pipes are used to fill a water-storage tank. The first pipe can fill the tank in 4 hours, and the two pipes together can fill the tank in 2 hours less time than the second pipe alone. How long would it take for the second pipe to fill the tank?

22. *Filling a swimming pool*   A hose can fill a swimming pool in 6 hours. Another hose needs 3 more hours to fill the pool than the two hoses combined. How long would it take the second hose to fill the pool?

23. *Mowing lawns*   Kristy can mow a lawn in 1 hour less time than her brother Steven. Together they can finish the job in 5 hours. How long would it take Kristy if she worked alone?

24. *Milking cows*   Working together, Sarah and Heidi can milk the cows in 2 hours. If they work alone, it takes Heidi 3 hours longer than it takes Sarah. How long would it take Heidi to milk the cows alone?

25. *Geometric problem*   Is it possible for a rectangle to have a width that is 3 units shorter than its diagonal and a length that is 4 units longer than its diagonal?

26. *Geometric problem*   If two opposite sides of a square are increased by 10 meters and the other sides are decreased by 8 meters, the area of the rectangle that is formed is 63 square meters. Find the area of the original square.

27. *Investment problem*   Maude and Matilda each have a bank CD. Maude's is $1000 larger than Matilda's, but the interest rate is 1% less. Last year Maude received interest of $280, and Matilda received $240. Find the rate of interest for each CD.

28. *Investment problem* Scott and Laura have both invested some money. Scott invested $3000 more than Laura and at a 2% higher interest rate. If Scott received $800 annual interest and Laura received $400, how much did Scott invest?

29. *Buying a microwave oven* Some mathematics professors would like to purchase a $150 microwave oven for the department workroom. If four of the professors don't contribute, everyone's share will increase by $10. How many professors are in the department?

30. *Planting a windscreen* A farmer intends to construct a windscreen by planting trees in a quarter-mile row. His daughter points out that 44 fewer trees will be needed if they are planted 1 foot farther apart. If the farmer takes her advice, how many trees will be needed? A row starts and ends with a tree. (*Hint:* 1 mile = 5280 feet.)

31. *Puzzle problem* If a wagon wheel had 10 more spokes, the angle between the spokes would decrease by 6°. How many spokes does the wheel have?

32. *Puzzle problem* A merchant could sell all of his calculators at list price for $180. If he had 3 more calculators, he could sell each one for $10 less and still receive $180. Find the list price of each calculator.

33. *Geometric problem* If one leg of a right triangle is 14 meters shorter than the other leg and the hypotenuse is 26 meters, find the length of the two legs.

34. *Geometric problem* Find the dimensions of a rectangle whose area is 180 cm^2 and whose perimeter is 54 cm.

2.5 COMPLEX NUMBERS

■ Powers of *i*   ■ Complex Numbers   ■ The Arithmetic of Complex Numbers   ■ Absolute Value of a Complex Number   ■ Factoring the Sum of Two Squares

So far, every solution of a quadratic equation that we have solved has been a real number. This is not always true. For example, if we use the quadratic formula to solve the quadratic equation $x^2 + x + 2 = 0$, we get solutions that are nonreal.

$$x = \frac{-b \pm \sqrt{b^2 - 4ac}}{2a}$$

$$x = \frac{-1 \pm \sqrt{1^2 - 4(1)(2)}}{2(1)} \qquad \text{Substitute 1 for } a, \text{ 1 for } b, \text{ and 2 for } c.$$

$$x = \frac{-1 \pm \sqrt{1 - 8}}{2}$$

$$x = \frac{-1 \pm \sqrt{-7}}{2}$$

Each solution to this quadratic equation involves the number $\sqrt{-7}$. This number cannot be real, because the square of a real number cannot be negative.

For years, mathematicians believed that square roots of negative numbers, denoted by symbols such as $\sqrt{-1}$, $\sqrt{-3}$, and $\sqrt{-7}$, were nonsense. Even the great English mathematician Sir Isaac Newton (1642–1727) called them "impossible numbers." In the 17th century, these symbols were called **imaginary numbers** by René Descartes.

Although no real number can be a solution of the equation $x^2 = -1$, there is a new number, an imaginary number denoted by i, that is a solution. Because i is defined to be a solution of $x^2 = -1$, we have

1. $i^2 = -1$

Since $\sqrt{-1}$ also appears to represent a solution of the equation $x^2 = -1$, we make this definition:

$$i = \sqrt{-1}$$

■ Powers of *i*

The powers of i with natural-number exponents produce an interesting pattern.

$i^1 = i$

$i^2 = -1$ From Equation 1.

$i^3 = i^2 i = -i$ Remember, $i^2 = -1$.

$i^4 = i^2 i^2 = (-1)(-1) = 1$

$i^5 = i^4 i = 1 \cdot i = i$

The pattern continues: $i, -1, -i, 1$.

We can simplify powers of i that involve negative integer exponents by rationalizing denominators.

$$i^{-1} = \frac{1}{i} = \frac{1 \cdot i}{i \cdot i} = \frac{i}{-1} = -i$$

$$i^{-2} = \frac{1}{i^2} = \frac{1}{-1} = -1$$

$$i^{-3} = \frac{1}{i^3} = \frac{1 \cdot i}{i^3 \cdot i} = \frac{1}{i^4} = \frac{i}{1} = i$$

$$i^{-4} = \frac{1}{i^4} = \frac{1}{1} = 1$$

If we define i^0 to be 1, then the pattern $i, -1, -i, 1$ repeats endlessly.

EXAMPLE 1   Simplify i^{365}.

Solution   Because $i^4 = 1$, each occurrence of i^4 is just a factor of 1. To determine how many factors of i^4 appear in i^{365}, we divide 365 by 4. The quotient is 91, the remainder is 1, and we have

$$i^{365} = (i^4)^{91} \cdot i^1 = 1^{91} \cdot i = i$$ ■

In Example 1, the number 365 left a remainder of 1 when it was divided by 4, and $i^{365} = i^1$. In general, if n is a natural number that leaves a remainder r when it is divided by 4, then $i^n = i^r$.

For example, to simplify i^{479}, we divide 479 by 4, and determine that the remainder is 3. Thus, $i^{479} = i^3$, which is $-i$.

Numbers such as i, $7i$, $\sqrt{3}i$, or πi, that represent the product of a real number and i, are imaginary numbers. The square of an imaginary number is a real number. For example, the square of $3i$ is -9, because

$$(3i)^2 = 3^2 i^2 = 9(-1) = -9$$

By the definition of square root, $3i$ is a square root of -9, and we write $\sqrt{-9} = 3i$. Similarly, we have

$$\sqrt{-25} = \sqrt{25 \cdot (-1)} = \sqrt{25}\sqrt{-1} = 5i$$
$$\sqrt{-7} = \sqrt{7}\sqrt{-1} = \sqrt{7}i$$

and

$$\sqrt{-16}\sqrt{-49} = 4i \cdot 7i = 28i^2 = -28$$

We note that $\sqrt{-16}\sqrt{-49}$ is not equal to $\sqrt{(-16)(-49)}$, which is $\sqrt{16 \cdot 49} = 28$.

  Warning! If x and y are both negative, then $\sqrt{xy}$ is *not* equal to $\sqrt{x}\sqrt{y}$.

■ Complex Numbers

Expressions such as $3 + 4i$, $-5 + 7i$, and $-1 - 9i$, which indicate the sum of a real number and an imaginary number, are called **complex numbers**.

Complex Numbers	A **complex number** is any number that can be expressed in the form $a + bi$, where a and b are real numbers and $i = \sqrt{-1}$. The number a is called the **real part** of the complex number, and b is called the **imaginary part**.

If $b = 0$, the complex number $a + bi$ is the real number a. Thus, any real number is a complex number with a zero imaginary part. If $a = 0$ and $b \neq 0$, the complex number $a + bi$ is the imaginary number bi. Thus, any imaginary number is a complex number with a zero real part. It follows that the set of real numbers and the set of imaginary numbers are both subsets of the set of complex numbers.

Figure 2-6 illustrates how the various sets of numbers are related.

The next definition shows how to determine whether two complex numbers are equal.

Equality of Complex Numbers	Two complex numbers are equal if and only if their real parts are equal and their imaginary parts are equal. Thus, if $a + bi$ and $c + di$ are two complex numbers, then $a + bi = c + di$ if and only if $a = c$ and $b = d$

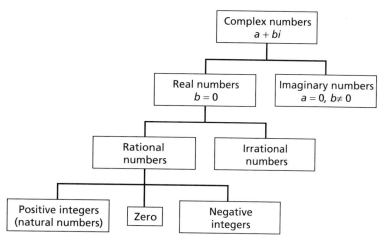

FIGURE 2-6

EXAMPLE 2   For what real values of x and y is $3x + 4i = (2y + x) + xi$?

Solution   Because the two complex numbers are equal, their imaginary parts must be equal: $x = 4$. Because the real parts of these two complex numbers must also be equal, $3x = 2y + x$. We solve the system of equations

$$\begin{cases} x = 4 \\ 3x = 2y + x \end{cases}$$

by substituting 4 for x in the second equation and solving for y. We find that $y = 4$. Hence, the solution is $x = 4$ and $y = 4$.   ∎

■ The Arithmetic of Complex Numbers

Complex numbers are added and subtracted as if they were binomials.

Addition and Subtraction of Complex Numbers	Two complex numbers such as $a + bi$ and $c + di$ are added and subtracted as binomials:

$$(a + bi) + (c + di) = (a + c) + (b + d)i$$
$$(a + bi) - (c + di) = (a - c) + (b - d)i$$

Because of the preceding definition, the sum or difference of two complex numbers is another complex number.

EXAMPLE 3   **a.** $(3 + 4i) + (2 + 7i) = 3 + 4i + 2 + 7i$
$$= 3 + 2 + 4i + 7i$$
$$= 5 + 11i$$

b. $(-5 + 8i) - (2 - 12i) = -5 + 8i - 2 + 12i$
$$= -5 - 2 + 8i + 12i$$
$$= -7 + 20i \qquad \blacksquare$$

Complex numbers are multiplied as if they were binomials.

Multiplication of Complex Numbers	Two complex numbers such as $a + bi$ and $c + di$ are multiplied as binomials, with $i^2 = -1$:
	$$(a + bi)(c + di) = (ac - bd) + (ad + bc)i$$

Because of the preceding definition, the product of two complex numbers is another complex number.

EXAMPLE 4   **a.** $(3 + 4i)(2 + 7i) = 6 + 21i + 8i + 28i^2$
$$= 6 + 21i + 8i + 28(-1)$$
$$= 6 - 28 + 29i$$
$$= -22 + 29i$$

b. $(5 - 7i)(1 + 3i) = 5 + 15i - 7i - 21i^2$
$$= 5 + 15i - 7i - 21(-1)$$
$$= 5 + 21 + 8i$$
$$= 26 + 8i \qquad \blacksquare$$

Warning! To avoid errors in determining the sign of the result, always express complex numbers in $a + bi$ form before attempting any algebra manipulations.

EXAMPLE 5   Find the product of $-2 + \sqrt{-16}$ and $4 - \sqrt{-9}$.

Solution   We first change each complex number to $a + bi$ form
$$-2 + \sqrt{-16} = -2 + \sqrt{16}\sqrt{-1} = -2 + 4i$$
$$4 - \sqrt{-9} = 4 - \sqrt{9}\sqrt{-1} = 4 - 3i$$

and then find the product.
$$(-2 + 4i)(4 - 3i) = -8 + 6i + 16i - 12i^2$$
$$= -8 + 6i + 16i - 12(-1)$$
$$= -8 + 12 + 22i$$
$$= 4 + 22i \qquad \blacksquare$$

We will use the next definition to find the quotient of two complex numbers.

Complex Conjugates	The complex numbers $a + bi$ and $a - bi$ are called **complex conjugates** of each other.

To find the product of $-\frac{1}{2} + 4i$ and its conjugate $-\frac{1}{2} - 4i$, for example, we proceed as follows:

$$\left(-\frac{1}{2} + 4i\right)\left(-\frac{1}{2} - 4i\right) = \frac{1}{4} + 2i - 2i - 16i^2$$

$$= \frac{1}{4} - 16(-1)$$

$$= \frac{1}{4} + 16$$

$$= \frac{65}{4}$$

In general, the product of the complex number $a + bi$ and its complex conjugate $a - bi$ is the real number $a^2 + b^2$, as the following work shows:

$$(a + bi)(a - bi) = a^2 - abi + abi - b^2i^2$$

$$= a^2 - b^2(-1)$$

$$= a^2 + b^2$$

This result enables us to divide one complex number by another and get another complex number.

EXAMPLE 6 Write the number $\dfrac{3}{2 + i}$ in $a + bi$ form.

Solution To write the number in $a + bi$ form, we multiply both the numerator and the denominator by the conjugate of the denominator and simplify.

$$\frac{3}{2 + i} = \frac{3(2 - i)}{(2 + i)(2 - i)}$$

$$= \frac{6 - 3i}{4 - 2i + 2i - i^2}$$

$$= \frac{6 - 3i}{4 + 1}$$

$$= \frac{6 - 3i}{5}$$

$$= \frac{6}{5} - \frac{3}{5}i$$

It is common practice to accept $\frac{6}{5} - \frac{3}{5}i$ as a substitute for $\frac{6}{5} + \left(-\frac{3}{5}\right)i$. ■

EXAMPLE 7   Write the number $\dfrac{2 - \sqrt{-16}}{3 + \sqrt{-1}}$ in $a + bi$ form.

Solution

$$\dfrac{2 - \sqrt{-16}}{3 + \sqrt{-1}} = \dfrac{2 - 4i}{3 + i}$$

$$= \dfrac{(2 - 4i)(3 - i)}{(3 + i)(3 - i)} \qquad \text{Multiply numerator and denominator by } 3 - i.$$

$$= \dfrac{6 - 2i - 12i + 4i^2}{9 - 3i + 3i - i^2} \qquad \text{Remove parentheses.}$$

$$= \dfrac{2 - 14i}{9 + 1} \qquad \text{Combine terms.}$$

$$= \dfrac{2}{10} - \dfrac{14i}{10}$$

$$= \dfrac{1}{5} - \dfrac{7}{5}i$$ ∎

Absolute Value of a Complex Number

We can also find absolute values of complex numbers.

Absolute Value of a Complex Number	If $a + bi$ is a complex number, then $$\lvert a + bi \rvert = \sqrt{a^2 + b^2}$$

Because of the previous definition, the absolute value of a complex number is a real number.

EXAMPLE 8   **a.** $\lvert 3 + 4i \rvert = \sqrt{3^2 + 4^2} = \sqrt{25} = 5$
b. $\lvert 2 - 5i \rvert = \sqrt{2^2 + (-5)^2} = \sqrt{29}$ ∎

EXAMPLE 9   Find **a.** $\left\lvert \dfrac{2i}{3 + i} \right\rvert$ and **b.** $\lvert a + 0i \rvert$, where a is unrestricted.

a. To find $\left\lvert \dfrac{2i}{3 + i} \right\rvert$, we first express $\dfrac{2i}{3 + i}$ in $a + bi$ form:

$$\dfrac{2i}{3 + i} = \dfrac{2i(3 - i)}{(3 + i)(3 - i)} = \dfrac{6i - 2i^2}{9 - i^2} = \dfrac{6i + 2}{10} = \dfrac{1}{5} + \dfrac{3}{5}i$$

We then find the absolute value of $\dfrac{1}{5} + \dfrac{3}{5}i$.

$$\left\lvert \dfrac{2i}{3 + i} \right\rvert = \left\lvert \dfrac{1}{5} + \dfrac{3}{5}i \right\rvert = \sqrt{\left(\dfrac{1}{5}\right)^2 + \left(\dfrac{3}{5}\right)^2} = \sqrt{\dfrac{10}{25}} = \dfrac{\sqrt{10}}{5}$$

b. If a is unrestricted, $|a + 0i| = \sqrt{a^2 + 0^2} = \sqrt{a^2} = |a|$. Thus, we have

$$|a| = \sqrt{a^2}$$
■

Part **b.** of Example 9 is an important result. We can guarantee that $|a| \geq 0$ by defining $|a|$ to be $\sqrt{a^2}$.

EXAMPLE 10 Solve the quadratic equation $x^2 - 4x + 5 = 0$.

Solution In this equation, $a = 1$, $b = -4$, and $c = 5$.

$$x = \frac{-b \pm \sqrt{b^2 - 4ac}}{2a}$$

$$= \frac{-(-4) \pm \sqrt{(-4)^2 - 4(1)(5)}}{2(1)} \qquad \text{Substitute 1 for } a, -4 \text{ for } b, \text{ and 5 for } c.$$

$$= \frac{4 \pm \sqrt{16 - 20}}{2}$$

$$= \frac{4 \pm \sqrt{-4}}{2}$$

$$= \frac{4 \pm 2i}{2} \qquad\qquad \sqrt{-4} = \sqrt{4}\sqrt{-1} = 2i$$

$$= 2 \pm i \qquad\qquad \frac{4 \pm 2i}{2} = \frac{2(2 \pm i)}{2} = 2 \pm i$$

The two roots $x = 2 + i$ and $x = 2 - i$ both satisfy the original equation. We note that the roots are complex conjugates. ■

■ Factoring the Sum of Two Squares

Because $i^2 = -1$, it is possible to factor the sum of two squares over the set of complex numbers. For example, to factor $9x^2 + 16y^2$, we proceed as follows:

$$9x^2 + 16y^2 = 9x^2 - (-1)16y^2$$
$$= 9x^2 - i^2(16y^2)$$
$$= (3x + 4yi)(3x - 4yi) \qquad \text{Factor the difference of two squares.}$$

2.5 EXERCISES

In Exercises 1–8, simplify each expression.

1. i^9

2. i^{27}

3. i^{38}

4. i^{99}

5. i^{-6}

6. i^0

7. i^{-10}

8. i^{-31}

In Exercises 9–12, find the values of x and y.

9. $x + (x + y)i = 3 + 8i$

10. $x + 5i = y - yi$

11. $3x - 2yi = 2 + (x + y)i$

12. $\begin{cases} 2 + (x + y)i = 2 - i \\ x + 3i = 2 + 3i \end{cases}$

In Exercises 13–44, perform all operations and express all answers in a + bi form.

13. $(2 - 7i) + (3 + i)$ **14.** $(-7 + 2i) + (2 - 8i)$ **15.** $(5 - 6i) - (7 + 4i)$ **16.** $(11 + 2i) - (13 - 5i)$

17. $(14i + 2) + (2 - \sqrt{-16})$ **18.** $(5 + \sqrt{-64}) - (23i - 32)$ **19.** $(3 + \sqrt{-4}) - (2 + \sqrt{-9})$

20. $(7 - \sqrt{-25}) + (-8 + \sqrt{-1})$ **21.** $(2 + 3i)(3 + 5i)$ **22.** $(5 - 7i)(2 + i)$

23. $(2 + 3i)^2$ **24.** $(3 - 4i)^2$ **25.** $(11 + \sqrt{-25})(2 - \sqrt{-36})$

26. $(6 + \sqrt{-49})(6 - \sqrt{-49})$ **27.** $(\sqrt{-16} + 3)(2 + \sqrt{-9})$ **28.** $(12 - \sqrt{-4})(-7 + \sqrt{-25})$

29. $\dfrac{1}{i^3}$ **30.** $\dfrac{3}{i^5}$ **31.** $\dfrac{-4}{i^{10}}$ **32.** $\dfrac{-10}{i^{24}}$

33. $\dfrac{1}{2 + i}$ **34.** $\dfrac{-2}{3 - i}$ **35.** $\dfrac{2i}{7 + i}$ **36.** $\dfrac{-3i}{2 + 5i}$

37. $\dfrac{2 + i}{3 - i}$ **38.** $\dfrac{3 - i}{1 + i}$ **39.** $\dfrac{4 - 5i}{2 + 3i}$ **40.** $\dfrac{34 + 2i}{2 - 4i}$

41. $\dfrac{5 - \sqrt{-16}}{-8 + \sqrt{-4}}$ **42.** $\dfrac{3 - \sqrt{-9}}{2 - \sqrt{-1}}$ **43.** $\dfrac{2 + i\sqrt{3}}{3 + i}$ **44.** $\dfrac{3 + i}{4 - i\sqrt{2}}$

In Exercises 45–60, find each absolute value.

45. $|3 + 4i|$ **46.** $|5 + 12i|$ **47.** $|2 + 3i|$ **48.** $|5 - i|$

49. $|-7 + \sqrt{-49}|$ **50.** $|-2 - \sqrt{16}|$ **51.** $\left| \dfrac{1}{2} + \dfrac{1}{2}i \right|$ **52.** $\left| \dfrac{1}{2} - \dfrac{1}{4}i \right|$

53. $|-6i|$ **54.** $|5i|$ **55.** $\left| \dfrac{2}{1 + i} \right|$ **56.** $\left| \dfrac{3}{3 + i} \right|$

57. $\left| \dfrac{-3i}{2 + i} \right|$ **58.** $\left| \dfrac{5i}{i - 2} \right|$ **59.** $\left| \dfrac{i + 2}{i - 2} \right|$ **60.** $\left| \dfrac{2 + i}{2 - i} \right|$

In Exercises 61–68, use the quadratic formula to solve each equation. Simplify all solutions and write them in a + bi form.

61. $x^2 + 2x + 2 = 0$ **62.** $a^2 + 4a + 8 = 0$ **63.** $y^2 + 4y + 5 = 0$ **64.** $x^2 + 2x + 5 = 0$

65. $x^2 - 2x = -5$ **66.** $z^2 - 3z = -8$ **67.** $x^2 - \dfrac{2}{3}x = -\dfrac{2}{9}$ **68.** $x^2 + \dfrac{5}{4} = x$

In Exercises 69–76, factor each expression over the set of complex numbers.

69. $x^2 + 4$ **70.** $16a^2 + 9$ **71.** $25p^2 + 36q^2$ **72.** $100r^2 + 49s^2$

73. $2y^2 + 8z^2$ **74.** $12b^2 + 75c^2$ **75.** $50m^2 + 2n^2$ **76.** $64a^4 + 4b^2$

77. Show that the addition of two complex numbers is commutative by adding the complex numbers $a + bi$ and $c + di$ in both orders and observing that the sums are equal.

78. Show that the multiplication of two complex numbers is commutative by multiplying the complex numbers $a + bi$ and $c + di$ in both orders and observing that the products are equal.

79. Show that the addition of complex numbers is associative.

80. Find three examples of complex numbers that are reciprocals of their own conjugates.

2.6   POLYNOMIAL AND RADICAL EQUATIONS

- Other Types of Equations That Can Be Solved by Factoring
- Radical Equations

The quadratic equation $ax^2 + bx + c = 0$ is an example of a polynomial equation of second degree, because its left-hand side contains a second-degree polynomial. Many polynomial equations of higher degree can be solved by factoring.

EXAMPLE 1   Solve the equations   **a.** $6x^3 - x^2 - 2x = 0$ and   **b.** $x^4 - 5x^2 + 4 = 0$.

Solution   **a.**

$$6x^3 - x^2 - 2x = 0$$
$$x(6x^2 - x - 2) = 0 \qquad \text{Factor out } x.$$
$$x(3x - 2)(2x + 1) = 0 \qquad \text{Factor } 6x^2 - x - 2.$$

$$x = 0 \quad \text{or} \quad 3x - 2 = 0 \quad \text{or} \quad 2x + 1 = 0 \qquad \text{Set each factor equal to 0.}$$
$$x = \frac{2}{3} \qquad\qquad x = -\frac{1}{2}$$

Verify that each solution satisfies the original equation.

b.

$$x^4 - 5x^2 + 4 = 0$$
$$(x^2 - 4)(x^2 - 1) = 0 \qquad \text{Factor } x^4 - 5x^2 + 4 = 0.$$
$$(x + 2)(x - 2)(x + 1)(x - 1) = 0 \qquad \text{Factor each difference of two squares.}$$

$$x + 2 = 0 \quad \text{or} \quad x - 2 = 0 \quad \text{or} \quad x + 1 = 0 \quad \text{or} \quad x - 1 = 0 \qquad \text{Set each factor equal to 0.}$$
$$x = -2 \qquad\qquad x = 2 \qquad\qquad x = -1 \qquad\qquad x = 1$$

Verify that each solution satisfies the original equation.   ■

Other Types of Equations That Can Be Solved by Factoring

There are many nonpolynomial equations that can be solved with factoring techniques. To solve such equations, we often use a theorem that states that equal powers of equal numbers are equal.

Theorem	If a and b are numbers, n is an integer, and $a = b$, then $$a^n = b^n$$

When we raise both sides of an equation to the same power, the resulting equation might not be equivalent to the original equation. For example, if we raise both sides of the equation

1. $x = 4$ with a solution set of $\{4\}$

to the second power, we obtain the equation

2. $x^2 = 16$ with a solution set of $\{4, -4\}$

Equations 1 and 2 are not equivalent, because they have different solution sets, and the solution -4 of Equation 2 does not satisfy Equation 1. Because raising both sides of an equation to the same power often introduces extraneous solutions (false solutions that don't satisfy the original equation), we must check all suspected solutions to be certain that they satisfy the original equation.

The following equation has an extraneous solution.

$$x - x^{1/2} - 6 = 0$$
$$(x^{1/2} - 3)(x^{1/2} + 2) = 0 \qquad\qquad\qquad \text{Factor } x - x^{1/2} - 6 = 0.$$
$$x^{1/2} - 3 = 0 \quad\text{or}\quad x^{1/2} + 2 = 0 \qquad \text{Set each factor equal to 0.}$$
$$x^{1/2} = 3 \qquad\qquad x^{1/2} = -2$$

Because equal powers of equal numbers are equal, we can square both sides of the previous equations to get

$$(x^{1/2})^2 = (3)^2 \quad\text{or}\quad (x^{1/2})^2 = (-2)^2$$
$$x = 9 \qquad\qquad x = 4$$

The root $x = 9$ satisfies the original equation, but the root $x = 4$ does not, as the following check shows:

If $x = 9$	**If $x = 4$**
$x - x^{1/2} - 6 = 0$	$x - x^{1/2} - 6 = 0$
$9 - 9^{1/2} - 6 \stackrel{?}{=} 0$	$4 - 4^{1/2} - 6 \stackrel{?}{=} 0$
$9 - 3 - 6 \stackrel{?}{=} 0$	$4 - 2 - 6 \stackrel{?}{=} 0$
$0 = 0$	$-4 \neq 0$

Thus, 9 is the only solution of the equation.

EXAMPLE 2   Solve the equation $2x^{2/5} - 5x^{1/5} - 3 = 0$.

Solution
$$2x^{2/5} - 5x^{1/5} - 3 = 0$$
$$(2x^{1/5} + 1)(x^{1/5} - 3) = 0 \qquad\qquad \text{Factor } 2x^{2/5} - 5x^{1/5} - 3 = 0.$$
$$2x^{1/5} + 1 = 0 \quad\text{or}\quad x^{1/5} - 3 = 0 \qquad \text{Set each factor equal to 0.}$$
$$2x^{1/5} = -1 \qquad\qquad x^{1/5} = 3$$
$$x^{1/5} = -\frac{1}{2}$$

Because equal powers of equal numbers are equal, we can raise both sides of each of the previous equations to the fifth power to get

$$(x^{1/5})^5 = \left(-\frac{1}{2}\right)^5 \quad\text{or}\quad (x^{1/5})^5 = (3)^5$$
$$x = -\frac{1}{32} \qquad\qquad x = 243$$

Verify that each solution satisfies the given equation.   ∎

Radical Equations

Equations containing radicals with variables in the radicand are called **radical equations**. To solve such equations, we will use the theorem stating that equal numbers raised to equal powers are equal.

 Warning!   Once again, we must check all suspected roots, because raising both sides of an equation to a power might introduce extraneous roots.

EXAMPLE 3   Solve the equation $\sqrt{x + 3} - 4 = 7$.

Solution   We begin by isolating the radical on the left-hand side.

$$\sqrt{x + 3} - 4 = 7$$
$$\sqrt{x + 3} = 11 \qquad \text{Add 4 to both sides.}$$
$$(\sqrt{x + 3})^2 = (11)^2 \qquad \text{Square both sides.}$$
$$x + 3 = 121 \qquad \text{Simplify.}$$
$$x = 118 \qquad \text{Subtract 3 from both sides.}$$

Since squaring both sides might introduce extraneous roots, we must check the suspected solution of 118.

$$\sqrt{x + 3} - 4 = 7$$
$$\sqrt{\mathbf{118} + 3} - 4 \overset{?}{=} 7$$
$$\sqrt{121} - 4 \overset{?}{=} 7$$
$$11 - 4 \overset{?}{=} 7$$
$$7 = 7$$

Because it checks, 118 is a root of the equation.   ■

EXAMPLE 4   Solve the equation $\sqrt{x + 3} = 3x - 1$.

Solution

$$\sqrt{x + 3} = 3x - 1$$
$$(\sqrt{x + 3})^2 = (3x - 1)^2 \qquad \text{Square both sides.}$$
$$x + 3 = 9x^2 - 6x + 1 \qquad \text{Remove parentheses.}$$
$$0 = 9x^2 - 7x - 2 \qquad \text{Add } -x - 3 \text{ to both sides.}$$
$$0 = (9x + 2)(x - 1) \qquad \text{Factor } 9x^2 - 7x - 2.$$

$$9x + 2 = 0 \qquad \text{or} \qquad x - 1 = 0 \qquad \text{Set each factor equal to 0.}$$
$$x = -\frac{2}{9} \qquad\qquad x = 1$$

Since squaring both sides might introduce extraneous roots, we must check each suspected solution.

$$\text{If } x = -\frac{2}{9}$$

$$\sqrt{x + 3} = 3x - 1$$

$$\sqrt{-\frac{2}{9} + 3} \overset{?}{=} 3\left(-\frac{2}{9}\right) - 1$$

$$\sqrt{\frac{25}{9}} \overset{?}{=} -\frac{2}{3} - 1$$

$$\frac{5}{3} \neq -\frac{5}{3}$$

$$\text{If } x = 1$$

$$\sqrt{x + 3} = 3x - 1$$

$$\sqrt{1 + 3} \overset{?}{=} 3(1) - 1$$

$$\sqrt{4} \overset{?}{=} 3 - 1$$

$$2 = 2$$

Since $-\frac{2}{9}$ does not satisfy the equation, it is extraneous; 1 is the only solution. ∎

EXAMPLE 5 Solve the equation $\sqrt[3]{x^3 + 56} = x + 2$.

Solution To eliminate the radical, we cube both sides of the equation. Then we proceed as follows:

$$\sqrt[3]{x^3 + 56} = x + 2$$

$$(\sqrt[3]{x^3 + 56})^3 = (x + 2)^3 \qquad \text{Square both sides.}$$

$$x^3 + 56 = x^3 + 6x^2 + 12x + 8 \qquad \text{Remove parentheses.}$$

$$0 = 6x^2 + 12x - 48 \qquad \text{Simplify.}$$

$$0 = x^2 + 2x - 8 \qquad \text{Divide both sides by 6.}$$

$$0 = (x + 4)(x - 2) \qquad \text{Factor } x^2 + 2x - 8.$$

$$x + 4 = 0 \quad \text{or} \quad x - 2 = 0 \qquad \text{Set each factor equal to 0.}$$

$$x = -4 \qquad \qquad x = 2$$

We check each suspected solution to see if either is extraneous.

$$\text{If } x = -4$$

$$\sqrt[3]{x^3 + 56} = x + 2$$

$$\sqrt[3]{(-4)^3 + 56} \overset{?}{=} -4 + 2$$

$$\sqrt[3]{-64 + 56} \overset{?}{=} -2$$

$$\sqrt[3]{-8} \overset{?}{=} -2$$

$$-2 = -2$$

$$\text{If } x = 2$$

$$\sqrt[3]{x^3 + 56} = x + 2$$

$$\sqrt[3]{2^3 + 56} \overset{?}{=} 2 + 2$$

$$\sqrt[3]{8 + 56} \overset{?}{=} 4$$

$$\sqrt[3]{64} \overset{?}{=} 4$$

$$4 = 4$$

Since both values satisfy the equation, both -4 and 2 are roots. ∎

EXAMPLE 6 Solve the equation $\sqrt{2x + 3} + \sqrt{x - 2} = 4$.

Solution To simplify the work, we write the equation in the form

$$\sqrt{2x + 3} = 4 - \sqrt{x - 2} \qquad \text{Subtract } \sqrt{x - 2} \text{ from both sides.}$$

so that the left-hand side contains only one radical. We can then square both sides to get

$$(\sqrt{2x + 3})^2 = (4 - \sqrt{x - 2})^2$$
$$2x + 3 = 16 - 8\sqrt{x - 2} + (x - 2)$$

After combining terms, we subtract x and 14 from both sides to leave a single radical on the right-hand side of the equation.

$$2x + 3 = 14 - 8\sqrt{x - 2} + x$$
$$x - 11 = -8\sqrt{x - 2}$$

We can square both sides again to eliminate the radical.

$$(x - 11)^2 = (-8\sqrt{x - 2})^2$$
$$x^2 - 22x + 121 = 64(x - 2)$$
$$x^2 - 86x + 249 = 0$$
$$(x - 3)(x - 83) = 0$$

$$x - 3 = 0 \quad \text{or} \quad x - 83 = 0 \qquad \text{Set each factor equal to 0.}$$
$$x = 3 \qquad\qquad\quad x = 83$$

Substituting these suspected roots into the given equation will show that 83 does not satisfy the equation. Thus, it is extraneous. However, 3 does satisfy the equation and is a root. ∎

2.6   EXERCISES

In Exercises 1-20, solve each equation for real values of the variable by factoring.

1. $x^3 + 9x^2 + 20x = 0$ **2.** $x^3 + 4x^2 - 21x = 0$ **3.** $6a^3 - 5a^2 - 4a = 0$ **4.** $8b^3 - 10b^2 + 3b = 0$

5. $y^4 - 26y^2 + 25 = 0$ **6.** $y^4 - 13y^2 + 36 = 0$ **7.** $x^4 - 37x^2 + 36 = 0$ **8.** $x^4 - 50x^2 + 49 = 0$

9. $2y^4 - 46y^2 = -180$ **10.** $2x^4 - 102x^2 = -196$ **11.** $z^{3/2} - z^{1/2} = 0$ **12.** $r^{5/2} - r^{3/2} = 0$

13. $2m^{2/3} + 3m^{1/3} - 2 = 0$ **14.** $6t^{2/5} + 11t^{1/5} + 3 = 0$ **15.** $x - 13x^{1/2} + 12 = 0$ **16.** $p + p^{1/2} - 20 = 0$

17. $2t^{1/3} + 3t^{1/6} - 2 = 0$ **18.** $z^3 - 7z^{3/2} - 8 = 0$ **19.** $6p + p^{1/2} - 1 = 0$ **20.** $3r - r^{1/2} - 2 = 0$

In Exercises 21–52, find all real solutions of each equation.

21. $\sqrt{x - 2} = 5$

22. $3\sqrt{x + 1} = \sqrt{6}$

23. $\sqrt{x + 3} = 2\sqrt{x}$

24. $\sqrt{16x + 4} = \sqrt{x + 4}$

25. $2\sqrt{x^2 + 3} = \sqrt{-16x - 3}$

26. $\sqrt{x^2 + 1} = \dfrac{\sqrt{-7x + 11}}{\sqrt{6}}$

27. $\sqrt[3]{7x + 1} = 4$

28. $\sqrt[3]{11a - 40} = 5$

29. $\sqrt[4]{30t + 25} = 5$

30. $\sqrt[4]{3z + 1} = 2$

31. $\sqrt{x^2 + 21} = x + 3$

32. $\sqrt{5 - x^2} = -(x + 1)$

33. $\sqrt{y + 2} = 4 - y$

34. $\sqrt{3z + 1} = z - 1$

35. $x - \sqrt{7x - 12} = 0$

36. $x - \sqrt{4x - 4} = 0$

37. $x + 4 = \sqrt{\dfrac{6x + 6}{5}} + 3$

38. $\sqrt{\dfrac{8x + 43}{3}} - 1 = x$

39. $\sqrt{\dfrac{x^2 - 1}{x - 2}} = 2\sqrt{2}$

40. $\dfrac{\sqrt{x^2 - 1}}{\sqrt{3x - 5}} = \sqrt{2}$

41. $\sqrt[3]{x^3 + 7} = x + 1$

42. $\sqrt[3]{x^3 - 7} + 1 = x$

43. $\sqrt[3]{8x^3 + 61} = 2x + 1$

44. $\sqrt[3]{8x^3 - 37} = 2x - 1$

45. $\sqrt{x + 3} = \sqrt{2x + 8} - 1$

46. $\sqrt{x + 2} + 1 = \sqrt{2x + 5}$

47. $\sqrt{y + 8} - \sqrt{y - 4} = -2$

48. $\sqrt{r} + \sqrt{r + 2} = 2$

49. $\sqrt{2b + 3} - \sqrt{b + 1} = \sqrt{b - 2}$

50. $\sqrt{a + 1} + \sqrt{3a} = \sqrt{5a + 1}$

51. $\sqrt{\sqrt{b} + \sqrt{b + 8}} = 2$

52. $\sqrt{\sqrt{x + 19} - \sqrt{x - 2}} = \sqrt{3}$

2.7 INEQUALITIES

■ Linear Inequalities ■ Compound Inequalities ■ Quadratic Inequalities
■ Rational Inequalities

In this section, we will develop techniques for solving inequalities. We begin by defining some terms.

Definition of $>$ and $<$	If a and b are real numbers, then
	$a > b$ if and only if $a - b$ is positive
	$a < b$ if and only if $b - a$ is positive

We note that if $a < b$, then $b > a$. If a is greater than or equal to b, we write $a \geq b$. Similarly, if a is less than or equal to b, we write $a \leq b$.

EXAMPLE 1   **a.** $5 > 3$ because $5 - 3$ is the positive real number 2.

b. $2 < 7$ because $7 - 2$ is the positive real number 5.

c. $3 \leq 3$ because $3 = 3$.

d. $2 \leq 3$ because $2 < 3$.

e. $x < x + 1$ because $(x + 1) - x = 1$ and 1 is positive. ■

We now state without proof several theorems involving $<$. Similar theorems exist for the other relations.

The first theorem states that any real number can be added to (or subtracted from) both sides of an inequality to obtain another inequality with the same order (direction). For example, since $2 < 5$, $2 + 3 < 5 + 3$ and $2 - 3 < 5 - 3$.

Theorem	Let a, b, and c represent real numbers.
	If $a < b$, then $a + c < b + c$.
	If $a < b$, then $a - c < b - c$.

The second theorem states that both sides of an inequality can be multiplied (or divided) by a positive number to obtain another inequality with the same order. For example, since $2 < 5$, $3(2) < 3(5)$ and $\frac{2}{3} < \frac{5}{3}$.

Theorem	Let a, b, and c represent real numbers.
	If $a < b$ and $c > 0$, then $ca < cb$.
	If $a < b$ and $c > 0$, then $\dfrac{a}{c} < \dfrac{b}{c}$.

The third theorem states that both sides of an inequality can be multiplied (or divided) by a negative number to obtain another inequality in the opposite order. For example, since $2 < 5$, $-3(2) > -3(5)$ and $\frac{2}{-3} > \frac{5}{-3}$.

Theorem	Let a, b, and c represent real numbers.
	If $a < b$ and $c < 0$, then $ca > cb$.
	If $a < b$ and $c < 0$, then $\dfrac{a}{c} > \dfrac{b}{c}$.

The previous theorems indicate that inequalities are solved like equations. However, we must always remember to change the order of an inequality when multiplying (or dividing) both sides of an inequality by a negative number.

The $<$ relation is neither reflexive (a number cannot be less than itself) nor symmetric (if a is less than b, then b is not less than a). However, it is transitive.

Theorem	If a, b, and c represent real numbers with $a < b$ and $b < c$, then $a < c$.

The relations $>$, $\leq$, and $\geq$ are also transitive.

■ Linear Inequalities

Inequalities such as $ax + c < 0$ or $ax - c \geq 0$, where $a \neq 0$, are called **linear inequalities**. Numbers that make an inequality true when substituted for the variable are solutions of the inequality. Two inequalities with the same solutions are called **equivalent inequalities**.

EXAMPLE 2   Solve the inequality $3(x + 2) < 8$.

Solution   Proceed as with equations.

$$3(x + 2) < 8$$
$$3x + 6 < 8 \qquad \text{Remove parentheses.}$$
$$3x < 2 \qquad \text{Subtract 6 from both sides.}$$
$$x < \frac{2}{3} \qquad \text{Divide both sides by 3.}$$

All numbers that are less than $\frac{2}{3}$ are solutions of the inequality. The solution set can be expressed in interval notation as $\left(-\infty, \frac{2}{3}\right)$ and can be graphed on the number line as in Figure 2-7. The symbol $-\infty$, read as "negative infinity," indicates that the interval is unbounded to the left of the origin.

FIGURE 2-7   ■

EXAMPLE 3   Solve the inequality $-5(x - 2) < 20 + x$.

Solution   Proceed as with equations.

$$-5(x - 2) < 20 + x$$
$$-5x + 10 < 20 + x \qquad \text{Remove parentheses.}$$
$$-6x + 10 < 20 \qquad \text{Subtract } x \text{ from both sides.}$$
$$-6x < 10 \qquad \text{Subtract 10 from both sides.}$$

We now divide both sides of the inequality by -6, which changes the order of the inequality.

$$x > \frac{10}{-6} \qquad \text{Divide both sides by } -6.$$
$$x > -\frac{5}{3} \qquad \text{Simplify the fraction.}$$

The graph of the solution set of the inequality is shown in Figure 2-8. The interval represented by the graph can be expressed in interval notation as $\left(-\frac{5}{3}, \infty\right)$.

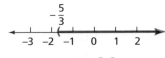

FIGURE 2-8   ■

EXAMPLE 4   An empty truck with driver weighs 4350 pounds. It is loaded with feed corn weighing 31 pounds per bushel. Between farm and market is a bridge with a 10,000-pound load limit. How many bushels can the truck legally carry?

Solution   We can let b represent the number of bushels in a legal load. Because the combined weight of the truck, driver, and cargo cannot exceed 10,000 pounds, we solve the inequality

$$4350 + 31b \leq 10,000$$
$$31b \leq 5650 \qquad \text{Subtract 4350 from each side.}$$
$$b \leq 182.2580645 \qquad \text{Divide both sides by 31.}$$

The truck cannot carry more than $182\frac{1}{4}$ bushels.   ■

Compound Inequalities

The statement that x is between 2 and 5 implies two inequalities

$$x > 2 \qquad \text{and} \qquad x < 5$$

It is customary to write both inequalities as the single **compound inequality**

$$2 < x < 5 \qquad \text{Read as ''2 is less than } x \text{ and } x \text{ is less than 5.''}$$

   Warning!   Remember that $2 < x < 5$ means that $x > 2$ *and* $x < 5$. The word *and* indicates that two inequalities must be true simultaneously.

To express that x is not between 2 and 5, we must convey the idea that either x is greater than or equal to 5 or x is less than or equal to 2. This is equivalent to the compound inequality

$$x \geq 5 \qquad \text{or} \qquad x \leq 2$$

This compound inequality is satisfied by all numbers x that satisfy either or both of its parts.

   Warning!   It is incorrect to write $x \geq 5$ or $x \leq 2$ as $2 \geq x \geq 5$, because this would mean that $2 \geq 5$, which is false.

EXAMPLE 5   Solve the inequality $5 < 3x - 7 \leq 8$.

Solution   We can isolate x between the inequality symbols by adding 7 to each part of the compound inequality to get

$$5 + 7 < 3x - 7 + 7 \leq 8 + 7 \qquad \text{Add 7 to each part.}$$
$$12 < 3x \leq 15 \qquad \text{Combine terms.}$$

and dividing all parts by 3 to get

$$4 < x \leq 5$$

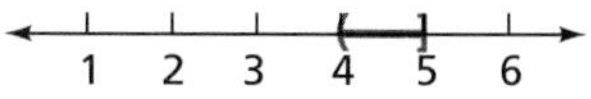

FIGURE 2-9

The solution set is the interval $(4, 5]$, whose graph appears in Figure 2-9.   ■

EXAMPLE 6   Solve the inequality $3 + x \leq 3x + 1 < 7x - 2$.

Solution   Because it is impossible to isolate x between the inequality symbols, we must solve each part of the compound inequality separately.

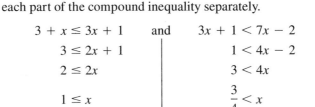

Because the connective in this compound inequality is "and," the solution set is the intersection of the intervals $[1, \infty)$ and $\left(\frac{3}{4}, \infty\right)$, which is $[1, \infty)$. The graph of the solution set is shown in Figure 2-10.

FIGURE 2-10   ■

It is possible for an inequality to be true for all values of its variable. For example,

$$x < x + 1$$

is true for all numbers x. It is also possible for an inequality to have no solutions. The inequality

$$x > x + 1$$

is not true for any number x.

*Hypatia
(370 A.D.–415 A.D.)
Hypatia is the earliest
known woman in the
history of mathematics. She
was a professor at the
University of Alexandria.
Because of her scientific
beliefs, she was considered
to be a heretic. At the age
of 45, she was attacked by
a mob and murdered for
her beliefs.*

■ Quadratic Inequalities

If $a \neq 0$, expressions such as $ax^2 + bx + c < 0$ and $ax^2 + bx + c > 0$ are called **quadratic inequalities**. We will discuss two methods for solving these inequalities.

EXAMPLE 7   Solve the quadratic inequality $x^2 - x - 6 > 0$.

Solution 1   First we solve the equation $x^2 - x - 6 = 0$:

$$x^2 - x - 6 = 0$$
$$(x + 2)(x - 3) = 0$$
$$x + 2 = 0 \quad \text{or} \quad x - 3 = 0$$
$$x = -2 \quad \quad x = 3$$

The graphs of these solutions on the number line establish the three intervals shown in Figure 2-11.

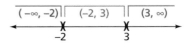

FIGURE 2-11

The solutions of the inequality $x^2 - x - 6 > 0$ will be the numbers in one or more of these intervals. To decide which intervals are solutions, we test a number in each one and see whether it satisfies the inequality.

$-6 \in (-\infty, -2)$	$0 \in (-2, 3)$	$5 \in (3, \infty)$
$x^2 - x - 6 > 0$	$x^2 - x - 6 > 0$	$x^2 - x - 6 > 0$
$(-6)^2 - (-6) - 6 > 0$	$0^2 - 0 - 6 > 0$	$5^2 - 5 - 6 > 0$
$36 + 6 - 6 > 0$	$0 - 0 - 6 > 0$	$25 - 5 - 6 > 0$
$36 > 0$	$-6 > 0$	$14 > 0$
This is true.	This is not true.	This is true.

The solutions are in the intervals $(-\infty, -2)$ or $(3, \infty)$. The graph of the solution set $(-\infty, -2) \cup (3, \infty)$ is shown in Figure 2-12.

FIGURE 2-12

Solution 2   Another method relies exclusively on the number line and a notation that keeps track of the signs of the factors of the quadratic inequality $x^2 - x - 6 > 0$.

When $x = 3$, the factor $x - 3$ is 0. When $x < 3$, the factor $x - 3$ is negative, and when $x > 3$, the factor $x - 3$ is positive.

When $x = -2$, the factor $x + 2$ is 0. When $x < -2$, the factor $x + 2$ is negative, and when $x > -2$, the factor $x + 2$ is positive.

Because the positiveness of the product $(x - 3)(x + 2)$ depends on the signs of $x - 3$ and $x + 2$, we keep track of the positiveness or negativeness of the factors by using $+$ and $-$ signs in the **sign graph** shown in Figure 2-13. Only to the left of -2 and to the right of $+3$ do the signs of both factors agree. Only there is the product positive. The graph appears in Figure 2-13.

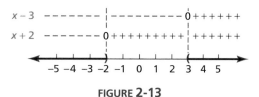

FIGURE 2-13   ■

EXAMPLE 8   Solve the inequality $x(x + 3) < -2$.

To get ready, we remove parentheses and add 2 to both sides of the inequality to make the right-hand side 0. We then solve the inequality $x^2 + 3x + 2 < 0$.

Solution 1   First we solve the quadratic equation $x^2 + 3x + 2 = 0$.

$$x^2 + 3x + 2 = 0$$
$$(x + 2)(x + 1) = 0$$
$$x + 2 = 0 \quad \text{or} \quad x + 1 = 0$$
$$x = -2 \quad \quad \quad x = -1$$

The graphs of these solutions on the number line establish the three intervals shown in Figure 2-14.

$$\overline{(-\infty, -2)} \quad \overline{(-2, -1)} \quad \overline{(-1, \infty)}$$

FIGURE **2-14**

The solutions of the inequality $x^2 + 3x + 2 < 0$ will be the numbers in one or more of these intervals. To decide which intervals are solutions, we test a number in each one and see whether it satisfies the inequality.

$-7 \in (-\infty, -2)$	$-\dfrac{3}{2} \in (-2, -1)$	$0 \in (-1, \infty)$
$x^2 + 3x + 2 < 0$	$x^2 + 3x + 2 < 0$	$x^2 + 3x + 2 < 0$
$(-7)^2 + 3(-7) + 2 < 0$	$\left(-\dfrac{3}{2}\right)^2 + 3\left(-\dfrac{3}{2}\right) + 2 < 0$	$0^2 + 3(0) + 2 < 0$
$49 - 21 + 2 < 0$	$\dfrac{9}{4} + \left(-\dfrac{9}{2}\right) + 2 < 0$	$0 + 0 + 2 < 0$
$30 < 0$	$-\dfrac{1}{4} < 0$	$2 < 0$
This is not true.	This is true.	This is not true.

The solution set is the interval $(-2, -1)$, whose graph appears in Figure 2-15.

FIGURE **2-15**

Solution 2   We can keep track of the positiveness or negativeness of the factors $x + 2$ and $x + 1$ by constructing the sign graph shown in Figure 2-16. Only between -2 and -1 do the factors have opposite signs. Here, the product is negative. The graph of the solution set is shown in Figure 2-16.

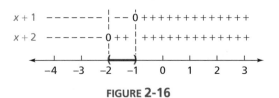

FIGURE **2-16**   ■

■ Rational Inequalities

An inequality that contains a fraction with a polynomial numerator and denominator is called a **rational inequality**. To solve such inequalities, we can use the same techniques used to solve quadratic inequalities.

EXAMPLE 9 Solve the rational inequality $\dfrac{x^2 - x - 2}{x^2 - 4x + 3} \leq 0$.

Solution 1 The intervals are found by solving the two quadratic equations $x^2 - x - 2 = 0$ and $x^2 - 4x + 3 = 0$. The solutions of the first equation are -1 and 2, and the solutions of the second equation are 1 and 3. The graphs of these solutions establish the five intervals shown in Figure 2-17.

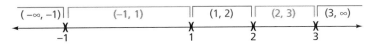

FIGURE 2-17

To decide which of these intervals are solutions, we test a number in each one to see whether it satisfies the inequality. Numbers in the intervals $(-1, 1)$ and $(2, 3)$ satisfy the rational inequality, but numbers in the intervals $(-\infty, -1)$, $(1, 2)$, and $(3, \infty)$ do not. The graph of the solution set is shown in Figure 2-18.

FIGURE 2-18

Because the numbers $x = -1$ and $x = 2$ make the numerator 0, they satisfy the inequality

$$\frac{x^2 - x - 2}{x^2 - 4x + 3} \leq 0$$

Thus, their graphs are drawn with brackets to show that -1 and 2 are included in the solution set. Because the numbers 1 and 3 give 0's in the denominator of the fraction, parentheses are placed at $x = 1$ and $x = 3$ to show that 1 and 3 are not in the solution set. The solution set is $[-1, 1) \cup [2, 3)$.

Solution 2 We can factor each of the trinomials and write the inequality in the form

$$\frac{(x - 2)(x + 1)}{(x - 3)(x - 1)} \leq 0$$

and then construct the sign graph shown in Figure 2-19. The value of the rational expression must be 0 or be negative. The only way it can be 0 is for its numerator to be 0. This happens when $x = 2$ and $x = -1$. The only way it can be negative is

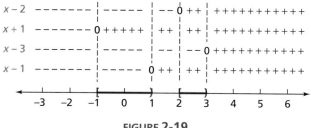

FIGURE **2-19**

to have an odd number of negative factors. This happens in the intervals between -1 and 1 and between 2 and 3.

The graph of the solution set appears in Figure 2-19. The brackets at $x = -1$ and $x = 2$ show that these are included in the solution set. The parentheses at $x = 1$ and $x = 3$ show that these two numbers are not included in the solution set. If they were included, the denominator of the fraction would be 0. ∎

EXAMPLE 10   Solve the inequality $\dfrac{6}{x} > 2$.

Solution   To get a 0 on the right-hand side, we add -2 to both sides of the inequality. Then we combine terms on the left-hand side.

$$\frac{6}{x} > 2$$

$$\frac{6}{x} - 2 > 0$$

$$\frac{6}{x} - \frac{2x}{x} > 0$$

$$\frac{6 - 2x}{x} > 0$$

The given inequality now has the form of a rational inequality. The intervals are found by solving the equations $6 - 2x = 0$ and $x = 0$. The solution of the first equation is 3, and the solution of the second equation is 0. This determines the intervals $(-\infty, 0)$, $(0, 3)$, and $(3, \infty)$. Because only the numbers in the interval $(0, 3)$ satisfy the original inequality, the solution set is $(0, 3)$. The graph is shown in Figure 2-20.

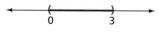

FIGURE **2-20**

To use the sign-graph method, we construct a sign graph as in Figure 2-21 to determine that the solution set is $(0, 3)$.

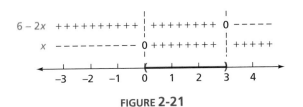

FIGURE 2-21   ■

> **Warning!**   It is tempting to solve Example 10 by multiplying both sides of
> the inequality by x and solving the inequality $6 > 2x$. However, this is not
> correct. Multiplying both sides of the inequality by x gives $6 > 2x$ only when x is
> positive. If x is negative, multiplying both sides of the inequality by x will reverse
> the direction of the $>$ symbol, and the inequality $\frac{6}{x} > 2$ will be equivalent to $6 < 2x$.
> If you fail to consider both cases, you will obtain a wrong answer.

2.7   EXERCISES

In Exercises 1–20, solve each inequality, graph the solution set, and write the answer in interval notation.

1. $3x + 2 < 5$ **2.** $-2x + 4 < 6$ **3.** $3x + 2 \geq 5$ **4.** $-2x + 4 \geq 6$

5. $-5x + 3 > -2$ **6.** $4x - 3 > -4$ **7.** $-5x + 3 \leq -2$ **8.** $4x - 3 \leq -4$

9. $2(x - 3) \leq -2(x - 3)$ **10.** $3(x + 2) \leq 2(x + 5)$

11. $\frac{3}{5}x + 4 > 2$ **12.** $\frac{1}{4}x - 3 > 5$ **13.** $\frac{x + 3}{4} < \frac{2x - 4}{3}$ **14.** $\frac{x + 2}{5} > \frac{x - 1}{2}$

15. $\frac{6(x - 4)}{5} \geq \frac{3(x + 2)}{4}$ **16.** $\frac{3(x + 3)}{2} < \frac{2(x + 7)}{3}$ **17.** $\frac{5}{9}(a + 3) - a \geq \frac{4}{3}(a - 3) - 1$

18. $\frac{2}{3}y - y \leq -\frac{3}{2}(y - 5)$ **19.** $\frac{2}{3}a - \frac{3}{4}a < \frac{3}{5}\left(a + \frac{2}{3}\right) + \frac{1}{3}$ **20.** $\frac{1}{4}b + \frac{2}{3}b - \frac{1}{2} > \frac{1}{2}(b + 1) + b$

In Exercises 21–40, solve each double inequality, write the answer in interval notation, and graph the solution set.

21. $4 < 2x - 8 \leq 10$ **22.** $3 \leq 2x + 2 < 6$ **23.** $9 \geq \frac{x - 4}{2} > 2$ **24.** $5 < \frac{x - 2}{6} < 6$

25. $0 \leq \frac{4 - x}{3} \leq 5$ **26.** $0 \geq \frac{5 - x}{2} \geq -10$ **27.** $-2 \geq \frac{1 - x}{2} \geq -10$ **28.** $-2 \leq \frac{1 - x}{2} < 10$

29. $-3x > -2x > -x$ **30.** $-3x < -2x < -x$ **31.** $x < 2x < 3x$ **32.** $x > 2x > 3x$

33. $2x + 1 < 3x - 2 < 12$ **34.** $2 - x < 3x + 5 < 18$

35. $2 + x < 3x - 2 < 5x + 2$ **36.** $x > 2x + 3 > 4x - 7$

37. $3 + x > 7x - 2 > 5x - 10$ **38.** $2 - x < 3x + 1 < 10x$

39. $x \leq x + 1 \leq 2x + 3$ **40.** $-x \geq -2x + 1 \geq -3x + 1$

In Exercises 41–68, solve each inequality, write the answer in interval notation, and graph the solution set.

41. $x^2 + 7x + 12 < 0$ **42.** $x^2 - 13x + 12 \leq 0$ **43.** $x^2 - 5x + 6 \geq 0$ **44.** $6x^2 + 5x - 6 > 0$

45. $x^2 + 5x + 6 < 0$

46. $x^2 + 9x + 20 \geq 0$

47. $6x^2 + 5x + 1 \geq 0$

48. $x^2 + 9x + 20 < 0$

49. $6x^2 - 5x < -1$

50. $9x^2 + 24x > -16$

51. $2x^2 \geq 3 - x$

52. $9x^2 \leq 24x - 16$

53. $\dfrac{x + 3}{x - 2} < 0$

54. $\dfrac{x + 3}{x - 2} > 0$

55. $\dfrac{x^2 + x}{x^2 - 1} > 0$

56. $\dfrac{x^2 - 4}{x^2 - 9} < 0$

57. $\dfrac{x^2 + 5x + 6}{x^2 + x - 6} \geq 0$

58. $\dfrac{x^2 + 10x + 25}{x^2 - x - 12} \leq 0$

59. $\dfrac{6x^2 - x - 1}{x^2 + 4x + 4} > 0$

60. $\dfrac{6x^2 - 3x - 3}{x^2 - 2x - 8} < 0$

61. $\dfrac{3}{x} > 2$

62. $\dfrac{3}{x} < 2$

63. $\dfrac{6}{x} < 4$

64. $\dfrac{6}{x} > 4$

65. $\dfrac{3}{x - 2} \leq 5$

66. $\dfrac{3}{x + 2} \leq 4$

67. $\dfrac{6}{x^2 - 1} < 1$

68. $\dfrac{6}{x^2 - 1} > 1$

In Exercises 69–76, solve each word problem.

69. Altitudes of an airplane A plane flies at altitudes from 27,720 feet to 33,000 feet. Express this range in miles. (*Hint:* 1 mile = 5280 feet.)

70. Geometry problem The perimeter of a rectangle is to be between 180 inches and 200 inches. Find the range of values for its length when its width is 40 inches.

71. Geometry problem The perimeter of an equilateral triangle is to be between 50 centimeters and 60 centimeters. Find the range of lengths of one side.

72. Geometry problem The perimeter of a square is to be from 25 meters to 60 meters. Find the range of values for its area.

73. Mathematics application Express the relationship $20 < l < 30$ in terms of P, where $P = 2l + 2w$.

74. Changing temperature scales Express the relationship $10 < C < 20$ in terms of F, where $F = \frac{9}{5}C + 32$.

75. Buying compact disks Andy can spend up to $275 on a compact disk player and some compact disks. If he can buy a disk player for $150 and disks for $9.75, what is the greatest number of disks that he can buy?

76. Buying videotapes Mary wants to spend less than $600 for a VCR and some videotapes. If the VCR of her choice costs $425 and videotapes cost $7.50, how many videotapes can she buy?

In Exercises 77–84, find the restrictions on x in each equation. (Hint: A radicand cannot be negative.)

77. $4 = \sqrt{3x - 4}$

78. $3 = \sqrt{6 - 4x}$

79. $5 = \dfrac{1}{\sqrt{3x + 2}}$

80. $2 = \dfrac{5}{\sqrt{5 - 7x}}$

81. $6 = \sqrt{x^2 - 9}$

82. $8 = \sqrt{25 - x^2}$

83. $2 = \dfrac{\sqrt{x^2 + 4x + 3}}{x}$

84. $4 = \dfrac{x^2 - 9}{\sqrt{x^2 - 5x + 6}}$

2.8 ABSOLUTE VALUE

■ Absolute Value Equations   ■ Absolute Value Inequalities

In this section, we review the definition of absolute value and examine its consequences in greater detail.

Absolute Value	The **absolute value** of the real number x, denoted as $|x|$, is defined as follows: If $x \geq 0$, then $|x| = x$. If $x < 0$, then $|x| = -x$.

Whether x is positive, 0, or negative, the above definition enables us to find $|x|$. If $x > 0$, the absolute value of x is the positive number x. If $x < 0$, the absolute value of x is the positive number $-x$. Since $|0| = 0$, we have that for all real numbers,

$$|x| \geq 0$$

EXAMPLE 1 Find   **a.** $|0|$,   **b.** $|7|$,   **c.** $|-3|$,   **d.** $-|-7|$,   and   **e.** $|x - 2|$.

Solution **a.** By the first part of the definition, $|0| = 0$.

b. Because 7 is positive, by the first part of the definition, $|7| = 7$.

c. Because -3 is negative, by the second part of the definition, $|-3| = -(-3) = 3$.

d. The expression $-|-7|$ means "the negative of the absolute value of -7." Thus,

$$-|-7| = -(7) = -7$$

e. To denote the absolute value of a variable quantity, we must give a conditional answer. Either

$$|x - 2| = x - 2 \qquad \text{when } x - 2 \geq 0 \text{ (or when } x \geq 2)$$

or

$$|x - 2| = -(x - 2) = 2 - x \qquad \text{when } x - 2 < 0 \text{ (or when } x < 2) \qquad ∎$$

■ Absolute Value Equations

To develop a strategy for solving equations that contain absolute value symbols, we consider the equation $|x| = a$, where a is positive. When $x \geq 0$, then $|x| = x$ and

$$x = a$$

However, when x is negative, then $|x| = -x$ and $-x = a$ and

$$x = -a$$

and we have the following theorem.

Theorem	If $a \geq 0$, the equation $|x| = a$ is equivalent to the two equations $$x = a \qquad \text{or} \qquad x = -a$$

The absolute value of x can be interpreted as the distance on a number line that a point is from the origin. For example, the solutions of the equation $|x| = a$ are represented by the two points that lie exactly a units from the origin, as shown in Figure 2-22.

FIGURE 2-22

EXAMPLE 2 Solve the equation $|3x - 5| = 7$.

Solution The equation $|3x - 5| = 7$ is equivalent to the two equations

$$3x - 5 = 7 \quad \text{or} \quad 3x - 5 = -7$$

which can be solved separately as follows:

$$3x - 5 = 7 \quad \text{or} \quad 3x - 5 = -7$$
$$3x = 12 \qquad\qquad 3x = -2$$
$$x = 4 \qquad\qquad x = -\frac{2}{3}$$

The graph of the solution set consists of the two points shown in Figure 2-23.

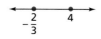

FIGURE 2-23 ∎

EXAMPLE 3 Solve the equation $|2x| = |x - 3|$.

Solution The equation $|2x| = |x - 3|$ will be satisfied when the quantities $2x$ and $x - 3$ are equal to each other or when they are the negatives of each other. This idea determines two equations, which can be solved separately:

$$2x = x - 3 \quad \text{or} \quad 2x = -(x - 3)$$
$$x = -3 \qquad\qquad 2x = -x + 3$$
$$\qquad\qquad 3x = 3$$
$$\qquad\qquad x = 1$$

Verify that both -3 and 1 satisfy the given equation. ∎

■ Absolute Value Inequalities

To solve an inequality that contains absolute values and the symbol $<$, we consider the inequality $|x| < a$. If x is positive or 0, then $|x| = x$, and

$$x < a$$

If x is negative, then $|x| = -x$ and $-x < a$ or

$$x > -a$$

Thus, if x is positive or 0, then x is less than a, and if x is negative, then x is greater than $-a$. This implies that x must be between $-a$ and a. The two inequalities

$$x < a \quad \text{and} \quad x > -a$$

are equivalent to the double inequality

$$-a < x < a$$

and we have the following theorem.

Theorem	If $a > 0$, the equation $|x| < a$ is equivalent to the double inequality $-a < x < a$ The symbol $<$ can be replaced with $\leq$.

The solutions to the inequality $|x| < a$ are the coordinates of the points on a number line that are less than a units from the origin. (See Figure 2-24.)

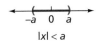

$$|x| < a$$

FIGURE 2-24

EXAMPLE 4   Solve the inequality $|x - 2| < 7$ and graph its solution set on a number line.

Solution   The inequality $|x - 2| < 7$ is equivalent to the double inequality

$$-7 < x - 2 < 7$$

We can add 2 to each part of this double inequality to get

$$-5 < x < 9$$

Thus, the solution set is the interval $(-5, 9)$. The graph appears in Figure 2-25.

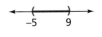

FIGURE 2-25   ■

To solve an inequality that contains absolute values and an "is greater than" symbol, we consider the inequality $|x| > a$ where a is positive. If x is positive or 0, then $|x| = x$ and

$$x > a$$

If x is negative, then $|x| = -x$ and

$$-x > a \quad \text{or} \quad x < -a$$

Thus, if x is positive or 0, then x is greater than a, and if x is negative, then x is less than $-a$. This implies that we can write the solutions of the inequality $|x| > a$ as

$$x < -a \quad \text{or} \quad x > a$$

and we have the following theorem.

Theorem	If $a > 0$, the equation $|x| > a$ is equivalent to the statement
	$$x < -a \quad \text{or} \quad x > a$$
	The symbols $<$ and $>$ can be replaced with $\leq$ and $\geq$.

The solutions of the inequality $|x| > a$ are the coordinates of the points on a number line that are greater than a units from the origin. (See Figure 2-26.)

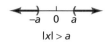

$$|x| > a$$

FIGURE 2-26

EXAMPLE 5 Solve the inequality $\left| \dfrac{2x + 3}{2} \right| + 7 \geq 12$ and graph its solution set on a number line.

Solution We begin by subtracting 7 from both sides of the inequality to isolate the absolute value expression on the left-hand side.

$$\left| \frac{2x + 3}{2} \right| \geq 5$$

This result is equivalent to the following two inequalities, which can be solved separately.

$$\frac{2x + 3}{2} \geq 5 \quad \text{or} \quad \frac{2x + 3}{2} \leq -5$$
$$2x + 3 \geq 10 \qquad\qquad 2x + 3 \leq -10$$
$$2x \geq 7 \qquad\qquad 2x \leq -13$$
$$x \geq \frac{7}{2} \qquad\qquad x \leq -\frac{13}{2}$$

Thus, the solution set is $\left(-\infty, -\frac{13}{2} \right] \cup \left[\frac{7}{2}, \infty \right)$. The graph appears in Figure 2-27.

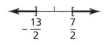

$$-\frac{13}{2} \qquad \frac{7}{2}$$

FIGURE 2-27

EXAMPLE 6 Solve the inequality $0 < |x - 5| \leq 3$ and graph its solution set on a number line.

Solution The double inequality $0 < |x - 5| \leq 3$ consists of two inequalities that can be solved separately. The first inequality, $0 < |x - 5|$, is true for all numbers x except 5. The second inequality, $|x - 5| \leq 3$, is equivalent to the double inequality

$$-3 \leq x - 5 \leq 3$$

or

$$2 \leq x \leq 8 \qquad \text{Add 5 to each part.}$$

Thus, the solution set consists of all numbers in the interval [2, 8] except 5. We can express this result as the union of the intervals [2, 5) and (5, 8). The graph appears in Figure 2-28.

FIGURE 2-28 ■

In Example 9 of Section 2.5, we saw that $|a|$ could be defined as

$$|a| = \sqrt{a^2}$$

We can use this fact in the next example.

EXAMPLE 7   Solve the inequality $|x + 2| > |x + 1|$.

Solution

$$|x + 2| > |x + 1|$$

$$\sqrt{(x + 2)^2} > \sqrt{(x + 1)^2} \qquad \text{Use } |a| = \sqrt{a^2}.$$

$$(x + 2)^2 > (x + 1)^2 \qquad \text{Square both sides.}$$

$$x^2 + 4x + 4 > x^2 + 2x + 1 \qquad \text{Expand each binomial.}$$

$$4x > 2x - 3 \qquad \text{Subtract } x^2 \text{ and 4 from both sides.}$$

$$2x > -3 \qquad \text{Subtract } 2x \text{ from both sides.}$$

$$x > -\frac{3}{2} \qquad \text{Divide both sides by 2.}$$

Thus, the solution set is the interval $\left(-\frac{3}{2}, \infty\right)$. Check several solutions to verify that this interval does not contain any extraneous solutions. ■

There are three other properties of absolute value that are sometimes useful.

Properties of Absolute Value	If a and b are real numbers, then
	1. $\|ab\| = \|a\|\|b\|$ **2.** $\left\|\dfrac{a}{b}\right\| = \dfrac{\|a\|}{\|b\|}$ $(b \neq 0)$
	3. $\|a + b\| \leq \|a\| + \|b\|$

Properties 1 and 2 state that the absolute value of a product (or a quotient) is the product (or the quotient) of the absolute values.

Property 3, called the **triangular inequality**, states that the absolute value of a sum is either equal to, or less than, the sum of the absolute values.

2.8 EXERCISES

In Exercises 1–12, write each expression without using absolute value symbols.

1. $|7|$ **2.** $|-9|$ **3.** $|0|$ **4.** $|3-5|$

5. $|5|-|-3|$ **6.** $|-3|+|5|$ **7.** $|\pi-2|$ **8.** $|\pi-4|$

9. $|x-5|$ and $x>5$ **10.** $|x-5|$ and $x<5$ **11.** $|x^3|$ **12.** $|2x|$

In Exercises 13–36, solve each equation for x.

13. $|x+2|=2$ **14.** $|2x+5|=3$ **15.** $|3x-1|=5$ **16.** $|7x-5|=3$

17. $\left|\dfrac{3x-4}{2}\right|=5$ **18.** $\left|\dfrac{10x+1}{2}\right|=\dfrac{9}{2}$ **19.** $\left|\dfrac{2x-4}{5}\right|=2$ **20.** $\left|\dfrac{3x+11}{7}\right|=1$

21. $\left|\dfrac{x-3}{4}\right|=-2$ **22.** $\left|\dfrac{x-5}{3}\right|=0$ **23.** $\left|\dfrac{4x-2}{x}\right|=3$ **24.** $\left|\dfrac{2(x-3)}{3x}\right|=6$

25. $|x|=x$ **26.** $|x|+x=2$ **27.** $|x+3|=|x|$ **28.** $|x+5|=|5-x|$

29. $|x-3|=|2x+3|$ **30.** $|x-2|=|3x+8|$ **31.** $|x+2|=|x-2|$ **32.** $|2x-3|=|3x-5|$

33. $\left|\dfrac{x+3}{2}\right|=|2x-3|$ **34.** $\left|\dfrac{x-2}{3}\right|=|6-x|$ **35.** $\left|\dfrac{3x-1}{2}\right|=\left|\dfrac{2x+3}{3}\right|$ **36.** $\left|\dfrac{5x+2}{3}\right|=\left|\dfrac{x-1}{4}\right|$

In Exercises 37–52, solve each inequality, express the solution set in interval notation, and graph it.

37. $|x-3|<6$ **38.** $|x-2|\geq 4$ **39.** $|x+3|>6$ **40.** $|x+2|\leq 4$

41. $|2x+4|\geq 10$ **42.** $|5x-2|<7$ **43.** $|3x+5|+1\leq 9$ **44.** $|2x-7|-3>2$

45. $|x+3|>0$ **46.** $|x-3|\leq 0$ **47.** $\left|\dfrac{5x+2}{3}\right|<1$ **48.** $\left|\dfrac{3x+2}{4}\right|>2$

49. $3\left|\dfrac{3x-1}{2}\right|>5$ **50.** $2\left|\dfrac{8x+2}{5}\right|\leq 1$ **51.** $\dfrac{|x-1|}{-2}>-3$ **52.** $\dfrac{|2x-3|}{-3}<-1$

In Exercises 53–62, solve each double inequality, express the solution set in interval notation, and graph it.

53. $0<|2x+1|<3$ **54.** $0<|2x-3|<1$ **55.** $8>|3x-1|>3$ **56.** $8>|4x-1|>5$

57. $2<\left|\dfrac{x-5}{3}\right|<4$ **58.** $3<\left|\dfrac{x-3}{2}\right|<5$ **59.** $10>\left|\dfrac{x-2}{2}\right|>4$ **60.** $5\geq\left|\dfrac{x+2}{3}\right|>1$

61. $2\leq\left|\dfrac{x+1}{3}\right|<3$ **62.** $8>\left|\dfrac{3x+1}{2}\right|>2$

In Exercises 63–70, solve each inequality and express the solution using interval notation.

63. $|x+1|\geq|x|$ **64.** $|x+1|<|x+2|$ **65.** $|2x+1|<|2x-1|$ **66.** $|3x-2|\geq|3x+1|$

67. $|x+1|<|x|$ **68.** $|x+2|\leq|x+1|$ **69.** $|2x+1|\geq|2x-1|$ **70.** $|3x-2|<|3x+1|$

2 CHAPTER SUMMARY

Key Words

break point (2.2)
completing the square (2.3)
complex conjugates (2.5)
complex numbers (2.5)
compound inequalities (2.7)
conditional equations (2.1)
contradictions (2.1)
discriminant (2.3)
equation (2.1)
equivalent equations (2.1)
equivalent inequalities (2.7)
extraneous solution (2.1)

first-degree equations (2.1)
formulas (2.1)
identities (2.1)
imaginary numbers (2.5)
impossible equations (2.1)
linear equations (2.1)
linear inequalities (2.7)
quadratic equations (2.3)
quadratic formula (2.3)
quadratic inequalities (2.7)
radical equations (2.6)

rational equations (2.1)
rational inequality (2.7)
root (2.1)
second-degree equation (2.3)
setup cost (2.2)
sign graph (2.7)
solution set (2.1)
square root property (2.3)
unit cost (2.2)
variable (2.1)
zero-factor theorem (2.3)

Key Ideas

(2.1) Any number can be added to or subtracted from both sides of an equation. If $a = b$ and c is a number, then

$$a + c = b + c \quad \text{and} \quad a - c = b - c$$

Both sides of an equation can be multiplied or divided by any nonzero number. If $a = b$ and $c \neq 0$, then

$$ac = bc \quad \text{and} \quad \frac{a}{c} = \frac{b}{c}$$

Any quantity can be substituted for its equal without changing the truth of a mathematical expression.

(2.2) Use the following steps to solve a word problem:

1. Read the problem.
2. Pick a variable to represent the quantity to be found.
3. Form an equation.
4. Solve the equation.
5. Check the solution in the words of the problem.

(2.3) **The Zero-Factor Theorem:** If $ab = 0$, then $a = 0$ or $b = 0$.

The Square Root Property: If $c > 0$, the equation $x^2 = c$ has two real roots:

$$x = \sqrt{c} \quad \text{and} \quad x = -\sqrt{c}$$

The Quadratic Formula:

$$x = \frac{-b \pm \sqrt{b^2 - 4ac}}{2a} \qquad (a \neq 0)$$

The Discriminant:

If $b^2 - 4ac > 0$, the roots of $ax^2 + bx + c = 0$ are unequal real numbers.

If $b^2 - 4ac = 0$, the roots of $ax^2 + bx + c = 0$ are equal real numbers.

If $b^2 - 4ac < 0$, the roots of $ax^2 + bx + c = 0$ are nonreal numbers.

(2.5) The complex numbers $a + bi$ and $c + di$ are equal if and only if $a = c$ and $b = d$.

$$(a + bi) + (c + di) = (a + c) + (b + d)i$$
$$(a + bi) - (c + di) = (a - c) + (b - d)i$$
$$(a + bi)(c + di) = (ac - bd) + (ad + bc)i$$
$$|a + bi| = \sqrt{a^2 + b^2}$$

(2.6) If $a = b$, then $a^2 = b^2$.

(2.7) If a, b, and c are real numbers:

If $a < b$, then $a + c < b + c$ and $a - c < b - c$.

If $a < b$ and $c > 0$, then $ac < bc$ and $\dfrac{a}{c} < \dfrac{b}{c}$.

If $a < b$ and $c < 0$, then $ac > bc$ and $\dfrac{a}{c} > \dfrac{b}{c}$.

If $a < b$ and $b < c$, then $a < c$.

Similar relationships hold for $>$, $\leq$, and $\geq$.

$(2.8)\ |x| = \begin{cases} x \text{ when } x \geq 0 \\ -x \text{ when } x < 0 \end{cases}$

If $a \geq 0$, then $|x| = a$ is equivalent to
$$x = a \text{ or } x = -a$$

If $a > 0$, then $|x| < a$ is equivalent to
$$-a < x < -a$$

If $a > 0$, then $|x| > a$ is equivalent to
$$x > a \text{ or } x < -a$$
$$|a| = \sqrt{a^2}$$

Properties of Absolute Value:

1. $|ab| = |a||b|$

2. $\left|\dfrac{a}{b}\right| = \dfrac{|a|}{|b|}$

3. $|a + b| \leq |a| + |b|$

---

2 CHAPTER REVIEW EXERCISES

In Review Exercises 1–4, find the restrictions on x.

1. $3x + 7 = 4$

2. $x + \dfrac{1}{x} = 2$

3. $\sqrt{x} = 4$

4. $\dfrac{1}{x - 2} = \dfrac{2}{x - 3}$

In Review Exercises 5–12, solve each equation and classify it as an identity, a conditional equation, or an equation with no solution.

5. $3(9x + 4) = 28$

6. $\dfrac{3}{2}a = 7(a + 11)$

7. $8(3x - 5) - 4(x + 3) = 12$

8. $\dfrac{x + 3}{x + 4} + \dfrac{x + 3}{x + 2} = 2$

9. $\dfrac{3}{x - 1} = \dfrac{1}{2}$

10. $\dfrac{8x^2 + 72x}{9 + x} = 8x$

11. $\dfrac{3x}{x - 1} - \dfrac{5}{x + 3} = 3$

12. $x + \dfrac{1}{2x - 3} = \dfrac{2x^2}{2x - 3}$

In Review Exercises 13–16, solve each formula for the indicated variable.

13. $C = \dfrac{5}{9}(F - 32);\ F$

14. $P_n = l + \dfrac{si}{f};\ f$

15. $\dfrac{1}{f} = \dfrac{1}{f_1} + \dfrac{1}{f_2};\ f_1$

16. $S = \dfrac{a - lr}{1 - r};\ l$

In Review Exercises 17–22, solve each word problem.

17. *Preparing a solution*   A liter of fluid is 50% alcohol. How much water must be added to dilute it to a 20% solution?

18. *Washing windows*   Scott can wash 37 windows in three hours, while Bill can wash 27 windows in 2 hours. How long will it take the two of them to wash 100 windows?

19. *Filling a tank*   A tank can be filled in 9 hours by one pipe and 12 hours by another. How long will it take both pipes to fill the empty tank?

20. *Producing brass*   How many ounces of pure zinc must be alloyed with 20 ounces of brass that is 30% zinc and 70% copper to produce brass that is 40% zinc?

21. *Lending money* A bank lent $10,000, part of it at 11% annual interest and the rest at 14%. If the annual income is $1265, how much was lent at each rate?

22. *Producing oriental rugs* A manufacturer of oriental rugs can use one loom with a setup cost of $750 that can weave a rug for $115. Another loom, with a setup cost of $950, can produce a rug for $95. How many rugs are produced if the costs are the same on each loom?

In Review Exercises 23–26, solve each equation by factoring.

23. $2x^2 - x - 6 = 0$

24. $12x^2 + 13x = 4$

25. $5x^2 - 8x = 0$

26. $27x^2 = 30x - 8$

In Review Exercises 27–28, solve each word problem.

27. *Fencing a field* A farmer wishes to enclose a rectangular garden with 300 yards of fencing. A river runs along one side of the garden, so no fencing is needed there. What will be the dimensions of the rectangle if the area is 10,450 square yards?

28. *Flying rates* A jet plane, flying 120 miles per hour faster than a propeller-driven plane, travels 3520 miles in 3 hours less time than the propeller plane requires to fly the same distance. How fast does each plane fly?

In Review Exercises 29–32, solve each equation by completing the square.

29. $x^2 - 8x + 15 = 0$

30. $3x^2 + 18x = -24$

31. $5x^2 - x - 1 = 0$

32. $5x^2 - x = 0$

In Review Exercises 33–36, use the quadratic formula to solve each equation.

33. $x^2 + 5x - 14 = 0$

34. $3x^2 - 25x = 18$

35. $5x^2 = 1 - x$

36. $-5 = a^2 + 2a$

37. Find the value of k that will make the roots of $kx^2 + 4x + 12 = 0$ equal.

38. Find the values of k that will make the roots of $4y^2 + (k + 2)y = 1 - k$ equal.

39. *Flight of a ball* A ball thrown into the air reaches a height h (in feet) according to the formula $h = -16t^2 + 64t$, where t is the time elapsed since the ball was thrown. Find the shortest time it will take the ball to reach a height of 48 feet.

40. *Finding the width of a border* A man plans to have a tile border of uniform width around a rectangular pool. If the area of the border is 160 square feet and the dimensions of the pool are 16 feet by 20 feet, how wide is the border?

In Review Exercises 41–54, perform all indicated operations and express all answers in $a + bi$ form.

41. Simplify i^{53}.

42. Simplify i^{103}.

43. $(2 - 3i) + (-4 + 2i)$

44. $(2 - 3i) - (4 + 2i)$

45. $(3 - \sqrt{-36}) + (\sqrt{-16} + 2)$

46. $(3 + \sqrt{-9})(2 - \sqrt{-25})$

47. $\dfrac{3}{i}$

48. $-\dfrac{2}{i^3}$

49. $\dfrac{3}{1 + i}$

50. $\dfrac{2i}{2 - i}$

51. $\dfrac{3 + i}{3 - i}$

52. $\dfrac{3 - 2i}{1 + i}$

53. $|3 - i|$

54. $\left|\dfrac{1 + i}{1 - i}\right|$

In Review Exercises 55–64, solve each equation.

55. $\dfrac{3x}{2} - \dfrac{2x}{x - 1} = x - 3$

56. $\dfrac{12}{x} - \dfrac{x}{2} = x - 3$

57. $x^4 - 2x^2 + 1 = 0$

58. $x^4 + 36 = 37x^2$

59. $a - a^{1/2} - 6 = 0$

60. $x^{2/3} + x^{1/3} - 6 = 0$

61. $\sqrt{x - 1} + x = 7$

62. $\sqrt{a + 9} - \sqrt{a} = 3$

63. $\sqrt{5 - x} + \sqrt{5 + x} = 4$

64. $\sqrt{y + 5} + \sqrt{y} = 1$

In Review Exercises 65–76, solve each inequality and graph the solution set.

65. $2x - 9 < 5$

66. $5x + 3 \geq 2$

67. $\dfrac{5(x - 1)}{2} < x$

68. $0 \leq \dfrac{3 + x}{2} < 4$

69. $(x + 2)(x - 4) > 0$

70. $(x - 1)(x + 4) < 0$

71. $x^2 - 2x - 3 < 0$

72. $2x^2 + x - 3 > 0$

73. $\dfrac{x + 2}{x - 3} \geq 0$

74. $\dfrac{x - 1}{x + 4} \leq 0$

75. $\dfrac{x^2 + x - 2}{x - 3} \geq 0$

76. $\dfrac{5}{x} < 2$

In Review Exercises 77–84, solve each equation or inequality. Graph the solution sets of the inequalities.

77. $|x + 1| = 6$

78. $|2x - 1| = |2x + 1|$

79. $|x + 3| < 3$

80. $|3x - 7| \geq 1$

81. $\left|\dfrac{x + 2}{3}\right| < 1$

82. $\left|\dfrac{x - 3}{4}\right| > 8$

83. $1 < |2x + 3| < 4$

84. $0 < |3x - 4| < 7$

2 CHAPTER TEST

In Questions 1–2, find the domain of x.

1. $\dfrac{x}{x(x - 1)}$

2. $\sqrt{x}$

In Questions 3–5, solve each equation.

3. $7(2a + 5) - 7 = 6(a + 8)$

4. $\dfrac{3}{y - 2} = \dfrac{3}{y - 2} - \dfrac{1}{y}$

5. $\dfrac{3}{x^2 - 5x - 14} = \dfrac{4}{x^2 + 5x + 6}$

6. Solve for x: $z = \dfrac{x - \mu}{\sigma}$.

7. A student's average grade on three chapter tests in algebra is 75. If the final exam is to count as two one-hour tests, what grade must the student make to bring his grade up to 80%?

8. A woman invests part of $20,000 at 6% interest and the rest at 7%. If her annual interest is $1260, how much did she invest at 6%?

In Questions 9–10, solve each equation.

9. $4x^2 - 8x + 3 = 0$

10. $2b^2 - 12 = -5b$

11. Write the quadratic formula.

12. Use the quadratic formula to solve

$$3x^2 - 5x - 9 = 0$$

13. Find k so that $x^2 + (k + 1)x + k + 4 = 0$ will have two equal roots.

14. The height of a projectile shot up into the air is given by the formula $h = -16t^2 + 128t$. Find the time t required for the projectile to return to its starting point.

In Questions 15–16, simplify each expression.

15. i^{13}

16. i^0

In Questions 17–20, perform each operation and write all answers in $a + bi$ form.

17. $(4 - 5i) - (-3 + 7i)$

18. $(4 - 5i)(3 - 7i)$

19. $\dfrac{2}{2 - i}$

20. $\dfrac{1 + i}{1 - i}$

In Questions 21–22, find each absolute value.

21. $|5 - 12i|$

22. $\left| \dfrac{1}{3 + i} \right|$

In Questions 23–26, solve each equation.

23. $z^4 - 13z^2 + 36 = 0$

24. $2p^{2/5} - p^{1/5} - 1 = 0$

25. $\sqrt{x + 5} = 12$

26. $\sqrt{2z + 3} = 1 - \sqrt{z + 1}$

In Questions 27–30, solve each inequality and graph the solution set.

27. $5x - 3 \leq 7$

28. $\dfrac{x + 3}{4} > \dfrac{2x - 4}{3}$

29. $1 + x < 3x - 3 < 4x - 2$

30. $\dfrac{x + 2}{x - 1} \leq 0$

In Questions 31–32, solve each equation.

31. $\left| \dfrac{3x + 2}{2} \right| = 4$

32. $|x + 3| = |x - 3|$

In Questions 33–35, solve each inequality and graph the solution set.

33. $|2x - 5| > 2$

34. $\left| \dfrac{2x + 3}{3} \right| \leq 5$

35. $0 < |x - 2| < 5$

CHAPTER

3

THE RECTANGULAR COORDINATE SYSTEM AND GRAPHS OF EQUATIONS

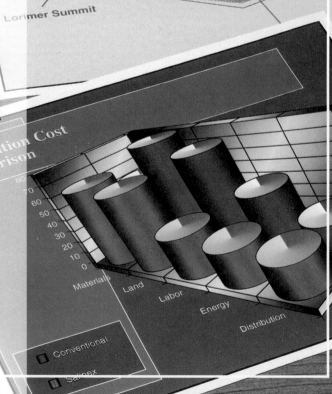

Mathematical expressions often indicate relationships between several quantities. For example, the formula $A = s^2$ describes the relationship between the area of a square and the length of one of its sides. Similarly, the formula $C = \pi d$ indicates the relationship between the diameter of the circle and the distance around the circle (its circumference). To visualize relationships such as these, we can draw graphs of their equations.

3.1 THE RECTANGULAR COORDINATE SYSTEM

- The Graph of an Equation
- The Distance Formula
- Graphing Lines Parallel to the *x*- and *y*-Axes
- The Midpoint Formula

René Descartes (1596–1650)

Descartes is famous for his work in philosophy as well as for his work in mathematics. His philosophy is expressed in the words "I think, therefore I am." He is best known in mathematics for his invention of a coordinate system and his work with conic sections.

We have seen that each point on a number line is associated with a single real number, its **coordinate**. The point is called the **graph** of the number. For example, the real number 3 is the coordinate of point *A* on the number line in Figure 3-1, and point *A* is the graph of the number 3.

FIGURE 3-1

The French mathematician René Descartes (1596–1650) gets the credit for associating points in the plane with pairs of real numbers. Descartes' idea is based on

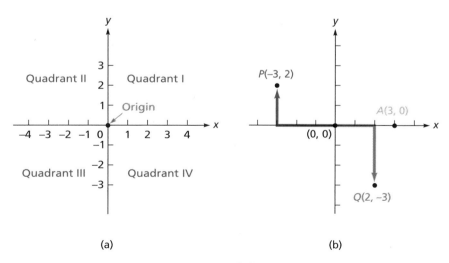

(a)   (b)

FIGURE 3-2

two perpendicular number lines, called the **x-axis** and the **y-axis**, which divide the plane into four **quadrants**, numbered with Roman numerals as shown in Figure 3-2(a). The axes intersect at a point called the **origin**, which is the zero point on each number line. The positive direction on the x-axis is to the right, the positive direction on the y-axis is upward, and the same unit distance is used on both axes. These axes form a **rectangular coordinate system**, or a **Cartesian coordinate system** (in honor of its inventor).

To **plot** the point associated with the pair of real numbers $(-3, 2)$, for example, we start at the origin, count 3 units to the left, and then two units up, as in Figure 3-2(b). The point P, called the **graph** of the pair $(-3, 2)$, lies in the second quadrant. The pair $(-3, 2)$ gives the **coordinates** of the point P. Point Q with coordinates $(2, -3)$ lies in the fourth quadrant. The coordinates of the origin are $(0, 0)$. The coordinate of point A on the x-axis is no longer just 3; it is now the pair $(3, 0)$.

Note that the pairs $(-3, 2)$ and $(2, -3)$ represent different points. Because the order of the real numbers in the pair (a, b) is important, such pairs are called **ordered pairs**. The first coordinate, a, of the ordered pair (a, b) is called the **x-coordinate** or the **abscissa**. The second coordinate, b, is the **y-coordinate**, or the **ordinate**. It is proper to say "the point P with coordinates (a, b)," but it is acceptable to say, simply, "point $P(a, b)$." The set of all points (x, y) is called the **xy-plane**.

EXAMPLE 1   In Figure 3-3, identify the quadrant or the axes where each point lies: **a.** $P(2, 3)$, **b.** $Q(-4, -2)$,   **c.** $R(-5, 0)$,   **d.** $S(0, 4)$,   and   **e.** $T(3, -2)$.

Solution   **a.** $P(2, 3)$ lies in quadrant I.

b. $Q(-4, -2)$ lies in quadrant III.

c. $R(-5, 0)$ lies on the x-axis. It is not in any quadrant.

d. $S(0, 4)$ lies on the y-axis. It is not in any quadrant.

e. $T(3, -2)$ lies in quadrant IV.

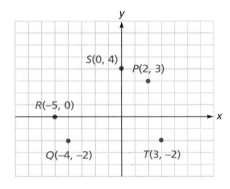

FIGURE 3-3   ■

■ ## The Graph of an Equation

Equations such as $x + 2y = 5$ define a relationship between the values of x and y. The **graph** of the equation is the set of all points $P(x, y)$ whose coordinates (x, y) satisfy the equation. To graph an equation, we choose a value for one of its variables (say, x), calculate the corresponding value of y, and plot the point (x, y). We do this for several points until we see a pattern and then join the points with a line or a curve. If the graph is a curve, the more points we plot, the more accurate the graph will be.

EXAMPLE 2 Graph the equation $x + 2y = 5$.

Solution We can pick values for either x or y, substitute them into the equation, and solve for the other variable. For example, if $y = -1$, we find x as follows.

$$x + 2y = 5$$
$$x + 2(-1) = 5 \qquad \text{Substitute } -1 \text{ for } y.$$
$$x - 2 = 5 \qquad \text{Simplify.}$$
$$x = 7 \qquad \text{Add 2 to both sides.}$$

Thus, $(7, -1)$ is one ordered pair that satisfies the equation. To find another, we let $x = 0$, and find y.

$$x + 2y = 5$$
$$\mathbf{0} + 2y = 5 \qquad \text{Substitute 0 for } x.$$
$$2y = 5 \qquad \text{Simplify.}$$
$$y = \frac{5}{2} \qquad \text{Divide both sides by 2.}$$

Thus, $\left(0, \frac{5}{2}\right)$ is another ordered pair that satisfies the equation.

These two ordered pairs and others that satisfy the equation $x + 2y = 5$ are shown in Figure 3-4. We plot these points on a rectangular coordinate system, as in the figure, and notice that they lie on a line. The line passing through the five points is the graph of the equation. Later, we will show that all points (x, y) that satisfy the equation will lie on a line.

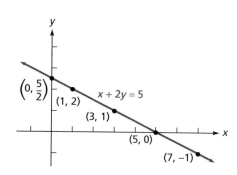

FIGURE 3-4   ∎

Because the line in Example 2 intersects the y-axis at the point $\left(0, \frac{5}{2}\right)$, the number $\frac{5}{2}$ is called the **y-intercept** of the line. Similarly, 5 is the **x-intercept** of the line.

EXAMPLE 3 The graph of the equation $4x - 3y = 12$ is a line. Use the x- and y-intercepts to graph it.

Solution   To find the y-intercept, we substitute 0 for x and solve for y.

$$4x - 3y = 12$$
$$4(0) - 3y = 12 \qquad \text{Substitute 0 for } x.$$
$$-3y = 12 \qquad \text{Simplify.}$$
$$y = -4 \qquad \text{Divide both sides by } -3.$$

Because the y-intercept is -4, the line intersects the y-axis at the point $(0, -4)$. To find the x-intercept, we substitute 0 for y and solve for x.

$$4x - 3y = 12$$
$$4x - 3(0) = 12 \qquad \text{Substitute 0 for } y.$$
$$4x = 12 \qquad \text{Simplify.}$$
$$x = 3 \qquad \text{Divide both sides by 4.}$$

Because the x-intercept is 3, the line intersects the x-axis at the point $(3, 0)$.

Although these two points are sufficient to draw the line, we find one other point to act as a check. If $x = 6$, for example, then $y = 4$, and the point $(6, 4)$ lies on the line. The graph appears in Figure 3-5.

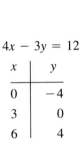

$$4x - 3y = 12$$

x	y
0	-4
3	0
6	4

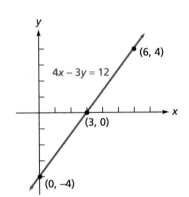

FIGURE 3-5

EXAMPLE 4   Graph the equation $3(y + 2) = 2x - 3$.

Solution   To make it easier to find pairs (x, y) that satisfy the equation, we first solve the equation for y.

$$3(y + 2) = 2x - 3$$
$$3y + 6 = 2x - 3 \qquad \text{Remove parentheses.}$$
$$3y = 2x - 9 \qquad \text{Subtract 6 from both sides.}$$
$$y = \frac{2}{3}x - 3 \qquad \text{Divide both sides by 3.}$$

We can substitute numbers for x to find the corresponding values of y. For example, we can let $x = 3$:

$$y = \frac{2}{3}x - 3$$

$$y = \frac{2}{3}(3) - 3 \qquad \text{Substitute 3 for } x.$$

$$y = 2 - 3 \qquad \text{Simplify.}$$

$$y = -1$$

Thus, the point $(3, -1)$ lies on the graph. To find the graph's y-intercept, we let $x = 0$ and find y:

$$y = \frac{2}{3}x - 3$$

$$y = \frac{2}{3}(0) - 3 \qquad \text{Substitute 0 for } x.$$

$$y = -3 \qquad \text{Simplify.}$$

Thus, the point $(0, -3)$ also lies on the graph. We plot these two points and a few others in Figure 3-6. The graph of the equation is the line that passes through these points.

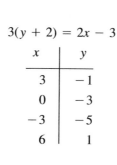

$$3(y + 2) = 2x - 3$$

x	y
3	-1
0	-3
-3	-5
6	1

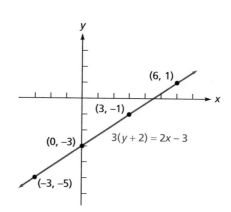

FIGURE 3-6 ■

■ Graphing Lines Parallel to the x- and y-Axes

EXAMPLE 5 Graph the equations **a.** $y = 2$ and **b.** $x = -3$

Solution **a.** Because x does not appear in the equation $y = 2$, the numbers assigned to x have no effect on y. The value of y is always 2. A table of values in Figure 3-7(a) gives several ordered pairs that satisfy the equation $y = 2$. After plotting these pairs (x, y) and joining them with a straight line, we see that the graph of $y = 2$ is a horizontal line, parallel to the x-axis and intersecting the y-axis at 2. The line $y = 2$ has a y-intercept of 2.

b. Because y does not appear in the equation $x = -3$, y can be any number, as long as $x = -3$. A table of values and the graph appear in Figure 3-7(b). The graph of $x = -3$ is a vertical line parallel to the y-axis and intersecting the x-axis at -3. The x-intercept of the line is -3; the line has no y-intercept.

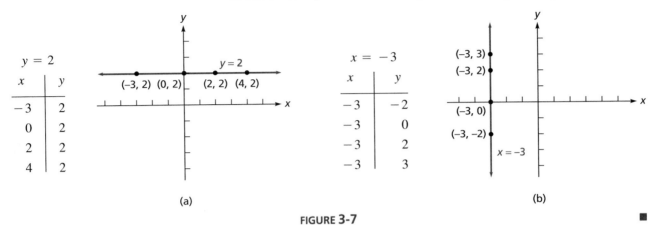

(a)

(b)

FIGURE 3-7

The results of Example 5 suggest the following facts.

Equations of Lines Parallel to the Coordinate Axes	If a and b are real numbers, then • The graph of the equation $x = a$ is a line parallel to the y-axis that intersects the x-axis at a. If $a = 0$, the line is the y-axis. • The graph of the equation $y = b$ is a line parallel to the x-axis that intersects the y-axis at b. If $b = 0$, the line is the x-axis.

■ The Distance Formula

Because the rectangular coordinate system uses the same unit of measure on the x- and the y-axes, we can derive a formula for finding the distance between any two points in the xy-plane. In deriving the formula, we use **subscript notation**, and denote one point as $P(x_1, y_1)$, read as "point P with coordinates of x sub 1 and y sub 1," and the second point as $Q(x_2, y_2)$.

If $P(x_1, y_1)$ and $Q(x_2, y_2)$ are two points in Figure 3-8 and point R has coordinates (x_2, y_1), triangle PQR is a right triangle. By the Pythagorean theorem, the square of

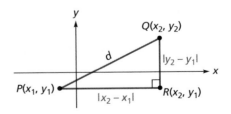

FIGURE 3-8

the hypotenuse of right triangle PQR is equal to the sum of the squares of the two legs. Because leg RQ is vertical, the square of its length is $(y_2 - y_1)^2$. Since leg PR is horizontal, the square of its length is $(x_2 - x_1)^2$. Thus, we have

1. $d^2 = (x_2 - x_1)^2 + (y_2 - y_1)^2$

Because equal positive numbers have equal positive square roots, we can take the positive square root of both sides of Equation 1 to obtain the **distance formula**.

The Distance Formula	The distance d between points (x_1, y_1) and (x_2, y_2) is given by $$d = \sqrt{(x_2 - x_1)^2 + (y_2 - y_1)^2}$$

EXAMPLE 6   Find the distance $d(PQ)$ between the points $P(-1, -2)$ and $Q(-7, 8)$.

Solution   We can let (x_1, y_1) be $P(-1, -2)$, let (x_2, y_2) be $Q(-7, 8)$, and substitute -1 for x_1, -2 for y_1, -7 for x_2, and 8 for y_2 into the formula and simplify.

$$d(PQ) = \sqrt{(x_2 - x_1)^2 + (y_2 - y_1)^2}$$
$$d(PQ) = \sqrt{[-7 - (-1)]^2 + [8 - (-2)]^2}$$
$$= \sqrt{(-6)^2 + (10)^2}$$
$$= \sqrt{36 + 100}$$
$$= \sqrt{136}$$
$$= \sqrt{4 \cdot 34}$$
$$= 2\sqrt{34}$$

■

■ The Midpoint Formula

The **midpoint** of a line segment is the point on the segment that is midway between its endpoints. The x-coordinate of the midpoint is the average of the x-coordinates of the segment's endpoints, and the y-coordinate of the midpoint is the average of the y-coordinates of the endpoints. For example, the x-coordinate of the midpoint M of the segment joining $P(3, 4)$ and $Q(5, 7)$ is the average of 3 and 5. The y-coordinate of the midpoint of the segment joining $P(3, 4)$ and $Q(5, 7)$ is the average of 4 and 7.

$$M\left(\frac{3 + 5}{2}, \frac{4 + 7}{2}\right) \quad \text{or} \quad M\left(4, \frac{11}{2}\right)$$

The Midpoint Formula	The midpoint of the line segment joining points $P(x_1, y_1)$ and $Q(x_2, y_2)$ is the point M: $$M\left(\frac{x_1 + x_2}{2}, \frac{y_1 + y_2}{2}\right)$$

Proof To prove that $M\left(\dfrac{x_1 + x_2}{2}, \dfrac{y_1 + y_2}{2}\right)$ is the midpoint of segment PQ, we must show that M is the same distance from each endpoint: that $d(PM) = d(MQ)$. We must also show that P, Q, and M all lie on the same line. See Figure 3-9.

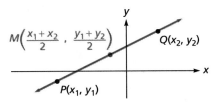

FIGURE 3-9

To show that $d(PM) = d(MQ)$, we use the distance formula and observe that the two distances are equal.

$$d(PM) = \sqrt{\left(\frac{x_1 + x_2}{2} - x_1\right)^2 + \left(\frac{y_1 + y_2}{2} - y_1\right)^2}$$

$$= \sqrt{\left(\frac{x_1 + x_2}{2} - \frac{2x_1}{2}\right)^2 + \left(\frac{y_1 + y_2}{2} - \frac{2y_1}{2}\right)^2} \qquad \text{To combine the fractions, find a common denominator.}$$

$$= \sqrt{\left(\frac{x_2 - x_1}{2}\right)^2 + \left(\frac{y_2 - y_1}{2}\right)^2} \qquad \text{Combine terms.}$$

$$= \sqrt{\frac{(x_2 - x_1)^2}{4} + \frac{(y_2 - y_1)^2}{4}} \qquad \text{Simplify.}$$

$$d(PM) = \frac{1}{2}\sqrt{(x_2 - x_1)^2 + (y_2 - y_1)^2}$$

$$d(MQ) = \sqrt{\left(x_2 - \frac{x_1 + x_2}{2}\right)^2 + \left(y_2 - \frac{y_1 + y_2}{2}\right)^2}$$

$$= \sqrt{\left(\frac{x_2 - x_1}{2}\right)^2 + \left(\frac{y_2 - y_1}{2}\right)^2} \qquad \text{Combine terms.}$$

$$= \sqrt{\frac{(x_2 - x_1)^2}{4} + \frac{(y_2 - y_1)^2}{4}} \qquad \text{Simplify.}$$

$$d(MQ) = \frac{1}{2}\sqrt{(x_2 - x_1)^2 + (y_2 - y_1)^2}$$

Since $d(PM) = d(MQ)$, point M is the same distance from each endpoint. The total distance between P and Q is

$$d(PQ) = \sqrt{(x_2 - x_1)^2 + (y_2 - y_1)^2}$$

Since $d(PM)$ and $d(MQ)$ are both $\frac{1}{2}[d(PQ)]$, points P, M, and Q must lie on the same line. The theorem is proved.   □

EXAMPLE 7   Find the midpoint of the line segment joining $P(-7, 2)$ and $Q(1, -4)$.

Solution   We use the midpoint formula and substitute $P(-7, 2)$ for $P(x_1, y_1)$ and $Q(1, -4)$ for $Q(x_2, y_2)$.

$$x_M = \frac{x_1 + x_2}{2} \quad \text{and} \quad y_M = \frac{y_1 + y_2}{2}$$

$$= \frac{-7 + 1}{2} \qquad\qquad = \frac{2 + (-4)}{2}$$

$$= \frac{-6}{2} \qquad\qquad = \frac{-2}{2}$$

$$= -3 \qquad\qquad = -1$$

The midpoint is $M(-3, -1)$.   ∎

EXAMPLE 8   The midpoint of the line segment joining $P(-3, 2)$ and $Q(x_2, y_2)$ is $M(1, 4)$. Find the coordinates of Q.

Solution   We can let $P(x_1, y_1)$ be $P(-3, 2)$, let $M(x_M, y_M)$ be $M(1, 4)$, and find the coordinates x_2 and y_2 of point $Q(x_2, y_2)$ as follows.

$$x_M = \frac{x_1 + x_2}{2} \quad \text{and} \quad y_M = \frac{y_1 + y_2}{2}$$

$$1 = \frac{-3 + x_2}{2} \qquad\qquad 4 = \frac{2 + y_2}{2}$$

$$2 = -3 + x_2 \qquad\qquad 8 = 2 + y_2 \qquad \text{Multiply both sides by 2.}$$

$$5 = x_2 \qquad\qquad\qquad 6 = y_2$$

The coordinates of point Q are $(5, 6)$.   ∎

3.1   EXERCISES

In Exercises 1–8, refer to Illustration 1 on page 146 and find the coordinates of each point.

1. A 2. B 3. C 4. D
5. E 6. F 7. G 8. H

In Exercises 9–20, graph each point in the xy-plane. Indicate the quadrant in which the point lies, or the axis on which it lies.

9. $(2, 5)$ 10. $(-3, 4)$ 11. $(-4, -5)$ 12. $(6, 2)$
13. $(5, 2)$ 14. $(3, -4)$ 15. $(4, 0)$ 16. $(-2, 2)$
17. $(0, 2)$ 18. $(3, 4)$ 19. $(-7, 0)$ 20. $(0, -5)$

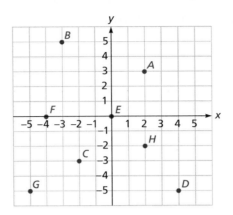

ILLUSTRATION 1

In Exercises 21–32, find the x- and y-intercepts and use them to graph each equation.

21. $x + y = 5$ **22.** $x - y = 3$ **23.** $2x - y = 4$ **24.** $3x + y = 9$

25. $3x + 2y = 6$ **26.** $2x - 3y = 6$ **27.** $4x - 5y = 20$ **28.** $3x - 5y = 15$

29. $3x + 4y = 12$ **30.** $4x - 3y = 12$ **31.** $2x + y = 5$ **32.** $x + 3y = 7$

In Exercises 33–44, solve each equation for y and graph the equation.

33. $y - 2x = 7$ **34.** $y + 3 = -4x$ **35.** $y + 5x = 5$ **36.** $y - 3x = 6$

37. $6x - 3y = 10$ **38.** $4x + 8y - 1 = 0$ **39.** $3x = 6y - 1$ **40.** $2x + 1 = 4y$

41. $2(x - y) = 3x + 2$ **42.** $5(x + 2) = 3y - x$ **43.** $3x + y = 3(x - 1)$ **44.** $2(y - x) = 3(x + 2)$

In Exercises 45–52, graph each equation.

45. $y = 3$ **46.** $x = -4$ **47.** $3x + 5 = -1$ **48.** $7y - 1 = 6$

49. $3(y + 2) = y$ **50.** $4 + 3y = 3(x + y)$ **51.** $3(y + 2x) = 6x + y$ **52.** $5(y - x) = x + 5y$

In Exercises 53–60, find the distance between P and O(0, 0).

53. $P(4, -3)$ **54.** $P(-5, 12)$ **55.** $P(-3, 2)$ **56.** $P(5, 0)$

57. $P(1, 1)$ **58.** $P(6, -8)$ **59.** $P(\sqrt{3}, 1)$ **60.** $P(\sqrt{7}, \sqrt{2})$

In Exercises 61–72, find the distance between P and Q.

61. $P(3, 7); Q(6, 3)$ **62.** $P(4, 9); Q(9, 21)$ **63.** $P(4, -6); Q(-1, 6)$ **64.** $P(0, 5); Q(6, -3)$

65. $P(-2, -15); Q(-9, -39)$ **66.** $P(-7, 11); Q(3, -13)$

67. $P(3, -3); Q(-5, 5)$ **68.** $P(6, -3); Q(-3, 2)$

69. $P(3, 7); Q(5, 7)$ **70.** $P(4, -6); Q(4, -8)$ **71.** $P(\pi, -2); Q(\pi, 5)$ **72.** $P(\sqrt{5}, 0); Q(0, 2)$

In Exercises 73–80, find the midpoint of the line segment PQ.

73. $P(2, 4); Q(6, 8)$ **74.** $P(3, -6); Q(-1, -6)$ **75.** $P(2, -5); Q(-2, 7)$ **76.** $P(0, 3); Q(-10, -13)$

77. $P(-8, 5); Q(8, -5)$ **78.** $P(3, -2); Q(2, -3)$ **79.** $P(0, 0); Q(\sqrt{5}, \sqrt{5})$ **80.** $P(\sqrt{3}, 0); Q(0, -\sqrt{5})$

In Exercises 81–84, find the distance between points P and Q. Assume that all variables represent real numbers.

81. $P(a, 0)$; $Q(0, b)$ **82.** $P(a, b)$; $Q(b, a)$ **83.** $P(a, b)$; $Q(0, 0)$ **84.** $P(a, -a)$; $Q(0, 0)$

In Exercises 85–88, find the midpoint of the line segment PQ. Assume that all variables represent real numbers.

85. $P(a, b)$; $Q(0, 0)$ **86.** $P(a, b)$; $Q(-a, -b)$ **87.** $P(a, b)$; $Q(b, a)$ **88.** $P(a, a)$; $Q(b, b)$

In Exercises 89–92, one endpoint P and the midpoint M of line segment PQ are given. Find the coordinates of the other endpoint, Q.

89. $P(1, 4)$; $M(3, 5)$ **90.** $P(2, -7)$; $M(-5, 6)$ **91.** $P(5, -5)$; $M(5, 5)$ **92.** $P(-7, 3)$; $M(0, 0)$

93. Show that a triangle with vertices at $(13, -2), (9, -8)$, and $(5, -2)$ is isosceles.

94. Show that a triangle with vertices at $(-1, 2), (3, 1)$, and $(4, 5)$ is isosceles.

95. Show that the points $(-3, -1), (3, 1), (1, 7)$, and $(-5, 5)$ are the vertices of a rhombus by showing that its four sides are equal.

96. Use the distance formula to find the radius of a circle with center at $(4, -2)$ and passing through $(6, -9)$.

97. Find the center of a circle if the endpoints of a diameter are $(-17, 6)$ and $(9, -12)$.

98. Find the coordinates of the two points on the *y*-axis that are $\sqrt{74}$ units from the point $(5, -3)$.

99. The diagonals of a square with area of 50 square units lie on the *x*- and the *y*-axes. Find the coordinates of the vertices.

100. Show that the distance between the points (a, b) and (c, d) is equal to the distance between the point $(a - c, b - d)$ and the origin.

101. In Illustration 2, points *M* and *N* are the midpoints of *AC* and *BC*, respectively. Find the length of *MN*.

102. In Illustration 3, points *M* and *N* are the midpoints of *AC* and *BC*, respectively. Show that
$$d(MN) = \tfrac{1}{2}[d(AB)]$$

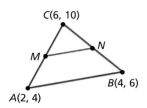

ILLUSTRATION **2**

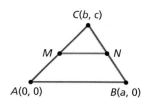

ILLUSTRATION **3**

103. In Illustration 4, point *M* is the midpoint of the hypotenuse of right triangle *AOB*. Show that the area of rectangle *OLMN* is one-half of the area of triangle *AOB*.

104. Rectangle *ABCD* in Illustration 5 is twice as long as it is wide, and its sides are parallel to the coordinate axes. If the perimeter is 42, find the coordinates of point *C*.

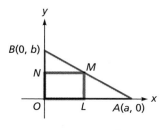

ILLUSTRATION **4**

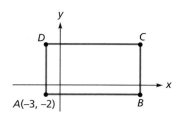

ILLUSTRATION **5**

105. 🔢   An ocean liner is located 23 miles east and 72 miles north of Pigeon Cove Lighthouse, and its home port is 47 miles west and 84 miles south of the lighthouse. How far is the ship from port? (See Illustration 6.)

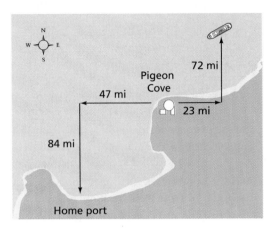

ILLUSTRATION 6

106. 🔢   Two holes are to be drilled at locations specified by the engineering drawings shown in Illustration 7. Find the distance between the centers of the holes.

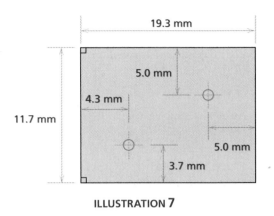

ILLUSTRATION 7

3.2   THE SLOPE OF A NONVERTICAL LINE

■ Interpretation of Slope   ■ Slopes of Parallel Lines   ■ Slopes of Perpendicular Lines

The **slope of a nonvertical line** drawn in the xy-plane is a measure of its tilt or inclination. Consider the line l in Figure 3-10(a), which passes through $P(x_1, y_1)$ and $Q(x_2, y_2)$. If point R is positioned so that line RQ is vertical and line PR is horizontal, then triangle PQR is a right triangle, and point R has coordinates (x_2, y_1).

The distance from R to Q, called the **rise**, is the change in the y-coordinates of points P and Q: $y_2 - y_1$. The horizontal distance from P to R, called the **run**, is the change in the x-coordinates of points P and Q: $x_2 - x_1$. The slope of the line passing through P and Q is the rise divided by the run.

$$m = \frac{\text{rise}}{\text{run}} = \frac{y_2 - y_1}{x_2 - x_1}$$

Because the y-coordinates of two points on a horizontal line are equal, $y_2 - y_1$ would be 0. Therefore, the slope of a horizontal line is 0. The slope of a vertical line does not exist, because $x_2 - x_1$ would be 0, and division by 0 is not defined. Therefore, *a vertical line has no defined slope.*

The definition of slope uses the coordinates of points P and Q to determine the slope of a line. However, the slope can be found by using any two points on the line. In Figure 3-10(b), we let point S be an arbitrary third point on the line l. Because

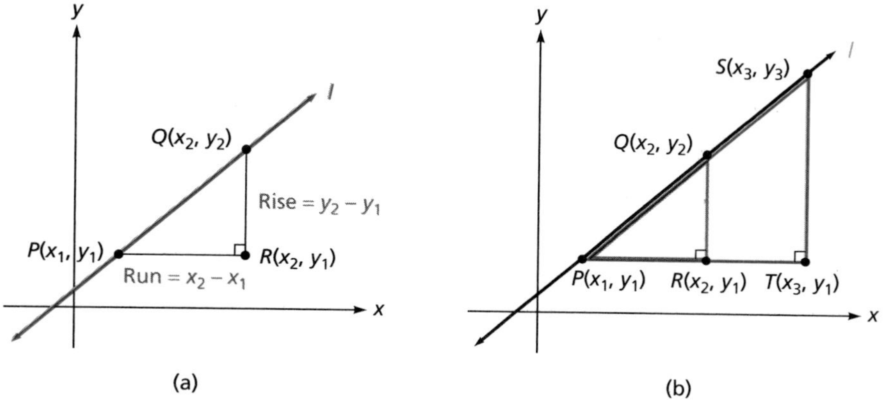

FIGURE 3-10

triangles PRQ and PTS are similar, their corresponding sides are in proportion, and the ratio of the rise to the run in either triangle can be used to find the slope:

$$m = \frac{y_2 - y_1}{x_2 - x_1} = \frac{y_3 - y_1}{x_3 - x_1}$$

Thus, the slope of a nonvertical line is a constant that we can find by using any two points on the line. Furthermore, if point P is on a line with slope m, and the slope of a segment PQ is also m, then point Q must also be on the line.

Slope	If $P(x_1, y_1)$ and $Q(x_2, y_2)$ are two points on a nonvertical line l in the xy-plane, then the slope of l is given by $$m = \frac{y_2 - y_1}{x_2 - x_1} \quad \text{provided } x_1 \neq x_2$$ If $x_1 = x_2$, then l is a vertical line and has no defined slope. The slope of a horizontal line is 0.

EXAMPLE 1   Find the slope of the line passing through the points $P(-1, -2)$ and $Q(7, 8)$.

Solution   We can substitute -1 for x_1, -2 for y_1, 7 for x_2, and 8 for y_2 in the formula for slope.

$$m = \frac{y_2 - y_1}{x_2 - x_1}$$

$$= \frac{8 - (-2)}{7 - (-1)}$$

$$= \frac{10}{8}$$

$$= \frac{5}{4}$$

Thus, the slope of the line is $\frac{5}{4}$. (See Figure 3-11.)

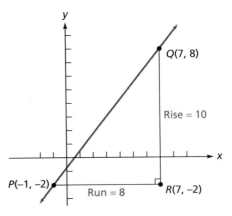

FIGURE 3-11

EXAMPLE 2 The graph of the equation $5x + 2y = 10$ is a line. Find its slope.

Solution We can find the coordinates of two points on the line by finding two ordered pairs (x, y) that satisfy the equation. Two convenient points are the x- and y-intercepts. To find the x-intercept, we let $y = 0$ and find that $x = 2$. Thus, the point $(x_1, y_1) = (2, 0)$ lies on the line. To find the y-intercept, we let $x = 0$ and find that $y = 5$. Thus, $(x_2, y_2) = (0, 5)$ is another point on the line. We substitute 2 for x_1, 0 for y_1, 0 for x_2, and 5 for y_2 into the formula for slope, and simplify.

$$m = \frac{y_2 - y_1}{x_2 - x_1} = \frac{5 - 0}{0 - 2} = -\frac{5}{2}$$

The slope of the line is $-\frac{5}{2}$. ■

As a point moves along a nonvertical line l from $P(x_1, y_1)$ to $Q(x_2, y_2)$, its y-coordinate changes by an amount $y_2 - y_1$. This change in the value of y is often denoted by Δy (read as "the change in y" or "delta y"). The difference in the x-coordinates is $x_2 - x_1$, often denoted by Δx. (See Figure 3-12.) Thus,

$$m = \frac{y_2 - y_1}{x_2 - x_1} = \frac{\Delta y}{\Delta x}$$

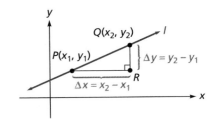

FIGURE 3-12

- If a line rises as it moves to the right—that is, if increasing x-values result in increasing y-values—then the slope of the line is positive. See Figure 3-13(a).
- If a line drops as it moves to the right—that is, if increasing x-values result in decreasing y-values—then the slope of the line is negative. See Figure 3-13(b).

- If a line is horizontal, then the slope of the line is 0, because the difference between the *y*-coordinates of any two points on the line is 0. See Figure 3-13(c).
- If a line is vertical, the slope is undefined. See Figure 3-13(d).

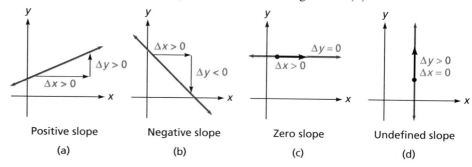

| Positive slope | Negative slope | Zero slope | Undefined slope |
| (a) | (b) | (c) | (d) |

FIGURE 3-13

■ Interpretation of Slope

Several applications of mathematics involve equations of lines and their slopes.

EXAMPLE 3 A trucker maintains a constant speed of 55 miles per hour during a trip from Chicago to Milwaukee. **a.** Write an equation that expresses the relationship between *d*, the distance traveled, and *t*, the number of hours of driving time, **b.** graph the equation, and **c.** interpret the slope of the line.

Solution **a.** The distance, rate of travel, and time of travel are related by the equation

$$\text{Distance} = \text{rate} \cdot \text{time} \quad \text{or} \quad d = r \cdot t$$

Since the trucker drives 55 miles per hour, $r = 55$, and we can substitute 55 for r in the equation $d = r \cdot t$ to obtain

$$d = 55t$$

b. We will graph the equation on a rectangular coordinate system with a vertical *d*-axis and a horizontal *t*-axis. We begin by finding a few points: If $t = 0$, then $d = 0$, and the point $(t, d) = (0, 0)$ lies on the graph. If $t = 1$, then $d = 55$, and the point $(t, d) = (1, 55)$ also lies on the graph. Because the *d*-coordinate of this last point is much greater than the *t*-coordinate, we will not use identical scales on the two axes. A table of values and the straight-line graph are shown in Figure 3-14.

$$d = 55t$$

t	d
0	0
1	55
2	110

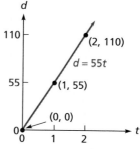

FIGURE 3-14

c. To find the slope of the line, we let $(t_1, d_1) = (0, 0)$, let $(t_2, d_2) = (1, 55)$, and use the formula for slope. In this example, the y-axis is now called the d-axis, and the x-axis is now the t-axis.

$$m = \frac{\Delta d}{\Delta t}$$

$$= \frac{d_2 - d_1}{t_2 - t_1}$$

$$= \frac{55 - 0}{1 - 0} \qquad \text{Substitute 0 for } t_1, \text{ 0 for } d_1, \text{ 1 for } t_2, \text{ and 55 for } d_2.$$

$$= 55$$

The slope of the line is 55, which is also the speed of the truck. The symbol Δd represents the number of miles driven in a time of Δt hours. The fraction $\frac{\Delta d \text{ miles}}{\Delta t \text{ hours}}$ represents the speed of the truck in miles per hour. ∎

Slope often represents a **rate of change**. In Example 3, the slope gives the rate of change of distance per unit of time. As another example, if temperature changes with time (measured in hours), the slope gives the rate of change of temperature in degrees per hour. If the amount of money in an account is increasing over the years, the slope gives the rate of growth, in dollars per year.

■ Slopes of Parallel Lines

A theorem relates parallel lines to their slopes.

Theorem	
	1. Nonvertical parallel lines have the same slope.
	2. Different lines with the same slope are parallel.

Proof Suppose that the nonvertical lines l_1 and l_2 in Figure 3-15 are parallel and have slopes of m_1 and m_2, respectively. Then right triangles ABC and DFE are similar, and it follows that

$$m_1 = \frac{\text{rise of } l_1}{\text{run of } l_1}$$

$$= \frac{\text{rise of } l_2}{\text{run of } l_2}$$

$$= m_2$$

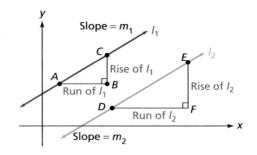

FIGURE 3-15

Thus, two nonvertical parallel lines have the same slope. □

In the exercises, you will be asked to prove that different lines with the same slope are parallel.

EXAMPLE 4 The line passing through $P(-2, 3)$ and $Q(3, -1)$ is parallel to the line passing through $R(4, -5)$ and $S(x, 7)$. Find x.

Solution Since the lines PQ and RS are parallel, their slopes are equal. We can find the slope of each line, set them equal, and solve the resulting equation for x:

Slope of PQ = slope of RS

$$\frac{-1 - 3}{3 - (-2)} = \frac{7 - (-5)}{x - 4}$$

$$\frac{-4}{5} = \frac{12}{x - 4} \qquad \text{Simplify.}$$

$$-4(x - 4) = 5 \cdot 12 \qquad \text{Multiply both sides by } 5(x - 4).$$

$$-4x + 16 = 60 \qquad \text{Remove parentheses and simplify.}$$

$$-4x = 44 \qquad \text{Subtract 16 from both sides.}$$

$$x = -11 \qquad \text{Divide both sides by } -4.$$

Thus, x is -11. The line passing through $P(-2, 3)$ and $Q(3, -1)$ is parallel to the line passing through $R(4, -5)$ and $S(-11, 7)$. ∎

Slopes of Perpendicular Lines

If the product of two numbers is -1, they are called **negative reciprocals** of each other. The following theorem describes the relationship between the slopes of perpendicular lines.

Theorem

1. If two nonvertical lines are perpendicular, their slopes are negative reciprocals of each other.

2. If the slopes of two lines are negative reciprocals, the lines are perpendicular.

Proof Suppose that l_1 and l_2 are lines with slopes of m_1 and m_2 that intersect at the origin. Let $P(a, b)$ be a point on l_1, and let $Q(c, d)$ be a point on l_2. Neither point P nor point Q can be the origin. (See Figure 3-16.)

First, we suppose that l_1 and l_2 are perpendicular. Then triangle POQ is a right triangle with its right angle at O, and by the Pythagorean theorem,

$$d(OP)^2 + d(OQ)^2 = d(PQ)^2$$

$$(a - 0)^2 + (b - 0)^2 + (c - 0)^2 + (d - 0)^2 = (a - c)^2 + (b - d)^2$$

$$a^2 + b^2 + c^2 + d^2 = a^2 - 2ac + c^2 + b^2 - 2bd + d^2$$

$$0 = -2ac - 2bd$$

$$bd = -ac$$

1.

$$\frac{b}{a} \cdot \frac{d}{c} = -1 \qquad \text{Divide both sides by } ac.$$

The coordinates of P are (a, b), and the coordinates of O are $(0, 0)$. Using the definition of slope, we have

$$m_1 = \frac{b - 0}{a - 0} = \frac{b}{a}$$

Similarly, we have

$$m_2 = \frac{d}{c}$$

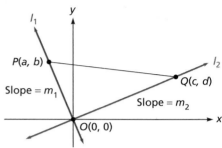

FIGURE 3-16

We substitute m_1 for $\frac{b}{a}$ and m_2 for $\frac{d}{c}$ in Equation 1 to obtain

$$m_1 m_2 = -1$$

Hence, if lines l_1 and l_2 are perpendicular, they have slopes that are negative reciprocals of each other.

Conversely, we suppose that the slopes of lines l_1 and l_2 are negative reciprocals of each other. Because the steps of the previous proof are reversible, we have that $d(OP)^2 + d(OQ)^2 = d(PQ)^2$. By the Pythagorean theorem, triangle POQ is a right triangle. Thus, l_1 and l_2 are perpendicular. ☐

It is also true that a line with a slope of 0 is horizontal and thus is perpendicular to a vertical line that has no defined slope.

EXAMPLE 5   Two lines intersect at the point $P(-5, 3)$. One passes through the point $Q(-1, -3)$, and the other passes through $R(1, 7)$. Are the lines perpendicular?

Solution   We first find the slopes of lines PQ and QR:

$$\text{Slope of } PQ = \frac{\Delta y}{\Delta x} = \frac{-3 - 3}{-1 - (-5)} = -\frac{6}{4} = -\frac{3}{2}$$

$$\text{Slope of } PR = \frac{\Delta y}{\Delta x} = \frac{7 - 3}{1 - (-5)} = \frac{4}{6} = \frac{2}{3}$$

Since the slopes of these lines are negative reiprocals of each other, the lines are perpendicular. ■

3.2   EXERCISES

In Exercises 1–16, find, if possible, the slope of the line passing through each pair of points.

1. $P(2, 5)$; $Q(3, 10)$

2. $P(3, -1)$; $Q(5, 3)$

3. $P(3, -2)$; $Q(-1, 5)$

4. $P(3, 7)$; $Q(6, 16)$

5. $P(8, -7)$; $Q(4, 1)$

6. $P(5, 17)$; $Q(17, 17)$

7. $P(-4, 3)$; $Q(-4, -3)$

8. $P(2, \sqrt{7})$; $Q(\sqrt{7}, 2)$

9. $P\left(\frac{3}{2}, \frac{2}{3}\right)$; $Q\left(\frac{5}{2}, \frac{7}{3}\right)$

10. $P\left(-\frac{2}{5}, \frac{1}{3}\right)$; $Q\left(\frac{3}{5}, -\frac{5}{3}\right)$

11. $P\left(\frac{3}{7}, \frac{3}{2}\right)$; $Q\left(\frac{5}{7}, -\frac{1}{2}\right)$

12. $P\left(\frac{1}{2}, -\frac{1}{2}\right)$; $Q\left(\frac{1}{2}, -\frac{7}{2}\right)$

13. $P(a, a - b)$; $Q(b, 0)$
   (Assume $a \neq b$)

14. $P(a, b - a)$; $Q(b, 0)$
   (Assume $a \neq b$)

15. $P(a + b, c)$; $Q(b + c, a)$
   (Assume $a \neq c$)

16. $P(b, 0)$; $Q(a + b, a)$
   (Assume $a \neq 0$)

In Exercises 17–24, find two points on the line and find the slope of the line.

17. $y = 3x + 2$ **18.** $y = 5x - 8$ **19.** $5x - 10y = 3$ **20.** $8y + 2x = 5$

21. $3(y + 2) = 2x - 3$ **22.** $4(x - 2) = 3y + 2$ **23.** $3(y + x) = 3(x - 2)$ **24.** $2x + 5 = 2(y + x)$

In Exercises 25–34, determine whether the lines with the given slope are parallel, perpendicular, or neither.

25. $m_1 = 3; m_2 = -\dfrac{1}{3}$ **26.** $m_1 = \dfrac{2}{3}; m_2 = \dfrac{3}{2}$

27. $m_1 = \sqrt{8}; m_2 = 2\sqrt{2}$ **28.** $m_1 = 1; m_2 = -1$

29. $m_1 = -\sqrt{2}; m_2 = \dfrac{\sqrt{2}}{2}$ **30.** $m_1 = 2\sqrt{7}; m_2 = \sqrt{28}$

31. $m_1 = -0.125; m_2 = 8$ **32.** $m_1 = 0.125; m_2 = \dfrac{1}{8}$

33. $m_1 = ab^{-1}; m_2 = -a^{-1}b \quad a, b \neq 0$ **34.** $m_1 = \left(\dfrac{a}{b}\right)^{-1}; m_2 = -\dfrac{b}{a} \quad a, b \neq 0$

In Exercises 35–40, determine if the line through the given points and the line through $R(-3, 5)$ and $S(2, 7)$ are parallel, perpendicular, or neither.

35. $P(2, 4); Q(7, 6)$ **36.** $P(-3, 8); Q(-13, 4)$ **37.** $P(-4, 6); Q(-2, 1)$ **38.** $P(0, -9); Q(4, 1)$

39. $P(a, a); Q(3a, 6a)$ and $a \neq 0$ **40.** $P(b, b); Q(-b, 6b)$ and $b \neq 0$

In Exercises 41–44, find the slopes of lines PQ and PR, and determine whether points P, Q, and R lie on the same line.

41. $P(-2, 8); Q(-6, 9); R(2, 5)$ **42.** $P(1, -1); Q(3, -2); R(-3, 0)$

43. $P(-a, a); Q(0, 0); R(a, -a)$ **44.** $P(a, a + b); Q(a + b, b); R(a - b, a)$

In Exercises 45–50, determine which, if any, of the three lines, PQ, PR, and QR are perpendicular.

45. $P(5, 4); Q(2, -5); R(8, -3)$ **46.** $P(8, -2); Q(4, 6); R(6, 7)$ **47.** $P(1, 3); Q(1, 9); R(7, 3)$

48. $P(2, -3); Q(-3, 2); R(3, 8)$ **49.** $P(0, 0); Q(a, b); R(-b, a)$ **50.** $P(a, b); Q(-b, a); R(a - b, a + b)$

51. *Charting temperature changes*   The following temperature readings were recorded over a four-hour period.

Time	12:00	1:00	2:00	3:00	4:00
Temperature	47°	53°	59°	65°	71°

Let t represent the time in hours, with 12:00 corresponding to $t = 0$. Let T represent the temperature. Plot the points (t, T) and draw the line through those points. Explain the meaning of $\frac{\Delta T}{\Delta t}$.

52. *Tracking the Dow*   The Dow Jones averages at the close of trade on three consecutive days were as follows:

Day	Monday	Tuesday	Wednesday
Closing Dow	2981	2964	2947

Let d represent the day, with $d = 0$ corresponding to Monday, and let D represent the Dow Jones average. Plot the points (d, D) and draw the graph. Explain the meaning of $\frac{\Delta D}{\Delta d}$.

53. *The speed of an airplane*   An airline pilot files a flight plan indicating her intention to fly at a constant speed of 590 miles per hour. Write an equation that expresses the distance traveled in terms of the flying time. Then graph the equation and interpret the slope of the line.

54. *Growth of savings*   A college student deposits $25 each month into a Holiday Club account at her bank. The account pays no interest. Write an equation that expresses the amount in her account in terms of the number of deposits. Then graph the line and interpret its slope.

55. *Geometry: right triangles* Show that the points $A(-1, -1)$, $B(-3, 4)$, and $C(4, 1)$ are the vertices of a right triangle.

56. *Geometry: right triangles* Show that the points $D(0, 1)$, $E(-1, 3)$, and $F(3, 5)$ are the vertices of a right triangle.

57. *Geometry: squares* Show that the points $A(1, -1)$, $B(3, 0)$, $C(2, 2)$, and $D(0, 1)$ are the vertices of a square.

58. *Geometry: squares* Show that the points $E(-1, -1)$, $F(3, 0)$, $G(2, 4)$, and $H(-2, 3)$ are the vertices of a square.

59. *Geometry: parallelograms* Show that the points $A(-2, -2)$, $B(3, 3)$, $C(2, 6)$, and $D(-3, 1)$ are the vertices of a parallelogram. (Show that both pairs of opposite sides are parallel.)

60. *Geometry: trapezoids* Show that points $E(1, -2)$, $F(5, 1)$, $G(3, 4)$, and $H(-3, 4)$ are the vertices of a trapezoid. (Show that only one pair of opposite sides are parallel.)

61. In Illustration 1, points M and N are midpoints of CB and BA, respectively. Show that MN is parallel to AC.

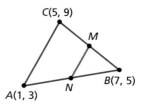

ILLUSTRATION 1

62. In Illustration 2, $d(AB) = d(AC)$. Show that AD is perpendicular to BC.

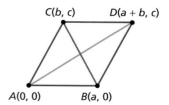

ILLUSTRATION 2

63. Prove that if two different lines have the same slope, they are parallel.

64. Prove: If the lines joining points $P(a, b)$ and $Q(c, d)$ to the origin are perpendicular, then $ac + bd = 0$.

3.3

EQUATIONS OF LINES

■ Point–Slope Form of the Equation of a Line ■ Slope–Intercept Form of the Equation of a Line ■ Using Mixed Forms ■ General Form of the Equation of a Line ■ Straight-Line Depreciation

We have seen that if we have an equation in two variables, we can draw its graph. In this section, we will begin with the graph and find the equation. More specifically, if we are given sufficient information about the equation's straight-line graph, we will be able to write the equation itself.

■ Point–Slope Form of the Equation of a Line

If we know a point through which a line passes and also know its slope, we can determine the equation of the line. Suppose that the nonvertical line *l* shown in Figure 3-17 has a slope of *m* and passes through the point $P(x_1, y_1)$. If $Q(x, y)$ is any other point on that line, we have

$$m = \frac{y - y_1}{x - x_1}$$ Use the definition of slope.

or, by clearing the equation of fractions,

$$y - y_1 = m(x - x_1)$$

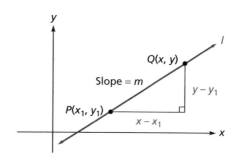

FIGURE **3-17**

This is the equation of a line that passes through the point (x_1, y_1) and has a slope of m. Because the equation displays the coordinates of a fixed point on the line and the slope of the line, it is called the **point–slope form of the equation of a line**.

Point–Slope Form of the Equation of a Line	The equation of the line passing through $P(x_1, y_1)$ with a slope m is $$y - y_1 = m(x - x_1)$$

EXAMPLE 1   Find the equation of the line passing through $P(3, -1)$ with a slope of $-\frac{5}{3}$. Then solve the equation for y.

Solution   We can substitute 3 for x_1, -1 for y_1, and $-\frac{5}{3}$ for m in the point–slope form of the equation of a line:

$$y - y_1 = m(x - x_1)$$

$$y - (-1) = -\frac{5}{3}(x - 3)$$

$$y + 1 = -\frac{5}{3}x + 5 \qquad \text{Remove parentheses.}$$

$$y = -\frac{5}{3}x + 4 \qquad \text{Subtract 1 from both sides.} \qquad \blacksquare$$

We can use the concept of slope to help graph the equation obtained in Example 1. We first plot the point $P(3, -1)$, as in Figure 3-18. Because the slope is $-\frac{5}{3}$, Δy is -5 provided that Δx is 3. Thus, we can locate another point Q on the line by starting at P, moving 3 units to the right, and then moving 5 units down. The graph of the equation is the line PQ.

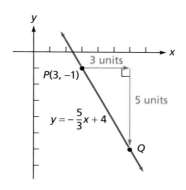

FIGURE 3-18

EXAMPLE 2   Find the equation of the line passing through the points $P(3, 7)$ and $Q(-5, 3)$. Solve the equation for y.

Solution   We begin by using points P and Q to find the slope of the line. We then use that slope and either of the points to find the equation of the line.

We let $P(x_1, y_1)$ be the point $P(3, 7)$ and $Q(x_2, y_2)$ be the point $Q(-5, 3)$.

$$m = \frac{y_2 - y_1}{x_2 - x_1} = \frac{3 - 7}{-5 - 3} = \frac{-4}{-8} = \frac{1}{2}$$

Finally, we substitute the coordinates of $P(3, 7)$ and $\frac{1}{2}$ for m in the point–slope form of the equation of a line and solve for y:

$$y - y_1 = m(x - x_1)$$

$$y - 7 = \frac{1}{2}(x - 3)$$

$$y = \frac{1}{2}x - \frac{3}{2} + 7 \qquad \text{Remove parentheses and add 7 to both sides.}$$

$$y = \frac{1}{2}x + \frac{11}{2}$$

We would have obtained the same result if we had used the coordinates of point Q instead of those of point P.   ∎

EXAMPLE 3   Find the equation of the line passing through the points $P(-3, 4)$ and $Q(-3, -2)$.

Solution   Because the x-coordinates of points P and Q are equal, the line has no defined slope. It is a vertical line, with equation $x = -3$. (See Figure 3-19.)

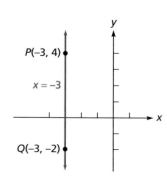

FIGURE 3-19   ∎

■ Slope–Intercept Form of the Equation of a Line

When we say that the *y*-intercept of a line is *b*, we mean that the line passes through the point $P(0, b)$. (See Figure 3-20.) Thus, if we know the slope and the *y*-intercept of a line, we can use the point–slope form to write the equation of the line.

We can write the equation of a line with slope *m* and *y*-intercept *b* by substituting 0 for x_1 and *b* for y_1 into the point–slope form of the equation of a line and simplifying:

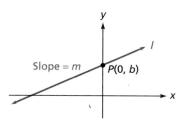

FIGURE 3-20

$$y - y_1 = m(x - x_1)$$
$$y - b = m(x - 0)$$
$$y = mx + b \qquad \text{Simplify and add } b \text{ to both sides.}$$

Because the equation $y = mx + b$ displays both the slope and the *y*-intercept of a line, it is called the **slope–intercept form of the equation of a line**.

Slope–Intercept Form of the Equation of a Line	The equation of the line with slope *m* and *y*-intercept *b* is $$y = mx + b$$

EXAMPLE 4 Use the slope–intercept form to find the equation of the line with slope of $\frac{7}{3}$ and *y*-intercept of -6.

Solution We substitute $\frac{7}{3}$ for *m* and -6 for *b* into the slope–intercept form of the equation of a line and simplify:

$$y = mx + b$$
$$y = \frac{7}{3}x + (-6)$$
$$y = \frac{7}{3}x - 6$$

■

EXAMPLE 5 Find the slope and the *y*-intercept of the line $3(y + 2) = 6x - 1$.

Solution We write the equation in the form $y = mx + b$ to determine the slope *m* and the *y*-intercept *b*.

$$3(y + 2) = 6x - 1$$
$$3y + 6 = 6x - 1 \qquad \text{Remove parentheses.}$$
$$3y = 6x - 7 \qquad \text{Subtract 6 from both sides.}$$
$$y = 2x - \frac{7}{3} \qquad \text{Divide both sides by 3.}$$

The slope of the line is 2, and the *y*-intercept is $-\frac{7}{3}$.

■

EXAMPLE 6 Find the *y*-intercept of the line having a slope of 2 and passing through the point $P(3, -5)$.

Solution   Because the line passes through the point $P(3, -5)$, the coordinates $x = 3$ and $y = -5$ must satisfy the equation. To find the y-intercept, we substitute 3 for x, -5 for y, and 2 for m into the slope–intercept form and solve for b.

$$y = mx + b$$
$$-5 = 2(3) + b$$
$$-5 = 6 + b$$
$$-11 = b$$

The y-intercept is -11.   ■

■ Using Mixed Forms

Depending on the information available, some forms of the equation of the line are more convenient to use than others. In the next examples, we use more than one form to determine the final results.

EXAMPLE 7   Find the equation of a line l that passes through the point $P(-2, 9)$ and is parallel to the line with equation $y = 4x - 73$.

Solution   The equation $y = 4x - 73$ is in slope–intercept form. From the equation, we read that the slope of its graph is 4. Because line l is parallel to the given line, it also has a slope of 4. Thus, we need the equation of a line passing through the point $(-2, 9)$ and having a slope of 4. We use the point–slope form:

$$y - y_1 = m(x - x_1)$$
$$y - 9 = 4[x - (-2)] \qquad \text{Let } x_1 = -2, y_1 = 9, \text{ and } m = 4.$$
$$y - 9 = 4x + 8 \qquad \text{Remove parentheses.}$$
$$y = 4x + 17$$

In slope–intercept form, the equation of line l is $y = 4x + 17$.   ■

EXAMPLE 8   Find the equation of the line l passing through the point $(2, 3)$ and perpendicular to the line $y = \frac{1}{3}x + 7$.

Solution   The slope of the given line $y = \frac{1}{3}x + 7$ is $\frac{1}{3}$. The slope of the required line must be -3, the negative reciprocal of $\frac{1}{3}$. We use the point–slope form to write the equation of a line with slope of -3 and passing through the point $(2, 3)$:

$$y - y_1 = m(x - x_1)$$
$$y - 3 = -3(x - 2)$$
$$y - 3 = -3x + 6$$
$$y = -3x + 9$$

In slope–intercept form the equation of the line is $y = -3x + 9$.   ■

■ General Form of the Equation of a Line

We have shown that the graph of any equation of the form

$$y = mx + b$$

is a line with slope m and y-intercept b. The graph of any equation of the form $Ax + By = C$ (where A and B are not *both* zero) is also a line. To see why, we look at three possibilities.

- If $B \neq 0$, the equation $Ax + By = 0$ can be rewritten in slope–intercept form as follows.

$$Ax + By = C$$
$$By = -Ax + C \qquad \text{Subtract } Ax \text{ from both sides.}$$
$$y = -\frac{A}{B}x + \frac{C}{B} \qquad \text{Divide both sides by } B.$$

This is the equation of a line with slope $-\frac{A}{B}$ and y-intercept $\frac{C}{B}$.

- If $A = 0$ and $B \neq 0$, then the equation can be written in the form $y = \frac{C}{B}$. This is the equation of a horizontal line with y-intercept $\frac{C}{B}$.

- If $A \neq 0$ and $B = 0$, then the equation can be written in the form $x = \frac{C}{A}$. This is the equation of a vertical line.

In general, for any real numbers A, B, and C (where A and B are not both 0), the equation $Ax + By = C$ represents a line. The equation is called the **general form of the equation of a line**. Whenever possible, we will write the general form $Ax + By = C$ so that A, B, and C are integers and $A \geq 0$.

Any equation that can be written in general form is called a **linear equation in x and y**.

General Form of the Equation of a Line	If A, B, and C are real numbers and A and B are not both 0, the graph of the equation $Ax + By = C$ is a line. The equation is called the **general form of the equation of a line**. If $B \neq 0$, the graph is a nonvertical line with slope of $-\dfrac{A}{B}$ and y-intercept of $\dfrac{C}{B}$. If $B = 0$, the graph is a vertical line with x-intercept of $\dfrac{C}{A}$.

EXAMPLE 9   Show that the two lines $3x - 2y = 5$ and $-6x + 4y = 7$ are parallel.

Solution   To show that two lines are parallel, we show that their slopes are equal. The first equation, $3x - 2y = 5$, is in general form, with $A = 3$, $B = -2$, and $C = 5$. By the previous theorem, the slope of the line is

$$m_1 = -\frac{A}{B} = -\frac{3}{-2} = \frac{3}{2}$$

The second equation, $-6x + 4y = 7$, is also in general form, with $A = -6$, $B = 4$, and $C = 7$. The slope of this line is

$$m_2 = -\frac{A}{B} = -\frac{-6}{4} = \frac{3}{2}$$

Since the slopes of the two lines are equal, the lines are parallel. ∎

Straight-Line Depreciation

Business and industry recognize that equipment loses value with age. For tax purposes, many businesses use **straight-line depreciation** to determine the current value of aging equipment.

To find the equation for depreciation, we let y represent the value of the equipment after x years of use, and we also assume that y is related to x by the equation of a line. We also need to know the equipment's purchase price, its expected useful life, and the value at the end of its life (called the **salvage value**).

EXAMPLE 10   A mail order distributor buys a copy machine for $3450 and expects it to last for 6 years. Then the copier can be sold for an estimated salvage value of $150. Let y represent the value of the copier after x years of service.

a. Find the equation of the line.

b. Determine the value of the copier after $3\frac{1}{2}$ years.

c. Explain the economic meaning of the y-intercept of the line.

d. Explain the economic meaning of the slope of the line.

Solution   **a.** To find the equation of the line, we first find two points on the line and find the line's slope. Then we use the point–slope form to find the equation of the line.

When the copier is new, its age, x, is 0, and its value, y, is 3450 dollars. Thus, the line passes through the point (0, 3450), as shown in Figure 3-21. When the copier is $x = 6$ years old, its value is $y = 150$ dollars. Thus, the line also passes through the point (6, 150). We use the points $(x_1, y_1) = (0, 3450)$ and $(x_2, y_2) = (6, 150)$ to find the slope of the line.

$$m = \frac{y_2 - y_1}{x_2 - x_1}$$

$$= \frac{150 - 3450}{6 - 0}$$

$$= \frac{-3300}{6}$$

$$= -550$$

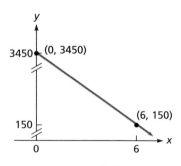

FIGURE **3-21**

To find the equation of the line, we substitute -550 for m, 0 for x_1, and 3450 for y_1 into the point–slope form of the equation of the line, and simplify.

$$y - y_1 = m(x - x_1)$$
$$y - 3450 = -550(x - 0)$$
$$y = -550x + 3450$$

Thus, the current value, y, of the copier is related to its age, x, by the equation $y = -550x + 3450$.

b. To determine the copier's value after $3\frac{1}{2}$ years, we substitute 3.5 for x in the equation and calculate y.

$$y = 3450 - 550x = 3450 - 550(3.5) = 3450 - 1925 = 1525$$

After $3\frac{1}{2}$ years, the copier is worth $\$1525$.

c. The y-intercept of the graph is the value of y found by letting $x = 0$. The corresponding value of y is the value of a zero-year-old copier. The y-intercept of 3450 is the copier's original cost.

d. Each year, the value of the copier decreases by $\$550$, because the slope of the line is -550. The slope of the depreciation line is called the **annual depreciation rate**. ■

3.3 EXERCISES

In Exercises 1–8, use the point–slope form to write the equation of the line passing through the given point and having the given slope. Leave the answer in point–slope form.

1. $P(2, 4)$; $m = 2$ **2.** $P(3, 5)$; $m = -3$ **3.** $P\left(-\frac{3}{2}, \frac{1}{2}\right)$; $m = 2$ **4.** $P\left(\frac{1}{4}, -2\right)$; $m = -6$

5. $P(5, -5)$; $m = -5$ **6.** $P(0, -8)$; $m = -3$ **7.** $P(\pi, 0)$; $m = \pi$ **8.** $P(0, \pi)$; $m = \pi$

In Exercises 9–16, use the point–slope form to write the equation of the line passing through the given point and having the given slope. Express the answer in slope–intercept form.

9. $P(1, 4)$; $m = 3$ **10.** $P(-4, 3)$; $m = -\frac{1}{2}$ **11.** $P\left(\frac{5}{2}, \frac{3}{2}\right)$; $m = -2$ **12.** $P(3, 8)$; $m = -3$

13. $P(0, 8)$; $m = -4$ **14.** $P(0, -8)$; $m = 4$ **15.** $P(0, 0)$; $m = \frac{1}{2}$ **16.** $P\left(\frac{1}{2}, -\frac{1}{2}\right)$; $m = \frac{1}{2}$

In Exercises 17–24, use the slope–intercept form to write the equation of the line with the given slope and y-intercept.

17. $m = 3$; $b = -2$ **18.** $m = -\frac{1}{3}$; $b = \frac{2}{3}$ **19.** $m = 5$; $b = -\frac{1}{5}$ **20.** $m = \sqrt{2}$; $b = \sqrt{2}$

21. $m = a$; $b = \frac{1}{a}$ **22.** $m = a$; $b = 2a$ **23.** $m = a$; $b = a$ **24.** $m = \frac{1}{a}$; $b = a$

In Exercises 25–32, use the slope–intercept form to write the equation of a line passing through the given point and having the given slope. Express the answer in general form.

25. $P(0, 0)$; $m = \frac{3}{2}$ **26.** $P(-3, -7)$; $m = -\frac{2}{3}$ **27.** $P(-3, 5)$; $m = -3$ **28.** $P(-5, 1)$; $m = 1$

29. $P(0, \sqrt{2})$; $m = \sqrt{2}$ **30.** $P(-\sqrt{3}, 0)$; $m = 2\sqrt{3}$ **31.** $P(r, s)$; $m = \frac{s}{r}$ **32.** $P(a, \frac{1}{a})$; $m = \frac{1}{a}$

In Exercises 33–40, write the equation of the line passing through the two given points. Express the answer in general form.

33. $P(3, 2)$; $Q(2, 3)$ **34.** $P(0, 5)$; $Q(-5, 0)$ **35.** $P(3, -2)$; $Q(4, -8)$ **36.** $P(-7, 5)$; $Q(-5, -7)$

37. $P(0, 0)$; $Q(-9, 13)$ **38.** $P(-9, -5)$; $Q(5, -9)$ **39.** $P(\frac{1}{2}, \frac{1}{3})$; $Q(\frac{3}{2}, -\frac{1}{3})$ **40.** $P(0, -\frac{5}{2})$; $Q(-1, \frac{1}{2})$

In Exercises 41–48, find, if possible, the slope and the y-intercept of each line.

41. $y = -7x + 12$ **42.** $y = \frac{7}{5}x - \frac{5}{2}$ **43.** $2y = 7x - 3$ **44.** $5y = -7 + x$

45. $3(x - 1) = y$ **46.** $3x + 6 = -2y + 15$

47. $2(y - 3) + x = 2y - 7$ **48.** $3(y + 2) + 2x = 2(y + x)$

In Exercises 49–58, determine the slope of the line with the given properties.

49. passes through $P(-3, 5)$ and $Q(7, -5)$ **50.** passes through $P(15, 3)$ and $Q(-3, -6)$

51. is parallel to the line $y = 3(x - 7)$ **52.** is parallel to the line $3y + 1 = 6(x - 2)$

53. is perpendicular to the line $y + 3x = 8$ **54.** is perpendicular to the line $x = 2y - 7$

55. is parallel to the x-axis   **56.** is perpendicular to the y-axis

57. is perpendicular to the line joining $P(3, -5)$ and $Q(-3, 5)$ **58.** is perpendicular to the line joining $P(-\frac{5}{2}, \frac{3}{2})$ and the origin.

In Exercises 59–76, write the equation of the line with the given properties. Write the answer in general form.

59. passes through $(3, 4)$ with slope of $\frac{1}{2}$ **60.** passes through $(-2, 6)$ with slope of $\frac{3}{5}$

61. passes through $(-2, 4)$ and $(5, -4)$ **62.** passes through $(3, 5)$ and $(-5, 3)$

63. passes through the origin and $(-2, 11)$ **64.** passes through the origin and $(5, -9)$

65. has y-intercept of 7 and passes through $(-3, 4)$ **66.** has y-intercept of -2 and passes through $(4, -3)$

67. has slope of $-\frac{2}{3}$ and y-intercept of 10   **68.** has slope of $-\frac{1}{2}$ and x-intercept of 10

69. passes through $(3, -5)$ and is parallel to the x-axis   **70.** has a y-intercept of 7 and is parallel to the x-axis

71. is parallel to the line with equation $y = 3x - 17$ and passes through $(0, -5)$ **72.** is perpendicular to the line with equation $y = 3x - 17$ and passes through $(0, -5)$

73. has an x-intercept of -4 and is parallel to the y-axis   **74.** is perpendicular to the line $y = -\frac{1}{3}x + 5$ and has the same y-intercept as the line $3x - 2y = 6$

75. is parallel to the line $3y - 5x = 0$ and has the same x-intercept as the line $5x - 2y = 15$ **76.** is parallel to the line $5x + 7y = 8$ and has the same y-intercept as the line $x = 7y - 5$

In Exercises 77–82, write the equation of the line with the given properties. Write the answer in slope–intercept form.

77. passes through the origin and is parallel to the line $3x + 2y = 6$ **78.** passes through the point $(2, 8)$ and is parallel to the line $2x - 3y = 12$

79. passes through the point $(-2, 3)$ and is perpendicular to the line $3x + 5y = 25$ **80.** passes through the origin and is perpendicular to the line $4x - y = 12$

81. passes through the origin and the point $P(r, s)$ **82.** passes through the points $P(r, s)$ and $Q(s, r)$

83. *Converting temperatures*   Water freezes at 32° Fahrenheit, or 0° Celsius. Water boils at 212°F, or 100°C. Find a formula for converting a temperature from Fahrenheit to Celsius.

84. *Converting temperatures*   A temperature of 0° Celsius is equal to a temperature of 273° Kelvin, and 100°C is 373°K. Find a formula for converting from Celsius to Kelvin.

85. *Atmospheric temperature*   The temperature T is related to the altitude x by an equation of the form $T = mx + b$. If $T = 30°$ at an altitude of 12,000 feet, and $T = -30°$ at 30,000 feet, find the equation.

86. *Converting units of speed*   A speed of 1 mile per hour is equal to 88 feet per minute, and (of course) 0 miles per hour is 0 feet per minute. Find an equation for converting a speed x, in miles per hour, to the corresponding speed y, in feet per minute.

87. *Liquid measure*   One U.S. gallon is equal to 3.785 liters. Find an equation for converting gallons to liters.

88. *Predicting stock prices*   The value of the stock of ABC Corporation has been increasing by the same fixed dollar amount each year. The pattern is expected to continue. Let 1980 be the base year corresponding to $x = 0$, with $x = 1, 2, 3, \ldots$ corresponding to later years. ABC stock was selling at $\$37\frac{1}{2}$ in 1985 and at $\$45$ in 1988. If y represents the price of ABC stock, find the equation $y = mx + b$ that relates x and y, and predict the price in the year 2000.

89. *Purchasing school supplies*   Spring Creek Elementary School began the school year with 300 reams of paper. The faculty uses the paper at the rate of 9 reams per week. Determine an equation of the form $y = mx + b$ that relates x, the number of weeks since school began, to y, the number of unused reams of paper.

90. *Purchasing school supplies*   How many reams of paper remain after 23 weeks? See Exercise 89.

91. *Investing in antiques*   An early American four-poster bed is expected to appreciate a constant amount each year. If the bed will be worth $\$2750$ three years from now, and $\$3298$ seven years from now, what is its present value?

92. *Library fines*   After 7 days, the library fine on several overdue books is $\$7.35$. After 9 days, the fine is $\$9.45$. What would be the fine after 11 days?

93. *Estimating inventory*   Inventory of unsold goods showed a surplus of 375 units in January and 264 in April. Assume that the relationship between inventory and time is given by the equation of a line. Estimate the expected inventory in March. Because March lies between January and April, this estimation is called **interpolation**.

94. *Estimating inventory*   Refer to Exercise 93 and estimate the inventory in July. Because July lies outside of the January–April interval, this estimation is called **extrapolation**.

95. *Oil depletion*   When a Petroland oil well was first brought on line, it produced 1900 barrels of crude oil per day. In each later year, owners expect its daily production to drop by 70 barrels. Find the daily production after $3\frac{1}{2}$ years.

96. *Oil depletion*   Refer to Exercise 95 and determine when the oil well can be expected to run dry.

97. *Waste management*   The corrosive waste in industrial sewage limits the useful life of the piping in a waste-processing plant to 12 years. The piping system was originally worth $\$137,000$, and it will cost the company $\$33,000$ to remove it at the end of its 12-year useful life. Find the depreciation equation.

98. *Real estate investment*   A two-family apartment complex was purchased for $\$585,000$, including the cost of the land (appraised at $\$175,000$). Tax law does not allow the depreciation of land. The owners expect the value of the land to triple during the 35-year useful life of the property, and they expect then to sell the property for $\$550,000$. Find the depreciation equation.

99. Prove that the equation of a line with x-intercept of a and y-intercept of b can be written in the form

$$\frac{x}{a} + \frac{y}{b} = 1$$

100. Prove that if $B \neq 0$, the graph of $Ax + By = C$ has a slope of $-\frac{A}{B}$ and a y-intercept of $\frac{C}{B}$.

101. Prove that if $B = 0$, the graph of $Ax + By = C$ is a vertical line with x-intercept of $\frac{C}{A}$.

102. Show that the graphs of $Ax + By = C$ and $kAx + kBy = D$, where $k \neq 0$, are parallel.

103. Show that the lines $Ax + By = C$ and $Bx - Ay = C$ are perpendicular.

104. Use the distance formula to find the equation of the perpendicular bisector of the line segment joining $A(3, 5)$ and $B(5, -3)$. (*Hint:* First find the midpoint of *AB*.)

105. Find the *x*-intercept of the line $y = mx + b$.

106. Find the equation of the *y*-axis.

107. Find the equation of the *x*-axis.

108. Find the *x*- and *y*-intercepts of the line $bx + ay = ab$.

3.4 GRAPHS OF OTHER EQUATIONS

■ Intercepts of Graphs ■ Symmetries of Graphs ■ Miscellaneous
Graphs ■ Circles ■ Graphing Equations of Circles

The graphs of many equations are curves. If we plot several points (x, y) that satisfy such an equation, the shape of the graph will usually become evident. We can sketch the graph by joining these points with a smooth curve.

■ Intercepts of Graphs

The **x-intercepts** of the graph of an equation in the variables *x* and *y* are the *x*-coordinates of the points where the graph intersects the *x*-axis. The **y-intercepts** of a graph are the *y*-coordinates of the points where the graph intersects the *y*-axis.

Intercepts of a Graph

The *x*-intercepts are $x = a$ and $x = b$.

To find the *x*-intercepts, let $y = 0$ and solve for *x*.

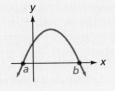

The *y*-intercept is $y = c$.

To find the *y*-intercepts, let $x = 0$ and solve for *y*.

To graph an equation such as $y = x^2 - 4$, we first find the x- and y-intercepts. To find the x-intercepts, we let $y = 0$ and solve for x.

$$y = x^2 - 4$$
$$0 = x^2 - 4 \qquad \text{Substitute 0 for } y.$$
$$0 = (x + 2)(x - 2) \qquad \text{Factor } x^2 - 4.$$

$$x + 2 = 0 \qquad \text{or} \qquad x - 2 = 0 \qquad \text{Set each factor equal to 0.}$$
$$x = -2 \qquad \qquad x = 2$$

Since the x-intercepts are -2 and 2, the graph intersects the x-axis at $(-2, 0)$ and $(2, 0)$.

To find the y-intercepts, we let $x = 0$ and solve for y.

$$y = x^2 - 4$$
$$y = 0^2 - 4 \qquad \text{Substitute 0 for } x.$$
$$y = -4$$

Since the only y-intercept is -4, the graph intersects the y-axis only at $(0, -4)$.

We can find several other pairs (x, y) that satisfy the equation by substituting numbers for x and finding the corresponding values of y. For example, if $x = -3$, then $y = (-3)^2 - 4$, or 5, and the point $(-3, 5)$ lies on the graph. If $x = 1$, then $y = 1^2 - 4$, or -3, and the point $(1, -3)$ lies on the graph.

The coordinates of these points and others appear in the table in Figure 3-22. We plot the points and draw a smooth curve through them, as in the figure.

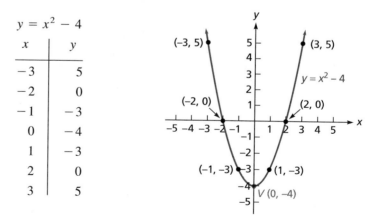

$y = x^2 - 4$

x	y
-3	5
-2	0
-1	-3
0	-4
1	-3
2	0
3	5

FIGURE 3-22

The graph in Figure 3-22 is called a **parabola**. The lowest point on the parabola, $V(0, -4)$ is called the **vertex**. Because the y-axis divides the parabola into two congruent halves, it is called an **axis of symmetry**, and we say that the parabola is **symmetric about the y-axis**.

■ Symmetries of Graphs

There are several ways a graph can exhibit symmetry. If the point $(-x, y)$ lies on a graph whenever the point (x, y) does, as in Figure 3-23(a), the graph is **symmetric about the y-axis**. If the point $(-x, -y)$ lies on the graph whenever the point (x, y) does, as in Figure 3-23(b), the graph is **symmetric about the origin**. If the point $(x, -y)$ lies on the graph whenever the point (x, y) does, as in Figure 3-23(c), the graph is **symmetric about the x-axis**.

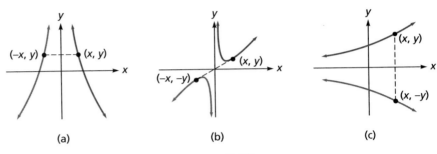

(a)   (b)   (c)

FIGURE 3-23

Tests for Symmetry for Equations in the Variables x and y	• To test the graph of an equation for y-axis symmetry, replace x with $-x$. If the new equation is equivalent to the original equation, the graph is symmetric about the y-axis. • To test the graph of an equation for symmetry about the origin, replace x with $-x$ and y with $-y$. If the resulting equation is equivalent to the original equation, the graph is symmetric about the origin. • To test the graph of an equation for x-axis symmetry, replace y with $-y$. If the resulting equation is equivalent to the original equation, the graph is symmetric about the x-axis.

EXAMPLE 1   Find the intercepts and the symmetries of the graph of $y = x^3 - x$. Then graph the equation.

Solution   *x-intercepts*   To find the x-intercepts, we let $y = 0$ and solve for x.

$$y = x^3 - x$$
$$0 = x^3 - x \qquad \text{Substitute 0 for } y.$$
$$0 = x(x^2 - 1) \qquad \text{Factor out } x.$$
$$0 = x(x + 1)(x - 1) \qquad \text{Factor } x^2 - 1.$$

$x = 0$ or $x + 1 = 0$ or $x - 1 = 0$ Set each factor equal to 0.
$\qquad\qquad\qquad x = -1 \qquad\qquad x = 1$

Because the x-intercepts are 0, -1, and 1, the graph intersects the x-axis at $(0, 0)$, $(-1, 0)$, and $(1, 0)$.

y-intercepts To find the *y*-intercepts, we let $x = 0$ and solve for *y*.

$$y = x^3 - x$$
$$y = 0^3 - 0 \qquad \text{Substitute 0 for } x.$$
$$y = 0$$

Because the *y*-intercept is 0, the graph intersects the *y*-axis at (0, 0).

Symmetry We test for symmetry about the *y*-axis by replacing *x* with $-x$, simplifying, and comparing the result to the original equation.

1. $y = x^3 - x$ The original equation.

      $y = (-x)^3 - (-x)$ Replace *x* with $-x$.

2. $y = -x^3 + x$ Simplify.

Because Equation 2 is not equivalent to Equation 1, the graph is not symmetric about the *y*-axis.

We test for symmetry about the origin by replacing *x* and *y* with $-x$ and $-y$, respectively, and comparing the result to the original equation.

1. $y = x^3 - x$ The original equation.

      $-y = (-x)^3 - (-x)$ Replace *x* with $-x$, and *y* with $-y$.

      $-y = -x^3 + x$ Simplify.

3. $y = x^3 - x$ Multiply both sides by -1.

Because Equation 3 is equivalent to Equation 1, the graph is symmetric about the origin.

We test for symmetry about the *x*-axis by replacing *y* with $-y$ and comparing the result to the original equation.

1. $y = x^3 - x$ The original equation.

      $-y = x^3 - x$ Replace *y* with $-y$.

4. $y = -x^3 + x$ Multiply both sides by -1.

Because Equation 4 is not equivalent to Equation 1, the graph is not symmetric about the *x*-axis.

To graph the equation, we plot the *x*-intercepts of $x = -1$, 0, and 1 and the *y*-intercept of $y = 0$. We also plot other points for positive values of *x* and use the fact that the curve is symmetric about the origin to draw the rest of the graph, as in Figure 3-24. ■

$y = x^3 - x$

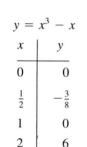

x	*y*
0	0
$\frac{1}{2}$	$-\frac{3}{8}$
1	0
2	6

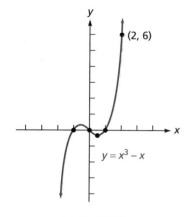

FIGURE 3-24

Miscellaneous Graphs

We now consider graphs of other equations. To graph each of these equations, we find any *x*- and *y*-intercepts, test for symmetries, plot points, and join them with a smooth curve.

EXAMPLE 2   Graph the equation $y = |x|$.

Solution   To find the x-intercepts, we substitute 0 for y and solve the equation

$$|x| = 0$$

Since 0 is the only solution, 0 is the only x-intercept of the graph.

To find the y-intercepts, we substitute 0 for x and solve the equation

$$y = |0|$$

Since 0 is the only solution, 0 is the only y-intercept of the graph. Because the x- and y-intercepts are both 0, the graph passes through the origin.

To test for y-axis symmetry, we replace x with $-x$.

5. $y = |x|$ The original equation.

  $y = |-x|$ Replace x with $-x$.

6. $y = |x|$ $|-x| = |x|$.

Because Equation 6 is equivalent to Equation 5, the graph is symmetric about the y-axis. The graph has no other symmetries.

We plot the x- and y-intercepts and several other pairs (x, y) and then use the y-axis symmetry to obtain the graph shown in Figure 3-25.

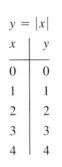

$y = |x|$

x	y
0	0
1	1
2	2
3	3
4	4

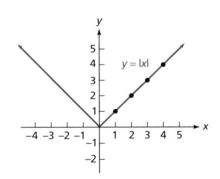

FIGURE 3-25

EXAMPLE 3   Graph the equation $y^2 = x$.

Solution   Because both the x- and y-intercepts are 0, the graph passes through the origin.

To test for x-axis symmetry, we replace y with $-y$.

7. $y^2 = x$ The original equation.

  $(-y)^2 = x$ Replace y with $-y$.

8. $y^2 = x$ Simplify.

Because Equation 8 is equivalent to Equation 7, the graph of the equation is symmetric about the x-axis. There are no other symmetries.

We can find other points on the graph by substituting numbers for y into the equation and finding x. We join the points with a smooth curve and use the x-axis symmetry to obtain the graph in Figure 3-26. It is a parabola opening to the right, with its vertex at the origin.

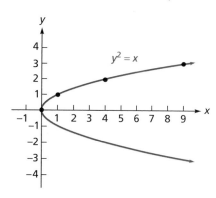

$y^2 = x$

x	y
0	0
1	1
4	2
9	3

FIGURE 3-26

EXAMPLE 4   Graph the equation $y = \sqrt{x}$.

Solution   Because both the x- and y-intercepts are 0, the graph passes through the origin.

We now test for symmetries. Because $\sqrt{x}$ is a real number only when $x \geq 0$, there is no point in replacing x with $-x$. Because $\sqrt{x} \geq 0$, and $y = \sqrt{x}$, there is also no point in replacing y with $-y$. The graph of the equation has no symmetries.

We plot several points to obtain the graph in Figure 3-27.

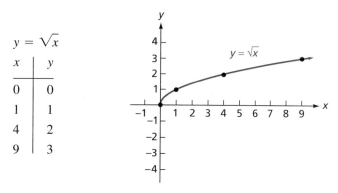

$y = \sqrt{x}$

x	y
0	0
1	1
4	2
9	3

FIGURE 3-27

The graphs in Examples 3 and 4 are related. If the equation $y^2 = x$ in Example 3 is solved for y, two equations result:

$$y = \sqrt{x} \qquad \text{and} \qquad y = -\sqrt{x}$$

The first equation, $y = \sqrt{x}$, was graphed in Example 4. It is the top half of the parabola shown in Example 3. The second equation, $y = -\sqrt{x}$, is the bottom half.

EXAMPLE 5   Graph the equation $xy = 1$.

Solution   We first look for intercepts. If $y = 0$, there is no value of x that will make the product xy equal to 1. Thus, the graph has no x-intercept. Similarly, the graph has no y-intercept.

We test for symmetry about the origin by replacing x and y with $-x$ and $-y$, respectively.

9. $xy = 1$ The original equation.

          $(-x)(-y) = 1$ Substitute $-x$ and $-y$ for x and y, respectively.

10. $xy = 1$ Simplify.

Because Equation 10 is equivalent to Equation 9, the graph is symmetric about the origin. There are no other symmetries.

To make it easier to find pairs (x, y) that satisfy the equation, we solve the equation for y to obtain $y = \frac{1}{x}$. We can then substitute positive values for x, find the corresponding values of y, and rely on the symmetry about the origin to draw the graph shown in Figure 3-28. This graph is called a **hyperbola**.

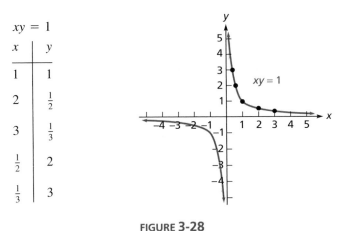

$xy = 1$

x	y
1	1
2	$\frac{1}{2}$
3	$\frac{1}{3}$
$\frac{1}{2}$	2
$\frac{1}{3}$	3

FIGURE 3-28

■

■ Circles

Because of the distance formula, we can develop the equation of a circle.

A Circle	A **circle** is the set of all points in a plane that are a fixed distance from a point called its **center**. The fixed distance is the **radius of the circle**.

To find the equation of a circle with radius r and center at point $C(h, k)$, we must find all points $P(x, y)$ in the xy-plane such that the length of line segment PC is r. (See Figure 3-29.) We can use the distance formula to find the length of CP, which is r:

$$r = \sqrt{(x - h)^2 + (y - k)^2}$$

After squaring both sides, we get

$$r^2 = (x - h)^2 + (y - k)^2$$

This equation is called the **standard equation of a circle**.

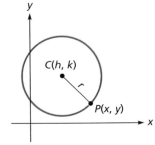

FIGURE 3-29

The Standard Equation of a Circle with Center at (h, k)	The graph of any equation that can be written in the form $$(x - h)^2 + (y - k)^2 = r^2$$ is a circle with radius r and center at point (h, k).

If $r = 0$, the circle is a single point called a **point circle**. If the center of the circle is the origin, then $(h, k) = (0, 0)$, and we have the following result.

The Standard Equation of a Circle with Center at $(0, 0)$	The graph of any equation that can be written in the form $$x^2 + y^2 = r^2$$ is a circle with radius r and center at the origin.

If we square the binomials in the equation $(x - h)^2 + (y - k)^2 = r^2$, we obtain an equation of the form

$$x^2 + y^2 + cx + dy + e = 0$$

where c, d, and e are real numbers. This form is called the **general form of the equation of a circle**.

EXAMPLE 6   Find the general form of the equation of the circle with radius 5 and center $(3, 2)$.

Solution   We can substitute 5 for r, 3 for h, and 2 for k in the standard equation of a circle and simplify:

$$(x - h)^2 + (y - k)^2 = r^2$$
$$(x - 3)^2 + (y - 2)^2 = 5^2$$
$$x^2 - 6x + 9 + y^2 - 4y + 4 = 25 \qquad \text{Remove parentheses.}$$
$$x^2 + y^2 - 6x - 4y - 12 = 0 \qquad \text{Subtract 25 from both sides and simplify.}$$

The general form is $x^2 + y^2 - 6x - 4y - 12 = 0$.   ■

EXAMPLE 7   Find the general form of the equation of the circle with endpoints of its diameter at $(8, -3)$ and $(-4, 13)$.

Solution   We can find the center $O(h, k)$ of the circle by finding the midpoint of its diameter. By the midpoint formulas and $(x_1, y_1) = (8, -3)$ and $(x_2, y_2) = (-4, 13)$, we have

$$h = \frac{x_1 + x_2}{2} \qquad k = \frac{y_1 + y_2}{2}$$

$$h = \frac{8 + (-4)}{2} \qquad k = \frac{-3 + 13}{2}$$

$$= \frac{4}{2} \qquad\qquad = \frac{10}{2}$$

$$= 2 \qquad\qquad\quad = 5$$

The center of the circle is $O(h, k) = O(2, 5)$.

To find the radius of the circle, we use the distance formula to find the distance between the center and one endpoint of the diameter. Since the center is $O(2, 5)$ and one endpoint is $(8, -3)$, we substitute 8 for x_1, -3 for y_1, 2 for x_2, and 5 for y_2 in the distance formula and simplify:

$$r = \sqrt{(x_2 - x_1)^2 + (y_2 - y_1)^2}$$
$$r = \sqrt{(2 - 8)^2 + [5 - (-3)]^2}$$
$$= \sqrt{(-6)^2 + (8)^2}$$
$$= \sqrt{36 + 64}$$
$$= \sqrt{100}$$
$$= 10$$

The radius of the circle is 10.

To find the equation of a circle with radius 10 and center at $(2, 5)$, we substitute 2 for h, 5 for k, and 10 for r in the standard equation of the circle and simplify:

$$(x - h)^2 + (y - k)^2 = r^2$$
$$(x - 2)^2 + (y - 5)^2 = 10^2$$
$$x^2 - 4x + 4 + y^2 - 10y + 25 = 100 \qquad \text{Remove parentheses.}$$
$$x^2 + y^2 - 4x - 10y - 71 = 0 \qquad \text{Subtract 100 from both sides and simplify.} \qquad ■$$

■ Graphing Equations of Circles

EXAMPLE 8   Graph the circle whose equation is $x^2 + y^2 - 4x + 2y = 20$.

Solution   To find the coordinates of the center and the radius, we write the equation in standard form by completing the square on both x and y:

$$x^2 + y^2 - 4x + 2y = 20$$
$$x^2 - 4x + y^2 + 2y = 20$$
$$x^2 - 4x + 4 + y^2 + 2y + 1 = 20 + 4 + 1 \qquad \begin{array}{l}\text{Add 4 and 1 to both sides} \\ \text{to complete the square.}\end{array}$$

$$(x - 2)^2 + (y + 1)^2 = 25 \qquad \begin{array}{l}\text{Factor } x^2 - 4x + 4 \text{ and} \\ y^2 + 2y + 1.\end{array}$$

$$(x - 2)^2 + [y - (-1)]^2 = 5^2$$

From the equation of the circle, we see that its radius is 5 and that the coordinates of its center are $h = 2$ and $k = -1$. The graph is shown in Figure 3-30. ■

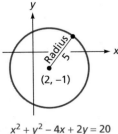

$x^2 + y^2 - 4x + 2y = 20$

FIGURE 3-30

3.4 EXERCISES

In Exercises 1–12, find the x- and y-intercepts of each graph. **Do not graph the equation.**

1. $y = x^2 - 4$ **2.** $y = x^2 - 9$ **3.** $y = 4x^2 - 2x$ **4.** $y = 2x - 4x^2$

5. $y = x^2 - 4x - 5$ **6.** $y = x^2 - 10x + 21$ **7.** $y = x^2 + x - 2$ **8.** $y = x^2 + 2x - 3$

9. $y = x^3 - 9x$ **10.** $y = x^3 + x$ **11.** $y = x^4 - 1$ **12.** $y = x^4 - 25x^2$

In Exercises 13–20, graph each parabola.

13. $y = x^2$ **14.** $y = -x^2$ **15.** $y = -x^2 + 2$ **16.** $y = x^2 - 1$

17. $y = x^2 - 4x$ **18.** $y = x^2 + 2x$ **19.** $y = \frac{1}{2}x^2 - 2x$ **20.** $y = \frac{1}{2}x^2 + 3$

In Exercises 21–36, find the symmetries, if any, of the graph of each equation. **Do not graph the equation.**

21. $y = x^2 + 2$ **22.** $y = 3x + 2$ **23.** $y^2 + 1 = x$ **24.** $y^2 + y = x$

25. $y^2 = x^2$ **26.** $y = 3x + 7$ **27.** $y = 3x^2 + 7$ **28.** $x^2 + y^2 = 1$

29. $y = 3x^3 + 7$ **30.** $y = 3x^3 + 7x$ **31.** $y^2 = 3x$ **32.** $y = 3x^4 + 7$

33. $y = |x|$ **34.** $y = |x + 1|$ **35.** $|y| = x$ **36.** $|y| = |x|$

In Exercises 37–52, graph each equation. Be sure to find any intercepts and symmetries.

37. $y = x^2 + 4x$ **38.** $y = x^2 - 6x$ **39.** $y = x^3$ **40.** $y = x^3 + x$

41. $y = |x - 2|$ **42.** $y = |x| - 2$ **43.** $y = 3 - |x|$ **44.** $y = 3|x|$

45. $y^2 = -x$ **46.** $y^2 = 4x$ **47.** $y^2 = 9x$ **48.** $y^2 = -4x$

49. $y = \sqrt{x} - 1$ **50.** $y = 1 - \sqrt{x}$ **51.** $xy = 4$ **52.** $xy = -9$

In Exercises 53–62, find the equation in general form of each circle with the given properties.

53. Center at the origin; $r = 1$ **54.** Center at the origin; $r = 4$

55. Center at $(6, 8)$; $r = 4$ **56.** Center at $(5, 3)$; $r = 2$

57. Center at $(3, -4)$; $r = \sqrt{2}$ **58.** Center at $(-9, 8)$; $r = 2\sqrt{3}$

59. Ends of diameter at $(3, -2)$ and $(3, 8)$ **60.** Ends of diameter at $(5, 9)$ and $(-5, -9)$

61. Center at $(-3, 4)$ and passing through the origin   **62.** Center at $(-2, 6)$ and passing through the origin

In Exercises 63–72, graph each equation.

63. $x^2 + y^2 - 25 = 0$ **64.** $x^2 + y^2 - 8 = 0$

65. $(x - 1)^2 + (y + 2)^2 = 4$ **66.** $(x + 1)^2 + (y - 2)^2 = 9$

67. $x^2 + y^2 + 2x - 24 = 0$ **68.** $x^2 + y^2 - 4y = 12$

69. $9x^2 + 9y^2 - 12y = 5$ **70.** $4x^2 + 4y^2 + 4y = 15$

71. $4x^2 + 4y^2 - 4x + 8y + 1 = 0$ **72.** $9x^2 + 9y^2 - 6x + 18y + 1 = 0$

3.5   GRAPHING DEVICES

- ■ Graphing Calculators with a Graphing Calculator
- ■ Using the ZOOM Feature
- ■ Solving Equations

FIGURE 3-31

In the previous section, we graphed equations in two variables by calculating and plotting several points (x, y) and joining them with a smooth curve. Plotting more points would have produced more accurate graphs, but plotting points is tedious work. Five or six points gave a fair compromise between accuracy and tedium. In this section, we will also graph by plotting points, but the graph will be very accurate, because we will plot hundreds of points. This is an impossible task by hand, but very easy if we use a graphing calculator. (See Figure 3-31.)

Several brands of graphing calculators and many computer graphing software packages are available. All of them have similar features, implemented in different ways. We will discuss how to use this technology to graph equations, but we will not show the keystrokes of any individual brand. For such details, consult the owner's manual.

■ Graphing Calculators

All graphing calculators have a **viewing window** to display graphs and other results. To see the proper picture of a graph, we must often set the minimum and the maximum values for the x- and y-axes, as well as the scale to be displayed. On a graphing calculator, we set the range of values for the coordinate axes to the numbers

$$\text{Xmin} = -10 \qquad \text{Xmax} = 10 \qquad \text{Ymin} = -10 \qquad \text{Ymax} = 10$$

These settings indicate that -10 is the minimum x-coordinate and the minimum y-coordinate that will be used in the graph, and that 10 is the maximum x- and y-coordinate to be used. In making these settings, we can move the cursor around by pressing the cursor keys, ◁ , ▷ , △ , and ▽ . To delete any unwanted digits, we press the DEL key.

To graph the equation $y = x^2 - 4$, for example, we enter the right-hand side of the equation after the symbol $Y_1 =$. The display will show the equation

$$Y_1 = X \wedge 2 - 4$$

We press the GRAPH key to produce the graph shown in Figure 3-32(a). To show more detail, we can redraw the graph after setting other range values. For example, the settings

$$\text{Xmin} = -4 \qquad \text{Xmax} = 4 \qquad \text{Ymin} = -4 \qquad \text{Ymax} = 4$$

produce the graph shown in Figure 3-32(b).

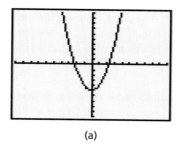 (a)

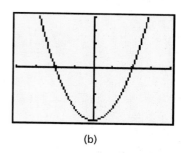 (b)

FIGURE **3-32**

EXAMPLE 1   Graph the equation $y = x^3 - 9x$ in these viewing windows:

a. $10 \leq x \leq 20$ and $10 \leq y \leq 20$

b. $-5 \leq x \leq 5$ and $-5 \leq y \leq 5$

c. $-10 \leq x \leq 10$ and $-12 \leq y \leq 12$

Solution   **a.** We enter the appropriate x- and y-values and enter the right-hand side of the equation. The viewing window will display

$$Y_1 = X \wedge 3 - 9 * X \qquad \text{Some calculators will accept 9X instead of } 9 * X.$$

Finally, we press the $\boxed{\text{GRAPH}}$ key and obtain a blank window. The blank window in Figure 3-33(a) shows that no portion of the graph appears in the coordinate grid seen in the viewing window.

b. As in part **a**, we set the range values. There is no need to re-enter the equation. We press the $\boxed{\text{GRAPH}}$ key to obtain the graph shown in Figure 3-33(b). In this case, only part of the graph appears, indicating that we need a slightly larger viewing window.

c. As in part **a**, we set the range values and press the $\boxed{\text{GRAPH}}$ key to obtain the more complete graph shown in Figure 3-33(c).

Graphing calculators have a TRACE function to help find the coordinates of points on a graph. To find the leftmost turning point of the graph shown in Figure 3-33(c), we enter TRACE mode and notice the flashing cursor in the viewing

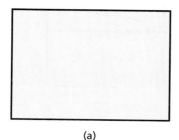 (a)

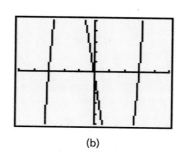 (b)

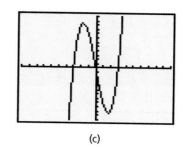 (c)

FIGURE **3-33**

window. The coordinates of the point marked by the cursor appear at the bottom of the viewing window; as the cursor moves along the curve, the coordinates change. If we move the cursor to the leftmost turning point of the graph, we read its coordinates as approximately $(-1.789, 10.375)$. ■

As Example 1 indicates, the choice of viewing window makes a big difference in the appearance of a graph. One of the challenges of using graphing calculators is finding an appropriate viewing window.

EXAMPLE 2   Find an appropriate viewing window for the graph of the equation $y = |x + 4| - 2$.

Solution   The smallest that an absolute value can be is 0, so the minimum value of y is $0 - 2$, or -2. We will set Ymin a bit lower: Ymin $= -3$. We also set Ymax to be 10 greater than Ymin: Ymax $= 7$.

The minimum value of y occurs when x is -4. To center the graph in the window, we will set Xmin and Xmax to 5 on either side of -4: Xmin $= -9$ and Xmax $= 1$.

We enter the right-hand side of the equation. Consult the owner's manual to find how to enter an absolute value. The graph is shown in Figure 3-34.

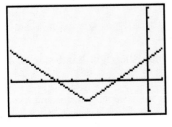

FIGURE 3-34

■

EXAMPLE 3   Graph the equation $y = \sqrt[3]{x - 2}$.

Solution   We can enter the equation by using a fractional exponent

$$Y_1 = (X - 2) \wedge (1/3)$$

and press $\boxed{\text{GRAPH}}$ to display the graph. Experiment with the range of values for the coordinate axes to produce an acceptable window. The values Xmin $= -5$, Xmax $= 15$, Ymin $= -5$, and Ymax $= 5$ produce the graph shown in Figure 3-35.

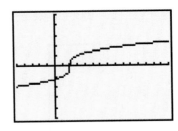

FIGURE 3-35   ■

Graphing calculators can display several graphs on the same axes. This is useful for graphing equations such as the circle $x^2 + y^2 = 25$. To do so, we must solve the equation for y:

$$x^2 + y^2 = 25$$
$$y^2 = 25 - x^2$$
$$y = \pm \sqrt{25 - x^2}$$

This last expression represents two equations: $y = \sqrt{25 - x^2}$ and $y = -\sqrt{25 - x^2}$. We graph both of these equations separately on the same coordinate axes by entering the first equation as Y_1 and the second as Y_2:

$$Y_1 = (25 - X \wedge 2) \wedge (1/2)$$
$$Y_2 = -Y_1$$

To define Y_2, we have reused the value of Y_1 to save time and avoid errors.

Depending on the setting of the maximum and minimum values of x and y, the graph may not appear to be a circle; it may look like Figure 3-36(a). Most graphing calculators can be set to display equal-sized divisions on the x- and y-axes, producing a better graph such as that shown in Figure 3-36(b).

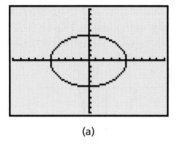

(a)

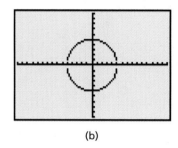

(b)

FIGURE 3-36

Using the ZOOM Feature

We have set appropriate viewing windows by setting individual range values. The ZOOM capabilities of a graphing calculator provide another way of establishing a viewing window. Suppose, for example, that we are interested in the behavior of the graph of $y = x^4$ close to the origin. We might begin by graphing the equation using the default range values to obtain the graph similar to that in Figure 3-37(a). We can zoom in on the graph near the origin and mark the corners of a new viewing window, as in Figure 3-37(b). The graph appears in the new window, as shown in Figure 3-37(c).

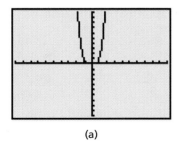

(a)

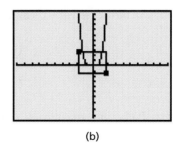

(b)

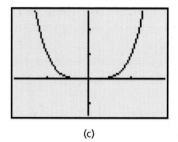

(c)

FIGURE 3-37

■ Solving Equations with a Graphing Calculator

It is easy to verify that the two solutions of the equation $x^2 - 3 = 0$, for example, are $\sqrt{3}$ and $-\sqrt{3}$, because either of these numbers make the left-hand side of the equation equal to 0. In the equation $y = x^2 - 3$, those same two values of x will cause y to equal 0. The solutions of the equation $x^2 - 3 = 0$ are the x-intercepts of the graph of the equation $y = x^2 - 3$.

This idea provides a method for using a graphing calculator to solve an equation such as $x^2 - 3 = 0$. We will use the TRACE and ZOOM features to close in on the x-intercepts of the graph of $y = x^2 - 3$.

EXAMPLE 4   Find the positive solution of the equation $x^2 - 3 = 0$.

Solution   We use a graphing calculator to graph $y = x^2 - 3$ and zoom in on the positive x-intercept, as in Figure 3-38(a). We then graph the equation again in the new window and enter the TRACE mode. We press the $\boxed{\triangleright}$ and $\boxed{\triangleleft}$ keys to bring the cursor as close to the intercept as possible, as in Figure 3-38(b). A value of x similar to that shown in Figure 3-38(b) provides an approximation to the solution. The solution provided in Figure 3-38(b) is $x = 1.734072$, which agrees to two decimal places with the actual solution, $\sqrt{3} = 1.732051$. Repeated zooms would provide more accurate results.

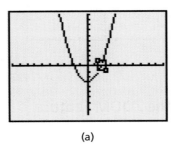

(a)

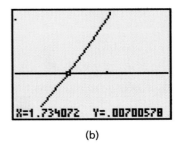

(b)

FIGURE 3-38   ■

3.5   EXERCISES

In Exercises 1–8, graph each equation. Use the range values Xmin $= -10$, Xmax $= 10$, Ymin $= -10$, *and* Ymax $= 10$ *to define the viewing window.*

1. $y = x^2 + 1$ **2.** $y = x^2 - 5$ **3.** $y = 3 - x^2$ **4.** $y = 7 - x^2$

5. $y = x^3 - x$ **6.** $y = x^2 - x^3$ **7.** $y = \dfrac{x^2 - x}{5}$ **8.** $y = \dfrac{x^2}{5} - 5$

In Exercises 9–12, graph each equation. Pick a viewing window that adequately displays the graph.

9. $y = x^2 - 10x$ **10.** $y = \dfrac{1}{x}$ **11.** $y = 9 + x^2$ **12.** $y = \sqrt{1 - x^2}$

In Exercises 13–16, graph each absolute value equation in a viewing window of x = −6 to 6 and y = −6 to 6.

13. $y = |x|$ **14.** $y = |x - 3|$ **15.** $y = |x| - 3$ **16.** $y = 3 - |x - 2|$

In Exercises 17–20, graph each equation by solving for y and graphing two separate equations. Use an appropriate viewing window.

17. $y^2 = x$ **18.** $x^2 + y^2 = 16$ **19.** $x^2 - y^2 = 4$ **20.** $6 - y^2 = x$

In Exercises 21–24, graph each equation and use the $\boxed{\text{TRACE}}$ *feature to find the coordinates of the vertex.*

21. $y = 2x^2 - x + 1$ **22.** $y = x^2 + 5x - 6$ **23.** $y = 7 + x - x^2$ **24.** $y = 2x^2 - 3x + 2$

In Exercises 25–28, solve each equation by using the $\boxed{\text{ZOOM}}$ *and* $\boxed{\text{TRACE}}$ *keys to find (to two decimal places) the x-intercepts of an appropriate graph.*

25. $x^2 - 7 = 0$ **26.** $x^2 - 3x + 2 = 0$ **27.** $x^3 - 3 = 0$ **28.** $3x^3 - x^2 - x = 0$

A graphing calculator can be used to solve inequalities. The solution of the inequality P(x) < 0 consists of those numbers x for which the graph of y = P(x) lies below the x-axis. To solve P(x) < 0, we graph y = P(x) and use the $\boxed{\text{TRACE}}$ *mode to find numbers x that produce negative values of y. In Exercises 29–30, solve each inequality.*

29. $x^2 + x - 6 < 0$ **30.** $x^2 - 3x - 10 > 0$

3.6 PROPORTION AND VARIATION

■ Solving Proportions   ■ Direct Variation   ■ Inverse Variation
■ Joint Variation   ■ Combined Variation

Quantities can be compared by using the concept of **ratio**. We might say, for example, that the ratio of men to women is 3 to 2, or that gasoline and oil are to be mixed in a ratio of 50 to 1.

Ratio

A **ratio** is the comparison of two numbers by their indicated quotient.

This definition implies that a ratio is a fraction. Thus, the denominator of a ratio cannot be 0. Some examples of ratios are

$$\frac{3}{5}, \quad \frac{x + 1}{9}, \quad \frac{a}{b}, \quad \text{and} \quad \frac{x^2 - 4}{x + 5}$$

Proportion

A **proportion** is an equation indicating that two ratios are equal.

Some examples of proportions are

$$\frac{2}{3} = \frac{4}{6}, \qquad \frac{x}{y} = \frac{3}{5}, \qquad \text{and} \qquad \frac{x^2 + 8}{2(x + 3)} = \frac{17(x + 3)}{2}$$

In the proportion $\frac{a}{b} = \frac{c}{d}$, the numbers b and c are called the **means** of the proportion, and the numbers a and d are called the **extremes**. Because the equation $\frac{a}{b} = \frac{c}{d}$ is equivalent to the equation $ad = bc$, we have the following theorem.

Theorem	In any proportion, the product of the means is equal to the product of the extremes.

This important theorem enables us to solve proportions.

■ Solving Proportions

EXAMPLE 1   Solve the proportion $\dfrac{x}{5} = \dfrac{2}{x + 3}$.

Solution

$$\frac{x}{5} = \frac{2}{x + 3}$$

$$x(x + 3) = 5 \cdot 2 \qquad \text{The product of the means equals the product of the extremes.}$$

$$x^2 + 3x = 10 \qquad \text{Remove parentheses and simplify.}$$

$$x^2 + 3x - 10 = 0 \qquad \text{Subtract 10 from both sides.}$$

$$(x - 2)(x + 5) = 0 \qquad \text{Factor the trinomial.}$$

$$x - 2 = 0 \quad \text{or} \quad x + 5 = 0 \qquad \text{Set each factor equal to 0.}$$

$$x = 2 \qquad \qquad x = -5$$

Thus, $x = 2$ or $x = -5$.   ■

EXAMPLE 2   Gasoline and oil for a lawn mower are to be mixed in a 50-to-1 ratio. How many ounces of oil should be mixed with 6 gallons of gasoline?

Solution   We first express 6 gallons as $6(128 \text{ ounces}) = 768$ ounces. We then let x represent the number of ounces of oil needed, set up the proportion, and solve it.

$$\frac{50}{1} = \frac{768}{x}$$

$$50x = 768 \qquad \text{The product of the means equals the product of the extremes.}$$

$$x = \frac{768}{50} \qquad \text{Divide both sides by 50.}$$

$$x \approx 15 \qquad \text{Read } \approx \text{ as "is approximately equal to."}$$

Approximately 15 ounces of oil should be added to 6 gallons of gasoline.   ■

■ **Direct Variation**

When scientists use functions to describe physical applications, they often use terminology involving the concept of variation.

Two variables are said to **vary directly** or be **directly proportional** if their ratio is a constant. The variables x and y vary directly when

$$\frac{y}{x} = k \qquad \text{or equivalently} \qquad y = kx$$

where k is a constant.

Direct Variation	The words "**y varies directly with x**," or "**y is directly proportional to x**," mean that $y = kx$ for some real-number constant k. The number k is called the **constant of proportionality**.

EXAMPLE 3   The distance traveled in a given time varies directly with the speed. If a car travels 70 miles at 30 miles per hour, how far will it travel in the same time at 45 miles per hour?

Solution   We let d represent distance traveled and let s represent the speed. "Distance varies directly with the speed" translates into the formula $d = ks$. The constant of proportionality k can be found by substituting 70 for d and 30 for s into the equation $d = ks$:

$$d = ks$$
$$\mathbf{70} = k(\mathbf{30})$$
$$k = \frac{7}{3}$$

To evaluate the distance d traveled at 45 miles per hour, we can now substitute $\frac{7}{3}$ for k and 45 for s into the formula $d = ks$.

$$d = \frac{7}{3}s$$
$$= \frac{7}{3}(\mathbf{45})$$
$$= 105$$

In the time it takes to go 70 miles at 30 miles per hour, the car could travel 105 miles at 45 miles per hour.   ■

EXAMPLE 4   For positive values of x, graph a few of the equations determined by the statement "y varies directly with x."

Solution   The statement "*y* varies directly with *x*" is equivalent to the equation $y = mx$, where *m* is the constant of proportionality. Because the equation is in the form $y = mx + b$ with $b = 0$, its graph is a line with slope *m* and *y*-intercept 0. The graphs of $y = mx$ for several values of *m* are shown in Figure 3-39.

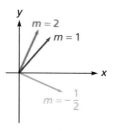

FIGURE 3-39

The graph of the relationship of direct variation is always a line that passes through the origin.   ■

■ Inverse Variation

Inverse Variation	The words "**y varies inversely with x,**" or "**y is inversely proportional to x,**" mean that $y = \frac{k}{x}$ for some real-number constant *k*.

EXAMPLE 5   The intensity of illumination from a light source varies inversely with the square of the distance from the source. If the intensity is 100 lumens at a distance of 20 feet, what will be the intensity at 30 feet?

Solution   If *I* is the intensity and *d* is the distance from the light source, the phrase "intensity varies inversely with the square of the distance" translates into the formula

1. $I = \dfrac{k}{d^2}$

We can evaluate *k* by substituting 100 for *I* and 20 for *d* into Equation 1 and solving for *k*.

$$I = \frac{k}{d^2}$$

$$100 = \frac{k}{20^2}$$

$$k = 40,000$$

We can substitute 40,000 for *k* and 30 for *d* into the formula to find the intensity at a distance of 30 feet:

$$I = \frac{k}{d^2}$$

$$I = \frac{40{,}000}{30^2}$$

$$= \frac{400}{9}$$

At 30 feet, the intensity of light would be $\frac{400}{9}$ lumens.   ∎

EXAMPLE 6   For positive values of x, graph a few of the equations determined by the statement "y varies inversely with x."

Solution   The statement is equivalent to the equation $y = \frac{k}{x}$ where k is a constant. For each choice of k, the equation determines a hyperbola similar to the one graphed in Example 7 in Section 3.4. Figure 3-40 shows the graphs of $y = \frac{k}{x}$ for three values of k.

The graphs of the relationship of inverse variation always look like the graphs shown in Figure 3-40.

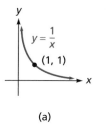
(a)

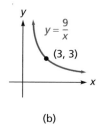
(b)

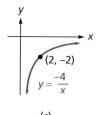
(c)

FIGURE **3-40**   ∎

■ Joint Variation

Joint Variation	The word "**y varies jointly with w and x**" mean that $y = kwx$ for some real-number constant k.

EXAMPLE 7   The energy an object has because of its motion is called **kinetic energy**. The kinetic energy of an object varies jointly with its mass and the square of its velocity. A 25-gram mass moving at the rate of 30 centimeters per second has a kinetic energy of 11,250 dyne-centimeters. Find the kinetic energy of a 10-gram mass that is moving at 40 centimeters per second.

Solution   We can let E, m, and v represent the kinetic energy, mass, and velocity, respectively. The phrase "energy varies jointly with its mass and the square of its velocity" translates into the formula

2. $E = kmv^2$

The constant k can be evaluated by substituting 11,250 for E, 25 for m, and 30 for v into Equation 2.

$$E = kmv^2$$
$$11{,}250 = k(25)(30)^2$$
$$11{,}250 = 22{,}500k$$
$$k = \frac{1}{2}$$

We can now substitute $\frac{1}{2}$ for k, 10 for m, and 40 for v into the formula to evaluate E.

$$E = kmv^2$$
$$= \frac{1}{2}mv^2$$
$$= \frac{1}{2}(10)(40)^2$$
$$= 8000$$

A 10-gram mass that is moving at 40 centimeters per second has a kinetic energy of 8000 dyne-centimeters. ∎

■ Combined Variation

The preceding terminology can be used in various combinations. In each of the following statements, the words on the left translate into the formula on the right:

y varies directly with x and inversely with z	$y = \dfrac{kx}{z}$
y varies jointly with the square of x and the cube root of z	$y = kx^2 \sqrt[3]{z}$
y varies jointly with x and the square root of z and inversely with the cube root of t	$y = \dfrac{kx\sqrt{z}}{\sqrt[3]{t}}$
y varies inversely with the product of x and z	$y = \dfrac{k}{xz}$

3.6 EXERCISES

In Exercises 1–4, solve each proportion.

1. $\dfrac{4}{x} = \dfrac{2}{7}$

2. $\dfrac{5}{2} = \dfrac{x}{6}$

3. $\dfrac{x}{2} = \dfrac{3}{x+1}$

4. $\dfrac{x+5}{6} = \dfrac{7}{8-x}$

In Exercises 5–6, set up and solve a proportion to answer each question.

5. The ratio of women to men in a mathematics class is 3 to 5. How many women are in the class if there are 30 men?

6. The ratio of lime to sand in mortar is 3 to 7. How much lime must be mixed with 21 bags of sand to make mortar?

In Exercises 7–12, find the constant of proportionality.

7. y is directly proportional to x. If $x = 30$, then $y = 15$.

8. z is directly proportional to t. If $t = 7$, then $z = 21$.

9. I is inversely proportional to R. If $R = 20$, then $I = 50$.

10. R is inversely proportional to the square of I. If $I = 25$, then $R = 100$.

11. E varies jointly with I and R. If $R = 25$ and $I = 5$, then $E = 125$.

12. z is directly proportional to the sum of x and y. If $x = 2$ and $y = 5$, then $z = 28$.

In Exercises 13–16, solve each problem.

13. y is directly proportional to x. If $y = 15$ when $x = 4$, find y when $x = \frac{7}{5}$.

14. w is directly proportional to z. If $w = -6$ when $z = 2$, find w when $z = -3$.

15. P varies jointly with r and s. If $P = 16$ when $r = 5$ and $s = -8$, find P when $r = 2$ and $s = 10$.

16. m varies jointly with the square of n and the square root of q. If $m = 24$ when $n = 2$ and $q = 4$, find m when $n = 5$ and $q = 9$.

In Exercises 17–20, decide if the graph could represent direct variation, inverse variation, or neither.

17.

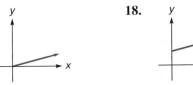

18.

19.

20.

21. *Gas laws* The volume of a gas varies directly with the temperature and inversely with the pressure. When the temperature of a certain gas is 330°, the pressure is 40 pounds per square inch, and the volume is 20 cubic feet. Find the volume when the pressure increases 10 pounds per square inch and the temperature decreases to 300°.

22. *Hooke's law* The force required to stretch a spring a distance d is directly proportional to d. A force of 5 newtons stretches a spring 0.2 meter. What force will stretch the spring 0.35 meter?

23. *Free-falling objects* The distance that an object will fall in t seconds varies directly with the square of t. An object falls 16 feet in 1 second. How long will it take to fall 144 feet?

24. *Heat dissipation* The power (in watts) dissipated as heat in a resistor varies jointly with the resistance (in ohms) and the square of the current (in amperes). A 10-ohm resistor carrying a current of 1 ampere dissipates 10 watts. How much power is dissipated in a 5-ohm resistor carrying a current of 3 amperes?

25. *Heat dissipation* The power (in watts) dissipated as heat in a resistor varies directly with the square of the voltage and inversely with the resistance. If 20 volts are placed across a 20-ohm resistor, it will dissipate 20 watts. What voltage across a 10-ohm resistor will dissipate 40 watts?

26. *Period of a pendulum* The time required for one complete swing of a pendulum is called the **period** of the pendulum. The period varies directly with the square of its length. If a 1-meter pendulum has a period of 1 second, find the length of a pendulum with a period of 2 seconds.

27. *Frequency of vibration* The **pitch**, or **frequency**, of a vibrating string varies directly with the square root of the tension. If a string vibrates at a frequency of 144 hertz due to a tension of 2 pounds, find the frequency when the tension is 18 pounds.

28. *Gravitational attraction* The gravitational attraction between two massive objects varies jointly with their masses and inversely with the square of the distance between them. What happens to this force if each mass is tripled and the distance between them is doubled?

29. *Plane geometry* The area of an equilateral triangle varies directly with the square of the length of a side. Find the constant of proportionality.

30. *Solid geometry* The diagonal of a cube varies directly with the length of a side. Find the constant of proportionality.

3 CHAPTER SUMMARY

Key Words

abscissa (3.1)
axis of symmetry (3.4)
center of a circle (3.4)
circle (3.4)
combined variation (3.6)
constant of proportionality (3.6)
coordinates (3.1)
direct variation (3.6)
graph (3.1)
inverse variation (3.6)

joint variation (3.6)
ordered pair (3.1)
ordinate (3.1)
origin (3.1)
parabola (3.4)
proportion (3.6)
quadrant (3.1)
radius of a circle (3.4)
ratio (3.6)

rectangular coordinate system (3.1)
slope of a nonvertical line (3.2)
symmetry (3.4)
vertex of a parabola (3.4)
x-axis (3.1)
x-intercept (3.1, 3.4)
xy-plane (3.1)
y-axis (3.1)
y-intercept (3.1, 3.4)

Key Ideas

(3.1) **The distance formula:**
$$d = \sqrt{(x_2 - x_1)^2 + (y_2 - y_1)^2}$$

The midpoint formula: The midpoint of the line segment joining (x_1, y_1) and (x_2, y_2) is the point M with coordinates

$$\left(\frac{x_1 + x_2}{2}, \frac{y_1 + y_2}{2} \right)$$

(3.2) The slope, m, of a nonvertical line passing through (x_1, y_1) and (x_2, y_2) is given by

$$m = \frac{y_2 - y_1}{x_2 - x_1} \qquad (x_2 \neq x_1)$$

A vertical line has no defined slope.
Nonvertical parallel lines have the same slope.
Perpendicular lines have slopes that are negative reciprocals of each other, provided neither line is vertical.

(3.3) **Point–slope form of the equation of a line:**
$$y - y_1 = m(x - x_1)$$
Slope–intercept form of the equation of a line:
$$y = mx + b$$
General form of the equation of a line: $Ax + By = C$
Equation of a horizontal line through the point (a, b):
$$y = b$$

Equation of a vertical line through the point (a, b):
$$x = a$$

(3.4) **Locating the intercepts of a graph:**
To find the **x-intercepts** of a graph, let $y = 0$ and solve for x.
To find the **y-intercepts** of a graph, let $x = 0$ and solve for y.

Tests for symmetry:
If $(-x, y)$ lies on the graph whenever (x, y) does, the graph is symmetric about the y-axis.
If $(-x, -y)$ lies on the graph whenever (x, y) does, the graph is symmetric about the origin.
If $(x, -y)$ lies on the graph whenever (x, y) does, the graph is symmetric about the x-axis.
Equations of a circle:
$(x - h)^2 + (y - k)^2 = r^2$; center at (h, k), radius r
$x^2 + y^2 = r^2$; center at $(0, 0)$, radius r
$x^2 + y^2 + cx + dy + e = 0$

(3.6) In any proportion, the product of the means is equal to the product of the extremes.

y varies directly with x: $\qquad y = kx$

y varies inversely with x: $\qquad y = \dfrac{k}{x}$

y varies jointly with w and x: $\qquad y = kwx$

3 CHAPTER REVIEW EXERCISES

In Review Exercises 1–4, refer to Illustration 1 and determine the coordinates of each point.

1. *A*   **2.** *B*

3. *C*   **4.** *D*

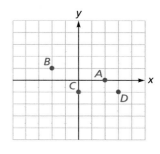

ILLUSTRATION 1

In Review Exercises 5–8, graph each point. Indicate the quadrant in which the point lies or the axis on which it lies.

5. $(-3, 5)$ **6.** $(5, -3)$ **7.** $(0, -7)$ **8.** $\left(-\frac{1}{2}, 0\right)$

In Review Exercises 9–12, use the x and y-intercepts to graph each equation.

9. $3x - 5y = 15$ **10.** $x + y = 7$ **11.** $x + y = -7$ **12.** $x - 5y = 5$

In Review Exercises 13–16, find the length and the midpoint of the line segment PQ.

13. $P(-3, 7); Q(3, -1)$ **14.** $P(0, 5); Q(-12, 10)$ **15.** $P(\sqrt{3}, 9); Q(\sqrt{3}, 7)$ **16.** $P(a, -a); Q(-a, a)$

In Review Exercises 17–20, find the slope of the line PQ, if possible.

17. $P(3, -5); Q(1, 7)$ **18.** $P(2, 7); Q(-5, -7)$ **19.** $P(b, a); Q(a, b)$ **20.** $P(a + b, b); Q(b, b - a)$

In Review Exercises 21–30, write the equation of the line with the given properties.

21. The line passes through the origin and the point $(-5, 7)$.

22. The line passes through $(-2, 1)$ and has a slope of -4.

23. The line passes through $(7, -5)$ and $(4, 1)$.

24. The line has a slope of $\frac{2}{3}$ and a *y*-intercept of 3.

25. The line has a slope of 0 and passes through $(-5, 17)$.

26. The line has no defined slope and passes through $(-5, 17)$.

27. The line passes through $(8, -2)$ and is parallel to the line segment joining $(2, 4)$ and $(4, -10)$.

28. The line passes through $(8, -2)$ and is perpendicular to the line segment joining $(2, 4)$ and $(4, -10)$.

29. The line is parallel to $3x - 4y = 7$ and passes through $(2, 0)$.

30. The line passes through $(7, 0)$ and is perpendicular to the line $3y + x - 4 = 0$.

In Review Exercises 31–38, graph each equation. Find all intercepts and symmetries.

31. $y = x^2 + 2$ **32.** $y = x^3 - 2$ **33.** $y = \frac{1}{2}|x|$ **34.** $y = -\sqrt{x - 4}$

35. $y = \sqrt{x} + 2$ **36.** $y = |x + 1| + 2$ **37.** $y = x^3 + 1$ **38.** $xy = 25$

In Review Exercises 39–40, write the equation of each circle.

39. Center at $(-3, 4)$; radius 12   **40.** Ends of diameter at $(-6, -3)$ and $(5, 8)$

In Review Exercises 41–42, graph each equation.

41. $x^2 + y^2 - 2y = 15$

42. $x^2 + y^2 - 4x + 2y = 4$

In Review Exercises 43–46, graph each equation on a graphing calculator. Some exercises might require graphing two equations.

43. $y = (x - 2)(x + 5)$ **44.** $y = x + 2|x|$ **45.** $y^2 = x - 3$ **46.** $x^2 - y^2 = 4$

In Review Exercises 47–50, solve each equation on a graphing calculator and use $\boxed{\text{ZOOM}}$ *and* $\boxed{\text{TRACE}}$ *to find the x-intercepts of the graph.*

47. $x^2 - 11 = 0$ **48.** $x^3 - x = 0$ **49.** $|x^2 - 2| - 1 = 0$ **50.** $x^2 - 3x = 5$

In Review Exercises 51–54, solve each proportion.

51. $\dfrac{x + 3}{10} = \dfrac{x - 1}{x}$ **52.** $\dfrac{x - 1}{2} = \dfrac{12}{x + 1}$ **53.** $\dfrac{3x - 5}{x} = \dfrac{x - 3}{1}$ **54.** $\dfrac{x + 7}{8} = \dfrac{x + 4}{x + 5}$

55. *Hooke's law*   The force required to stretch a spring is proportional to the amount of stretch. If a 3-pound force stretches a spring 5 inches, what force would stretch the spring 3 inches?

56. *Gas laws*   The volume of gas in a balloon varies directly as the temperature and inversely as the pressure. If the volume is 400 cubic centimeters when the temperature is 300°K and the pressure is 25 dynes per square centimeter, find the volume when the temperature is 200°K and the pressure is 20 dynes per square centimeter.

57. *Kinetic energy*   A moving body has a kinetic energy proportional to the square of its velocity. By what factor does the kinetic energy of an automobile increase if its speed increases from 30 miles per hour to 50 miles per hour?

58. 🖩 *Electrical resistance*   The resistance of a wire varies directly as the length of the wire and inversely as the square of its diameter. A 1000-foot length of wire, 0.05 inches in diameter, has a resistance of 200 ohms. What would be the resistance of a 1500-foot length of wire that is 0.08 inches in diameter?

59. *Billing for services*   Angie's Painting and Decorating Service charges a fixed amount for accepting a wallpapering job, then adds a fixed dollar amount for each roll of wallpaper that is hung. If Angie bills a customer $177 to hang 11 rolls and $294 to add 20 rolls, find the cost to hang 27 rolls.

60. *Paying for college*   Rolf must earn $5040 for next semester's tuition. Assume he works x hours tutoring algebra at $14 per hour and y hours tutoring Spanish at $18 per hour, and achieves his goal. Write an equation expressing the relationship between x and y and graph the equation. If Rolf tutors algebra for 180 hours, how long must he tutor Spanish?

3   CHAPTER TEST

In Questions 1–2, indicate the quadrant in which the point lies or the axis on which it lies.

1. $(-3, \pi)$

2. $(0, -8)$

In Questions 3–4, find the x- and y-intercepts and use them to graph the equation.

3. $x + 3y = 6$

4. $2x - 5y = 10$

In Questions 5–8, graph each equation.

5. $2(x + y) = 3x + 5$

6. $3x - 5y = 3(x - 5)$

7. $\frac{1}{2}(x - 2y) = y - 1$

8. $\dfrac{x + y - 5}{7} = 3x$

In Questions 9–10, find the distance between points P and Q.

9. $P(1, -1); Q(-3, 4)$

10. $P(0, \pi); Q(-\pi, 0)$

In Questions 11–12, find the midpoint of the line segment PQ.

11. $P(3, -7); Q(-3, 7)$

12. $P(0, \sqrt{2}); Q(\sqrt{8}, \sqrt{18})$

In Questions 13–14, find the slope of the line PQ.

13. $P(3, -9); Q(-5, 1)$

14. $P(\sqrt{3}, 3); Q(-\sqrt{12}, 0)$

In Questions 15–16, determine if the two lines are parallel, perpendicular, or neither.

15. $y = 3x - 2; y = 2x - 3$

16. $2x - 3y = 5; 3x + 2y = 7$

In Questions 17–22, write the equation of the line with the given properties.

17. passing through $(3, -5); m = 2$

18. $m = 3; b = \frac{1}{2}$

19. parallel to $2x - y = 3; b = 5$

20. perpendicular to $2x - y = 3; b = 5$

21. passing through $\left(2, -\frac{3}{2}\right)$ and $\left(3, \frac{1}{2}\right)$

22. parallel to the y-axis and passing through $(3, -4)$

In Questions 23–24, find the x- and y-intercepts of each graph.

23. $y = x^3 - 16x$

24. $y = |x - 4|$

In Questions 25–26, find the symmetries of each graph.

25. $y^2 = x - 1$

26. $y = x^4 + 1$

In Questions 27–30, graph each equation.

27. $y = x^2 - 9$

28. $x = |y|$

29. $y = 2\sqrt{x}$

30. $x = y^3$

In Questions 31–32, write the equation of each circle.

31. Center at $(5, 7)$; radius of 8

32. Center at $(2, 4)$; passing through $(6, 8)$

In Questions 33–34, graph each equation.

33. $x^2 + y^2 = 9$

34. $x^2 - 4x + y^2 + 3 = 0$

In Questions 35–36, write each statement as an equation.

35. y varies directly as the square of z.

36. w varies jointly with r and the square of s.

37. P varies directly with Q. $P = 7$ when $Q = 2$. Find P when $Q = 5$.

38. y is directly proportional to x and inversely proportional to the square of z. $y = 16$ when $x = 3$ and $z = 2$. Find x when $y = 2$ and $z = 3$.

 In Questions 39–40, use a graphing calculator to find the positive root of each equation.

39. $x^2 - 7 = 0$

40. $x^2 - 5x - 5 = 0$

CUMULATIVE REVIEW EXERCISES

In Exercises 1–2, assume that $\mathbf{A} = \{3, 4, 5, 6\}$, $\mathbf{B} = \{2, 3\}$, *and* $\mathbf{C} = \{6, 7\}$. *Find each set.*

1. $\mathbf{A} \cap \mathbf{B}$

2. $\mathbf{A} \cup (\mathbf{B} \cap \emptyset)$

In Exercises 3–4, graph each interval on the number line.

3. $\{x \mid -4 \le x < 7\}$

4. $\{x \mid x \ge 2 \text{ or } x < 0\}$

In Exercises 5–6, tell which property of the real numbers justifies each expression.

5. $(a + b) + c = c + (a + b)$

6. If $x < 3$ and $3 < y$, then $x < y$.

In Exercises 7–14, simplify each expression. Assume that all variables represent positive numbers. Give all answers with positive exponents.

7. $(81a^4)^{1/2}$

8. $81(a^4)^{1/2}$

9. $(a^{-3}b^{-2})^{-2}$

10. $\left(\dfrac{4x^4}{12x^2y}\right)^{-2}$

11. $\left(\dfrac{4x^0y^2}{x^2y}\right)^{-2}$

12. $\left(\dfrac{4x^{-5}y^2}{6x^{-2}y^{-3}}\right)^2$

13. $(a^{1/2}b)^2(ab^{1/2})^2$

14. $(a^{1/2}b^{1/2}c)^2$

In Exercises 15–18, rationalize each denominator and simplify. Assume that all variables represent positive numbers.

15. $\dfrac{3}{\sqrt{3}}$

16. $\dfrac{2}{\sqrt[3]{4x}}$

17. $\dfrac{3}{y - \sqrt{3}}$

18. $\dfrac{3x}{\sqrt{x} - 1}$

In Exercises 19–22, simplify each expression and combine similar terms.

19. $\sqrt{75} - 3\sqrt{5}$

20. $\sqrt{18} + \sqrt{8} - 2\sqrt{2}$

21. $(\sqrt{2} - \sqrt{3})^2$

22. $(3 - \sqrt{5})(3 + \sqrt{5})$

In Exercises 23–28, perform the indicated operations and simplify, when necessary.

23. $(3x^2 - 2x + 5) - 3(x^2 + 2x - 1)$

24. $5x^2(2x^2 - x) + x(x^2 - x^3)$

25. $(3x - 5)(2x + 7)$

26. $(z + 2)(z^2 - z + 2)$

27. $3x + 2\overline{)6x^3 + x^2 + x + 2}$

28. $x^2 + 2\overline{)3x^4 + 7x^2 - x + 2}$

In Exercises 29–32, factor each polynomial.

29. $3t^2 - 6t$

30. $3x^2 - 10x - 8$

31. $x^8 + x^4 + 1$

32. $x^6 - 1$

In Exercises 33–38, perform the indicated operations and simplify.

33. $\dfrac{x^2 - 4}{x^2 + 5x + 6} \cdot \dfrac{x^2 - 2x - 15}{x^2 + 3x - 10}$

34. $\dfrac{6x^3 + x^2 - x}{x + 2} \div \dfrac{3x^2 - x}{x^2 + 4x + 4}$

35. $\dfrac{2}{x + 3} + \dfrac{5x}{x - 3}$

36. $\dfrac{x - 2}{x + 3}\left(\dfrac{x + 3}{x^2 - 4} - 1\right)$

37. $\dfrac{\dfrac{1}{a} + \dfrac{1}{b}}{\dfrac{1}{ab}}$

38. $\dfrac{x^{-1} - y^{-1}}{x - y}$

In Exercises 39–40, solve each equation.

39. $\dfrac{3x}{x + 5} = \dfrac{x}{x - 5}$

40. $8(2x - 3) - 3(5x + 2) = 4$

In Exercises 41–42, solve each formula for the indicated variable.

41. $\dfrac{1}{R} = \dfrac{1}{R_1} + \dfrac{1}{R_2}; R$

42. $S = \dfrac{a - lr}{1 - r}; r$

43. *Gardening* A gardener wishes to enclose her rectangular raspberry patch with 40 feet of fencing. The raspberry bushes are planted along the garage, so no fencing is needed on that side. What will be the dimensions if the total area is 192 square feet?

44. *Financial planning* A college student invests part of a $25,000 inheritance at 7% interest and the rest at 6%. If his annual interest is $1670, how much did he invest at 6%?

In Exercises 45–48, perform the indicated operations and express all answers in a + bi form.

45. $\dfrac{2 + i}{2 - i}$

46. $\dfrac{i(3 - i)}{(1 + i)(1 + i)}$

47. $|3 + 4i|$

48. $\dfrac{5}{i^7} + 5i$

In Exercises 49–52, solve each equation.

49. $\dfrac{x + 3}{x - 1} - \dfrac{6}{x} = 1$

50. $x^4 + 36 = 13x^2$

51. $\sqrt{y + 2} + \sqrt{11 - y} = 5$

52. $z^{2/3} - 13z^{1/3} + 36 = 0$

In Exercises 53–58, solve each inequality. Graph the solution on the number line.

53. $5x - 7 \le 4$

54. $x^2 - 8x + 15 > 0$

55. $\dfrac{x^2 + 4x + 3}{x - 2} \ge 0$

56. $\dfrac{9}{x} > x$

57. $|2x - 3| \ge 5$

58. $\left|\dfrac{3x - 5}{2}\right| < 2$

In Exercises 59–60, use the x- and y-intercepts to graph each equation.

59. $5x - 3y = 15$

60. $3x + 2y = 12$

In Exercises 61–62, find the length, the midpoint, and the slope of the line segment PQ.

61. $P\left(-2, \tfrac{7}{2}\right); Q\left(3, -\tfrac{1}{2}\right)$

62. $P(3, 7); Q(-7, 3)$

In Exercises 63–66, write the equation of the line with the given properties. Give the answer in slope–intercept form.

63. The line passes through $(-3, 5)$ and $(3, -7)$.

64. The line passes through $\left(\frac{3}{2}, \frac{5}{2}\right)$ and has a slope of $\frac{7}{2}$.

65. The line is parallel to $3x - 5y = 7$ and passes through $(-5, 3)$.

66. The line is perpendicular to $x - 4y = 12$ and passes through the origin.

In Exercises 67–70, graph each equation. Make use of intercepts and symmetries.

67. $x^2 = y - 2$

68. $y^2 = x - 2$

69. $x^2 + y^2 = 100$

70. $x^2 - 2x + y^2 = 8$

In Exercises 71–72, solve each proportion.

71. $\dfrac{x - 2}{x} = \dfrac{x - 6}{5}$

72. $\dfrac{x + 2}{x - 6} = \dfrac{3x + 1}{2x - 11}$

73. *Dental billing*   The billing schedule for dental x-rays specifies a fixed amount for the office visit plus a fixed amount for each x-ray exposure. If having 2 x-rays costs \$37 and having 4 costs \$54, find the cost of 5 exposures.

74. *Automobile collisions*   The energy dissipated in an automobile collision varies directly with the square of the speed. By what factor does the energy increase in a 50-mile-per-hour collision compared with a 20-mph collision?

CHAPTER 4

FUNCTIONS

In Chapter 3, we saw that an equation in x and y sets up a correspondence between numbers x and values y. Correspondence between the elements of two sets is a common occurrence in everyday life. For example,

- To every house, there corresponds an address.
- To every car, there corresponds a license plate.
- To every state, there correspond two United States senators.

This idea of correspondence is basic in mathematics and is the topic of this chapter.

4.1 FUNCTIONS AND FUNCTION NOTATION

■ Relations   ■ Functions   ■ The Vertical Line Test   ■ Function Notation   ■ Linear Functions

An equation, a table of ordered pairs of values, or a graph can describe a correspondence between two variables. The equation $y = x^2 - 1$, for example, sets up a correspondence in which each number x determines a value y. For example, if $x = 2$, the value

$$y = 2^2 - 1 = 3$$

is produced. The equation $y = x^2 - 1$ determines the table of ordered pairs and the graph shown in Figure 4-1.

To see how the graph determines the correspondence, we draw a vertical and a horizontal line through any point, such as point P, on the graph shown in Figure 4-1. Because these lines intersect the x-axis at 2 and the y-axis at 3, the point $P(2, 3)$ associates the number 3 on the y-axis with the number 2 on the x-axis.

$y = x^2 - 1$

x	y
-2	3
-1	0
0	-1
1	0
2	3

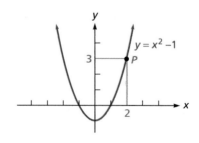

FIGURE **4-1**

■ Relations

Any correspondence between the elements of two sets is called a **relation**. In this section, we will consider relations between two sets of numbers.

Relation	A **relation** is a correspondence that assigns to each number x in a set **X** one or more values of y in a set **Y**.
	The set **X** is called the **domain** of the relation. The set of all y-values that correspond to numbers x in the domain is called the **range**.

Since the domain and range of a relation are sets of real numbers, the relation defines a set of ordered pairs (x, y), where x is an element of the domain and y is the corresponding value in the range. The **graph of the relation** is the graph of all of these ordered pairs in the xy-plane.

EXAMPLE 1 Graph the relation described by the equation $x = |y|$ and find the domain and range.

Solution We construct a table of ordered pairs and draw the graph, as in Figure 4-2. In this relation, the value $y = 0$ corresponds to the number $x = 0$. There are two values of y that correspond to each positive number x. For example, to the number $x = 2$, there correspond the y-values $y = 2$ and $y = -2$.

From the graph, we can see that any number x in the interval $[0, \infty)$ produces one or more values of y. This interval is the domain of the relation. Since y can be any value in the interval $(-\infty, \infty)$, this interval is the range of the relation.

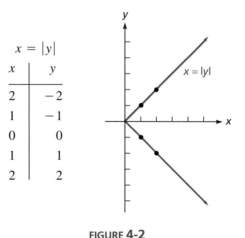

FIGURE **4-2**

As Example 1 suggests, the graph of a relation illustrates the domain and range of the relation, as well as the correspondence. For example, the interval shown on the x-axis in Figure 4-3(a) is the domain of the relation whose graph is shown, because it contains all numbers x for which the relation is defined.

To the particular number x shown in Figure 4-3(b), there corresponds a y-value on the y-axis. The set of all possible values of y that correspond to numbers x in the domain is the range of the relation. The range is the interval shown on the y-axis in Figure 4-3(c).

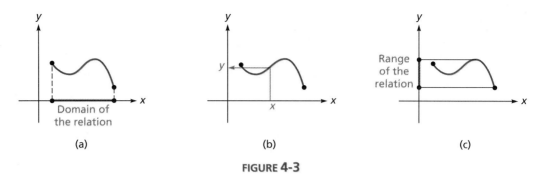

FIGURE 4-3

Functions

Any relation that assigns *exactly one value of y* to each number x is called a **function**. In a function, the value of y that corresponds to a particular number x is called the **image of x**, and we say that the function is **defined at x**. For the function determined by the equation $y = x^2 - 1$, the image of 2 is 3.

Because the variable y depends on x in a function, y is called the **dependent variable**. The variable x is called the **independent variable**.

Function	A **function** is a correspondence that assigns to each number x in some set **X** exactly one value y in some set **Y**.
	The set **X** is the **domain of the function**. The value y that corresponds to a particular x is called the **image of x** under the function. The set of all images of x is the **range** of the function.

Unless otherwise indicated, we will assume that the domain of a function is its implied domain—the set of all real numbers x for which the function is defined.

By definition, a function is a correspondence that assigns to each element of some set **X** exactly one element of a set **Y**. We can visualize this correspondence with the diagram of Figure 4-4(a). The function f that assigns the element y to the element x

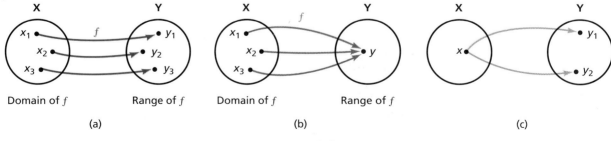

FIGURE 4-4

is represented by an arrow leaving x and pointing to y. The set of all those elements in **X** from which arrows originate is the domain of the function. The set of all elements of **Y** to which arrows point is the range.

To be a function, each element of the domain must have exactly one image in the range. However, the same value of y could be the image of several numbers x. In the function shown in Figure 4-4(b), the single image y corresponds to the three numbers x_1, x_2, and x_3 in the domain.

The correspondence shown in Figure 4-4(c) is not a function, because two values of y correspond to the same number x.

 Warning!   From the definitions of function and relation, a function is a relation, but a relation is not necessarily a function.

To determine whether an equation determines y to be a function of x, we first solve the equation for y and then see whether one value of y corresponds to each number x. If this is so, the equation does determine y to be a function of x.

EXAMPLE 2   Decide whether the relations defined by   **a.** $7x + y = 5$ and   **b.** $y = \dfrac{3}{x - 2}$ are functions.

Solution   **a.** We first solve the equation $7x + y = 5$ for y to get

$$y = 5 - 7x$$

Since $5 - 7x$ has exactly one value for each number we substitute for x, each number x does determine exactly one value y. Therefore, the equation does determine y to be a function of x.

Because x can represent any real number, the domain of the function is $\mathfrak{R}$, the set of real numbers. Because the resulting value of y can be any real number, the range of the function is also $\mathfrak{R}$.

b. The equation $y = \dfrac{3}{x-2}$ also defines y to be a function of x, because each number we substitute for x, except 2, gives exactly one value y. We cannot substitute 2 for x, because division by 0 is undefined. The domain of this function is the set of all real numbers except 2. Since the numerator of the fraction is the constant 3, y cannot be 0. Therefore, the range of this function is the set of all real numbers except 0.   ■

The Vertical Line Test

A test, called the **vertical line test**, can be applied to the graph of a relation to determine whether the relation is a function. If each vertical line that intersects a graph does so exactly once, then each number x determines exactly one value y, and the graph represents a function. See Figure 4-5(a).

If any vertical line intersects the graph more than once, then to some numbers x there correspond more than one value y, and the relation is not a function. See Figure 4-5(b).

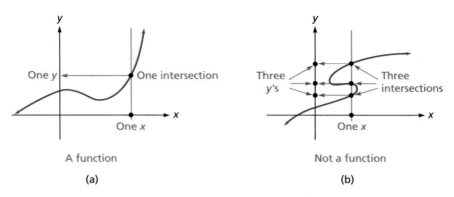

FIGURE 4-5

EXAMPLE 3 Graph the points (x, y) for which **a.** $x = y^2$ and **b.** $y = \sqrt{x}$. From the graphs, decide whether these relations are functions.

Solution **a.** Several ordered pairs (x, y) that satisfy the equation are listed in the table shown in Figure 4-6(a). We plot these points and draw the graph as shown in the figure. Because some vertical lines intersect the curve twice, there are some numbers x that determine two values of y. Thus, the graph fails the vertical line test, and the relation is not a function.

b. Because $\sqrt{x} \geq 0$, the values of y defined by the equation $y = \sqrt{x}$ are never negative. We can plot ordered pairs (x, y) that satisfy the equation and draw the graph shown in Figure 4-6(b). Because every vertical line that intersects the curve does so exactly once, each number x determines exactly one value of y. Thus, the graph passes the vertical line test. The relation is a function.

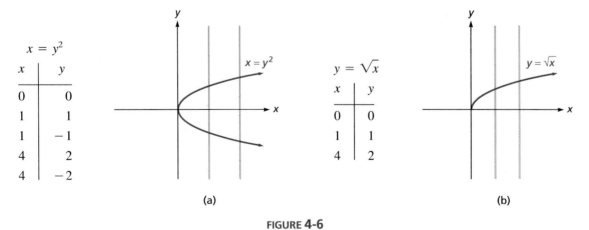

FIGURE 4-6

■

Function Notation

To indicate that y is a function of x, we can use **function notation** and write

$$y = f(x)$$

Strictly speaking, y is the image of x under the function f, and the correspondence itself is the function. However, it is common to read $y = f(x)$ as "y is a function of x." Functions are often denoted by letters other than f.

Function notation provides a way of indicating the image of a particular number x. The symbol $f(2)$, read as "f of 2" or "f at 2," indicates the image of 2 under the function f.

$$f(x) = 5 - 7x$$
$$f(2) = 5 - 7(2) \qquad \text{Substitute 2 for } x.$$
$$= -9$$

Therefore, if $x = 2$, then $y = f(2) = -9$. Similarly, the image of -5 is $f(-5)$ because

$$f(x) = 5 - 7x$$
$$f(-5) = 5 - 7(-5) \qquad \text{Substitute } -5 \text{ for } x.$$
$$= 40$$

Thus, $f(-5) = 40$.

EXAMPLE 4   Let $g(x) = 3x^2 + x - 4$. Find **a.** $g(-3)$, **b.** $g(k)$, **c.** $g(-t^3)$, and **d.** $g(k + 1)$.

Solution   **a.** $g(x) = 3x^2 + x - 4$ **b.** $g(x) = 3x^2 + x - 4$
$\qquad\qquad\quad g(-3) = 3(-3)^2 + (-3) - 4$ $\qquad\quad g(k) = 3k^2 + k - 4$
$\qquad\qquad\qquad\quad = 3(9) - 3 - 4$
$\qquad\qquad\qquad\quad = 20$

c. $g(x) = 3x^2 + x - 4$ **d.** $g(x) = 3x^2 + x - 4$
$\quad g(-t^3) = 3(-t^3)^2 + (-t^3) - 4$ $\quad g(k + 1) = 3(k + 1)^2 + (k + 1) - 4$
$\qquad\qquad = 3t^6 - t^3 - 4$ $\qquad\qquad\qquad = 3(k^2 + 2k + 1) + k + 1 - 4$
$\qquad\qquad\qquad\qquad\qquad\qquad\qquad = 3k^2 + 6k + 3 + k + 1 - 4$
$\qquad\qquad\qquad\qquad\qquad\qquad\qquad = 3k^2 + 7k$ ■

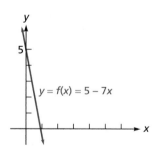

FIGURE 4-7

For most functions of algebra, the domains and ranges will be sets of real numbers. If f is such a function, then its graph is the set of all points $(x, f(x))$ in the xy-plane, where x is an element of the domain of f. In other words, the graph of f is the graph of the equation $y = f(x)$. The graph of the function $f(x) = 5 - 7x$, for example, is a line with a slope -7 and y-intercept 5. (See Figure 4-7.)

EXAMPLE 5   Let $f(x) = |x|$. Graph f and find its domain and range.

Solution   If $x = 1$, then $f(1) = 1$, and the point $(1, 1)$ lies on the graph. If $x = -2$, then $f(-2) = 2$, and the point $(-2, 2)$ lies on the graph. These points and others are plotted in Figure 4-8(a), which shows the complete graph.

The domain of f is the set of all real numbers, and the range is the set of non-negative real numbers, as shown in Figure 4-8(b).

$y = |x|$

x	y
2	2
1	1
0	0
−1	1
−2	2

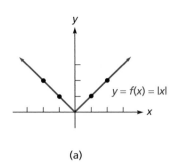

(a)

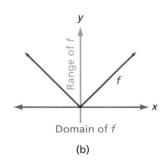

(b)

FIGURE **4-8**

Linear Functions

The equation of a nonvertical line defines a special function called a linear function.

Linear Function	A **linear function** is a function determined by an equation of the form $$f(x) = mx + b$$

EXAMPLE 6 The cost of electricity in a midwestern city is a linear function of x, the number of kilowatt-hours used. If the cost of 100 kilowatt-hours is \$17 and the cost of 500 kilowatt-hours is \$57, determine the formula that expresses the function.

Solution Since c (the cost of electricity) is given to be a linear function of x, there are constants m and b such that

1. $c = mx + b$

Because $c = 17$ when $x = 100$, the point $(x_1, c_1) = (100, 17)$ lies on the straight-line graph of this function. Furthermore, $c = 57$ when $x = 500$, and the point $(x_2, c_2) = (500, 57)$ also lies on that line. The slope of the line is

$$m = \frac{c_2 - c_1}{x_2 - x_1}$$

$$= \frac{57 - 17}{500 - 100}$$

$$= \frac{40}{400}$$

$$= 0.10$$

Thus, $m = 0.10$. To determine b, we can substitute 0.10 for m and the coordinates of point $P(100, 17)$ into the equation $c = mx + b$ and solve for b.

$$c = mx + b$$
$$17 = 0.10(100) + b$$
$$17 = 10 + b$$
$$7 = b$$

Thus, $c = 0.10x + 7$. The company charges \$7, plus 10¢ per kilowatt-hour used. ∎

The next example illustrates how the concept of function can be used in geometry.

EXAMPLE 7 Express the area of a square as a function of its perimeter.

Solution The area of a square is a function of the length s of a side. This can be expressed as

2. $A = f(s) = s^2$

The perimeter P is also a function of the length s of a side, given by

3. $P = g(s) = 4s$

To express the area A as a function of the perimeter P, we must determine a function h such that $A = h(P)$. To do so, we solve Equation 3 for s and substitute into Equation 2.

$$P = 4s$$

$$s = \frac{P}{4} \qquad \text{Divide both sides by 4.}$$

$$A = \left(\frac{P}{4}\right)^2 \qquad \text{Substitute } \frac{P}{4} \text{ for } s \text{ in Equation 2.}$$

$$= \frac{P^2}{16}$$

Therefore,

$$A = h(P) = \frac{P^2}{16}$$

In the context of this example, the domain of h is not the implied domain of the set of real numbers. Because the perimeter of a square is a positive number, the domain of h is the set of positive real numbers. ∎

4.1 EXERCISES

In Exercises 1–4, graph each relation and find its domain and range.

1. $2x + 3y = 6$ **2.** $y^2 = -x$ **3.** $x^2 + y^2 = 16$ **4.** $x = \sqrt{y}$

In Exercises 5–6, indicate the domain and range of each relation as intervals on the x- and y-axes.

5.
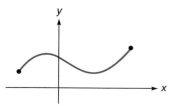

6.
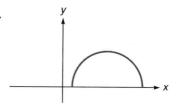

In Exercises 7–18, assume that all variables represent real numbers. Indicate whether each equation determines y to be a function of x.

7. $y = x$

8. $y - 2x = 0$

9. $y^2 = x$

10. $|y| = x$

11. $y = x^2$

12. $y - 7 = 7$

13. $y^2 - 4x = 1$

14. $|x - 2| = y$

15. $|x| = |y|$

16. $x = 7$

17. $y = 7$

18. $|x + y| = 7$

In Exercises 19–26, let the function f be defined by the equation $y = f(x)$, where x and $f(x)$ are real numbers. Find the domain and range of each function.

19. $f(x) = 3x + 5$

20. $f(x) = -5x + 2$

21. $f(x) = x^2$

22. $f(x) = x^2 + 3$

23. $f(x) = \dfrac{3}{x + 1}$

24. $f(x) = \dfrac{-7}{x + 3}$

25. $f(x) = \sqrt{x}$

26. $f(x) = \sqrt{x^2}$

In Exercises 27–32, some correspondences are illustrated. If the correspondence could represent y as a function of x, so indicate. If the illustration could not represent a function, explain why.

27.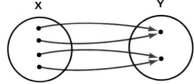

28.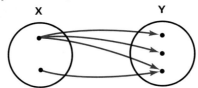

29.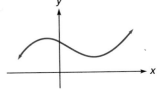

30.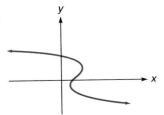

31.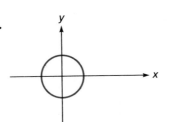

32.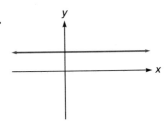

In Exercises 33–44, let the function f be defined by the equation $y = f(x)$, where x and $f(x)$ are real numbers. Find $f(2)$, $f(-3)$, $f(k)$, and $f(k^2 - 1)$.

33. $f(x) = 3x - 2$

34. $f(x) = 5x + 7$

35. $f(x) = \dfrac{1}{2}x + 3$

36. $f(x) = \dfrac{2}{3}x + 5$

37. $f(x) = x^2$

38. $f(x) = 3 - x^2$

39. $f(x) = \dfrac{2}{x + 4}$

40. $f(x) = \dfrac{3}{x - 5}$

41. $f(x) = \dfrac{1}{x^2 - 1}$

42. $f(x) = \dfrac{3}{x^2 + 3}$

43. $f(x) = \sqrt{x^2 + 1}$

44. $f(x) = \sqrt{x^2 - 1}$

In Exercises 45–60, graph each equation. Use the vertical line test to decide whether the equation defines y to be a function of x.

45. $y = 2x + 3$

46. $y = 3x + 2$

47. $2x = 3y - 3$

48. $3x = 2(y + 1)$

49. $y = -x^2$

50. $y^2 = 1 - x$

51. $y^2 = x + 1$

52. $|y| = -x$

53. $y = x^3$

54. $x = y^3$

55. $x = \sqrt{y}$

56. $y = \sqrt{x + 2}$

57. $xy = 1$

58. $xy = -4$

59. $x^2 + y^2 = 25$

60. $(x - 1)^2 + y^2 = 4$

61. Express the radius of a circle as a function of its circumference.

62. Express the radius of a circle as a function of its area.

63. Express the perimeter of a square as a function of its area.

64. Express the area of a circle as a function of its circumference.

65. Express the volume of a cube as a function of the area of one of its faces.

66. Express the length of the diagonal of a square as a function of the length of one edge.

67. Express the area of a square as a function of the length of a diagonal.

68. Express the circumference of a circle as a function of its area.

69. *Temperature conversion* The Fahrenheit temperature reading F is a linear function of the Celsius reading C. If $C = 0$ when $F = 32$ and the readings are the same at $-40°$, express F as a function of C.

70. *Free-falling objects* The velocity of a falling object is a linear function of the time t it has been falling. If $v = 15$ when $t = 0$ and $v = 79$ when $t = 2$, express v as a function of t.

71. *Water billing* The cost c of water is a linear function of n, the number of gallons used. If 1000 gallons cost $4.70 and 9000 gallons cost $14.30, express c as a function of n.

72. *Simple interest* The amount A of money on deposit for t years in an account earning simple interest is a linear function of t. Express that function as an equation if $A = \$272$ when $t = 3$ and $A = \$320$ when $t = 5$.

In Exercises 73–76, use a graphing calculator to graph each relation. Then give the domain and range of the relation.

73. $y = \sqrt{2x - 5}$ **74.** $y = |3x + 2|$ **75.** $y = \sqrt[3]{5x - 1}$ **76.** $y = -\sqrt[3]{3x + 2}$

4.2 QUADRATIC FUNCTIONS

■ Quadratic Functions ■ Finding the Vertex of a Parabola
■ Optimization Problems

The linear function defined by the equation $y = f(x) = mx + b$ is a first-degree polynomial function because its right-hand side is a first-degree polynomial in the variable x. In this section, we will discuss functions involving polynomials of second degree.

■ Quadratic Functions

A function defined by a polynomial of second degree is called a **quadratic function**.

Quadratic Function	A **quadratic function** is a second-degree polynomial function in one variable. It is defined by an equation of the form $y = f(x) = ax^2 + bx + c$, where a, b, and c are constants and $a \neq 0$.

EXAMPLE 1   Graph the functions **a.** $y = f(x) = x^2 - 2x - 3$ and **b.** $y = f(x) = -\dfrac{1}{3}x^2 + 3$.

Solution   We can plot several points with coordinates that satisfy the equations and join them with smooth curves to obtain the graphs shown in Figure 4-9.

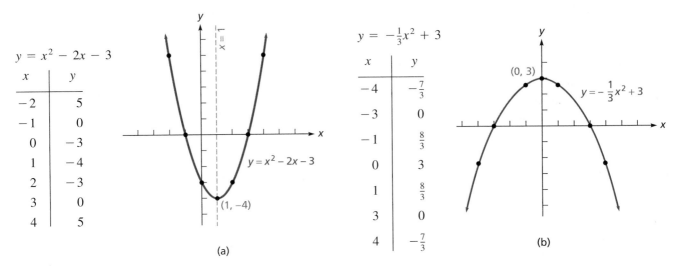

$y = x^2 - 2x - 3$

x	y
−2	5
−1	0
0	−3
1	−4
2	−3
3	0
4	5

$y = -\dfrac{1}{3}x^2 + 3$

x	y
−4	$-\dfrac{7}{3}$
−3	0
−1	$\dfrac{8}{3}$
0	3
1	$\dfrac{8}{3}$
3	0
4	$-\dfrac{7}{3}$

(a)   (b)

FIGURE 4-9   ■

The vertex of the parabola shown in Figure 4-9(a) is the point $(1, -4)$. The axis of symmetry is the line $x = 1$, the vertical line that passes through the vertex.

The vertex of the parabola in Figure 4-9(b) is the point $(0, 3)$. Its axis is the y-axis, the line $x = 0$.

■ Finding the Vertex of a Parabola

If a, h, and k are constants and $a \ne 0$, the graph of the equation

$$y - k = a(x - h)^2$$

is a parabola, because if the parentheses were removed and the equation solved for y, the right-hand side would be a second-degree polynomial. If $a > 0$, the parabola opens upward, as in Figure 4-9(a). If $a < 0$, the parabola opens downward, as in Figure 4-9(b).

The equation $y - k = a(x - h)^2$ displays the coordinates (h, k) of the **vertex** of its parabolic graph, as the following discussion will show.

If $a > 0$, the graph of $y - k = a(x - h)^2$ is a parabola opening upward. The vertex of this parabola is the point on the graph that has the least possible y-coordinate. Since a is positive and $(x - h)^2$ is never negative, the minimum value of $a(x - h)^2$ is 0, and this occurs when $x = h$.

When the right-hand side of the equation is 0, the left-hand side is also 0, and the corresponding value of y is k. Therefore, the parabola's vertex is the point (h, k).

A similar argument will show that when $a < 0$, the vertex is the highest point on the graph, which is the point (h, k).

Theorem	The graph of the equation

$$y - k = a(x - h)^2 \qquad (a \neq 0)$$

is a parabola with its vertex at the point (h, k). The parabola opens upward if $a > 0$ and downward if $a < 0$.

We can use the process of completing the square to transform an equation from the form $y = ax^2 + bx + c$ to the form $y - k = a(x - h)^2$. In that second form, we can read the coordinates (h, k) of the vertex directly.

EXAMPLE 2 Use completing the square to find the vertex of the parabola $y = 2x^2 - 5x - 3$.

Solution We complete the square on the right-hand side of the equation:

$$y = 2x^2 - 5x - 3$$

$$y + 3 = 2x^2 - 5x \qquad \text{Add 3 to both sides.}$$

$$y + 3 = 2\left(x^2 - \frac{5}{2}x \qquad\right) \qquad \text{Factor a 2 from the terms on the right-hand side.}$$

To complete the square within the parentheses, we add the square of one-half of the coefficient of x.

- First, we identify the coefficient of x, which is $-\dfrac{5}{2}$.

- We then find one-half of it, which is $\dfrac{1}{2}\left(-\dfrac{5}{2}\right) = -\dfrac{5}{4}$.

- Finally, we square that result, which is $\left(-\dfrac{5}{4}\right)^2 = \dfrac{25}{16}$.

We add $\frac{25}{16}$ within the parentheses. Because of the factor of 2, however, we are really adding twice $\frac{25}{16}$, or $\frac{25}{8}$, to the right-hand side of the equation. We must also add $\frac{25}{8}$ to the left-hand side.

$$y + 3 = 2\left(x^2 - \frac{5}{2}x \qquad\right)$$

$$y + 3 + \frac{\mathbf{25}}{\mathbf{8}} = 2\left(x^2 - \frac{5}{2}x + \frac{\mathbf{25}}{\mathbf{16}}\right) \qquad \text{Add } \frac{25}{8} \text{ to both sides.}$$

$$y + \frac{49}{8} = 2\left(x - \frac{5}{4}\right)^2 \qquad \text{Combine terms and factor.}$$

$$y - \left(-\frac{49}{8}\right) = 2\left(x - \frac{5}{4}\right)^2$$

The equation is now in the form

$$y - k = a(x - h)^2$$

with $h = \frac{5}{4}$ and $k = -\frac{49}{8}$. The vertex is the point $(h, k) = \left(\frac{5}{4}, -\frac{49}{8}\right)$. ∎

To find the vertex of any parabola defined by $y = ax^2 + bx + c$, we write the equation in the form $y - k = a(x - h)^2$ by completing the square:

$$y = ax^2 + bx + c$$

$$y = a\left(x^2 + \frac{b}{a}x \qquad\right) + c \qquad \text{Factor out } a.$$

$$y + \frac{b^2}{4a} = a\left(x^2 + \frac{b}{a}x + \frac{b^2}{4a^2}\right) + c \qquad \text{Add } \frac{b^2}{4a} \text{ to both sides.}$$

$$y + \frac{b^2}{4a} - c = a\left(x^2 + \frac{b}{a}x + \frac{b^2}{4a^2}\right) \qquad \text{Subtract } c \text{ from both sides.}$$

$$y - \left(c - \frac{b^2}{4a}\right) = a\left(x - \frac{-b}{2a}\right)^2 \qquad \text{Factor.}$$

We compare this equation to the standard form $y - k = a(x - h)^2$ to determine that $h = -\dfrac{b}{2a}$ and $k = c - \dfrac{b^2}{4a}$. This proves the following theorem.

Theorem	The graph of the equation
	$$y = ax^2 + bx + c \qquad (a \neq 0)$$ is a parabola with vertex $\left(-\dfrac{b}{2a}, c - \dfrac{b^2}{4a}\right)$.

■ Optimization Problems

EXAMPLE 3   A farmer has 400 feet of fencing to enclose a rectangular corral. To save money and fencing, he intends to use one bank of the river as one boundary of the corral, as in Figure 4-10. Find the dimensions that would enclose the largest area.

FIGURE 4-10

Solution   We let x represent the width of the fenced area. Then $400 - 2x$ represents the length. Because the area A of a rectangle is the product of the length and the width, we have

$$A = (400 - 2x)x \qquad \text{or} \qquad A = -2x^2 + 400x$$

The graph of this equation is a parabola, and since the coefficient of x^2 is negative, the parabola opens downward. Since its vertex is its highest point, the A-coordinate of the vertex represents the maximum area, and the x-coordinate represents the width of the corral with that maximum area. We compare the equations

$$A = -2x^2 + 400x \qquad \text{and} \qquad y = ax^2 + bx + c$$

to determine that $a = -2$, $b = 400$, and $c = 0$. The vertex of the parabola is the point

$$\left(-\frac{b}{2a}, c - \frac{b^2}{4a} \right) = \left(-\frac{400}{2(-2)}, 0 - \frac{400^2}{4(-2)} \right) = (100, 20{,}000)$$

If the farmer's fence runs 100 feet out from the river, 200 feet parallel to the river, and 100 feet back to the river, it will enclose the largest possible area, 20,000 square feet.   ■

To determine a selling price that will maximize revenue, manufacturers must consider the economic principle of supply and demand: Increasing the number of units manufactured decreases the price that can be charged for each unit.

EXAMPLE 4   A manufacturer of automobile air bag modules has determined that x units can be manufactured and sold each week at $\$(384 - 0.1\,x)$ each. Find the weekly production level that will maximize the revenue from sales, and then find that revenue.

Solution   We let y represent the revenue from sales. Because the revenue is the product of the number of units sold and the price charged for each unit, we have

$$y = x(384 - 0.1\,x) \qquad \text{or} \qquad y = -0.1x^2 + 384x$$

The graph of this equation is a parabola. Since the coefficient of x^2 is negative, it opens downward, and its vertex represents its highest point. The x-coordinate of the vertex is the production level that will maximize revenue, and the y-coordinate is that maximum revenue. We compare the equations

$$y = 0.1x^2 + 384x \qquad \text{and} \qquad y = ax^2 + bx + c$$

to see that $a = -0.1$, $b = 384$, and $c = 0$. Thus, the vertex of the parabola is the point

$$\left(-\frac{b}{2a}, c - \frac{b^2}{4a} \right) = \left(-\frac{384}{2(-0.1)}, 0 - \frac{384^2}{4(-0.1)} \right) = (1920, 368{,}640)$$

The greatest possible revenue from sales is $\$368{,}640$, attained at a weekly production level of 1920 units.   ■

4.2 EXERCISES

In Exercises 1–8, graph each quadratic equation.

1. $y = x^2 - x$

2. $y = x^2 + 2x$

3. $y = -3x^2 + 2$

4. $y = -3x^2 + 4$

5. $y = -\dfrac{1}{2}x^2 + 3$

6. $y = \dfrac{1}{2}x^2 - 2$

7. $y = x^2 - 4x + 1$

8. $y = -x^2 - 4x + 1$

In Exercises 9–16, find the vertex of each parabola.

9. $y = x^2 - 1$

10. $y = -x^2 + 2$

11. $y = x^2 - 4x + 4$

12. $y = x^2 - 10x + 25$

13. $y = x^2 + 6x - 3$

14. $y = -x^2 + 9x - 2$

15. $y = -2x^2 + 12x - 17$

16. $y = 2x^2 + 16x + 33$

In Exercises 17–30, find each maximum.

17. *Architecture*   A parabolic arch has an equation of $x^2 + 20y - 400 = 0$, where x is measured in feet. Find the maximum height of the arch.

18. *Ballistics*   An object is thrown from the origin of a coordinate system with the x-axis along the ground and the y-axis vertical. Its path, or **trajectory**, is given by the equation $y = 400x - 16x^2$. Find the object's maximum height.

19. *Ballistics*   A child throws a ball up a hill that makes an angle of 45° with the horizontal. The ball lands 100 feet up the hill. Its trajectory is a parabola with equation $y = -x^2 + ax$ for some number a. Find a. (See Illustration 1.)

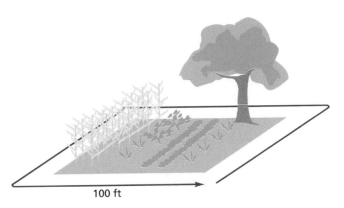

100 ft

ILLUSTRATION 2

21. *Maximizing storage area*   A farmer wants to partition a rectangular feed-storage area in a corner of his barn. The barn walls form two sides of the stall, and the farmer has 50 feet of partition for the remaining two sides. What dimensions will maximize the area? (See Illustration 3.)

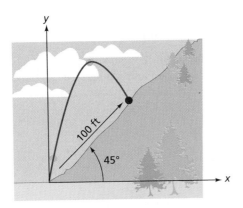

ILLUSTRATION 1

20. *Maximizing area*   The rectangular garden in Illustration 2 has a width of x and a perimeter of 100 feet. Find x so that the area of the rectangle is maximum.

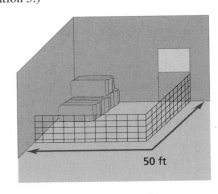

50 ft

ILLUSTRATION 3

22. *Maximizing grazing area* A rancher wishes to enclose a rectangular, partitioned corral with 1800 feet of fencing. (See Illustration 4.) What dimensions of the corral would enclose the largest possible area? Find the maximum area.

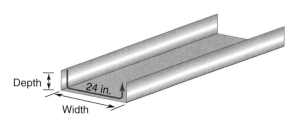

Depth   24 in.

Width

ILLUSTRATION 5

ILLUSTRATION 4

23. *Sheet-metal fabrication* A 24-inch-wide sheet of metal is to be bent into a rectangular trough with the cross section shown in Illustration 5. Find the dimensions that will maximize the amount of water the trough can carry. That is, find the dimensions that will maximize the cross-sectional area.

24. *Landscape design* A gardener will use D feet of edging to border a rectangular plot of ground. Show that she will enclose the maximum area if the rectangle is a square.

25. *Puzzle problem* The sum of two numbers is 6, and the sum of the squares of those two numbers is as small as possible. What are the two numbers?

26. *Puzzle problem* What number most exceeds its square?

27. *Selling television sets* A wholesaler of appliances finds that she can sell $(1200 - p)$ television sets each week when the price is p dollars. What price will maximize revenue?

28. *Finding mass transit fares* The Municipal Transit Authority serves 150,000 commuters daily when the fare is $1.80. Market research has determined that every penny decrease in the fare will result in 1000 new riders. What fare will maximize revenue?

29. *Finding hotel rates* A 300-room hotel is two-thirds filled when the nightly room rate is $90. Experience has shown that each $5 increase in cost results in ten fewer occupied rooms. Find the nightly rate that will maximize income.

30. *Selling concert tickets* Tickets for a concert are cheaper when purchased in quantity. The first 100 tickets are priced at $10 each, but each additional block of 100 tickets purchased decreases the cost of each ticket by 50¢. How many blocks of tickets should be sold to maximize the revenue?

In Exercises 31–34, use this information: At a time t seconds after an object is tossed vertically upward, it reaches a height s in feet given by the equation $s = 80t - 16t^2$.

31. In how many seconds does the object reach its maximum height?

32. In how many seconds does the object return to the point from which it was thrown?

33. What is the maximum height reached by the object?

34. Show that it takes the same amount of time for the object to reach its maximum height as it does to return from that height to the point from which it was thrown.

35. *Preparing for calculus* Find the dimensions of the largest rectangle that can be inscribed in the right triangle *ABC* shown in Illustration 6.

36. *Preparing for calculus* Point *P* lies in the first quadrant and on the line $x + y = 1$ in such position that the area of triangle *OPA* is maximum. Find the coordinates of *P*. (See Illustration 7.)

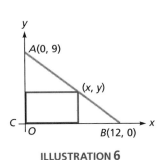

ILLUSTRATION 6

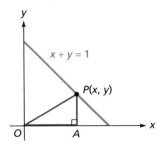

ILLUSTRATION 7

In Exercises 37–40, use a graphing calculator to determine the coordinates of the vertex of each parabola. You will have to select appropriate viewing windows.

37. $y = 2x^2 + 9x - 56$ **38.** $y = 14x - \dfrac{x^2}{5}$ **39.** $y = (x - 7)(5x + 2)$ **40.** $y = -x(0.2 + 0.1x)$

4.3 POLYNOMIAL AND MISCELLANEOUS FUNCTIONS

■ Graphing Polynomial Functions   ■ Even and Odd Functions
■ Increasing and Decreasing Functions   ■ Piecewise-Defined Functions
■ The Greatest Integer Function

So far, we have discussed two types of polynomial functions—first-degree (or linear functions) and second-degree (or quadratic functions). In this section, we will discuss polynomial functions of higher degree.

Polynomial Functions

A **polynomial function in one independent variable (say, *x*)** is defined by an equation of the form $y = P(x)$, where $P(x)$ is a polynomial in the variable *x*.
The **degree of the polynomial function** $y = P(x)$ is the degree of $P(x)$.

■ Graphing Polynomial Functions

EXAMPLE 1 Graph the function *f*, where $y = f(x) = x^3 - 4x$.

Solution To check the graph of $y = f(x) = x^3 - 4x$ for symmetry about the *y*-axis, we replace *x* with $-x$, simplify, and compare the result to the original equation.

1. $y = x^3 - 4x$ The original equation.

$y = (-x)^3 - 4(-x)$ Replace x with $-x$.

2. $y = -x^3 + 4x$ Simplify.

Because Equation 2 is not equivalent to Equation 1, the graph is not symmetric about the y-axis.

Next, we test for symmetry about the origin by replacing x and y with $-x$ and $-y$, respectively.

1. $y = x^3 - 4x$ The original equation.

$-y = (-x)^3 - 4(-x)$ Replace x with $-x$, and y with $-y$.

$-y = -x^3 + 4x$ Simplify.

3. $y = x^3 - 4x$ Multiply both sides by -1.

Because Equation 3 is identical to Equation 1, the graph is symmetric about the origin. Since the equation is given to be a function, there can be no symmetry about the x-axis.

We can also find the x-intercepts of the graph—the points at which the graph intersects the x-axis. The x-intercepts are the numbers x for which $y = 0$. To find them, we solve the equation $x^3 - 4x = 0$.

$$x^3 - 4x = 0$$
$$x(x^2 - 4) = 0$$
$$x(x + 2)(x - 2) = 0$$

$x = 0$ or $x + 2 = 0$ or $x - 2 = 0$

$x = -2$ $x = 2$

The x-intercepts are 0, -2, and 2.

The y-intercept of the graph is the value of y obtained when $x = 0$. The y-intercept is 0. Since the y-intercept is 0, the graph passes through the origin.

Finally, we plot the intercepts and a few points for positive x and use the symmetry to draw the rest of the graph, as shown in Figure 4-11.

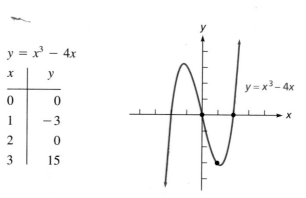

$y = x^3 - 4x$

x	y
0	0
1	-3
2	0
3	15

$y = x^3 - 4x$

FIGURE 4-11

EXAMPLE 2   Graph the function f, where $y = f(x) = x^4 - 5x^2 + 4$.

Solution   Because the variable x appears with only even exponents, $y = f(x)$ is equivalent to $y = f(-x)$, and the graph is symmetric about the y-axis.

$$y = x^4 - 5x^2 + 4 \qquad \text{The original equation.}$$
$$y = (-x)^4 - 5(-x)^2 + 4 \qquad \text{Replace } x \text{ with } -x.$$
$$y = x^4 - 5x^2 + 4 \qquad \text{Remove parentheses.}$$

The graph is not symmetric about the origin or the x-axis, however.

The x-intercepts of the graph are those numbers x for which $y = 0$. We set y equal to 0 and solve for x.

$$y = x^4 - 5x^2 + 4$$
$$0 = x^4 - 5x^2 + 4$$
$$0 = (x^2 - 4)(x^2 - 1)$$
$$0 = (x + 2)(x - 2)(x + 1)(x - 1)$$

$x + 2 = 0$	or	$x - 2 = 0$	or	$x + 1 = 0$	or	$x - 1 = 0$
$x = -2$		$x = 2$		$x = -1$		$x = 1$

The graph has four x-intercepts: $x = -2$, $x = 2$, $x = -1$, and $x = 1$. However, it has only one y-intercept: $y = 4$, the value of y produced when $x = 0$.

To graph the equation, we plot the intercepts and several other points to the right of the y-axis. We then rely on the graph's symmetry about the y-axis to graph the other half. A table of values and the graph appear in Figure 4-12.

$y = x^4 - 5x^2 + 4$

x	y
0	4
1	0
$\frac{3}{2}$	$-\frac{35}{16}$
2	0
3	40

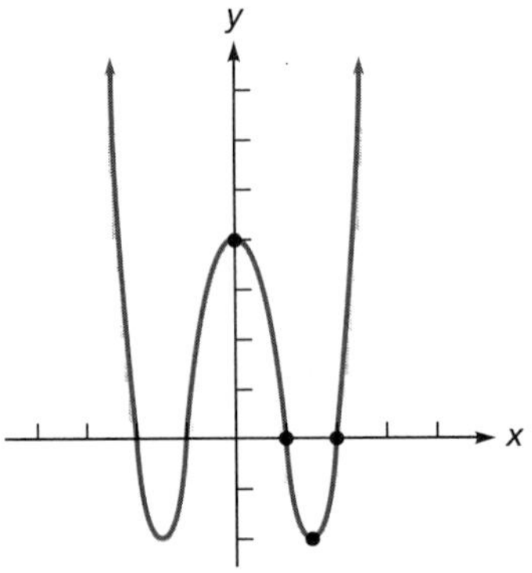

$y = x^4 - 5x^2 + 4$

FIGURE 4-12

Even and Odd Functions

A function with a graph that is symmetric about the y-axis is called an **even function**. A function with a graph that is symmetric about the origin is called an **odd function**.

Even and Odd Functions

A function f with the property that $f(-x) = f(x)$ for all x in the domain of f is called an **even function**.

A function f with the property that $f(-x) = -f(x)$ for all x in the domain of f is called an **odd function**.

Since the graph in Example 1 is symmetric about the origin, it represents an odd function. Since the graph in Example 2 is symmetric about the y-axis, it represents an even function.

EXAMPLE 3 Decide whether the functions are even or odd: **a.** $y = f(x) = x^2$ and **b.** $y = f(x) = x^3$.

Solution **a.** We find $f(-x)$ and compare it to $f(x)$:

$$y = f(-x) = (-x)^2 = x^2 \qquad \text{which is equivalent to } y = f(x) = x^2$$

Since $f(-x) = f(x)$, the function is an even function.

b. We find $f(-x)$ and compare it to $f(x)$:

$$y = f(-x) = (-x)^3 = -x^3 \qquad \text{which is not equivalent to } y = f(x) = x^3$$

Since $f(-x) \neq f(x)$, the function is not an even function.

To determine whether the function is an odd function, we find $f(-x)$ and compare it to $-f(x)$:

$$y = f(-x) = (-x)^3 = -x^3 \qquad \text{which is equivalent to } -f(x) = -x^3$$

Since $f(-x) = -f(x)$, the function is an odd function. ■

Increasing and Decreasing Functions

If the values $f(x)$ increase as x increases on an interval, we say that the function is **increasing on the interval**. See Figure 4-13(a). If the values $f(x)$ decrease as x increases on an interval, we say that the function is **decreasing on the interval**. See Figure 4-13(b). If the values $f(x)$ remain unchanged as x increases on an interval, we say that the function is **constant on the interval**. See Figure 4-13(c).

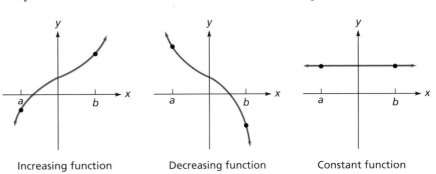

Increasing function   Decreasing function   Constant function
    (a)   (b)   (c)

FIGURE **4-13**

Piecewise-Defined Functions

Some functions, called **piecewise-defined functions**, are defined by using different equations for different intervals in their domains. To illustrate, we will graph the piecewise-defined function f given by

$$y = f(x) = \begin{cases} -2 & \text{if } x \le 0 \\ x + 1 & \text{if } x > 0 \end{cases}$$

For each number x, we decide which part of the definition to use. If $x \le 0$, the corresponding value of y is -2. In the interval $(-\infty, 0]$, the function is constant.

If $x > 0$, the corresponding value of y is $x + 1$. In the interval $(0, \infty)$, the function is increasing. The graph appears in Figure 4-14.

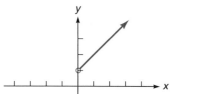

$$y = f(x) = \begin{cases} -2 \text{ if } x \le 0 \\ x + 1 \text{ if } x > 0 \end{cases}$$

FIGURE 4-14

EXAMPLE 4 Graph the function defined by $y = f(x) = \begin{cases} -x & \text{if } x < 0 \\ x^2 & \text{if } 0 \le x \le 1. \\ 1 & \text{if } x > 1 \end{cases}$

Solution For each number x, we decide which part of the definition to use. If $x < 0$, the corresponding value of y is determined by the equation $y = -x$. In the interval $(-\infty, 0)$, the function is decreasing.

If $0 \le x \le 1$, the corresponding value of y is x^2. In the interval $(0, 1)$, the function is increasing.

If $x > 1$, the corresponding value of y is the constant 1. In the interval $(1, \infty)$, the function is constant.

The graph appears in Figure 4-15.

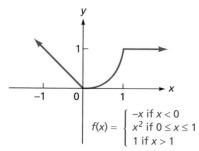

$$f(x) = \begin{cases} -x \text{ if } x < 0 \\ x^2 \text{ if } 0 \le x \le 1 \\ 1 \text{ if } x > 1 \end{cases}$$

FIGURE 4-15

The Greatest Integer Function

Another type of function, called the **greatest integer function**, is important in computer applications and others. This function is determined by the equation $y = [\![x]\!]$, where the value of y that corresponds to x is the greatest integer that is less than or equal to x. For example,

$$[\![2.71]\!] = 2, \quad \left[\!\left[23\tfrac{1}{2}\right]\!\right] = 23, \quad [\![10]\!] = 10, \quad [\![\pi]\!] = 3 \quad [\![-2.5]\!] = -3$$

EXAMPLE 5 Graph $y = [\![x]\!]$.

Solution We list several intervals and the corresponding values of the greatest integer function:

$$[0, 1) \qquad y = [\![x]\!] = 0 \qquad \text{For numbers from 0 to 1, not including 1,}$$
the greatest integer in the interval is 0.

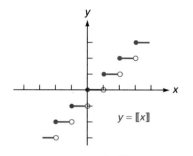

FIGURE 4-16

[1, 2) $y = [\![x]\!] = 1$ For numbers from 1 to 2, not including 2, the greatest integer in the interval is 1.

[2, 3) $y = [\![x]\!] = 2$ For numbers from 2 to 3, not including 3, the greatest integer in the interval is 2.

Within each interval, the values of y are constant, but they jump by 1 at integer values of x. The graph is shown in Figure 4-16. From the graph, we can see that the domain of the greatest integer function is the interval $(-\infty, \infty)$. The range is the set of integers, $\{\ldots, -3, -2, -1, 0, 1, 2, 3, \ldots\}$. ∎

Since the greatest integer function is made up of a series of horizontal line segments, it is an example of a group of functions that are called **step functions**.

EXAMPLE 6   To print a business form, a printer charges $10 for the order, plus $20 for each box containing 500 forms. The printer counts any portion of a box as a full box. Graph this step function.

Solution   If we order the forms and then change our mind before the forms are printed, the cost will be $10. Thus, the ordered pair (0, 10) will be on the graph. If we purchase one full box, the cost will be $10 for the order and $20 for the printing, for a total of $30. Thus, the ordered pair (1, 30) will be on the graph. The cost for $1\frac{1}{2}$ boxes will be the same as the cost for 2 full boxes, or $50. Thus, the ordered pairs (1.5, 50) and (2, 50) are on the graph.

The complete graph is shown in Figure 4-17.

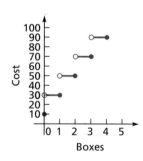

FIGURE 4-17   ∎

4.3   EXERCISES

In Exercises 1–8, graph each polynomial function.

1. $y = x^3$ **2.** $y = x^4$ **3.** $y = x^3 + x^2$ **4.** $y = x^3 - x$

5. $y = -x^3$ **6.** $y = -x^3 + 1$ **7.** $y = x^4 - 2x^2 + 1$ **8.** $y = x^4 - 5x^2 + 4$

In Exercises 9–16, tell whether each function is even or odd. If it is neither, so indicate.

9. $y = x^4 + x^2$ **10.** $y = x^3 - 2x$ **11.** $y = x^3 + x^2$ **12.** $y = x^6 - x^2$

13. $y = x^5 + x^3$ **14.** $y = x^3 - x^2$ **15.** $y = 2x^3 - 3x$ **16.** $y = 4x^2 - 5$

In Exercises 17–22, tell where each function is increasing, decreasing, or constant.

17.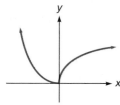

18.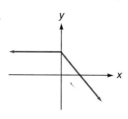

19.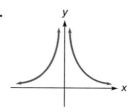

20.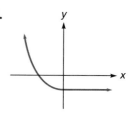

21. $y = f(x) = x^2 - 4x + 4$

22. $y = f(x) = 4 - x^2$

In Exercises 23–28, graph each piecewise-defined function.

23. $y = f(x) = \begin{cases} x + 2 & \text{if } x < 0 \\ 2 & \text{if } x \geq 0 \end{cases}$

24. $y = f(x) = \begin{cases} 2x & \text{if } x < 0 \\ -2x & \text{if } x \geq 0 \end{cases}$

25. $y = f(x) = \begin{cases} -x & \text{if } x < 0 \\ x^2 & \text{if } x \geq 0 \end{cases}$

26. $y = f(x) = \begin{cases} |x| & \text{if } x < 0 \\ \sqrt{x} & \text{if } x \geq 0 \end{cases}$

27. $y = f(x) = \begin{cases} 0 & \text{if } x < 0 \\ x^2 & \text{if } 0 \leq x \leq 2 \\ 4 - 2x & \text{if } x > 2 \end{cases}$

28. $y = f(x) = \begin{cases} 2 & \text{if } x < 0 \\ 2 - x & \text{if } 0 \leq x < 2 \\ x & \text{if } x \geq 2 \end{cases}$

In Exercises 29–32, graph each function.

29. $y = [\![2x]\!]$

30. $y = \left[\!\left[\dfrac{1}{3}x + 3\right]\!\right]$

31. $y = [\![x]\!] - 1$

32. $y = [\![x + 2]\!]$

33. *Renting a car*   A rental company charges $20 to rent a car for one day, plus $2 for every 100 miles, or portion of 100 miles, that it is driven. Graph the ordered pairs (m, c), where m represents the miles driven and c represents the cost. Find the cost if the car is driven 275 miles.

34. *Riding in a taxi*   A taxi cab company charges $3 for a trip up to 1 mile, and $2 for every extra mile (or portion of a mile). Graph the ordered pairs (m, c), where m represents the miles traveled and c represents the cost. Find the cost to ride $10\frac{1}{4}$ miles.

35. *Computer communications*   A computer information network charges for connect time at a rate of $12 per hour, computed for every minute or fraction of a minute. Graph the points (t, c), where c is the cost of t minutes of connect time. Find the cost of $7\frac{1}{2}$ minutes.

36. *Rounding numbers*   Measurements are rarely exact. They are often *rounded* to an appropriate precision. Graph the points (x, y), where y is the result of rounding the number x to the nearest ten.

37. *Signum function*   Computer programmers often use the following function, denoted by $y = \text{sgn } x$. Graph this function and find its domain and range.

$$y = \begin{cases} -1 & \text{if } x < 0 \\ 0 & \text{if } x = 0 \\ 1 & \text{if } x > 0 \end{cases}$$

38. Graph the function defined by $y = \frac{|x|}{x}$ and compare it to the graph in Exercise 37. Are the graphs the same?

In Exercises 39–42, use a graphing calculator to explore the properties of graphs of polynomial functions, and write a brief paragraph summarizing your observations.

39. Graph the function $y = x^2 + ax$ for several values of a. How does the graph change?

40. Graph the function $y = x^3 + ax$ for several values of a. How does the graph change?

41. Graph the function $y = (x - a)(x - b)$ for several values of a and b. What is the relationship between the x-intercepts and the equation?

42. Use the insight you gained in Exercise 41 to factor $x^3 - 3x^2 - 4x + 12$.

4.4 TRANSLATING AND STRETCHING GRAPHS

■ Vertical and Horizontal Translations   ■ Vertical and Horizontal Stretchings

■ Vertical and Horizontal Translations

The graphs of different equations may be identical except for their position in the xy-plane. For example, Figure 4-18(a) shows the graph of the function $y = x^2 + k$ for three different values of k. The graph of $y = x^2 + 2$ is identical to the graph of $y = x^2$ except that it is shifted 2 units upward. Similarly, the graph of $y = x^2 - 3$ is identical to the graph of $y = x^2$ except that it is shifted 3 units downward. Such shifts are called **vertical translations**.

Figure 4-18(b) shows the graph of the equation $y = (x + h)^2$ for three different values of h. The graph of $y = (x - 2)^2$ is identical to the graph of $y = x^2$ except that it is shifted 2 units to the right. The graph of $y = (x + 3)^2$ is identical to the graph of $y = x^2$ except that it is shifted 3 units to the left. Such shifts are called **horizontal translations**.

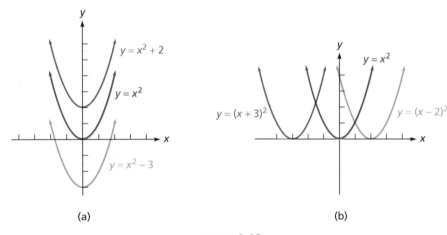

(a)   (b)

FIGURE 4-18

In general, we can make the following observations.

Vertical Translations	If f is a function and k is a positive number, then
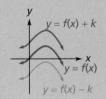	• The graph of $y = f(x) + k$ is identical to the graph of $y = f(x)$ except that it is translated k units upward. • The graph of $y = f(x) - k$ is identical to the graph of $y = f(x)$ except that it is translated k units downward.

Horizontal Translations

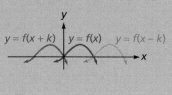

If f is a function and k is a positive number, then

• The graph of $y = f(x - k)$ is identical to the graph of $y = f(x)$ except that it is translated k units to the right.

• The graph of $y = f(x + k)$ is identical to the graph of $y = f(x)$ except that it is translated k units to the left.

EXAMPLE 1   Graph the function $y = (x - 5)^3 + 4$.

Solution   The graph of $y = (x - 5)^3 + 4$ is identical to the graph of $y = (x - 5)^3$ except that it has been translated 4 units upward. The graph of $y = (x - 5)^3$ is identical to the graph of $y = x^3$ except that it has been translated 5 units to the right. Thus, to graph $y = (x - 5)^3 + 4$, graph $y = x^3$ and translate it 4 units upward and 5 units to the right. (See Figure 4-19.)

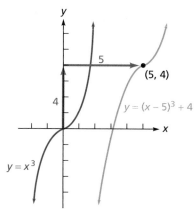

FIGURE **4-19**

EXAMPLE 2   Use a graphing calculator to show the effect of vertical and horizontal translations by graphing $y = x^2$ and $y = (x - 3)^2 + 2$.

Solution   The graph of $y = x^2$ is a parabola opening upward with vertex at the origin. The graph of $y = (x - 3)^2 + 2$ should be that same parabola translated 3 units to the right and 2 units upward. Figure 4-20 shows the result of plotting these equations on a graphing calculator. The TRACE function shows (within the accuracy of the graphing process) that the vertex of the translated graph is the point (3, 2), as expected.

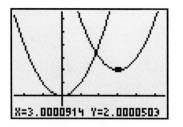

FIGURE **4-20**

Vertical and Horizontal Stretchings

Figure 4-21(a) shows the graph of $y = f(x) = x^2 - 1$ with intercepts of 1 and -1. Because each y-value on the graph of $y = 3f(x) = 3(x^2 - 1)$ is 3 times the y-value on the graph of $y = x^2 - 1$, the graph of $y = x^2 - 1$ has been stretched vertically by a factor of 3 to become the graph of $y = 3(x^2 - 1)$. This stretching does not change the x-intercepts of the graph.

Figure 4-21(b) shows the graphs of $y = f(x) = x^2 - 1$ and $y = f(3x) = (3x)^2 - 1$. The graph of $y = x^2 - 1$ has been stretched horizontally by a factor of $\frac{1}{3}$ to become the graph of $y = (3x)^2 - 1$. This stretching does not change the y-intercept of the graph.

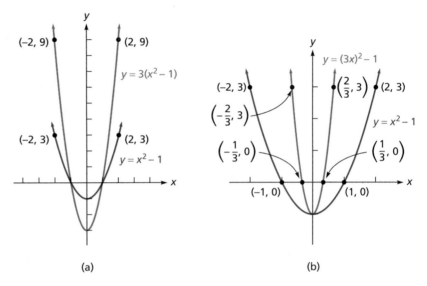

(a)   (b)

FIGURE 4-21

The effect of vertically stretching a graph by a factor of -1 is to reflect the graph in the x-axis. In Figure 4-22(a), the graph $y = -f(x) = -x^2$ can be obtained by reflecting the graph of $y = f(x) = x^2$ in the x-axis. The effect of horizontally

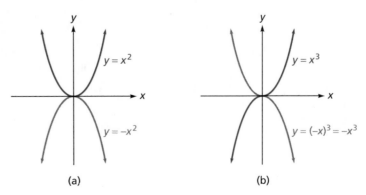

(a)   (b)

FIGURE 4-22

stretching a graph by a factor of -1 is to reflect the graph in the y-axis. In Figure 4-22(b), the graph of $y = f(-x) = (-x)^3$ can be obtained by reflecting the graph of $y = f(x) = x^3$ in the y-axis. Because $(-x)^3 = -x^3$, the reflection of the graph of $y = x^3$ in the y-axis is also the reflection in the x-axis.

EXAMPLE 3   Use a graphing calculator to show the effect of a vertical stretching and reflection by graphing $y = f(x) = x^2 - 4$ and $y = -2f(x) = -2(x^2 - 4)$.

Solution   The graph of $y = x^2 - 4$ is a parabola opening upward with vertex at $(0, -4)$. If that graph is stretched vertically by a factor of 2 and then reflected in the x-axis, the graph of $y = -2(x^2 - 4)$ should result. That is what happens, as Figure 4-23 shows. Notice that the x-intercepts of the graphs are the same.

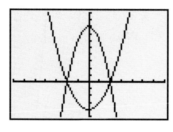

FIGURE **4-23**   ■

In general, we can make the following observations.

Vertical Stretching

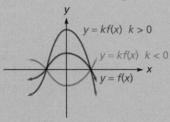

- If $k > 0$, the graph of $y = kf(x)$ can be obtained by stretching the graph of $y = f(x)$ vertically by a factor of k.

- If $k < 0$, the graph of $y = kf(x)$ can be obtained by stretching the reflection of the graph of $y = f(x)$ in the x-axis by a factor of $|k|$.

Horizontal Stretching

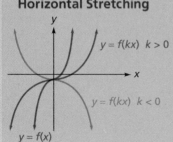

- If $k > 0$, the graph of $y = f(kx)$ can be obtained by stretching the graph of $y = f(x)$ horizontally by a factor of $\frac{1}{k}$.

- If $k < 0$, the graph of $y = f(kx)$ can be obtained by stretching the reflection of $y = f(x)$ in the y-axis by a factor of $\left|\frac{1}{k}\right|$.

4.4   EXERCISES

In Exercises 1–8, the graph of each equation is a translation of the graph of $y = f(x) = x^2$. Graph each equation.

1. $y = x^2 - 2$

2. $y = (x - 2)^2$

3. $y = (x + 3)^2$

4. $y = x^2 + 3$

5. $y = (x + 1)^2 + 2$

6. $y = (x - 3)^2 - 1$

7. $y = \left(x + \dfrac{1}{2}\right)^2 - \dfrac{1}{2}$

8. $y = \left(x - \dfrac{3}{2}\right)^2 + \dfrac{5}{2}$

In Exercises 9–16, the graph of each equation is a translation of the graph of $y = f(x) = x^3$. Graph each equation.

9. $y = x^3 + 1$

10. $y = x^3 - 3$

11. $y = (x - 2)^3$

12. $y = (x + 3)^3$

13. $y = (x - 2)^3 - 3$

14. $y = (x + 1)^3 + 4$

15. $y + 2 = x^3$

16. $y - 7 = (x - 5)^3$

In Exercises 17–20, the graph of each equation is a stretching of the graph of $y = f(x) = x^2$. Graph each equation.

17. $y = 2x^2$

18. $y = \dfrac{1}{2}x^2$

19. $y = -3x^2$

20. $y = -\dfrac{1}{3}x^2$

In Exercises 21–24, the graph of each equation is a stretching of the graph of $y = f(x) = x^3$. Graph each equation.

21. $y = \left(\dfrac{1}{2}x\right)^3$

22. $y = \dfrac{1}{8}x^3$

23. $y = -8x^3$

24. $y = (-2x)^3$

In Exercises 25–34, graph each equation.

25. $y = |x - 2| + 1$

26. $y = |x + 5| - 2$

27. $y = \sqrt{x - 2} + 1$ $(x \geq 2)$

28. $y = \sqrt{x + 5} - 2$ $(x \geq -5)$

29. $y = |3x|$

30. $y = 3|x|$

31. $y = 2\sqrt{x} + 3$ $(x \geq 0)$

32. $y = 2\sqrt{x + 3}$ $(x \geq -3)$

33. $y = -2|x| + 3$

34. $y = -2|x + 3|$

In Exercises 35–38, use a graphing calculator to perform each experiment. Write a brief paragraph describing your findings.

35. Investigate the translations of the graph of a function by graphing the parabola $y = (x - k)^2 + k$ for several values of k. What do you observe about successive positions of the vertex?

36. Investigate the translations of the graph of a function by graphing the parabola $y = (x - k)^2 + k^2$ for several values of k. What do you observe about successive positions of the vertex?

37. Investigate the horizontal stretching of the graph of a function by graphing $y = \sqrt{ax}$ for several values of a. What do you observe?

38. Investigate the vertical stretching of the graph of a function by graphing $y = b\sqrt{x}$ for several values of b. What do you observe? Are these graphs different than the graphs in Exercise 37?

4.5   RATIONAL FUNCTIONS

■ Horizontal and Vertical Asymptotes   ■ Graphing Rational Functions
■ Graphs with Discontinuities

Certain functions, called **rational functions**, are defined by equations of the form

$$y = \frac{P(x)}{Q(x)}$$

where $P(x)$ and $Q(x)$ are polynomials. Because $Q(x)$ appears in the denominator of a fraction, it cannot equal 0. Thus, the domain of the rational function must exclude all values of x for which $Q(x) = 0$.

Here are some examples of rational functions and their domains:

$$y = f(x) = \frac{3}{x - 2}$$ Domain: $\{x \mid x \text{ is a real number and } x \neq 2\}$

$$y = f(x) = \frac{5x + 2}{x^2 - 4}$$ Domain: $\{x \mid x \text{ is a real number and } x \neq 2 \text{ and } x \neq -2\}$

$$y = f(x) = \frac{3x^2 - 2}{(2x + 1)(x - 3)}$$ Domain: $\left\{x \mid x \text{ is a real number and } x \neq -\frac{1}{2} \text{ and } x \neq 3\right\}$

■ Horizontal and Vertical Asymptotes

It is useful to study the behavior of rational functions near values that are not in their domains. For example, the domain of the rational function $y = f(x) = \frac{2}{x}$ is the set of all real numbers except 0. To discover the behavior of its graph near $x = 0$, we construct the following tables. Table 1 shows x approaching 0 from the left, and Table 2 shows x approaching 0 from the right.

TABLE 1

x	-1	-0.1	-0.01	-0.001
$f(x)$	-2	-20	-200	-2000

TABLE 2

x	1	0.1	0.01	0.001
$f(x)$	2	20	200	2000

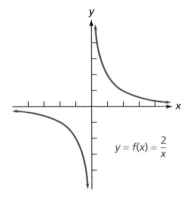

$$y = f(x) = \frac{2}{x}$$

FIGURE 4-24

Table 1 illustrates that as x approaches 0 from the left, the values of $f(x)$ decrease without bound. Table 2 shows that as x approaches 0 from the right, the values of $f(x)$ increase without bound. These results are shown in the graph of Figure 4-24. The rational function $y = \frac{2}{x}$ behaves in this manner because as x approaches 0, $(x \to 0)$, the denominator of the fraction approaches 0, but the numerator does not.

The graph in the figure also shows the behavior of the rational function $y = f(x) = \frac{2}{x}$ as x approaches ∞ ($x \to \infty$) and as x approaches $-\infty$ ($x \to -\infty$).

From the figure, we see that as x gets closer and closer to 0, the graph of $y = f(x) = \frac{2}{x}$ gets closer and closer to the y-axis. In this case, the y-axis is called a **vertical asymptote**.

Furthermore, if $x \to \infty$ or $x \to -\infty$, the graph approaches the x-axis. Here, the x-axis is called a **horizontal asymptote**.

More formally, we make the following definition.

Vertical and Horizontal Asymptotes	The line $x = a$ is a **vertical asymptote** for the graph of a function f when $$f(x) \to \infty \qquad \text{or} \qquad f(x) \to -\infty$$ as $x \to a$ from either the left or the right. The line $y = b$ is a **horizontal asymptote** for the graph of a function f when $$f(x) \to b$$ as $x \to \infty$ or as $x \to -\infty$.

Figure 4-25 shows several graphs and their asymptotes.

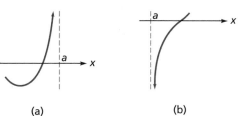

Vertical asymptote at $x = a$

$f(x) \to \infty$ as $x \to a$ from the left

$f(x) \to -\infty$ as $x \to a$ from the right

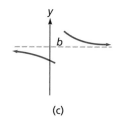

Horizontal asymptote at $y = b$

$f(x) \to b$ as $x \to \infty$ or $x \to -\infty$

(a)   (b)   (c)

FIGURE 4-25

■ Graphing Rational Functions

EXAMPLE 1   Graph the rational function defined by $y = f(x) = \dfrac{x^2 - 4}{x^2 - 1}$.

Solution   We first check for symmetries. Because x appears to even powers only, $y = f(x) = \dfrac{x^2 - 4}{x^2 - 1}$ is equivalent to $y = f(-x) = \dfrac{(-x)^2 - 4}{(-x)^2 - 1}$, and the graph will be symmetric about the y-axis. There is no symmetry about the origin.

To find the numbers excluded from the domain of this function, we set its denominator equal to 0, solve for x, and find the domain to be the set of all real numbers x except $x = 1$ and $x = -1$.

$$x^2 - 1 = 0$$
$$(x + 1)(x - 1) = 0$$

$$x + 1 = 0 \qquad \text{or} \qquad x - 1 = 0$$
$$x = -1 \qquad \qquad x = 1$$

As $x \to -1$ or as $x \to 1$, the denominator of the fraction $\dfrac{x^2 - 4}{x^2 - 1}$ approaches 0, but the numerator does not. Thus, $y \to \infty$, and the lines $x = 1$ and $x = -1$ are vertical asymptotes of the graph.

We can find the y-intercept of the graph by setting x equal to 0 and solving for y:

$$y = \frac{x^2 - 4}{x^2 - 1} = \frac{0^2 - 4}{0^2 - 1} = \frac{-4}{-1} = 4$$

The y-intercept is 4.

We can find the x-intercepts of the graph by setting the numerator of the rational expression equal to 0 and solving for x:

$$x^2 - 4 = 0$$
$$(x + 2)(x - 2) = 0$$

$$x + 2 = 0 \qquad \text{or} \qquad x - 2 = 0$$
$$x = -2 \qquad \qquad x = 2$$

Thus, the x-intercepts of the curve are 2 and -2.

To find any horizontal asymptotes, we divide $x^2 - 4$ by $x^2 - 1$ and write the answer in quotient $+ \frac{\text{remainder}}{\text{divisor}}$ form:

$$
\begin{array}{r}
1 \\
x^2 - 1 \overline{\smash{)}\, x^2 - 4} \\
\underline{x^2 - 1} \\
-3
\end{array}
$$

Hence,

$$y = \frac{x^2 - 4}{x^2 - 1} = 1 + \frac{-3}{x^2 - 1}$$

If $x \to \infty$, the denominator $(x^2 - 1) \to \infty$ also, and the value of the fraction $\dfrac{-3}{x^2 - 1}$ approaches 0. For large values of x, the fraction is negative, and the corresponding values of y are all less than 1. Because the fraction approaches 0 as $x \to \infty$, then $y \to 1$. Therefore, as $x \to \infty$, the graph of the function

$$y = \frac{x^2 - 4}{x^2 - 1}$$

approaches the line $y = 1$ from below. Because of symmetry, the curve also approaches the line $y = 1$ as $x \to -\infty$. The graph of the rational function is shown in Figure 4-26.   ∎

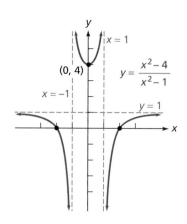

FIGURE 4-26

EXAMPLE 2 Graph the rational function defined by $y = f(x) = \dfrac{3x}{x - 2}$.

Solution We first look for symmetries. Because x appears to an odd power, the function is not symmetric about the y-axis. There is no symmetry about the origin either.

- To locate the vertical asymptotes, we find the numbers excluded from the domain by setting the denominator equal to 0 and solving for x. We find the domain to be the set of all real numbers except $x = 2$. The line $x = 2$ is a vertical asymptote.
- To find the y-intercept, we let $x = 0$, and find the y-intercept to be 0. Because $y = 0$ when $x = 0$, the graph passes through the origin.
- To find the x-intercepts, we set the numerator equal to 0 and solve for x:

$$3x = 0$$
$$x = 0$$

Thus, the only x-intercept is 0.

- We can find the horizontal asymptotes, if any, by dividing $3x$ by $x - 2$ and expressing the answer in quotient $+ \frac{\text{remainder}}{\text{divisor}}$ form.

$$y = \frac{3x}{x - 2} = 3 + \frac{6}{x - 2}$$

As $x \to \infty$, the fraction $\frac{6}{x - 2} \to 0$, so the graph approaches the line $y = 3$. This line is a horizontal asymptote.

To find what happens to the graph when x is greater than 2, we pick a value of x such that $x > 2$ and find the corresponding value of y. If $x = 3$, for example, $y = 9$. After plotting the point $(3, 9)$, we can use the intercepts and asymptotes to sketch the graph shown in Figure 4-27. ∎

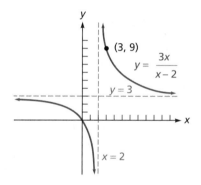

FIGURE 4-27

EXAMPLE 3 Graph the rational function defined by $y = f(x) = \dfrac{1}{x(x - 1)^2}$.

Solution The expansion of $x(x - 1)^2$ gives a third-degree polynomial in which x appears to be an odd power. Thus, there is no symmetry about the y-axis. Because $y = f(x)$ and $-y = f(-x)$ are not equivalent, there is no symmetry about the origin either.

- To find any vertical asymptotes, we note that 0 and 1 make the denominator 0. Thus, the vertical asymptotes of the graph are the lines $x = 0$ and $x = 1$.
- Since x cannot be 0, the graph has no y-intercept, and since y cannot be 0, the graph has no x-intercept.
- As $x \to \infty$, the denominator of the fraction becomes large, and the corresponding values of y approach 0. Thus, the line $y = 0$ is a horizontal asymptote.

The table in Figure 4-28 lists the coordinates of several points lying in regions between the asymptotes. The values of y change sign at $x = 0$: To the left of the y-axis, $y < 0$, and to the right of the y-axis, $y > 0$. This happens because x appears to an odd power as a factor in the denominator. The value of y does not change sign at $x = 1$: To the left and to the right of the asymptote $x = 1$, y is positive. The function behaves this way because $(x - 1)$ appears to an even power in the denominator. The graph of the rational function appears in Figure 4-28.

$$y = \frac{1}{x(x-1)^2}$$

x	y
-1	$-\frac{1}{4}$
$\frac{1}{2}$	8
2	$\frac{1}{2}$

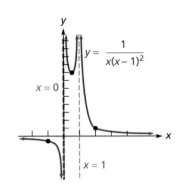

FIGURE 4-28

EXAMPLE 4 Graph the rational function $y = f(x) = \dfrac{1}{x^2 + 1}$.

Solution Because x appears only to an even power, the graph will be symmetric about the y-axis.

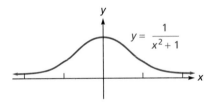

- Since no numbers x will make the denominator 0, the graph has no vertical asymptotes.
- Since $f(0) = 1$, the y-intercept of the graph is 1. Since y cannot be 0, there is no x-intercept.
- Since $f(x) \to 0$ as $x \to \infty$, the graph has $y = 0$ as a horizontal asymptote. However, because the denominator is always positive, the graph lies entirely above the x-axis, as shown in Figure 4-29.

FIGURE 4-29

EXAMPLE 5 Graph the rational function defined by $y = f(x) = \dfrac{x^2 + x - 2}{x - 3}$.

Solution We first factor the numerator of the rational expression

$$y = \frac{(x - 1)(x + 2)}{x - 3}$$

Most of the information is straightforward:

- The function is not symmetric about the y-axis or the origin.
- A vertical asymptote is $x = 3$.
- The y-intercept is $\frac{2}{3}$.
- The x-intercepts are 1 and -2.
- We perform the long division and write the rational expression as

$$y = \frac{x^2 + x - 2}{x - 3} = x + 4 + \frac{10}{x - 3}$$

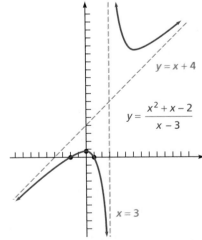

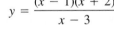

As before, the fraction $\frac{10}{x - 3} \to 0$ as $x \to \infty$. This time, however, the graph does not approach a constant. Instead, it approaches the line $y = x + 4$. Because this line is not horizontal, the graph has no horizontal asymptote. However, it does have a **slant asymptote** or an **oblique asymptote**: the line $y = x + 4$. Putting this information together and graphing the rational function, we obtain the graph shown in Figure 4-30.

FIGURE 4-30

EXAMPLE 6   Discuss the nature of the asymptotes of the following rational functions:

a. $y = \dfrac{x + 2}{x^2 - 1}$ **b.** $y = \dfrac{x^2 + x + 2}{x^2 - 1}$

c. $y = \dfrac{3x^3 + 2x^2 + 2}{x^2 - 1}$ **d.** $y = \dfrac{x^4 + x + 2}{x^2 - 1}$

Solution   All four functions have vertical asymptotes at $x = 1$ and $x = -1$ because at these values, their denominators are 0. The nature of the remaining asymptotes must be considered.

a. Because the degree of the numerator is less than the degree of the denominator, we can find the horizontal asymptote by dividing both the numerator and the denominator by x^2, which is the largest power of x in the denominator.

$$y = \frac{x + 2}{x^2 - 1} = \frac{\dfrac{x}{x^2} + \dfrac{2}{x^2}}{\dfrac{x^2}{x^2} - \dfrac{1}{x^2}} = \frac{\dfrac{1}{x} + \dfrac{2}{x^2}}{1 - \dfrac{1}{x^2}}$$

The three fractions in the final result, $\dfrac{1}{x}$, $\dfrac{2}{x^2}$, and $\dfrac{1}{x^2}$ all approach 0 as $x \to \infty$. Thus, y approaches

$$\frac{0 + 0}{1 - 0} = 0$$

The graph of this function has a horizontal asymptote of $y = 0$.

In the next three parts, we can perform a long division.

b. $y = \dfrac{x^2 + x + 2}{x^2 - 1} = 1 + \dfrac{x + 3}{x^2 - 1}$

As $x \to \infty$, the fraction $\dfrac{x + 3}{x^2 - 1} \to 0$ (for reasons discussed in part **a**), and $y \to 1$. This graph has a horizontal asymptote of $y = 1$.

c. $y = \dfrac{3x^3 + 2x^2 + 2}{x^2 - 1} = 3x + 2 + \dfrac{3x + 4}{x^2 - 1}$

Here again the last fraction approaches 0 as $x \to \infty$, and the graph approaches the slant asymptote $y = 3x + 2$.

d. $y = \dfrac{x^4 + x + 2}{x^2 - 1} = x^2 + 1 + \dfrac{x + 3}{x^2 - 1}$

As $x \to \infty$, the fractional part again approaches 0, and the curve approaches $y = x^2 + 1$. However, $y = x^2 + 1$ does not represent a line. The graph of this function has no horizontal or slant asymptotes.   ∎

We generalize the results of Example 6 and summarize the techniques discussed in this section as follows.

Steps for Graphing Rational Functions	Perform the following steps to graph the rational function $y = \dfrac{P(x)}{Q(x)}$ where $\dfrac{P(x)}{Q(x)}$ is in simplest form.

Check for symmetry.   If the polynomials $P(x)$ and $Q(x)$ involve only even powers of x, the graph is symmetric about the y-axis. Otherwise, y-axis symmetry does not exist. Check for symmetry about the origin.

Look for vertical asymptotes.   The real roots of $Q(x) = 0$, if any, determine the vertical asymptotes of the graph.

Look for y-intercepts.   Let $x = 0$. The resulting value of y, if any, is the y-intercept of the graph.

Look for x-intercepts.   The real roots of $P(x) = 0$, if any, are the x-intercepts of the graph.

Look for horizontal asymptotes.

- If the degree of $P(x)$ is less than the degree of $Q(x)$, the line $y = 0$ is a horizontal asymptote.

- If the degrees of $P(x)$ and $Q(x)$ are equal, the line $y = \dfrac{p}{q}$, where p and q are the lead coefficients of $P(x)$ and $Q(x)$, is a horizontal asymptote.

   Warning!   Be sure that $P(x)$ and $Q(x)$ are written in descending powers of x before applying this rule.

- If the degree of $P(x)$ is greater than the degree of $Q(x)$, there is no horizontal asymptote.

   Warning!   A graph might intersect a horizontal asymptote when $|x|$ is small. However, it will approach but never touch a horizontal asymptote as $x \to \infty$ or as $x \to -\infty$.

Look for slant asymptotes.   If the degree of $P(x)$ is one greater than the degree of $Q(x)$, there is a slant asymptote. To find it, divide $P(x)$ by $Q(x)$ and ignore the remainder.

■ Graphs with Discontinuities

We have discussed rational functions where the fraction is in simplified form. We now consider a rational function $y = \dfrac{P(x)}{Q(x)}$ where $P(x)$ and $Q(x)$ have a common factor. Graphs of such functions contain gaps or **discontinuities** that are not the result of vertical asymptotes.

EXAMPLE 7   Find the domain of the rational function defined by $y = f(x) = \dfrac{x^2 - x - 12}{x - 4}$ and graph the function.

Solution   Because a denominator cannot be 0, $x \neq 4$. Thus, the domain of the function is the set of all real numbers except 4.

When we factor the numerator of the rational expression, we see that its numerator and denominator have a common factor of $x - 4$.

$$y = \frac{x^2 - x - 12}{x - 4}$$

$$= \frac{(x + 3)(x - 4)}{x - 4}$$

For $x \neq 4$, the common factor of $x - 4$ can be divided out. The resulting function is equivalent to the original function only when we keep the restriction that $x \neq 4$. Thus,

$$y = \frac{(x + 3)(x - 4)}{x - 4} = x + 3 \qquad \text{provided that } x \neq 4$$

When $x = 4$, the function is not defined. The graph of the given rational function appears in Figure 4-31. It is a line with the point with x-coordinate of 4 missing. ∎

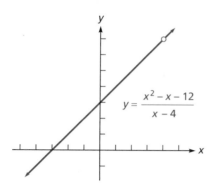

$y = \dfrac{x^2 - x - 12}{x - 4}$

FIGURE 4-31

EXAMPLE 8   Use a graphing calculator to graph the rational function $y = \dfrac{x^2 - 4}{x^2 + 2x - 35}$.

Solution   We begin by setting a viewing window of $-10 \leq x \leq 10$ and $-10 \leq y \leq 10$, and enter and graph the function. The result is shown in Figure 4-32.

The calculator graphs by drawing short lines between points with slightly different x-coordinates. When two such points straddle a vertical asymptote and their y-coordinates are far apart, the graphing calculator draws a line between them anyway, producing the vertical asymptotes that appear in the figure. Graphing calculators do not show all asymptotes, however. For example, the horizontal asymptote $y = 1$ is not displayed. ∎

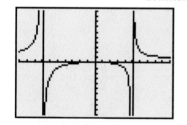

FIGURE 4-32

4.5   EXERCISES

In Exercises 1–8, find the domain of each rational function. **Do not graph the function.**

1. $y = \dfrac{x^2}{x - 2}$

2. $y = \dfrac{x^3 - 3x^2 + 1}{x + 3}$

3. $y = \dfrac{2x^2 + 7x - 2}{x^2 - 25}$

4. $y = \dfrac{5x^2 + 1}{x^2 + 5}$

5. $y = \dfrac{x - 1}{x^3 - x}$

6. $y = \dfrac{x + 2}{2x^2 - 9x + 9}$

7. $y = \dfrac{3x^2 + 5}{x^2 + 1}$

8. $y = \dfrac{7x^2 - x + 2}{x^4 + 4}$

In Exercises 9–36, find all vertical, horizontal, and slant asymptotes, x- and y-intercepts, and symmetries, and then graph each function.

9. $y = \dfrac{1}{x - 2}$

10. $y = \dfrac{3}{x + 3}$

11. $y = \dfrac{x}{x - 1}$

12. $y = \dfrac{x}{x + 2}$

13. $y = \dfrac{x + 1}{x + 2}$

14. $y = \dfrac{x - 1}{x - 2}$

15. $y = \dfrac{2x - 1}{x - 1}$

16. $y = \dfrac{3x + 2}{x^2 - 4}$

17. $y = \dfrac{x^2 - 9}{x^2 - 4}$

18. $y = \dfrac{x^2 - 4}{x^2 - 9}$

19. $y = \dfrac{x^2 - x - 2}{x^2 - 4x + 3}$

20. $y = \dfrac{x^2 + 7x + 12}{x^2 - 7x + 12}$

21. $y = \dfrac{x^2 + 2x - 3}{x^3 - 4x}$

22. $y = \dfrac{3x^2 - 4x + 1}{2x^3 + 3x^2 + x}$

23. $y = \dfrac{x^2 - 9}{x^2}$

24. $y = \dfrac{3x^2 - 12}{x^2}$

25. $y = \dfrac{x}{(x + 3)^2}$

26. $y = \dfrac{x}{(x - 1)^2}$

27. $y = \dfrac{x + 1}{x^2(x - 2)}$

28. $y = \dfrac{x - 1}{x^2(x + 2)^2}$

29. $y = \dfrac{x}{x^2 + 1}$

30. $y = \dfrac{x - 1}{x^2 + 2}$

31. $y = \dfrac{3x^2}{x^2 + 1}$

32. $y = \dfrac{x^2 - 9}{2x^2 + 1}$

33. $y = \dfrac{x^2 - 2x - 8}{x - 1}$

34. $y = \dfrac{x^2 + x - 6}{x + 2}$

35. $y = \dfrac{x^3 + x^2 + 6x}{x^2 - 1}$

36. $y = \dfrac{x^3 - 2x^2 + x}{x^2 - 4}$

In Exercises 37–44, the numerator and denominator of the fraction share a common factor. Graph each rational function.

37. $y = \dfrac{x^2}{x}$

38. $y = \dfrac{x^2 - 1}{x - 1}$

39. $y = \dfrac{x^3 + x}{x}$

40. $y = \dfrac{x^3 - x^2}{x - 1}$

41. $y = \dfrac{x^2 - 2x + 1}{x - 1}$

42. $y = \dfrac{2x^2 + 3x - 2}{x + 2}$

43. $y = \dfrac{x^3 - 1}{x - 1}$

44. $y = \dfrac{x^2 - x}{x^2}$

45. Can a rational function have two horizontal asymptotes? Explain.

46. Can a rational function have two slant asymptotes? Explain.

In Exercises 47–49, a, b, c, and d are nonzero constants.

47. Show that the graph of $y = \dfrac{ax + b}{cx^2 + d}$ has the horizontal asymptote $y = 0$.

48. Show that the graph of $y = \dfrac{ax^2 + b}{cx^2 + d}$ has the horizontal asymptote $y = \dfrac{a}{c}$.

49. Show that the graph of $y = \dfrac{ax^3 + b}{cx^2 + d}$ has the slant asymptote $y = \dfrac{a}{c}x$.

50. Graph the rational function $y = \dfrac{x^3 + 1}{x}$ and explain why the curve is said to have a parabolic asymptote.

In Exercises 51–54, use a graphing calculator to perform each experiment. Write a brief paragraph describing your findings.

51. Investigate the positioning of the vertical asymptotes of a rational function by graphing $y = \dfrac{x}{x - k}$ for several values of k. What do you observe?

52. Investigate the positioning of the vertical asymptotes of a rational function by graphing $y = \dfrac{x}{x^2 - k}$ for $k = 4$, 1, -1, and 0. What do you observe?

53. Find the range of the rational function $y = \dfrac{kx^2}{x^2 + 1}$ for several values of k. What do you observe?

54. Investigate the positioning of the x-intercepts of a rational function by graphing $y = \dfrac{x^2 - k}{x}$ for $k = 1$, -1, and 0. What do you observe?

4.6 OPERATIONS ON FUNCTIONS

■ Algebra of Functions   ■ Composition of Functions

■ Algebra of Functions

With the following definition, it is possible to perform some arithmetic on algebraic functions.

Operations on Functions

If the ranges of the functions f and g are subsets of the real numbers, then

1. The sum of f and g, denoted as $f + g$, is defined by
$$(f + g)(x) = f(x) + g(x)$$

2. The difference of f and g, denoted as $f - g$, is defined by
$$(f - g)(x) = f(x) - g(x)$$

3. The product of f and g, denoted as $f \cdot g$, is defined by
$$(f \cdot g)(x) = f(x)g(x)$$

4. The quotient of f and g, denoted as f/g, is defined by
$$(f/g)(x) = \frac{f(x)}{g(x)}$$

The domain of each function, unless otherwise restricted, is the set of real numbers x that are in the domain of both f and g. In the case of the quotient f/g, there is the further restriction that $g(x) \neq 0$.

EXAMPLE 1   Let $f(x) = 3x + 1$ and $g(x) = 2x - 3$. Find   **a.** $f + g$ and   **b.** $f - g$ and give the domain of each function.

Solution   **a.** $(f + g)(x) = f(x) + g(x)$

$\qquad\qquad\qquad = (3x + 1) + (2x - 3)$ Substitute $3x + 1$ for $f(x)$ and $2x - 3$ for $g(x)$.

$\qquad\qquad\qquad = 5x - 2$ Combine terms.

The domain of $f + g$ is the set of real numbers that are in the domain of both f and g. Since the domain of both f and g is the set of real numbers, the domain of $f + g$ is the set of real numbers.

b. $(f - g)(x) = f(x) - g(x)$

$$= (3x + 1) - (2x - 3) \qquad \text{Substitute } 3x + 1 \text{ for } f(x)$$
$$\text{and } 2x - 3 \text{ for } g(x).$$

$$= x + 4 \qquad \text{Combine terms.}$$

Since the domain of both f and g is the set of real numbers, the domain of $f - g$ is the set of real numbers. ∎

EXAMPLE 2 Let $f(x) = 3x + 1$ and $g(x) = 2x - 3$. Find **a.** $f \cdot g$ and **b.** f/g and give the domain of each function.

Solution **a.** $(f \cdot g)(x) = f(x) \cdot g(x)$

$$= (3x + 1)(2x - 3) \qquad \text{Substitute } 3x + 1 \text{ for } f(x)$$
$$\text{and } 2x - 3 \text{ for } g(x).$$

$$= 6x^2 - 7x - 3 \qquad \text{Multiply the binomials.}$$

Since the domain of both f and g is the set of real numbers, the domain of $f \cdot g$ is the set of real numbers.

b. $(f/g)(x) = \dfrac{f(x)}{g(x)}$

$$= \frac{3x + 1}{2x - 3} \qquad \text{Substitute } 3x + 1 \text{ for } f(x) \text{ and } 2x - 3 \text{ for } g(x).$$

Since $2x - 3$ cannot equal 0, the domain of f/g is the set of real numbers except $\frac{3}{2}$. ∎

EXAMPLE 3 Let $f(x) = x^2 - 4$ and $g(x) = \sqrt{x}$. If all expressions represent real numbers, find the functions **a.** $f + g$, **b.** $f \cdot g$, **c.** f/g, and **d.** g/f. Give the domain of each result.

Solution Because all real numbers can be squared, the domain of f is the set of real numbers. Because all expressions must represent real numbers, the domain of g is $\{x \mid x \geq 0\}$.

a. $(f + g)(x) = f(x) + g(x)$
$$= x^2 - 4 + \sqrt{x}$$

The domain consists of the numbers x that are in both the domain of f and the domain of g. The domain of $f + g$ is $\{x \mid x \geq 0\}$.

b. $(f \cdot g)(x) = f(x)g(x)$
$$= (x^2 - 4)\sqrt{x}$$
$$= x^2\sqrt{x} - 4\sqrt{x}$$

The domain consists of the numbers x that are in both the domain of f and the domain of g. The domain of $f \cdot g$ is $\{x \mid x \geq 0\}$.

c. $(f/g)(x) = \dfrac{f(x)}{g(x)}$

$\qquad\quad = \dfrac{x^2 - 4}{\sqrt{x}}$

The domain consists of the numbers x that are in both the domain of f and the domain of g, except 0, because division by 0 is undefined. The domain of f/g is $\{x \mid x > 0\}$.

d. $(g/f)(x) = \dfrac{g(x)}{f(x)}$

$\qquad\quad = \dfrac{\sqrt{x}}{x^2 - 4}$

The domain consists of the numbers x that are in both the domain of f and the domain of g, except 2, because division by 0 is undefined. The domain of f/g is $\{x \mid x \geq 0 \text{ and } x \neq 2\}$. ∎

EXAMPLE 4   Find $(f + g)(3)$ when $f(x) = x^2 + 1$ and $g(x) = 2x + 1$.

Solution   We first find $(f + g)(x)$.

$$(f + g)(x) = f(x) + g(x)$$
$$= x^2 + 1 + 2x + 1$$
$$= x^2 + 2x + 2$$

We then find $(f + g)(3)$.

$$(f + g)(x) = x^2 + 2(x) + 2$$
$$(f + g)(3) = 3^2 + 2(3) + 2$$
$$= 9 + 6 + 2$$
$$= 17$$

∎

EXAMPLE 5   Let $h(x) = x^2 + 3x + 2$. Find two functions f and g such that   **a.** $h = f + g$ and   **b.** $h = f \cdot g$.

Solution   **a.** Several answers are possible. One possibility is $f(x) = x^2$ and $g(x) = 3x + 2$, for then

$$h(x) = x^2 + 3x + 2$$
$$= (x^2) + (3x + 2)$$
$$= f(x) + g(x)$$
$$= (f + g)(x)$$

Another possibility is $f(x) = x^2 + 2x$ and $g(x) = x + 2$.

b. Again, several answers are possible. One possibility is suggested by factoring the trinomial $x^2 + 3x + 2$:

$$h(x) = x^2 + 3x + 2 = (x + 1)(x + 2)$$

Let $f(x) = x + 1$ and $g(x) = x + 2$. Then

$$
\begin{aligned}
h(x) &= x^2 + 3x + 2 \\
&= (x + 1)(x + 2) \\
&= f(x) \cdot g(x) \\
&= (f \cdot g)(x)
\end{aligned}
$$

Another possibility is $f(x) = 3$ and $g(x) = \dfrac{x^2}{3} + x + \dfrac{2}{3}$.   ■

Composition of Functions

Often one quantity is a function of a second quantity that depends, in turn, on a third quantity. A farmer's income, for example, is a function of the number of bushels of grain harvested. But the harvest is a function of the number of inches of rainfall received. Such chains of dependence can be analyzed mathematically by considering the **composition of functions**.

We suppose that $y = f(x)$ and $y = g(x)$ define two functions. Any number x in the domain of g will produce a corresponding value $g(x)$ in the range of g. If this value $g(x)$ is in the domain of the function f, then $g(x)$ is an acceptable input into the function f, and a corresponding value $f(g(x))$ is determined. This two-step process defines a new function, called a **composite function**, denoted by $f \circ g$.

Figure 4-33(a) represents a function f as a machine that takes numbers x and turns them into values $f(x)$. Figure 4-33(b) represents a function machine for $y = f(x) = \sqrt{x}$. If we put numbers in the domain of f into the machine, the values $f(x)$ come out. For example, the machine turns 4 into 2.

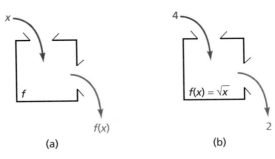

(a)   (b)

FIGURE 4-33

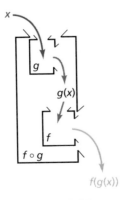

FIGURE 4-34

The two function machines shown in Figure 4-34 illustrate the composition $f \circ g$. Function g receives a number x into its input hopper and turns it into $g(x)$. The value $g(x)$ falls into the input of function f, and function f turns $g(x)$ into $f(g(x))$. If functions g and f are thought of as the two-stage internal workings of a single function machine, that machine would be named $f \circ g$.

To be in the domain of the composite function $f \circ g$, a number x has to be acceptable input into g. That is, x must be in the domain of g. Also, the output of g

must be acceptable input into f. That is, $g(x)$ must be in the domain of f. Thus, the domain of $f \circ g$ consists of those numbers x that are in the domain of g, and for which $g(x)$ is in the domain of f.

We can also visualize the domain of the composite function $f \circ g$ by referring to Figure 4-35 and noting that x must be in the domain of the function g to produce a value $g(x)$. We also note that $g(x)$ must be in the domain of function f to produce a value $f(g(x))$. Hence, the domain of $f \circ g$ consists of all those numbers x that are possible inputs into g, and for which each $g(x)$ is a permissible input into f.

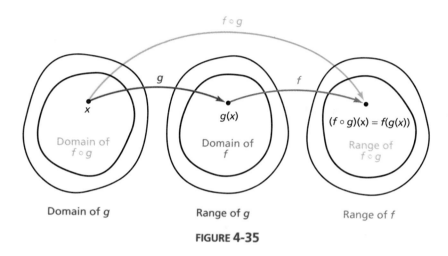

FIGURE **4-35**

Composite Functions	The **composite function** $f \circ g$ is defined by
	$$(f \circ g)(x) = f(g(x))$$
	The **domain** of the composite function $f \circ g$ consists of all those elements in the domain of g for which $g(x)$ is an element in the domain of f.

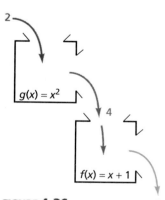

FIGURE **4-36**

To illustrate the composition of functions, we suppose that $f(x) = x + 1$ and $g(x) = x^2$ and then calculate $(f \circ g)(2)$. We do so by calculating $f(g(2))$. Because $g(x) = x^2$, $g(2) = 2^2 = 4$. Because 4 is in the domain of f, we calculate $f(4)$, which is $4 + 1 = 5$. Thus,

$$f \circ g = f(g(2))$$
$$= f(4)$$
$$= 5$$

To visualize this work with function machines, we refer to Figure 4-36. When we put 2 into the machine, it first falls into the g machine, and $g(x)$ turns the 2 into 4. Then 4 falls into the f machine, and $f(x)$ turns the 4 into 5. Thus, $f(g(x)) = 5$.

EXAMPLE 6   If $f(x) = 2x + 7$ and $g(x) = 4x + 1$, find   **a.** $(f \circ g)(x)$ and   **b.** $(g \circ f)(x)$.

Solution   **a.**
$$(f \circ g) = f(g(x)) = f(4x + 1)$$
$$= 2(4x + 1) + 7$$
$$= 8x + 9$$

b.
$$(g \circ f)(x) = g(f(x)) = g(2x + 7)$$
$$= 4(2x + 7) + 1$$
$$= 8x + 29$$ ■

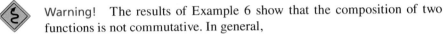

Warning!   The results of Example 6 show that the composition of two functions is not commutative. In general,

$$(f \circ g)(x) \neq (g \circ f)(x)$$

EXAMPLE 7   Let $f(x) = \sqrt{x}$ and $g(x) = x - 3$. Find the domain of functions   **a.** f,   **b.** g, **c.** $f \circ g$,   and   **d.** $g \circ f$.

Solution   **a.** If the range is to be a subset of the real numbers, the domain of f is the set of nonnegative real numbers: $\{x \mid \geq 0\}$.

b. The domain of g is the set of real numbers.

c. Because $g(x)$ must be in the domain of f, the value $g(x)$ must be nonnegative. Thus,

$$g(x) \geq 0$$
$$x - 3 \geq 0 \qquad \text{Because } g(x) = x - 3.$$
$$x \geq 3$$

Hence, the domain of $f \circ g$ is $\{x \mid x \geq 3\}$.

d. The domain of $g \circ f$ contains those numbers x for which $f(x)$ is also in the domain of g. Because the domain of g includes all real numbers, the domain of $g \circ f$ is just the domain of f itself: $\{x \mid x \geq 0\}$.   ■

EXAMPLE 8   Let $f(x) = \sqrt{1 - x}$ and $g(x) = x^2$. If all expressions represent real numbers, find   **a.** $(f \circ g)(x)$ and   **b.** $(g \circ f)(x)$, and give the domain of each function.

Solution   We begin by finding the domain of f. Because all expressions represent real numbers, the radicand $1 - x$ cannot be negative, so x must be less than or equal to 1. Thus, the domain of f is the set $\{x \mid x \leq 1\}$.

Now we find the domain of g. Because the square of any real number is still a real number, the domain of g is the set of real numbers.

a. $(f \circ g)(x) = f(g(x))$
$$= f(x^2)$$
$$= \sqrt{1 - x^2}$$

The domain of $f \circ g$ consists of those numbers x for which $g(x)$ is in the domain of f. Thus, $g(x)$ must be less than or equal to 1.

$$g(x) \leq 1$$
$$x^2 \leq 1 \qquad \text{Because } g(x) = x^2.$$

Thus, the domain of $f \circ g$ is $\{x \mid -1 \leq x \leq 1\}$.

b. $(g \circ f)(x) = g(f(x))$
$$= g(\sqrt{1-x})$$
$$= (\sqrt{1-x})^2$$
$$= 1 - x$$

The domain of $g \circ f$ consists of those numbers x for which $f(x)$ is in the domain of g. Because any real number is acceptable to g, the domain of $g \circ f$ is just the domain of f: $\{x \mid x \leq 1\}$. ∎

4.6   EXERCISES

In Exercises 1–4, let $f(x) = 2x + 1$ and $g(x) = 3x - 2$. Find each function and determine its domain.

1. $f + g$ **2.** $f - g$ **3.** $f \cdot g$ **4.** f/g

In Exercises 5–8, let $f(x) = x^2 + x$ and $g(x) = x^2 - 1$. Find each function and determine its domain.

5. $f - g$ **6.** $f + g$ **7.** f/g **8.** $f \cdot g$

In Exercises 9–16, let $f(x) = x^2 - 1$ and $g(x) = 3x - 2$. Find each value, if possible.

9. $(f + g)(2)$ **10.** $(f + g)(-3)$ **11.** $(f - g)(0)$ **12.** $(f - g)(-5)$

13. $(f \cdot g)(2)$ **14.** $(f \cdot g)(-1)$ **15.** $(f/g)\left(\dfrac{2}{3}\right)$ **16.** $(f/g)(t)$

In Exercises 17–24, find two functions f and g such that the given correspondence can be expressed as the function indicated. Several answers are possible.

17. $y = 3x^2 + 2x; f + g$ **18.** $y = 3x^2; f \cdot g$ **19.** $y = \dfrac{3x^2}{x^2 - 1}; f/g$ **20.** $y = 5x + x^2; f - g$

21. $y = x(3x^2 + 1); f - g$ **22.** $y = (3x - 2)(3x + 2); f + g$

23. $y = x^2 + 7x - 18; f \cdot g$ **24.** $y = 5x^5; f/g$

In Exercises 25–28, let $f(x) = 2x - 5$ and $g(x) = 5x - 2$. Find each value.

25. $(f \circ g)(2)$ **26.** $(g \circ f)(-3)$ **27.** $(f \circ f)\left(-\dfrac{1}{2}\right)$ **28.** $(g \circ g)\left(\dfrac{3}{5}\right)$

In Exercises 29–32, let $f(x) = 3x^2 - 2$ and $g(x) = 4x + 4$. Find each value.

29. $(f \circ g)(-3)$ **30.** $(g \circ f)(3)$ **31.** $(f \circ f)(\sqrt{3})$ **32.** $(g \circ g)(-4)$

In Exercises 33–36, let $f(x) = 3x$ and $g(x) = x + 1$. Find each composite function and determine its domain.

33. $f \circ g$ **34.** $g \circ f$ **35.** $f \circ f$ **36.** $g \circ g$

In Exercises 37–40, let $f(x) = x^2$ and $g(x) = 2x$. Find each composite function and determine its domain.

37. $g \circ f$ **38.** $f \circ g$ **39.** $g \circ g$ **40.** $f \circ f$

In Exercises 41–44, let $f(x) = \sqrt{x}$ and $g(x) = x^2 + 1$. Find each composite function and determine its domain.

41. $f \circ g$ **42.** $g \circ f$ **43.** $f \circ f$ **44.** $g \circ g$

In Exercises 45–48, let $f(x) = \sqrt{x + 1}$ and $g(x) = x^2 - 1$. Find each composite function and determine its domain.

45. $g \circ f$ **46.** $f \circ g$ **47.** $g \circ g$ **48.** $f \circ f$

In Exercises 49–60, find two functions f and g such that the composition $f \circ g$ expresses the given correspondence. Several answers are possible.

49. $y = 3x - 2$ **50.** $y = 7x - 5$ **51.** $y = x^2 - 2$ **52.** $y = x^3 - 3$

53. $y = (x - 2)^2$ **54.** $y = (x - 3)^3$ **55.** $y = \sqrt{x + 2}$ **56.** $y = \dfrac{1}{x - 5}$

57. $y = \sqrt{x} + 2$ **58.** $y = \dfrac{1}{x} - 5$ **59.** $y = x$ **60.** $y = 3$

61. Let $f(x) = 3x$. Show that $(f + f)(x) = f(x + x)$.

62. Let $g(x) = x^2$. Show that $(g + g)(x) \neq g(x + x)$.

63. Let $f(x) = \dfrac{x - 1}{x + 1}$. Find $(f \circ f)(x)$.

64. Let $g(x) = \dfrac{x}{x - 1}$. Find $(g \circ g)(x)$.

In Exercises 65–68, let $y = f(x) = x^2 - x$, $g(x) = x - 3$, and $h(x) = 3x$. Use a graphing calculator to graph both functions on the same axis. Write a brief paragraph summarizing your observations.

65. f and $f \circ g$ **66.** f and $g \circ f$ **67.** f and $f \circ h$ **68.** f and $h \circ f$

4.7 INVERSE FUNCTIONS

- The Horizontal Line Test
- The Relationship between the Graphs f and f^{-1}
- Finding the Inverse of a Function

By definition, each element x in the domain of a function has a single image y. For some functions, different numbers x in the domain can have the same image. See Figure 4-37(a). For other functions, called **one-to-one functions**, different numbers x have different images. See Figure 4-37(b).

One-to-One Functions   A function f from a set **X** to a set **Y** is called **one-to-one** if and only if different numbers in the domain of f have different images in the range of f. That is, if x_1 and x_2 are two numbers in the domain of f and $x_1 \neq x_2$, then $f(x_1) \neq f(x_2)$.

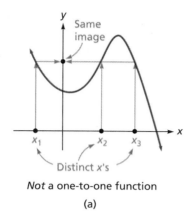

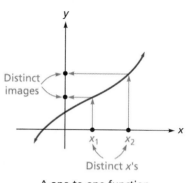

FIGURE **4-37**

EXAMPLE 1   Determine whether the functions   **a.** $f(x) = x^2$ and   **b.** $f(x) = x^3$ are one-to-one.

Solution   **a.** The function $f(x) = x^2$ is not one-to-one, because different numbers in the domain have the same image. For example, 2 and -2 have the same image: $f(2) = f(-2) = 4$.

b. The function $f(x) = x^3$ is one-to-one, because different numbers x produce different images $f(x)$. This is because different numbers have different cubes.   ■

The Horizontal Line Test

The **horizontal line test** can be used to determine whether the graph of a function represents a one-to-one function. If every horizontal line that intersects the graph does so only once, the function is one-to-one. See Figure 4-38(a). However, if any horizontal line intersects the graph of a function more than once, the function is not one-to-one. See Figure 4-38(b).

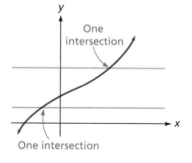

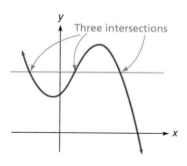

FIGURE **4-38**

Figure 4-39 illustrates a function f from a set $\mathbf{X}$ to a set $\mathbf{Y}$. Since several arrows point to a single y, the function f is not one-to-one. If the arrows in Figure 4-39(a) were reversed, however, the diagram would not represent a function, because to some y of set $\mathbf{Y}$ would correspond several values x in set $\mathbf{X}$.

If the arrows of the one-to-one function f in Figure 4-39(b) were reversed, the diagram would still represent a function. This backward function is called the **inverse of function** f; it is denoted by the symbol f^{-1}.

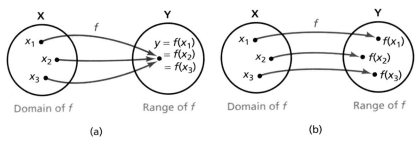

(a)   (b)

FIGURE **4-39**

Inverse Function	If f is a one-to-one function with domain $\mathbf{X}$ and range $\mathbf{Y}$, then f^{-1}, called the **inverse function of** f, has domain $\mathbf{Y}$ and range $\mathbf{X}$. The function f^{-1} is defined by

$$f^{-1}(y) = x \qquad \text{is equivalent to} \qquad f(x) = y$$

for all y in $\mathbf{Y}$.

 Warning! The -1 in the notation for inverse function is not an exponent. Remember that

$$f^{-1}(x) \neq \frac{1}{f(x)}$$

We refer to the one-to-one function f and its inverse f^{-1} shown in Figure 4-40. To the number x in the domain of f, there corresponds its image $f(x)$ in the range of f. However, the value $f(x)$ is also in the domain of f^{-1}, and the image of $f(x)$ under the function f^{-1} is $f^{-1}(f(x)) = x$. Thus,

$$(f^{-1} \circ f)(x) = f^{-1}(f(x)) = x$$

Furthermore, if y is any number in the domain of f^{-1}, then

$$(f \circ f^{-1})(x) = f(f^{-1}(x)) = y$$

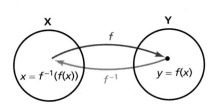

FIGURE **4-40**

Theorem	If f is a one-to-one function with domain $\mathbf{X}$ and range $\mathbf{Y}$, there is a one-to-one function f^{-1} with domain $\mathbf{Y}$ and range $\mathbf{X}$ such that

$$(f^{-1} \circ f)(x) = x \qquad \text{and} \qquad (f \circ f^{-1})(y) = y$$

To show that one function is the inverse of another, we must show that their composition is the **identity function**, the function that turns every real number x into x.

EXAMPLE 2 Show that $f(x) = x^3$ is the inverse function of $g(x) = \sqrt[3]{x}$.

Solution To show that f is the inverse function of g, we must show that $f \circ g$ and $g \circ f$ are the identity function.

$$(f \circ g)(x) = f(g(x)) = f(\sqrt[3]{x}) = (\sqrt[3]{x})^3 = x$$

$$(g \circ f)(x) = g(f(x)) = g(x^3) = \sqrt[3]{x^3} = x$$ ∎

Finding the Inverse of a Function

If f is a one-to-one function $y = f(x)$, Then f^{-1} reverses the correspondence of f. That is, if $f(a) = b$, then $f^{-1}(b) = a$. To determine f^{-1}, we perform the following steps.

Method for Finding f^{-1}

1. Write the function in the form $y = f(x)$.
2. Interchange the positions of variables x and y in the equation $y = f(x)$. The resulting equation, $x = f(y)$, defines the inverse function f^{-1}.
3. Solve the equation $x = f(y)$ for y, if possible. The result is $f^{-1}(x)$.

EXAMPLE 3 Find the inverse of the function $y = f(x) = \frac{3}{2}x + 2$ and verify the result.

Solution Since the function is written in the form $y = f(x)$, we can find the inverse function by interchanging the positions of x and y.

$$x = \frac{3}{2}y + 2$$

To write the inverse function in the form $y = f^{-1}(x)$, we can solve the equation for y.

$$x = \frac{3}{2}y + 2$$
$$2x = 3y + 4 \qquad \text{Multiply both sides by 2.}$$
$$2x - 4 = 3y \qquad \text{Subtract 4 from both sides.}$$
$$y = \frac{2x - 4}{3} \qquad \text{Divide both sides by 3.}$$

The inverse of the function $y = f(x) = \frac{3}{2}x + 2$ is $y = f^{-1}(x) = \frac{2x - 4}{3}$. To verify that

$$f(x) = \frac{3}{2}x + 2 \qquad \text{and} \qquad f^{-1}(x) = \frac{2x - 4}{3}$$

are inverses of each other, we must show that $f \circ g$ and $g \circ f$ are the identity function.

$$(f \circ f^{-1})(x) = f(f^{-1}(x)) = f\left(\frac{2x - 4}{3}\right) = \frac{3}{2}\left(\frac{2x - 4}{3}\right) + 2 = x - 2 + 2 = x$$

$$(f^{-1} \circ f)(x) = f^{-1}(f(x)) = f^{-1}\left(\frac{3}{2}x + 2\right) = \frac{2\left(\frac{3}{2}x + 2\right) - 4}{3} = \frac{3x + 4 - 4}{3} = x$$

■

■ The Relationship between the Graphs of *f* and *f*⁻¹

Because we interchange the positions of x and y to find the inverse of a function, the point (b, a) lies on the graph of $y = f^{-1}(x)$ whenever the point (a, b) lies on the graph of $y = f(x)$. Thus, the graph of a function and its inverse are reflections of each other about the line $y = x$.

EXAMPLE 4　Find the inverse of $y = f(x) = x^3 + 3$. Graph both the function and its inverse on the same set of coordinate axes.

Solution　We find the inverse of the function $y = f(x) = x^3 + 3$ by interchanging x and y. Thus, the inverse is

$$x = y^3 + 3$$

To write the inverse in the form $f^{-1}(x)$, we solve the equation for y to get

$$x - 3 = y^3$$
$$y = \sqrt[3]{x - 3}$$

Thus, $f^{-1}(x) = \sqrt[3]{x - 3}$. In the graph that appears in Figure 4-41, the line $y = x$ is the axis of symmetry.

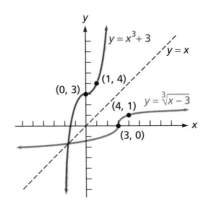

FIGURE 4-41

■

EXAMPLE 5   The function defined by $y = f(x) = x^2 + 3$ is not one-to-one. However, the function becomes one-to-one if the domain is restricted to a subset of the real numbers, such as the set $\{x \mid x \le 0\}$.

a. Find the range of f.

b. Find the inverse of f along with its domain and range.

c. Graph each function.

Solution   **a.** The function f is defined by $y = f(x) = x^2 + 3$, with domain $\{x \mid x \le 0\}$. If x is replaced with numbers from this domain, y ranges over the values 3 and above. Thus, the range of f is the set $\{y \mid y \ge 3\}$.

b. To find the inverse of f, we interchange x and y in the equation that defines f and solve for y.

$$y = x^2 + 3 \qquad \text{where } x \le 0$$
$$x = y^2 + 3 \qquad \text{where } y \le 0 \qquad \text{Interchange } x \text{ and } y.$$
$$x - 3 = y^2 \qquad \text{where } y \le 0 \qquad \text{Subtract 3 from both sides.}$$

To solve this equation for y, we take the square root of both sides. Because $y \le 0$, we have

$$-\sqrt{x - 3} = y \qquad \text{where } y \le 0$$

Thus, the inverse of f is defined by

$$y = f^{-1}(x) = -\sqrt{x - 3}$$

It has domain $\{x \mid x \ge 3\}$ and range $\{y \mid y \le 0\}$.

c. The graphs of these two functions appear in Figure 4-42. Note that the line of symmetry is $y = x$.

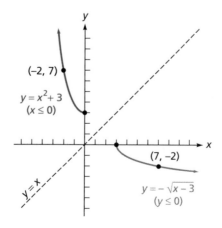

FIGURE 4-42

If a function is defined by the equation $y = f(x)$, we can often find the domain of f by inspection. Finding the range can be more difficult. One way to find the range of f is to find the domain of f^{-1}.

EXAMPLE 6   Find the domain and range of the function $y = f(x) = \dfrac{2}{x} + 3$.

Solution   Because x cannot be 0, the domain of f is

$$\{x \mid x \text{ is a real number and } x \neq 0\}$$

To find the range of f, we find the domain of f^{-1}. To find f^{-1}, we proceed as follows:

$$y = \frac{2}{x} + 3$$

$$x = \frac{2}{y} + 3 \qquad \text{Interchange } x \text{ and } y.$$

$$xy = 2 + 3y \qquad \text{Multiply both sides by } y.$$

$$xy - 3y = 2 \qquad \text{Subtract } 3y \text{ from both sides.}$$

$$y(x - 3) = 2 \qquad \text{Factor out } y.$$

$$y = \frac{2}{x - 3} \qquad \text{Divide both sides by } x - 3.$$

This final equation defines f^{-1}, whose domain is $\{x \mid x \text{ is a real number and } x \neq 3\}$. Because the range of f is equal to the domain of f^{-1}, the range is

$$\{y \mid y \text{ is a real number and } y \neq 3\}$$ ∎

4.7   EXERCISES

In Exercises 1–12, determine whether each function is one-to-one.

1. $y = 3x$ **2.** $y = \dfrac{1}{2}x$ **3.** $y = x^2 + 3$ **4.** $y = x^4 - x^2$

5. $y = x^3 - x$ **6.** $y = x^2 - x$ **7.** $y = |x|$ **8.** $y = |x - 3|$

9. $y = 5$ **10.** $y = \sqrt{x - 5};\ x \geq 5$ **11.** $y = (x - 2)^2;\ x \geq 2$ **12.** $y = \dfrac{1}{x}$

In Exercises 13–16, use the horizontal line test to determine whether each graph represents a one-to-one function.

13. 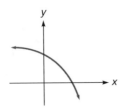   **14.** 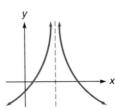   **15.** 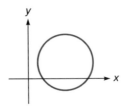   **16.**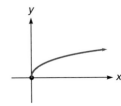

In Exercises 17–20, verify that the functions are inverses by showing that f ∘ g and g ∘ f are identity functions.

17. $f(x) = 5x$ and $g(x) = \dfrac{1}{5}x$

18. $f(x) = 3x + 2$ and $g(x) = \dfrac{x - 2}{3}$

19. $f(x) = \dfrac{x + 1}{x}$ and $g(x) = \dfrac{1}{x - 1}$

20. $f(x) = \dfrac{x + 1}{x - 1}$ and $g(x) = \dfrac{x + 1}{x - 1}$

In Exercises 21–28, each equation defines a one-to-one function f. Determine f^{-1} and verify that $f \circ f^{-1}$ and $f^{-1} \circ f$ are the identity function.

21. $y = 3x$

22. $y = \dfrac{1}{3}x$

23. $y = 3x + 2$

24. $y = 2x - 5$

25. $y = \dfrac{1}{x + 3}$

26. $y = \dfrac{1}{x - 2}$

27. $y = \dfrac{1}{2x}$

28. $y = \dfrac{1}{x^3}$

In Exercises 29–40, find the inverse of each one-to-one function and graph both the function and its inverse on the same set of coordinate axes.

29. $y = 5x$

30. $y = \dfrac{3}{2}x$

31. $y = 2x - 4$

32. $y = \dfrac{3}{2}x - 2$

33. $x - y = 2$

34. $x + y = 0$

35. $2x + y = 4$

36. $3x + 2y = 6$

37. $y = \dfrac{1}{2x}$

38. $y = \dfrac{1}{x - 3}$

39. $y = \dfrac{x + 1}{x - 1}$

40. $y = \dfrac{x - 1}{x}$

In Exercises 41–46, the function f defined by the given equation is one-to-one on the given domain. Find $f^{-1}(x)$.

41. $f(x) = x^2 - 3, \{x \mid x \le 0\}$

42. $f(x) = \dfrac{1}{x^2}, \{x \mid x > 0\}$

43. $f(x) = x^4 - 8, \{x \mid x \ge 0\}$

44. $f(x) = \dfrac{-1}{x^4}, \{x \mid x < 0\}$

45. $f(x) = \sqrt{4 - x^2}, \{x \mid 0 \le x \le 2\}$

46. $f(x) = \sqrt{x^2 - 1}, \{x \mid x \le -1\}$

In Exercises 47–50, find the domain and the range of f. Find the range by finding the domain of f^{-1}.

47. $f(x) = \dfrac{x}{x - 1}$

48. $f(x) = \dfrac{x - 2}{x + 3}$

49. $f(x) = \dfrac{1}{x} - 2$

50. $f(x) = \dfrac{3}{x} - \dfrac{1}{2}$

4   CHAPTER SUMMARY

Key Words

asymptote (4.5)

composition of functions (4.6)

decreasing function (4.3)

degree of a polynomial function (4.3)

domain of a function (4.1)

domain of a relation (4.1)

even function (4.3)

function (4.1)

function notation (4.1)

graph of a function (4.1)

graph of a relation (4.1)

greatest integer function (4.3)

horizontal asymptote (4.5)

horizontal line test (4.7)

identity function (4.7)

image of x (4.1)

increasing function (4.3)

inverse functions (4.7)

linear function (4.1)
odd function (4.3)
one-to-one function (4.7)
piecewise-defined function (4.3)
polynomial function (4.3)

quadratic function (4.2)
range of a function (4.1)
range of a relation (4.1)
rational function (4.5)
relation (4.1)

step function (4.3)
vertex of a parabola (4.2)
vertical asymptotes (4.5)
vertical line test (4.1)

Key Ideas

(4.1) A **function** is a correspondence that assigns to each number x in some set **X** a single value y in some set **Y**. A **relation** is a correspondence that assigns to each number x in a set **X** one or more values y in some set **Y**.

(4.2) The graph of the equation $y - k = a(x - h)^2$ (with $a \neq 0$) is a parabola with its vertex at the point (h, k). The parabola opens upward if $a > 0$ and downward if $a < 0$. The graph of the equation $y = ax^2 + bx + c$ (with $a \neq 0$) is a parabola with vertex $\left(-\dfrac{b}{2a}, c - \dfrac{b^2}{4a} \right)$.

(4.3) The graph of an **even function** is symmetric about the y-axis.
The graph of an **odd function** is symmetric about the origin.

(4.4) **Vertical translations:**

If $k > 0$, the graph of $\begin{cases} y = f(x) + k \\ y = f(x) - k \end{cases}$ is identical to

the graph of $y = f(x)$ except that it is translated

k units $\begin{cases} \text{upward} \\ \text{downward} \end{cases}$.

Horizontal translations:

If $k > 0$, the graph of $\begin{cases} y = f(x - k) \\ y = f(x + k) \end{cases}$ is identical to

the graph of $y = f(x)$ except that it is translated k units

to the $\begin{cases} \text{right} \\ \text{left} \end{cases}$.

Vertical stretchings:
If $k > 0$, the graph of $y = kf(x)$ can be obtained by stretching the graph of $y = f(x)$ vertically by a factor of k.
If $k < 0$, the graph of $y = kf(x)$ can be obtained by stretching the reflection of the graph of $y = f(x)$ in the x-axis vertically by a factor of $|k|$.

Horizontal stretchings:
If $k > 0$, the graph of $y = f(kx)$ can be obtained by stretching the graph of $y = f(x)$ horizontally by a factor of k.

If $k < 0$, the graph of $y = f(kx)$ can be obtained by stretching the reflection of the graph of $y = f(x)$ in the y-axis horizontally by a factor of $\left| \frac{1}{k} \right|$.

(4.5) To graph the rational function $y = \dfrac{P(x)}{Q(x)}$:

1. Check symmetries.
2. Look for vertical asymptotes.
3. Look for y-intercepts.
4. Look for x-intercepts.
5. Look for horizontal asymptotes.
6. Look for slant asymptotes.

(4.6) If the ranges of the functions f and g are subsets of the real numbers, then

1. $(f + g)(x) = f(x) + g(x)$
2. $(f - g)(x) = f(x) - g(x)$
3. $(f \cdot g)(x) = f(x) \cdot g(x)$
4. $(f/g)(x) = \dfrac{f(x)}{g(x)}$

The domain of each function, unless otherwise restricted, is the set of real numbers x that are in the domain of both f and g. In the case of the quotient f/g, there is the further restriction that $g(x) \neq 0$.

Composition of functions:

$$(f \circ g)(x) = f(g(x))$$

The domain of $f \circ g$ is the set of all x in the domain of g for which $g(x)$ is in the domain of f.

(4.7) **Inverse functions:**
If f is a one-to-one function with domain **X** and range **Y**, there is a one-to-one function f^{-1} with domain **Y** and range **X** such that

$$(f^{-1} \circ f)(x) = x \quad \text{and} \quad (f \circ f^{-1})(y) = y$$

The graph of a function is symmetric to the graph of its inverse. The axis of symmetry is the line $y = x$.

4   CHAPTER REVIEW EXERCISES

In Review Exercises 1–4, indicate whether the equation defines y as a function of x. Assume that all variables represent real numbers.

1. $y = 3$ **2.** $y + 5x^2 = 2$ **3.** $y^2 - x = 5$ **4.** $y = |x| + x$

In Review Exercises 5–8, find the domain and the range of the function defined by each equation.

5. $f(x) = 3x^2 - 5x$ **6.** $f(x) = \dfrac{3x^2}{x - 5}$ **7.** $f(x) = \sqrt{x - 1}$ **8.** $f(x) = \sqrt{x^2 + 1}$

In Review Exercises 9–12, find $f(2), f(-3),$ and $f(0)$.

9. $f(x) = 5x - 2$ **10.** $f(x) = \dfrac{6}{x - 5}$ **11.** $f(x) = |x - 2|$ **12.** $f(x) = \dfrac{x^2 - 3}{x^2 + 3}$

In Review Exercises 13–16, graph each parabola and find its vertex.

13. $y = x^2 - x$ **14.** $y = x - x^2$ **15.** $y = x^2 - 3x - 4$ **16.** $y = 3x^2 - 8x - 3$

In Review Exercises 17–18, find each maximum or minimum.

17. *Architecture*   A parabolic arch has an equation of $3x^2 + y - 300 = 0$. Find the maximum height of the arch.

18. *Puzzle problem*   The sum of two numbers is 1, and their product is as large as possible. Find the numbers.

In Review Exercises 19–22, graph each polynomial function and tell whether it is even, odd, or neither.

19. $y = x^3 - x$ **20.** $y = x^2 - 4x$ **21.** $y = x^3 - x^2$ **22.** $y = 1 - x^4$

In Review Exercises 23–24, graph each piecewise-defined function and tell when it is increasing, decreasing, or constant.

23. $y = f(x) = \begin{cases} x + 5 & \text{if } x \le 0 \\ 5 - x & \text{if } x > 0 \end{cases}$

24. $y = f(x) = \begin{cases} x + 3 & \text{if } x \le 0 \\ 3 & \text{if } x > 0 \end{cases}$

In Review Exercises 25–26, graph each step function.

25. $y = f(x) = [\![x - 2]\!]$

26. $y = f(x) = [\![x]\!] - 2$

In Review Exercises 27–28, each function is a translation of a simpler function. Graph both on one set of coordinate axes.

27. $y = \sqrt{x + 2} + 3$

28. $y = |x - 4| + 2$

In Review Exercises 29–30, each function is a stretching of a simpler function. Graph both on one set of coordinate axes.

29. $y = \dfrac{1}{3}x^3$

30. $y = (-5x)^3$

In Review Exercises 31–34, graph each rational function.

31. $y = \dfrac{x}{(x - 1)^2}$ **32.** $y = \dfrac{(x - 1)^2}{x}$ **33.** $y = \dfrac{x^2 - x - 2}{x^2 + x - 2}$ **34.** $y = \dfrac{x^3 + x}{x^2 - 4}$

In Review Exercises 35–38, let $f(x) = x^2 - 1$ and $g(x) = 2x + 1$. Find each function.

35. $f + g$ **36.** $f \cdot g$ **37.** $f \circ g$ **38.** $g \circ f$

In Review Exercises 39–42, each equation defines a one-to-one function. Determine f^{-1}.

39. $y = 7x - 1$ **40.** $y = \dfrac{1}{2 - x}$ **41.** $y = \dfrac{x}{1 - x}$ **42.** $y = \dfrac{3}{x^3}$

---

4 CHAPTER TEST

In Questions 1–2, find the domain and range of each function.

1. $f(x) = \dfrac{3}{x - 5}$ **2.** $f(x) = \sqrt{x + 3}$

In Questions 3–4, find $f(-1)$ and $f(2)$.

3. $f(x) = \dfrac{x}{x - 1}$ **4.** $f(x) = \sqrt{x + 7}$

In Questions 5–8, find the vertex of each parabola.

5. $y = 3(x - 7)^2 - 3$ **6.** $y = x^2 - 2x - 3$
7. $y = 3x^2 - 24x + 38$ **8.** $y = 5 - 4x - x^2$

In Questions 9–10, graph each polynomial function.

9. $y = x^4 - x^2$ **10.** $y = x^5 - x^3$

In Questions 11–12, assume that an object tossed vertically upward reaches a height of h feet after t seconds, where $h = 100t - 16t^2$.

11. In how many seconds does the object reach its maximum height?

12. What is that maximum height?

13. *Suspension bridge*   The cable of a suspension bridge is in the shape of the parabola $x^2 - 2500y + 25,000 = 0$ in the coordinate system shown in Illustration 1. Distances are in feet. How far above the roadway is the cable's lowest point?

14. Refer to Question 13. How far above the roadway does the cable attach to the vertical pillars?

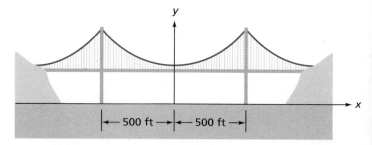

ILLUSTRATION 1

In Questions 15–16, find all asymptotes of the graph of each rational function. **Do not graph the function.**

15. $y = \dfrac{x - 1}{x^2 - 9}$

16. $y = \dfrac{x^2 - 5x - 14}{x - 3}$

In Questions 17–18, graph each rational function. Check for asymptotes, intercepts, and symmetry.

17. $y = \dfrac{x^2}{x^2 - 9}$

18. $y = \dfrac{x}{x^2 + 1}$

In Questions 19–20, graph each rational function. The numerator and denominator share a common factor.

19. $y = \dfrac{2x^2 - 3x - 2}{x - 2}$

20. $y = \dfrac{x}{x^2 - x}$

In Questions 21–24, let $f(x) = 3x$ and $g(x) = x^2 + 2$. Find each function.

21. $f + g$

22. $g \circ f$

23. f/g

24. $f \circ g$

In Questions 25–26, $f(x)$ is one-to-one. Find f^{-1}.

25. $f(x) = \dfrac{x + 1}{x - 1}$

26. $f(x) = x^3 - 3$

In Questions 27–28, graph each function.

27. $y = (x - 3)^2 + 1$

28. $y = \sqrt{x - 1} + 5$

In Questions 29–30, find the range of f by finding the domain of f^{-1}.

29. $y = \dfrac{3}{x} - 2$

30. $y = \dfrac{3x - 1}{x - 3}$

5

SOLVING POLYNOMIAL EQUATIONS

We have seen that many polynomial equations can be solved by factoring. For example, to solve the equation $x^3 - 3x^2 + 2x = 0$, we proceed as follows:

$$x^3 - 3x^2 + 2x = 0$$

$$x(x^2 - 3x + 2) = 0 \qquad \text{Factor out } x.$$

$$x(x - 1)(x - 2) = 0 \qquad \text{Factor } x^2 - 3x + 2.$$

$$x = 0 \quad \text{or} \quad x - 1 = 0 \quad \text{or} \quad x - 2 = 0 \qquad \text{Set each factor equal to 0.}$$
$$x = 1 \qquad\qquad x = 2$$

The solution set of this equation is $\{0, 1, 2\}$. Verify that each root checks.

Not all polynomials factor as easily as this example. In this chapter we will discuss how to solve more complicated polynomial equations.

THE REMAINDER AND FACTOR THEOREMS

■ The Remainder Theorem   ■ The Factor Theorem

A **polynomial equation** is an equation that can be written in the form $P(x) = 0$, where

$$P(x) = a_n x^n + a_{n-1} x^{n-1} + \cdots + a_1 x + a_0$$

is a polynomial of degree n, where n is a natural number. A **zero of the polynomial** $P(x)$ is any number r for which $P(r) = 0$. It follows that a zero of $P(x)$ is a root of the polynomial equation $P(x) = 0$. For example, 2 is a zero of the polynomial

$$P(x) = x^2 - 3x + 2$$

because

$$P(2) = 2^2 - 3(2) + 2$$
$$= 4 - 6 + 2$$
$$= 0$$

Since 2 is a zero of $x^2 - 3x + 2$, it is also a root of the polynomial equation

$$x^2 - 3x + 2 = 0$$

■ The Remainder Theorem

There is a relationship between a zero r of a polynomial $P(x)$ and the results of a long division of $P(x)$ by the binomial $x - r$. This relationship is best illustrated with an example.

EXAMPLE 1   Let $P(x) = 3x^3 - 5x^2 + 3x - 10$.   **a.** Find $P(1)$ and $P(-2)$.   **b.** Divide $P(x)$ by $x - 1$ and by $x + 2$.

Solution

a. $P(1) = 3(1)^3 - 5(1)^2 + 3(1) - 10$ $\qquad$ $P(-2) = 3(-2)^3 - 5(-2)^2 + 3(-2) - 10$

$\qquad\qquad = 3 - 5 + 3 - 10$ $\qquad\qquad\qquad\qquad = 3(-8) - 5(4) + 3(-2) - 10$

$\qquad\qquad = -9$ $\qquad\qquad\qquad\qquad\qquad\qquad = -60$

b.

$$
\begin{array}{r}
3x^2 - 2x + 1 \\
x - 1\overline{)3x^3 - 5x^2 + 3x - 10} \\
\underline{3x^3 - 3x^2} \\
-2x^2 + 3x \\
\underline{-2x^2 + 2x} \\
+ x - 10 \\
\underline{x - 1} \\
- 9
\end{array}
\qquad
\begin{array}{r}
3x^2 - 11x + 25 \\
x + 2\overline{)3x^3 - 5x^2 + 3x - 10} \\
\underline{3x^3 + 6x^2} \\
-11x^2 + 3x \\
\underline{-11x^2 - 22x} \\
25x - 10 \\
\underline{25x + 50} \\
-60
\end{array}
$$ ∎

We note that when $P(x)$ was divided by $x - 1$ in Example 1, the remainder was $P(1)$, or -9. Furthermore, when $P(x)$ was divided by $x - (-2)$, or $x + 2$, the remainder was $P(-2)$, or -60. These results are not coincidental. The following theorem, called the **remainder theorem**, asserts that the division of any polynomial $P(x)$ by the binomial $x - r$ gives $P(r)$ as the remainder.

The Remainder Theorem	If $P(x)$ is a polynomial, r is a real or nonreal complex number, and $P(x)$ is divided by $x - r$, then the remainder is $P(r)$.

Proof To divide $P(x)$ by $x - r$, we must find a **quotient** $q(x)$ and a remainder $r(x)$ such that

$$\textbf{dividend} = \textbf{divisor} \;\cdot\; \textbf{quotient} + \textbf{remainder}$$
$$\downarrow \qquad\qquad \downarrow \qquad\qquad \downarrow \qquad\qquad \downarrow$$
$$P(x) \;=\; (x - r) \;\cdot\; q(x) \;+\; r(x)$$

Furthermore, by the division algorithm, the degree of the remainder $r(x)$ must be less than the degree of the divisor $x - r$. Since the divisor, $x - r$, is of degree 1, the remainder, $r(x)$, must be some constant R.

In the equation $P(x) = (x - r)q(x) + R$, the polynomial on the left-hand side is the same as the polynomial on the right-hand side. In particular, the values that these polynomials assume for any number x must be equal. If we replace x with r, we have

$$P(r) = (r - r)q(r) + R$$
$$= (0)q(r) + R$$
$$= R$$

Thus, $P(r) = R$. $\qquad\qquad$ □

EXAMPLE 2 Without performing a long division, find the remainder that will occur when $2x^4 - 10x^3 + 17x^2 - 14x - 3$ is divided by $x - 3$.

Solution To find the remainder without long division, we can use the remainder theorem and find $P(3)$.

$$P(x) = 2x^4 - 10x^3 + 17x^2 - 14x - 3$$
$$P(3) = 2(3)^4 - 10(3)^3 + 17(3)^2 - 14(3) - 3$$
$$= 0$$

The remainder will be 0. Although this calculation is tedious to do by hand, it is easy to do with a calculator. ∎

■ The Factor Theorem

If $R = 0$ in the equation $P(x) = (x - r)Q(x) + R$, then $P(x)$ factors as $(x - r)Q(x)$. This fact will help us factor polynomials.

The Factor Theorem	Let $P(x)$ be any polynomial and let r be a real or nonreal complex number. Then $$P(r) = 0 \text{ if and only if } x - r \text{ is a factor of } P(x).$$

Proof *Part 1:* First we assume that $P(r) = 0$ and prove that $x - r$ is a factor of $P(x)$. To do so, we divide $P(x)$ by $x - r$. By the remainder theorem, the remainder will be $P(r)$. But, by assumption, $P(r) = 0$. Hence, $x - r$ divides $P(x)$ evenly and $x - r$ is a factor of $P(x)$.

Part 2: Conversely, we assume that $x - r$ is a factor of $P(x)$ and prove that $P(r) = 0$. Because, by assumption, $x - r$ is a factor of $P(x)$, $x - r$ divides $P(x)$ evenly, and the division has a remainder of 0. By the remainder theorem, this remainder is $P(r)$. Hence, $P(r) = 0$. □

EXAMPLE 3 Determine whether $x + 2$ is a factor of $P(x) = x^4 - 7x^2 - 6x$. If it is, find the other factor by long division.

Solution By the factor theorem, $x + 2$, or $x - (-2)$, is a factor of $P(x)$ if -2 is a zero of $P(x)$. So we evaluate $P(-2)$ to see if $P(-2) = 0$.

$$P(x) = x^4 - 7x^2 - 6x$$
$$P(-2) = (-2)^4 - 7(-2)^2 - 6(-2)$$
$$= 16 - 28 + 12$$
$$= 0$$

Since -2 is a zero of $P(x)$, we know that $x - (-2)$, or $x + 2$, is a factor of $P(x)$. To find the other factor, we divide $x^4 - 7x^2 - 6x$ by $x + 2$.

$$\begin{array}{r}
x^3 - 2x^2 - 3x \\
x + 2\overline{)x^4 - 7x^2 - 6x} \\
\underline{x^4 + 2x^3} \\
-2x^3 - 7x^2 \\
\underline{-2x^3 - 4x^2} \\
-3x^2 - 6x \\
\underline{-3x^2 - 6x} \\
0
\end{array}$$

Thus, $x^4 - 7x^2 - 6x$ factors as $(x + 2)(x^3 - 2x^2 - 3x)$. ∎

EXAMPLE 4 Let $P(x) = 3x^3 - 5x^2 + 3x - 10$. Show that $P(2) = 0$ and use this fact to factor $P(x)$.

Solution We calculate $P(2)$.

$$\begin{aligned}
P(x) &= 3x^3 - 5x^2 + 3x - 10 \\
P(2) &= 3(2)^3 - 5(2)^2 + 3(2) - 10 \\
&= 3(8) - 5(4) + 6 - 10 \\
&= 0
\end{aligned}$$

Because $P(2) = 0$, it follows from the factor theorem that $x - 2$ is a factor of $P(x)$. To determine the other factor, we divide $P(x)$ by $x - 2$.

$$\begin{array}{r}
3x^2 + x + 5 \\
x - 2\overline{)3x^3 - 5x^2 + 3x - 10} \\
\underline{3x^3 - 6x^2} \\
x^2 + 3x \\
\underline{x^2 - 2x} \\
5x - 10 \\
\underline{5x - 10} \\
0
\end{array}$$

Thus, $3x^3 - 5x^2 + 3x - 10$ factors as $(x - 2)(3x^2 + x + 5)$. ∎

EXAMPLE 5 Solve the equation $3x^3 - 5x^2 + 3x - 10 = 0$.

Solution From Example 4, we have

$$\begin{aligned}
3x^3 - 5x^2 + 3x - 10 &= 0 \\
(x - 2)(3x^2 + x + 5) &= 0
\end{aligned}$$

To solve for x, we set each factor equal to 0 and apply the quadratic formula to the equation $3x^2 + x + 5 = 0$.

$$
\begin{array}{c|c}
\begin{aligned}
x - 2 &= 0 \\
x &= 2
\end{aligned}
&
\text{or} \quad
\begin{aligned}
3x^2 + x + 5 &= 0 \\
x &= \frac{-1 \pm \sqrt{1^2 - 4(3)(5)}}{2(3)} \\
x &= \frac{-1 \pm i\sqrt{59}}{6}
\end{aligned}
\end{array}
$$

The complete solution set is

$$\left\{ 2, -\frac{1}{6} + \frac{\sqrt{59}}{6}i, -\frac{1}{6} - \frac{\sqrt{59}}{6}i \right\}$$

∎

EXAMPLE 6   Find a polynomial $P(x)$ with zeros of 3, 3, and -5.

Solution   By the factor theorem, if 3, 3, and -5 are zeros of $P(x)$, then $x - 3$, $x - 3$, and $x - (-5)$ are factors of $P(x)$. Thus,

$$P(x) = (x - 3)(x - 3)(x + 5)$$
$$P(x) = (x^2 - 6x + 9)(x + 5) \qquad \text{Multiply } x - 3 \text{ and } x - 3.$$
$$P(x) = x^3 - x^2 - 21x + 45$$

The polynomial $P(x) = x^3 - x^2 - 21x + 45$ has zeros of 3, 3, and -5.

∎

Because 3 occurs twice as a zero in Example 6, we say that 3 is a **zero of multiplicity 2**.

EXAMPLE 7   Find the three cube roots of -1.

Solution   The three cube roots of -1 are the solutions of the equation $x^3 = -1$, which can be written in the form

1. $x^3 + 1 = 0$

By inspection, we know that one of the cube roots of -1 is -1 and that -1 is a root of Equation 1. Because of the factor theorem, we know that $x - (-1)$, or $x + 1$, is a factor of $x^3 + 1$. Thus, $x + 1$ divides $x^3 + 1$. To obtain the other factor, we can use long division.

$$
\begin{array}{r}
x^2 - x + 1 \\
x + 1 \overline{)x^3 \qquad\qquad + 1} \\
\underline{x^3 + x^2} \\
-x^2 \\
\underline{-x^2 - x} \\
x + 1 \\
\underline{x + 1} \\
0
\end{array}
$$

Thus, the equation $x^3 + 1 = 0$ factors as

$$(x + 1)(x^2 - x + 1) = 0$$

Setting each factor equal to 0 and solving for x gives the three cube roots of -1. We can use the quadratic formula to solve the quadratic equation.

$$x + 1 = 0 \quad \text{or} \quad x^2 - x + 1 = 0$$

$$x = -1 \qquad\qquad x = \frac{-(-1) \pm \sqrt{(-1)^2 - 4(1)(1)}}{2(1)}$$

$$x = \frac{1 \pm \sqrt{-3}}{2}$$

$$x = \frac{1 \pm i\sqrt{3}}{2}$$

The three cube roots of -1 are

$$-1, \frac{1}{2} + \frac{\sqrt{3}}{2}i, \frac{1}{2} - \frac{\sqrt{3}}{2}i$$

Verify that cube of each number is -1. ∎

5.1 EXERCISES

In Exercises 1–8, let $P(x) = 2x^4 - 2x^3 + 5x^2 - 1$. Find each value by substituting the given value of x into the polynomial and simplifying. Then find the value by performing a long division and finding the remainder.

1. $P(1)$ **2.** $P(2)$ **3.** $P(-2)$ **4.** $P(-1)$

5. $P(-3)$ **6.** $P(3)$ **7.** $P\left(\dfrac{1}{2}\right)$ **8.** $P\left(-\dfrac{1}{2}\right)$

In Exercises 9–16, find the remainder that will occur when $P(x) = 3x^4 + 5x^3 - 4x^2 - 2x + 1$ is divided by the given binomial. **Do not use long division.**

9. $x - 1$ **10.** $x - 2$ **11.** $x + 2$ **12.** $x + 1$

13. $x - 12$ **14.** $x + 15$ **15.** $x + 3.25$ **16.** $x - 7.12$

In Exercises 17–20, a polynomial is given. Use the remainder theorem to find each value.

17. $P(x) = x^5 - 1; P(2)$ **18.** $P(x) = x^5 + 4x^2 - 1; P(-3)$

19. $P(x) = -x^5 + 6x^3 - 20x + 1; P(-3)$ **20.** $P(x) = 3x^4 - 2x^3 + 7x^2 - 5x + 3; P(2)$

In Exercises 21–28, use the factor theorem to decide whether each statement is true. If the statement is not true, so indicate.

21. $x - 1$ is a factor of $x^7 - 1$.

22. $x - 2$ is a factor of $x^3 - x^2 + 2x - 8$.

23. $x - 1$ is a factor of $3x^5 + 4x^2 - 7$.

24. $x + 1$ is a factor of $3x^5 + 4x^2 - 7$.

25. $x + 3$ is a factor of $2x^3 - 2x^2 + 1$.

26. $x - 3$ is a factor of $3x^5 - 3x^4 + 5x^2 - 13x - 6$.

27. $x - 1$ is a factor of $x^{1984} - x^{1776} + x^{1492} - x^{1066}$.

28. $x + 1$ is a factor of $x^{1984} + x^{1776} - x^{1492} - x^{1066}$.

In Exercises 29–34, a partial solution set for each equation is given. Find the complete solution set.

29. $x^3 + 3x^2 - 13x - 15 = 0; \{-1\}$

30. $x^3 + 6x^2 + 5x - 12 = 0; \{1\}$

31. $x^4 - 2x^3 - 2x^2 + 6x - 3 = 0; \{1, 1\}$

32. $x^5 + 4x^4 + 4x^3 - x^2 - 4x - 4 = 0; \{-2, -2\}$

33. $x^4 - 5x^3 + 7x^2 - 5x + 6 = 0; \{2, 3\}$

34. $x^4 + 2x^3 - 3x^2 - 4x + 4 = 0; \{1, -2\}$

In Exercises 35–46, find a polynomial with the given zeros.

35. $4, 5$ **36.** $-3, 5$ **37.** $1, 1, 1$ **38.** $1, 0, -1$

39. $2, 4, 5$ **40.** $7, 6, 3$ **41.** $1, -1, \sqrt{2}, -\sqrt{2}$ **42.** $0, 0, 0, \sqrt{3}, -\sqrt{3}$

43. $\sqrt{2}, i, -i$ **44.** $i, i, 2$ **45.** $0, 1 + i, 1 - i$ **46.** $i, 2 + i, 2 - i$

In Exercises 47–50, find the three cube roots of each number.

47. 1 **48.** 64 **49.** -125 **50.** -216

51. Find a_0 if 0 is a zero of
$$P(x) = a_n x^n + a_{n-1} x^{n-1} + \cdots + a_1 x + a_0$$

52. Find a_1 if 0 occurs twice as a zero of
$$P(x) = a_n x^n + a_{n-1} x^{n-1} + \cdots + a_1 x + a_0$$

5.2   SYNTHETIC DIVISION

■ Using Synthetic Division to Evaluate Polynomials   ■ Nested Forms of a Polynomial   ■ Computer Application of Nested Forms

Synthetic division is a shortcut method used to divide higher-degree polynomials by binomials of the form $x - r$. To see how synthetic division works, we consider the following long division. On the left is the complete long division. On the right is a modified version of the long division, in which the variables have been removed.

$$
\begin{array}{r}
2x^2 + 10x + 27 \\
x - 3 \overline{) 2x^3 + 4x^2 - 3x + 10} \\
\underline{2x^3 - 6x^2} \\
10x^2 - 3x \\
\underline{10x^2 - 30x} \\
27x + 10 \\
\underline{27x - 81} \\
(\text{remainder})\ 91
\end{array}
\qquad
\begin{array}{r}
2 + 10 + 27 \\
1 - 3 \overline{) 2 + 4 - 3 + 10} \\
\underline{2 - 6} \\
10 - 3 \\
\underline{10 - 30} \\
27 + 10 \\
\underline{27 - 81} \\
(\text{remainder})\ 91
\end{array}
$$

We can shorten the work on the right even further by omitting the numbers printed in color.

$$
\begin{array}{r}
2 + 10 + 27 \\
-3 \overline{) 2 + 4 - 3 + 10} \\
\underline{- 6} \\
10 \\
\underline{- 30} \\
27 \\
\underline{- 81} \\
(\text{remainder})\ 91
\end{array}
$$

We can then compress the work vertically to get

$$
\begin{array}{r}
2 + 10 + 27 \\
-3)\overline{2 + 4 - 3 + 10} \\
- 6 - 30 - 81 \\
\hline
10 \quad 27 \quad 91
\end{array}
$$

There is no reason why the quotient (represented by the numbers 2, 10, and 27) must appear above the long division symbol. If we write the 2 on the bottom line, the bottom line gives both the coefficients of the quotient and the remainder. The top line can now be eliminated, and the division appears as

$$
\begin{array}{r|l}
-3 & 2 + 4 - 3 + 10 \\
& - 6 - 30 - 81 \\
\hline
& 2 \quad 10 \quad 27 \quad 91
\end{array}
$$

The bottom line was obtained by subtracting the middle line from the top line. If we replace the -3 in the divisor with $+3$, the signs of each number in the middle line will be reversed in the division process. Then, the bottom line can be obtained by addition. Thus, we have the final form of the synthetic division.

$$
\begin{array}{r|l}
+3 & 2 + 4 - 3 + 10 \\
& + 6 + 30 + 81 \\
\hline
& 2 \quad 10 \quad 27 \,\big|\, 91
\end{array}
$$

The coefficients of the dividend.

The coefficients of the quotient and the remainder.

Thus, $\dfrac{2x^3 + 4x^2 - 3x + 10}{x - 3} = 2x^2 + 10x + 27 + \dfrac{91}{x - 3}$.

EXAMPLE 1   Use synthetic division to divide $10x + 3x^4 - 8x^3 + 3$ by $x - 2$.

Solution   We must first write the terms of $P(x)$ in descending powers of x:

$$3x^4 - 8x^3 + 10x + 3$$

We then write the coefficients of the dividend with its terms in descending powers of x and the 2 from the divisor in the following form:

$$
\begin{array}{r|l}
2 & 3 \ -8 \quad \mathbf{0} \quad 10 \quad 3 \\
& \\
\hline
& \\
\end{array}
$$

Write 0 for the coefficient of the missing x^2 term.

Then we follow these steps:

$$
\begin{array}{r|l}
2 & 3 \ -8 \quad 0 \quad 10 \quad 3 \\
& \downarrow \\
\hline
& 3
\end{array}
$$

Bring down the 3.

$$
\begin{array}{r|l}
2 & 3 \ -8 \quad 0 \quad 10 \quad 3 \\
& 6 \\
\hline
& 3 \ -2
\end{array}
$$

Multiply 2 and 3 together to get 6, and add 6 and -8 to get -2.

$$\begin{array}{r|rrrrr} 2 & 3 & -8 & 0 & 10 & 3 \\ & & 6 & -4 & & \\ \hline & 3 & -2 & -4 & & \end{array}$$

Multiply 2 and -2 together to get -4, and add -4 and 0 to get -4.

$$\begin{array}{r|rrrrr} 2 & 3 & -8 & 0 & 10 & 3 \\ & & 6 & -4 & -8 & \\ \hline & 3 & -2 & -4 & 2 & \end{array}$$

Multiply 2 and -4 together to get -8, and add -8 and 10 to get 2.

$$\begin{array}{r|rrrr|r} 2 & 3 & -8 & 0 & 10 & 3 \\ & & 6 & -4 & -8 & 4 \\ \hline & 3 & -2 & -4 & 2 & 7 \end{array}$$

Multiply 2 and 2 together to get 4, and add 4 and 3 to get 7.

The coefficients of the quotient and the remainder.

Thus,

$$\frac{10x + 3x^4 - 8x^3 + 3}{x - 2} = 3x^3 - 2x^2 - 4x + 2 + \frac{7}{x - 2}$$ ∎

◼ Using Synthetic Division to Evaluate Polynomials

EXAMPLE 2   Use synthetic division to find $P(-2)$ when $P(x) = 5x^3 + 3x^2 - 21x - 1$.

Solution   Because of the remainder theorem, $P(-2)$ will be the remainder when $P(x)$ is divided by $x - (-2)$. We use synthetic division to find the remainder.

$$\begin{array}{r|rrrr} -2 & 5 & 3 & -21 & -1 \\ & & 10 & & \\ \hline & 5 & -7 & & \end{array} \quad \begin{array}{r|rrrr} -2 & 5 & 3 & -21 & -1 \\ & & -10 & 14 & \\ \hline & 5 & -7 & -7 & \end{array} \quad \begin{array}{r|rrr|r} -2 & 5 & 3 & -21 & -1 \\ & & -10 & 14 & 14 \\ \hline & 5 & -7 & -7 & 13 \end{array}$$

Because the remainder is 13, $P(-2) = 13$.   ∎

EXAMPLE 3   If $P(x) = x^3 - x^2 + x - 1$, find $P(i)$, where $i = \sqrt{-1}$.

Solution   We can use synthetic division.

$$\begin{array}{r|rrrr} i & 1 & -1 & +1 & -1 \\ & & i & -1-i & +1 \\ \hline & 1 & i-1 & -i & 0 \end{array}$$

Since the remainder is 0, $P(i) = 0$ and i is a zero of $P(x)$.   ∎

◼ Nested Forms of a Polynomial

Synthetic division provides an easy way to evaluate polynomials when a calculator is available. To illustrate, we recall the polynomial discussed in Example 2.

$$P(x) = 5x^3 + 3x^2 - 21x - 1$$

If we divide $P(x)$ by $x - k$, we get a remainder that is a polynomial in k.

$$
\begin{array}{c|cccc}
k & 5 & 3 & -21 & -1 \\
 & & 5k & (5k + 3)k & [(5k + 3)k - 21]k \\
\hline
 & 5 & 5k + 3 & (5k + 3)k - 21 & [(5k + 3)k - 21]k - 1
\end{array}
$$

The remainder shows that $P(k) = [(5k + 3)k - 21]k - 1$, which is called the **nested form**, or **Horner's method**, of writing a polynomial. Because the value of $P(x)$ at any number x is equal to the remainder, we can evaluate $P(x)$ by evaluating $P(k)$. This is easy to do if we press these keys on a calculator:

For $P(k)$

5 ☒ k ＋ 3 ＝ ☒ k － 21 ＝ ☒ k － 1 ＝

For $P(-2)$

5 ☒ –2 ＋ 3 ＝ ☒ –2 － 21 ＝ ☒ –2 － 1 ＝

After pressing the keys on the right for $P(-2)$, we obtain $P(-2) = 13$, which agrees with the result obtained in Example 2.

When k is a large number or a decimal, we can save keystrokes by saving the value of k in memory and then recalling it each time we need to enter the value of k.

We can more easily write a polynomial in nested form by successively factoring the variable from those terms that contain the variable. For example, we can write

$$P(x) = 5x^3 + 3x^2 - 21x - 1$$

in nested form by factoring x from the first three terms to get

$$P(x) = (5x^2 + 3x - 21)x - 1$$

and then factoring x from $5x^2 + 3x$ to get

$$P(x) = [(5x + 3)x - 21]x - 1$$

■ Computer Application of Nested Forms

Computers take more time to perform multiplications than additions. Since time is money, it makes sense to enter problems into a computer in a way that will allow the computer to operate efficiently, especially if there is a large amount of calculating to be done.

Writing polynomials in nested form provides a way to evaluate polynomials efficiently with a computer. For example, the polynomial $P(x) = 3x^4 + 2x^3 + 5x^2 + 7x + 1$ can be written in expanded form as

$$P(x) = 3 \cdot x \cdot x \cdot x \cdot x + 2 \cdot x \cdot x \cdot x + 5 \cdot x \cdot x + 7 \cdot x + 1$$

to illustrate that the evaluation of this polynomial will involve 10 multiplications and 4 additions. By writing the polynomial in nested form, we can reduce the number of time-consuming multiplications.

$$
\begin{aligned}
P(x) &= 3x^4 + 2x^3 + 5x^2 + 7x + 1 \\
&= (3x^3 + 2x^2 + 5x + 7)x + 1 \qquad \text{Factor } x \text{ from the first four terms.}
\end{aligned}
$$

$$= [(3x^2 + 2x + 5)x + 7]x + 1 \qquad \text{Factor } x \text{ from the first three terms.}$$
$$= \{[(3x + 2)x + 5]x + 7\}x + 1 \qquad \text{Factor } x \text{ from the first two terms.}$$

To emphasize the number of multiplications, we write the previous equation as

$$P(x) = \{[(3 \cdot x + 2) \cdot x + 5] \cdot x + 7\} \cdot x + 1$$

There are still four additions but only four multiplications. If the polynomial were to be evaluated for many different values of x, the computer time saved by using the nested form would be substantial.

5.2 EXERCISES

In Exercises 1–8, $P(x) = 5x^3 + 2x^2 - x + 1$. *Use synthetic division to find each value of* $P(x)$.

1. $P(2)$ **2.** $P(-2)$ **3.** $P(-5)$ **4.** $P(3)$

5. $P(0)$ **6.** $P(5)$ **7.** $P(-i)$ **8.** $P(i)$

In Exercises 9–16, $P(x) = 2x^4 - x^2 + 2$. *Use synthetic division to find each value of* $P(x)$.

9. $P(1)$ **10.** $P(-1)$ **11.** $P(-2)$ **12.** $P(3)$

13. $P\left(\dfrac{1}{2}\right)$ **14.** $P\left(\dfrac{1}{3}\right)$ **15.** $P(i)$ **16.** $P(-i)$

In Exercises 17–24, $P(x) = x^4 - 8x^3 + 8x + 14x^2 - 15$. *Write the terms of* $P(x)$ *in descending powers of* x *and use synthetic division to find each value of* $P(x)$.

17. $P(1)$ **18.** $P(0)$ **19.** $P(-3)$ **20.** $P(-1)$

21. $P(3)$ **22.** $P(5)$ **23.** $P(2)$ **24.** $P(-5)$

In Exercises 25–32, $P(x) = 8 - 8x^2 + x^5 - x^3$. *Write the terms of* $P(x)$ *in descending powers of* x *and use synthetic division to find each value of* $P(x)$.

25. $P(0)$ **26.** $P(1)$ **27.** $P(2)$ **28.** $P(-2)$

29. $P(i)$ **30.** $P(-i)$ **31.** $P(-2i)$ **32.** $P(2i)$

In Exercises 33–40, use synthetic division to express $P(x) = 3x^3 - 2x^2 - 6x - 4$ *in the form* (*divisor*)(*quotient*) + *remainder for each of the given divisors.*

33. $x - 1$ **34.** $x - 2$ **35.** $x - 3$ **36.** $x - 4$

37. $x + 1$ **38.** $x + 2$ **39.** $x + 3$ **40.** $x + 4$

In Exercises 41–52, use synthetic division to perform each division.

41. $\dfrac{x^3 + x^2 + x - 3}{x - 1}$ **42.** $\dfrac{x^3 - x^2 - 5x + 6}{x - 2}$ **43.** $\dfrac{7x^3 - 3x^2 - 5x + 1}{x + 1}$ **44.** $\dfrac{2x^3 + 4x^2 - 3x + 8}{x - 3}$

45. $\dfrac{4x^4 - 3x^3 - x + 5}{x - 3}$ **46.** $\dfrac{x^4 + 5x^3 - 2x^2 + x - 1}{x + 1}$ **47.** $\dfrac{3x^5 - 768x}{x - 4}$ **48.** $\dfrac{x^5 - 4x^2 + 4x + 4}{x + 3}$

49. $\dfrac{4x^3 + x^4 + x^2 + 5}{x - 2}$ **50.** $\dfrac{x^3 + 3x^4 - x + 2}{x - 3}$ **51.** $\dfrac{3x^3 + 5x^5 + x^2 + 5}{x + 2}$ **52.** $\dfrac{4x^2 + 2x^5 - 2x - 3}{x + 3}$

In Exercises 53–58, write each polynomial in nested form.

53. $P(x) = 3x^3 + 4x^2 + 5x + 12$ **54.** $P(x) = 7x^3 - 5x^2 - 2x + 8$

55. $P(x) = 4x^4 - 2x^3 - 8x^2 + 3x + 5$ **56.** $P(x) = 5x^4 + 3x^3 + x^2 - x - 2$

57. $P(x) = 5x^5 + 3x^3 + 9x + 2$ **58.** $P(x) = 6x^5 - 2x^4 + 5x^3 + 2x + 3$

In Exercises 59–60, use a calculator and synthetic division to find each value.

59. $P(1.3)$ where $P(x) = 2.5x^3 - 0.78x^2 - 2.7x + 4.3$ **60.** $P(0.13)$ where $P(x) = 2.1x^3 - 1.2x^2 + 3.5x - 1.8$

5.3 DESCARTES' RULE OF SIGNS AND BOUNDS ON ROOTS

■ Descartes' Rule of Signs   ■ Bounds on Roots

The remainder theorem and synthetic division provide a way of verifying that a particular number is a root of a polynomial equation, but they do not provide the roots. Selecting numbers at random and checking to see whether they work is not an efficient technique. Some guidelines are needed to indicate how many roots to expect, the kind of roots to expect, and where they are located. This section develops several theorems that provide such guidelines.

Before attempting to find the roots of a complicated polynomial equation, however, we need to know whether or not it even has a root. This question was answered by Carl Friedrich Gauss (1777–1855) when he proved the **fundamental theorem of algebra**.

The Fundamental Theorem of Algebra	If $P(x)$ is a polynomial with positive degree, then $P(x)$ has at least one zero.

The fundamental theorem of algebra guarantees that polynomials such as

$$2x^3 + 3 \qquad \text{and} \qquad 32.75x^{1984} + ix^3 - (2 + i)x - 5$$

all have zeros. Since all polynomials with positive degree have a zero, their corresponding polynomial equations all have a root. It is the zero of the polynomial.

The next theorem will help us show that every nth-degree polynomial equation has exactly n roots.

The Polynomial Factorization Theorem	If $n > 0$ and $P(x)$ is an nth-degree polynomial, then $P(x)$ has exactly n linear factors: $$P(x) = a_n(x - r_1)(x - r_2)(x - r_3) \cdots (x - r_n)$$ where $r_1, r_2, r_3, \ldots, r_n$ are either real or nonreal complex numbers and a_n is the leading coefficient of $P(x)$.

Proof Let $P(x)$ be a polynomial of degree n, where $n > 0$. Because of the fundamental theorem of algebra, we know that $P(x)$ has a zero r_1 and that the equation $P(x) = 0$ has r_1 for a root. Furthermore, by the factor theorem, we know that $x - r_1$ is a factor of $P(x)$. Thus,

$$P(x) = (x - r_1)Q_1(x)$$

If the lead coefficient of the nth-degree polynomial $P(x)$ is a_n, then $Q_1(x)$ is a polynomial of degree $n - 1$ whose lead coefficient is also a_n. By the fundamental theorem of algebra, we know that $Q_1(x)$ also has a zero, r_2. According to the factor theorem, $x - r_2$ is a factor of $Q_1(x)$, and

$$P(x) = (x - r_1)(x - r_2)Q_2(x)$$

where $Q_2(x)$ is a polynomial of degree $n - 2$ with lead coefficient a_n. This process can continue only to n factors of the form $x - r_i$ until the final quotient $Q_n(x)$ is a polynomial of degree $n - n$, or degree 0. Thus, the polynomial $P(x)$ factors completely as

1. $P(x) = a_n(x - r_1)(x - r_2)(x - r_3) \cdots (x - r_n)$ □

If we substitute any one of the numbers $r_1, r_2, r_3, \ldots, r_n$ for x in Equation 1, $P(x)$ will equal 0. Thus, each value of r is a zero of $P(x)$ and a root of the equation $P(x) = 0$. There can be no other roots, because no single factor in Equation 1 is 0 for any value of x not included in the list $r_1, r_2, r_3, \ldots, r_n$.

The values of r in the previous list need not be distinct. Any number r_i that occurs k times as a root of a polynomial equation is a root of multiplicity k.

The following theorem summarizes the previous discussion.

Theorem	If multiple roots are counted individually, the polynomial equation $P(x) = 0$ with degree n ($n > 0$) has exactly n roots among the complex numbers.

The next theorem points out a pattern in the nonreal complex roots of polynomial equations with real-number coefficients.

Conjugate Pairs Theorem	If a polynomial equation $P(x) = 0$ with real-number coefficients has a complex root $a + bi$ with $b \neq 0$, then its conjugate $a - bi$ is also a root.

This theorem is often stated as *"Complex roots of real polynomial equations occur in complex conjugate pairs."* The proof is omitted.

EXAMPLE 1   Find a fourth-degree polynomial equation with real coefficients and i as a root of multiplicity 2.

Solution   Because i is a root twice and a fourth-degree polynomial equation has four roots, we must find the other two roots. According to the conjugate pairs theorem, the missing roots are the conjugates of the given roots. Thus, the complete solution set is

$$\{i, i, -i, -i\}$$

The equation is

$$(x - i)(x + i)(x - i)(x + i) = 0$$
$$(x^2 + 1)(x^2 + 1) = 0$$
$$x^4 + 2x^2 + 1 = 0$$ ■

EXAMPLE 2   Find a quadratic equation with a double root of i.

Solution   If i is a root twice in a quadratic equation, the equation is

$$(x - i)(x - i) = 0$$
$$x^2 - 2ix - 1 = 0$$

In this polynomial equation, the roots are not in complex conjugate pairs. This is not surprising, because the coefficient $2i$ is not a real number.

 Warning!   The theorem "Complex roots of real polynomial equations occur in complex conjugate pairs" applies only to polynomial equations with real coefficients. ■

■ Descartes' Rule of Signs

René Descartes is credited with a theorem known as **Descartes' rule of signs**, which enables us to estimate the number of positive, negative, and nonreal roots of a polynomial equation.

If a polynomial is written in descending powers of x and we scan it from left to right, we say that a variation in signs occurs whenever successive terms have opposite signs. For example, the polynomial

$$P(x) = \overbrace{3x^5 - 2x^4}\ \overbrace{- 5x^3 + x^2}\ \overbrace{- x - 9}$$

has three variations in sign, and the polynomial

$$P(-x) = 3(-x)^5 - 2(-x)^4 - 5(-x)^3 + (-x)^2 - (-x) - 9$$

$$= -3x^5 - 2x^4 + 5x^3 + x^2 + x - 9$$

has two variations in sign.

Descartes' Rule of Signs	If $P(x)$ is a polynomial with real coefficients, the number of positive roots of $P(x) = 0$ is either equal to the number of variations in sign of $P(x)$ or less than that by an even number. The number of negative roots of $P(x) = 0$ is either equal to the number of variations in sign of $P(-x)$ or less than that by an even number.

The proof of this theorem is omitted.

As an example of Descartes' rule of signs, we consider the polynomial equation

$$P(x) = x^8 + x^6 + x^4 + x^2 + 1 = 0$$

Since $P(x)$ is of eighth degree, the equation has 8 roots. Since there are no variations in sign of $P(x)$, there are 0 positive roots.

Because

$$P(-x) = (-x)^8 + (-x)^6 + (-x)^4 + (-x)^2 + 1 = 0$$
$$P(-x) = x^8 + x^6 + x^4 + x^2 + 1 = 0$$

has no variations in sign either, there are 0 negative roots. Thus, all 8 roots are nonreal complex numbers, and they will occur in complex conjugate pairs.

EXAMPLE 3 Discuss the possibilities for the roots of $3x^3 - 2x^2 + x - 5 = 0$.

Solution Since there are three variations of sign in $P(x) = 3x^3 - 2x^2 + x - 5 = 0$, there can be either 3 positive roots or only 1 (1 is less than 3 by the even number 2).

Because

$$P(-x) = 3(-x)^3 - 2(-x)^2 + (-x) - 5$$
$$= -3x^3 - 2x^2 - x - 5$$

has no variations in sign, there are 0 negative roots. Furthermore, 0 is not a root, because the polynomial does not have a common factor of x.

If there are 3 positive roots, then all of the roots are accounted for. If there is 1 positive root, the remaining two roots must be nonreal complex numbers. The following chart indicates these possibilities.

Number of positive roots	Number of negative roots	Number of nonreal roots
3	0	0
1	0	2

The number of nonreal complex roots is the number needed to bring the total number of roots up to three.   ■

EXAMPLE 4   Discuss the possibilities for the roots of $P(x) = 5x^5 - 3x^3 - 2x^2 + x - 1 = 0$.

Solution   Since there are three variations of sign in $P(x)$, the number of positive roots is either 3 or 1. Because $P(-x) = -5x^5 + 3x^3 - 2x^2 - x - 1$ has two variations in sign, there are 2 or 0 negative roots. The possibilities are shown in the following chart.

Number of positive roots	Number of negative roots	Number of nonreal roots
1	0	4
3	0	2
1	2	2
3	2	0

In each case, the number of nonreal complex roots is an even number. This is expected, because this polynomial has real coefficients and its nonreal complex roots will occur in conjugate pairs.   ■

■ Bounds on Roots

A final theorem, presented without proof, provides a way to find **bounds** on the roots of a polynomial equation, enabling us to look for roots where they can be found.

Theorem   Let the lead coefficient of a polynomial $P(x)$ with real coefficients be positive and do a synthetic division of the coefficients by a positive number c. If each term in the last row of the division is nonnegative, then no number greater than c can be a root of $P(x) = 0$. (c is an **upper bound** of the real roots.)

If $P(x)$ is synthetically divided by a negative number d and the signs in the last row alternate,* then no value less than d can be a root of $P(x) = 0$. (d is a **lower bound** of the real roots.)

*If 0 appears in the third row, that 0 can be assigned either a $+$ or $-$ sign to help the signs alternate.

EXAMPLE 5   Establish integer bounds for the roots of $2x^3 + 3x^2 - 5x - 7 = 0$.

Solution   We will perform several synthetic divisions by positive integers, looking for non-negative values in the last row. Then we will divide by several negative integers, looking for alternating signs in the last row. Trying 1 first gives

$$
\begin{array}{r|rrrr}
1 & 2 & 3 & -5 & -7 \\
  &   & 2 & 5 & 0 \\
\hline
  & +2 & +5 & 0 & -7
\end{array}
$$

Because one of the signs in the last row is negative, we cannot claim that 1 is an upper bound of the roots of the equation. We now try 2.

$$
\begin{array}{r|rrrr}
2 & 2 & 3 & -5 & -7 \\
  &   & 4 & 14 & 18 \\
\hline
  & +2 & +7 & +9 & +11
\end{array}
$$

Because the last row is entirely nonnegative, 2 is an upper bound. That is, no number greater than 2 can be a root of the equation.

Now we try some negative divisors, beginning with -3.

$$
\begin{array}{r|rrrr}
-3 & 2 & 3 & -5 & -7 \\
   &   & -6 & 9 & -12 \\
\hline
   & +2 & -3 & +4 & -19
\end{array}
$$

Since the signs in the last row alternate, -3 is a lower bound. That is, no number less than -3 can be a root. To see if there is a greater lower bound, we try -2.

$$
\begin{array}{r|rrrr}
-2 & 2 & 3 & -5 & -7 \\
   &   & -4 & 2 & 6 \\
\hline
   & +2 & -1 & -3 & -1
\end{array}
$$

Since the signs in the last row do not alternate, we do not know if -2 is a lower bound.

All of the real roots of the equation lie in the interval $(-3, 2)$. ■

It is important to understand what the theorem on the bounds of roots says and what it doesn't say. If we divide synthetically by a positive number c and the last row of the synthetic division is entirely nonnegative, then the theorem guarantees that c is an upper bound of the roots. If, however, the last row contains some negative values, c may be an upper bound anyway. Similarly, if we divide by a negative number d and the last row does not alternate, then d might still be a lower bound. In these situations, the theorem does not help us.

This is illustrated by Example 5. It can be shown that the smallest negative root of the equation is approximately -1.81. Thus, -2 is a lower bound for the roots of the equation. However, when we checked -2, the last row of the synthetic division did not alternate. Thus, the theorem does not always determine the best bounds for the roots of the equation.

5.3 EXERCISES

In Exercises 1–4, tell how many roots each equation has.

1. $x^{10} = 1$

2. $x^{40} = 1$

3. $3x^4 - 4x^2 - 2x = -7$ **4.** $-32x^{111} - x^5 = 1$

5. One root of $x(3x^4 - 2) = 12x$ is 0. How many other roots are there?

6. One root of $3x^2(x^7 - 14x + 3) = 0$ is 0. How many other roots are there?

In Exercises 7–10, write a third-degree polynomial equation with real coefficients and the given roots.

7. $3, -i$ **8.** $1, i$ **9.** $2, 2 + i$ **10.** $-2, 3 - i$

In Exercises 11–14, write a fourth-degree polynomial equation with real coefficients and the given roots.

11. $3, 2, i$ **12.** $1, 2, -i$ **13.** $i, 1 - i$ **14.** $i, 2 - i$

In Exercises 15–28, use Descartes' rule of signs to find the number of possible positive, negative, and nonreal roots of each equation. **Do not try to find the roots.**

15. $3x^3 + 5x^2 - 4x + 3 = 0$ **16.** $3x^3 - 5x^2 - 4x - 3 = 0$

17. $2x^3 + 7x^2 + 5x + 5 = 0$ **18.** $-2x^3 - 7x^2 - 5x - 4 = 0$

19. $8x^4 = -5$ **20.** $-3x^3 = -5$

21. $x^4 + 8x^2 - 5x - 10 = 0$ **22.** $5x^7 + 3x^6 - 2x^5 + 3x^4 + 9x^3 + x^2 + 1 = 0$

23. $-x^{10} - x^8 - x^6 - x^4 - x^2 - 1 = 0$ **24.** $x^{10} + x^8 + x^6 + x^4 + x^2 + 1 = 0$

25. $x^9 + x^7 + x^5 + x^3 + x = 0$ (Is 0 a root?)   **26.** $-x^9 - x^7 - x^5 - x^3 - x = 0$ (Is 0 a root?)

27. $-2x^4 - 3x^2 + 2x + 3 = 0$ **28.** $-7x^5 - 6x^4 + 3x^3 - 2x^2 + 7x - 4 = 0$

In Exercises 29–38, find integer bounds for the roots of each equation.

29. $x^2 - 2x - 4 = 0$ **30.** $9x^2 - 6x - 1 = 0$ **31.** $18x^2 - 6x - 1 = 0$ **32.** $2x^2 - 10x - 9 = 0$

33. $6x^3 - 13x^2 - 110x = 0$ **34.** $12x^3 + 20x^2 - x - 6 = 0$

35. $x^5 + x^4 - 8x^3 - 8x^2 + 15x + 15 = 0$ **36.** $3x^4 - 5x^3 - 9x^2 + 15x = 0$

37. $3x^5 - 11x^4 - 2x^3 + 38x^2 - 21x - 15 = 0$ **38.** $3x^6 - 4x^5 - 21x^4 + 4x^3 + 8x^2 + 8x + 32 = 0$

39. Explain why the fundamental theorem of algebra guarantees that every polynomial equation of positive degree has at least one root.

40. Explain why the fundamental theorem of algebra and the factor theorem guarantee that an nth-degree polynomial equation has n roots.

41. Prove that any odd-degree polynomial equation with real coefficients must have at least one real root.

42. If a, b, c, and d are positive numbers, prove that $ax^4 + bx^2 + cx - d = 0$ has exactly two nonreal roots.

5.4   RATIONAL ROOTS OF POLYNOMIAL EQUATIONS

■ Finding Possible Rational Roots   ■ Finding Rational Roots

In this section, we will actually find the rational roots of polynomial equations with integer coefficients. The following theorem enables us to list the possible rational roots of a polynomial equation with integer coefficients.

Theorem

Let the polynomial equation

$$P(x) = a_n x^n + a_{n-1} x^{n-1} + a_{n-2} x^{n-2} + \cdots + a_1 x + a_0 = 0$$

have integer coefficients. If the rational number $\frac{p}{q}$ (written in lowest terms) is a root of $P(x) = 0$, then p is a factor of the constant a_0, and q is a factor of the lead coefficient a_n.

Proof Let $\frac{p}{q}$ be a rational root (written in lowest terms) of the equation $P(x) = 0$. Then the equation is satisfied by $\frac{p}{q}$.

1. $\quad a_n \left(\dfrac{p}{q}\right)^n + a_{n-1} \left(\dfrac{p}{q}\right)^{n-1} + a_{n-2} \left(\dfrac{p}{q}\right)^{n-2} + \cdots + a_1 \left(\dfrac{p}{q}\right) + a_0 = 0$

We can clear Equation 1 of fractions by multiplying both sides by the least common denominator q^n to get

2. $\quad a_n p^n + a_{n-1} p^{n-1} q + a_{n-2} p^{n-2} q^2 + \cdots + a_1 p q^{n-1} + a_0 q^n = 0$

We can factor p from all but the last term and subtract $a_0 q^n$ from both sides to get

$$p(a_n p^{n-1} + a_{n-1} p^{n-2} q + a_{n-2} p^{n-3} q^2 + \cdots + a_1 q^{n-1}) = -a_0 q^n$$

Since p is a factor of the left-hand side, it is also a factor of the right-hand side. Thus, p is a factor of $-a_0 q^n$, but because $\frac{p}{q}$ was written in lowest terms, p cannot be a factor of q^n. Hence, p is a factor of a_0.

We can factor q from all but the first term of Equation 2 and subtract $a_n p^n$ from both sides to get

$$q(a_{n-1} p^{n-1} + a_{n-2} p^{n-2} q + a_{n-3} p^{n-3} q^2 + \cdots + a_0 q^{n-1}) = -a_n p^n$$

Since q is a factor of the left-hand side, it is also a factor of the right-hand side. Because q is not a factor of p^n, it must be a factor of a_n. $\qquad\square$

Finding Possible Rational Roots

To illustrate the previous theorem, we consider the polynomial equation

$$\frac{1}{2} x^4 + \frac{2}{3} x^3 + 3x^2 - \frac{3}{2} x + 3 = 0$$

Because the previous theorem requires integer coefficients, we multiply both sides of the equation by 6 to clear it of fractions.

$$3x^4 + 4x^3 + 18x^2 - 9x + 18 = 0$$

By the previous theorem, the only possible numerators for the rational roots of the equation are the factors of the constant term 18:

$$\pm 1, \ \pm 2, \ \pm 3, \ \pm 6, \ \pm 9, \text{ and } \pm 18$$

The only possible denominators are the factors of the lead coefficient 3:

$$\pm 1 \text{ and } \pm 3$$

We can form a list of all possible rational solutions by listing the combinations of possible numerators and denominators:

$$\pm \frac{1}{1}, \ \pm \frac{2}{1}, \ \pm \frac{3}{1}, \ \pm \frac{6}{1}, \ \pm \frac{9}{1}, \ \pm \frac{18}{1}, \ \pm \frac{1}{3}, \ \pm \frac{2}{3}, \ \pm \frac{3}{3}, \ \pm \frac{6}{3}, \ \pm \frac{9}{3}, \ \pm \frac{18}{3}$$

Since several of these possibilities are duplicates, we can condense the list to get

> *Possible Rational Roots*
>
> $$\pm 1, \ \pm 2, \ \pm 3, \ \pm 6, \ \pm 9, \ \pm 18, \ \pm \frac{1}{3}, \ \pm \frac{2}{3}$$

EXAMPLE 1   Prove that $\sqrt{2}$ is an irrational number.

Solution   We know that $\sqrt{2}$ is a real root of the polynomial equation $x^2 - 2 = 0$. Any rational root of this equation must have a numerator that is a factor of the constant term -2:

$$\pm 1 \text{ or } \pm 2$$

Furthermore, a rational root must have a denominator that is a factor of the lead coefficient 1:

$$\pm 1$$

Thus, the only possible rational roots of the equation are $\pm \frac{1}{1}$ and $\pm \frac{2}{1}$. Since none of these possibilities satisfies the equation, the root, $\sqrt{2}$, must be irrational.   ∎

■ Finding Rational Roots

EXAMPLE 2   Solve the polynomial equation $P(x) = 2x^3 + 3x^2 - 8x + 3 = 0$.

Solution   Because the equation is of third degree, it has 3 roots. According to Descartes' rule of signs, there are two possible combinations of positive, negative, and nonreal roots. They are summarized as follows:

Number of positive roots	Number of negative roots	Number of nonreal roots
2	1	0
0	1	2

The only possible rational roots are

$$\pm \frac{3}{1}, \ \pm \frac{1}{1}, \ \pm \frac{3}{2}, \ \pm \frac{1}{2}$$

or

$$-3, -\frac{3}{2}, -1, -\frac{1}{2}, \frac{1}{2}, 1, \frac{3}{2}, 3$$

We can check each one of the possibilities to see whether it is a root. We can start with $\frac{3}{2}$, for example,

$$
\frac{3}{2} \begin{array}{|rrrr} 2 & 3 & -8 & 3 \\ & 3 & 9 & \frac{3}{2} \\ \hline 2 & 6 & 1 & \frac{9}{2} \end{array}
$$

Since the remainder is not 0, the number $\frac{3}{2}$ is not a root and can be crossed off the list. Because every number in the last row of the synthetic division is positive, $\frac{3}{2}$ is an upper bound. Thus, 3 cannot be a root either, and we can cross it off the list as well.

$$-3, -\frac{3}{2}, -1, -\frac{1}{2}, \frac{1}{2}, 1, \cancel{\frac{3}{2}}, \cancel{3}$$

We now try another possible solution such as $\frac{1}{2}$.

$$
\frac{1}{2} \begin{array}{|rrrr} 2 & 3 & -8 & 3 \\ & 1 & 2 & -3 \\ \hline 2 & 4 & -6 & 0 \end{array}
$$

Since the remainder is 0, the number $\frac{1}{2}$ is a root. Thus, the binomial $x - \frac{1}{2}$ is a factor of $P(x)$, and any remaining roots must be supplied by the remaining factor, which is the quotient $2x^2 + 4x - 6$. We can find the other roots by solving the equation $2x^2 + 4x - 6 = 0$ or the equation

$$x^2 + 2x - 3 = 0 \qquad \text{Divide both sides by 2.}$$

This equation, called the **depressed equation**, is a quadratic equation that can be solved by factoring:

$$
\begin{aligned}
x^2 + 2x - 3 &= 0 \\
(x - 1)(x + 3) &= 0 \\
x - 1 = 0 \qquad &\text{or} \qquad x + 3 = 0 \\
x = 1 \qquad &\qquad\qquad x = -3
\end{aligned}
$$

The solution set of the given equation is $\left\{\frac{1}{2}, 1, -3\right\}$. ∎

EXAMPLE 3 Solve the equation $P(x) = x^7 - 2x^6 - 5x^5 + 6x^4 - x^3 + 2x^2 + 5x - 6 = 0$.

Solution Because the equation is of seventh degree, it has 7 roots. According to Descartes' rule of signs, there are six possible combinations of positive, negative, and nonreal roots.

Number of positive roots	Number of negative roots	Number of nonreal roots
5	2	0
3	2	2
1	2	4
5	0	2
3	0	4
1	0	6

The only possible rational roots are

$$-6, -3, -2, -1, 1, 2, 3, 6$$

We check each one, crossing off those that do not satisfy the equation. We begin with -3.

$$
\begin{array}{r|rrrrrrr}
-3 & 1 & -2 & -5 & 6 & -1 & 2 & 5 & -6 \\
   &   & -3 & 15 & -30 & 72 & -213 & 633 & -1914 \\
\hline
   & 1 & -5 & 10 & -24 & 71 & -211 & 638 & -1920
\end{array}
$$

Since the last number in the synthetic division is not 0, -3 is not a root and can be crossed off the list. Furthermore, because the signs in the last row alternate, -3 is a lower bound, and we can cross off -6 as well.

$$\cancel{-6}, \cancel{-3}, -2, -1, 1, 2, 3, 6$$

We now try -2:

$$
\begin{array}{r|rrrrrrr}
-2 & 1 & -2 & -5 & 6 & -1 & 2 & 5 & -6 \\
   &   & -2 & 8 & -6 & 0 & 2 & -8 & 6 \\
\hline
   & 1 & -4 & 3 & 0 & -1 & 4 & -3 & 0
\end{array}
$$

Since the remainder is 0, -2 is a root.

Because this root is negative, we can revise the chart of possibilities to eliminate the possibility that there are 0 negative roots.

Number of positive roots	Number of negative roots	Number of nonreal roots
5	2	0
3	2	2
1	2	4

Because -2 is a root, the factor theorem states that $x - (-2)$, or $x + 2$, is a factor of $P(x)$. Any remaining roots can be found by solving the depressed equation

$$x^6 - 4x^5 + 3x^4 - x^2 + 4x - 3 = 0$$

Because the constant term of this equation is different from the constant term of the original equation, we can cross off some other possible rational roots. The number -2 cannot be a root a second time, because it is not a factor of -3. Furthermore, the numbers 2 and 6 are no longer possible roots, because neither is a factor of -3.

The list of possible roots is now

$$\cancel{-6},\ \cancel{-3},\ \cancel{-2},\ -1,\ 1,\ 2,\ 3,\ \cancel{6}$$

Since we know that there is one more negative root, we will now synthetically divide the coefficients of the depressed equation by -1.

$$
\begin{array}{r|rrrrrr}
-1 & 1 & -4 & 3 & 0 & -1 & 4 & -3 \\
   &   & -1 & 5 & -8 & 8 & -7 & 3 \\
\hline
   & 1 & -5 & 8 & -8 & 7 & -3 & 0
\end{array}
$$

Since the remainder is 0, -1 is a root, and the solution set so far is $\{-2, -1, \ldots\}$. The number -1 cannot be a root again, because we have found both negative roots. When we cross off -1, we have only two possibilities left.

$$\cancel{-6},\ \cancel{-3},\ \cancel{-2},\ \cancel{-1},\ 1,\ 2,\ 3,\ \cancel{6}$$

The depressed equation is now $x^5 - 5x^4 + 8x^3 - 8x^2 + 7x - 3 = 0$. We can synthetically divide the coefficients of this equation by 1 to get

$$
\begin{array}{r|rrrrrr}
1 & 1 & -5 & 8 & -8 & 7 & -3 \\
  &   & 1 & -4 & 4 & -4 & 3 \\
\hline
  & 1 & -4 & 4 & -4 & 3 & 0
\end{array}
$$

Thus, 1 joins the solution set $\{-2, -1, 1, \ldots\}$. To see if 1 is a root a second time, we synthetically divide the coefficients of the new depressed equation by 1.

$$
\begin{array}{r|rrrrr}
1 & 1 & -4 & 4 & -4 & 3 \\
  &   & 1 & -3 & 1 & -3 \\
\hline
  & 1 & -3 & 1 & -3 & 0
\end{array}
$$

Again, 1 is a root, and the solution set is now $\{-2, -1, 1, 1, \ldots\}$. To see if 1 is a root a third time, we synthetically divide the coefficients of the new depressed equation by 1.

$$
\begin{array}{r|rrrr}
1 & 1 & -3 & 1 & -3 \\
  &   & 1 & -2 & -1 \\
\hline
  & 1 & -2 & -1 & -4
\end{array}
$$

Since the remainder is not 0, the number 1 is not a root three times, and we can cross off 1 from the list of possibilities, leaving only 3.

$$\cancel{-6},\ \cancel{-3},\ \cancel{-2},\ \cancel{-1},\ \cancel{1},\ 2,\ 3,\ \cancel{6}$$

To see if 3 is a root, we synthetically divide the coefficients of the new depressed equation by 3.

$$\underline{3} \begin{array}{rrrr} 1 & -3 & 1 & -3 \\ & 3 & 0 & 3 \\ \hline 1 & 0 & 1 & 0 \end{array}$$

Thus, 3 joins the solution set, which is now $\{-2, -1, 1, 1, 3\ldots\}$.

The depressed equation is now $x^2 + 1 = 0$, which can be solved as a quadratic equation.

$$x^2 + 1 = 0$$
$$x^2 = -1$$
$$x = i \quad \text{or} \quad x = -i$$

Thus, the complete solution set is $\{-2, -1, 1, 1, 3, i, -i\}$. The solution set contains 3 positive roots, 2 negative roots, and 2 nonreal roots that are complex conjugates. This combination was one of the predicted possibilities. ∎

EXAMPLE 4 To protect cranberry crops from the damage of early freezes, growers flood the cranberry bogs. Three irrigation sources, used together, can flood a cranberry bog in one day. If the sources were used one at a time, the second source would require one day longer to flood the bog than the first, and the third would require four days longer than the first. If the bog must be flooded before a freeze that is predicted in three days, can the water in the second two sources be diverted to other bogs?

Solution Let x represent the number of days it would take the first irrigation source to flood the cranberry bog. Then $x + 1$ and $x + 4$ represent the number of days it would take the second and the third sources, respectively, to flood the bog.

Because the first source, alone, requires x days to flood the bog, that source could fill $\frac{1}{x}$ of the bog in one day. Similarly, in one day's time, the remaining sources could flood $\frac{1}{x+1}$ and $\frac{1}{x+4}$ of the bog. However, in one day, the three sources together would flood 1 cranberry bog. Thus, we have the equation

The part of the cranberry bog the first source can flood in one day		the part the second source can flood in one day		the part the third source can flood in one day		one bog.
	+		+		=	

$$\frac{1}{x} \quad + \quad \frac{1}{x+1} \quad + \quad \frac{1}{x+4} \quad = \quad 1$$

We multiply both sides of the equation by $x(x + 1)(x + 4)$ to clear it of fractions, and then we simplify.

$$x(x + 1)(x + 4)\left(\frac{1}{x} + \frac{1}{x+1} + \frac{1}{x+4}\right) = 1 \cdot x(x + 1)(x + 4)$$
$$(x + 1)(x + 4) + x(x + 4) + x(x + 1) = x(x + 1)(x + 4)$$
$$x^2 + 5x + 4 + x^2 + 4x + x^2 + x = x^3 + 5x^2 + 4x$$
$$0 = x^3 + 2x^2 - 6x - 4$$

To solve the cubic equation $x^3 + 2x^2 - 6x - 4 = 0$, we first list its possible rational solutions, which are the factors of the constant term, -4:

$$-4, -2, -1, 1, 2, \text{ and } 4$$

One solution of this equation is $x = 2$, because when we divide synthetically by 2, the remainder is 0:

$$
\begin{array}{r|rrrr}
2 & 1 & 2 & -6 & -4 \\
  &   & 2 & 8 & 4 \\
\hline
  & 1 & 4 & 2 & 0
\end{array}
$$

We find the remaining solutions by using the quadratic formula to solve the depressed equation, $x^2 + 4x + 2 = 0$. These two solutions, $-2 + \sqrt{2}$ and $-2 - \sqrt{2}$, are both negative numbers. Because the time to flood a bog cannot be negative, we discard these values and conclude that $x = 2$ is the only meaningful solution.

Because the first source, alone, could flood the bog in 2 days, and it is 3 days until the freeze, the water from the second two sources can be diverted to flood other cranberry bogs. ■

5.4   EXERCISES

In Exercises 1–14, find all rational roots for each equation.

1. $x^3 - 5x^2 - x + 5 = 0$

2. $x^3 + 7x^2 - x - 7 = 0$

3. $x^3 - 2x^2 - 9x + 18 = 0$

4. $x^3 + 3x^2 - 4x - 12 = 0$

5. $x^3 - 2x^2 - x + 2 = 0$

6. $x^3 + 2x^2 - x - 2 = 0$

7. $x^4 - 10x^3 + 35x^2 - 50x + 24 = 0$

8. $x^4 + 4x^3 + 6x^2 + 4x + 1 = 0$

9. $x^4 + 3x^3 - 13x^2 - 9x + 30 = 0$

10. $x^4 - 8x^3 + 14x^2 + 8x - 15 = 0$

11. $x^5 + 3x^4 - 5x^3 - 15x^2 + 4x + 12 = 0$

12. $x^5 - 3x^4 - 5x^3 + 15x^2 + 4x - 12 = 0$

13. $x^7 - 12x^5 + 48x^3 - 64x = 0$

14. $x^7 + 7x^6 + 21x^5 + 35x^4 + 35x^3 + 21x^2 + 7x + 1 = 0$

In Exercises 15–28, find all roots for each equation.

15. $3x^3 - 2x^2 + 12x - 8 = 0$

16. $4x^4 - 8x^3 - x^2 + 8x - 3 = 0$

17. $3x^4 - 14x^3 + 11x^2 + 16x - 12 = 0$

18. $2x^4 - x^3 - 2x^2 - 4x - 40 = 0$

19. $12x^4 + 20x^3 - 41x^2 + 20x - 3 = 0$

20. $4x^5 - 12x^4 + 15x^3 - 45x^2 - 4x + 12 = 0$

21. $6x^5 - 7x^4 - 48x^3 + 81x^2 - 4x - 12 = 0$

22. $36x^4 - x^2 + 2x - 1 = 0$

23. $30x^3 - 47x^2 - 9x + 18 = 0$

24. $20x^3 - 53x^2 - 27x + 18 = 0$

25. $15x^3 - 61x^2 - 2x + 24 = 0$

26. $12x^4 + x^3 + 42x^2 + 4x - 24 = 0$

27. $20x^3 - 44x^2 + 9x + 18 = 0$

28. $24x^3 - 82x^2 + 89x - 30 = 0$

In Exercises 29–32, $1 + i$ is a root of each equation. Find all of the roots.

29. $x^3 - 5x^2 + 8x - 6 = 0$

30. $x^3 - 2x + 4 = 0$

31. $x^4 - 2x^3 - 7x^2 + 18x - 18 = 0$

32. $x^4 - 2x^3 - 2x^2 + 8x - 8 = 0$

In Exercises 33–36, find all rational roots for each equation.

33. $x^3 - \dfrac{4}{3}x^2 - \dfrac{13}{3}x - 2 = 0$

34. $x^3 - \dfrac{19}{6}x^2 + \dfrac{1}{6}x + 1 = 0$

35. $x^{-5} - 8x^{-4} + 25x^{-3} - 38x^{-2} + 28x^{-1} - 8 = 0$

36. $1 - x^{-1} - x^{-2} - 2x^{-3} = 0$

37. If n is an even integer and c is a positive constant, show that $x^n + c = 0$ has no real roots.

38. If n is an even positive integer and c is a positive constant, show that $x^n - c = 0$ has two real roots.

39. *Parallel resistance*   If three resistors with resistances R_1, R_2, and R_3 are wired in parallel, their combined resistance R is given by the formula $\dfrac{1}{R} = \dfrac{1}{R_1} + \dfrac{1}{R_2} + \dfrac{1}{R_3}$. The design of a laboratory voltmeter requires that resistance R_2 be 10 ohms greater than resistance R_1, that R_3 be 50 ohms greater than R_1, and that their combined resistance be 6 ohms. Find the exact value of each resistance.

40. *Fabricating sheet metal*   The open tray in Illustration 1 is to be manufactured from a 12-by-14-inch rectangular sheet of metal by cutting squares from the four corners and folding up the sides. If the volume of the tray is to be 160 cubic inches, what size squares should be cut from each corner?

41. *Precalculus application*   A rectangle is inscribed in the parabola $y = 16 - x^2$, as shown in Illustration 2. Find the point (x, y) if the area of the rectangle is 42 square units.

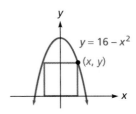

ILLUSTRATION 2

42. *Precalculus application*   One corner of the rectangle in Illustration 3 is at the origin, and the opposite corner (x, y) lies in the first quadrant on the curve $y = x^3 - 2x^2$. Find the point (x, y) if the area of the rectangle is 27 square units.

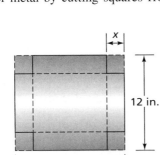

ILLUSTRATION 1

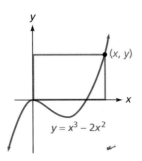

ILLUSTRATION 3

5.5   IRRATIONAL ROOTS OF POLYNOMIAL EQUATIONS

■ The Bisection Method   ■ Finding Roots with a Graphing Calculator

First-degree equations are easy to solve, and all quadratic equations can be solved with the quadratic formula. There are formulas for solving third- and fourth-degree polynomial equations, although they are complicated.

There are no formulas for solving polynomial equations of degree 5 or greater. This fact was proved for fifth-degree equations by the Norwegian mathematician Niels Henrik Abel (1802–1829) and for equations of degree greater than 5 by the French mathematician Evariste Galois (1811–1832).

To solve a high-degree polynomial equation with integer coefficients, we can use the methods of the previous section to find its rational roots. Once they are found, however, the remaining depressed equation would have to be a first- or second-degree equation for us to complete the solution.

The purpose of this section is to discuss ways of approximating irrational roots of these higher-degree polynomial equations. The following theorem, called the **intermediate value theorem**, will lead to a way of locating an interval that contains a root.

The Intermediate Value Theorem	Let $P(x)$ be a polynomial with real coefficients. If $P(a) \neq P(b)$ for $a < b$, then $P(x)$ takes on all values between $P(a)$ and $P(b)$ on the closed interval $[a, b]$.

Justification   The intermediate value theorem becomes meaningful if we consider the graph of the polynomial $y = P(x)$, shown in Figure 5-1. We have seen that graphs of polynomials are *continuous* curves, a technical term that means, roughly, that they can be drawn without lifting the pencil from the paper. If $P(a) \neq P(b)$, then the continuous curve joining the points $A(a, P(a))$ and $B(b, P(b))$ must take on all values between $P(a)$ and $P(b)$ in the interval $[a, b]$, because the curve has no gaps in it.

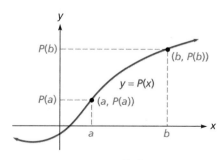

FIGURE 5-1

The next theorem is an immediate consequence of the intermediate value theorem.

| **Location Theorem** | Let $P(x)$ be a polynomial with real coefficients. If $P(a)$ and $P(b)$ have opposite signs, then there is at least one number r in the interval (a, b) for which $P(r) = 0$. |

Proof   See Figure 5-2. By the intermediate value theorem, $P(x)$ takes on all values between $P(a)$ and $P(b)$. Since $P(a)$ and $P(b)$ have opposite signs, the number 0 lies between them. Thus, there is a number r between a and b for which $P(r) = 0$. This number r is a zero of $P(x)$ and a root of the equation $P(x) = 0$.

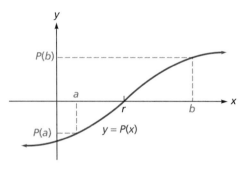

FIGURE 5-2   □

The Bisection Method

The previous theorem provides a method for finding the roots of $P(x) = 0$ to any degree of accuracy required.

Suppose we find, by trial and error, that the numbers x_l and x_r (for left and right) straddle a root; that is, $x_l < x_r$ and $P(x_l)$ and $P(x_r)$ have opposite signs. Further suppose that $P(x_l) < 0$ and $P(x_r) > 0$. We can compute the number c that is halfway between x_l and x_r and then compute $P(c)$. If $P(c) = 0$, then we've found a root. If $P(c)$ is not 0, then we proceed in one of two ways:

1. If $P(c) < 0$, the root r lies between c and x_r, as shown in Figure 5-3(a). In this case, we can let c become a new x_l and repeat the procedure.
2. If $P(c) > 0$, the root r lies between x_l and c, as shown in Figure 5-3(b). In this case, we can let c become a new x_r and repeat the procedure.

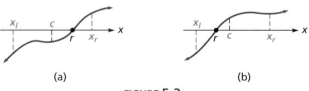

(a)   (b)

FIGURE 5-3

At any stage in this procedure, the root is contained between the current values of x_l and x_r. If the original bounds were 1 unit apart, after ten repetitions of this procedure, the root would be between bounds that were 2^{-10} units apart. After 20 repetitions, the bounds would be only 2^{-20} units apart, and the actual zero of $P(x)$ would be within 0.000001 of either x_l or x_r. This procedure is called the **bisection method**.

EXAMPLE 1   Solve the equation $P(x) = x^2 - 2 = 0$ for x and express the roots to the nearest tenth.

Solution   We first note that $P(1) = -1$ and $P(2) = 2$ have opposite signs. Thus, there is a root between 1 and 2. We can set $x_l = 1$ and $x_r = 2$ and compute the midpoint c:

$$c = \frac{x_l + x_r}{2} = \frac{1 + 2}{2} = 1.5$$

Because $P(c) = P(1.5) = 0.25$ is a positive number, we let c become our new x_r, and we calculate a new midpoint, which we will call c_1.

$$c_1 = \frac{x_l + x_r}{2} = \frac{1 + \mathbf{1.5}}{2} = 1.25$$

Because $P(c_1) = P(1.25) = -0.4375$ is a negative number, we let c_1 become our new x_l, and we calculate a new midpoint, which we will call c_2.

$$c_2 = \frac{x_l + x_r}{2} = \frac{\mathbf{1.25} + 1.5}{2} = 1.375$$

Because $P(c_2) = P(1.375) = -0.109375$ is a negative number, we let c_2 become our new x_l, and we calculate a new midpoint, which we will call c_3.

$$c_3 = \frac{x_l + x_r}{2} = \frac{\mathbf{1.375} + 1.5}{2} = 1.4375$$

Because $P(c_3) = P(1.4375) = 0.066406$ is a positive number, we let c_3 become our new x_r, and we calculate a new midpoint, which we will call c_4.

$$c_4 = \frac{x_l + x_r}{2} = \frac{1.375 + \mathbf{1.4375}}{2} = 1.40625$$

From here on, the first two digits of the midpoints will remain 1.4. Thus, the root r of the equation (to the nearest tenth) is 1.4. Of course, you already knew that the correct root was $\sqrt{2} \approx 1.414$.   ∎

■ Finding Roots with a Graphing Calculator

We can approximate the roots of an equation by using the TRACE and ZOOM capabilities of a graphing calculator. The next example illustrates the method and also finds a root that the bisection method would have missed.

EXAMPLE 2   Use a graphing calculator to solve the equation $x^4 - 6x^2 + 9 = 0$.

Solution   The solutions of an equation of the form $P(x) = 0$ are the x-intercepts of the graph of $y = P(x)$. Thus, we can use a graphing calculator to approximate the solutions of the equation $x^4 - 6x^2 + 9 = 0$ by graphing the function $y = x^4 - 6x^2 + 9$ in a suitable viewing window and using TRACE to approximate the x-intercepts.

We begin by observing that the function $y = x^4 - 6x^2 + 9$ is symmetric about the y-axis, so that if r is a positive x-intercept, then $-r$ is also an x-intercept. We graph the function in a viewing window set to display values of x and y between -5 and 5, and zoom in on the area of the graph shown in Figure 5-4(a). The result of graphing in the new window appears in Figure 5-4(b). We zoom in on the intercept again, producing the graph in Figure 5-4(c). There, we enter the TRACE mode and find that the y-coordinate of the intercept is about 7.5×10^{-8}, which is very close to 0. The displayed x-coordinate is a good approximation to the actual positive solution: $x \approx 1.732$. By symmetry, we know that the negative solution is $x \approx -1.732$.

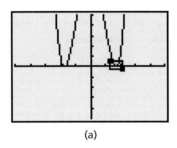
(a)

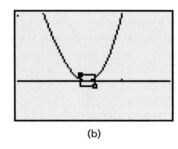
(b)

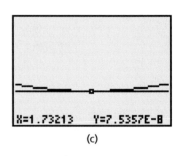
(c)

FIGURE **5-4**

These graphs illustrate a situation in which the bisection method fails. Since the graph does not cross the axis, values of the function on the opposite sides of the intercept are not of opposite sign, and the bisection method is useless. ■

5.5 EXERCISES

 In Exercises 1–10, show that each equation has at least one real root between the specified numbers.

1. $2x^2 + x - 3 = 0$; -2 and -1

2. $2x^3 + 17x^2 + 31x - 20 = 0$; -1 and 2

3. $3x^3 - 11x^2 - 14x = 0$; 4 and 5

4. $2x^3 - 3x^2 + 2x - 3 = 0$; 1 and 2

5. $x^4 - 8x^2 + 15 = 0$; 1 and 2

6. $x^4 - 8x^2 + 15 = 0$; 2 and 3

7. $30x^3 + 10 = 61x^2 + 39x$; 2 and 3

8. $30x^3 + 10 = 61x^2 + 39x$; -1 and 0

9. $30x^3 + 10 = 61x^2 + 39x$; 0 and 1

10. $5x^3 - 9x^2 - 4x + 9 = 0$; -1 and 2

 In Exercises 11–18, use the bisection method to find the following values to the nearest tenth.

11. The positive root of $x^2 - 3 = 0$.

12. The negative root of $x^2 - 3 = 0$.

13. The negative root of $x^2 - 5 = 0$.

14. The positive root of $x^2 - 5 = 0$.

15. The positive root of $x^3 - x^2 - 2 = 0$.

16. The negative root of $x^3 - x + 2 = 0$.

17. The negative root of $3x^4 + 3x^3 - x^2 - 4x - 4 = 0$.

18. The positive root of
$$x^5 + x^4 - 4x^3 - 4x^2 - 5x - 5 = 0.$$

In Exercises 19–22, use a graphing calculator to find the distinct real solutions of each equation to the nearest tenth. Which roots, if any, would the bisection method fail to find?

19. $x^2 - 5 = 0$

20. $x^2 - 10x + 25 = 0$

21. $x^3 - 5x^2 + 8x - 4 = 0$

22. $x^3 - 5x^2 - 2x + 10 = 0$

23. *Precalculus application*   Use the bisection method or a graphing calculator to find (to the nearest hundredth) the coordinates of the two points on the graph of $y = x^3$ that lie 1 unit from the origin. (See Illustration 1.)

24. *Fabricating a shipping crate*   The width of the shipping crate in Illustration 2 is to be 2 feet greater than its height and 5 feet longer than its width, and its volume is to be 170 cubic feet. Use the bisection method or a graphing calculator to find the height of the crate to the nearest tenth of a foot.

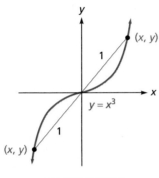

ILLUSTRATION 1

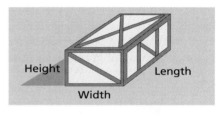

ILLUSTRATION 2

5 CHAPTER SUMMARY

Key Words

bisection method (5.5)
bounds on roots (5.3)
depressed equation (5.4)
Descartes' rule of signs (5.3)

factor theorem (5.1)
fundamental theorem of algebra (5.3)
nested forms (5.2)
polynomial equation (5.1)

quotient (5.1)
remainder (5.1)
synthetic division (5.2)
zeros of a polynomial (5.1)

Key Ideas

(5.1) If $n > 0$, every nth degree polynomial has at least one zero.

The Remainder Theorem. If $P(x)$ is a polynomial, r is any number, and $P(x)$ is divided by $x - r$, then the remainder is $P(r)$.

The Factor Theorem. Let $P(x)$ be any polynomial and r be any number. Then, $P(r) = 0$ if and only if $x - r$ is a factor of $P(x)$.

(5.2) Synthetic division can be used to find the quotient and the remainder when a polynomial is to be divided by a binomial of the form $x - r$.

(5.3) **The Fundamental Theorem of Algebra.** If $P(x)$ is a polynomial with positive degree, then $P(x)$ has at least one zero.

The Polynomial Factorization Theorem. If $n > 0$ and $P(x)$ is an nth-degree polynomial, then $P(x)$ has exactly n linear factors.

If multiple roots are counted individually, a polynomial equation $P(x) = 0$ with degree $n > 0$ has exactly n roots among the complex numbers.

Nonreal complex roots of polynomial equations with real coefficients occur in complex conjugate pairs.

Descartes' Rule of Signs. If $P(x)$ is a polynomial with real coefficients, the number of positive roots of $P(x) = 0$ is equal to the number of variations in sign of $P(x)$, or less than that by an even number.

The number of negative roots of $P(x) = 0$ is equal to the number of variations in sign of $P(-x)$, or less than that by an even number.

Upper Bounds. Let the lead coefficient of the polynomial $P(x)$ with real coefficients be positive and do a synthetic division of the coefficients of $P(x)$ by the positive number c. If none of the terms in the bottom row is negative, then c is an upper bound for the real roots of $P(x) = 0$.

Lower Bounds. Let the lead coefficient of the polynomial $P(x)$ with real coefficients be positive and do a synthetic division of the coefficients of $P(x)$ by the negative number

d. If the signs in the bottom row alternate, then d is a lower bound for the real roots of $P(x) = 0$.

(5.4) If the polynomial equation $P(x) = a_n + a_{n-1}x^{n-1} + a_{n-2}x^{n-2} + \cdots + a_1x + a_0 = 0$ has integer coefficients and the rational number $\frac{p}{q}$ (written in lowest terms) is a root of $P(x) = 0$, then p is a factor of a_0, and q is a factor of a_n.

(5.5) **The Intermediate Value Theorem.** Let $P(x)$ be a polynomial with real coefficients. If $P(a) \neq P(b)$ for $a < b$, then $P(x)$ takes on all values between $P(a)$ and $P(b)$ on the closed interval $[a, b]$.

Let $P(x)$ be a polynomial with real coefficients. If $P(a)$ and $P(b)$ have opposite signs, then there is at least one number r between a and b for which $P(r) = 0$.

5   CHAPTER REVIEW EXERCISES

In Review Exercises 1–4, let $P(x) = 4x^4 + 2x^3 - 3x^2 - 2$. *Use synthetic division to evaluate* $P(x)$ *for the given value.*

1. $P(0)$ **2.** $P(2)$ **3.** $P(-3)$ **4.** $P\left(\dfrac{1}{2}\right)$

In Review Exercises 5–8, use the factor theorem to decide whether each statement is true. If it is false, so indicate.

5. $x - 2$ is a factor of $x^3 + 4x^2 - 2x + 4$.

6. $x + 3$ is a factor of $2x^4 + 10x^3 + 4x^2 + 7x + 21$.

7. $x - 5$ is a factor of $x^5 - 3125$.

8. $x - 6$ is a factor of $x^5 - 6x^4 - 4x + 24$.

In Review Exercises 9–12, find the polynomial of lowest degree with the given zeros.

9. $-1, 2,$ and $\dfrac{3}{2}$ **10.** $1, -3,$ and $\dfrac{1}{2}$ **11.** $2, -5, i,$ and $-i$ **12.** $3, 2, i,$ and $-i$

In Review Exercises 13–16, use synthetic division to find the quotient and remainder when the given polynomial is divided by the given divisor.

13. $3x^4 + 2x^2 + 3x + 7; x - 3$ **14.** $2x^4 - 3x^2 + 3x - 1; x - 2$

15. $5x^5 - 4x^4 + 3x^3 - 2x^2 + x - 1; x + 2$ **16.** $4x^5 + 2x^4 - x^3 + 3x^2 + 2x + 1; x + 1$

In Review Exercises 17–20, tell how many roots each equation has.

17. $3x^6 - 4x^5 + 3x + 2 = 0$ **18.** $2x^6 - 5x^4 + 5x^3 - 4x + x - 12 = 0$

19. $3x^{65} - 4x^{50} + 3x^{17} + 2x = 0$ **20.** $x^{1984} - 12 = 0$

In Review Exercises 21–26, use Descartes' rule of signs to find the number of possible positive, negative, and nonreal roots for each equation. **Do not attempt to solve the equation.**

21. $3x^4 + 2x^3 - 4x + 2 = 0$

22. $2x^4 - 3x^3 + 5x^2 + x - 5 = 0$

23. $4x^5 + 3x^4 + 2x^3 + x^2 + x = 7$

24. $3x^7 - 4x^5 + 3x^3 + x - 4 = 0$

25. $x^4 + x^2 + 24{,}567 = 0$

26. $-x^7 - 5 = 0$

In Review Exercises 27–30, find all roots of each equation.

27. $2x^3 + 17x^2 + 41x + 30 = 0$

28. $3x^3 + 2x^2 + 2x - 1 = 0$

29. $4x^4 - 25x^2 + 36 = 0$

30. $2x^4 - 11x^3 - 6x^2 + 64x + 32 = 0$

In Review Exercises 31–32, show that each polynomial has a zero between the two given numbers.

31. $5x^3 + 37x^2 + 59x + 18 = 0$; -1 and 0

32. $6x^3 - x^2 - 10x - 3 = 0$; 1 and 2

In Review Exercises 33–34, use the bisection method to find the positive root of each equation to the nearest tenth.

33. $x^3 - 2x^2 - 9x - 2 = 0$

34. $6x^2 - 13x - 5 = 0$

In Review Exercises 35–36, use a graphing calculator to find the positive root of each equation to the nearest hundredth.

35. $6x^2 - 7x - 5 = 0$

36. $3x^2 + x - 2 = 0$

37. *Designing solar collectors* The space available for the installation of three solar collecting panels requires that their lengths differ by the amounts shown in Illustration 1, and that the total of their widths be 15 meters. To be equally effective, each panel must measure 60 square meters. Determine the dimensions of each panel.

38. *Designing a storage tank* The design specifications for the cylindrical storage tank in Illustration 2 require that its height be 3 feet greater than the radius of its circular base, and that the volume of the tank be 19,000 cubic feet. Use the bisection method or a graphing calculator to find the radius of the tank to the nearest hundredth of a foot.

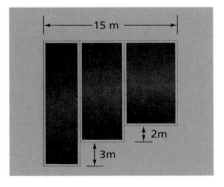

ILLUSTRATION 1

ILLUSTRATION 2

5   CHAPTER TEST

In Questions 1–2, a polynomial is given. Use the remainder theorem to find each value.

1. $P(x) = 3x^3 - 9x - 5$; $P(2)$

2. $P(x) = x^5 + 2$; $P(-2)$

In Questions 3–4, find a polynomial with the given zeros.

3. $5, -1, 0$

4. $i, -i, \sqrt{3}, -\sqrt{3}$

In Questions 5–8, let $P(x) = 3x^3 - 2x^2 + 4$. Use synthetic division to find each value of $P(x)$.

5. $P(1)$

6. $P(-2)$

7. $P\left(-\dfrac{1}{3}\right)$

8. $P(i)$

In Questions 9–10, use synthetic division to express $P(x) = 2x^3 - 3x^2 - 4x - 1$ in the form (divisor)(quotient) + remainder, for each divisor.

9. $x - 2$

10. $x + 1$

In Questions 11–12, use synthetic division to perform each division.

11. $\dfrac{2x^2 - 7x - 15}{x - 5}$

12. $\dfrac{3x^3 + 7x^2 + 2x}{x + 2}$

In Questions 13–14, write a third-degree polynomial equation with real coefficients and the given roots.

13. $2, i$

14. $1, 2 + i$

*In Questions 15–16, use Descartes' rule of signs to find the number of possible positive, negative, and nonreal roots of each equation. **Do not find the roots.***

15. $3x^5 - 2x^4 + 2x^2 - x - 3 = 0$

16. $2x^3 - 5x^2 - 2x - 1 = 0$

In Questions 17–18, find integer bounds for the roots of each equation.

17. $x^5 - x^4 - 5x^3 + 5x^2 + 4x - 5 = 0$

18. $2x^3 - 11x^2 + 10x + 3 = 0$

In Question 19, find all roots of the equation.

19. $2x^3 + 3x^2 - 11x - 6 = 0$

In Question 20, use the bisection method to find the positive root of the equation to the nearest tenth.

20. $x^2 - 11 = 0$

C H A P T E R

EXPONENTIAL AND LOGARITHMIC FUNCTIONS

In this chapter, we discuss two functions that are important in many applications of mathematics. The first type, the **exponential functions**, are used to compute compound interest and to model population growth and radioactive decay. The second type, **logarithmic functions**, are used to measure the acidity of solutions, the intensity of earthquakes, and safe noise levels in factories.

6.1 EXPONENTIAL FUNCTIONS

■ Graphing Exponential Functions   ■ Applications of Exponential Functions
■ Watching Money Grow

*Sir Isaac Newton
(1642–1727)
Newton made contributions
in many of the sciences.
He is best known in
mathematics for his
development of calculus
and in physics for his
discovery of the laws
of motion.*

In this section, we will consider exponential expressions such as 3^x, where x is a real number. Since we have defined 3^x only when x is a rational number, we must now define the symbol 3^x where x is an irrational number.

To define irrational exponents, we consider the expression $3^{\sqrt{2}}$, where $\sqrt{2}$ is the irrational number $1.414213562\ldots$. Because $1 < \sqrt{2} < 2$, it can be shown that

$$3^1 < 3^{\sqrt{2}} < 3^2$$

and because $1.4 < \sqrt{2} < 1.5$, it can be shown that

$$3^{1.4} < 3^{\sqrt{2}} < 3^{1.5}$$

The value of $3^{\sqrt{2}}$ is bounded by two numbers involving only rational powers of 3, as shown in the following list. As the list continues, its value gets squeezed into a smaller and smaller interval:

$$3^1 = \mathbf{3} < 3^{\sqrt{2}} < \mathbf{9} = 3^2$$
$$3^{1.4} \approx \mathbf{4.656} < 3^{\sqrt{2}} < \mathbf{5.196} \approx 3^{1.5}$$
$$3^{1.41} \approx \mathbf{4.7070} < 3^{\sqrt{2}} < \mathbf{4.7590} \approx 3^{1.42}$$
$$3^{1.414} \approx \mathbf{4.727695} < 3^{\sqrt{2}} < \mathbf{4.732892} \approx 3^{1.415}$$

There is exactly one real number that is greater than any of the increasing numbers on the left-hand side of the previous list and less than all of the decreasing numbers on the right-hand side. We will define this number to be $3^{\sqrt{2}}$.

To find an approximation for $3^{\sqrt{2}}$, we can press these keys on a scientific calculator:

$$3 \;\boxed{y^x}\; 2 \;\boxed{\sqrt{x}}\; \boxed{=} \qquad \text{(You may have to press } \boxed{\text{2nd}} \text{ before } \boxed{\sqrt{x}} \text{.)}$$

The display will show 4.7288044. Thus, $3^{\sqrt{2}} \approx 4.7288044$.

In general, if b is a positive number and x is any real number, the exponential expression b^x represents a single positive number. It can be shown that all of the familiar properties of exponents hold for irrational exponents as well.

If $b > 0$ and $b \neq 1$, the function defined by the equation $y = f(x) = b^x$ is called an **exponential function**.

Exponential Function	The **exponential function with base *b*** is defined by the equation

$$y = f(x) = b^x \qquad \text{where } b > 0, b \neq 1, \text{ and } x \text{ is a real number}$$

The **domain of the exponential function** is the interval $(-\infty, \infty)$ of real numbers. The **range of the exponential function** is the interval $(0, \infty)$ of positive numbers.

■ Graphing Exponential Functions

Because the domain and range of the exponential function $y = f(x) = b^x$ are sets of real numbers, we can graph the exponential functions on a rectangular coordinate system.

EXAMPLE 1 Graph the exponential functions   **a.** $y = 2^x$ and   **b.** $y = 4^x$.

Solution   **a.** To construct the graph of $y = 2^x$, shown in Figure 6-1(a), we calculate several points (x, y) whose coordinates satisfy the equation, plot the points, and join them with a smooth curve.

b. To construct the graph of $y = 4^x$, shown in Figure 6-1(b), we calculate several points (x, y) whose coordinates satisfy the equation, plot the points, and join them with a smooth curve.

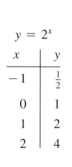

$y = 2^x$

x	y
-1	$\frac{1}{2}$
0	1
1	2
2	4

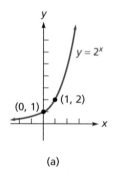

(a)

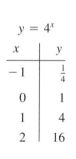

$y = 4^x$

x	y
-1	$\frac{1}{4}$
0	1
1	4
2	16

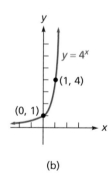

(b)

FIGURE 6-1

By looking at the graphs, we see that the domain of each function is the interval $(-\infty, \infty)$, and the range is the interval $(0, \infty)$.   ■

In each graph in Example 1, the values of y increase as the values of x increase. For this reason, the functions defined by $y = 2^x$ and $y = 4^x$ are increasing functions. Furthermore, each graph approaches the x-axis as x gets smaller, and each graph passes through the point $(0, 1)$. The graph of $y = 2^x$ passes through the point $(1, 2)$, and the graph of $y = 4^x$ passes through the point $(1, 4)$.

EXAMPLE 2   Graph the exponential functions $y = \left(\dfrac{1}{2}\right)^{x}$ and $y = \left(\dfrac{1}{4}\right)^{x}$.

Solution   We can calculate and plot several pairs (x, y) that satisfy each equation. The graph of $y = \left(\frac{1}{2}\right)^{x}$ appears in Figure 6-2(a), and the graph of $y = \left(\frac{1}{4}\right)^{x}$ appears in Figure 6-2(b). The domain of each function is the interval $(-\infty, \infty)$, and the range is the interval $(0, \infty)$.

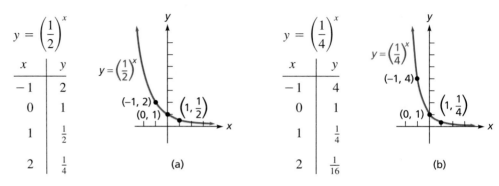

FIGURE 6-2

In each graph in Example 2, the values of y decrease as the values of x increase. For this reason, the functions defined by $y = \left(\frac{1}{2}\right)^{x}$ and $y = \left(\frac{1}{4}\right)^{x}$ are decreasing functions. Furthermore, each graph approaches the x-axis as x gets larger, and each graph passes through the point $(0, 1)$. The graph of $y = \left(\frac{1}{2}\right)^{x}$ passes through the point $\left(1, \frac{1}{2}\right)$, and the graph of $y = \left(\frac{1}{4}\right)^{x}$ passes through the point $\left(1, \frac{1}{4}\right)$.

Examples 1 and 2 illustrate the following fact.

Theorem	The graph of the exponential function defined by $y = f(x) = b^{x}$ passes through the points $(0, 1)$ and $(1, b)$.

An exponential function with base b is either increasing (for $b > 1$) or decreasing (for $0 < b < 1$). Because distinct real numbers x will determine distinct values b^{x}, exponential functions are one-to-one.

Theorem	The exponential function defined by $$y = f(x) = b^{x} \qquad \text{where } b > 0 \text{ and } b \neq 1$$ is one-to-one, which means that **1.** If $b^{r} = b^{s}$, then $r = s$. **2.** If $r \neq s$, then $b^{r} \neq b^{s}$.

EXAMPLE 3 On the same set of coordinate axes, graph $y = \left(\dfrac{3}{2}\right)^x$ and $y = \left(\dfrac{2}{3}\right)^x$.

Solution We plot several pairs (x, y) that satisfy each equation and draw each graph as in Figure 6-3.

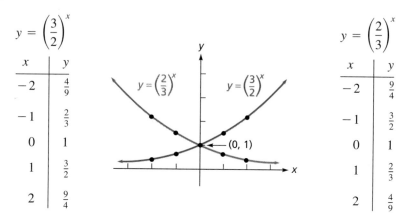

$y = \left(\dfrac{3}{2}\right)^x$

x	y
-2	$\frac{4}{9}$
-1	$\frac{2}{3}$
0	1
1	$\frac{3}{2}$
2	$\frac{9}{4}$

$y = \left(\dfrac{2}{3}\right)^x$

x	y
-2	$\frac{9}{4}$
-1	$\frac{3}{2}$
0	1
1	$\frac{2}{3}$
2	$\frac{4}{9}$

FIGURE 6-3

We note that $\frac{3}{2}$ and $\frac{2}{3}$ are reciprocals of each other and that the graphs are reflections of each other about the y-axis. This follows from the properties of exponents. If the number x in the first equation $y = \left(\frac{3}{2}\right)^x$ is replaced with $-x$, the result is the second equation, $y = \left(\frac{2}{3}\right)^x$.

$$y = \left(\frac{3}{2}\right)^x$$

$$y = \left(\frac{3}{2}\right)^{-x} \qquad \text{Replace } x \text{ with } -x.$$

$$= \left[\left(\frac{3}{2}\right)^{-1}\right]^x$$

$$= \left(\frac{2}{3}\right)^x$$

■

We summarize the properties of the exponential function with base b as follows.

Properties of Exponential Functions

The domain of an exponential function is the interval $(-\infty, \infty)$ of real numbers. The range is the interval $(0, \infty)$ of positive numbers.

If $b > 1$, then $y = b^x$ defines an increasing function.
If $0 < b < 1$, then $y = b^x$ defines a decreasing function.
The graph of $y = b^x$ passes through the points $(0, 1)$ and $(1, b)$.
The graphs of $y = b^x$ and $y = b^{-x}$ are reflections of each other about the y-axis.
The x-axis is an asymptote of the graph of $y = b^x$.
The exponential function defined by $y = b^x$ is one-to-one.

In Section 4.3, we saw that when $k > 0$, the graph of

1. $y = f(x) + k$ **2.** $y = f(x) - k$
3. $y = f(x - k)$ **4.** $y = f(x + k)$

is identical to the graph of $y = f(x)$, except that the graph is translated k units

1. upward   **2.** downward
3. to the right   **4.** to the left

The same principles apply to the graphs of exponential functions.

EXAMPLE 4   Graph the functions defined by **a.** $y = 2^x - 3$, **b.** $y = 2^{x+2}$, and **c.** $y = 2(3^{x/2})$.

Solution   **a.** The graph of $y = 2^x - 3$, shown in Figure 6-4(a), is identical to the graph of $y = 2^x$, except that it is translated 3 units downward.

b. The graph of $y = 2^{x+2}$, shown in Figure 6-4(b), is identical to the graph of $y = 2^x$, except that it is translated 2 units to the left.

c. The graph of $y = 2(3^{x/2})$, shown in Figure 6-4(c), has the same shape as an exponential function. To determine its graph, we plot several pairs (x, y) that satisfy the equation and join them with a smooth curve.

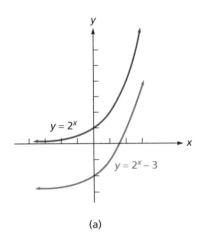
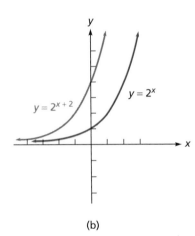
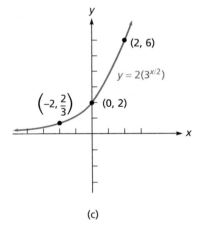

(a)   (b)   (c)

FIGURE 6-4

Applications of Exponential Functions

A mathematical description of an observed event is called a **model** of that event. Many events that change with time can be modeled by functions defined by equations of the form

$$y = f(t) = ab^{kt}$$ **Warning!** Remember that ab^{kt} means $a(b^{kt})$.

where a, b, and k are constants and t represents time. If f is an increasing function such as the one in Example 4(c), then y is said to **grow exponentially**. If f is a decreasing function, then y is said to **decay exponentially**.

Radioactive Decay

A decreasing function determined by the equation of the form $y = ab^{kt}$ models a process called **radioactive decay**. The atomic structure of a radioactive material changes as the material emits radiation. Uranium, for example, changes (decays) into thorium, then into radium, and eventually into lead.

Experiments have determined the time it takes for half of a sample of a given radioactive material to decompose. This time is a constant, called the material's **half-life**. The amount of radioactive material present in a radioactive material decays exponentially according to the following model.

Radioactive Decay Formula	If A is the amount of radioactive material present at time t, A_0 was the amount present at $t = 0$, and h is the material's half-life, then $$A = A_0 2^{-t/h}$$

EXAMPLE 5 The half-life of radium is approximately 1600 years. How much of a 1-gram sample will remain after 660 years?

Solution In this example, $A_0 = 1$, $h = 1600$, and $t = 660$. We substitute these values into the formula for radioactive decay and simplify.

$$A = A_0 2^{-t/h}$$
$$A = 1 \cdot 2^{-660/1600}$$
$$\approx 0.751320306 \qquad \text{Use a calculator.}$$

After 660 years, approximately 0.75 gram of radium will remain. ■

Compound Interest

An example of exponential growth is **compound interest**. If interest earned on money in a bank account is allowed to accumulate in the account, that interest will also earn interest. The balance in such an account will grow exponentially according to the following model.

Compound Interest Formula	If A_0 dollars are deposited in an account earning an annual rate r, compounded k times per year, then the amount A in the account after t years is given by $$A = A_0\left(1 + \frac{r}{k}\right)^{kt}$$

EXAMPLE 6 In the name of a newborn child, a parent deposits $8000 in a savings plan that earns 9% interest, compounded quarterly. If the money is left untouched, how much will the child have in 55 years?

Solution We can substitute 8000 for A_0, 0.09 for r, and 55 for t into the formula for compound interest and calculate A. Because interest is paid quarterly, $k = 4$.

$$A = A_0\left(1 + \frac{r}{k}\right)^{kt}$$

$$A = 8000\left(1 + \frac{0.09}{4}\right)^{4(55)}$$

$$= 8000(1 + 0.0225)^{220}$$

$$= 1,069,103.27 \qquad \text{Use a calculator.}$$

In 55 years, the account will be worth $1,069,103.27—more than a million dollars.   ■

In financial calculations, the initial amount deposited is often called the **present value**, denoted by *PV*. The amount to which the account will grow is called the **future value**, denoted by *FV*. The interest rate used for each compounding period is the **periodic interest rate**, *i*, and the number of times interest is compounded is the **number of compounding periods**, *n*. Using these definitions, an alternate formula for the compound interest formula is as follows.

Compound Interest Formula	$FV = PV(1 + i)^n$

This alternate formula appears on many business calculators. To use this formula to solve Example 6, we proceed as follows:

$$FV = PV(1 + i)^n$$

$$FV = 8000(1 + 0.0225)^{220} \qquad i = \frac{0.09}{4} = 0.0225 \text{ and } n = 4(55) = 220.$$

$$= 8000(1.0225)^{220}$$

$$= 1,069,103.27 \qquad \text{Use a calculator.}$$

Oceanography

EXAMPLE 7   The intensity *I* of light at a distance *x* meters below the surface of a body of water decreases exponentially according to the formula

$$I = I_0 k^x$$

where I_0 is the intensity of light above the water and *k* is a constant that depends on the clarity of the water. For a certain area of the Atlantic Ocean, $I_0 = 12$ and $k = 0.6$. Find the intensity of light at a depth of 5 meters.

Solution   After substituting 12 for I_0 and 0.6 for *k*, we have the formula

$$I = 12(0.6)^x$$

We substitute 5 for x and calculate I.

$$I = 12(0.6)^5$$
$$I \approx 0.93312$$

At a depth of 5 meters, the intensity of the light is slightly less than 1, or about one-twelfth of the intensity at the surface. ∎

■ Watching Money Grow

EXAMPLE 8   If $1 is deposited in an account earning 9% annual interest, compounded monthly, estimate how much will be in the account in 70 years.

Solution   We can substitute 1 for A_0, .09 for r, and 12 for k into the formula

$$A = A_0\left(1 + \frac{r}{k}\right)^{kt}$$

and simplify to get

$$A = (1.0075)^{12t}$$

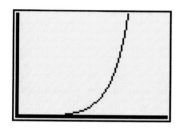

FIGURE 6-5

We now use a graphing calculator to see how the money grows year by year. We graph the function $A = (1.0075)^{12t}$ in the viewing window $0 \le t \le 100$ and $0 \le A \le 750$ to obtain the graph shown in Figure 6-5. We can then use the TRACE and ZOOM features to estimate that $1 grows to the surprising amount of approximately $532 in 70 years. ∎

6.1   EXERCISES

   In Exercises 1–4, find each value to four decimal places.

1. $4^{\sqrt{3}}$ **2.** $5^{\sqrt{2}}$ **3.** 7^{π} **4.** $3^{-\pi}$

In Exercises 5–12, graph each exponential function.

5. $y = 3^x$ **6.** $y = 5^x$ **7.** $y = \left(\frac{1}{5}\right)^x$ **8.** $y = \left(\frac{1}{3}\right)^x$

9. $y = -2^x$ **10.** $y = -3^x$ **11.** $y = \left(\frac{3}{4}\right)^x$ **12.** $y = \left(\frac{4}{3}\right)^x$

In Exercises 13–28, graph each function.

13. $y = 3^x - 1$ **14.** $y = 2^x + 3$ **15.** $y = 2^x + 1$ **16.** $y = 4^x - 4$

17. $y = 3^{x-1}$ **18.** $y = 2^{x+3}$ **19.** $y = 3^{x+1}$ **20.** $y = 2^{x-3}$

21. $y = 2^{x+1} - 2$ **22.** $y = 3^{x-1} + 2$ **23.** $y = 3^{x-2} + 1$ **24.** $y = 3^{x+2} - 1$

25. $y = 5(2^x)$ **26.** $y = 2(5^x)$ **27.** $y = 3^{-x}$ **28.** $y = 2^{-x}$

In Exercises 29–36, find the value of b, if any, that would cause the graph of $y = b^x$ to look like the graph indicated.

29.

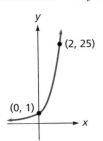

30.

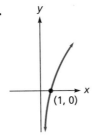

31.

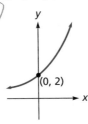

32.

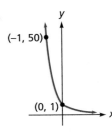

33.

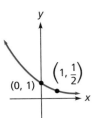

34.

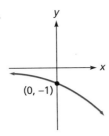

35.

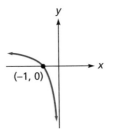

36.

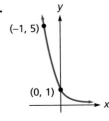

In Exercises 37–42, solve each problem.

37. Tritium decay Tritium, a radioactive isotope of hydrogen, has a half-life of 12.4 years. Of an initial sample of 50 grams, how much will remain after 100 years?

38. Carbon-14 decay The half-life of radioactive carbon-14 is 5700 years. How much of an initial sample will remain after 3000 years?

39. Radioactive decay A radioactive material decays according to the formula $A = A_0\left(\frac{2}{3}\right)^t$, where A_0 is the amount present initially and t is the time in years. Find the amount that will be present in 5 years.

40. Plutonium decay One of the isotopes of plutonium, ^{237}Pu, decays with a half-life of 40 days. How much of an initial sample will remain after 60 days?

41. Californium decay The half-life of one isotope of californium, ^{253}Cf, is 18 days. The half-life of another, ^{254}Cf, is 60 days. If 1 gram of each is present initially, find the amount of ^{254}Cf present when 0.5 gram of ^{253}Cf remains. (*Hint:* One-half of the ^{253}Cf remains after one half-life.)

42. Comparing radioactive decay One isotope of holonium, ^{162}Ho, has a half-life of 22 minutes. The half-life of a second isotope, ^{164}Ho, is 37 minutes. Starting with a sample containing equal amounts, find the ratio of the amounts of ^{162}Ho to ^{164}Ho after one hour.

In Exercises 43–48, assume there are no deposits or withdrawals.

43. Compound interest An initial deposit of $500 earns 8% interest, compounded quarterly. How much will be in the account in 10 years?

44. Compound interest An initial deposit of $1000 earns 9% interest, compounded monthly. How much will be in the account in $4\frac{1}{2}$ years?

45. Comparing interest rates How much more interest could $500 earn in 5 years, compounded quarterly, if the annual interest rate were $5\frac{1}{2}\%$ instead of 5%?

46. Comparing savings plans One bank guarantees to pay interest at 7.25%, compounded monthly. Another bank offers 7.35%, compounded annually. Which bank provides the better investment?

47. Compound interest If $1 had been invested on July 4, 1776, at 5% interest, compounded annually, what would it be worth on July 4, 2076?

48. *Frequency of compounding* $1000 is invested in each of two accounts, both paying 9% annual interest. In the first account, interest compounds quarterly; in the second account, interest compounds daily. Find the difference between the accounts after 20 years.

In Exercises 49–70, solve each problem.

49. *360/365 method* Some financial institutions pay daily interest, compounded by the 360/365 method, by using the formula

$$A = A_0\left(1 + \frac{r}{360}\right)^{365t} \qquad (t \text{ is in years})$$

Using this method, what will an initial investment of $1000 be worth in 5 years, assuming a 7% annual interest rate?

50. *360/365 method* See Exercise 49 and find what an initial investment of $5000 will be worth in 25 years, assuming a 7% annual interest rate.

51. *Oceanography* In one location in the Arctic Ocean, the intensity I, in lumens, of light at a depth of x meters is given by $I = 8\left(\frac{1}{2}\right)^x$. Find the intensity at a depth of 2 meters.

52. *Comparing light intensities* At location A in the Pacific Ocean, the intensity, I, at a depth of x meters is given by the equation $I = 7(0.8)^x$. At location B, the equation is $I = 9(0.7)^x$. At which location will the light be brighter at a depth of 5 meters?

53. *Comparing light intensities* At location A in the Atlantic Ocean, the intensity I, in lumens, at a depth of x meters is given by $I = 6(0.5)^x$. At location B, the equation is $I = 8(0.55)^x$. At which location will the light be brighter at a depth of 4.5 meters?

54. *Drug absorption in smokers* The biological half-life of the asthma medication theophylline is 4.5 hours for smokers. Twelve hours after a dose of 1 unit is administered, find the amount of the drug retained in a smoker's system.

55. *Drug absorption in nonsmokers* For a nonsmoker, the biological half-life of theophylline is 8 hours. Twelve hours after a one-unit dose is administered, find the amount of the drug retained in a nonsmoker's system.

56. *Relative drug absorption* Determine the ratio of theophylline retained in the systems of a smoker and a nonsmoker, twelve hours after they receive equal doses of the drug. See Exercises 54 and 55.

57. *Heart disease* Digoxin, a form of digitalis, is used in the treatment of congestive heart disease and atrial fibrillation. The typical half-life of digoxin in patients with normal kidney function is $1\frac{1}{2}$ days. What part of a unit dose of digoxin is retained in the system after $2\frac{1}{2}$ days?

58. *Bacterial growth* A colony of 6 million bacteria is growing in a culture medium. The population P after t hours is modeled by the formula $P = (6 \times 10^6)(2.3)^t$. Find the population after 4 hours.

59. *Population growth* The population of Eagle River is growing exponentially according to the model $P = 375(1.3)^t$, where t is measured in years from the present date. Find the population in 3 years.

60. *Bluegill population* A northern Wisconsin lake is stocked with a population of 10,000 bluegill, which is expected to grow exponentially according to the model $P = P_0 2^{t/2}$. How many bluegill will be in the lake in 5 years?

61. *Battery charge* The charge remaining in a battery is decreasing exponentially according to the formula $C = C_0(0.7)^t$, where C is the charge remaining after t days and C_0 is the initial charge. If a charge of 2.471×10^{-5} coulombs remains after 7 days, find the battery's initial charge.

62. *Carrying charge* A college student takes advantage of the ad shown in Illustration 1 and buys a bedroom set for $1100. He plans to pay the $1100 plus interest when his income tax refund comes in 8 months. At that time, what will he need to pay?

BUY NOW, PAY LATER!

Only $1\frac{3}{4}$% interest per month.

ILLUSTRATION 1

63. *Credit-card interest* A major bank credit card charges interest at the rate of 21% per year, compounded monthly. If a college senior charges her last tuition bill of $1500 and intends to pay it in one year, what will she have to pay?

64. *Real estate appreciation*   Rico buys property in a neighborhood where houses are expected to appreciate at the rate of 6% per year. If he pays $87,000, how much capital gain will he have to pay tax on if he sells the house in 15 years?

65. *Boat depreciation*   Carol buys an 18-foot ski boat for $10,500. She estimates that it will depreciate 8% per year, provided that she takes good care of it. How much does she expect to get for the boat if she sells it in 9 years?

66. *Trade-in value*   The credit union tells Frank that the $18,500 automobile he has chosen will depreciate 15% per year. What will the car be worth when he trades it for a new car in 7 years, 3 months, when he turns 40?

67. *Car loan*   A high school graduate purchases a car to use for transportation to college. He promises to pay

$4400 for the car, plus $7\frac{1}{2}$% annual interest, compounded monthly, after completing his summer job 5 months from now. What will he have to pay?

68. *Financial planning*   To have P available in n years, A can be invested now in an account paying interest at an annual rate i, compounded annually. Show that

$$A = P(1 + i)^{-n}$$

69. *Atmospheric pressure*   Atmospheric pressure P (in pounds per square inch) is an exponential function of the altitude a (in feet above sea level) given by

$$P = 14.7(2^{-0.000056a})$$

Find the pressure at 5.0 miles.

70. *Newton's law of cooling*   A bucket of water, initially at 100°C, is placed in a room at temperature 40°C. The temperature, T, of the water after t hours is given by

$$T = 40 + 60(0.75)^t$$

Find the temperature of the water in $3\frac{1}{2}$ hours.

    In Exercises 71–74, use a graphing calculator to answer each question.

71. If $1 is deposited in an account earning 8% interest, compounded monthly, how much will be in the account in 60 years?

72. If $1 is deposited in an account earning 9% interest, compounded quarterly, how much will be in the account in 60 years?

73. One gram of radium decays according to the formula

$$A = 2^{-0.000625t}$$

where t is in years. How much of a 1-gram sample will be left in 660 years, in 1600 years, and in 3000

years? (Set the viewing window to $0 \le t \le 3000$ and $0 \le A \le 2$.)

74. One gram of plutonium decays according to the formula

$$A = 2^{-0.025t}$$

where t is in days. How much of a 1-gram sample will be left in 30 days, in 40 days, and in 80 days? (Set the viewing window to $0 \le t \le 100$ and $0 \le A \le 2$.)

    In Exercises 75–80, use a graphing calculator to perform each experiment. Write a brief paragraph describing what you find.

75. Graph $y = 2^{x+k}$ for many values of k.

76. Graph $y = 2^x + k$ for many values of k.

77. Graph $y = 2^{kx}$ for many values of k.

78. Graph $y = k2^x$ for many values of k.

79. Graph $y = 2^{kx}$ and $y = 2^{-kx}$ for many values of k.

80. Graph $y = k2^x$ and $y = k2^{-x}$ for many values of k.

6.2   BASE-e EXPONENTIAL FUNCTIONS

■ Graphing the Exponential Function ■ Applications of Exponential
Functions   ■ The Malthusian Theory

Leonhard Euler
(1707–1783)
Euler first used the letter i to
represent $\sqrt{-1}$, the letter e
for the base of natural
logarithms, and the symbol
Σ for summation. Euler was
one of the most prolific
mathematicians of all time,
contributing to almost all
areas of mathematics.
Much of his work was
accomplished after he
became blind.

In mathematical models of natural events, the number $e = 2.71828182845904\ldots$
appears often as the base of an exponential function. We introduce this important
number by recalling the formula for compound interest,

$$A = A_0\left(1 + \frac{r}{k}\right)^{kt}$$

and allowing k, representing the number of compounding periods per year, to
become very large. To see what happens, we let $k = rp$, where p is a new variable.

$$A = A_0\left(1 + \frac{r}{k}\right)^{kt}$$

$$A = A_0\left(1 + \frac{r}{rp}\right)^{rpt} \qquad \text{Substitute } rp \text{ for } k.$$

$$A = A_0\left(1 + \frac{1}{p}\right)^{rpt} \qquad \text{Simplify } \frac{r}{rp}.$$

$$A = A_0\left[\left(1 + \frac{1}{p}\right)^{p}\right]^{rt} \qquad \text{Remember that } (x^m)^n = x^{mn}.$$

Because the annual rate r is a positive constant and $k = rp$, it follows that as k
becomes very large, then so does p. The question of what happens to the value of A
becomes tied to the question: "What happens to the value of

$$\left(1 + \frac{1}{p}\right)^{p}$$

as p becomes very large?"

Some results calculated for increasing values of p appear in Table 6-1.

p	$\left(1+\frac{1}{p}\right)^{p}$
1	2
10	2.5937425
100	2.7048138
1,000	2.7169239
1,000,000	2.7182805
1,000,000,000	2.7182818

TABLE 6-1

As the results in this table suggest, as p increases, the value of

$$\left(1 + \frac{1}{p}\right)^p$$

approaches the number e.

If interest earned on an amount A_0 is compounded more and more often, the number p grows large without bound, and the formula

$$A = A_0\left[\left(1 + \frac{1}{p}\right)^p\right]^{rt}$$

becomes

$$A = A_0 e^{rt} \qquad \text{Substitute } e \text{ for } \left(1 + \frac{1}{p}\right)^p.$$

When the amount invested grows exponentially according to the formula $A = A_0 e^{rt}$, interest is said to be **compounded continuously**.

Continuous Compound Interest Formula	$A = A_0 e^{rt}$

EXAMPLE 1   If $25,000 accumulates interest at an annual rate of 8%, compounded continuously, find the balance in the account in 50 years.

Solution $A = A_0 e^{rt}$

$A = 25{,}000 e^{(0.08)(50)}$

$\quad = 25{,}000 e^4$

$\quad \approx 1{,}364{,}953.75 \qquad$ Use a calculator.

In 50 years, the balance will be $1,364,953.75—more than a million dollars. ∎

The exponential function $y = e^x$ is so important that it is often called **the exponential function**. The exponential function is often denoted as exp. Thus,

$$\exp(x) = e^x$$

■ **Graphing the Exponential Function**

EXAMPLE 2   Graph the exponential function defined by $y = f(x) = e^x$.

Solution   We can use a calculator to find several points with coordinates (x, y) that satisfy the equation, plot each point, and join them with a smooth curve. The graph appears in Figure 6-6. The domain is the interval $(-\infty, \infty)$, and the range is the interval $(0, \infty)$.

$y = e^x$

x	y
-1	0.37
0	1
1	2.7
2	7.4

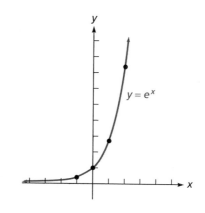

FIGURE **6-6**

EXAMPLE 3   Graph the exponential function defined by $y = f(x) = 3e^{-x/2}$.

Solution   We can use a calculator to find several points with coordinates (x, y) that satisfy the equation, plot each point, and join them with a smooth curve. The graph appears in Figure 6-7.

$y = 3e^{-x/2}$

x	y
-2	8.2
-1	4.9
0	3
1	1.8
2	1.1

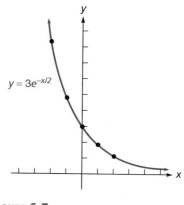

FIGURE **6-7**

■ Applications of Exponential Functions

Malthusian Population Growth

An equation based on the exponential function provides a model for population growth. One such model, called the **Malthusian model of population growth,** assumes a constant birth rate and a constant death rate and incorporates no other factors. In this model, the population P grows exponentially according to the following formula.

Malthusian Model of Population Growth	If b is the annual birth rate, d is the annual death rate, t is the time in years, P_0 is the initial population at $t = 0$, and P is the current population, then $$P = P_0 e^{kt}$$ where $k = b - d$ is the **annual growth rate**, the difference between the annual birth rate and death rate.

EXAMPLE 4   The population of the United States is approximately 253 million people. Assuming that the annual birth rate is 19 per 1000 and the annual death rate is 7 per 1000, what does the Malthusian model predict the U.S. population will be in 50 years?

Solution   Since k is the difference between the birth and death rates, we have

$$k = b - d$$

$$k = \frac{19}{1000} - \frac{7}{1000} \qquad \text{Substitute } \frac{19}{1000} \text{ for } b \text{ and } \frac{7}{1000} \text{ for } d.$$

$$k = 0.019 - 0.007$$

$$= 0.012$$

We can now substitute 253,000,000 for P_0, 50 for t, and 0.012 for k in the formula for the Malthusian model of population growth and simplify.

$$P = P_0 e^{kt}$$

$$P = (253{,}000{,}000)e^{(0.012)(50)}$$

$$= (253{,}000{,}000)e^{0.6}$$

$$\approx 460{,}996{,}057 \qquad \text{Use a calculator.}$$

After 50 years, the population of the United States will exceed 460 million people.

∎

Biology

EXAMPLE 5   A population of 1000 bacteria doubles every 8 hours. Assuming the Malthusian model, find the population in 12 hours.

Solution   We can substitute 1000 for P_0, 2000 for P, and 8 for t into the formula for the Malthusian model of population growth to get

$$P = P_0 e^{kt}$$

$$2000 = 1000 e^{k(8)}$$

$$2 = e^{k(8)} \qquad \text{Divide both sides by 1000.}$$

$$2^{1/8} = [e^{k(8)}]^{1/8} \qquad \text{Raise both sides to the } \frac{1}{8} \text{ power.}$$

1. $\qquad 2^{1/8} = e^{k} \qquad \text{Simplify.}$

We know that the population grows according to the formula

$$P = 1000e^{kt}$$
$$= 1000(2^{1/8})^t \qquad \text{See Equation 1 and substitute } 2^{1/8} \text{ for } e^k.$$
$$= 1000(2^{t/8})$$

To find the population after 12 hours, we substitute 12 for t and simplify.

$$P = 1000(2^{t/8})$$
$$= 1000(2^{12/8})$$
$$= 1000(2^{3/2})$$
$$\approx 2828.427125 \qquad \text{Use a calculator.}$$

After 12 hours, there are approximately 2800 bacteria.   ∎

Epidemiology

Many infectious diseases, including some caused by viruses, spread most rapidly when they first appear in a population, but then more slowly as the number of uninfected individuals decreases. These situations are often modeled by a function, called a **logistic function**, of the form

$$P = \frac{M}{1 + \left(\dfrac{M}{P_0} - 1\right)e^{-kt}}$$

where P is the size of the infected population at any time t, P_0 was the infected population size at $t = 0$, and k is a constant determined by how easily the virus spreads in a given environment. M is the theoretical maximum size of the population P.

EXAMPLE 6   In a certain city with population of 1,200,000, there are currently 1000 cases of infection with the HIV virus. If the spread of the virus is projected by the formula

$$P = \frac{1,200,000}{1 + (1200 - 1)e^{-0.4t}}$$

how many people will have the HIV virus in 3 years?

Solution   We can substitute 3 for t in the given formula and calculate P.

$$P = \frac{1,200,000}{1 + (1200 - 1)e^{-0.4t}}$$
$$P = \frac{1,200,000}{1 + (1199)e^{-0.4(3)}}$$
$$\approx 3314$$

In 3 years, approximately 3300 people are expected to have the virus.   ∎

■ The Malthusian Theory

The English economist Thomas Robert Malthus (1766–1834) was a pioneer in population study. He believed that poverty and starvation were unavoidable because the human population tends to grow exponentially, but the food supply tends to grow linearly.

EXAMPLE 7   Suppose that a country with a population of 1000 people is growing exponentially according to the formula

$$P = 1000e^{0.02t}$$

where *t* is in years. Furthermore, assume that the food supply (measured in adequate food per day per person) is growing linearly according to the formula

$$y = 30.625x + 2000$$

In how many years will the population outstrip the food supply?

Solution   We can set a viewing window of a graphing calculator to $0 \le x \le 100$ and $0 \le y \le 10{,}000$, graph each function as in Figure 6-8, and use the TRACE and ZOOM features to find the point where the two graphs intersect. We see that the food supply will be adequate for approximately 71.7 years. After that, the population will exceed 4196 people, and not all of them will have enough to eat.

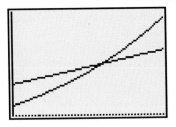

FIGURE 6-8

■

6.2   EXERCISES

 In Exercises 1–8, graph each function.

1. $y = -e^x$ **2.** $y = e^{-x}$ **3.** $y = e^{-0.5x}$ **4.** $y = -e^{2x}$

5. $y = 2e^{-x}$ **6.** $y = -3e^x$ **7.** $y = e^x + 1$ **8.** $y = e^x - 2$

In Exercises 9–16, tell whether the graph of $y = e^x$ could look like the graph indicated.

9.
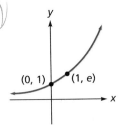

10.
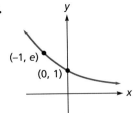

11.
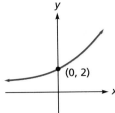

12.
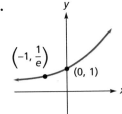

13.

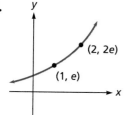

14.

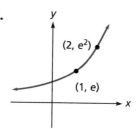

15.

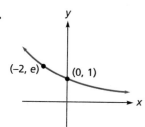

16.

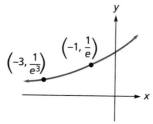

In Exercises 17–22, assume that there are no deposits or withdrawals.

17. **Continuous compound interest** An initial investment of $5000 earns 8.2% interest, compounded continuously. What will the investment be worth in 12 years?

18. **Continuous compound interest** An initial investment of $2000 earns 8% interest, compounded continuously. What will the investment be worth in 15 years?

19. **Determining the initial deposit** An account now contains $11,180 and has been accumulating interest at 7% annual interest, compounded continuously, for 7 years. Find the initial deposit.

20. **Determining a previous balance** An account now contains $3610 and has been accumulating interest at 8% annual interest, compounded continuously. How much was in the account 1 year ago?

21. **Comparison of compounding methods** An initial deposit of $5000 grows at an annual rate of 8.5% for 5 years. Compare the final balances resulting from continuous compounding and annual compounding.

22. **Comparison of compounding methods** An initial deposit of $30,000 grows at an annual rate of 8% for 20 years. Compare the final balances resulting from continuous compounding and annual compounding.

In Exercises 23–40, solve each problem.

23. **Population growth** The growth of a population is modeled by

$$P = 173e^{0.03t}$$

How large will the population be when $t = 20$?

24. **Population decline** The decline of a population is modeled by

$$P = 1.2 \times 10^6 e^{-0.008t}$$

How large will the population be when $t = 30$?

25. **Epidemics** The spread of ungulate fever through a herd of cattle can be modeled by the formula

$$P = P_0 e^{0.27t}$$

If a rancher does not act quickly to treat 2 cases, how many cattle will have the disease in 12 days?

26. **Alcohol absorption** In one individual, the percent alcohol level t minutes after drinking two shots of whiskey is given by $P = 0.3(1 - e^{-0.05t})$. Find the blood alcohol level after 20 minutes.

27. **World population growth** The population of the earth is approximately 5.2 billion people and is growing at an annual rate of 1.9%. Assuming a Malthusian growth model, find the world population in 30 years.

28. **World population growth** Assuming a Malthusian growth model, find the world population in 40 years. (See Exercise 27.)

29. **World population growth** Assuming a Malthusian growth model, by what factor will the current population of the earth increase in 50 years? (See Exercise 27.)

30. **World population growth** Assuming a Malthusian growth model, by what factor will the current population of the earth increase in 100 years? (See Exercise 27.)

31. **Growth of a nation** A country with a population of 2×10^5 people is expected to double every 20 years. Assuming a Malthusian model, find the population in 35 years.

32. **Town planning** The population of a small town is now 140 persons and is expected to grow exponentially, tripling every 15 years. Assuming a Malthusian model, what do the city planners project the population to be in 5 years?

33. **Medicine** The concentration, x, of a certain drug in an organ after t minutes is given by $x = 0.08(1 - e^{-0.1t})$. Find the concentration of the drug after 30 minutes.

34. **Medicine** Refer to Exercise 33. Find the initial concentration of the drug.

35. **Epidemics** Refer to Example 6. How many people will have the HIV virus in 5 years?

36. **Epidemics** Refer to Example 6. How many people will have the HIV virus in 10 years?

37. **Drug absorption** The amount A of a drug remaining in a person's bloodstream after t hours is given by the formula

$$A = A_0 e^{kt}$$

where A_0 is the initial dose. After 2.3 hours, one-half of an initial dose of triazolam (a drug for treating insomnia) will remain. What percent will remain after 24 hours?

38. **Skydiving** Before the parachute opens, a skydiver's velocity v (in meters per second) is given by $v = 50(1 - e^{-0.2t})$. Find the initial velocity.

39. **Skydiving** Refer to Exercise 38 and find the velocity after 20 seconds.

40. **Free-falling objects** After t seconds, a certain falling object has a velocity v given by $v = 50(1 - e^{-0.3t})$. Which is falling faster after 2 seconds, this object or the skydiver in Exercise 38?

41. If $e^{t+3} = ke^t$, find k.

42. If $e^{3t} = k^t$, find k.

In Exercises 43–44, use a graphing calculator to solve each problem.

43. In Example 7, suppose that better farming methods change the formula for food growth to $y = 31x + 2000$. How long will the food supply be adequate?

44. In Example 7, suppose that a birth-control program changed the formula for population growth to $P = 1000e^{0.01t}$. How long will the food supply be adequate?

45. The value of e can be calculated to any degree of accuracy by adding the first several terms of the following list:

$$1, 1, \frac{1}{2}, \frac{1}{2 \cdot 3}, \frac{1}{2 \cdot 3 \cdot 4}, \frac{1}{2 \cdot 3 \cdot 4 \cdot 5}, \text{ and so on}$$

The more terms that are added, the closer the sum will be to e. Add the first eight numbers in the preceding list. To how many decimal places is the sum accurate?

46. Graph the function defined by the equation

$$y = f(x) = \frac{e^x + e^{-x}}{2}$$

from $x = -2$ to $x = 2$. The graph will look like a parabola, but it is not. The graph, called a **catenary**, is important in the design of power distribution networks, because it represents the shape of a uniform flexible cable whose ends are suspended from the same height. The function is called the **hyperbolic cosine function**.

47.    Graph the logistic function of Example 6:

$$P = \frac{1,200,000}{1 + (1199)e^{-0.4t}}$$

You will need to set appropriate RANGE values: Let x vary from 0 to 40, and y from 0 to 1,500,000.

48. Use the TRACE capabilities of a graphing calculator to explore the logistic function of Exercise 47. As time passes, what value does P approach? How many years does it take for 20% of the population to become infected? For 80%?

6.3   LOGARITHMIC FUNCTIONS

■ Common Logarithms   ■ Natural Logarithms   ■ Graphing Logarithmic Functions

Because an exponential function defined by $y = b^x$ is one-to-one, it has an inverse function that is defined by the equation $x = b^y$. To express this inverse function in the form $y = f^{-1}(x)$, we must solve the equation $x = b^y$ for y. To do so, we need the following definition.

Logarithmic Function

If $b > 0$ and $b \neq 1$, the **logarithmic function with base b** is defined by

$$y = \log_b x \qquad \text{if and only if} \qquad x = b^y$$

The **domain of the logarithmic function** is the interval $(0, \infty)$ of positive numbers. The **range of the logarithmic function** is the interval $(-\infty, \infty)$ of real numbers.

Since the function $y = \log_b x$ is the inverse of the one-to-one exponential function $y = b^x$, the logarithmic function is also a one-to-one function.

Warning!   Because the domain of the logarithmic function is the set of positive numbers, it is impossible to find the logarithm of 0 or the logarithm of a negative number.

The previous definition guarantees that any pair (x, y) that satisfies the equation $y = \log_b x$ also satisfies the equation $x = b^y$. Thus,

$$\log_b x = y \qquad \text{because} \qquad x = b^y$$
$$\log_5 25 = 2 \qquad \text{because} \qquad 25 = 5^2$$
$$\log_7 1 = 0 \qquad \text{because} \qquad 1 = 7^0$$
$$\log_{16} 4 = \frac{1}{2} \qquad \text{because} \qquad 4 = 16^{1/2}$$
$$\log_2 \frac{1}{8} = -3 \qquad \text{because} \qquad \frac{1}{8} = 2^{-3}$$

In each of the previous examples, the logarithm of a number is an exponent. This suggests the following important fact: log_b *x* **is the exponent to which *b* is raised to get *x*.** To say this with an equation, we write

$$b^{\log_b x} = x$$

EXAMPLE 1   Find *y* in each equation: **a.** $\log_5 1 = y$,   **b.** $\log_2 8 = y$,   and   **c.** $\log_7 \dfrac{1}{7} = y$.

Solution   **a.** We can change the equation $\log_5 1 = y$ into the equivalent exponential form $1 = 5^y$. Since $1 = 5^0$, it follows that $y = 0$. Thus,

$$\log_5 1 = 0$$

b. $\log_2 8 = y$ is equivalent to $8 = 2^y$. Since $8 = 2^3$, it follows that $y = 3$. Thus,

$$\log_2 8 = 3$$

c. $\log_7 \dfrac{1}{7} = y$ is equivalent to $\dfrac{1}{7} = 7^y$. Since $\dfrac{1}{7} = 7^{-1}$, it follows that $y = -1$. Thus,

$$\log_7 \dfrac{1}{7} = -1$$

■

EXAMPLE 2   Find *a* in each equation: **a.** $\log_3 \dfrac{1}{9} = a$,   **b.** $\log_a 32 = 5$,   and   **c.** $\log_9 a = -\dfrac{1}{2}$.

Solution   **a.** $\log_3 \dfrac{1}{9} = a$ is equivalent to $\dfrac{1}{9} = 3^a$. Since $\dfrac{1}{9} = 3^{-2}$, it follows that $a = -2$.

b. $\log_a 32 = 5$ is equivalent to $32 = a^5$. Since $32 = 2^5$, it follows that $a = 2$.

c. $\log_9 a = -\dfrac{1}{2}$ is equivalent to $a = 9^{-1/2}$. Since $\dfrac{1}{3} = 9^{-1/2}$, it follows that $a = \dfrac{1}{3}$.

■

Common Logarithms

For computational purposes, base-10 logarithms are very convenient. For this reason, base-10 logarithms are called **common logarithms**. When the base *b* is not indicated in the notation $\log x$, we assume that $b = 10$:

$$\log x \qquad \text{means} \qquad \log_{10} x$$

Because base-10 logarithms appear often in mathematics and science, it is a good idea to become familiar with the following base-10 logarithms:

$$\log_{10} \dfrac{1}{100} = -2 \qquad \text{because} \qquad 10^{-2} = \dfrac{1}{100}$$

$$\log_{10} \dfrac{1}{10} = -1 \qquad \text{because} \qquad 10^{-1} = \dfrac{1}{10}$$

$$\log_{10} 1 = 0 \qquad \text{because} \qquad 10^0 = 1$$

$$\log_{10} 10 = 1 \qquad \text{because} \qquad 10^1 = 10$$
$$\log_{10} 100 = 2 \qquad \text{because} \qquad 10^2 = 100$$
$$\log_{10} 1000 = 3 \qquad \text{because} \qquad 10^3 = 1000$$

In general, we have

$$\log_{10} 10^x = x$$

In the past, we had to rely on extensive tables to find common logarithms of numbers whose logarithms were not obvious. Today, however, logarithms are easy to find with a calculator. For example, to find log 2.34, we press these keys on a scientific calculator:

2.34 $\boxed{\text{LOG}}$ (You may have to press a $\boxed{\text{2nd}}$ function key before $\boxed{\text{LOG}}$.)

The display will read .369215857. Thus, to four decimal places,

$$\log 2.34 = 0.3692$$

For the keys to press on a graphing calculator, consult the owner's manual.

EXAMPLE 3 Find x in the equation log $x = 0.7482$.

Solution The equation log $x = 0.7482$ is equivalent to $10^{0.7482} = x$. To find x with a scientific calculator, we press these keys:

10 $\boxed{y^x}$ .7482 $\boxed{=}$

The display will read 5.600154388. Thus, to four decimal places,

$$x = 5.6002$$

If your calculator has a $\boxed{10^x}$ key, simply enter .7482 and press it to get the same result. (You might have to press a $\boxed{\text{2nd}}$ function key first.) For the keys to press on a graphing calculator, consult the owner's manual. ∎

Natural Logarithms

Because the number e appears often in mathematical models of events in nature, base-e logarithms are called **natural logarithms**. They are also called **Napierian logarithms** after John Napier (1550–1617). Natural logarithms are usually denoted by the symbol ln x, rather than $\log_e x$:

ln x means **$\log_e x$**

We can also find natural logarithms with a calculator. For example, to find ln 2.34 with a scientific calculator, we press these keys:

2.34 $\boxed{\text{LN}}$ (You may have to press a $\boxed{\text{2nd}}$ function key before $\boxed{\text{LN}}$.)

The display will read .850150929. Thus, to four decimal places,

$$\ln 2.34 = 0.8502$$

For the keys to press on a graphing calculator, consult the owner's manual.

John Napier (1550–1617) Napier is famous for his work with natural logarithms. In fact, natural logarithms are often called Napierian logarithms. He also invented a device, called Napier's rods, that did multiplications mechanically. His device was a forerunner of modern-day computers.

EXAMPLE 4   Find x in each equation: **a.** $\ln x = 1.335$ and   **b.** $\ln x = \log 5.5$

Solution   **a.** The equation $\ln x = 1.335$ is equivalent to $e^{1.335} = x$. To use a scientific calculator to find x, we press these keys:

$$1.335 \boxed{e^x} \qquad \text{(You may have to press a } \boxed{\text{2nd}} \text{ function key before } \boxed{e^x} \text{ .)}$$

The display will read 3.799995946. Thus, to four decimal places

$$x = 3.8000$$

For the keys to press on a graphing calculator, consult the owner's manual.

b. The equation $\ln x = \log 5.5$ is equivalent to $e^{\log 5.5} = x$. To use a scientific calculator to find x, we press these keys:

$$5.5 \boxed{\text{LOG}} \boxed{e^x} \text{ (You may have to press a } \boxed{\text{2nd}} \text{ function key before } \boxed{e^x} \text{ .)}$$

The display will read 2.096695826. Thus, to four decimal places,

$$x = 2.0967$$

For the keys to press on a graphing calculator, consult the owner's manual.   ■

Graphing Logarithmic Functions

To graph the logarithmic function $y = \log_2 x$, we can calculate and plot several points with coordinates (x, y) that satisfy the equation $x = 2^y$. After joining these points with a smooth curve, we have the graph shown in Figure 6-9(a).

To graph $y = \log_{1/2} x$, we can calculate and plot several points with coordinates (x, y) that satisfy the equation $x = \left(\frac{1}{2}\right)^y$. After joining these points with a smooth curve, we have the graph shown in Figure 6-9(b).

We can see from the graphs that the domain of each function is the interval $(0, \infty)$, and the range is the interval $(-\infty, \infty)$.

$y = \log_2 x$

x	y
$\frac{1}{4}$	-2
$\frac{1}{2}$	-1
1	0
2	1
4	2
8	3

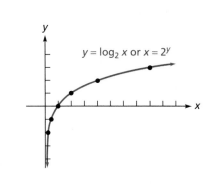

$y = \log_{1/2} x$

x	y
8	-3
4	-2
2	-1
1	0
$\frac{1}{2}$	1
$\frac{1}{4}$	2

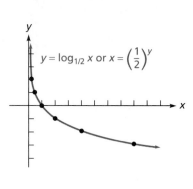

FIGURE 6-9

EXAMPLE 5   Graph the functions defined by   **a.** $y = 3 + \log_2 x$ and   **b.** $y = \log_2 (x - 1)$.

Solution   **a.** The graph of $y = 3 + \log_2 x$ is identical to the graph of $y = \log_2 x$, except that it is translated 3 units upward. See Figure 6-10(a).

b. The graph of $y = \log_2 (x - 1)$ is identical to the graph of $y = \log_2 x$, except that it is translated 1 unit to the right. See Figure 6-10(b).

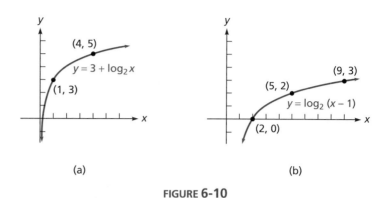

FIGURE **6-10** ■

EXAMPLE 6 Graph the functions determined by **a.** $y = \log x$ and **b.** $y = \ln x$.

Solution **a.** The equation $y = \log x$ is equivalent to the equation $x = 10^y$. To get the graph of $y = \log x$, we can plot points that satisfy the equation $x = 10^y$ and join them with a smooth curve. The graph appears in Figure 6-11(a).

b. The equation $y = \ln x$ is equivalent to the equation $x = e^y$. To get the graph of $y = \ln x$, we can plot points that satisfy the equation $x = e^y$ and join them with a smooth curve. The graph appears in Figure 6-11(b).

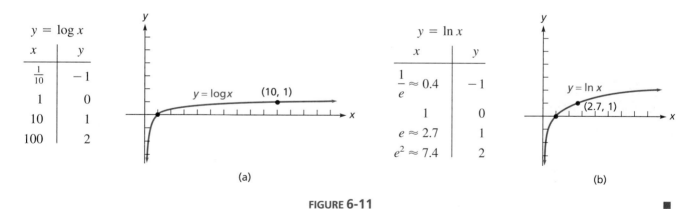

$y = \log x$

x	y
$\frac{1}{10}$	-1
1	0
10	1
100	2

$y = \ln x$

x	y
$\frac{1}{e} \approx 0.4$	-1
1	0
$e \approx 2.7$	1
$e^2 \approx 7.4$	2

FIGURE **6-11** ■

The previous examples suggest that the graphs of logarithmic functions are similar to those in Figure 6-12. If $b > 1$, the logarithmic function is increasing, as in Figure 6-12(a). If $0 < b < 1$, the logarithmic function is decreasing, as in Figure 6-12(b). Each graph of $y = \log_b x$ passes through the points $(1, 0)$ and $(b, 1)$ and has the y-axis as an asymptote.

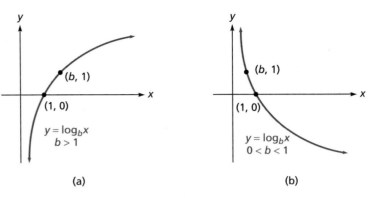

FIGURE **6-12**

The exponential and logarithmic functions are inverses of each other and, there-fore, have symmetry about the line $y = x$. The graphs of $y = \log_b x$ and $y = b^x$ are shown in Figure 6-13(a) when $b > 1$, and in Figure 6-13(b) when $0 < b < 1$.

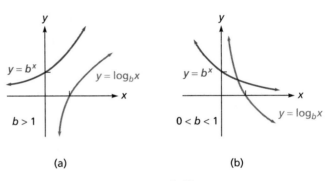

FIGURE **6-13**

6.3 EXERCISES

In Exercises 1–8, write each equation in exponential form.

1. $\log_3 81 = 4$

2. $\log_7 7 = 1$

3. $\log_{1/2} \dfrac{1}{8} = 3$

4. $\log_{1/5} 1 = 0$

5. $\log_4 \dfrac{1}{64} = -3$

6. $\log_6 \dfrac{1}{36} = -2$

7. $\log_x y = z$

8. $\log_m n = \dfrac{1}{z}$

In Exercises 9–16, write each equation in logarithmic form.

9. $8^2 = 64$

10. $10^3 = 1000$

11. $4^{-2} = \dfrac{1}{16}$

12. $3^{-4} = \dfrac{1}{81}$

13. $\left(\dfrac{1}{2}\right)^{-5} = 32$

14. $\left(\dfrac{1}{3}\right)^{-3} = 27$

15. $x^y = z$

16. $m^n = p$

In Exercises 17–56, find each value of x.

17. $\log_2 8 = x$

18. $\log_3 9 = x$

19. $\log_4 64 = x$

20. $\log_6 216 = x$

21. $\log_{1/2} \dfrac{1}{8} = x$

22. $\log_{1/3} \dfrac{1}{81} = x$

23. $\log_9 3 = x$

24. $\log_{125} 5 = x$

25. $\log_{1/2} 8 = x$

26. $\log_{1/2} 16 = x$

27. $\log_8 x = 2$

28. $\log_7 x = 0$

29. $\log_7 x = 1$

30. $\log_2 x = 8$

31. $\log_{25} x = \dfrac{1}{2}$

32. $\log_4 x = \dfrac{1}{2}$

33. $\log_5 x = -2$

34. $\log_3 x = -4$

35. $\log_{36} x = -\dfrac{1}{2}$

36. $\log_{27} x = -\dfrac{1}{3}$

37. $\log_{100} \dfrac{1}{1000} = x$

38. $\log_{5/2} \dfrac{4}{25} = x$

39. $\log_{27} 9 = x$

40. $\log_{12} x = 0$

41. $\log_x 5^3 = 3$

42. $\log_x 5 = 1$

43. $\log_x \dfrac{9}{4} = 2$

44. $\log_x \dfrac{\sqrt{3}}{3} = \dfrac{1}{2}$

45. $\log_x \dfrac{1}{64} = -3$

46. $\log_x \dfrac{1}{100} = -2$

47. $\log_{2\sqrt{2}} x = 2$

48. $\log_4 8 = x$

49. $2^{\log_2 5} = x$

50. $3^{\log_3 4} = x$

51. $x^{\log_4 6} = 6$

52. $x^{\log_3 8} = 8$

53. $\log_{10} 10^3 = x$

54. $\log_{10} 10^{-2} = x$

55. $10^{\log_{10} x} = 100$

56. $10^{\log_{10} x} = \dfrac{1}{10}$

In Exercises 57–68, use a calculator to find each value, if possible. Express all answers to four decimal places.

57. $\log 3.25$

58. $\log 0.57$

59. $\log 0.00467$

60. $\log 375.876$

61. $\ln 0.93$

62. $\ln 7.39$

63. $\ln 37.896$

64. $\ln 0.00465$

65. $\log (\ln 1.7)$

66. $\ln (\log 9.8)$

67. $\ln (\log 0.1)$

68. $\log (\ln 0.01)$

In Exercises 69–76, use a calculator to find each value of y, if possible. Express all answers to two decimal places.

69. $\log y = 1.4023$

70. $\ln y = 2.6490$

71. $\ln y = 4.24$

72. $\log y = 0.926$

73. $\log y = -3.71$

74. $\ln y = -0.28$

75. $\log y = \ln 8$

76. $\ln y = \log 7$

In Exercises 77–88, graph the function defined by each equation.

77. $y = \log_3 x$

78. $y = \log_4 x$

79. $y = \log_{1/3} x$

80. $y = \log_{1/4} x$

81. $y = 2 + \log_2 x$

82. $y = \log_2 (x - 1)$

83. $y = \log_3 (x + 2)$

84. $y = -3 + \log_3 x$

85. $y = 1 + \ln x$

86. $y = \ln (x - 1)$

87. $y = 2 \ln x$

88. $y = \ln 2x$

In Exercises 89–94, graph each pair of equations on one set of coordinate axes.

89. $y = \log_3 x$ and $y = \log_3 (3x)$

90. $y = \log_3 x$ and $y = \log_3 \left(\dfrac{x}{3}\right)$

91. $y = \log_2 x$ and $y = \log_2 (x + 1)$

92. $y = \log_2 x$ and $y = \log_2 (-x)$

93. $y = \log_5 x$ and $y = 5^x$

94. $y = \ln x$ and $y = e^x$

In Exercises 95–98, find the value of b, if any, that would cause the graph of y = $\log_b x$ to look like the graph indicated.

95.

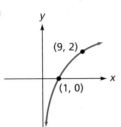

96.

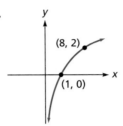

97.

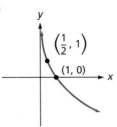

98.

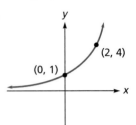

In Exercises 99–102, tell whether the graph of y = ln x could look like the graph indicated.

99.

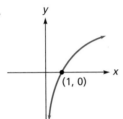

100.

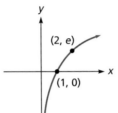

101.

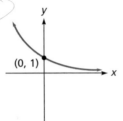

102.

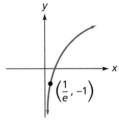

In Exercises 103–106, use a graphing calculator to perform each experiment. Write a brief paragraph describing what you find.

103. Graph $y = k + \log x$ for many values of k.

104. Graph $y = \log (x + k)$ for many values of k.

105. Graph $y = k \log x$ for many values of k.

106. Graph $y = \log kx$ for many values of k.

6.4 PROPERTIES OF LOGARITHMS

◼ The Change-of-Base Formula

Because logarithms are exponents, it is not surprising that the laws of exponents have counterparts in logarithmic notation. We will use the properties of exponents to develop many of the properties of logarithms discussed in this section.

Theorem	If b is a positive number and $b \neq 1$, then
	1. $\log_b 1 = 0$ **2.** $\log_b b = 1$
	3. $\log_b b^x = x$ **4.** $b^{\log_b x} = x$ $(x > 0)$

Properties 1 through 4 follow directly from the definition of logarithm:

1. $\log_b 1 = 0$ because $b^0 = 1$.

2. $\log_b b = 1$ because $b^1 = b$.

3. $\log_b b^x = x$ because $b^x = b^x$.

4. $b^{\log_b x} = x$ because $\log_b x$ is the exponent to which b is raised to get x.

EXAMPLE 1   Simplify each expression: **a.** $\log_3 1$, **b.** $\log_4 4$, **c.** $\log_7 7^3$, and **d.** $b^{\log_b 3}$.

Solution   **a.** By Property 1, $\log_3 1 = 0$ because $3^0 = 1$.

b. By Property 2, $\log_4 4 = 1$ because $4^1 = 4$.

c. By Property 3, $\log_7 7^3 = 3$ because $7^3 = 7^3$.

d. By Property 4, $b^{\log_b 3} = 3$ because $\log_b 3$ is the power to which b is raised to get 3.

■

Properties 3 and 4 also indicate that the composition of the exponential and logarithmic functions (in both directions) is the identity function. This is expected, because the exponential and logarithmic functions are inverse functions.

Theorem	If M, N, and b are positive numbers and $b \neq 1$, then
	5. $\log_b MN = \log_b M + \log_b N$ **6.** $\log_b \dfrac{M}{N} = \log_b M - \log_b N$

Proof   To prove Property 5, we let $x = \log_b M$ and $y = \log_b N$. Because of the definition of logarithms, these equations can be written in the form

$$M = b^x \qquad \text{and} \qquad N = b^y$$

Then $MN = b^x b^y$ and the properties of exponents give

$$MN = b^{x+y}$$

Using the definition of logarithms gives

$$\log_b MN = x + y$$

Substituting the values of x and y completes the proof.

$$\log_b MN = \log_b M + \log_b N$$

□

The proof of Property 6 is similar to the proof of Property 5 and is left as an exercise.

Warning!   Property 5 of logarithms asserts that the logarithm of the product of two numbers is equal to the sum of their logarithms. The logarithm of a sum or a difference usually does not simplify. In general,

$$\log_b (M + N) \neq \log_b M + \log_b N \quad \text{and} \quad \log_b (M - N) \neq \log_b M - \log_b N$$

Property 6 of logarithms asserts that the logarithm of the quotient of two numbers is equal to the difference of their logarithms. The logarithm of a quotient is not the quotient of the logarithms:

$$\log_b \frac{M}{N} \neq \frac{\log_b M}{\log_b N}$$

EXAMPLE 2   Use a calculator to verify Property 5 of logarithms by showing that

$$\ln[(3.7)(15.9)] = \ln 3.7 + \ln 15.9$$

Solution   We can calculate the left- and right-hand sides of the equation separately and compare the results. To use a scientific calculator to find $\ln[(3.7)(15.9)]$, we press these keys:

3.7 $\boxed{x}$ 15.9 $\boxed{=}$ $\boxed{\text{LN}}$

The display will read 4.074651929.
   To find $\ln 3.7 + \ln 15.9$, we press these keys:

3.7 $\boxed{\text{LN}}$ $\boxed{+}$ 15.9 $\boxed{\text{LN}}$ $\boxed{=}$

The display will read 4.074651929. Since the left- and right-hand sides are equal, the equation is verified. For the keys to press on a graphing calculator, consult the owner's manual. ∎

Theorem   If M, p, and b are real numbers, M and b are positive, and $b \neq 1$, then

7. $\log_b M^p = p \log_b M$ **8.** If $\log_b x = \log_b y$, then $x = y$.

Proof   To prove Property 7, we let $x = \log_b M$, write the expression in exponential form, and raise both sides to the pth power:

$$M = b^x$$
$$(M)^p = (b^x)^p$$
$$M^p = b^{px}$$

Using the definition of logarithms gives

$$\log_b M^p = px$$

Substituting the value for x completes the proof:

$$\log_b M^p = p \log_b M$$

□

Property 8 follows from the fact that the logarithmic function is a one-to-one function. We will use Property 8 extensively in Section 6.6, when we are solving logarithmic equations.

Properties 5, 6, and 7 of logarithms can be used to expand or condense logarithmic expressions.

EXAMPLE 3   Assume that x, y, and z are positive numbers. Use the properties of logarithms to write each expression in terms of the logarithms of x, y, and z:

a. $\log_b xyz$ and   **b.** $\log_b \dfrac{x}{yz}$.

Solution   **a.** $\log_b xyz = \log_b (xy)z$

$\qquad\qquad = \log_b (xy) + \log_b z$ $\qquad$ Use Property 5 of logarithms.

$\qquad\qquad = \log_b x + \log_b y + \log_b z$ $\qquad$ Use Property 5 of logarithms.

b. $\log_b \dfrac{x}{yz} = \log_b x - \log_b (yz)$ $\qquad$ Use Property 6 of logarithms.

$\qquad\qquad = \log_b x - (\log_b y + \log_b z)$ $\qquad$ Use Property 5 of logarithms.

$\qquad\qquad = \log_b x - \log_b y - \log_b z$ $\qquad$ Remove parentheses. ∎

EXAMPLE 4   Assume that x, y, and z are positive numbers. Use the properties of logarithms to write each expression in terms of the logarithms of x, y, and z:

a. $\log_b (x^3 y^2 z)$ and   **b.** $\log_b \dfrac{y^2 \sqrt{z}}{x}$.

Solution   **a.** $\log_b (x^3 y^2 z) = \log_b x^3 + \log_b y^2 + \log_b z$ $\qquad$ Use Property 5 of logarithms.

$\qquad\qquad = 3 \log_b x + 2 \log_b y + \log_b z$ $\qquad$ Use Property 7 of logarithms.

b. $\log_b \dfrac{y^2 \sqrt{z}}{x} = \log_b (y^2 \sqrt{z}) - \log_b x$ $\qquad$ Use Property 6 of logarithms.

$\qquad\qquad = \log_b y^2 + \log_b z^{1/2} - \log_b x$ $\qquad$ Use Property 5 of logarithms.

$\qquad\qquad = 2 \log_b y + \dfrac{1}{2} \log_b z - \log_b x$ $\qquad$ Use Property 7 of logarithms. ∎

EXAMPLE 5   Assume that x, y, z, and b are positive numbers and $b \neq 0$. Use the properties of logarithms to write each expression as the logarithm of a single quantity:

a. $2 \log_b x + \dfrac{1}{3} \log_b y$ and   **b.** $\dfrac{1}{2} \log_b (x - 2) - \log_b y + 3 \log_b z$.

Solution   **a.** $2 \log_b x + \dfrac{1}{3} \log_b y = \log_b x^2 + \log_b y^{1/3}$ $\qquad$ Use Property 7 of logarithms.

$\qquad\qquad = \log_b (x^2 y^{1/3})$ $\qquad$ Use Property 5 of logarithms.

$\qquad\qquad = \log_b (x^2 \sqrt[3]{y})$ $\qquad$ Change $y^{1/3}$ to $\sqrt[3]{y}$.

b. $\dfrac{1}{2} \log_b (x - 2) - \log_b y + 3 \log_b z = \log_b (x - 2)^{1/2} - \log_b y + \log_b z^3$ Use Property 7 of logarithms.

$$= \log_b \dfrac{(x - 2)^{1/2}}{y} + \log_b z^3 \qquad \text{Use Property 6 of logarithms.}$$

$$= \log_b \dfrac{z^3 \sqrt{x - 2}}{y} \qquad \text{Use Property 5 of logarithms.} \qquad \blacksquare$$

We summarize the eight properties of logarithms in the following box.

Properties of Logarithms	If b, M, and N are positive numbers and $b \neq 1$, then

1. $\log_b 1 = 0$ 　　　　　　　　　　　　 2. $\log_b b = 1$

3. $\log_b b^x = x$ 　　　　　　　　　　　 4. $b^{\log_b x} = x$ 　　$(x > 0)$

5. $\log_b MN = \log_b M + \log_b N$ 　　6. $\log_b \dfrac{M}{N} = \log_b M - \log_b N$

7. $\log_b M^p = p \log_b M$ 　　　　　　 8. If $\log_b x = \log_b y$, then $x = y$.

EXAMPLE 6   Given that $\log_{10} 2 \approx 0.3010$ and $\log_{10} 3 \approx 0.4771$, find approximations for **a.** $\log_{10} 18$　and　**b.** $\log_{10} 2.5$.

Solution　**a.** $\log_{10} 18 = \log_{10} (2 \cdot 3^2)$ 　　　　**b.** $\log_{10} 2.5 = \log_{10} \left(\dfrac{5}{2} \right)$

$\qquad\qquad\quad = \log_{10} 2 + \log_{10} 3^2 \qquad\qquad\qquad\quad = \log_{10} 5 - \log_{10} 2$

$\qquad\qquad\quad = \log_{10} 2 + 2 \log_{10} 3 \qquad\qquad\qquad\quad = \log_{10} \dfrac{10}{2} - \log_{10} 2$

$\qquad\qquad\quad \approx 0.3010 + 2(0.4771) \qquad\qquad\qquad\quad = \log_{10} 10 - \log_{10} 2 - \log_{10} 2$

$\qquad\qquad\quad \approx 1.2552 \qquad\qquad\qquad\qquad\qquad\qquad\quad = 1 - 2 \log_{10} 2 \qquad \log_{10} 10 = 1$

$\qquad\qquad\qquad\qquad\qquad\qquad\qquad\qquad\qquad\qquad\qquad\;\; \approx 1 - 2(0.3010)$

$\qquad\qquad\qquad\qquad\qquad\qquad\qquad\qquad\qquad\qquad\qquad\;\; \approx 0.3980 \qquad\qquad\qquad \blacksquare$

■ **The Change-of-Base Formula**

If we know the base-a logarithm of a number, we can find its logarithm to some other base b by using a formula called the **change-of-base formula**.

The Change-of-Base Formula	If a, b, and x are real numbers, then

$$\log_b x = \dfrac{\log_a x}{\log_a b}$$

Proof We begin with the equation $\log_b x = y$ and proceed as follows:

1.

$$\log_b x = y$$

$$b^y = x \qquad \text{Change the equation from logarithmic to exponential form.}$$

$$\log_a b^y = \log_a x \qquad \text{Take the base-}a \text{ logarithm of both sides.}$$

$$y \log_a b = \log_a x \qquad \text{Use Property 7 of logarithms.}$$

$$y = \frac{\log_a x}{\log_a b} \qquad \text{Divide both sides by } \log_a b.$$

$$\log_b x = \frac{\log_a x}{\log_a b} \qquad \text{Refer to Equation 1 and substitute } \log_b x \text{ for } y.$$

$\square$

If we know logarithms to base a (for example, $a = 10$), we can find the logarithm of x to a new base b. We simply divide the base-a logarithm of x by the base-a logarithm of b.

EXAMPLE 7 Use the change-of-base formula to find $\log_3 5$.

Solution We can substitute 3 for b, 10 for a, and 5 for x into the change-of-base formula:

$$\log_b x = \frac{\log_a x}{\log_a b}$$

$$\log_3 5 = \frac{\log_{10} 5}{\log_{10} 3}$$

$$\approx 1.464973521$$

Thus, to four decimal places, $\log_3 5 = 1.4650$. ∎

6.4 EXERCISES

In Exercises 1–8, simplify each expression.

1. $\log_4 1$

2. $\log_4 4$

3. $\log_4 4^7$

4. $4^{\log_4 8}$

5. $5^{\log_5 10}$

6. $\log_5 5^2$

7. $\log_5 5$

8. $\log_5 1$

 In Exercises 9–14, use a calculator to verify each equation.

9. $\log[(3.7)(2.9)] = \log 3.7 + \log 2.9$

10. $\ln \dfrac{9.3}{2.1} = \ln 9.3 - \ln 2.1$

11. $\ln (3.7)^3 = 3 \ln 3.7$

12. $\log 3.2 = \dfrac{\ln 3.2}{\ln 10}$

13. $\log \sqrt{14.1} = \dfrac{1}{2} \log 14.1$

14. $\ln 9.7 = \dfrac{\log 9.7}{\log e}$

In Exercises 15–26, assume that x, y, z, and b are positive numbers. Use the properties of logarithms to write each expression in terms of the logarithms of x, y, and z.

15. $\log_b 2xy$

16. $\log_b 3xz$

17. $\log_b \dfrac{2x}{y}$

18. $\log_b \dfrac{x}{yz}$

19. $\log_b x^2y^3$

20. $\log_b x^3y^2z$

21. $\log_b (xy)^{1/3}$

22. $\log_b x^{1/2}y^3$

23. $\log_b x\sqrt{z}$

24. $\log_b \sqrt{xy}$

25. $\log_b \dfrac{\sqrt[3]{x}}{\sqrt[3]{yz}}$

26. $\log_b \sqrt[4]{\dfrac{x^3y^2}{z^4}}$

In Exercises 27–34, assume that x, y, and z are positive numbers. Use the properties of logarithms to write each expression as the logarithm of a single quantity.

27. $\log_b (x + 1) - \log_b x$

28. $\log_b x + \log_b (x + 2) - \log_b 8$

29. $2 \log_b x + \dfrac{1}{3} \log_b y$

30. $-2 \log_b x - 3 \log_b y + \log_b z$

31. $-3 \log_b x - 2 \log_b y + \dfrac{1}{2} \log_b z$

32. $3 \log_b (x + 1) - 2 \log_b (x + 2) + \log_b x$

33. $\log_b \left(\dfrac{x}{z} + x\right) - \log_b \left(\dfrac{y}{z} + y\right)$

34. $\log_b (xy + y^2) - \log_b (xz + yz) + \log_b z$

In Exercises 35–54, tell whether the given statement is true. If it is not true, so indicate.

35. $\log_b ab = \log_b a + 1$

36. $\log_b \dfrac{1}{a} = -\log_b a$

37. $\log_b 0 = 1$

38. $\log_b 2 = \log_2 b$

39. $\log_b (x + y) \neq \log_b x + \log_b y$

40. $\log_a xy = (\log_b x)(\log_b y)$

41. If $\log_a b = c$, then $\log_b a = c$.

42. If $\log_a b = c$, then $\log_b a = \dfrac{1}{c}$.

43. $\log_7 7^7 = 7$

44. $7^{\log_7 7} = 7$

45. $\log_b (-x) = -\log_b x$

46. If $\log_b a = c$, then $\log_b a^p = pc$.

47. $\dfrac{\log_b A}{\log_b B} = \log_b A - \log_b B$

48. $\log_b (A - B) = \dfrac{\log_b A}{\log_b B}$

49. $\log_b \dfrac{1}{5} = -\log_b 5$

50. $3 \log_b \sqrt[3]{a} = \log_b a$

51. $\dfrac{1}{3} \log_b a^3 = \log_b a$

52. $\log_{4/3} y = -\log_{3/4} y$

53. $\log_b y + \log_{1/b} y = 0$

54. $\log_{10} 10^3 = 3(10^{\log_{10} 3})$

In Exercises 55–66, assume that $\log_{10} 4 = 0.6021$, $\log_{10} 7 = 0.8451$, and $\log_{10} 9 = 0.9542$. Use these values and the properties of logarithms to find each value.

55. $\log_{10} 28$

56. $\log_{10} \dfrac{7}{4}$

57. $\log_{10} 2.25$

58. $\log_{10} 36$

59. $\log_{10} \dfrac{63}{4}$

60. $\log_{10} \dfrac{4}{63}$

61. $\log_{10} 252$

62. $\log_{10} 49$

63. $\log_{10} 112$ **64.** $\log_{10} 324$ **65.** $\log_{10} \dfrac{144}{49}$ **66.** $\log_{10} \dfrac{324}{63}$

In Exercises 67–74, use the change-of-base formula to find each logarithm.

67. $\log_3 7$ **68.** $\log_7 3$ **69.** $\log_\pi 3$ **70.** $\log_3 \pi$

71. $\log_3 8$ **72.** $\log_5 10$ **73.** $\log_{\sqrt 2} \sqrt 5$ **74.** $\log_\pi e$

75. Prove Property 6 of logarithms:

$$\log_b \frac{M}{N} = \log_b M - \log_b N$$

76. Show that $-\log_b x = \log_{1/b} x$.

77. Show that $e^{x \ln a} = a^x$.

78. Show that $e^{\ln x} = x$.

79. Show that $\ln(e^x) = x$.

80. If $\log_b 3x = 1 + \log_b x$, find b.

81. Explain why $\ln (\log 0.9)$ is undefined.

82. Explain why $\log_b (\ln 1)$ is undefined.

6.5   APPLICATIONS OF LOGARITHMS

■ Applications of Base-10 Logarithms   ■ Applications of Base-e Logarithms

In this section, we consider many applications of logarithmic functions.

■ Applications of Base-10 Logarithms

Chemistry

In chemistry, common logarithms are used to express the acidity of solutions. The more acidic a solution, the greater the concentration of hydrogen ions. This concentration is indicated indirectly by the **pH scale**, or **hydrogen-ion index**. The pH of a solution is defined as follows.

---

pH of a Solution	If $[\text{H}^+]$ is the hydrogen-ion concentration in gram-ions per liter, then $$\text{pH} = -\log[\text{H}^+]$$

EXAMPLE 1   Since pure water has approximately 10^{-7} gram-ions per liter, its pH is

$$\begin{aligned}
\text{pH} &= -\log [\text{H}^+] \\
&= -\log 10^{-7} \\
&= -(-7) \log 10 &&\text{Use Property 7 of logarithms.} \\
&= -(-7) \cdot 1 &&\text{Use Property 2 of logarithms.} \\
&= 7
\end{aligned}$$

Seawater has a pH of approximately 8.5. To find its hydrogen-ion concentration, we can solve the following equation for $[H^+]$.

$$8.5 = -\log[H^+]$$
$$-8.5 = \log[H^+]$$
$$[H^+] = 10^{-8.5} \qquad \text{Change the equation from logarithmic form to exponential form.}$$

We can use a calculator to find that

$$[H^+] \approx 3.2 \times 10^{-9} \text{ gram-ions per liter} \qquad ■$$

Electrical Engineering

In electrical engineering, common logarithms are used to express the voltage gain (or loss) of an electronic device such as an amplifier or a length of transmission line. The unit of gain (or loss), called the **decibel**, is defined by the following logarithmic function.

Decibel Voltage Gain	If E_O is the output voltage of a device and E_I is the input voltage, the decibel voltage gain is given by
	$$\text{db voltage gain} = 20 \log \frac{E_O}{E_I}$$

EXAMPLE 2   If the input to an amplifier is 0.5 volt and the output is 40 volts, its decibel voltage gain is found by substituting these values into the formula for db voltage gain:

$$\text{db voltage gain} = 20 \log \frac{E_O}{E_I}$$

$$\text{db voltage gain} = 20 \log \frac{40}{0.5}$$

$$= 20 \log 80$$

$$\approx 38 \qquad \text{Use a calculator.}$$

The amplifier provides a 38-decibel voltage gain. $\qquad ■$

Geology

In seismology, common logarithms are used to measure the intensity of earthquakes on the **Richter scale**. The intensity of an earthquake is given by the following logarithmic function.

Richter Scale	If R is the intensity of an earthquake, A is the amplitude (measured in micrometers), and P is the period (the time of one oscillation of the earth's surface, measured in seconds), then
	$$R = \log \frac{A}{P}$$

EXAMPLE 3   To calculate the intensity of an earthquake with an amplitude of 10,000 micrometers (1 centimeter) and a period of 0.1 second, we substitute 10,000 for A and 0.1 for P into the Richter scale formula and simplify:

$$R = \log \frac{A}{P}$$

$$R = \log \frac{10{,}000}{0.1}$$

$$= \log 100{,}000$$

$$= \log 10^5$$

$$= 5 \log 10 \qquad \text{Use Property 7 of logarithms.}$$

$$= 5 \qquad \text{Use Property 2 of logarithms.}$$

The earthquake measures 5 on the Richter scale.   ■

Applications of Base-e Logarithms

Electronics

A battery charges at a rate that depends on how close it is to being fully charged. It charges fastest when it is most discharged. The charge C at any instant is modeled as follows.

Formula for Charging Batteries	If M is the theoretical maximum charge that a battery can hold and k is a positive constant that depends on the battery and charger, then $$C = M(1 - e^{-kt})$$

EXAMPLE 4   Plotting the variable C against the variable t in the formula $C = M(1 - e^{-kt})$ gives a curve like that shown in Figure 6-14. The graph shows that a full charge, M, is never attained. However, the actual charge can come very close to M, provided the battery is charged long enough.

To find out how long it will take a battery to reach a given charge C, we solve the equation $C = M(1 - e^{-kt})$ for t:

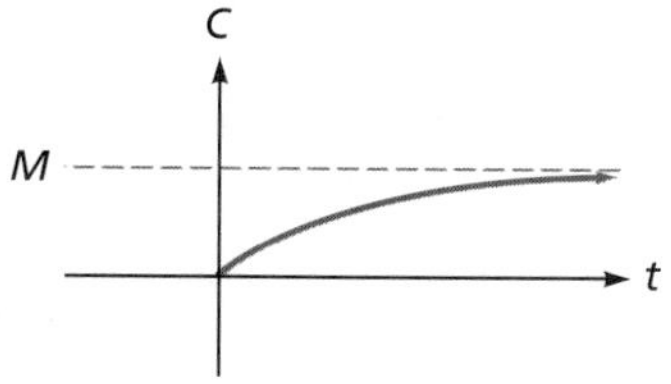

FIGURE 6-14

$$C = M(1 - e^{-kt})$$

$$\frac{C}{M} = 1 - e^{-kt} \qquad \text{Divide both sides by } M.$$

$$\frac{C}{M} - 1 = -e^{-kt} \qquad \text{Subtract 1 from both sides.}$$

$$1 - \frac{C}{M} = e^{-kt} \qquad \text{Multiply both sides by } -1.$$

$$\ln\left(1 - \frac{C}{M}\right) = -kt \qquad \begin{array}{l}\text{Change the exponential equation} \\ \text{to logarithmic form.}\end{array}$$

$$-\frac{1}{k} \ln\left(1 - \frac{C}{M}\right) = t \qquad \text{Multiply both sides by } -\frac{1}{k}.$$

The formula that determines the time t required to charge a battery to a given level C is

$$t = -\frac{1}{k} \ln \left(1 - \frac{C}{M} \right)$$

  ■

Physiology

In physiology, experiments suggest that the relationship between the loudness and the intensity of sound is a logarithmic one known as the **Weber-Fechner Law**.

Weber-Fechner Law	If L is the apparent loudness of a sound and I is the actual intensity, then $$L = k \ln I$$

EXAMPLE 5   To find the increase in intensity of a sound that will cause a doubling of the apparent loudness, we consider the equation

$$L_0 = k \ln I_0$$

To double the apparent loudness, we multiply both sides of the equation by 2 and use Property 7 of logarithms:

$$2L_0 = 2k \ln I_0$$
$$= k \ln (I_0)^2$$

Thus, to double the apparent loudness of a sound, the intensity must be squared.   ■

Population Growth

If a population grows exponentially at a certain annual rate, the time required for the population to double is called the **doubling time**, which is calculated as follows.

Formula for Doubling Time	If r is the annual rate and t is the time required for a population to double, then $$t = \frac{\ln 2}{r}$$

EXAMPLE 6   The population of the earth is growing at the rate of approximately 2% per year. If this rate continues, when will the population double?

Solution   Because the population is growing at the rate of 2% per year, we substitute 0.02 for r into the formula for doubling time and simplify.

$$t = \frac{\ln 2}{r}$$

$$t = \frac{\ln 2}{0.02}$$

$$\approx 34.65735903$$

At the current rate of growth, the population of the earth will double in about $34\frac{1}{2}$ years.

∎

Isothermal Expansion

When energy is added to a gas, its temperature and volume could increase. In **isothermal expansion**, the temperature remains constant, and only the volume changes. The energy required is calculated as follows.

Formula for Isothermal Expansion	If the temperature T is constant, the energy E required to increase the volume of one mole of gas from an initial volume V_i to a final volume V_f is given by $$E = RT \ln \left(\frac{V_f}{V_i} \right)$$ E is measured in joules and T in degrees Kelvin, and R is the **universal gas constant**, 8.314 joules/mole/K.

EXAMPLE 7 Find the amount of energy that must be supplied to triple the volume of one mole of gas at a constant temperature of 300K.

Solution We can substitute 8.314 for R and 300 for T directly into the formula. Because the final volume is to be three times the initial volume, we can also substitute $3V_i$ for V_f.

$$E = RT \ln \left(\frac{V_f}{V_i} \right)$$

$$E = (8.314)(300) \ln \left(\frac{3V_i}{V_i} \right)$$

$$\approx 2494.2 \ln 3$$

$$\approx 2740$$

Approximately 2740 joules of energy must be added to the gas to triple its volume.

∎

6.5 EXERCISES

In Exercises 1–26, solve each problem.

1. *pH of a solution* Find the pH of a solution with a hydrogen-ion concentration of 1.7×10^{-5} gram-ions per liter.

2. *pH of calcium hydroxide* Find the hydrogen-ion concentration of a saturated solution of calcium hydroxide whose pH is 13.2.

3. *pH of apples* The pH of apples can range from 2.9 to 3.3. Find the range in the hydrogen-ion concentration.

4. *pH of sour pickles* The hydrogen-ion concentration of sour pickles is 6.31×10^{-4}. Find the pH.

5. *db voltage gain of an amplifier* The db gain of an amplifier is 29. Find the input voltage when the output voltage is 20 volts.

6. *db voltage gain of an amplifier* The db gain of an amplifier is 35. Find the output voltage when the input voltage is 0.05 volt.

7. Power output of an amplifier Power (measured in watts) is directly proportional to the square of the voltage: $P = kE^2$. If P_O is the power output of an amplifier and P_I is the power input, show that the formula for db voltage gain is

$$\text{db voltage gain} = 10 \log \frac{P_O}{P_I}$$

8. Voltage gain of an amplifier An amplifier produces power output of 30 watts when driven by an input signal of 0.1 watt. Find the amplifier's voltage gain. (See Exercise 7.)

9. Earthquakes An earthquake has amplitude of 5000 micrometers and a period of 0.2 second. Find its measure on the Richter scale.

10. Earthquakes An earthquake with amplitude of 8000 micrometers measures 6 on the Richter scale. Find its period.

11. Earthquakes An earthquake with a period of $\frac{1}{4}$ second measures 4 on the Richter scale. Find its amplitude.

12. Earthquakes By what factor must the amplitude of an earthquake change to increase its severity by 1 point on the Richter scale? Assume that the period remains constant.

13. Battery charge If a battery can reach half of its full charge in 6 hours, how long will it take the battery to reach a 90% charge? Assume that the battery was fully discharged when it began charging.

14. Battery charge A battery reaches 80% of a full charge in 8 hours. If it started charging when it was fully discharged, how long did it take to reach a 40% charge?

15. Change in loudness If the intensity of a sound is doubled, what is the apparent change in loudness?

16. Change in loudness What change in intensity of sound will cause an apparent tripling of the loudness?

17. Population growth A town's population grows at the rate of 12% per year. If this growth rate remains constant, how long will it take the population to double?

18. Population growth A population growing at an annual rate r will triple in a time t given by the formula

$$t = \frac{\ln 3}{r}$$

How long will it take the population of the town in Exercise 17 to triple?

19. Isothermal expansion One mole of gas expands isothermally to triple its volume. If the gas temperature is 400K, what energy is absorbed?

20. Isothermal expansion One mole of gas expands isothermally to double its volume. If the gas temperature is 300K, what energy is absorbed?

21. Depreciation In business, equipment is often depreciated using the double declining-balance method. In this method, a piece of equipment with a life expectancy of N years, costing $\$C$, will depreciate to a value of $\$V$ in n years, where n is given by the formula

$$n = \frac{\log V - \log C}{\log \left(1 - \dfrac{2}{N} \right)}$$

A computer that cost \$37,000 has a life expectancy of 5 years. If it has depreciated to a value of \$8000, how old is it?

22. Depreciation A typewriter worth \$470 when new had a life expectancy of 12 years. If it is now worth \$189, how old is it? (See Exercise 21.)

23. Time for money to grow twentyfold If $\$P$ is invested at the end of each year in an annuity earning $r\%$ annual interest, then the amount in the account will be $\$A$ after n years, where

$$n = \frac{\log \left[\dfrac{Ar}{P} + 1 \right]}{\log (1 + r)}$$

If \$1000 is invested each year in an annuity earning 12% annual interest, when will the account be worth \$20,000?

24. Time for money to grow tenfold If \$5000 is invested each year in an annuity earning 8% annual interest, when will the account be worth \$50,000? (See Exercise 23.)

25. Breakdown voltage The coaxial power cable shown in Illustration 1 has a central wire with radius $R_1 = 0.25$ centimeter. It is insulated from a surrounding shield with inside radius $R_2 = 2$ centimeters. The maximum voltage the cable can withstand is called the **breakdown voltage** V of the insulation. V is given by the formula

$$V = ER_1 \ln \frac{R_2}{R_1}$$

where E is the **dielectric strength** of the insulation. If $E = 400{,}000$ volts/centimeter, find V.

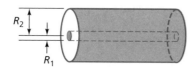

ILLUSTRATION 1

26. *Breakdown voltage* In Exercise 25, if the inside diameter of the shield were doubled, what voltage could the cable withstand?

27. Use the formula $P = P_0 e^{rt}$ to verify that P will be twice P_0 when $t = \dfrac{\ln 2}{r}$.

28. Use the formula $P = P_0 e^{rt}$ to verify that P will be three times as large as P_0 when $t = \dfrac{\ln 3}{r}$.

29. One form of the logistic function is given by the equation

$$y = \frac{1}{1 + e^{-2x}}$$

Find the y-intercept of its graph.

30.  Graph the logistic function in Exercise 29.

6.6

EXPONENTIAL AND LOGARITHMIC EQUATIONS

■ Solving Exponential Equations   ■ Solving Logarithmic Equations
■ Applications of Exponential and Logarithmic Equations
■ Solving Equations with a Graphing Calculator

An **exponential equation** is an equation that contains a variable in one of its exponents. Some examples of exponential equations are

$$3^x = 5, \qquad 6^{x-3} = 2^x, \qquad \text{and} \qquad 3^{2x+1} - 10(3^x) + 3 = 0$$

A **logarithmic equation** is an equation with a logarithmic expression that contains a variable. Some examples of logarithmic equations are

$$\log 2x = 25, \qquad \ln x - \ln (x - 12) = 24, \qquad \text{and} \qquad \log x = \log \frac{1}{x} + 4$$

In this section, we will learn how to solve many of these equations.

■ Solving Exponential Equations

EXAMPLE 1   Solve the exponential equation $3^x = 5$.

Solution   Since logarithms of equal numbers are equal, we can take the common logarithm of each side of the equation. Property 7 of logarithms then provides a way of moving the variable x from its position as an exponent to a position as a coefficient.

$$3^x = 5$$

$\log 3^x = \log 5$	Take the common logarithm of each side.
$x \log 3 = \log 5$	Use Property 7 of logarithms.

1. $x = \dfrac{\log 5}{\log 3}$ Divide both sides by log 3.

≈ 1.464973521 Use a calculator.

To four decimal places, $x = 1.4650$.   ■

 Warning!   A careless reading of Equation 1 leads to a common error. The right-hand side of Equation 1 calls for a division, not a subtraction.

$$\dfrac{\log 5}{\log 3} \qquad \text{means} \qquad (\log 5) \div (\log 3)$$

It is the expression $\log \dfrac{5}{3}$ that means $\log 5 - \log 3$.

EXAMPLE 2   Solve the exponential equation $6^{x-3} = 2^x$.

Solution

$$6^{x-3} = 2^x$$

$\log 6^{x-3} = \log 2^x$	Take the common logarithm of each side.
$(x - 3) \log 6 = x \log 2$	Use Property 7 of logarithms.
$x \log 6 - 3 \log 6 = x \log 2$	Use the distributive property.
$x \log 6 - x \log 2 = 3 \log 6$	Add 3 log 6 and subtract x log 2 from both sides.
$x(\log 6 - \log 2) = 3 \log 6$	Factor out x on the left-hand side.

$x = \dfrac{3 \log 6}{\log 6 - \log 2}$ Divide both sides by log 6 $-$ log 2.

$x \approx 4.892789261$ Use a calculator.   ■

EXAMPLE 3   Solve the exponential equation $2^{x^2+2x} = \dfrac{1}{2}$.

Solution   Since $\dfrac{1}{2} = 2^{-1}$, we can write the equation in the form

$$2^{x^2+2x} = 2^{-1}$$

Since equal quantities with equal bases have equal exponents, we have

$x^2 + 2x = -1$	
$x^2 + 2x + 1 = 0$	Add 1 to both sides.
$(x + 1)(x + 1) = 0$	Factor the trinomial.
$x + 1 = 0 \qquad \text{or} \qquad x + 1 = 0$	Set each factor equal to 0.
$x = -1 \qquad\qquad\qquad x = -1$	

Verify that -1 satisfies the equation.   ■

■ Solving Logarithmic Equations

EXAMPLE 4   Solve the logarithmic equation $\log x + \log (x - 3) = 1$.

Solution
$$\log x + \log (x - 3) = 1$$

$$\log x(x - 3) = 1 \qquad \text{Use Property 5 of logarithms.}$$

$$x(x - 3) = 10^1 \qquad \begin{array}{l}\text{Use the definition of logarithms to change}\\ \text{the equation to exponential form.}\end{array}$$

$$x^2 - 3x - 10 = 0 \qquad \begin{array}{l}\text{Remove parentheses and subtract 10}\\ \text{from both sides.}\end{array}$$

$$(x + 2)(x - 5) = 0 \qquad \text{Factor the trinomial.}$$

$$\begin{array}{ccc} x + 2 = 0 & \text{or} & x - 5 = 0 \\ x = -2 & & x = 5 \end{array}$$

Check: The number -2 is not a solution, because it does not satisfy the equation (a negative number does not have a logarithm). We will check the remaining number 5.

$$\log x + \log (x - 3) = 1$$

$$\log 5 + \log (5 - 3) \stackrel{?}{=} 1 \qquad \text{Substitute 5 for } x.$$

$$\log 5 + \log (2) \stackrel{?}{=} 1$$

$$\log 10 \stackrel{?}{=} 1 \qquad \text{Use Property 5 of logarithms.}$$

$$1 = 1 \qquad \text{Use Property 2 of logarithms.}$$

Since 5 does check, it is a root of the equation.   ■

EXAMPLE 5   Solve the logarithmic equation $\log_b (3x + 2) - \log_b (2x - 3) = 0$.

Solution
$$\log_b (3x + 2) - \log_b (2x - 3) = 0$$

$$\log_b (3x + 2) = \log_b (2x - 3) \qquad \text{Add } \log_b (2x - 3) \text{ to both sides.}$$

$$(3x + 2) = (2x - 3) \qquad \text{If } \log_b r = \log_b s, \text{ then } r = s.$$

$$x = -5 \qquad \text{Subtract } 2x \text{ and 2 from both sides.}$$

Check: $\log_b (3x + 2) - \log_b (2x - 3) = 0$

$$\log_b [3(-5) + 2] - \log_b [2(-5) - 3] \stackrel{?}{=} 0$$

$$\log_b (-13) - \log_b (-13) \stackrel{?}{=} 0$$

Since the logarithm of a negative number does not exist, -5 is an extraneous root and must be discarded. This equation has no roots.   ■

EXAMPLE 6   Solve the logarithmic equation $\dfrac{\log (5x - 6)}{\log x} = 2$.

Solution   We can multiply both sides of the equation to get

$$\log (5x - 6) = 2 \log x$$

and apply Property 7 of logarithms to get

$$\log (5x - 6) = \log x^2$$

By Property 8 of logarithms, $5x - 6 = x^2$, because they have equal logarithms. Thus,

$$5x - 6 = x^2$$
$$0 = x^2 - 5x + 6$$
$$0 = (x - 3)(x - 2)$$
$$x - 3 = 0 \quad \text{or} \quad x - 2 = 0$$
$$x = 3 \quad \mid \quad x = 2$$

Verify that both 2 and 3 satisfy the equation.   ■

■ Applications of Exponential and Logarithmic Equations

Carbon-14 Dating

When a living organism dies, the oxygen/carbon dioxide cycle common to all living things ceases, and carbon-14, a radioactive isotope with a half-life of 5700 years, is no longer absorbed. By measuring the amount of carbon-14 present in an ancient object, archeologists can estimate the object's age.

The amount A of radioactive material present at time t is given by the model

$$A = A_0 2^{-t/h}$$

where A_0 is the amount present initially and h is the half-life of the material.

EXAMPLE 7   How old is a wooden statue that contains only one-third of its original carbon-14 content?

Solution   To find the time t when $A = \dfrac{1}{3}A_0$, we substitute $\dfrac{A_0}{3}$ for A and 5700 for h into the radioactive decay formula and solve for t:

$$A = A_0 2^{-t/h}$$
$$\frac{A_0}{3} = A_0 2^{-t/5700}$$

$1 = 3(2^{-t/5700})$	Divide both sides by A_0.
$\log 1 = \log 3(2^{-t/5700})$	Take the common logarithm of each side.
$0 = \log 3 + \log 2^{-t/5700}$	Use Properties 1 and 5 of logarithms.
$-\log 3 = -\dfrac{t}{5700} \log 2$	Subtract log 3 from both sides and use Property 7 of logarithms.
$5700\left(\dfrac{\log 3}{\log 2}\right) = t$	Multiply both sides by $-\dfrac{5700}{\log 2}$.
$t \approx 9034.286254$	Use a calculator.

The wooden statue is approximately 9000 years old.   ■

Population Growth

When there is sufficient food and space, populations of living organisms tend to increase exponentially according to the Malthusian growth model

$$P = P_0 e^{kt}$$

where P_0 is the initial population at $t = 0$, and k depends on the rate of growth.

EXAMPLE 8   The bacteria in a laboratory culture increased from an initial population of 500 to 1500 in 3 hours. Find the time it will take for the population to reach 10,000.

Solution

$P = P_0 e^{kt}$	
$1500 = 500(e^{k3})$	Substitute 1500 for P, 500 for P_0, and 3 for t.
$3 = e^{3k}$	Divide both sides by 500.
$3k = \ln 3$	Change the equation from exponential to logarithmic form.
$k = \dfrac{\ln 3}{3}$	Divide both sides by 3.

To find when the population will reach 10,000, we substitute 10,000 for P, 500 for P_0, and $\frac{\ln 3}{3}$ for k in the equation $P = P_0 e^{kt}$ and solve for t:

$P = P_0 e^{kt}$	
$10,000 = 500 e^{[\ln 3/3]t}$	
$20 = e^{[\ln 3/3]t}$	Divide both sides by 500.
$\left(\dfrac{\ln 3}{3}\right)t = \ln 20$	Change the equation to logarithmic form.
$t = \dfrac{3 \ln 20}{\ln 3}$	Multiply both sides by $\dfrac{3}{\ln 3}$.
≈ 8.180499084	Use a calculator.

The population will reach 10,000 in a little more than 8 hours.   ■

■ Solving Equations with a Graphing Calculator

EXAMPLE 9   Use a graphing calculator to solve the equation $\log x + \log (x - 3) = 1$.

Solution   We can subtract 1 from both sides of the equation to get

$$\log x + \log (x - 3) - 1 = 0$$

and graph the corresponding function

$$y = \log x + \log (x - 3) - 1$$

in the viewing window $0 \le x \le 20$ and $-2 \le y \le 2$ to obtain the graph shown in Figure 6-15.

Since the root of the equation is the x-intercept, we can find the root by zooming in on the value of the x-intercept. The root is $x = 5$.   ■

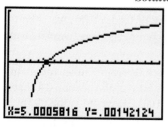

X=5.0005816   Y=.00142124

FIGURE 6-15

6.6   EXERCISES

 In Exercises 1–24, solve each exponential equation. Give all answers to four decimal places.

1. $4^x = 5$

2. $7^x = 12$

3. $13^{x-1} = 2$

4. $5^{x+1} = 3$

5. $2^{x+1} = 3^x$

6. $5^{x-3} = 3^{2x}$

7. $2^x = 3^x$

8. $3^{2x} = 4^x$

9. $7^{x^2} = 10$

10. $8^{x^2} = 11$

11. $8^{x^2} = 9^x$

12. $5^{x^2} = 2^{5x}$

13. $2^{x^2-2x} = 8$

14. $5^{x^2-3x} = 625$

15. $3^{x^2+4x} = \dfrac{1}{81}$

16. $7^{x^2+3x} = \dfrac{1}{49}$

17. $4^{x+2} - 4^x = 15$ (*Hint:* $4^{x+2} = 4^x 4^2$.)

18. $3^{x+3} + 3^x = 84$ (*Hint:* $3^{x+3} = 3^x 3^3$.)

19. $2(3^x) = 6^{2x}$

20. $2(3^{x+1}) = 3(2^{x-1})$

21. $2^{2x} - 10(2^x) + 16 = 0$ (*Hint:* Let $y = 2^x$.)

22. $3^{2x} - 10(3^x) + 9 = 0$ (*Hint:* Let $y = 3^x$.)

23. $2^{2x+1} - 2^x = 1$ (*Hint:* $2^{a+b} = 2^a 2^b$.)

24. $3^{2x+1} - 10(3^x) + 3 = 0$ (*Hint:* $3^{a+b} = 3^a 3^b$.)

In Exercises 25–50, solve each logarithmic equation, if possible.

25. $\log (2x - 3) = \log (x + 4)$

26. $\log (3x + 5) - \log (2x + 6) = 0$

27. $\log \dfrac{4x + 1}{2x + 9} = 0$

28. $\log \dfrac{5x + 2}{2(x + 7)} = 0$

29. $\log x^2 = 2$

30. $\log x^3 = 3$

31. $\log x + \log (x - 48) = 2$

32. $\log x + \log (x + 9) = 1$

33. $\log x + \log (x - 15) = 2$

34. $\log x + \log (x + 21) = 2$

35. $\log (x + 90) = 3 - \log x$

36. $\log (x - 6) - \log (x - 2) = \log \dfrac{5}{x}$

37. $\log (x - 1) - \log 6 = \log (x - 2) - \log x$

38. $\log (2x - 3) - \log (x - 1) = 0$

39. $\log x^2 = (\log x)^2$

40. $\log (\log x) = 1$

41. $\dfrac{\log (3x - 4)}{\log x} = 2$

42. $\dfrac{\log (8x - 7)}{\log x} = 2$

43. $\dfrac{\log (5x + 6)}{2} = \log x$

44. $\dfrac{1}{2} \log (4x + 5) = \log x$

45. $\log_3 x = \log_3 \left(\dfrac{1}{x} \right) + 4$

46. $\log_5 (7 + x) + \log_5 (8 - x) - \log_5 2 = 2$

47. $2 \log_2 x = 3 + \log_2 (x - 2)$

48. $2 \log_3 x - \log_3 (x - 4) = 2 + \log_3 2$

49. $\log (7y + 1) = 2 \log (y + 3) - \log 2$

50. $2 \log (y + 2) = \log (y + 2) - \log 12$

In Exercises 51–68, solve each problem.

51. Tritium decay   The half-life of tritium is 12.4 years. How long will it take for 25% of a sample of tritium to decompose?

52. Radioactive decay   In two years, 20% of a radioactive element decays. Find its half-life.

53. Thorium decay   An isotope of thorium, ^{227}Th, has a half-life of 18.4 days. How long will it take 80% of the sample to decompose?

54. Lead decay   An isotope of lead, ^{201}Pb, has a half-life of 8.4 hours. How many hours ago was there 30% more of the substance?

55. Carbon-14 dating   A cloth fragment is found in an ancient tomb. It contains 60% of the carbon-14 that it is assumed to have had initially. How old is the cloth?

56. Carbon-14 dating   Only 10% of the carbon-14 in a small wooden bowl remains. How old is the bowl?

57. *Compound interest* If $500 is deposited in an account paying 8.5% annual interest, compounded semiannually, how long will it take for the account to increase to $800?

58. *Continuous compound interest* In Exercise 57, how long will it take if the interest is compounded continuously?

59. *Compound interest* If $1300 is deposited in a savings account paying 9% interest, compounded quarterly, how long will it take the account to increase to $2100?

60. *Compound interest* A sum of $5000 deposited in an account grows to $7000 in 5 years. Assuming annual compounding, what interest rate is being paid?

61. *Rule of seventy* A rule of thumb for finding how long it takes an investment to double is called the **rule of seventy**. To apply the rule, divide 70 by the interest rate written as a percent. At 5%, it requires $\frac{70}{5} = 14$ years to double the investment. At 7%, it takes $\frac{70}{7} = 10$ years. Explain why this formula works.

62. *Bacterial growth* A bacteria culture grows according to the formula

$$P = P_0 a^t$$

If it takes 5 days for the culture to triple in size, how long will it take to double in size?

63. *Oceanography* The intensity I of a light a distance x meters beneath the surface of a lake decreases exponentially. If the light intensity at 6 meters is 70% of the intensity at the surface, at what depth will the intensity be 20%?

64. *Rodent control* The rodent population in a city is currently estimated at 30,000. If it is expected to double every 5 years, when will the population reach 1 million?

65. *Newton's law of cooling* Water initially at 100°C is left to cool in a chamber at temperature 60°C. After 3 minutes, the water temperature is 90°C. If the water temperature T is a function of time t given by $T = 60 + 40e^{kt}$, find k.

66. *Newton's law of cooling* Refer to Exercise 65 and find the time for the water temperature to reach 70°C.

67. *Newton's law of cooling* A block of steel, initially at 0°C, is placed in an oven heated to 300°C. After 5 minutes, the temperature of the steel is 100°C. If the steel temperature T is a function of time t given by $T = 300 - 300e^{kt}$, find k.

68. *Newton's law of cooling* Refer to Exercise 67 and find the time for the steel temperature to reach 200°C.

In Exercises 69–72, use a graphing calculator to solve each equation.

69. $\log x + \log (x - 15) = 2$

70. $\log x + \log (x + 3) = 1$

71. $2^{x+1} = 7$

72. $\ln (2x + 5) - \ln 3 = \ln (x - 1)$

6   CHAPTER SUMMARY

Key Words

carbon dating (6.6)
change-of-base formula (6.4)
common logarithms (6.3)
compound interest (6.1)
continuous compound interest (6.2)
decibel voltage gain (6.5)
domain of an exponential function (6.1)
domain of a logarithmic function (6.3)
e (6.2)

exponential decay (6.1)
exponential equation (6.6)
exponential function (6.1)
exponential growth (6.1)
half-life (6.1)
logarithmic equation (6.6)
logarithmic function (6.3)
logistic function (6.2)
Malthusian model of population growth (6.2)

model (6.1)
Napierian logarithms (6.3)
natural logarithms (6.3)
pH scale (6.5)
radioactive decay (6.1)
range of an exponential function (6.1)
range of a logarithmic function (6.3)
Richter scale (6.5)
Weber-Fechner law (6.5)

Key Ideas

(6.1) The exponential function $y = b^x$, where $b > 0$, $b \neq 1$ and x is a real number, is one-to-one. Its domain is the set of real numbers, and its range is the set of positive numbers.

(6.2) $e = 2.718281828 \ldots$

The exponential function $y = e^x$ is one-to-one. Its domain is the set of real numbers, and its range is the set of positive numbers.

(6.3) The logarithmic function $y = \log_b x$, where $b > 0$, $b \neq 1$, and x is a positive number, is one-to-one. Its domain is the set of positive numbers, and its range is the set of real numbers.

The equation $y = \log_b x$ is equivalent to the equation $x = b^y$.

Logarithms of negative numbers do not exist.

The functions defined by $y = \log_b x$ and $y = b^x$ are inverse functions.

Common logarithms are base-10 logarithms.

Natural logarithms are base-e logarithms.

(6.4) **Properties of Logarithms:** If b, M, and N are positive numbers and $b \neq 1$, then

1. $\log_b 1 = 0$
2. $\log_b b = 1$
3. $\log_b b^x = x$
4. $b^{\log_b x} = x$ $\quad (x > 0)$
5. $\log_b MN = \log_b M + \log_b N$
6. $\log_b \dfrac{M}{N} = \log_b M - \log_b N$
7. $\log_b M^p = p \log_b M$
8. If $\log_b x = \log_b y$, then $x = y$.

The Change-of-Base Formula: $\log_b y = \dfrac{\log_a y}{\log_a b}$

(6.5) Many applications can be modeled by logarithmic functions.

(6.6) Many applications lead to exponential and logarithmic equations.

6 CHAPTER REVIEW EXERCISES

In Review Exercises 1–4, graph the function defined by each equation.

1. $y = \left(\dfrac{6}{5}\right)^x$ **2.** $y = e^x$ **3.** $y = \log x$ **4.** $y = \ln x$

In Review Exercises 5–8, graph each pair of equations on one set of coordinate axes.

5. $y = \left(\dfrac{1}{3}\right)^x$ and $y = \log_{1/3} x$ **6.** $y = \left(\dfrac{2}{5}\right)^x$ and $y = \log_{2/5} x$

7. $y = 4^x$ and $y = \log_4 x$ **8.** $y = 3^x$ and $y = \log_3 x$

In Review Exercises 9–20, find each value.

9. $\log_3 9$ **10.** $\log_9 \dfrac{1}{3}$ **11.** $\log_\pi 1$ **12.** $\log_5 0.04$

13. $\log_a \sqrt{a}$ **14.** $\log_a \sqrt[3]{a}$ **15.** $\ln e^4$ **16.** $\ln 1$

17. $10^{\log_{10} 7}$ **18.** $e^{\log_e 3}$ **19.** $\log_b b^4$ **20.** $\ln e^9$

In Review Exercises 21–44, solve each equation for x.

21. $\log_2 x = 3$ **22.** $\log_3 x = -2$ **23.** $\log_x 9 = 2$ **24.** $\log_x 0.125 = -3$

25. $\log_7 7 = x$ **26.** $\log_3 \sqrt{3} = x$ **27.** $\log_8 \sqrt{2} = x$ **28.** $\log_6 36 = x$

29. $\log_{1/3} 9 = x$ **30.** $\log_{1/2} 1 = x$ **31.** $\log_x 3 = \dfrac{1}{3}$ **32.** $\log_x 25 = -2$

33. $\log_2 x = 5$ **34.** $\log_{\sqrt{3}} x = 4$ **35.** $\log_{\sqrt{3}} x = 6$ **36.** $\log_{0.1} 10 = x$

37. $\log_x 2 = -\dfrac{1}{3}$ **38.** $\log_x 32 = 5$ **39.** $\log_{0.25} x = -1$ **40.** $\log_{0.125} x = -\dfrac{1}{3}$

41. $\log_{\sqrt{2}} 32 = x$ **42.** $\log_{\sqrt{5}} x = -4$ **43.** $\log_{\sqrt{3}} 9\sqrt{3} = x$ **44.** $\log_{\sqrt{5}} 5\sqrt{5} = x$

In Review Exercises 45–46, write each expression in terms of the logarithms of x, y, and z.

45. $\log_b \dfrac{x^2 y^3}{z^4}$ **46.** $\log_b \sqrt{\dfrac{x}{yz^2}}$

In Review Exercises 47–48, write each expression as the logarithm of a single quantity.

47. $3 \log_b x - 5 \log_b y + 7 \log_b z$ **48.** $\dfrac{1}{2}\log_b x + 3 \log_b y - 7 \log_b z$

In Review Exercises 49–52, assume that $\log a = 0.60$, $\log b = 0.36$, and $\log c = 2.40$. Find the value of each expression.

49. $\log abc$ **50.** $\log a^2 b$ **51.** $\log \dfrac{ac}{b}$ **52.** $\log \dfrac{a^2}{c^3 b^2}$

In Review Exercises 53–66, solve for x, where possible.

53. $3^x = 7$ **54.** $5^{x+2} = 625$ **55.** $2^x = 3^{x-1}$ **56.** $2^{x^2+4x} = \dfrac{1}{8}$

57. $\log x + \log (29 - x) = 2$ **58.** $\log_2 x + \log_2 (x - 2) = 3$

59. $\log_2 (x + 2) + \log_2 (x - 1) = 2$ **60.** $\dfrac{\log (7x - 12)}{\log x} = 2$

61. $\log x + \log (x - 5) = \log 6$ **62.** $\log 3 - \log (x - 1) = -1$
63. $e^{x \ln 2} = 9$ **64.** $\ln x = \ln (x - 1)$
65. $\ln x = \ln (x - 1) + 1$ **66.** $\ln x = \log_{10} x$ (*Hint:* Use the change-of-base formula.)

▦ *In Review Exercises 67–70, solve each problem.*

67. Carbon-14 dating A wooden statue excavated in Egypt has a carbon-14 content that is two-thirds of that found in living wood. If the half-life of carbon-14 is 5700 years, how old is the statue?

68. pH of grapefruit The pH of grapefruit juice is approximately 3.1. Find its hydrogen-ion concentration.

69. Radioactive decay One-third of a radioactive material decays in 20 years. Find its half-life.

70. Formula for pH Some chemistry textbooks define the pH of a solution by the formula

$$pH = \log_{10} \dfrac{1}{[H^+]}$$

Show that this definition is equivalent to the one given in this book.

6 CHAPTER TEST

In Questions 1–2, graph each function.

1. $y = 2^x + 1$

2. $y = 2^{-x}$

In Questions 3–4, solve each problem.

3. A radioactive material decays according to the formula $A = A_0(2)^{-t}$. How much of a 3-gram sample will be left in 6 years?

4. An initial deposit of \$1000 earns 6% interest, compounded twice a year. How much will be in the account in one year?

In Questions 5–6, graph each function.

5. $y = e^x$

6. $y = 2e^x$

7. An account contains \$2000 and has been earning 8% interest, compounded continuously. How much will be in the account in 10 years?

In Questions 8–12, find each value.

8. $\log_7 343$

9. $\log_3 \dfrac{1}{27}$

10. $\log_{10} 10^{12} + 10^{\log_{10} 5}$

11. $\log_{3/2} \dfrac{9}{4}$

12. $\log_{2/3} \dfrac{27}{8}$

In Questions 13–14, graph each function.

13. $y = \log (x - 1)$

14. $y = 2 + \ln x$

In Questions 15–16, write each expression in terms of the logarithms of a, b, and c.

15. $\log a^2 b c^3$

16. $\ln \sqrt{\dfrac{a}{b^2 c}}$

In Questions 17–18, write each expression as a logarithm of a single quantity.

17. $\dfrac{1}{2} \log (a + 2) + \log b - 2 \log c$

18. $\dfrac{1}{3}(\log a - 2 \log b) - \log c$

In Questions 19–20, assume that $\log 2 = 0.3010$ *and* $\log 3 = 0.4771$. *Find each value.*

19. $\log 24$

20. $\log \dfrac{8}{3}$

*In Questions 21–22, use the change-of-base formula to find each logarithm. **Do not attempt to simplify the answer.***

21. $\log_7 3$

22. $\log_\pi e$

In Questions 23–26, tell whether each statement is true. If it is not true, answer false.

23. $\log_a ab = 1 + \log_a b$

24. $\dfrac{\log a}{\log b} = \log a - \log b$

25. $\log a^{-3} = \dfrac{1}{3 \log a}$

26. $\ln(-x) = -\ln x$

27. Find the pH of a solution with a hydrogen-ion concentration of 3.7×10^{-7}. (*Hint:* $pH = -\log[H^+]$.)

28. Find the db voltage gain of an amplifier when $E_O = 60$ volts and $E_I = 0.3$ volt.
(*Hint:* db gain $= 20 \log (E_O/E_I)$.)

In Questions 29–32, solve each equation.

29. $3^{x-1} = 100^x$

30. $3^{x^2-2x} = 27$

31. $\log(5x + 2) = \log(2x + 5)$

32. $\log x + \log(x - 9) = 1$

CUMULATIVE REVIEW EXERCISES

In Cumulative Review Exercises 1–4, decide whether each equation defines a function.

1. $y = 3x - 1$

2. $y = x^2 + 3$

3. $y = \dfrac{1}{x - 2}$

4. $y^2 = 4x$

In Cumulative Review Exercises 5–8, find the domain and range of each function.

5. $y = f(x) = x^2 + 5$

6. $y = f(x) = \dfrac{7}{x + 2}$

7. $y = f(x) = -\sqrt{x - 2}$

8. $y = f(x) = \sqrt{x + 4}$

In Cumulative Review Exercises 9–10, find the vertex of the parabolic graph of each equation.

9. $y = x^2 + 5x - 6$

10. $y = -x^2 + 5x + 6$

In Cumulative Review Exercises 11–14, graph the function defined by each equation.

11. $y = x^2 - 4$

12. $y = -x^2 + 4$

13. $y = x^3 + x$

14. $y = -x^4 + 2x^2 + 1$

In Cumulative Review Exercises 15–16, graph the function defined by each equation.

15. $y = \dfrac{x}{x - 3}$

16. $y = \dfrac{x^2 - 1}{x^2 - 9}$

In Cumulative Review Exercises 17–20, $f(x) = 3x - 4$ and $g(x) = x^2 + 1$. Find each function and its domain.

17. $(f + g)(x)$

18. $(f - g)(x)$

19. $(f \cdot g)(x)$

20. $(g/f)(x)$

In Cumulative Review Exercises 21–24, $f(x) = 3x - 4$ and $g(x) = x^2 + 1$. Find each value.

21. $(f \circ g)(2)$

22. $(g \circ f)(2)$

23. $(f \circ g)(x)$

24. $(g \circ f)(x)$

In Cumulative Review Exercises 25–28, find the inverse of the function defined by each equation.

25. $y = 3x + 2$ **26.** $y = \dfrac{1}{x - 3}$ **27.** $y = x^2 + 5$ **28.** $3x - y = 1$

In Cumulative Review Exercises 29–30, write each sentence as an equation.

29. y varies directly with the product of w and z.

30. y varies directly with x but inversely with t^2.

In Cumulative Review Exercises 31–34, let $P(x) = 4x^3 + 3x + 2$. Use synthetic division to find each value.

31. $P(1)$ **32.** $P(-2)$ **33.** $P\!\left(\dfrac{1}{2}\right)$ **34.** $P(i)$ $(i = \sqrt{-1})$

In Cumulative Review Exercises 35–38, tell whether each binomial is a factor of $P(x) = x^3 + 2x^2 - x - 2$. Use synthetic division.

35. $(x + 1)$ **36.** $(x - 2)$ **37.** $(x - 1)$ **38.** $(x + 2)$

In Cumulative Review Exercises 39–40, tell how many roots each polynomial equation has.

39. $x^{12} - 4x^8 + 2x^4 + 12 = 0$ **40.** $x^{2000} - 1 = 0$

In Cumulative Review Exercises 41–42, tell the maximum number of possible positive roots and the maximum number of negative roots for each equation.

41. $x^4 + 2x^3 - 3x^2 + x + 2 = 0$ **42.** $x^4 - 3x^3 - 2x^2 - 3x - 5 = 0$

In Cumulative Review Exercises 43–44, solve each equation.

43. $x^3 + x^2 - 9x - 9 = 0$ **44.** $x^3 - 2x^2 - x + 2 = 0$

In Cumulative Review Exercises 45–48, graph the function defined by each equation.

45. $y = 3^x - 2$ **46.** $y = 2e^x$ **47.** $y = \log_3 x$ **48.** $y = \ln(x - 2)$

In Cumulative Review Exercises 49–52, find each value.

49. $\log_2 64$ **50.** $\log_{1/3} 27$ **51.** $\ln e^3$ **52.** $2^{\log_2 2}$

In Cumulative Review Exercises 53–56, write each expression in terms of the logarithms of a, b, and c.

53. $\log abc$ **54.** $\log \dfrac{a^2 b}{c}$ **55.** $\log \sqrt{\dfrac{ab}{c^2}}$ **56.** $\log \dfrac{\sqrt{ab^2}}{c}$

In Cumulative Review Exercises 57–58, write each expression as the logarithm of a single quantity.

57. $3 \log a - 3 \log b$ **58.** $\dfrac{1}{2} \log a + 3 \log b - \dfrac{1}{2} \log c$

In Cumulative Review Exercises 59–62, solve each equation, if possible.

59. $3^{x+1} = 8$ **60.** $3^{x-1} = 3^x$

61. $\log x + \log 2 = 3$ **62.** $\log(x + 1) + \log(x - 1) = 1$

CHAPTER 7

LINEAR SYSTEMS

Equations containing two variables have infinitely many solutions, although pairs of these equations often have a single common solution. For example, the equation $x + y = 5$ has infinitely many solutions, as does the equation $x - y = 1$. However, only the pair of numbers $x = 3$ and $y = 2$ is a solution of both equations, because $3 + 2 = 5$ and $3 - 2 = 1$. The pair of equations

$$\begin{cases} x + y = 5 \\ x - y = 1 \end{cases}$$

is called a **system of equations**, and the solution $x = 3$, $y = 2$ is called its **simultaneous solution**, or just its **solution**. The process of finding the solution of a system of equations is called **solving the system**.

A **linear equation** in the variables $x, y, z, \ldots$ is any equation of the form

$$ax + by + cz + \cdots = k$$

where $a, b, c, \ldots$ and k are constants. In this chapter, we will discuss methods for solving systems of m linear equations in n variables.

7.1   SYSTEMS OF LINEAR EQUATIONS

- The Graphing Method
- The Addition Method
- An Inconsistent System
- The Substitution Method
- A System with Infinitely Many Solutions

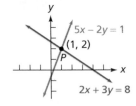

FIGURE 7-1

The graph of an equation in two variables represents the equation's infinitely many solutions. For example, the graph of one of the lines in Figure 7-1 represents the infinitely many solutions of the equation

$$5x - 2y = 1$$

The other line in the figure is the graph of the infinitely many solutions of the equation

$$2x + 3y = 8$$

Only point P lies on both lines. The coordinates of P, $x = 1$ and $y = 2$, are the only numbers that satisfy both equations. Thus, the simultaneous solution of the system of equations

$$\begin{cases} 5x - 2y = 1 \\ 2x + 3y = 8 \end{cases}$$

is the pair of numbers $x = 1$ and $y = 2$, or simply the pair $(1, 2)$.

This discussion suggests a graphical method of solving systems of equations in two variables.

■ The Graphing Method

We use the following steps to solve a system of two equations in two variables by graphing.

Steps to Follow When Using the Graphing Method	1. On a single coordinate grid, graph each equation. 2. Find the coordinates of the point or points where all of the graphs intersect. These coordinates give the solutions of the system. 3. If the graphs have no point in common, the system has no solution.

EXAMPLE 1   Use the graphing method to solve each system:

$$\textbf{a.} \begin{cases} 3x + y = 1 \\ -x + 2y = 9 \end{cases}, \quad \textbf{b.} \begin{cases} 2x - 3y = 4 \\ 4x = -4 + 6y \end{cases} \quad \text{and} \quad \textbf{c.} \begin{cases} y = 4 - x \\ 2x + 2y = 8 \end{cases}$$

Solution   **a.** The graphs of the equations are the lines shown in Figure 7-2(a). The solution of this system is given by the coordinates of the point where the lines intersect, the point $(-1, 4)$. Verify that the solution of the system is $x = -1$ and $y = 4$.

b. The graphs of the equations are the lines shown in Figure 7-2(b). In this case, the lines are parallel. Because parallel lines do not intersect, the system has no solutions.

c. The graphs of the equations are the lines shown in Figure 7-2(c). In this case, the lines are the same. Because the two lines have infinitely many points in common, the system has infinitely many solutions. All ordered pairs whose coordinates satisfy one of the equations satisfy the other also.

   To find some of these solutions, we substitute numbers for x in the first equation and solve for y. If $x = 3$, for example, then $y = 1$. Thus, one solution is the pair $(3, 1)$. Other solutions are $(0, 4)$ and $(5, -1)$.

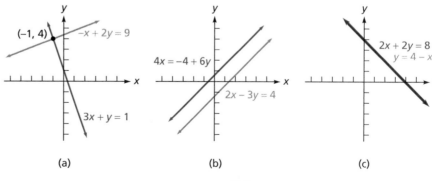

(a)   (b)   (c)

FIGURE **7-2**   ■

Example 1 illustrates three situations that can occur when we solve systems of equations. If a system of equations has at least one solution (as in parts **a** and **c** of Example 1), the system is called **consistent**. If it has no solutions (as in part **b**), it is called **inconsistent**.

If a system of two equations in two variables has exactly one solution (as in part **a**), or no solution (as in part **b**), the equations in the system are called **independent**. If a system of linear equations has infinitely many solutions (as in part **c**), the equations of the system are called **dependent**.

The following is a summary of the possible outcomes when two equations, each with two variables, are graphed.

Possible Outcomes When Graphing Two Equations, Each with Two Variables	Possible figure	If the	then
		lines are distinct and intersect,	the equations are independent, and the system is consistent. The system has one solution.
		lines are distinct and parallel,	the equations are independent, and the system is inconsistent. The system has no solutions.
		lines coincide,	the equations are dependent, and the system is consistent. The system has infinitely many solutions.

The graphing method has limitations. First, the method is limited to equations in two variables. Systems of equations with three or more variables cannot be solved graphically. Second, it is difficult to find exact solutions such as $\left(\frac{37}{29}, \frac{13}{7}\right)$ graphically. However, the graphing calculator's TRACE and ZOOM capabilities provide good approximations of such solutions.

EXAMPLE 2   Use a graphing calculator to solve the system $\begin{cases} x - 4y = 7 \\ x + 2y = 4 \end{cases}$.

Solution   To enter the equations into a graphing calculator, we must solve them for y. We do so and get the equivalent system

$$\begin{cases} y = \dfrac{x - 7}{4} \\ y = \dfrac{4 - x}{2} \end{cases}$$

Depending on the viewing window, the two graphs will be similar to those in Figure 7-3(a). We zoom in on the intersection as in Figure 7-3(b) and use TRACE to find an approximate solution of $x = 4.992$ and $y = .501$. Verify that the actual solution is $x = 5$, $y = -\frac{1}{2}$.

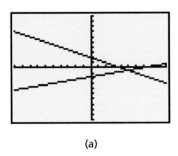

(a)

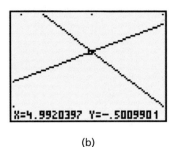

(b)

FIGURE **7-3**   ◼

We now consider two algebraic techniques that provide solutions for systems containing any number of variables. The goal of each of these methods is to produce simpler but **equivalent** systems—that is, systems with the same solutions as the original system—until it is easy to find the solution.

◼ The Substitution Method

We can take the following steps to use substitution to solve a system of two equations in two variables.

Steps to Follow When Using the Substitution Method to Solve Two Equations in *x* and *y*	**1.** Solve one equation for a variable—say, *y*. **2.** Substitute the expression obtained for *y* for every *y* in the other equation. **3.** Solve the equation that results. **4.** Substitute the solution found in Step 3 in the equation found in Step 1 and solve for *y*.

EXAMPLE 3   Use the substitution method to solve the system $\begin{cases} 3x + y = 1 \\ -x + 2y = 9 \end{cases}$.

Solution   We can solve the first equation for *y* and substitute the result for *y* in the second equation.

$$\begin{cases} 3x + y = 1 \rightarrow y = \boxed{1 - 3x} \\ -x + 2y = 9 \end{cases}$$

The substitution gives one linear equation with one variable, which we can solve for *x*:

$$-x + 2(1 - 3x) = 9$$
$$-x + 2 - 6x = 9$$
$$-7x = 7 \qquad \text{Combine terms and subtract 2 from both sides.}$$
$$x = -1 \qquad \text{Divide both sides by } -7.$$

To find y, we substitute -1 for x in the equation $y = 1 - 3x$ and simplify:

$$y = 1 - 3x$$
$$= 1 - 3(\mathbf{-1})$$
$$= 1 + 3$$
$$= 4$$

The solution to the system is the pair $(-1, 4)$. ■

■ The Addition Method

As with substitution, the purpose of the addition method is to combine the equations of the system to eliminate terms involving one of the variables. To do so, we follow these steps.

Steps to Follow When Using the Addition Method to Solve Two Equations in x and y	1. Write the equations of the system in general form so that terms with the same variable are aligned vertically.
	2. Multiply all the terms of one or both of the equations by constants chosen to make the coefficients of x (or y) differ only in sign.
	3. Add the equations. Solve the equation that results, if possible.
	4. Substitute the value obtained in Step 3 into either of the original equations and solve for the remaining variable.
	5. The results obtained in Steps 3 and 4 are the solution of the system. **Check the solutions in both of the equations of the system.**

EXAMPLE 4   Use the addition method to solve the system $\begin{cases} 3x + 2y = 8 \\ 2x - 5y = 18 \end{cases}$.

Solution   If we choose to eliminate the variable y, we multiply the terms of the first equation by 5 and the terms of the second equation by 2, to obtain an equivalent system in which the coefficients of y differ only in sign. Then we add the equations to eliminate y.

$$\begin{cases} 3x + 2y = 8 \\ 2x - 5y = 18 \end{cases} \quad \begin{matrix} \rightarrow \\ \rightarrow \end{matrix} \quad \begin{matrix} 15x + 10y = 40 \\ \underline{4x - 10y = 36} \\ 19x \qquad\quad = 76 \\ x = 4 \end{matrix} \quad \begin{matrix} \text{Multiply by 5.} \\ \text{Multiply by 2.} \\ \text{Add the equations.} \\ \text{Divide both sides by 19.} \end{matrix}$$

We now substitute 4 for x into either of the original equations and solve for y. For example, we use the first equation.

$$3x + 2y = 8$$
$$3(\mathbf{4}) + 2y = 8 \qquad \text{Substitute 4 for } x.$$
$$12 + 2y = 8 \qquad \text{Simplify.}$$

$$2y = -4 \qquad \text{Subtract 12 from both sides.}$$
$$y = -2 \qquad \text{Divide both sides by 2.}$$

Verify that the solution of the system is $(4, -2)$. ∎

■ A System with Infinitely Many Solutions

EXAMPLE 5 Use the addition method to solve the system $\begin{cases} x + 2y = 3 \\ 2x + 4y = 6 \end{cases}$.

Solution We can multiply both sides of the first equation by -2 and add the results to the second equation to get

$$\begin{array}{r} -2x - 4y = -6 \\ 2x + 4y = 6 \\ \hline 0 = 0 \end{array}$$

Although the result $0 = 0$ is true, it does not give the value of y.

Since the second equation in the original system is twice the first, the equations are equivalent. If we were to solve this system by graphing, the graphs of the two equations would coincide. The pairs (x, y) of coordinates of points on that one line make up the infinite set of solutions to the given system. The system is consistent, and the equations are dependent.

To find some of the solutions, we substitute values for x into either equation and solve for y. If $x = 5$, for example, then $y = -1$, and the pair $(5, -1)$ is a solution. Other solutions are $(3, 0)$, $\left(0, \frac{3}{2}\right)$, and $(-7, 5)$. ∎

■ An Inconsistent System

EXAMPLE 6 Solve the system $\begin{cases} x + y = 3 \\ x + y = 2 \end{cases}$.

Solution We can multiply both sides of the second equation by -1 and add the results to the first equation to get

$$\begin{array}{r} x + y = 3 \\ -x - y = -2 \\ \hline 0 = 1 \end{array}$$

Because $0 \neq 1$, the system has no solutions. If we had graphed each equation in this system, the graphs would be parallel lines. This system is inconsistent, and the equations are independent. ∎

EXAMPLE 7 Solve the system $\begin{cases} \dfrac{x + 2y}{4} + \dfrac{x - y}{5} = \dfrac{6}{5} \\ \dfrac{x + y}{7} - \dfrac{x - y}{3} = -\dfrac{12}{7} \end{cases}$

Solution We first clear the equations of fractions by multiplying both sides of the first equation by 20 and both sides of the second equation by 21.

$$\begin{cases} 20\left(\dfrac{x+2y}{4} + \dfrac{x-y}{5}\right) = 20\left(\dfrac{6}{5}\right) \\ 21\left(\dfrac{x+y}{7} - \dfrac{x-y}{3}\right) = 21\left(-\dfrac{12}{7}\right) \end{cases}$$

$$\begin{cases} 5(x+2y) + 4(x-y) = 4(6) \\ 3(x+y) - 7(x-y) = -3(12) \end{cases}$$

$$\begin{cases} 9x + 6y = 24 \\ -4x + 10y = -36 \end{cases} \qquad \text{Remove parentheses and combine terms.}$$

$$\begin{cases} 3x + 2y = 8 \\ 2x - 5y = 18 \end{cases} \qquad \begin{array}{l} \text{Divide both sides by 3.} \\ \text{Divide both sides by } -2. \end{array}$$

The solution is $(4, -2)$. (See Example 4.) ∎

In Example 8, we use both substitution and addition to solve a system of three equations in three variables.

EXAMPLE 8 Solve the system

1. $\begin{cases} x + 2y + z = 8 \\ 2x + y - z = 1 \\ x + y - 2z = -3 \end{cases}$
2.
3.

Solution We can add Equations 1 and 2 to eliminate the variable z

1. $x + 2y + z = 8$
2. $\underline{2x + y - z = 1}$
4. $3x + 3y = 9$

and divide both sides of Equation 4 by 3 to get

5. $x + y = 3$

We now choose a different pair of equations (say, Equations 1 and 3) and eliminate z again. If we multiply both sides of Equation 1 by 2 and add the result to Equation 3, we get

$2x + 4y + 2z = 16$
3. $\underline{x + y - 2z = -3}$
6. $3x + 5y = 13$

Equations 5 and 6 form a system of two equations in two variables, which we can solve by substitution.

5. $\begin{cases} x + y = 3 \\ 3x + 5y = 13 \end{cases} \qquad \rightarrow \qquad y = \boxed{3 - x}$
6.

$$3x + 5(3 - x) = 13$$
$$3x + 15 - 5x = 13 \qquad \text{Remove parentheses.}$$
$$-2x = -2 \qquad \text{Combine terms and subtract 15 from both sides.}$$
$$x = 1 \qquad \text{Divide both sides by } -2.$$

To find y, we substitute 1 for x in the equation $y = 3 - x$.

$$y = 3 - 1$$
$$y = 2$$

To find z, we substitute 1 for x and 2 for y in any one of the original equations and find that $z = 3$. The solution to the given system is the triple

$$(1, 2, 3)$$

Because the solution is unique, the system is consistent, and its equations are independent. Verify that $x = 1$, $y = 2$, and $z = 3$ satisfy each equation in the original system. ∎

EXAMPLE 9 An airplane flies 600 miles with the wind for 2 hours and returns against the wind in 3 hours. Find the speed of the wind and the air speed of the plane.

Solution If a represents the air speed and w represents wind speed, then the ground speed of the plane with the wind is the combined speed $a + w$. On the return trip, against a headwind, the ground speed is $a - w$. This and the other information is organized in the chart in Figure 7-4 to give a system of two equations in the variables a and w.

	d	$=$ r	$\cdot$ t
Outbound trip	600	$a+w$	2
Return trip	600	$a-w$	3

FIGURE 7-4

Since $d = rt$, we have $\begin{cases} 600 = 2(a + w) \\ 600 = 3(a - w) \end{cases}$ or

7. $\begin{cases} 300 = a + w \\ 200 = a - w \end{cases}$

We can add the equations to get

$$500 = 2a$$
$$a = 250 \qquad \text{Divide both sides by 2.}$$

To find w, we substitute 250 for a into one of the previous equations (such as Equation 7) and solve for w:

7. $\quad 300 = a + w$
$$300 = 250 + w$$
$$w = 50 \qquad \text{Subtract 250 from both sides.}$$

The air speed of the plane is 250 miles per hour. With a 50 mile-per-hour tailwind, the ground speed is $250 + 50$, or 300 miles per hour. At 300 miles per hour, the 600-mile trip will take 2 hours.

With a 50-mile-per-hour headwind, the ground speed is $250 - 50$, or 200 miles per hour. At 200 miles per hour, the 600-mile trip will take 3 hours. ∎

7.1 EXERCISES

In Exercises 1–4, solve each system of equations by the graphing method.

1. $\begin{cases} y = -3x + 5 \\ x - 2y = -3 \end{cases}$

2. $\begin{cases} x - 2y = -3 \\ 3x + y = -9 \end{cases}$

3. $\begin{cases} 3x + 2y = 2 \\ -2x + 3y = 16 \end{cases}$

4. $\begin{cases} x + y = 0 \\ 5x - 2y = 14 \end{cases}$

In Exercises 5–8, use a graphing calculator to approximate the solution of each system of equations. Give answers to the nearest tenth.

5. $\begin{cases} y = -5.7x + 7.8 \\ y = 37.2 - 19.1x \end{cases}$

6. $\begin{cases} y = 3.4x - 1 \\ x = -4.7y + 3 \end{cases}$

7. $\begin{cases} 5.3x - 9.2y = 6.0 \\ y = \dfrac{5.5 - 2.7x}{3.5} \end{cases}$

8. $\begin{cases} 29x + 17y = 7 \\ -17x + 23y = 19 \end{cases}$

In Exercises 9–16, solve each system of equations by the substitution method, if possible.

9. $\begin{cases} y = x - 1 \\ y = 2x \end{cases}$

10. $\begin{cases} y = 2x - 1 \\ x + y = 5 \end{cases}$

11. $\begin{cases} 2x + 3y = 0 \\ y = 3x - 11 \end{cases}$

12. $\begin{cases} 2x + y = 3 \\ y = 5x - 11 \end{cases}$

13. $\begin{cases} 4x + 3y = 3 \\ 2x - 6y = -1 \end{cases}$

14. $\begin{cases} 4x + 5y = 4 \\ 8x - 15y = 3 \end{cases}$

15. $\begin{cases} x + 3y = 1 \\ 2x + 6y = 3 \end{cases}$

16. $\begin{cases} x - 3y = 14 \\ 3(x - 12) = 9y \end{cases}$

In Exercises 17–32, solve each system of equations by the addition method, if possible.

17. $\begin{cases} 5x - 3y = 12 \\ 2x - 3y = 3 \end{cases}$

18. $\begin{cases} 2x + 3y = 8 \\ -5x + y = -3 \end{cases}$

19. $\begin{cases} x - 7y = -11 \\ 8x + 2y = 28 \end{cases}$

20. $\begin{cases} 3x + 9y = 9 \\ -x + 5y = -3 \end{cases}$

21. $\begin{cases} 3(x - y) = y - 9 \\ 5(x + y) = -15 \end{cases}$

22. $\begin{cases} 2(x + y) = y + 1 \\ 3(x + 1) = y - 3 \end{cases}$

23. $\begin{cases} 2 = \dfrac{1}{x + y} \\ 2 = \dfrac{3}{x - y} \end{cases}$

24. $\begin{cases} \dfrac{1}{x + y} = 12 \\ \dfrac{3x}{y} = -4 \end{cases}$

25. $\begin{cases} 0.5x = 0.1y \\ y - 0.5x = 1.5 \end{cases}$

26. $\begin{cases} -0.3x + 0.1y = -0.1 \\ 6x - 2y = 2 \end{cases}$

27. $\begin{cases} x + 2(x - y) = 2 \\ 3(y - x) - y = 5 \end{cases}$

28. $\begin{cases} x + \dfrac{y}{3} = \dfrac{5}{3} \\ \dfrac{x + y}{3} = 3 - x \end{cases}$

29. $\begin{cases} \dfrac{3}{2}x + \dfrac{1}{3}y = 2 \\ \dfrac{2}{3}x + \dfrac{1}{9}y = 1 \end{cases}$

30. $\begin{cases} \dfrac{x + y}{2} + \dfrac{x - y}{5} = 2 \\ x = \dfrac{y}{2} + 1 \end{cases}$

31. $\begin{cases} \dfrac{x - y}{5} + \dfrac{x + y}{2} = 6 \\ \dfrac{x - y}{2} - \dfrac{x + y}{4} = 3 \end{cases}$

32. $\begin{cases} \dfrac{x - 2}{5} + \dfrac{y + 3}{2} = 5 \\ \dfrac{x + 3}{2} + \dfrac{y - 2}{3} = 6 \end{cases}$

In Exercises 33–50, solve each system of equations, if possible.

33. $\begin{cases} x + y + z = 3 \\ 2x + y + z = 4 \\ 3x + y - z = 5 \end{cases}$

34. $\begin{cases} x - y - z = 0 \\ x + y - z = 0 \\ x - y + z = 2 \end{cases}$

35. $\begin{cases} x - y + z = 0 \\ x + y + 2z = -1 \\ -x - y + z = 0 \end{cases}$

36. $\begin{cases} 2x + y - z = 7 \\ x - y + z = 2 \\ x + y - 3z = 2 \end{cases}$

37. $\begin{cases} 2x + y = 4 \\ x - z = 2 \\ y + z = 1 \end{cases}$

38. $\begin{cases} 3x + y + z = 0 \\ 2x - y + z = 0 \\ 2x + y + z = 0 \end{cases}$

39. $\begin{cases} x + y + z = 6 \\ 2x + y + 3z = 17 \\ x + y + 2z = 11 \end{cases}$

40. $\begin{cases} x + y + z = 3 \\ 2x + y + z = 6 \\ x + 2y + 3z = 2 \end{cases}$

41. $\begin{cases} x + y + z = 3 \\ x + z = 2 \\ 2x + 2y + 2z = 3 \end{cases}$ **42.** $\begin{cases} x + y + z = 3 \\ x + z = 2 \\ 2x + y + 2z = 5 \end{cases}$ **43.** $\begin{cases} x + 2y - z = 2 \\ 2x - y = -1 \\ 3x + y + z = 1 \end{cases}$ **44.** $\begin{cases} x + y = 2 \\ y + z = 2 \\ 3x + 3y = 2 \end{cases}$

45. $\begin{cases} 3x + 4y + 2z = 4 \\ 6x - 2y + z = 4 \\ 3x - 8y - 6z = -3 \end{cases}$ **46.** $\begin{cases} x + y = 2 \\ y + z = 2 \\ x - z = 0 \end{cases}$ **47.** $\begin{cases} 2x - y - z = 0 \\ x - 2y - z = -1 \\ x - y - 2z = -1 \end{cases}$ **48.** $\begin{cases} x + 3y - z = 5 \\ 3x - y + z = 2 \\ 2x + y = 1 \end{cases}$

49. $\begin{cases} (x + y) + (y + z) + (z + x) = 6 \\ (x - y) + (y - z) + (z - x) = 0 \\ x + y + 2z = 4 \end{cases}$ **50.** $\begin{cases} (x + y) + (y + z) = 1 \\ (x + z) + (x + z) = 3 \\ (x - y) - (x - z) = -1 \end{cases}$

In Exercises 51–62, use systems of two equations in two variables to solve each word problem.

51. Planning for harvest A farmer raises corn and soybeans on 350 acres of land. Because of expected prices at harvest time, he thinks it would be wise to plant 100 more acres of corn than of soybeans. How many acres of each does he plant?

52. Resource allocation 120,000 gallons of fuel are to be divided between two airlines. Triple A Airways requires twice as much as UnityAir. How much fuel should be allocated to Triple A?

53. Country club membership There is an initiation fee to join the Pine River Country Club, as well as monthly dues. The total cost after 7 months of membership will be $3025, and after $1\frac{1}{2}$ years, $3850. Find both the initiation fee and the monthly dues.

54. Framing pictures A rectangular picture frame has a perimeter of 1900 centimeters and a width that is 250 centimeters less than its length. Find the area of the picture.

55. Boating in moving water A Mississippi riverboat can travel 30 kilometers downstream in three hours and can make the return trip in five hours. Find the speed of the boat in still water.

56. Making an alloy A metallurgist wants to make 60 grams of an alloy that is to be 34% copper. She has samples that are 9% copper and 84% copper. How many grams of each must she use?

57. Archimedes' law of the lever The two weights shown in Illustration 1 will be in balance if the product of one weight and its distance from the fulcrum is equal to the product of the other weight and its distance from the fulcrum. Two weights are in balance when one is 2 meters and the other 3 meters from the fulcrum. If the fulcrum remains in the same spot, and weights were interchanged, the closer weight would need to be increased by 5 pounds to maintain balance. Find the weights.

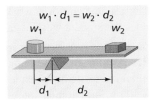

ILLUSTRATION 1

58. Lifting a weight A 112-pound force can lift the 448-pound load in Illustration 2. If the fulcrum is moved 1 foot away from the load, a 192-pound force is required. Find the length of the lever.

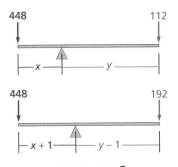

ILLUSTRATION 2

59. Flight range The speed of an airplane with a tailwind is 300 miles per hour, and with a headwind, 220 miles per hour. On a day with no wind, how far could the plane travel on a 5-hour fuel supply?

60. Writing test questions For a test question, a mathematics teacher wants to find two constants a and b such that the test item "*Simplify*: $a(x + 2y) - b(2x - y)$" will have an answer of $-3x + 9y$. What constants a and b should the teacher use?

61. ▦ **Break-even point** Rollowheel, Inc. can manufacture a pair of in-line roller skates for $43.53. Daily fixed costs attributable to their manufacture amount to $742.72. A pair of skates can be sold for $89.95. Find equations expressing the expenses, E, and the revenue, R, as functions of x, the number of pairs of skates manufactured and sold. At what production level will expenses equal revenues?

62. ▦ **Choosing salary options** For its sales staff, a marketing company offers two salary options. One is $326 per week and a commission of $3\frac{1}{2}\%$ of sales. The other is $200 per week and $4\frac{1}{4}\%$ of sales. Find equations that express incomes as functions of sales, and find the weekly sales level that produces equal salaries.

In Exercises 63–70, use systems of three equations in three variables to solve each word problem.

63. **Work schedule** A college student earns $99.25 per week by working three part-time jobs. Half of his 30-hour work week is spent cooking hamburgers at a fast-food chain, earning $2.85 per hour. In addition, the student earns $3.15 per hour working at a gas station and $5 per hour doing janitorial work. How many hours per week does the student work at each job?

64. **Investment income** A woman invested a $22,000 rollover IRA account in three banks paying 5%, 6%, and 7% annual interest. She invested $2000 more at 6% than at 5%. The total annual interest she earned was $1370. How much did she invest at each rate?

65. **Age distribution** Approximately 3 million people live in Costa Rica. 2.61 million are less than 50 years old, and 1.95 million are older than 14. How many people are in each of the categories 0–14 years, 15–49 years, and 50 years and older?

66. **Designing an arch** The engineer designing a parabolic arch knows that its equation has the form $y = ax^2 + bx + c$. Use the information in Illustration 3 to find a, b, and c. Assume the distances are given in feet. (*Hint:* The coordinates of points on the parabola satisfy its equation.)

67. **Geometry** The sum of the angles of a triangle is 180°. The largest angle is 20° greater than the sum of the other two, and the largest is also 10° greater than 3 times the smallest. How large is each angle?

68. **Ballistics** The path of a thrown object is a parabola with the equation $y = ax^2 + bx + c$. Use the information in Illustration 4 to find a, b, and c. (Distances are in feet.)

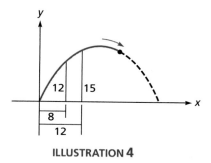

ILLUSTRATION 4

69. **Library shelving** To use space effectively, librarians wish to fill shelves completely. One 35-inch shelf can hold 3 dictionaries, 5 atlases, and 1 thesaurus; or 6 dictionaries and 2 thesauruses; or 2 dictionaries, 4 atlases, and 3 thesauruses. How wide is one copy of each kind of book?

70. **Copy machine productivity** When both copy machines A and B are working, secretaries can make 100 copies in one minute. In one minute's time, copiers A and C together produce 140 copies, and all three working together produce 180 copies. How many copies per minute can each machine produce separately?

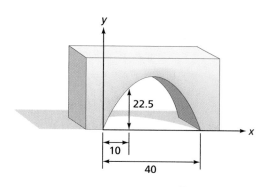

ILLUSTRATION 3

7.2   GAUSSIAN ELIMINATION AND MATRIX METHODS

- ■ Elementary Row Operations   ■ Row Echelon Form of a Matrix
- ■ An Inconsistent System   ■ A System of Dependent Equations
- ■ Gauss–Jordan Elimination

Carl Friedrich Gauss
(1777–1855)
Many people consider
Gauss to be the greatest
mathematician of all time.
He made contributions in
the areas of number theory,
solutions of equations,
geometry of curved
surfaces, and statistics. For
his efforts, he earned the
title "Prince of the
Mathematicians."

There is a method, called **Gaussian elimination** after the German mathematician Carl Friedrich Gauss (1777–1855), that can be used to solve systems of linear equations. In this method, a system of equations is transformed into an equivalent system that can be solved by a process called **back substitution**.

In Gaussian elimination, we will solve a system of equations by working only with the coefficients of the variables. All of the information needed to find a solution of

$$\begin{cases} x + 2y + z = 8 \\ 2x + y - z = 1 \\ x + y - 2z = -3 \end{cases}$$

for example, is contained in the following rectangular array of numbers, called a **matrix**:

$$\begin{bmatrix} 1 & 2 & 1 & \vdots & 8 \\ 2 & 1 & -1 & \vdots & 1 \\ 1 & 1 & -2 & \vdots & -3 \end{bmatrix}$$

Each row of the matrix represents one of the equations of the system.

- The first row represents the equation $x + 2y + z = 8$.
- The second row represents the equation $2x + y - z = 1$.
- The third row represents the equation $x + y - 2z = -3$.

Because the matrix has three rows and four columns, it is called a 3×4 matrix (read as "3 by 4"). The matrix to the left of the broken line is called the **coefficient matrix**. The entire matrix is called the **augmented matrix**, or the **system matrix**.

EXAMPLE 1   Use Gaussian elimination to solve the system

1.
2. $\begin{cases} x + 2y + z = 8 \\ 2x + y - z = 1 \\ x + y - 2z = -3 \end{cases}$
3.

Solution   We first multiply each term in Equation 1 by -2 and add the result to Equation 2 to obtain Equation 4. Then we multiply each term of Equation 1 by -1 and add the result to Equation 3 to obtain Equation 5. This gives an equivalent system with the same solution as the original system. The system is shown in both equation form and in matrix form. Note that the variable x does not appear in Equation 4 or 5.

$$
\begin{array}{l}
\textbf{1.} \\
\textbf{4.} \\
\textbf{5.}
\end{array}
\left\{
\begin{array}{rcr}
x + 2y + z & = & 8 \\
-3y - 3z & = & -15 \\
-y - 3z & = & -11
\end{array}
\right.
\qquad
\left[
\begin{array}{ccc|c}
1 & 2 & 1 & 8 \\
0 & -3 & -3 & -15 \\
0 & -1 & -3 & -11
\end{array}
\right]
$$

Now we divide both sides of Equation 4 by -3 to obtain Equation 6,

$$
\begin{array}{l}
\textbf{1.} \\
\textbf{6.} \\
\textbf{5.}
\end{array}
\left\{
\begin{array}{rcr}
x + 2y + z & = & 8 \\
y + z & = & 5 \\
-y - 3z & = & -11
\end{array}
\right.
\qquad
\left[
\begin{array}{ccc|c}
1 & 2 & 1 & 8 \\
0 & 1 & 1 & 5 \\
0 & -1 & -3 & -11
\end{array}
\right]
$$

and add Equation 6 to Equation 5 to obtain Equation 7.

$$
\begin{array}{l}
\textbf{1.} \\
\textbf{6.} \\
\textbf{7.}
\end{array}
\left\{
\begin{array}{rcr}
x + 2y + z & = & 8 \\
y + z & = & 5 \\
-2z & = & -6
\end{array}
\right.
\qquad
\left[
\begin{array}{ccc|c}
1 & 2 & 1 & 8 \\
0 & 1 & 1 & 5 \\
0 & 0 & -2 & -6
\end{array}
\right]
$$

Finally, we divide both sides of Equation 7 by -2 to obtain the system

$$
\begin{array}{l}
\textbf{1.} \\
\textbf{6.}
\end{array}
\left\{
\begin{array}{rcl}
x + 2y + z & = & 8 \\
y + z & = & 5 \\
z & = & 3
\end{array}
\right.
\qquad
\left[
\begin{array}{ccc|c}
1 & 2 & 1 & 8 \\
0 & 1 & 1 & 5 \\
0 & 0 & 1 & 3
\end{array}
\right]
$$

The system can now be solved by back substitution. Because $z = 3$, we can substitute 3 for z in Equation 6 and solve for y:

$$
\begin{array}{rl}
\textbf{6.} \quad & y + z = 5 \\
& y + 3 = 5 \\
& y = 2
\end{array}
$$

We can now substitute 2 for y and 3 for z in Equation 1 and solve for x:

$$
\begin{array}{rl}
\textbf{1.} \quad & x + 2y + z = 8 \\
& x + 2(2) + 3 = 8 \\
& x + 7 = 8 \\
& x = 1
\end{array}
$$

Verify that the solution of the system is the ordered triple $(1, 2, 3)$. ∎

■ Elementary Row Operations

In Example 1, we performed operations on the equations as well as on the rows of the corresponding matrices. These operations are called **elementary row operations**.

Elementary Row Operations	If any of the following operations are performed on the rows of a system matrix, the matrix of an equivalent system results:

Type 1 row operation: Two rows of a matrix can be interchanged.

Type 2 row operation: The elements of a row of a matrix can be multiplied by a nonzero constant.

Type 3 row operation: Any row can be changed by adding a multiple of another row to it.

Any matrix that can be obtained from another matrix by a sequence of elementary row operations is called **row equivalent** to the original matrix. Thus, row equivalent matrices represent equivalent systems of equations.

- A type 1 row operation is equivalent to writing the equations of a system in a different order.
- A type 2 row operation is equivalent to multiplying both sides of an equation by a nonzero constant.
- A type 3 row operation is equivalent to adding a multiple of one equation to another.

■ Row Echelon Form of a Matrix

We will use row operations to write matrices in a form called **row echelon form**.

Row Echelon Form of a Matrix	A matrix is in **row echelon form** if it has the following properties:

1. Any rows containing only 0's are at the bottom of the matrix.
2. The first nonzero entry in each row (called the **lead entry**) is 1.
3. Lead entries appear farther to the right as we move down the rows of the matrix.

The matrices

$$\begin{bmatrix} 1 & 3 & 0 & 7 \\ 0 & 0 & 1 & 8 \\ 0 & 0 & 0 & 0 \end{bmatrix}, \qquad \begin{bmatrix} 0 & 1 & 2 \\ 0 & 0 & 1 \end{bmatrix}, \qquad \text{and} \qquad \begin{bmatrix} 1 & 1 \\ 0 & 0 \\ 0 & 0 \\ 0 & 0 \end{bmatrix}$$

are in row echelon form, because all rows that consist entirely of 0's are at the bottom of the matrix, the first entry in each of the other rows is 1, and the lead entries appear farther to the right as we move down the rows. The matrices

$$\begin{bmatrix} 1 & 3 & 0 & 7 \\ 0 & 0 & 1 & 8 \\ 0 & 0 & 1 & 0 \end{bmatrix}, \quad \begin{bmatrix} 0 & 3 & 0 \\ 0 & 0 & 1 \end{bmatrix}, \quad \text{and} \quad \begin{bmatrix} 0 & 0 \\ 1 & 0 \\ 0 & 1 \\ 0 & 0 \end{bmatrix}$$

are not in row echelon form. In the first matrix, the lead entry in the last row is not to the right of the lead entry of the middle row. In the second matrix, the lead entry of the first row is not 1. In the third matrix, the rows of 0's are not last.

EXAMPLE 2   Solve the following system by transforming its system matrix into row echelon form and back substituting.

$$\begin{cases} x + 2y + 3z = 4 \\ 2x - y - 2z = 0 \\ x - 3y - 3z = -2 \end{cases}$$

Solution   This system is represented by the augmented matrix

$$\begin{bmatrix} 1 & 2 & 3 & \vdots & 4 \\ 2 & -1 & -2 & \vdots & 0 \\ 1 & -3 & -3 & \vdots & -2 \end{bmatrix}$$

We will perform elementary row operations to reduce the system matrix to row echelon form. We can use the 1 in the upper left-hand corner of the system matrix to zero out the rest of the first column. To do so, we multiply the first row by -2 and add the result to the second row. We indicate that operation with the notation $(-2)R1 + R2$.

$$\begin{bmatrix} 1 & 2 & 3 & \vdots & 4 \\ 2 & -1 & -2 & \vdots & 0 \\ 1 & -3 & -3 & \vdots & -2 \end{bmatrix} (-2)R1 + R2 \rightarrow \begin{bmatrix} 1 & 2 & 3 & \vdots & 4 \\ 0 & -5 & -8 & \vdots & -8 \\ 1 & -3 & -3 & \vdots & -2 \end{bmatrix}$$

Next, we multiply row one by -1 and add the result to row three to get a new row three. This fills the rest of the first column with 0's.

$$(-1)R1 + R3 \rightarrow \begin{bmatrix} 1 & 2 & 3 & \vdots & 4 \\ 0 & -5 & -8 & \vdots & -8 \\ 0 & -5 & -6 & \vdots & -6 \end{bmatrix}$$

We multiply row two by -1 and add the result to row three to get

$$(-1)R2 + R3 \rightarrow \begin{bmatrix} 1 & 2 & 3 & \vdots & 4 \\ 0 & -5 & -8 & \vdots & -8 \\ 0 & 0 & 2 & \vdots & 2 \end{bmatrix}$$

Finally, we multiply row two by $-\frac{1}{5}$ and row three by $\frac{1}{2}$.

$$\begin{array}{c} \\ \left(-\tfrac{1}{5}\right)R2 \to \\ \left(\tfrac{1}{2}\right)R3 \to \end{array} \left[\begin{array}{ccc|c} 1 & 2 & 3 & 4 \\ 0 & 1 & \dfrac{8}{5} & \dfrac{8}{5} \\ 0 & 0 & 1 & 1 \end{array}\right]$$

The final matrix is in row echelon form and represents the system

$$\begin{cases} x + 2y + 3z = 4 \\ \quad\quad y + \dfrac{8}{5}z = \dfrac{8}{5} \\ \quad\quad\quad\quad z = 1 \end{cases}$$

To find y, we substitute 1 for z in the second equation.

$$y + \frac{8}{5}z = \frac{8}{5}$$

$$y + \frac{8}{5}(1) = \frac{8}{5}$$

$$y = 0$$

To solve for x, we substitute 0 for y and 1 for z in the first equation.

$$x + 2y + 3z = 4$$
$$x + 2(0) + 3(1) = 4$$
$$x + 3 = 4$$
$$x = 1$$

Verify that the solution of the original system is the ordered triple $(x, y, z) = (1, 0, 1)$.

■

Since each of the two previous examples had solutions, each system was consistent. The next two examples illustrate a system that is inconsistent and a system whose equations are dependent.

■ An Inconsistent System

EXAMPLE 3   Use matrix methods to solve the system $\begin{cases} x + y + z = 3 \\ 2x - y + z = 2. \\ \quad\quad 3y + z = 1 \end{cases}$

Solution   We form the augmented matrix and use elementary row operations to write the augmented matrix in row echelon form. We keep the 1 in the top left position but use it to zero out the rest of the first column, and then proceed as follows:

$$\begin{bmatrix} 1 & 1 & 1 & \vdots & 3 \\ 2 & -1 & 1 & \vdots & 2 \\ 0 & 3 & 1 & \vdots & 1 \end{bmatrix} \xrightarrow{(-2)R1 + R2} \begin{bmatrix} 1 & 1 & 1 & \vdots & 3 \\ 0 & -3 & -1 & \vdots & -4 \\ 0 & 3 & 1 & \vdots & 1 \end{bmatrix} \xrightarrow[(1)R2 + R3]{} \begin{bmatrix} 1 & 1 & 1 & \vdots & 3 \\ 0 & -3 & -1 & \vdots & -4 \\ 0 & 0 & 0 & \vdots & -3 \end{bmatrix}$$

The final matrix is not in row echelon form, but there is no point in continuing. Because the last row of the matrix represents the equation

$$0x + 0y + 0z = -3$$

and no values of x, y, and z could make $0 = -3$, the given system has no solution and is inconsistent. ∎

■ A System of Dependent Equations

EXAMPLE 4 Use matrices to solve the system $\begin{cases} x + 2y + z = 8 \\ 2x + y - z = 1 \\ x - y - 2z = -7 \end{cases}$.

Solution We can set up the augmented matrix and use row operations to reduce it to row echelon form:

$$\begin{bmatrix} 1 & 2 & 1 & \vdots & 8 \\ 2 & 1 & -1 & \vdots & 1 \\ 1 & -1 & -2 & \vdots & -7 \end{bmatrix} \begin{matrix} \\ (-2)R1 + R2 \rightarrow \\ (-1)R1 + R3 \rightarrow \end{matrix} \begin{bmatrix} 1 & 2 & 1 & \vdots & 8 \\ 0 & -3 & -3 & \vdots & -15 \\ 0 & -3 & -3 & \vdots & -15 \end{bmatrix}$$

$$\begin{matrix} (-\frac{1}{3})R2 \rightarrow \\ (-\frac{1}{3})R3 \rightarrow \end{matrix} \begin{bmatrix} 1 & 2 & 1 & \vdots & 8 \\ 0 & 1 & 1 & \vdots & 5 \\ 0 & 1 & 1 & \vdots & 5 \end{bmatrix}$$

$$\begin{matrix} \\ \\ (-1)R2 + R3 \rightarrow \end{matrix} \begin{bmatrix} 1 & 2 & 1 & \vdots & 8 \\ 0 & 1 & 1 & \vdots & 5 \\ 0 & 0 & 0 & \vdots & 0 \end{bmatrix}$$

The final matrix is in echelon form and represents the system

1. $\begin{cases} x + 2y + z = 8 \\ y + z = 5 \\ 0x + 0y + 0z = 0 \end{cases}$
2.
3.

Since all coefficients in Equation 3 are 0, it can be ignored. To solve this system by back substitution, we solve Equation 2 for the variable y

$$y = 5 - z$$

and then substitute $5 - z$ for y in Equation 1.

1. $\qquad x + 2y + z = 8$

$$x + 2(5 - z) + z = 8$$

$$x + 10 - 2z + z = 8 \qquad \text{Remove parentheses.}$$
$$x + 10 - z = 8 \qquad \text{Combine terms.}$$
$$x = -2 + z \qquad \text{Solve for } x.$$

A general solution of this system is $(x, y, z) = (-2 + z, 5 - z, z)$.

There are many specific solutions to this system. The variable z can be any real number, but once we choose a number z, the numbers x and y are determined. For example, if $z = 3$, then $x = 1$ and $y = 2$. Thus, one possible solution of this system is $x = 1$, $y = 2$, and $z = 3$. Choosing $z = 2$ gives the solution $(0, 3, 2)$. Because this system has infinitely many solutions, it is consistent, but the equations are dependent. ∎

Gauss–Jordan Elimination

In Gaussian elimination, we perform row operations until the system matrix is in row echelon form. Then we convert the matrix back into equation form and solve by back substitution. A second method, called **Gauss–Jordan elimination**, uses row operations to produce a matrix in **reduced row echelon form**. With this method, we can obtain the solution of the system from that matrix directly and back substitution is not needed.

Reduced Row Echelon Form of a Matrix	A matrix is in **reduced row echelon form** if **1.** It is in row echelon form. **2.** The entries above each lead entry are also zero.

In the following example, we solve a system by Gauss–Jordan elimination.

EXAMPLE 5   Use Gauss–Jordan elimination to solve the system $\begin{cases} w + 2x + 3y + z = 4 \\ x + 4y - z = 0. \\ w - x - y + 2z = 2 \end{cases}$

Solution   We row reduce the augmented matrix as follows. We first use the 1 in the upper left corner to zero out the rest of the column:

$$\begin{bmatrix} 1 & 2 & 3 & 1 & | & 4 \\ 0 & 1 & 4 & -1 & | & 0 \\ 1 & -1 & -1 & 2 & | & 2 \end{bmatrix} \xrightarrow[(-1)R1 + R3]{} \begin{bmatrix} 1 & 2 & 3 & 1 & | & 4 \\ 0 & 1 & 4 & -1 & | & 0 \\ 0 & -3 & -4 & 1 & | & -2 \end{bmatrix}$$

The lead entry in the second row is 1. We use it to zero out the rest of the second column.

$$\begin{matrix} (-2)R2 + R1 \rightarrow \\ \\ (3)R2 + R3 \rightarrow \end{matrix} \begin{bmatrix} 1 & 0 & -5 & 3 & | & 4 \\ 0 & 1 & 4 & -1 & | & 0 \\ 0 & 0 & 8 & -2 & | & -2 \end{bmatrix}$$

To make the lead entry in the third row equal to 1, we multiply the third row by $\frac{1}{8}$ and then use it to zero out the rest of the third column.

$$\left(\tfrac{1}{8}\right)R3 \to \begin{bmatrix} 1 & 0 & -5 & 3 & \vdots & 4 \\ 0 & 1 & 4 & -1 & \vdots & 0 \\ 0 & 0 & 1 & -\dfrac{1}{4} & \vdots & -\dfrac{1}{4} \end{bmatrix}$$

$$\begin{matrix} (5)R3 + R1 \to \\ (-4)R3 + R2 \to \\ \\ \end{matrix} \begin{bmatrix} 1 & 0 & 0 & \dfrac{7}{4} & \vdots & \dfrac{11}{4} \\ 0 & 1 & 0 & 0 & \vdots & 1 \\ 0 & 0 & 1 & -\dfrac{1}{4} & \vdots & -\dfrac{1}{4} \end{bmatrix}$$

The final matrix is in reduced row echelon form. Note that each lead entry is 1, and each 1 is alone in its column. The matrix represents the system of equations

$$\begin{cases} w & + \dfrac{7}{4}z = \dfrac{11}{4} \\ \quad x & = 1 \\ \quad y - \dfrac{1}{4}z = -\dfrac{1}{4} \end{cases} \quad \text{or} \quad \begin{cases} w = \dfrac{11}{4} - \dfrac{7}{4}z \\ x = 1 \\ y = -\dfrac{1}{4} + \dfrac{1}{4}z \end{cases}$$

The system of equations has infinitely many solutions. To find some of them, we can choose any value for z in the general solution, and the corresponding values of w, x, and y will be determined. For example, if $z = 1$, then $w = 1$, $x = 1$, and $y = 0$, Thus, $(w, x, y, z) = (1, 1, 0, 1)$ is a solution. Similarly, if $z = -1$, then another solution is $(w, x, y, z) = \left(\frac{9}{2}, 1, -\frac{1}{2}, -1\right)$. ∎

EXAMPLE 6   Use Gauss–Jordan elimination to solve the system $\begin{cases} 2x + y = 4 \\ x - 3y = 9 \\ x + 4y = -5 \end{cases}$.

Solution   To get a 1 in the top left corner, we could multiply the first row by $\frac{1}{2}$, but that would introduce fractions. Instead, we can exchange the first two rows and proceed as follows:

$$\begin{bmatrix} 2 & 1 & \vdots & 4 \\ 1 & -3 & \vdots & 9 \\ 1 & 4 & \vdots & -5 \end{bmatrix} \xrightarrow{R1 \leftrightarrow R2} \begin{bmatrix} 1 & -3 & \vdots & 9 \\ 2 & 1 & \vdots & 4 \\ 1 & 4 & \vdots & -5 \end{bmatrix}$$

$$\begin{matrix} \\ (-2)R1 + R2 \to \\ (-1)R1 + R3 \to \end{matrix} \begin{bmatrix} 1 & -3 & \vdots & 9 \\ 0 & 7 & \vdots & -14 \\ 0 & 7 & \vdots & -14 \end{bmatrix}$$

$$\begin{array}{c} \\ \left(\tfrac{1}{7}\right)R2 \to \\ \\ \left(\tfrac{1}{7}\right)R3 \to \end{array} \left[\begin{array}{cc|c} 1 & -3 & 9 \\ 0 & 1 & -2 \\ 0 & 1 & -2 \end{array}\right]$$

$$\begin{array}{c} (3)R2 + R1 \to \\ \\ (-1)R2 + R3 \to \end{array} \left[\begin{array}{cc|c} 1 & 0 & 3 \\ 0 & 1 & -2 \\ 0 & 0 & 0 \end{array}\right]$$

This final matrix is in reduced echelon form, and it represents the system

$$\begin{cases} x = & 3 \\ y = & -2 \end{cases}$$

Verify that $x = 3$ and $y = -2$ satisfy the equations of the original system.   ∎

7.2   EXERCISES

In Exercises 1–8, use Gaussian elimination to solve each system. **Do not use matrices.**

1. $\begin{cases} x + y = 7 \\ x - 2y = -1 \end{cases}$
2. $\begin{cases} x + 3y = 8 \\ 2x - 5y = 5 \end{cases}$
3. $\begin{cases} x - y = 1 \\ 2x - y = 8 \end{cases}$
4. $\begin{cases} x - 5y = 4 \\ 2x + 3y = 21 \end{cases}$

5. $\begin{cases} x + 2y - z = 2 \\ x - 3y + 2z = 1 \\ x + y - 3z = -6 \end{cases}$
6. $\begin{cases} x + 5y - z = 2 \\ x + 2y + z = 3 \\ x + y + z = 2 \end{cases}$
7. $\begin{cases} x - y - z = -3 \\ 5x + y = 6 \\ y + z = 4 \end{cases}$
8. $\begin{cases} x + y = 1 \\ x + z = 3 \\ y + z = 2 \end{cases}$

In Exercises 9–12, indicate whether each matrix is in row echelon form, reduced row echelon form, or neither.

9. $\begin{bmatrix} 1 & 3 & 0 & 5 \\ 0 & 1 & 2 & 7 \\ 0 & 0 & 1 & 0 \end{bmatrix}$
10. $\begin{bmatrix} 1 & 3 & 0 & 5 \\ 0 & 1 & 2 & 7 \\ 0 & 0 & 0 & 0 \end{bmatrix}$
11. $\begin{bmatrix} 1 & 0 & 1 \\ 0 & 1 & 5 \\ 0 & 0 & 0 \\ 0 & 0 & 0 \end{bmatrix}$
12. $\begin{bmatrix} 1 & 0 & 1 \\ 0 & 1 & 5 \\ 0 & 0 & 1 \\ 0 & 0 & 0 \end{bmatrix}$

In Exercises 13–28, write each system of equations as a matrix and solve it by Gaussian elimination.

13. $\begin{cases} 2x + y = 3 \\ x - 3y = 5 \end{cases}$
14. $\begin{cases} x + 2y = -1 \\ 3x - 5y = 19 \end{cases}$
15. $\begin{cases} x - 7y = -2 \\ 5x - 2y = -10 \end{cases}$
16. $\begin{cases} 3x - y = 3 \\ 2x + y = -3 \end{cases}$

17. $\begin{cases} 2x - y = 5 \\ x + 3y = 6 \end{cases}$
18. $\begin{cases} 3x - 5y = -25 \\ 2x + y = 5 \end{cases}$
19. $\begin{cases} x - 2y = 3 \\ -2x + 4y = 6 \end{cases}$
20. $\begin{cases} 2x - y = 7 \\ -x + \tfrac{1}{3}y = -\tfrac{7}{3} \end{cases}$

21. $\begin{cases} x - y + z = 3 \\ 2x - y + z = 4 \\ x + 2y - z = -1 \end{cases}$
22. $\begin{cases} 2x + y - z = 1 \\ x + y - z = 0 \\ 3x + y + 2z = 2 \end{cases}$
23. $\begin{cases} x + y - z = -1 \\ 3x + y = 4 \\ y - 2z = -4 \end{cases}$
24. $\begin{cases} 3x + y = 7 \\ x - z = 0 \\ y - 2z = -8 \end{cases}$

25. $\begin{cases} x - y + z = 2 \\ 2x + y + z = 5 \\ 3x \quad\;\; - 4z = -5 \end{cases}$ **26.** $\begin{cases} x \quad\quad\; + z = -1 \\ 3x + y \quad\quad = 2 \\ 2x + y + 5z = 3 \end{cases}$ **27.** $\begin{cases} x + y + 2z = 4 \\ -x - y - 3z = -5 \\ 2x + y + z = 2 \end{cases}$ **28.** $\begin{cases} 2x - y + z = 6 \\ 3x + y - z = 2 \\ -x + 3y - 3z = 8 \end{cases}$

In Exercises 29–48, write each system of equations as a matrix and solve it by Gauss–Jordan elimination. If a system has infinitely many solutions, show a general solution.

29. $\begin{cases} x - 2y = 7 \\ \quad\quad y = 3 \end{cases}$ **30.** $\begin{cases} x - 2y = -11 \\ \quad\quad y = 8 \end{cases}$ **31.** $\begin{cases} x + 2y - z = 3 \\ \quad\quad y + 3z = 1 \\ \quad\quad\quad\; z = -2 \end{cases}$ **32.** $\begin{cases} x - 3y + 2z = -1 \\ \quad\quad y - 2z = 3 \\ \quad\quad\quad\; z = 5 \end{cases}$

33. $\begin{cases} x - y = 7 \\ x + y = 13 \end{cases}$ **34.** $\begin{cases} x + 2y = 7 \\ 2x - y = -1 \end{cases}$ **35.** $\begin{cases} x - \dfrac{1}{2}y = 0 \\ x + 2y = 0 \end{cases}$ **36.** $\begin{cases} x - y = 5 \\ -x + \dfrac{1}{5}y = -9 \end{cases}$

37. $\begin{cases} x + y + 2z = 0 \\ x + y + z = 2 \\ x \quad\;\; + z = 1 \end{cases}$ **38.** $\begin{cases} x + 2y = -3 \\ x + 4y = -2 \\ 2x + z = -8 \end{cases}$ **39.** $\begin{cases} 2x + y - 2z = 1 \\ -x + y - 3z = 0 \\ 4x + 3y = 4 \end{cases}$ **40.** $\begin{cases} 3x + y = 3 \\ 3x + y - z = 2 \\ 6x \quad\;\; + z = 5 \end{cases}$

41. $\begin{cases} 2x - 2y + 3z + t = 2 \\ x + y + z + t = 5 \\ -x + 2y - 3z + 2t = 2 \\ x + y + 2z - t = 4 \end{cases}$ **42.** $\begin{cases} x + y + 2z + t = 1 \\ x + 2y + z + t = 2 \\ 2x + y + z + t = 4 \\ x + y + z + 2t = 3 \end{cases}$

43. $\begin{cases} x + y + t = 4 \\ x \quad\;\; + z + t = 2 \\ 2x + 2y + z + 2t = 8 \\ x - y + z - t = -2 \end{cases}$ **44.** $\begin{cases} x - y + 2z + t = 3 \\ 3x - 2y - z - t = 4 \\ 2x + y + 2z - t = 10 \\ x + 2y + z - 3t = 8 \end{cases}$ **45.** $\begin{cases} \dfrac{1}{3}x + \dfrac{3}{4}y - \dfrac{2}{3}z = -2 \\ x + \dfrac{1}{2}y + \dfrac{1}{3}z = 1 \\ \dfrac{1}{6}x - \dfrac{1}{8}y - z = 0 \end{cases}$

46. $\begin{cases} \dfrac{1}{4}x + y + 3z = 1 \\ \dfrac{1}{2}x - 4y + 6z = -1 \\ \dfrac{1}{3}x - 2y - 2z = -1 \end{cases}$ **47.** $\begin{cases} \dfrac{1}{2}x + \dfrac{1}{4}y - z = 2 \\ \dfrac{2}{3}x + \dfrac{1}{4}y + \dfrac{1}{2}z = \dfrac{3}{2} \\ \dfrac{2}{3}x \quad\quad + z = -\dfrac{1}{3} \end{cases}$ **48.** $\begin{cases} \dfrac{5}{7}x - \dfrac{1}{3}y + z = 0 \\ \dfrac{2}{7}x + y + \dfrac{1}{8}z = 9 \\ 6x + 4y - \dfrac{27}{4}z = 20 \end{cases}$

In Exercises 49–56, each system does not contain the same number of equations as variables. Solve each system using Gauss–Jordan elimination. If a system has infinitely many solutions, show a general solution.

49. $\begin{cases} x + y = -2 \\ 3x - y = 6 \\ 2x + 2y = -4 \\ x - y = 4 \end{cases}$ **50.** $\begin{cases} x - y = -3 \\ 2x + y = -3 \\ 3x - y = -7 \\ 4x + y = -7 \end{cases}$ **51.** $\begin{cases} x + 2y + z = 4 \\ 3x - y - z = 2 \end{cases}$

52. $\begin{cases} x + 2y - 3z = -5 \\ 5x + y - z = -11 \end{cases}$ **53.** $\begin{cases} w + x = 1 \\ w + y = 0 \\ x + z = 0 \end{cases}$ **54.** $\begin{cases} w + x - y + z = 2 \\ 2w - x - 2y + z = 0 \\ w - 2x - y + z = -1 \end{cases}$

55. $\begin{cases} x + y = 3 \\ 2x + y = 1 \\ 3x + 2y = 2 \end{cases}$

56. $\begin{cases} x + 2y + z = 4 \\ x - y + z = 1 \\ 2x + y + 2z = 2 \\ 3x + 3z = 6 \end{cases}$

In Exercises 57–58, use matrix methods to solve each system of equations.

57. $\begin{cases} x^2 + y^2 + z^2 = 14 \\ 2x^2 + 3y^2 - 2z^2 = -7 \\ x^2 - 5y^2 + z^2 = 8 \end{cases}$

(*Hint:* Solve first as a linear system in x^2, y^2, and z^2.)

58. $\begin{cases} 5\sqrt{x} + 2\sqrt{y} + \sqrt{z} = 22 \\ \sqrt{x} + \sqrt{y} - \sqrt{z} = 5 \\ 3\sqrt{x} - 2\sqrt{y} - 3\sqrt{z} = 10 \end{cases}$

7.3 MATRIX ALGEBRA

- Adding and Subtracting Matrices
- Matrices on a Graphing Calculator
- The Identity Matrix
- Multiplying Matrices
- An Application of Matrices

In this section, we will discuss some topics of the algebra of matrices.

$m \times n$ Matrix

An **$m \times n$ matrix** is a rectangular array of mn numbers arranged in m rows and n columns. We say that the matrix is of **order $m \times n$**.

Matrices are often denoted by capital letters such as A, B, and C. To denote the entries in an $m \times n$ matrix A, we use double subscript notation: The entry in the first row, third column is a_{13}, and the entry in the ith row, jth column is a_{ij}. We can use any of the following notations to denote the $m \times n$ matrix A:

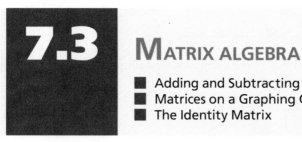

$$A, \qquad [a_{ij}], \qquad \left.\begin{bmatrix} a_{11} & a_{12} & a_{13} \cdots a_{1n} \\ a_{21} & a_{22} & a_{23} \cdots a_{2n} \\ \vdots & & \vdots \\ a_{m1} & a_{m2} & a_{m3} \cdots a_{mn} \end{bmatrix}\right\} m \text{ rows}$$

$$\underbrace{\qquad\qquad\qquad}_{n \text{ columns}}$$

Two matrices are equal if they are the same size, with the same entries in corresponding positions.

Equality of Matrices	If $A = [a_{ij}]$ and $B = [b_{ij}]$ are both $m \times n$ matrices, then
	$A = B$ if and only if $a_{ij} = b_{ij}$ for all i and j

We can multiply a matrix by a constant by multiplying each of its entries by that number. For example, if

$$A = \begin{bmatrix} 1 & 2 \\ 3 & 4 \end{bmatrix}, \quad \text{then} \quad 5A = 5 \begin{bmatrix} 1 & 2 \\ 3 & 4 \end{bmatrix} = \begin{bmatrix} 5 & 10 \\ 15 & 20 \end{bmatrix}$$

If A is a matrix, the real number k in the product kA is called a **scalar**.

Multiplying a Matrix by a Scalar	If $A = [a_{ij}]$ is an $m \times n$ matrix and k is a scalar, then
	$kA = k[a_{ij}] = [ka_{ij}]$ for all i and j

EXAMPLE 1 Let $\begin{bmatrix} 5 & y \\ 15 & z \end{bmatrix} = 5 \begin{bmatrix} x & 3 \\ 3 & y \end{bmatrix}$. Find x, y, and z.

Solution We simplify the right-hand side of the expression by multiplying each entry of the matrix by 5.

$$\begin{bmatrix} 5 & y \\ 15 & z \end{bmatrix} = \begin{bmatrix} 5x & 15 \\ 15 & 5y \end{bmatrix}$$

Because the matrices are equal, their corresponding entries are equal. Thus, $5 = 5x$, $y = 15$, and $z = 5y$. We conclude that $x = 1$, $y = 15$, and $z = 75$. ∎

■ Adding and Subtracting Matrices

We can add matrices of the same size by adding the entries in corresponding positions.

 Warning! Matrices of different sizes cannot be added.

The Sum of Two Matrices	Let $A = [a_{ij}]$ and $B = [b_{ij}]$ be two $m \times n$ matrices. The sum, $A + B$, is the $m \times n$ matrix found by adding the corresponding entries of matrices A and B:
	$A + B = [a_{ij} + b_{ij}]$ for all i and j

EXAMPLE 2 Add the matrices $\begin{bmatrix} 2 & 1 & 3 \\ 1 & -1 & 0 \end{bmatrix}$ and $\begin{bmatrix} 1 & -1 & 2 \\ -1 & 1 & 5 \end{bmatrix}$.

Solution Since each matrix is 2×3, we can find their sum by adding their corresponding entries.

$$\begin{bmatrix} 2 & 1 & 3 \\ 1 & -1 & 0 \end{bmatrix} + \begin{bmatrix} 1 & -1 & 2 \\ -1 & 1 & 5 \end{bmatrix} = \begin{bmatrix} 2+1 & 1-1 & 3+2 \\ 1-1 & -1+1 & 0+5 \end{bmatrix}$$
$$= \begin{bmatrix} 3 & 0 & 5 \\ 0 & 0 & 5 \end{bmatrix}$$

∎

The arithmetic of matrices is similar to the arithmetic of real numbers. The real number 0, for example, is the additive identity, because $a + 0 = 0 + a = a$ for any real number a. A matrix consisting entirely of 0's is an additive identity for matrices, because it behaves like the real number 0:

$$\begin{bmatrix} 1 & 2 \\ 3 & 4 \end{bmatrix} + \begin{bmatrix} \mathbf{0} & \mathbf{0} \\ \mathbf{0} & \mathbf{0} \end{bmatrix} = \begin{bmatrix} \mathbf{0} & \mathbf{0} \\ \mathbf{0} & \mathbf{0} \end{bmatrix} + \begin{bmatrix} 1 & 2 \\ 3 & 4 \end{bmatrix} = \begin{bmatrix} 1 & 2 \\ 3 & 4 \end{bmatrix}$$

The Additive Identity Matrix

Let A be any $m \times n$ matrix. There is an $m \times n$ matrix $\mathbf{0}$, called the **zero matrix** or the **additive identity matrix**, for which

$$A + \mathbf{0} = \mathbf{0} + A = A$$

The matrix $\mathbf{0}$ consists of m rows and n columns of 0's.

Every matrix also has an additive inverse.

The Additive Inverse of a Matrix

Any $m \times n$ matrix A has an **additive inverse**, an $m \times n$ matrix $-A$ with the property that the sum of A and $-A$ is the zero matrix:

$$A + (-A) = (-A) + A = \mathbf{0}$$

The entries of $-A$ are the negatives of the corresponding entries of A:

$$-A = (-1)A$$

The additive inverse of $A = \begin{bmatrix} 1 & -3 & 2 \\ 0 & 1 & -5 \end{bmatrix}$ is the matrix $-A = (-1)A = \begin{bmatrix} -1 & 3 & -2 \\ 0 & -1 & 5 \end{bmatrix}$ because their sum is the zero matrix:

$$A + (-A) = \begin{bmatrix} 1 & -3 & 2 \\ 0 & 1 & -5 \end{bmatrix} + \begin{bmatrix} -1 & 3 & -2 \\ 0 & -1 & 5 \end{bmatrix}$$
$$= \begin{bmatrix} 1-1 & -3+3 & 2-2 \\ 0+0 & 1-1 & -5+5 \end{bmatrix}$$
$$= \begin{bmatrix} 0 & 0 & 0 \\ 0 & 0 & 0 \end{bmatrix}$$

Subtraction of matrices is defined as follows.

The Difference of Two Matrices	If A and B are $m \times n$ matrices, their difference, $A - B$, is the sum of A and the additive inverse of B:

$$A - B = A + (-B)$$

For example, $\begin{bmatrix} 3 & 7 \\ -4 & 0 \end{bmatrix} - \begin{bmatrix} -1 & 4 \\ -5 & 1 \end{bmatrix} = \begin{bmatrix} 3 & 7 \\ -4 & 0 \end{bmatrix} + \begin{bmatrix} 1 & -4 \\ 5 & -1 \end{bmatrix} = \begin{bmatrix} 4 & 3 \\ 1 & -1 \end{bmatrix}.$

■ Multiplying Matrices

Unlike addition of matrices, we do not multiply matrices by multiplying corresponding positions. Instead, the product AB involves the rows of A and the columns of B. We illustrate how to find the product of two matrices by finding the product of a 2×3 matrix A and a 3×3 matrix B. The result is the 2×3 matrix C.

$$AB = \begin{bmatrix} 1 & 2 & 3 \\ 4 & 5 & 6 \end{bmatrix} \begin{bmatrix} a & b & c \\ d & e & f \\ g & h & i \end{bmatrix} = C$$

Each entry of matrix C is the result of a calculation that involves a row of A and a column of B. For example, the first-row, third-column entry of matrix C is the sum of the products of corresponding entries of the first row of A and the third column of B:

$$\begin{bmatrix} \mathbf{1} & \mathbf{2} & \mathbf{3} \\ 4 & 5 & 6 \end{bmatrix} \begin{bmatrix} a & b & c \\ d & e & f \\ g & h & i \end{bmatrix} = \begin{bmatrix} ? & ? & \mathbf{1c + 2f + 3i} \\ ? & ? & ? \end{bmatrix}$$

Similarly, the second-row, second-column entry of matrix C is the sum of the products of the second row of A and the second column of B.

$$\begin{bmatrix} 1 & 2 & 3 \\ \mathbf{4} & \mathbf{5} & \mathbf{6} \end{bmatrix} \begin{bmatrix} a & b & c \\ d & e & f \\ g & h & i \end{bmatrix} = \begin{bmatrix} ? & ? & 1c + 2f + 3i \\ ? & \mathbf{4b + 5e + 6h} & ? \end{bmatrix}$$

The other entries of the product are computed similarly.

For the product AB to exist, the number of columns of A must equal the number of rows of B. If the product exists, it will have as many rows as A and as many columns as B:

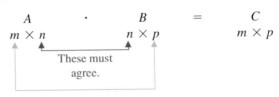

Arthur Cayley (1821–1895)
Cayley taught mathematics at Cambridge University. When he refused to take religious vows, he was fired and became a lawyer. After 14 years, he returned to mathematics and to Cambridge. Cayley was a major force in developing the theory of matrices.

More formally, we have the following definition.

The Product of Two Matrices	Let $A = [a_{ij}]$ be an $m \times n$ and $B = [b_{ij}]$ be an $n \times p$ matrix. The product, AB, is the $m \times p$ matrix C found as follows:

$AB = C = [c_{ij}]$, where c_{ij} is the sum of the products of the corresponding entries in the ith row of A and the jth column of B, where $i = 1, 2, 3, \ldots, m$ and $j = 1, 2, 3, \ldots, p$.

EXAMPLE 3   Find $C = AB$ if $A = \begin{bmatrix} 1 & 2 & 4 \\ -2 & 1 & -1 \end{bmatrix}$ and $B = \begin{bmatrix} 1 & 5 \\ -2 & 4 \\ 1 & -3 \end{bmatrix}$.

Solution   Because matrix A is 2×3 and B is 3×2, the product C is defined; it is 2×2. To find entry c_{11} of C, we find the total of the products of the entries in the first row of A and the first column of B:

$$c_{11} = 1 \cdot 1 + 2 \cdot (-2) + 4 \cdot 1 = 1$$

$$\begin{bmatrix} \mathbf{1} & \mathbf{2} & \mathbf{4} \\ -2 & 1 & -1 \end{bmatrix} \begin{bmatrix} \mathbf{1} & 5 \\ \mathbf{-2} & 4 \\ \mathbf{1} & -3 \end{bmatrix} = \begin{bmatrix} \mathbf{1} & ? \\ ? & ? \end{bmatrix}$$

To find entry c_{12}, we move across the first row of A and down the second column of B:

$$c_{12} = 1 \cdot 5 + 2 \cdot 4 + 4 \cdot (-3) = 1$$

$$\begin{bmatrix} \mathbf{1} & \mathbf{2} & \mathbf{4} \\ -2 & 1 & -1 \end{bmatrix} \begin{bmatrix} 1 & \mathbf{5} \\ -2 & \mathbf{4} \\ 1 & \mathbf{-3} \end{bmatrix} = \begin{bmatrix} 1 & \mathbf{1} \\ ? & ? \end{bmatrix}$$

To find entry c_{21}, we move across the second row of A and down the first column of B:

$$c_{21} = (-2) \cdot 1 + 1 \cdot (-2) + (-1) \cdot 1 = -5$$

$$\begin{bmatrix} 1 & 2 & 4 \\ \mathbf{-2} & \mathbf{1} & \mathbf{-1} \end{bmatrix} \begin{bmatrix} \mathbf{1} & 5 \\ \mathbf{-2} & 4 \\ \mathbf{1} & -3 \end{bmatrix} = \begin{bmatrix} 1 & 1 \\ \mathbf{-5} & ? \end{bmatrix}$$

Finally, we find entry c_{22}:

$$c_{22} = (-2) \cdot 5 + 1 \cdot 4 + (-1) \cdot (-3) = -3$$

$$\begin{bmatrix} 1 & 2 & 4 \\ \mathbf{-2} & \mathbf{1} & \mathbf{-1} \end{bmatrix} \begin{bmatrix} 1 & \mathbf{5} \\ -2 & \mathbf{4} \\ 1 & \mathbf{-3} \end{bmatrix} = \begin{bmatrix} 1 & 1 \\ \mathbf{-5} & \mathbf{-3} \end{bmatrix}$$

∎

EXAMPLE 4   Find the product $\begin{bmatrix} 1 & -1 & 2 \\ 1 & 3 & 0 \\ 0 & 1 & 1 \end{bmatrix} \begin{bmatrix} 2 & 1 \\ 1 & 3 \\ 0 & 1 \end{bmatrix}$.

Solution   Because the matrices are 3×3 and 3×2, the product is a 3×2 matrix.

$$\begin{bmatrix} 1 & -1 & 2 \\ 1 & 3 & 0 \\ 0 & 1 & 1 \end{bmatrix} \begin{bmatrix} 2 & 1 \\ 1 & 3 \\ 0 & 1 \end{bmatrix} = \begin{bmatrix} 1 \cdot 2 + (-1) \cdot 1 + 2 \cdot 0 & 1 \cdot 1 + (-1) \cdot 3 + 2 \cdot 1 \\ 1 \cdot 2 + 3 \cdot 1 + 0 \cdot 0 & 1 \cdot 1 + 3 \cdot 3 + 0 \cdot 1 \\ 0 \cdot 2 + 1 \cdot 1 + 1 \cdot 0 & 0 \cdot 1 + 1 \cdot 3 + 1 \cdot 1 \end{bmatrix}$$

$$= \begin{bmatrix} 1 & 0 \\ 5 & 10 \\ 1 & 4 \end{bmatrix}$$ ∎

EXAMPLE 5   Find the product $[1 \quad 2 \quad 3] \begin{bmatrix} 4 \\ 5 \\ 6 \end{bmatrix}$.

Solution   Because the first matrix is 1×3 and the second matrix is 3×1, the product is a 1×1 matrix:

$$[1 \quad 2 \quad 3] \begin{bmatrix} 4 \\ 5 \\ 6 \end{bmatrix} = [1 \cdot 4 + 2 \cdot 5 + 3 \cdot 6] = [32]$$ ∎

EXAMPLE 6   Find the product $\begin{bmatrix} 1 \\ 2 \\ 3 \end{bmatrix} [4 \quad 5 \quad 6]$.

Solution   Because the first matrix is 3×1 and the second matrix is 1×3, the product is a 3×3 matrix:

$$\begin{bmatrix} 1 \\ 2 \\ 3 \end{bmatrix} [4 \quad 5 \quad 6] = \begin{bmatrix} 1 \cdot 4 & 1 \cdot 5 & 1 \cdot 6 \\ 2 \cdot 4 & 2 \cdot 5 & 2 \cdot 6 \\ 3 \cdot 4 & 3 \cdot 5 & 3 \cdot 6 \end{bmatrix} = \begin{bmatrix} 4 & 5 & 6 \\ 8 & 10 & 12 \\ 12 & 15 & 18 \end{bmatrix}$$ ∎

EXAMPLE 7   If $A = \begin{bmatrix} 1 & 1 \\ 0 & 0 \end{bmatrix}$ and $B = \begin{bmatrix} 0 & 1 \\ 0 & 1 \end{bmatrix}$, find AB and BA to show that multiplication of matrices is not commutative.

Solution $AB = \begin{bmatrix} 1 & 1 \\ 0 & 0 \end{bmatrix} \begin{bmatrix} 0 & 1 \\ 0 & 1 \end{bmatrix} = \begin{bmatrix} 0 & 2 \\ 0 & 0 \end{bmatrix}$

$BA = \begin{bmatrix} 0 & 1 \\ 0 & 1 \end{bmatrix} \begin{bmatrix} 1 & 1 \\ 0 & 0 \end{bmatrix} = \begin{bmatrix} 0 & 0 \\ 0 & 0 \end{bmatrix}$

Since the products are not equal, matrix multiplication is not commutative. ∎

Matrices on a Graphing Calculator

Several models of graphing calculators are able to do matrix arithmetic. For example, to find the sum and the product of the matrices

$$A = \begin{bmatrix} 2 & 3.7 \\ -2.1 & 3 \end{bmatrix} \quad \text{and} \quad B = \begin{bmatrix} 2 & -1 \\ 0 & 0.3 \end{bmatrix}$$

on a Texas Instruments TI-82 graphing calculator, we press MATRIX , select EDIT, and enter the size and entries of matrix *A*, as shown in Figure 7-5(a). We press 2nd   CLEAR   MATRIX and enter matrix *B*. Pressing 2nd   1   +   2nd   2   ENTER gives *A* + *B* as shown in Figure 7-5(b). Figure 7-5(b) also shows the product *AB*.

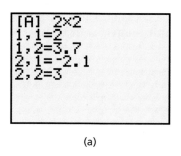

(a)

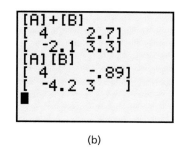

(b)

FIGURE 7-5

■ An Application of Matrices

EXAMPLE 8   Supplies must be purchased for a company's security officers. The quantities and prices of each item required for each shift are given in the following charts.

Quantities					Unit Prices	
	Uniforms	Badges	Whistles		Uniforms	47
Day shift	17	13	19		Badges	7
Night shift	14	24	27		Whistles	5

Use matrix multiplication to find the cost of supplies for each shift.

Solution   We can write the quantities and prices in matrix form and multiply them to get a cost matrix.

$$C = QP$$

$$= \begin{bmatrix} 17 & 13 & 19 \\ 14 & 24 & 27 \end{bmatrix} \begin{bmatrix} 47 \\ 7 \\ 5 \end{bmatrix}$$

$$= \begin{bmatrix} 17 \cdot 47 + 13 \cdot 7 + 19 \cdot 5 \\ 14 \cdot 47 + 24 \cdot 7 + 27 \cdot 5 \end{bmatrix} \quad \begin{array}{l} (17 \text{ uniforms})(\$47) + (13 \text{ badges})(\$7) + (19 \text{ whistles})(\$5) \\ (14 \text{ uniforms})(\$47) + (24 \text{ badges})(\$7) + (27 \text{ whistles})(\$5) \end{array}$$

$$= \begin{bmatrix} 985 \\ 961 \end{bmatrix}$$

It will cost \$985 to buy supplies for the day shift, and it will cost \$961 to equip the night shift.

Note that as in ordinary arithmetic, *Total cost* = (*Quantity*)(*Unit price*). ■

■ The Identity Matrix

The real number 1 is called the **identity for multiplication**, because multiplying a number by 1 does not change the number: $a \cdot 1 = 1 \cdot a = a$. There is a **multiplicative identity matrix** with a similar property.

Identity Matrix	Let A be an $n \times n$ matrix. There is an $n \times n$ **identity matrix** I for which

$$AI = IA = A$$

The matrix I consists of 1's on its diagonal and 0's elsewhere.

$$I = \begin{bmatrix} 1 & 0 & 0 \cdots 0 \\ 0 & 1 & 0 \cdots 0 \\ 0 & 0 & 1 \cdots 0 \\ \vdots & \vdots & \vdots & \vdots \\ 0 & 0 & 0 \cdots 1 \end{bmatrix}$$

 Warning! An identity matrix is always a square matrix—that is, it has the same number of rows and columns.

We illustrate the previous definition for the 3×3 identity matrix.

$$\begin{bmatrix} 1 & 0 & 0 \\ 0 & 1 & 0 \\ 0 & 0 & 1 \end{bmatrix}\begin{bmatrix} 1 & 2 & 3 \\ 4 & 5 & 6 \\ 7 & 8 & 9 \end{bmatrix} = \begin{bmatrix} 1 \cdot 1 + 0 \cdot 4 + 0 \cdot 7 & 1 \cdot 2 + 0 \cdot 5 + 0 \cdot 8 & 1 \cdot 3 + 0 \cdot 6 + 0 \cdot 9 \\ 0 \cdot 1 + 1 \cdot 4 + 0 \cdot 7 & 0 \cdot 2 + 1 \cdot 5 + 0 \cdot 8 & 0 \cdot 3 + 1 \cdot 6 + 0 \cdot 9 \\ 0 \cdot 1 + 0 \cdot 4 + 1 \cdot 7 & 0 \cdot 2 + 0 \cdot 5 + 1 \cdot 8 & 0 \cdot 3 + 0 \cdot 6 + 1 \cdot 9 \end{bmatrix}$$

$$= \begin{bmatrix} 1 & 2 & 3 \\ 4 & 5 & 6 \\ 7 & 8 & 9 \end{bmatrix}$$

$$\begin{bmatrix} 1 & 2 & 3 \\ 4 & 5 & 6 \\ 7 & 8 & 9 \end{bmatrix}\begin{bmatrix} 1 & 0 & 0 \\ 0 & 1 & 0 \\ 0 & 0 & 1 \end{bmatrix} = \begin{bmatrix} 1 & 2 & 3 \\ 4 & 5 & 6 \\ 7 & 8 & 9 \end{bmatrix}$$

In many ways, the properties of matrices are similar to the properties of real numbers. Some properties are listed below. Not all of the properties of real numbers carry over to matrices, however. We will consider some exceptions in the exercises.

Properties of Matrices	Let A, B, and C be matrices and a and b be scalars.

- **The Commutative Property of Addition.** $\quad A + B = B + A$
- **The Associative Property of Addition.** $\quad A + (B + C) = (A + B) + C$
- **The Associative Properties of Scalar Multiplication.** $\quad a(bA) = (ab)A$
  $\quad a(AB) = (aA)B$
- **Distributive Properties of Scalar Multiplication.** $\quad (a + b)A = aA + bA$
  $\quad a(A + B) = aA + aB$
- **The Associative Property of Multiplication.** $\quad A(BC) = (AB)C$
- **The Distributive Properties.** $\quad A(B + C) = AB + AC$
  $\quad (A + B)C = AC + BC$

EXAMPLE 9   Verify the distributive property $A(B + C) = AB + AC$ using the matrices

$$\begin{bmatrix} 2 & 1 \\ 3 & 1 \end{bmatrix}\left(\begin{bmatrix} 3 & -4 \\ 1 & -1 \end{bmatrix} + \begin{bmatrix} -1 & 3 \\ 0 & 1 \end{bmatrix}\right) = \begin{bmatrix} 2 & 1 \\ 3 & 1 \end{bmatrix}\begin{bmatrix} 3 & -4 \\ 1 & -1 \end{bmatrix} + \begin{bmatrix} 2 & 1 \\ 3 & 1 \end{bmatrix}\begin{bmatrix} -1 & 3 \\ 0 & 1 \end{bmatrix}$$

Solution   We perform the operations on the left-hand side and the right-hand side separately, and then compare the results.

$$\begin{bmatrix} 2 & 1 \\ 3 & 1 \end{bmatrix}\left(\begin{bmatrix} 3 & -4 \\ 1 & -1 \end{bmatrix} + \begin{bmatrix} -1 & 3 \\ 0 & 1 \end{bmatrix}\right) = \begin{bmatrix} 2 & 1 \\ 3 & 1 \end{bmatrix}\left(\begin{bmatrix} 2 & -1 \\ 1 & 0 \end{bmatrix}\right)$$ Perform the addition within the parentheses.

$$= \begin{bmatrix} 5 & -2 \\ 7 & -3 \end{bmatrix}$$ Multiply.

$$\begin{bmatrix} 2 & 1 \\ 3 & 1 \end{bmatrix}\begin{bmatrix} 3 & -4 \\ 1 & -1 \end{bmatrix} + \begin{bmatrix} 2 & 1 \\ 3 & 1 \end{bmatrix}\begin{bmatrix} -1 & 3 \\ 0 & 1 \end{bmatrix} = \begin{bmatrix} 7 & -9 \\ 10 & -13 \end{bmatrix} + \begin{bmatrix} -2 & 7 \\ -3 & 10 \end{bmatrix}$$ Perform the multiplications.

$$= \begin{bmatrix} 5 & -2 \\ 7 & -3 \end{bmatrix}$$ Add.

Because the results are the same, this example illustrates the distributive property. ∎

7.3   EXERCISES

In Exercises 1–4, find values of x and y, if any, that will make the two matrices equal.

1. $\begin{bmatrix} x & y \\ 1 & 3 \end{bmatrix} = \begin{bmatrix} 2 & 5 \\ 1 & 3 \end{bmatrix}$

2. $\begin{bmatrix} x & 5 \\ 3 & y \end{bmatrix} = \begin{bmatrix} 0 & 5 \\ 3 & 2 \end{bmatrix}$

3. $\begin{bmatrix} x + y & 3 + x \\ -2 & 5y \end{bmatrix} = \begin{bmatrix} 3 & 4 \\ -2 & 10 \end{bmatrix}$

4. $\begin{bmatrix} x + y & x - y \\ 2x & 3y \end{bmatrix} = \begin{bmatrix} -x & x - 2 \\ -y & 8 - y \end{bmatrix}$

In Exercises 5–8, find 5A.

5. $A = \begin{bmatrix} 3 & -3 \\ 0 & -2 \end{bmatrix}$

6. $A = \begin{bmatrix} 3 & \frac{3}{5} \\ 0 & -1 \end{bmatrix}$

7. $A = \begin{bmatrix} 5 & 5 & -2 \\ -2 & -5 & 1 \end{bmatrix}$

8. $A = \begin{bmatrix} -3 & 1 & 2 \\ -3 & -2 & -5 \end{bmatrix}$

In Exercises 9–10, find A + B.

9. $A = \begin{bmatrix} 2 & 1 & -1 \\ -3 & 2 & 5 \end{bmatrix}, B = \begin{bmatrix} -3 & 1 & 2 \\ -3 & -2 & -5 \end{bmatrix}$

10. $A = \begin{bmatrix} 3 & 2 & 1 \\ -2 & 3 & -3 \\ -4 & -2 & -1 \end{bmatrix}, B = \begin{bmatrix} -2 & 6 & -2 \\ 5 & 7 & -1 \\ -4 & -6 & 7 \end{bmatrix}$

In Exercises 11–12, find A − B.

11. $A = \begin{bmatrix} -3 & 2 & -2 \\ -1 & 4 & -5 \end{bmatrix}, B = \begin{bmatrix} 3 & -3 & -2 \\ -2 & 5 & -5 \end{bmatrix}$

12. $A = \begin{bmatrix} 2 & 2 & 0 \\ -2 & 8 & 1 \\ 3 & -3 & -8 \end{bmatrix}, B = \begin{bmatrix} -4 & 3 & 7 \\ -1 & 2 & 0 \\ 1 & 4 & -1 \end{bmatrix}$

In Exercises 13–14, find 5A + 3B.

13. $A = \begin{bmatrix} 3 & 1 & -2 \\ -4 & 3 & -2 \end{bmatrix}, B = \begin{bmatrix} 1 & -2 & 2 \\ -5 & -5 & 3 \end{bmatrix}$

14. $A = \begin{bmatrix} 2 & -5 \\ -5 & 2 \end{bmatrix}, B = \begin{bmatrix} 5 & -2 \\ 2 & -5 \end{bmatrix}$

In Exercises 15–16, find the additive inverse of each matrix.

15. $A = \begin{bmatrix} 5 & -2 & 7 \\ -5 & 0 & 3 \\ -2 & 3 & -5 \end{bmatrix}$

16. $A = \begin{bmatrix} 3 & -\dfrac{2}{3} & -5 & \dfrac{1}{2} \end{bmatrix}$

In Exercises 17–28, find each product, if possible.

17. $\begin{bmatrix} 2 & 3 \\ 3 & -2 \end{bmatrix} \begin{bmatrix} 1 & 2 \\ 0 & -2 \end{bmatrix}$

18. $\begin{bmatrix} -2 & 3 \\ 3 & -2 \end{bmatrix} \begin{bmatrix} 2 & 4 \\ -5 & 7 \end{bmatrix}$

19. $\begin{bmatrix} -4 & -2 \\ 21 & 0 \end{bmatrix} \begin{bmatrix} -5 & 6 \\ 21 & -1 \end{bmatrix}$

20. $\begin{bmatrix} -5 & 4 \\ 4 & -5 \end{bmatrix} \begin{bmatrix} 6 & -2 \\ 1 & 3 \end{bmatrix}$

21. $\begin{bmatrix} 2 & 1 & 3 \\ 1 & 2 & -1 \\ 0 & 1 & 0 \end{bmatrix} \begin{bmatrix} 1 & 2 & 3 \\ 2 & -2 & 1 \\ 0 & 0 & 1 \end{bmatrix}$

22. $\begin{bmatrix} 2 & 1 & 1 \\ 1 & 1 & 2 \\ 1 & -2 & -1 \end{bmatrix} \begin{bmatrix} 1 & 2 & 3 \\ 1 & 2 & -3 \\ -1 & -1 & 3 \end{bmatrix}$

23. $\begin{bmatrix} 1 & -2 & -3 \\ 2 & 0 & 1 \end{bmatrix} \begin{bmatrix} 4 \\ -5 \\ -6 \end{bmatrix}$

24. $\begin{bmatrix} 1 \\ -2 \\ -3 \end{bmatrix} [4 \quad -5 \quad -6]$

25. $[1 \quad 2 \quad 3] \begin{bmatrix} 4 & 5 & 6 \\ 7 & 8 & 9 \end{bmatrix}$

26. $\begin{bmatrix} 2 & 3 & 4 \\ 1 & 2 & 3 \\ -2 & 2 & 2 \end{bmatrix} \begin{bmatrix} -1 \\ 2 \\ 3 \end{bmatrix}$

27. $\begin{bmatrix} 2 & 5 \\ -3 & 1 \\ 0 & -2 \\ 1 & -5 \end{bmatrix} \begin{bmatrix} 3 & -2 & 4 \\ -2 & -3 & 1 \end{bmatrix}$

28. $\begin{bmatrix} 1 & 4 & 0 & 0 \\ -4 & 1 & 0 & -2 \\ 0 & 0 & 1 & 0 \\ 0 & 2 & 0 & 1 \end{bmatrix} \begin{bmatrix} 1 \\ 2 \\ -2 \\ -1 \end{bmatrix}$

In Exercises 29–36, let $A = \begin{bmatrix} 1 & 3 \\ 2 & 5 \end{bmatrix}$, $B = \begin{bmatrix} -1 \\ 3 \end{bmatrix}$, and $C = [3 \quad 2]$. Perform the indicated operations, if possible.

29. $A - BC$

30. $AB + B$

31. $CB - AB$

32. CAB

33. ABC

34. $CA + C$

35. A^2B

36. $(BC)^2$

 In Exercises 37–40, let $A = \begin{bmatrix} 2.3 & -1.7 & 3.1 \\ -2 & 3.5 & 1 \\ -8 & 4.7 & 9.1 \end{bmatrix}$, $B = \begin{bmatrix} -2.5 \\ 5.2 \\ -7 \end{bmatrix}$, and $C = \begin{bmatrix} -5.8 \\ 2.9 \\ 4.1 \end{bmatrix}$. Use a graphing calculator to find each result.

37. AB

38. $B + C$

39. A^2

40. $AB + C$

In Exercises 41–42, use a graphing calculator with matrix capabilities to solve each problem.

41. Beverage sales Beverages were sold to parents and children at the school basketball game in the quantities and prices given in the following charts. Find matrices Q and P that represent the quantities and prices, find the product QP, and interpret the result.

Quantities

	Coffee	Milk	Cola
Adult males	217	23	319
Adult females	347	24	340
Children	3	97	750

Price

Coffee	$ .75
Milk	$1.00
Cola	$1.25

42. Production costs Each of four factories manufactures three products in the daily quantities and unit costs given in the charts. Find a suitable matrix product to represent production costs.

Production Quantities

Factory	Product A	Product B	Product C
Ashtabula	19	23	27
Boston	17	21	22
Chicago	21	18	20
Denver	27	25	22

Unit Production Costs

	Day shift	Night shift
Product A	$1.20	$1.35
Product B	$ .75	$ .85
Product C	$3.50	$3.70

In Exercises 43–46, $A = \begin{bmatrix} 1 & 2 \\ 1 & 3 \end{bmatrix}$, $B = \begin{bmatrix} 2 & 1 & -5 \\ 1 & 1 & 2 \end{bmatrix}$, $C = \begin{bmatrix} -2 & -1 & 6 \\ 0 & -1 & -1 \end{bmatrix}$, $D = \begin{bmatrix} 1 & 2 \\ 1 & 3 \end{bmatrix}$, and $E = \begin{bmatrix} 1 & 2 \\ 1 & 3 \end{bmatrix}$. Verify

each property by performing the operations on each side of the equation and comparing the results.

43. Distributive property

$$A(B + C) = AB + AC$$

44. Associative property of scalar multiplication

$$5(6A) (5 \cdot 6)A$$

45. Associative property of scalar multiplication

$$3(AB) = (3A)B$$

46. Associative property of multiplication

$$A(DE) = (AD)E$$

47. Connectivity matrix An entry of 1 in the following **connectivity matrix** A indicates that the person associated with that row knows the address of the person associated with that column. For example, the 1 in Bill's row and Al's column indicates that Bill can write to Al. The 0 in Bill's row and Carl's column indicates that Bill cannot write to Carl. However, Bill could ask Al to forward his letter to Carl. The matrix A^2 indicates the numbers of ways that one person can write to another with a letter that is forwarded exactly once. Find A^2.

$$
\begin{array}{c}
\phantom{\text{Carl}}\text{Al}\ \ \text{Bill}\ \ \text{Carl} \\
\begin{array}{c}
\text{Al} \\
\text{Bill} \\
\text{Carl}
\end{array}
\left[\begin{array}{ccc}
0 & 1 & 1 \\
1 & 0 & 0 \\
0 & 1 & 0
\end{array}\right] = A
\end{array}
$$

48. **Communication routing**   Refer to Exercise 47. Find and interpret the matrix $A + A^2$. Can everyone receive a letter from everyone else with at most one forwarding?

49. **Routing telephone calls**   A new long-distance telephone carrier has established several direct microwave links among four cities. In the following connectivity matrix, entries a_{ij} and a_{ji} indicate the number of direct links between cities i and j. For example, cities 2 and 4 are not linked directly, but they could be connected through city 3. Find and interpret the matrices A^2 and $A + A^2$.

$$
A = \left[\begin{array}{cccc}
0 & 2 & 1 & 0 \\
2 & 0 & 1 & 0 \\
1 & 1 & 0 & 2 \\
0 & 0 & 2 & 0
\end{array}\right]
$$

50. **Communication on one-way channels**   Three communication centers are linked as indicated in Illustration 1, with communication only in the direction of the arrows. Thus, location 1 can send a message directly to location 2 along two paths, but location 2 can return a message directly on only one path. Entry c_{ij} of matrix C indicates the number of channels from i to j. Find and interpret C^2.

$$
C = \left[\begin{array}{ccc}
0 & 2 & 1 \\
1 & 0 & 1 \\
2 & 0 & 0
\end{array}\right]
$$

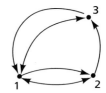

ILLUSTRATION 1

51. If A and B are 2×2 matrices, is $(AB)^2$ equal to A^2B^2? Support your answer.

52. Let a, b, and c be real numbers. If $ab = ac$ and $a \neq 0$, then $b = c$. Find 2×2 matrices, A, B, and C, where $A \neq 0$, to show that such a law does not hold for all matrices.

53. Another property of the real numbers is that, if $ab = 0$, then either $a = 0$ or $b = 0$. To show that this property is not true for matrices, find two nonzero 2×2 matrices, A and B, such that $AB = 0$.

54. Find 2×2 matrices to show that $(A + B)(A - B) \neq A^2 - B^2$.

7.4 MATRIX INVERSION

■ Finding an Inverse by Row Operations   ■ Finding Inverses on a Graphing Calculator   ■ Solving a System of Equations   ■ An Application of Matrix Inversion

Two real numbers are called **multiplicative inverses** if their product is the multiplicative identity 1. Some matrices have multiplicative inverses also.

Matrix Inverses

If A and B are $n \times n$ matrices, I is the $n \times n$ identity matrix, and

$$AB = BA = I$$

then A and B are called **multiplicative inverses**. Matrix A is the **inverse** of B, and B is the **inverse** of A.

It can be shown that if a matrix A has an inverse, it only has one inverse. The inverse of A is denoted by A^{-1}. Because of the definition of matrix inverses,

$$AA^{-1} = A^{-1}A = I$$

EXAMPLE 1   Show that $A = \begin{bmatrix} 1 & 1 & 0 \\ 4 & 3 & 0 \\ 2 & 1 & -1 \end{bmatrix}$ and $B = \begin{bmatrix} -3 & 1 & 0 \\ 4 & -1 & 0 \\ -2 & 1 & -1 \end{bmatrix}$ are inverses.

Solution $AB = \begin{bmatrix} 1 & 1 & 0 \\ 4 & 3 & 0 \\ 2 & 1 & -1 \end{bmatrix} \begin{bmatrix} -3 & 1 & 0 \\ 4 & -1 & 0 \\ -2 & 1 & -1 \end{bmatrix}$

$= \begin{bmatrix} -3 + 4 + 0 & 1 - 1 + 0 & 0 + 0 + 0 \\ -12 + 12 + 0 & 4 - 3 + 0 & 0 + 0 + 0 \\ -6 + 4 + 2 & 2 - 1 - 1 & 0 + 0 + 1 \end{bmatrix} = \begin{bmatrix} 1 & 0 & 0 \\ 0 & 1 & 0 \\ 0 & 0 & 1 \end{bmatrix}$

$BA = \begin{bmatrix} -3 & 1 & 0 \\ 4 & -1 & 0 \\ -2 & 1 & -1 \end{bmatrix} \begin{bmatrix} 1 & 1 & 0 \\ 4 & 3 & 0 \\ 2 & 1 & -1 \end{bmatrix} = \begin{bmatrix} 1 & 0 & 0 \\ 0 & 1 & 0 \\ 0 & 0 & 1 \end{bmatrix}$ ■

■ Finding an Inverse by Row Operations

A matrix that has an inverse is called a **nonsingular matrix** and is said to be **invertible**. Otherwise, it is called a **singular matrix** and is not invertible. The following theorem, stated without proof, provides a way to find the inverse of an invertible matrix.

Theorem

If a sequence of elementary row operations performed on the $n \times n$ matrix A reduces A to the $n \times n$ identity matrix I, then those same row operations, performed in the same order on the identity matrix I, will transform I into A^{-1}.

Furthermore, if no sequence of row operations will reduce A to I, then A is not invertible.

To use this theorem, we perform elementary row operations on matrix A to change it to the identity matrix I. At the same time, we perform the same elementary row operations on the identity matrix I. This changes I into A^{-1}. A notation for this process uses an n-row–by–$2n$-column matrix, with matrix A as the left half and matrix I as the right half. If A is invertible, the proper row operations performed on $[A \mid I]$ will transform it into $[I \mid A^{-1}]$.

EXAMPLE 2   Find the inverse of matrix A if $A = \begin{bmatrix} 2 & -4 \\ 4 & -7 \end{bmatrix}$.

Solution We can set up a 2×4 matrix with A on the left and I on the right of the broken line:

$$[A \mid I] = \begin{bmatrix} 2 & -4 & \vdots & 1 & 0 \\ 4 & -7 & \vdots & 0 & 1 \end{bmatrix}$$

We perform row operations on the entire matrix to transform the left half into I:

$$\begin{bmatrix} 2 & -4 & \vdots & 1 & 0 \\ 4 & -7 & \vdots & 0 & 1 \end{bmatrix} \quad \begin{array}{c} (\frac{1}{2})R1 \rightarrow \\ \\ (-2)R1 + R2 \rightarrow \end{array} \begin{bmatrix} 1 & -2 & \vdots & \frac{1}{2} & 0 \\ 0 & 1 & \vdots & -2 & 1 \end{bmatrix}$$

$$(2)R2 + R1 \rightarrow \begin{bmatrix} 1 & 0 & \vdots & -\frac{7}{2} & 2 \\ 0 & 1 & \vdots & -2 & 1 \end{bmatrix}$$

Matrix A has been transformed into I. Thus, the right side of the previous matrix is A^{-1}. Verify this by finding AA^{-1} and $A^{-1}A$ and showing that each product is I:

$$AA^{-1} = \begin{bmatrix} 2 & -4 \\ 4 & -7 \end{bmatrix} \begin{bmatrix} -\frac{7}{2} & 2 \\ -2 & 1 \end{bmatrix} = \begin{bmatrix} 1 & 0 \\ 0 & 1 \end{bmatrix}$$

$$A^{-1}A = \begin{bmatrix} -\frac{7}{2} & 2 \\ -2 & 1 \end{bmatrix} \begin{bmatrix} 2 & -4 \\ 4 & -7 \end{bmatrix} = \begin{bmatrix} 1 & 0 \\ 0 & 1 \end{bmatrix}$$

■

EXAMPLE 3 Find the inverse of matrix A if $A = \begin{bmatrix} 1 & 1 & 0 \\ 1 & 2 & 1 \\ 2 & 3 & 2 \end{bmatrix}$.

Solution We set up a 3×6 matrix with A on the left and I and on the right of the broken line,

$$[A \mid I] = \begin{bmatrix} 1 & 1 & 0 & \vdots & 1 & 0 & 0 \\ 1 & 2 & 1 & \vdots & 0 & 1 & 0 \\ 2 & 3 & 2 & \vdots & 0 & 0 & 1 \end{bmatrix}$$

and perform row operations on the entire matrix to transform the left half into I.

$$\begin{bmatrix} 1 & 1 & 0 & \vdots & 1 & 0 & 0 \\ 1 & 2 & 1 & \vdots & 0 & 1 & 0 \\ 2 & 3 & 2 & \vdots & 0 & 0 & 1 \end{bmatrix} \begin{array}{c} \\ (-1)R1 + R2 \rightarrow \\ (-2)R1 + R3 \rightarrow \end{array} \begin{bmatrix} 1 & 1 & 0 & \vdots & 1 & 0 & 0 \\ 0 & 1 & 1 & \vdots & -1 & 1 & 0 \\ 0 & 1 & 2 & \vdots & -2 & 0 & 1 \end{bmatrix}$$

$$\begin{array}{c} (-1)R2 + R1 \rightarrow \\ \\ (-1)R2 + R3 \rightarrow \end{array} \begin{bmatrix} 1 & 0 & -1 & \vdots & 2 & -1 & 0 \\ 0 & 1 & 1 & \vdots & -1 & 1 & 0 \\ 0 & 0 & 1 & \vdots & -1 & -1 & 1 \end{bmatrix}$$

$$\begin{array}{c} R3 + R1 \rightarrow \\ (-1)R3 + R2 \rightarrow \end{array} \begin{bmatrix} 1 & 0 & 0 & \vdots & 1 & -2 & 1 \\ 0 & 1 & 0 & \vdots & 0 & 2 & -1 \\ 0 & 0 & 1 & \vdots & -1 & -1 & 1 \end{bmatrix}$$

Since the left half has been transformed into the identity matrix, the right half has become A^{-1}. Thus,

$$A^{-1} = \begin{bmatrix} 1 & -2 & 1 \\ 0 & 2 & -1 \\ -1 & -1 & 1 \end{bmatrix}$$

∎

EXAMPLE 4 Find the inverse of $A = \begin{bmatrix} 1 & 2 \\ 2 & 4 \end{bmatrix}$, if possible.

Solution We form the 2×4 matrix

$$[A \mid I] = \begin{bmatrix} 1 & 2 & | & 1 & 0 \\ 2 & 4 & | & 0 & 1 \end{bmatrix}$$

and begin to transform the left side of the matrix into the identity matrix I:

$$\begin{bmatrix} 1 & 2 & | & 1 & 0 \\ 2 & 4 & | & 0 & 1 \end{bmatrix} (-2)R1 + R2 \rightarrow \begin{bmatrix} 1 & 2 & | & 1 & 0 \\ 0 & 0 & | & -2 & 1 \end{bmatrix}$$

In obtaining the second-row, first-column position of A, the entire second row of A is "zeroed out." Since we cannot transform matrix A into the identity, A is not invertible. ∎

Finding Inverses on a Graphing Calculator

We can use a graphing calculator that performs matrix operations to find the inverse of a matrix. For example, to find the inverse of

$$A = \begin{bmatrix} 2 & 2 & 3 \\ 1 & 2 & 3 \\ 1 & 0 & 1 \end{bmatrix}$$

on a Texas Instruments TI-82 graphing calculator, we press $\boxed{\text{MATRIX}}$ and select EDIT to enter matrix A. We exit entry mode by pressing $\boxed{\text{2nd}}$ $\boxed{\text{CLEAR}}$ and display A^{-1} by pressing $\boxed{\text{2nd}}$ $\boxed{1}$ $\boxed{x^{-1}}$ $\boxed{\text{ENTER}}$. The display appears in Figure 7-6(a). To verify that $AA^{-1} = I$, we press $\boxed{\text{2nd}}$ $\boxed{1}$ $\boxed{\text{2nd}}$ $\boxed{1}$ $\boxed{x^{-1}}$ $\boxed{\text{ENTER}}$ to obtain the result shown in Figure 7-6(b).

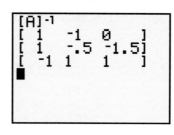

(a)

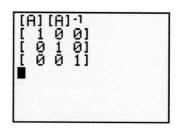

(b)

FIGURE 7-6

■ Solving a System of Equations

If we multiply the matrices on the left-hand side of the matrix equation

$$\begin{bmatrix} 1 & 1 & 0 \\ 1 & 2 & 1 \\ 2 & 3 & 2 \end{bmatrix} \begin{bmatrix} x \\ y \\ z \end{bmatrix} = \begin{bmatrix} 3 \\ -2 \\ 1 \end{bmatrix}$$

and set the corresponding entries equal, we get the system of equations

$$\begin{cases} x + y & = 3 \\ x + 2y + z = -2 \\ 2x + 3y + 2z = 1 \end{cases}$$

Thus, a system of equations can be written as a matrix equation $AX = B$, where A is the coefficient matrix of the system, X is a column matrix of variables, and B is the column matrix of constants. The next example shows that if matrix A is invertible, the matrix equation $AX = B$ is easy to solve.

EXAMPLE 5 Solve the system $\begin{bmatrix} 1 & 1 & 0 \\ 1 & 2 & 1 \\ 2 & 3 & 2 \end{bmatrix} \begin{bmatrix} x \\ y \\ z \end{bmatrix} = \begin{bmatrix} 20 \\ 30 \\ 55 \end{bmatrix}$.

Solution The 3×3 matrix on the left is the matrix whose inverse was found in Example 3. We can multiply each side of the equation on the left by this inverse to obtain an equivalent system of equations.

$$\begin{bmatrix} 1 & -2 & 1 \\ 0 & 2 & -1 \\ -1 & -1 & 1 \end{bmatrix} \begin{bmatrix} 1 & 1 & 0 \\ 1 & 2 & 1 \\ 2 & 3 & 2 \end{bmatrix} \begin{bmatrix} x \\ y \\ z \end{bmatrix} = \begin{bmatrix} 1 & -2 & 1 \\ 0 & 2 & -1 \\ -1 & -1 & 1 \end{bmatrix} \begin{bmatrix} 20 \\ 30 \\ 55 \end{bmatrix}$$

$$\begin{bmatrix} 1 & 0 & 0 \\ 0 & 1 & 0 \\ 0 & 0 & 1 \end{bmatrix} \begin{bmatrix} x \\ y \\ z \end{bmatrix} = \begin{bmatrix} 15 \\ 5 \\ 5 \end{bmatrix} \qquad \text{Multiply the matrices. On the left, remember that } A^{-1}A = I.$$

$$\begin{bmatrix} x \\ y \\ z \end{bmatrix} = \begin{bmatrix} 15 \\ 5 \\ 5 \end{bmatrix} \qquad IX = X.$$

The solution of this system can be read directly from the matrix on the right-hand side. Verify that the values $x = 15$, $y = 5$, $z = 5$ satisfy all three of the original equations. ■

Example 5 suggests the following result.

Solving Systems of Equations	If A is invertible, the solution of the matrix equation $AX = B$ is $$X = A^{-1}B$$

This method is especially useful for finding solutions of several systems of equations that differ from each other only in the column matrix B. If the coefficient matrix A remains unchanged from one system of equations to the next, then A^{-1} needs to be found only once. The solution of each system is found by a single matrix multiplication, $A^{-1}B$.

■ An Application of Matrix Inversion

EXAMPLE 6   A company that manufactures diagnostic medical equipment spends time on paperwork, manufacture, and testing for each of three versions of a crucial circuit board. The times spent on each and the total time available are given in the following tables.

Hours Required per Unit

	Product A	Product B	Product C
Paperwork	1	1	0
Manufacture	1	2	1
Testing	2	3	2

Hours Available

Paperwork	20
Manufacturing	30
Testing	55

Solution   We can let x, y, and z represent the number of units of products A, B, and C to be manufactured, respectively. We can then set up the following system of equations:

Paperwork:	$x + y = 20$	One hour is needed for every A, and one hour for every B.
Manufacturing:	$x + 2y + z = 30$	One hour is needed for every A and C, and two hours for every B.
Testing:	$2x + 3y + 2z = 55$	Two hours are needed for every A and C, and three hours for every B.

In matrix form, the system becomes

$$\begin{bmatrix} 1 & 1 & 0 \\ 1 & 2 & 1 \\ 2 & 3 & 2 \end{bmatrix} \begin{bmatrix} x \\ y \\ z \end{bmatrix} = \begin{bmatrix} 20 \\ 30 \\ 55 \end{bmatrix}$$

We solved this matrix equation in Example 5. To use all of the available time, the company should manufacture 15 units of product A and 5 units each of products B and C.   ■

7.4   EXERCISES

In Exercises 1–16, find the inverse of each matrix, if possible.

1. $\begin{bmatrix} 3 & -4 \\ -2 & 3 \end{bmatrix}$ **2.** $\begin{bmatrix} 2 & 3 \\ 3 & 5 \end{bmatrix}$ **3.** $\begin{bmatrix} 3 & 7 \\ 2 & 5 \end{bmatrix}$ **4.** $\begin{bmatrix} 1 & -2 \\ 2 & -5 \end{bmatrix}$

5. $\begin{bmatrix} 1 & 0 & 3 \\ -1 & 1 & 3 \\ -2 & 1 & 1 \end{bmatrix}$

6. $\begin{bmatrix} 2 & 1 & -1 \\ 2 & 2 & -1 \\ -1 & -1 & 1 \end{bmatrix}$

7. $\begin{bmatrix} 3 & 2 & 1 \\ 1 & 1 & -1 \\ 4 & 3 & 1 \end{bmatrix}$

8. $\begin{bmatrix} -2 & 1 & -3 \\ 2 & 3 & 0 \\ 1 & 0 & 1 \end{bmatrix}$

9. $\begin{bmatrix} 1 & 3 & 5 \\ 0 & 1 & 6 \\ 1 & 4 & 11 \end{bmatrix}$

10. $\begin{bmatrix} 1 & 2 & 3 \\ 4 & 5 & 6 \\ 7 & 8 & 9 \end{bmatrix}$

11. $\begin{bmatrix} 1 & 2 & 3 \\ 0 & 1 & 2 \\ 0 & 0 & 1 \end{bmatrix}$

12. $\begin{bmatrix} 1 & 2 & 3 \\ 0 & 1 & 1 \\ 0 & -1 & 0 \end{bmatrix}$

13. $\begin{bmatrix} 1 & 6 & 4 \\ 1 & -2 & -5 \\ 2 & 4 & -1 \end{bmatrix}$

14. $\begin{bmatrix} 1 & 1 & 1 \\ 1 & 0 & -1 \\ 1 & 2 & 3 \end{bmatrix}$

15. $\begin{bmatrix} 1 & 2 & 3 & 4 \\ 0 & 1 & 2 & 3 \\ 0 & 0 & 1 & 2 \\ 0 & 0 & 0 & 1 \end{bmatrix}$

16. $\begin{bmatrix} 1 & 0 & 0 & 0 \\ 1 & 1 & 0 & 0 \\ 1 & 1 & 1 & 0 \\ 1 & 2 & 2 & 1 \end{bmatrix}$

In Exercises 17–20, use a graphing calculator with matrix capabilities to find the inverse of each matrix.

17. $\begin{bmatrix} 1 & 1 & -1 \\ 0.5 & 1 & 0.5 \\ 1 & 1 & -1.5 \end{bmatrix}$

18. $\begin{bmatrix} -2 & -1 & 1 \\ 0.5 & -1.5 & -0.5 \\ 0 & 1 & 0.5 \end{bmatrix}$

19. $\begin{bmatrix} 3 & 3 & -3 & 2 \\ 1 & -4 & 3 & -5 \\ 3 & 0 & -2 & -1 \\ -1 & 5 & -3 & 6 \end{bmatrix}$

20. $\begin{bmatrix} 1 & 0 & 0 & 0 \\ 2 & 1 & 0 & 0 \\ 3 & 2 & 1 & 0 \\ 4 & 3 & 2 & 1 \end{bmatrix}$

In Exercises 21–28, use the method of Example 5 to solve each system of equations. Note that several systems have the same coefficient matrix.

21. $\begin{cases} 3x - 4y = 1 \\ -2x + 3y = 5 \end{cases}$

22. $\begin{cases} 3x - 4y = -1 \\ -2x + 3y = 3 \end{cases}$

23. $\begin{cases} 3x - 4y = 0 \\ -2x + 3y = 0 \end{cases}$

24. $\begin{cases} 3x - 4y = -3 \\ -2x + 3y = -2 \end{cases}$

25. $\begin{cases} 2x + y - z = 2 \\ 2x + 2y - z = 4 \\ -x - y + z = -1 \end{cases}$

26. $\begin{cases} 2x + y - z = 3 \\ 2x + 2y - z = -1 \\ -x - y + z = 4 \end{cases}$

27. $\begin{cases} -2x + y - 3z = 2 \\ 2x + 3y = -3 \\ x + z = 5 \end{cases}$

28. $\begin{cases} -2x + y - 3z = 5 \\ 2x + 3y = 1 \\ x + z = -2 \end{cases}$

In Exercises 29–32, use a graphing calculator to solve each system of equations. Use the method of Example 5.

29. $\begin{cases} 5x + 3y = 13 \\ -7x + 5y = -9 \end{cases}$

30. $\begin{cases} 8x - 3y = 7 \\ -3x + 2y = 0 \end{cases}$

31. $\begin{cases} 5x + 2y + 3z = 12 \\ 2x + 5z = 7 \\ 3x + z = 4 \end{cases}$

32. $\begin{cases} 3x + 2y - z = 0 \\ 5x - 2y = 5 \\ 3x + y + z = 6 \end{cases}$

33. **Manufacturing and testing** The numbers of hours required to manufacture and test each of two models of heart monitor are given in the first table, and the number of hours available each week for manufacturing and testing is given in the second table.

Hours Required per Unit

	Model A	Model B
Manufacturing	23	27
Testing	21	22

Hours Available

Manufacturing	127
Testing	108

Find the number of each model that can be manufactured each week.

34. **Cryptography** The letters of a message, called **plaintext**, are assigned values 1–26 and are written in groups of 3 as 3×1 matrices. To write a message in **cyphertext**, each 3×1 matrix is multiplied by matrix A, where

$$A = \begin{bmatrix} 1 & 1 & 0 \\ 2 & 3 & 3 \\ 1 & 1 & 1 \end{bmatrix}$$

The cyphertext of one message is $AY = \begin{bmatrix} 30 \\ 122 \\ 49 \end{bmatrix}$. Find the plaintext.

In Exercises 35–36, let $A = \begin{bmatrix} 3 & 0 & 0 \\ -2 & -1 & -2 \\ 3 & 6 & 3 \end{bmatrix}$ *and* $X = \begin{bmatrix} x \\ y \\ z \end{bmatrix}$. *Solve each equation. Each solution is called an eigenvector of the matrix A.*

35. $AX - 2IX = 0$ (*Hint:* Factor out *X*.)

36. $AX = 3IX$

37. If the $n \times n$ matrix *A* is invertible and if *B* and *C* are $n \times n$ matrices such that $AB = AC$, prove that $B = C$.

38. If *B* is an $n \times n$ matrix that behaves as an identity $(AB = BA = A$, for any $n \times n$ matrix *A*), prove that $B = I$. (*Hint:* If the equation is true for all *A*, is it true for $A = I$?)

39. Prove that $\begin{bmatrix} a & b \\ c & d \end{bmatrix}$ has an inverse if and only if $ad - bc \neq 0$. (*Hint:* Try to find the inverse and see what happens.)

40. Show that $\begin{bmatrix} x & 1 \\ -1 & x \end{bmatrix}$ is invertible for all numbers *x*.

(*Hint:* See Exercise 39.)

In Exercises 41–42, use examples chosen from 2×2 matrices to support each answer.

41. Does $(AB)^{-1} = A^{-1}B^{-1}$?

42. Does $(AB)^{-1} = B^{-1}A^{-1}$?

43. The **transpose** of a matrix *A*, denoted by A^T, is formed by exchanging the rows and columns of *A*: The first row of *A* becomes the first column of A^T, the second row becomes the second column, and so on. With an example chosen from 2×2 matrices, verify that $(A^{-1})^T = (A^T)^{-1}$.

In Exercises 44–45, let $A = \begin{bmatrix} -1 & -1 \\ 1 & 1 \end{bmatrix}$.

44. Show that $A^2 = 0$.

45. Show that the inverse of $I - A$ is $I + A$.

46. Suppose that *B* is any matrix for which $B^2 = 0$. Show that $I - B$ is invertible by showing that the inverse of $I - B$ is $I + B$.

47. Suppose that *C* is any matrix for which $C^3 = 0$. Show that $I - C$ is invertible by showing that the inverse of $I - C$ is $I + C + C^2$.

7.5 DETERMINANTS

■ Evaluating Determinants of Higher-Order Matrices ■ Properties of Determinants ■ Using Determinants to Solve Systems of Equations ■ Writing Equations of Lines ■ Finding Areas of Triangles

There is a function, called the **determinant function**, that associates a number with every square matrix of any order. The function is denoted by the symbols $\det(A)$ or $|A|$.

The Determinant of a 2 × 2 Matrix

If a, b, c, and d are numbers, the determinant of $A = \begin{bmatrix} a & b \\ c & d \end{bmatrix}$ is

$$\det(A) = \begin{vmatrix} a & b \\ c & d \end{vmatrix} = ad - bc$$

 Warning! Do not confuse the notation $|A|$ with absolute value symbols.

EXAMPLE 1 **a.** $\begin{vmatrix} 1 & 2 \\ 3 & 4 \end{vmatrix} = 1 \cdot 4 - 2 \cdot 3$ **b.** $\begin{vmatrix} -2 & 3 \\ -\pi & \frac{1}{2} \end{vmatrix} = (-2)\left(\frac{1}{2}\right) - (3)(-\pi)$

$$= 4 - 6 \qquad\qquad\qquad = -1 + 3\pi$$

$$= -2$$

■

■ Evaluating Determinants of Higher-Order Matrices

To evaluate determinants of higher-order matrices, we need to define the **minor** and the **cofactor** of an element in a matrix.

Minor and Cofactor

Let $A = [a_{ij}]$ be a square matrix of order $n \geq 2$.

1. The **minor** of a_{ij}, denoted as M_{ij}, is the determinant of the $n - 1 \times n - 1$ matrix formed by deleting the ith row and jth column of A.

2. The **cofactor** of a_{ij}, denoted as C_{ij}, is $\begin{cases} M_{ij} \text{ when } i + j \text{ is even} \\ -M_{ij} \text{ when } i + j \text{ is odd.} \end{cases}$

EXAMPLE 2 In matrix $A = \begin{bmatrix} 1 & 2 & 3 \\ 4 & 5 & 6 \\ 7 & 8 & 9 \end{bmatrix}$, find the cofactors of **a.** a_{31} and **b.** a_{12}.

Solution **a.** The minor M_{31} is the minor of $a_{31} = 7$ appearing in row 3, column 1. It is found by deleting row 3 and column 1:

$$M_{31} = \begin{vmatrix} 1 & 2 & 3 \\ 4 & 5 & 6 \\ 7 & 8 & 9 \end{vmatrix} = \begin{vmatrix} 2 & 3 \\ 5 & 6 \end{vmatrix}$$

Because $i + j$ is even ($3 + 1 = 4$), the cofactor of the minor M_{31} is M_{31}:

$$C_{31} = M_{31} = \begin{vmatrix} 2 & 3 \\ 5 & 6 \end{vmatrix} = 2 \cdot 6 - 3 \cdot 5 = 12 - 15 = -3$$

b. The minor M_{12} is the minor of $a_{12} = 2$ appearing in row 1, column 2. It is found by deleting row 1 and column 2.

$$M_{12} = \begin{vmatrix} 1 & 2 & 3 \\ 4 & 5 & 6 \\ 7 & 8 & 9 \end{vmatrix} = \begin{vmatrix} 4 & 6 \\ 7 & 9 \end{vmatrix}$$

Because $i + j$ is odd ($1 + 2 = 3$), the cofactor of the minor M_{12} is $-M_{12}$:

$$C_{12} = -M_{12} = -\begin{vmatrix} 4 & 6 \\ 7 & 9 \end{vmatrix} = -(4 \cdot 9 - 6 \cdot 7) = -(36 - 42) = -(-6) = 6 \quad \blacksquare$$

We are now ready to evaluate determinants of higher-order matrices.

Expanding a Determinant by Minors

If A is a square matrix of order $n \ge 2$, the value $|A|$ is the sum of the products of the elements in any row (or column) and the cofactors of those elements.

To illustrate the method of expanding a determinant by minors, we expand a 3×3 determinant by using different rows and columns.

EXAMPLE 3   Expanding on row 1:

$$\begin{vmatrix} 1 & 2 & -3 \\ -1 & 0 & 1 \\ -2 & 2 & 1 \end{vmatrix} = a_{11}C_{11} + a_{12}C_{12} + a_{13}C_{13}$$

$$= 1\begin{vmatrix} 0 & 1 \\ 2 & 1 \end{vmatrix} + 2\left[-\begin{vmatrix} -1 & 1 \\ -2 & 1 \end{vmatrix}\right] + (-3)\begin{vmatrix} -1 & 0 \\ -2 & 2 \end{vmatrix}$$
$$= 1(-2) - 2(-1) - 3(-2)$$
$$= -2 - 2 + 6$$
$$= 2$$

Expanding on row 3:

$$\begin{vmatrix} 1 & 2 & -3 \\ -1 & 0 & 1 \\ -2 & 2 & 1 \end{vmatrix} = a_{31}C_{31} + a_{32}C_{32} + a_{33}C_{33}$$

$$= -2\begin{vmatrix} 2 & -3 \\ 0 & 1 \end{vmatrix} + 2\left[-\begin{vmatrix} 1 & -3 \\ -1 & 1 \end{vmatrix}\right] + 1\begin{vmatrix} 1 & 2 \\ -1 & 0 \end{vmatrix}$$
$$= -2(2) + 2(+2) + 1(2)$$
$$= -4 + 4 + 2$$
$$= 2$$

Expanding on column 2:

$$\begin{vmatrix} 1 & 2 & -3 \\ -1 & 0 & 1 \\ -2 & 2 & 1 \end{vmatrix} = a_{12}C_{12} + a_{22}C_{22} + a_{32}C_{32}$$

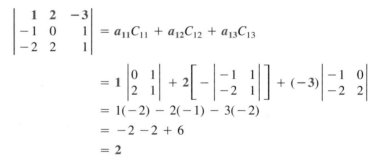

Gabriel Cramer
(1704–1752)
Although other
mathematicians had
worked with determinants,
it was the work of Cramer
that popularized them.

$$= 2\left[- \begin{vmatrix} -1 & 1 \\ -2 & 1 \end{vmatrix} \right] + 0\begin{vmatrix} 1 & -3 \\ -2 & 1 \end{vmatrix} + 2\left[- \begin{vmatrix} 1 & -3 \\ -1 & 1 \end{vmatrix} \right]$$
$$= 2(-1) + 0(-5) + 2(+2)$$
$$= -2 + 4$$
$$= 2$$

In each case, the result is 2. ∎

EXAMPLE 4   Evaluate $\begin{vmatrix} 0 & 0 & 2 & 0 \\ 1 & 2 & 17 & -3 \\ -1 & 0 & 28 & 1 \\ -2 & 2 & -37 & 1 \end{vmatrix}$.

Solution   Because row 1 contains three 0's, we expand the determinant along row 1. Then only one cofactor needs to be evaluated.

$$\begin{vmatrix} 0 & 0 & 2 & 0 \\ 1 & 2 & 17 & -3 \\ -1 & 0 & 28 & 1 \\ -2 & 2 & -37 & 1 \end{vmatrix} = 0\begin{vmatrix} ? \end{vmatrix} - 0\begin{vmatrix} ? \end{vmatrix} + 2\begin{vmatrix} 1 & 2 & -3 \\ -1 & 0 & 1 \\ -2 & 2 & 1 \end{vmatrix} - 0\begin{vmatrix} ? \end{vmatrix}$$
$$= 2(2) \qquad \text{See Example 3.}$$
$$= 4 \qquad\qquad\qquad\qquad ∎$$

Example 4 suggests the following theorem.

Zero Row or Column Theorem	If every entry in a row or column of a square matrix A is 0, then $$|A| = 0$$

Proof   If we evaluate a determinant along a row or column of 0's, the product of each entry and its cofactor will be 0. Since the value of the determinant is the sum of these products, its value will be 0. □

Many graphing calculators are able to find the determinant of a square matrix. If you own a graphing calculator with this capability, try evaluating many of the determinants in this section with your calculator.

■ Properties of Determinants

We have seen that there are elementary row operations for transforming matrices. There are similar row and column operations for transforming determinants.

Row and Column Operations for Determinants

Let A be a square matrix and k be a real number.

1. If a matrix B is obtained from matrix A by interchanging two rows (or columns), then $|B| = -|A|$.

2. If B is obtained from A by multiplying every element in a row (or column) of A by k, then $|B| = k|A|$.

3. If B is obtained from A by adding k times any row (or column) of A to another row (or column) of A, then $|B| = |A|$.

We will illustrate each row operation by showing that it is true for a 2×2 matrix. In the exercises, you will be asked to illustrate each column operation by showing that it is true for a 2×2 matrix.

$$\text{Let } |A| = \begin{vmatrix} a & b \\ c & d \end{vmatrix} = ad - bc.$$

Interchanging two rows. Let B be obtained from A by interchanging its two rows. Then

$$|B| = \begin{vmatrix} c & d \\ a & b \end{vmatrix} = cb - da = -(ad - bc) = -|A|$$

Multiplying every element in a row by k. Let B be obtained from A by multiplying its second row by k. Then

$$|B| = \begin{vmatrix} a & b \\ kc & kd \end{vmatrix} = akd - bkc = k(ad - bc) = k|A|$$

Adding k times any row of A to another row of A. Let B be obtained from A by adding k times its first row to its second row. Then

$$|B| = \begin{vmatrix} a & b \\ ka + c & kb + d \end{vmatrix} = a(kb + d) - b(ka + c)$$
$$= akb + ad - bka - bc = ad - bc = |A|$$

We will use several row and column operations in the next example.

EXAMPLE 5 Evaluate $\begin{vmatrix} 10 & 20 & -10 & 20 \\ 2 & 1 & 1 & 1 \\ 1 & 2 & -3 & 2 \\ 2 & -1 & -1 & 1 \end{vmatrix}$.

Solution To get smaller numbers in the first row, we can use a type 2 row operation and multiply each entry in the first row by $\frac{1}{10}$. However, we must then multiply the resulting determinant by 10 to retain the original value.

1.
$$\begin{vmatrix} 10 & 20 & -10 & 20 \\ 2 & 1 & 1 & 1 \\ 1 & 2 & -3 & 2 \\ 2 & -1 & -1 & 1 \end{vmatrix} \xrightarrow{\left(\frac{1}{10}\right)R1} = 10 \begin{vmatrix} 1 & 2 & -1 & 2 \\ 2 & 1 & 1 & 1 \\ 1 & 2 & -3 & 2 \\ 2 & -1 & -1 & 1 \end{vmatrix}$$

To get three 0's in the first row of the second determinant in Equation 1, we perform a type 3 row operation and expand the new determinant along its first row.

$$10 \begin{vmatrix} 1 & 2 & -1 & 2 \\ 2 & 1 & 1 & 1 \\ 1 & 2 & -3 & 2 \\ 2 & -1 & -1 & 1 \end{vmatrix} \xrightarrow{(-1)R3 + R1} = 10 \begin{vmatrix} 0 & 0 & 2 & 0 \\ 2 & 1 & 1 & 1 \\ 1 & 2 & -3 & 2 \\ 2 & -1 & -1 & 1 \end{vmatrix}$$

2.
$$= 10(2) \begin{vmatrix} 2 & 1 & 1 \\ 1 & 2 & 2 \\ 2 & -1 & 1 \end{vmatrix}$$

To introduce 0's into the 3×3 determinant in Equation 2, we perform a column operation on the 3×3 determinant and expand the result on its first column.

$$20 \begin{vmatrix} 2 & 1 & 1 \\ 1 & 2 & 2 \\ 2 & -1 & 1 \end{vmatrix} \overset{(-2)C3 + C1}{\underset{\downarrow}{=}} 20 \begin{vmatrix} 0 & 1 & 1 \\ -3 & 2 & 2 \\ 0 & -1 & 1 \end{vmatrix} = 20 \left[-(-3) \begin{vmatrix} 1 & 1 \\ -1 & 1 \end{vmatrix} \right]$$

$$= 20(3)[1 - (-1)] = 60(2) = 120 \qquad \blacksquare$$

We consider one final theorem.

Theorem	If A is a square matrix with two identical rows (or columns), then $$|A| = 0$$

Proof If the square matrix A has two identical rows (or columns), we can apply a type 3 row (or column) operation to "zero out" one of those rows (or columns). Since the matrix would then have an all-zero row (or column), its determinant would be 0, by the zero row or column theorem. $\qquad \square$

■ Using Determinants to Solve Systems of Equations

We can solve the general system $\begin{cases} ax + by = e \\ cx + dy = f \end{cases}$ by multiplying the first equation by d, multiplying the second equation by $-b$, and adding to get

$$\begin{array}{rcl} adx + bdy & = & ed \\ -bcx - bdy & = & -bf \\ \hline adx - bcx & = & ed - bf \end{array}$$

If $ad \neq bc$, we can solve the resulting equation for x:

$$adx - bcx = ed - bf$$
$$(ad - bc)x = ed - bf \qquad \text{Factor out } x.$$

3. $$x = \frac{ed - bf}{ad - bc} \qquad \text{Divide both sides by } ad - bc.$$

If $ad \neq bc$, we can also solve the system for y to get

4. $$y = \frac{af - ec}{ad - bc}$$

The values of x and y in Equations 3 and 4 can be expressed as determinants.

$$x = \frac{\begin{vmatrix} e & b \\ f & d \end{vmatrix}}{\begin{vmatrix} a & b \\ c & d \end{vmatrix}} = \frac{ed - bf}{ad - bc} \qquad y = \frac{\begin{vmatrix} a & e \\ c & f \end{vmatrix}}{\begin{vmatrix} a & b \\ c & d \end{vmatrix}} = \frac{af - ec}{ad - bc}$$

If we compare these formulas with the original system,

$$\begin{cases} ax + by = e \\ cx + dy = f \end{cases}$$

we see that the denominator determinants consist of the coefficients of the variables of each equation:

$$\text{Denominator determinants} = \begin{vmatrix} a & b \\ c & d \end{vmatrix}$$

To find the numerator determinant for x, we replace the first column of the denominator determinant with the column on the right-hand side of the equal sign.

To find the numerator determinant for y, we replace the second column of the denominator determinant with the column on the right-hand side of the equal sign.

$$x = \frac{\begin{vmatrix} e & b \\ f & d \end{vmatrix}}{\begin{vmatrix} a & b \\ c & d \end{vmatrix}} \qquad y = \frac{\begin{vmatrix} a & e \\ c & f \end{vmatrix}}{\begin{vmatrix} a & b \\ c & d \end{vmatrix}}$$

The method of using determinants to solve systems of equations is called **Cramer's rule**.

Cramer's Rule for Two Equations in Two Variables

If the system $\begin{cases} ax + by = e \\ cx + dy = f \end{cases}$ has a single solution, it is given by

$$x = \frac{D_x}{D} \quad \text{and} \quad y = \frac{D_y}{D}$$

where $D = \begin{vmatrix} a & b \\ c & d \end{vmatrix}$, $D_x = \begin{vmatrix} e & b \\ f & d \end{vmatrix}$, and $D_y = \begin{vmatrix} a & e \\ c & f \end{vmatrix}$

If D, D_x, and D_y are all 0, the system is consistent but the equations are dependent. If $D = 0$ and $D_x \neq 0$ or $D_y \neq 0$, the system is inconsistent.

EXAMPLE 6 Use determinants to solve the system $\begin{cases} 3x + 2y = 7 \\ -x + 5y = 9 \end{cases}$.

Solution

$$x = \frac{\begin{vmatrix} 7 & 2 \\ 9 & 5 \end{vmatrix}}{\begin{vmatrix} 3 & 2 \\ -1 & 5 \end{vmatrix}} = \frac{7 \cdot 5 - 2 \cdot 9}{3 \cdot 5 - 2(-1)} = \frac{35 - 18}{15 + 2} = \frac{17}{17} = 1$$

$$y = \frac{\begin{vmatrix} 3 & 7 \\ -1 & 9 \end{vmatrix}}{\begin{vmatrix} 3 & 2 \\ -1 & 5 \end{vmatrix}} = \frac{3 \cdot 9 - 7(-1)}{3 \cdot 5 - 2(-1)} = \frac{27 + 7}{15 + 2} = \frac{34}{17} = 2$$

Verify that the pair $(1, 2)$ satisfies both of the equations in the given system. ∎

We can use Cramer's rule to solve systems of n equations in n variables, where each equation has the form

$$a_1x_1 + a_2x_2 + \cdots + a_nx_n = c$$

To do so, we let D be the coefficient matrix of the system and let D_{x_i} be the matrix formed by replacing the ith column of D by the column of constants from the right of the equal signs. If $|D| \neq 0$, Cramer's rule provides the following unique solution.

Cramer's Rule for *n* Equations in *n* Variables

$$x_1 = \frac{|D_{x_1}|}{|D|}, \quad x_2 = \frac{|D_{x_2}|}{|D|}, \ldots, x_n = \frac{|D_{x_n}|}{|D|}$$

EXAMPLE 7 Use Cramer's rule to solve the system $\begin{cases} 2x - y + 2z = 3 \\ x - y + z = 2. \\ x + y + 2z = 3 \end{cases}$

Solution Each of the values x, y, and z is the quotient of two 3×3 determinants. The denominator of each quotient is the determinant consisting of the nine coefficients of the variables. The numerators for x, y, and z are modified copies of this denominator determinant. We substitute the column of constants for the coefficients of the variable for which we are solving.

$$\begin{cases} 2x - y + 2z = 3 \\ x - y + z = 2 \\ x + y + 2z = 3 \end{cases}$$

$$x = \frac{\begin{vmatrix} 3 & -1 & 2 \\ 2 & -1 & 1 \\ 3 & 1 & 2 \end{vmatrix}}{\begin{vmatrix} 2 & -1 & 2 \\ 1 & -1 & 1 \\ 1 & 1 & 2 \end{vmatrix}} = \frac{3\begin{vmatrix} -1 & 1 \\ 1 & 2 \end{vmatrix} - (-1)\begin{vmatrix} 2 & 1 \\ 3 & 2 \end{vmatrix} + 2\begin{vmatrix} 2 & -1 \\ 3 & 1 \end{vmatrix}}{2\begin{vmatrix} -1 & 1 \\ 1 & 2 \end{vmatrix} - (-1)\begin{vmatrix} 1 & 1 \\ 1 & 2 \end{vmatrix} + 2\begin{vmatrix} 1 & -1 \\ 1 & 1 \end{vmatrix}} = \frac{2}{-1} = -2$$

$$y = \frac{\begin{vmatrix} 2 & 3 & 2 \\ 1 & 2 & 1 \\ 1 & 3 & 2 \end{vmatrix}}{\begin{vmatrix} 2 & -1 & 2 \\ 1 & -1 & 1 \\ 1 & 1 & 2 \end{vmatrix}} = \frac{2\begin{vmatrix} 2 & 1 \\ 3 & 2 \end{vmatrix} - 3\begin{vmatrix} 1 & 1 \\ 1 & 2 \end{vmatrix} + 2\begin{vmatrix} 1 & 2 \\ 1 & 3 \end{vmatrix}}{-1} = \frac{1}{-1} = -1$$

$$z = \frac{\begin{vmatrix} 2 & -1 & 3 \\ 1 & -1 & 2 \\ 1 & 1 & 3 \end{vmatrix}}{\begin{vmatrix} 2 & -1 & 2 \\ 1 & -1 & 1 \\ 1 & 1 & 2 \end{vmatrix}} = \frac{2\begin{vmatrix} -1 & 2 \\ 1 & 3 \end{vmatrix} - (-1)\begin{vmatrix} 1 & 2 \\ 1 & 3 \end{vmatrix} + 3\begin{vmatrix} 1 & -1 \\ 1 & 1 \end{vmatrix}}{-1} = \frac{-3}{-1} = 3$$

Verify that the triple $(-2, -1, 3)$ satisfies each equation in the given system. ■

■ Writing Equations of Lines

If we are given the coordinates of two points in the xy-plane, we can use determinants to write the equation of a line.

Two-Point Form of the Equation of a Line	The equation of the line passing through points $P(x_1, y_1)$ and $Q(x_2, y_2)$ is given by
	$$\begin{vmatrix} x & y & 1 \\ x_1 & y_1 & 1 \\ x_2 & y_2 & 1 \end{vmatrix} = 0$$

EXAMPLE 8   Write the equation of the line passing through $P(-2, 3)$ and $Q(4, -5)$.

Solution   We set up the equation $\begin{vmatrix} x & y & 1 \\ -2 & 3 & 1 \\ 4 & -5 & 1 \end{vmatrix} = 0$ and expand along the first row to get

$$[3(1) - 1(-5)]\, x - [-2(1) - 1(4)]\, y + [(-2)(-5) - 3(4)]\, 1 = 0$$

$$8x + 6y - 2 = 0$$

Add 2 to both sides and divide both sides by 2. $4x + 3y = 1$

In general form, the equation of the line is $4x + 3y = 1$. Verify that the coordinates of each point satisfy the equation.   ∎

Finding Areas of Triangles

Area of a Triangle	If points $P(x_1, y_1)$, $Q(x_2, y_2)$, and $R(x_3, y_3)$ are the vertices of a triangle, the area of the triangle is given by
	$$A = \pm \frac{1}{2} \begin{vmatrix} x_1 & y_1 & 1 \\ x_2 & y_2 & 1 \\ x_3 & y_3 & 1 \end{vmatrix}$$ Pick either $+$ or $-$ to make the area positive.

EXAMPLE 9   Find the area of the triangle shown in Figure 7-7.

Solution   We set up the equation

$$A = \pm \frac{1}{2} \begin{vmatrix} 0 & 0 & 1 \\ 5 & 0 & 1 \\ 5 & 12 & 1 \end{vmatrix}$$

and expand the determinant along the first row to get

$$= \pm \frac{1}{2}\left[1 \begin{vmatrix} 5 & 0 \\ 5 & 12 \end{vmatrix} \right]$$

$$= \pm \frac{1}{2}(60 - 0)$$

$$= 30$$

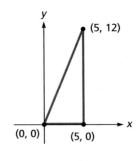

FIGURE 7-7

The area of the triangle is 30 square units. To check the answer, find the area by another method.   ∎

7.5 EXERCISES

In Exercises 1–4, evaluate each determinant.

1. $\begin{vmatrix} 2 & 1 \\ -2 & 3 \end{vmatrix}$

2. $\begin{vmatrix} -3 & -6 \\ 2 & -5 \end{vmatrix}$

3. $\begin{vmatrix} 2 & -3 \\ -3 & 5 \end{vmatrix}$

4. $\begin{vmatrix} 5 & 8 \\ -6 & -2 \end{vmatrix}$

*In Exercises 5–12, $A = \begin{vmatrix} 1 & -2 & 3 \\ 4 & 5 & -6 \\ -7 & 8 & 9 \end{vmatrix}$. Find each minor or cofactor. **Do not evaluate.***

5. M_{21}

6. M_{13}

7. M_{33}

8. M_{32}

9. C_{21}

10. C_{13}

11. C_{33}

12. M_{32}

In Exercises 13–24, evaluate each determinant.

13. $\begin{vmatrix} 2 & -3 & 5 \\ -2 & 1 & 3 \\ 1 & 3 & -2 \end{vmatrix}$

14. $\begin{vmatrix} 1 & 3 & 1 \\ -2 & 5 & 3 \\ 3 & -2 & -2 \end{vmatrix}$

15. $\begin{vmatrix} 1 & -1 & 2 \\ 2 & 1 & 3 \\ 1 & 1 & -1 \end{vmatrix}$

16. $\begin{vmatrix} 1 & 3 & 1 \\ 2 & 1 & -1 \\ 2 & -1 & 1 \end{vmatrix}$

17. $\begin{vmatrix} 2 & 1 & -1 \\ 1 & 3 & 5 \\ 2 & -5 & 3 \end{vmatrix}$

18. $\begin{vmatrix} 3 & 1 & -2 \\ -3 & 2 & 1 \\ 1 & 3 & 0 \end{vmatrix}$

19. $\begin{vmatrix} 0 & 1 & -3 \\ -3 & 5 & 2 \\ 2 & -5 & 3 \end{vmatrix}$

20. $\begin{vmatrix} 1 & -7 & -2 \\ -2 & 0 & 3 \\ -1 & 7 & 1 \end{vmatrix}$

21. $\begin{vmatrix} 0 & 0 & 1 & 0 \\ -2 & 1 & 0 & 1 \\ 1 & 0 & 1 & 2 \\ 2 & 0 & 1 & 2 \end{vmatrix}$

22. $\begin{vmatrix} 1 & 0 & -2 & 1 \\ 0 & 1 & 0 & 1 \\ 0 & 3 & -1 & 2 \\ 0 & -1 & 0 & 1 \end{vmatrix}$

23. $\begin{vmatrix} 1 & 2 & 1 & 3 \\ -2 & 1 & -3 & 1 \\ -1 & 0 & 1 & -2 \\ 2 & -1 & -1 & 3 \end{vmatrix}$

24. $\begin{vmatrix} -1 & 3 & -2 & 5 \\ 2 & 1 & 0 & 1 \\ 1 & 3 & -2 & 5 \\ 2 & -1 & 0 & -1 \end{vmatrix}$

In Exercises 25–28, decide whether each statement is true. Do not evaluate the determinants.

25. $\begin{vmatrix} 1 & 3 & -4 \\ -2 & 1 & 3 \\ 1 & 3 & 2 \end{vmatrix} = -\begin{vmatrix} -2 & 1 & 3 \\ 1 & 3 & -4 \\ 1 & 3 & 2 \end{vmatrix}$

26. $\begin{vmatrix} 4 & 6 & 8 \\ 10 & 5 & 15 \\ 20 & 5 & 10 \end{vmatrix} = \begin{vmatrix} 2 & 3 & 4 \\ 10 & 5 & 15 \\ 20 & 5 & 10 \end{vmatrix}$

27. $\begin{vmatrix} -2 & -3 & -4 \\ 5 & -1 & 2 \\ 1 & 2 & 3 \end{vmatrix} = \begin{vmatrix} 2 & 3 & 4 \\ -5 & 1 & -2 \\ 1 & 2 & 3 \end{vmatrix}$

28. $\begin{vmatrix} 1 & 2 & 3 \\ 4 & 5 & 6 \\ 7 & 8 & 9 \end{vmatrix} = \begin{vmatrix} 5 & 7 & 9 \\ 4 & 5 & 6 \\ 7 & 8 & 9 \end{vmatrix}$

In Exercises 29–32, assume that $\begin{vmatrix} a & b & c \\ d & e & f \\ g & h & i \end{vmatrix} = 3$ and find the value of each determinant.

29. $\begin{vmatrix} d & e & f \\ a & b & c \\ -g & -h & -i \end{vmatrix}$

30. $\begin{vmatrix} 5a & 5b & 5c \\ -d & -e & -f \\ 3g & 3h & 3i \end{vmatrix}$

31. $\begin{vmatrix} a+g & b+h & c+i \\ d & e & f \\ g & h & i \end{vmatrix}$

32. $\begin{vmatrix} g & h & i \\ a & b & c \\ d & e & f \end{vmatrix}$

In Exercises 33–44, use Cramer's rule to find the solution of each system of equations, if possible.

33. $\begin{cases} 3x + 2y = 7 \\ 2x - 3y = -4 \end{cases}$ **34.** $\begin{cases} x - 5y = -6 \\ 3x + 2y = -1 \end{cases}$ **35.** $\begin{cases} x - y = 3 \\ 3x - 7y = 9 \end{cases}$ **36.** $\begin{cases} 2x - y = -6 \\ x + y = 0 \end{cases}$

37. $\begin{cases} x + 2y + z = 2 \\ x - y + z = 2 \\ x + y + 3z = 4 \end{cases}$ **38.** $\begin{cases} x + 2y - z = -1 \\ 2x + y - z = 1 \\ x - 3y - 5z = 17 \end{cases}$ **39.** $\begin{cases} 2x - y + z = 5 \\ 3x - 3y + 2z = 10 \\ x + 3y + z = 0 \end{cases}$ **40.** $\begin{cases} x - y - z = 2 \\ x + y + z = 2 \\ -x - y + z = -4 \end{cases}$

41. $\begin{cases} \dfrac{x}{2} + \dfrac{y}{3} + \dfrac{z}{2} = 11 \\ \dfrac{x}{3} + y - \dfrac{z}{6} = 6 \\ \dfrac{x}{2} + \dfrac{y}{6} + z = 16 \end{cases}$ **42.** $\begin{cases} \dfrac{x}{2} + \dfrac{y}{5} + \dfrac{z}{3} = 17 \\ \dfrac{x}{5} + \dfrac{y}{2} + \dfrac{z}{5} = 32 \\ x + \dfrac{y}{3} + \dfrac{z}{2} = 30 \end{cases}$

43. $\begin{cases} 2p - q + 3r - s = 0 \\ p + q \quad - s = -1 \\ 3p \quad - r = 2 \\ p - 2q \quad + 3s = 7 \end{cases}$ **44.** $\begin{cases} a + b + c + d = 8 \\ a + b + c + 2d = 7 \\ a + b + 2c + 3d = 3 \\ a + 2b + 3c + 4d = 4 \end{cases}$

In Exercises 45–48, write the equation of the line that passes through the given points.

45. $P(0, 0)$, $Q(4, 6)$ **46.** $P(2, 3)$, $Q(6, 8)$ **47.** $P(-2, 3)$, $Q(5, -3)$ **48.** $P(1, -2)$, $Q(-4, 3)$

In Exercises 49–52, find the area of each triangle with vertices at the given points.

49. $P(0, 0)$, $Q(12, 0)$, $R(12, 5)$ **50.** $P(0, 0)$, $Q(0, 5)$, $R(12, 5)$

51. $P(2, 3)$, $Q(10, 8)$, $R(0, 20)$ **52.** $P(1, 1)$, $Q(6, 6)$, $R(2, 10)$

In Exercises 53–55, illustrate each column operation by showing that it is true for the determinant $\begin{vmatrix} a & b \\ c & d \end{vmatrix}$.

53. Interchanging two columns. **54.** Multiplying each element in a column by k.

55. Adding k times any column of A to another column of A.

56. Use the method of addition to solve $\begin{cases} ax + by = e \\ cx + dy = f \end{cases}$ for y, and thereby show that $y = \dfrac{af - ec}{ad - bc}$.

In Exercises 57–60, find each determinant. What do you discover?

57. $\begin{vmatrix} 1 & 3 & 4 \\ 0 & 5 & 2 \\ 0 & 0 & 2 \end{vmatrix}$ **58.** $\begin{vmatrix} 2 & 1 & -2 \\ 0 & 3 & 4 \\ 0 & 0 & -1 \end{vmatrix}$ **59.** $\begin{vmatrix} 1 & 2 & 4 & 3 \\ 0 & 2 & 2 & 1 \\ 0 & 0 & 3 & 2 \\ 0 & 0 & 0 & 4 \end{vmatrix}$ **60.** $\begin{vmatrix} 2 & 1 & -2 & 1 \\ 0 & 2 & 2 & -1 \\ 0 & 0 & 3 & 1 \\ 0 & 0 & 0 & 2 \end{vmatrix}$

In Exercises 61–64, expand the determinants and solve for x.

61. $\begin{vmatrix} 3 & x \\ 1 & 2 \end{vmatrix} = \begin{vmatrix} 2 & -1 \\ x & -5 \end{vmatrix}$ **62.** $\begin{vmatrix} 4 & x^2 \\ 1 & -1 \end{vmatrix} = \begin{vmatrix} x & 4 \\ 2 & 3 \end{vmatrix}$

63. $\begin{vmatrix} 3 & x & 1 \\ x & 0 & -2 \\ 4 & 0 & 1 \end{vmatrix} = \begin{vmatrix} 2 & x \\ x & 4 \end{vmatrix}$ **64.** $\begin{vmatrix} x & -1 & 2 \\ -2 & x & 3 \\ 4 & -3 & -1 \end{vmatrix} = \begin{vmatrix} 2 & 2 \\ 5 & x \end{vmatrix}$

65. Use an example chosen from 2×2 matrices to show that the determinant of the product of two matrices is the product of the determinants of those two matrices.

66. Find an example among 2×2 matrices to show that the determinant of a sum of two matrices is not equal to the sum of the determinants of those matrices.

67. A determinant is a function that associates a number with every square matrix. Give the domain and the range of that function.

68. Use an example chosen from 2×2 matrices to show that for $n \times n$ matrices A and B, $AB \neq BA$ but $|AB| = |BA|$.

69. If A and B are matrices and $|AB| = 0$, must $|A| = 0$ or $|B| = 0$? Support your answer.

70. If A and B are matrices and $|AB| = 0$, must $A = \mathbf{0}$ or $B = \mathbf{0}$? Support your answer.

71. For what value(s) of x does $\begin{vmatrix} x - 2 & -3 \\ -2 & x - 3 \end{vmatrix} = 0$?

72. Show that if $\det(A) \neq 0$, then the inverse of $\begin{bmatrix} a & b \\ c & d \end{bmatrix}$ is $\dfrac{1}{\det(A)} \begin{vmatrix} d & -b \\ -c & a \end{vmatrix}$.

 In Exercises 73–74, use a graphing calculator to evaluate each determinant.

73. $\begin{vmatrix} 2.3 & 5.7 & 6.1 \\ 3.4 & 6.2 & 8.3 \\ 5.8 & 8.2 & 9.2 \end{vmatrix}$

74. $\begin{vmatrix} .32 & -7.4 & -6.7 \\ 3.3 & 5.5 & -.27 \\ -8 & -.13 & 5.47 \end{vmatrix}$

7.6 PARTIAL FRACTIONS

■ When $Q(x)$ Has Distinct Linear Factors ■ When $Q(x)$ Has Repeated Linear Factors ■ When $Q(x)$ Has Distinct Quadratic Factors ■ When $Q(x)$ Has Repeated Quadratic Factors ■ When the Degree of $P(x)$ Is Equal to or Greater Than the Degree of $Q(x)$

*Johann Bernoulli
(1667–1748)
A Swiss mathematician and teacher of Euler. His method of decomposition by partial fractions was a major contribution to the calculus.*

In this section, we discuss how to break down fractions into sums of simpler fractions. We begin by reviewing how to add fractions. For example, to find the sum

$$\frac{2}{x} + \frac{6}{x + 1} + \frac{-1}{(x + 1)^2}$$

we write each fraction with an LCD of $x(x + 1)^2$, add the fractions by adding their numerators and keeping the common denominator, and simplify.

$$\frac{2}{x} + \frac{6}{x + 1} + \frac{-1}{(x + 1)^2} = \frac{2(x + 1)^2}{x(x + 1)^2} + \frac{6x(x + 1)}{(x + 1)x(x + 1)} + \frac{-1x}{(x + 1)^2 x}$$

$$= \frac{2x^2 + 4x + 2 + 6x^2 + 6x - x}{x(x + 1)^2}$$

$$= \frac{8x^2 + 9x + 2}{x(x + 1)^2}$$

To reverse the addition process and write a fraction as the sum of simpler fractions with denominators of smallest possible degree, we are **decomposing a fraction into partial fractions**. For example, to decompose the fraction

$$\frac{8x^2 + 9x + 2}{x(x + 1)^2}$$

into partial fractions, we will assume that there are constants A, B, and C such that

$$\frac{8x^2 + 9x + 2}{x(x + 1)^2} = \frac{A}{x} + \frac{B}{x + 1} + \frac{C}{(x + 1)^2}$$

After writing the terms on the right-hand side as fractions with an LCD of $x(x + 1)^2$, we add the fractions to get

$$\frac{8x^2 + 9x + 2}{x(x + 1)^2} = \frac{A(x + 1)^2}{x(x + 1)^2} + \frac{Bx(x + 1)}{x(x + 1)(x + 1)} + \frac{Cx}{(x + 1)^2 x}$$

$$\frac{8x^2 + 9x + 2}{x(x + 1)^2} = \frac{Ax^2 + 2Ax + A + Bx^2 + Bx + Cx}{x(x + 1)^2}$$

1. $\quad \dfrac{8x^2 + 9x + 2}{x(x + 1)^2} = \dfrac{(A + B)x^2 + (2A + B + C)x + A}{x(x + 1)^2} \qquad$ Factor x^2 from $Ax^2 + Bx^2$.
$\qquad\qquad\qquad\qquad\qquad\qquad\qquad\qquad\qquad\qquad\qquad\quad$ Factor x from $2Ax + Bx + Cx$.

Since the fractions on the left- and right-hand sides of Equation 1 are equal, the coefficients of their polynomial numerators are equal.

$$\begin{cases} A + B & = 8 & \text{The coefficients of } x^2. \\ 2A + B + C & = 9 & \text{The coefficients of } x. \\ A & = 2 & \text{The constants.} \end{cases}$$

We can solve this system to find that $A = 2$, $B = 6$, and $C = -1$. Hence,

$$\frac{8x^2 + 9x + 2}{x(x + 1)^2} = \frac{2}{x} + \frac{6}{x + 1} + \frac{-1}{(x + 1)^2}$$

Of course, these are the fractions we added at the beginning of the discussion.

The method of partial fractions is based on the following theorem, which is given without proof.

Polynomial Factorization Theorem	The factorization of any polynomial $Q(x)$ with real coefficients is the product of polynomials of the forms $\qquad (ax + b)^n \qquad$ and $\qquad (ax^2 + bx + c)^n$ where n is a positive integer and $ax^2 + bx + c$ is irreducible over the real numbers.

■ When $Q(x)$ Has Distinct Linear Factors

Let $\frac{P(x)}{Q(x)}$ be the quotient of two polynomials with real coefficients where the degree of $P(x)$ is less than the degree of $Q(x)$, and let the fraction $\frac{P(x)}{Q(x)}$ be in lowest terms. If $Q(x)$ has n distinct linear factors, the partial fraction decomposition of $\frac{P(x)}{Q(x)}$ is

$$\frac{P(x)}{Q(x)} = \frac{A_1}{a_1x + b_1} + \frac{A_2}{a_2x + b_2} + \frac{A_3}{a_3x + b_3} + \cdots + \frac{A_n}{a_nx + b_n}$$

for some constants $A_1, A_2, \ldots.$

EXAMPLE 1 Decompose $\dfrac{9x + 2}{(x + 2)(3x - 2)}$ into partial fractions.

Solution Since each factor in the denominator is linear, there are constants A and B such that

$$\frac{9x + 2}{(x + 2)(3x - 2)} = \frac{A}{x + 2} + \frac{B}{3x - 2}$$

After writing the terms on the right-hand side as fractions with an LCD of $(x + 2)(3x - 2)$, we add the fractions to get

$$\frac{9x + 2}{(x + 2)(3x - 2)} = \frac{A(3x - 2)}{(x + 2)(3x - 2)} + \frac{B(x + 2)}{(x + 2)(3x - 2)}$$

$$\frac{9x + 2}{(x + 2)(3x - 2)} = \frac{3Ax - 2A + Bx + 2B}{(x + 2)(3x - 2)}$$

2. $\dfrac{9x + 2}{(x + 2)(3x - 2)} = \dfrac{(3A + B)x - 2A + 2B}{(x + 2)(3x - 2)}$ Factor x from $3Ax + Bx.$

Since the fractions in Equation 2 are equal, the coefficients of their polynomial numerators are equal, and we have

$$\begin{cases} 9 = 3A + B & \text{The coefficients of } x. \\ 2 = -2A + 2B & \text{The constants.} \end{cases}$$

We can solve this system to find $A = 2$ and $B = 3$. Hence,

$$\frac{9x + 2}{(x + 2)(3x - 2)} = \frac{2}{x + 2} + \frac{3}{3x - 2}$$

Verify that the sum of $\dfrac{2}{x + 2}$ and $\dfrac{3}{3x - 2}$ is $\dfrac{9x + 2}{(x + 2)(3x - 2)}.$ ∎

When $Q(x)$ Has Repeated Linear Factors

If $Q(x)$ has n linear factors of $ax + b$, then $(ax + b)^n$ is a factor of $Q(x)$. Each factor of the form $(ax + b)^n$ generates the following sum of n partial fractions:

$$\frac{A_1}{ax + b} + \frac{A_2}{(ax + b)^2} + \frac{A_3}{(ax + b)^3} + \cdots + \frac{A_n}{(ax + b)^n}$$

EXAMPLE 2 Decompose $\dfrac{3x^2 - x + 1}{x(x - 1)^2}$ into partial fractions.

Solution In this example, each factor in the denominator is linear. The linear factor x appears once, and the linear factor $x - 1$ appears twice. Thus, there are constants A, B, and C such that

$$\frac{3x^2 - x + 1}{x(x - 1)^2} = \frac{A}{x} + \frac{B}{x - 1} + \frac{C}{(x - 1)^2}$$

After writing the terms on the right-hand side as fractions with an LCD of $x(x - 1)^2$, we combine them to get

$$\frac{3x^2 - x + 1}{x(x - 1)^2} = \frac{A(x - 1)^2}{x(x - 1)^2} + \frac{Bx(x - 1)}{x(x - 1)(x - 1)} + \frac{Cx}{(x - 1)^2 x}$$

$$= \frac{Ax^2 - 2Ax + A + Bx^2 - Bx + Cx}{x(x - 1)^2}$$

$$= \frac{(A + B)x^2 + (-2A - B + C)x + A}{x(x - 1)^2} \qquad \begin{array}{l} \text{Factor } x^2 \text{ from } Ax^2 + Bx^2. \\ \text{Factor } x \text{ from } -2Ax - Bx + Cx. \end{array}$$

Since the fractions are equal, the coefficients of the polynomial numerators are equal, and we have

$$\begin{cases} A + B & = 3 & \text{The coefficients of } x^2. \\ -2A - B + C = -1 & \text{The coefficients of } x. \\ A & = 1 & \text{The constants.} \end{cases}$$

We can solve this system to find that $A = 1$, $B = 2$, and $C = 3$. Hence,

$$\frac{3x^2 - x + 1}{x(x - 1)^2} = \frac{A}{x} + \frac{B}{x - 1} + \frac{C}{(x - 1)^2}$$

$$= \frac{1}{x} + \frac{2}{x - 1} + \frac{3}{(x - 1)^2} \qquad \blacksquare$$

■ When $Q(x)$ Has Distinct Quadratic Factors

If $Q(x)$ has an irreducible quadratic factor of the form $ax^2 + bx + c$, the partial fraction decomposition of $\frac{P(x)}{Q(x)}$ will have a corresponding term of the form

$$\frac{Bx + C}{ax^2 + bx + c}$$

EXAMPLE 3   Decompose $\dfrac{2x^2 + x + 1}{x^3 + x}$ into partial fractions.

Solution   Since the denominator can be written as a product of a linear factor and an irreducible quadratic factor, the partial fractions have the form

$$\frac{2x^2 + x + 1}{x(x^2 + 1)} = \frac{A}{x} + \frac{Bx + C}{x^2 + 1}$$

$$= \frac{A(x^2 + 1)}{x(x^2 + 1)} + \frac{(Bx + C)x}{x(x^2 + 1)}$$

$$= \frac{Ax^2 + A + Bx^2 + Cx}{x(x^2 + 1)}$$

$$= \frac{(A + B)x^2 + Cx + A}{x(x^2 + 1)} \qquad \text{Factor } x^2 \text{ from } Ax^2 + Bx^2.$$

We can equate the corresponding coefficients of the numerators $2x^2 + x + 1$ and $(A + B)x^2 + Cx + A$ to get the system

$$\begin{cases} A + B = 2 & \text{The coefficients of } x^2. \\ C = 1 & \text{The coefficients of } x. \\ A = 1 & \text{The constants.} \end{cases}$$

with solutions $A = 1$, $B = 1$, $C = 1$. Thus, the partial fraction decomposition is

$$\frac{2x^2 + x + 1}{x(x^2 + 1)} = \frac{A}{x} + \frac{Bx + C}{x^2 + 1}$$

$$= \frac{1}{x} + \frac{1x + 1}{x^2 + 1}$$

$$= \frac{1}{x} + \frac{x + 1}{x^2 + 1}$$

∎

■ When $Q(x)$ Has Repeated Quadratic Factors

If $Q(x)$ has n factors of $ax^2 + bx + c$, then $(ax^2 + bx + c)^n$ is a factor. Each factor of the form $(ax^2 + bx + c)^n$ generates a sum of n partial fractions of the form

$$\frac{B_1x + C_1}{ax^2 + bx + c} + \frac{B_2x + C_2}{(ax^2 + bx + c)^2} + \cdots + \frac{B_nx + C_n}{(ax^2 + bx + c)^n}$$

EXAMPLE 4   Find the partial fraction decomposition of $\dfrac{3x^2 + 5x + 5}{(x^2 + 1)^2}$.

Solution   Since the quadratic factor $x^2 + 1$ is used twice, we must find constants A, B, C, and D such that

$$\frac{3x^2 + 5x + 5}{(x^2 + 1)^2} = \frac{Ax + B}{x^2 + 1} + \frac{Cx + D}{(x^2 + 1)^2}$$

We add the fractions on the right-hand side to get

$$\frac{3x^2 + 5x + 5}{(x^2 + 1)^2} = \frac{(Ax + B)(x^2 + 1)}{(x^2 + 1)(x^2 + 1)} + \frac{Cx + D}{(x^2 + 1)^2}$$

$$= \frac{Ax^3 + Ax + Bx^2 + B + Cx + D}{(x^2 + 1)^2}$$

$$= \frac{Ax^3 + Bx^2 + (A + C)x + B + D}{(x^2 + 1)^2} \qquad \text{Factor } x \text{ from } Ax + Cx.$$

If we add the term $0x^3$ to the numerator on the left-hand side, we can equate the corresponding coefficients in the numerators to get

$$\begin{aligned} A &= 0 & \text{The coefficients of } x^3. \\ B &= 3 & \text{The coefficients of } x^2. \\ A + C &= 5 & \text{The coefficients of } x. \\ B + D &= 5 & \text{The constants.} \end{aligned}$$

with a solution of $A = 0$, $B = 3$, $C = 5$, and $D = 2$. Thus, the partial fraction decomposition is

$$\frac{3x^2 + 5x + 5}{(x^2 + 1)^2} = \frac{Ax + B}{x^2 + 1} + \frac{Cx + D}{(x^2 + 1)^2}$$

$$= \frac{0x + 3}{x^2 + 1} + \frac{5x + 2}{(x^2 + 1)^2}$$

$$= \frac{3}{x^2 + 1} + \frac{5x + 2}{(x^2 + 1)^2} \qquad \blacksquare$$

■ When the Degree of *P(x)* Is Equal to or Greater Than the Degree of *Q(x)*

When the degree of $P(x)$ is equal to or greater than the degree of $Q(x)$ in the fraction $\frac{P(x)}{Q(x)}$, we must first do a long division before decomposing the fraction into partial fractions. We illustrate this situation in the next example.

EXAMPLE 5   Decompose $\dfrac{x^2 + 4x + 2}{x^2 + x}$ into partial fractions.

Solution   Because the degree of the numerator and denominator are the same, we must do a long division and express the fraction in quotient $+ \frac{\text{remainder}}{\text{divisor}}$ form:

$$\begin{array}{r} 1 \\ x^2 + x \overline{)x^2 + 4x + 2} \\ \underline{x^2 + x} \\ 3x + 2 \end{array}$$

Hence,

3. $\dfrac{x^2 + 4x + 2}{x^2 + x} = 1 + \dfrac{3x + 2}{x^2 + x}$

Because the degree of the numerator of the fraction on the right-hand side of Equation 3 is less than the degree of the denominator, we can find its partial fraction decomposition:

$$\frac{3x + 2}{x^2 + x} = \frac{3x + 2}{x(x + 1)}$$

$$= \frac{A}{x} + \frac{B}{x + 1}$$

$$= \frac{A(x + 1) + Bx}{x(x + 1)}$$

$$= \frac{(A + B)x + A}{x(x + 1)}$$

We equate the corresponding coefficients in the numerator and solve the resulting system of equations to find that the solution is $A = 2, B = 1$. Thus,

$$\frac{x^2 + 4x + 2}{x^2 + x} = 1 + \frac{2}{x} + \frac{1}{x + 1}$$

∎

7.6   EXERCISES

In Exercises 1–44, decompose each fraction into partial fractions.

1. $\dfrac{3x - 1}{x(x - 1)}$

2. $\dfrac{4x + 6}{x(x + 2)}$

3. $\dfrac{2x - 15}{x(x - 3)}$

4. $\dfrac{5x + 21}{x(x + 7)}$

5. $\dfrac{3x + 1}{(x + 1)(x - 1)}$

6. $\dfrac{9x - 3}{(x + 1)(x - 2)}$

7. $\dfrac{-2x + 11}{x^2 - x - 6}$

8. $\dfrac{7x + 2}{x^2 + x - 2}$

9. $\dfrac{3x - 23}{x^2 + 2x - 3}$

10. $\dfrac{-x - 17}{x^2 - x - 6}$

11. $\dfrac{9x - 31}{2x^2 - 13x + 15}$

12. $\dfrac{-2x - 6}{3x^2 - 7x + 2}$

13. $\dfrac{5x^2 + 9x + 3}{x(x + 1)^2}$

14. $\dfrac{2x^2 - 7x + 2}{x(x - 1)^2}$

15. $\dfrac{-2x^2 + x - 2}{x^2(x - 1)}$

16. $\dfrac{x^2 + x + 1}{x^3}$

17. $\dfrac{4x^2 + 4x - 2}{x(x^2 - 1)}$

18. $\dfrac{x^2 - 6x - 13}{(x + 2)(x^2 - 1)}$

19. $\dfrac{3x^2 - 13x + 18}{x^3 - 6x^2 + 9x}$

20. $\dfrac{3x^2 + 13x + 20}{x^3 + 4x^2 + 4x}$

21. $\dfrac{x^2 - 2x - 3}{(x - 1)^3}$

22. $\dfrac{x^2 + 8x + 18}{(x + 3)^3}$

23. $\dfrac{x^2 + x + 3}{x(x^2 + 3)}$

24. $\dfrac{3x^2 + 8x + 11}{(x + 1)(x^2 + 2x + 3)}$

25. $\dfrac{5x^2 + 2x + 2}{x^3 + x}$

26. $\dfrac{x^3 + 4x^2 + 2x + 1}{x^4 + x^3 + x^2}$

27. $\dfrac{3x^3 + 5x^2 + 3x + 1}{x^2(x^2 + x + 1)}$

28. $\dfrac{4x^3 + 5x^2 + 3x + 4}{x^2(x^2 + 1)}$

29. $\dfrac{2x^2 + 1}{x^4 + x^2}$

30. $\dfrac{-3x^2 + x - 5}{(x + 1)(x^2 + 2)}$

31. $\dfrac{-x^2 - 3x - 5}{x^3 + x^2 + 2x + 2}$

32. $\dfrac{-2x^3 + 7x^2 + 6}{x^2(x^2 + 2)}$

33. $\dfrac{x^3 + 4x^2 + 3x + 6}{(x^2 + 2)(x^2 + x + 2)}$

34. $\dfrac{x^3 + 3x^2 + 2x + 4}{(x^2 + 1)(x^2 + x + 2)}$

35. $\dfrac{2x^4 + 6x^3 + 20x^2 + 22x + 25}{x(x^2 + 2x + 5)^2}$

36. $\dfrac{x^3 + 3x^2 + 6x + 6}{(x^2 + x + 5)(x^2 + 1)}$

37. $\dfrac{x^3}{x^2 + 3x + 2}$

38. $\dfrac{2x^3 + 6x^2 + 3x + 2}{x^3 + x^2}$

39. $\dfrac{3x^3 + 3x^2 + 6x + 4}{3x^3 + x^2 + 3x + 1}$

40. $\dfrac{x^4 + x^3 + 3x^2 + x + 4}{(x^2 + 1)^2}$

41. $\dfrac{x^3 + 3x^2 + 2x + 1}{x^3 + x^2 + x}$

42. $\dfrac{x^4 - x^3 + x^2 - x + 1}{(x^2 + 1)^2}$

43. $\dfrac{2x^4 + 2x^3 + 3x^2 - 1}{(x^2 - x)(x^2 + 1)}$

44. $\dfrac{x^4 - x^3 + 5x^2 + x + 6}{(x^2 + 3)(x^2 + 1)}$

7.7 GRAPHS OF LINEAR INEQUALITIES

■ Systems of Inequalities

The **graph of an inequality** in x and y is the graph of all ordered pairs (x, y) that satisfy the inequality. In this section, we will consider graphs of **linear inequalities**—inequalities that can be expressed in a form such as

$$Ax + By < C, \qquad Ax + By > C, \qquad Ax + By \leq C, \qquad \text{or} \qquad Ax + By \geq C$$

To graph the inequality $y > 3x + 2$, for example, we note that exactly one of the following statements is true:

$$y = 3x + 2, \qquad y < 3x + 2, \qquad \text{or} \qquad y > 3x + 2$$

The graph of $y = 3x + 2$ is a line, as shown in Figure 7-8(a). The graphs of the inequalities are half-planes, one on each side of that line. We can think of the graph of $y = 3x + 2$ as a boundary separating the two half-planes. The graph of $y = 3x + 2$ is drawn with a broken line to show that it is not part of the graph of $y > 3x + 2$.

To find which half-plane is the graph of $y > 3x + 2$, we can substitute the coordinates of any point on one side of the line (say, the origin $(0, 0)$) into the inequality and simplify:

$$y > 3x + 2$$
$$0 > 3(0) + 2$$
$$0 > 2$$

Since $0 > 2$ is false, the coordinates $(0, 0)$ do not satisfy the inequality, and the origin is not in the half-plane that is the graph of $y > 3x + 2$. Thus, the graph is the half-plane located on the other side of the broken line. The graph of the inequality $y > 3x + 2$ is shown in Figure 7-8(b).

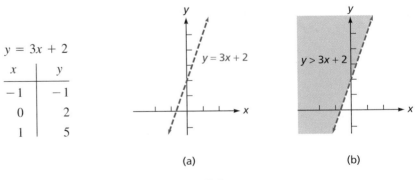

$y = 3x + 2$

x	y
-1	-1
0	2
1	5

(a)   (b)

FIGURE 7-8

EXAMPLE 1 Graph the inequality $2x - 3y \leq 6$.

Solution This inequality is the combination of the inequality $2x - 3y < 6$ and the equation $2x - 3y = 6$. We start by graphing the equation $2x - 3y = 6$ to establish the boundary that separates the two half-planes. However, this time we draw a solid line, because equality is permitted. See Figure 7-9(a). To decide which half-plane represents $2x - 3y \leq 6$, we check to see whether the coordinates of the origin satisfy the inequality:

$$2x - 3y \leq 6$$
$$2(0) - 3(0) \leq 6$$
$$0 \leq 6$$

Because $0 \leq 6$ is true, the origin lies in the graph of $2x - 3y \leq 6$. The complete graph is shown in Figure 7-9(b).

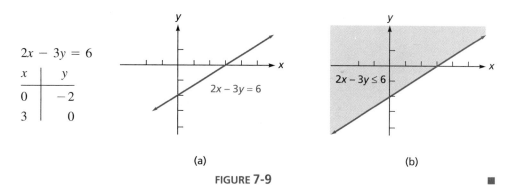

$2x - 3y = 6$

x	y
0	-2
3	0

(a)　　　　　　　　　　(b)

FIGURE 7-9

EXAMPLE 2 Graph the inequality $y < 2x$.

Solution Since the graph of $y = 2x$ is not part of the graph of the inequality, we graph the boundary with a broken line, as in Figure 7-10(a).

To decide which half-plane represents the graph of $y < 2x$, we check to see whether the coordinates of some fixed point satisfy the inequality. This time, however, we cannot use the origin as a test point, because the boundary line passes through the origin. Thus, we choose some other point—say (3, 1)—for a test point:

$$y < 2x$$
$$1 < 2(3)$$
$$1 < 6$$

Since $1 < 6$, the point (3, 1) lies in the graph, and we have the graph shown in Figure 7-10(b).

$$y = 2x$$

x	y
0	0
1	2

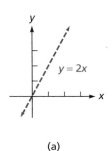

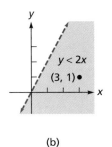

(a)   (b)

FIGURE 7-10

■

Systems of Inequalities

We now consider systems of inequalities. To graph the solution set of the system

$$\begin{cases} y < 5 \\ x \leq 6 \end{cases}$$

we graph each inequality on the same set of coordinate axes, as in Figure 7-11. The graph of the inequality $y < 5$ is the half-plane that lies below the line $y = 5$. The graph of the inequality $x \leq 6$ includes the half-plane that lies to the left of the line $x = 6$ together with the line $x = 6$.

The portion of the *xy*-plane where the two graphs intersect is the graph of the system. Any point that lies in the doubly-shaded region has coordinates that satisfy both inequalities in the system.

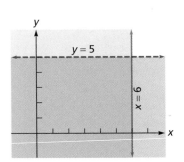

FIGURE 7-11

EXAMPLE 3 Graph the solution set of the system $\begin{cases} x + y \leq 1 \\ 2x - y > 2 \end{cases}$.

Solution On the same set of coordinate axes, we graph each inequality, as in Figure 7-12. The graph of $x + y \leq 1$ includes the line graph of the equation $x + y = 1$ and all points below it. Because the boundary line is included, we draw it as a solid line.

The graph of $2x - y > 2$ contains only those points below the line graph of the equation $2x - y = 2$. Because the boundary line is not included, we draw it as a broken line.

$x + y = 1$			$2x - y = 2$	
x	y		x	y
0	1		0	-2
1	0		1	0

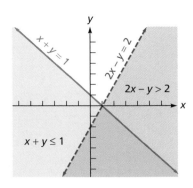

FIGURE 7-12

The area that is shaded twice represents the simultaneous solution of the system of inequalities. Any point in the doubly shaded region has coordinates that satisfy both inequalities. ∎

EXAMPLE 4 Graph the solution set of the system $\begin{cases} y < x^2 \\ y > \dfrac{x^2}{4} - 2 \end{cases}$.

Solution The graph of $y = x^2$ is a parabola opening upward with vertex at the origin, as shown in Figure 7-13. The points with coordinates that satisfy the inequality are the points below the parabola.

The graph of $y = \dfrac{x^2}{4} - 2$ is also a parabola opening upward. This time, however, the points that satisfy the inequality are the points above the parabola. Thus, the graph of the solution set is the shaded area between the two parabolas.

$y = x^2$	
x	y
0	0
1	1
−1	1
2	4
−2	4

$x = \dfrac{x^2}{4} - 2$	
x	y
0	−2
2	−1
−2	−1
4	2
−4	2

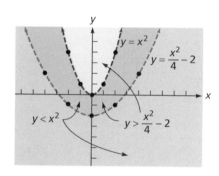

FIGURE 7-13 ∎

EXAMPLE 5 Graph the solution set of the system $\begin{cases} x + y \le 4 \\ x - y \le 6. \\ x \ge 0 \end{cases}$

Solution On the same set of coordinate axes, we graph each inequality, as in Figure 7-14. The graph of the inequality $x + y \le 4$ includes the line $x + y = 4$ and all points below it. Because the boundary line is included, we draw it as a solid line.

The graph of the inequality $x - y \le 6$ contains the line $x - y = 6$ and all points above it. The graph of the inequality $x \ge 0$ contains the y-axis and all points to the right of it. The solution of the system of inequalities is the shaded area in the figure.

The coordinates of the corner points of the shaded area are $(0, 4)$, $(0, -6)$, and $(5, -1)$.

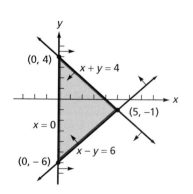

FIGURE 7-14

∎

EXAMPLE 6   Graph the solution set of the system $\begin{cases} x \geq 1 \\ y \geq x \\ 4x + 5y < 20 \end{cases}$.

Solution   The graph of the solution set of $x \geq 1$ includes those points on the graph of $x = 1$ and to the right. See Figure 7-15(a).

The graph of the solution set of $y \geq x$ includes those points on the graph of $y = x$ and above it. See Figure 7-15(b).

The graph of the solution set of $4x + 5y < 20$ includes those points below the graph of $4x + 5y = 20$. See Figure 7-15(c).

If these graphs are merged onto a single set of coordinate axes as in Figure 7-15(d), the graph of the original system of inequalities includes those points within the shaded triangle together with the points on the sides of the triangle drawn as solid lines. The coordinates of the corner points are $\left(1, \frac{16}{5}\right)$, $(1, 1)$ and $\left(\frac{20}{9}, \frac{20}{9}\right)$.

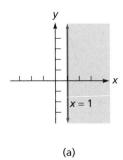

(a)

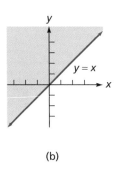

(b)

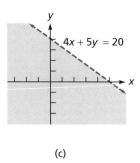

(c)

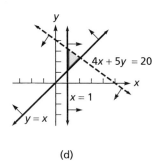

(d)

FIGURE 7-15

7.7   EXERCISES

In Exercises 1–12, graph each linear inequality.

1. $2x + 3y < 12$ **2.** $4x - 3y > 6$ **3.** $x < 3$ **4.** $y > -1$

5. $4x - y > 4$ **6.** $x - 2y < 5$ **7.** $y > 2x$ **8.** $y < 3x$

9. $y \leq \frac{1}{2}x + 1$ **10.** $y \geq \frac{1}{3}x - 1$ **11.** $2y \geq 3x - 2$ **12.** $3y \leq 2x + 3$

In Exercises 13–32, graph the solution set of each system.

13. $\begin{cases} y < 3 \\ x \geq 2 \end{cases}$ **14.** $\begin{cases} y \geq -2 \\ x < 0 \end{cases}$ **15.** $\begin{cases} y \geq 1 \\ x < 2 \end{cases}$ **16.** $\begin{cases} y \leq -1 \\ x > -1 \end{cases}$

17. $\begin{cases} y \leq x - 2 \\ y \geq 2x + 1 \end{cases}$ **18.** $\begin{cases} y < 3x + 2 \\ y < -2x + 3 \end{cases}$ **19.** $\begin{cases} x + y < 2 \\ x + y \leq 1 \end{cases}$ **20.** $\begin{cases} 3x + 2y > 6 \\ x + 3y \leq 2 \end{cases}$

21. $\begin{cases} x + 2y < 3 \\ 2x - 4y < 8 \end{cases}$ **22.** $\begin{cases} 3x + y \leq 1 \\ -x + 2y \geq 9 \end{cases}$ **23.** $\begin{cases} 2x - 3y \geq 6 \\ 3x + 2y < 6 \end{cases}$ **24.** $\begin{cases} 4x + 2y \leq 6 \\ 2x - 4y \geq 10 \end{cases}$

25. $\begin{cases} 2x - y \le 0 \\ x + 2y \le 10 \\ y \ge 0 \end{cases}$

26. $\begin{cases} 3x - 2y \ge 5 \\ 2x + y \ge 8 \\ x \le 5 \end{cases}$

27. $\begin{cases} x - 2y \ge 0 \\ x - y \le 2 \\ x \ge 0 \end{cases}$

28. $\begin{cases} 2x + 3y \le 6 \\ x - y \ge 4 \\ y \ge -4 \end{cases}$

29. $\begin{cases} x + y \le 4 \\ x - y \le 4 \\ x \ge 0 \\ y \ge 0 \end{cases}$

30. $\begin{cases} 2x + 3y \ge 12 \\ 2x - 3y \le 6 \\ x \ge 0 \\ y \le 4 \end{cases}$

31. $\begin{cases} 3x - 2y \le 6 \\ x + 2y \le 10 \\ x \ge 0 \\ y \ge 0 \end{cases}$

32. $\begin{cases} 3x + 2y \ge 12 \\ 5x - y \le 15 \\ x \ge 0 \\ y \le 4 \end{cases}$

7.8 LINEAR PROGRAMMING

■ Applications of Linear Programming

*Charles Babbage
(1792–1871)
In 1823, Babbage built a
steam-powered digital
calculator, which he called a
Difference Engine. Thought
to be a crackpot by his
London neighbors, Babbage
was a visionary. His machine
embodied principles still
used in modern computers.*

Systems of inequalities provide the basis for an area of applied mathematics known as **linear programming**. Linear programming helps answer questions such as "How can a business make as much money as possible?" or "How can I plan a nutritious menu at a school cafeteria at the least cost?"

Linear programming was developed during World War II when it became necessary to move huge quantities of men, materials, and supplies as efficiently and economically as possible. High-speed computers and linear programming routinely solve problems such as finding the best production and transportation schedules, efficiently routing telephone calls, and developing models of the world economy.

In such problems, the solution depends on certain **constraints**: The business has limited resources; the menu must contain sufficient vitamins and minerals; and so on. Any solution that satisfies the constraints is called a **feasible solution**. In linear programming, the constraints are expressed as a system of linear inequalities, and the quantity that is to be maximized (or minimized) is expressed as a linear function of several variables.

To solve a linear programming problem, we will maximize or minimize a function subject to constraints. For example, suppose that the profit P to be earned by a company is given by the function $P = y + 2x$, subject to the constraints

$$\begin{cases} x + y \ge 1 \\ x - y \le 1 \\ x - y \ge 0 \\ x \le 2 \end{cases}$$

We can find the solution to this system of inequalities and then find the coordinates of each corner of the region R shown in Figure 7-16(a). We then rewrite the equation

$$P = y + 2x$$

in the equivalent form

$$y = -2x + P$$

This is the equation for a set of parallel lines, each with a slope of -2 and a y-intercept of P. Many such lines pass through the region R. To decide which of these provides the maximum value of P, we refer to Figure 7-16(b) and locate the line with the greatest y-intercept. Since line l has the greatest y-intercept and crosses region R at the corner $(2, 2)$, the maximum value of P (subject to the given constraints) is

$$P = y + 2x = 2 + 2(2) = 6$$

Thus, the profit P has a maximum value of 6, subject to the given constraints. This profit occurs when $x = 2$ and $y = 2$.

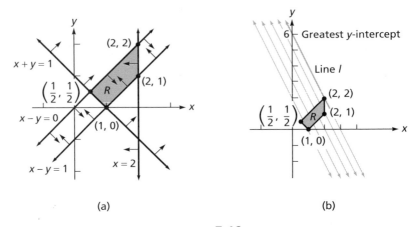

(a)   (b)

FIGURE 7-16

This example illustrates the following important theorem.

Theorem	If a linear function, subject to the constraints of a system of linear inequalities in two variables, attains a maximum or a minimum value, that value will occur at a corner or along an entire edge of the region R that represents the solution of the system.

In a linear programming problem, the function to be maximized (or minimized) is called the **objective function**. The solution region R for the set of constraints is called the **feasible region**.

EXAMPLE 1   If $P = 2x + 3y$, find the maximum value of P subject to the given constraints:

$$\begin{cases} x \geq 0 \\ y \geq 0 \\ x + y \leq 4 \\ 2x + y \leq 6 \end{cases}$$

Solution We solve the system of inequalities to find the feasible region R shown in Figure 7-17. The coordinates of its corners are $(0, 0)$, $(3, 0)$, $(0, 4)$, and $(2, 2)$.

Because the maximum value of P will occur at a corner of R, we substitute the coordinates of each corner point into the objective function

$$P = 2x + 3y$$

and find the one that gives the maximum value of P.

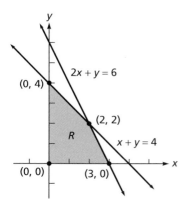

FIGURE 7-17

Point	$P = 2x + 3y$
$(0, 0)$	$P = 2(0) + 3(0) = 0$
$(3, 0)$	$P = 2(3) + 3(0) = 6$
$(2, 2)$	$P = 2(2) + 3(2) = 10$
$(0, 4)$	$P = 2(0) + 3(4) = 12$

The maximum value $P = 12$ occurs when $x = 0$ and $y = 4$. ■

EXAMPLE 2 If $P = 3x + 2y$, find the minimum value of P subject to the given constraints:

$$\begin{cases} x + y \geq 1 \\ x - y \leq 1 \\ x - y \geq 0 \\ x \leq 2 \end{cases}$$

Solution We refer to the feasible region shown in Figure 7-18, with corners at $\left(\frac{1}{2}, \frac{1}{2}\right)$, $(2, 2)$, $(2, 1)$ and $(1, 0)$. Because the minimum value of P occurs at a corner point of R, we substitute the coordinates of each corner point into the objective function

$$P = 3x + 2y$$

and find which one gives the minimum value of P.

Point	$P = 3x + 2y$
$\left(\frac{1}{2}, \frac{1}{2}\right)$	$P = 3\left(\frac{1}{2}\right) + 2\left(\frac{1}{2}\right) = \frac{5}{2}$
$(2, 2)$	$P = 3(2) + 2(2) = 10$
$(2, 1)$	$P = 3(2) + 2(1) = 8$
$(1, 0)$	$P = 3(1) + 2(2) = 3$

FIGURE 7-18

The minimum value $P = \frac{5}{2}$ occurs when $x = \frac{1}{2}$ and $y = \frac{1}{2}$. ■

EXAMPLE 3 An accountant earns a profit of $60 for preparing an individual tax return and $150 for preparing a commercial tax return. On average, an individual return requires 3 hours of the accountant's time, of which 1 hour is spent working on a computer. Each commercial return requires 4 hours of the accountant's time, of which 2 hours is spent working on a computer. If the accountant's time is limited to 240 hours and the available computer time is limited to 100 hours, how should the accountant divide his time between individual and commercial returns to maximize his income?

Solution   Suppose that x represents the number of individual returns completed and y represents the number of commercial returns completed. Because each of the x individual returns earns a profit of $60 and each of the y commercial returns earns a profit of $150, the profit function P is given by the equation

$$P = 60x + 150y$$

The following table provides the information about time requirements.

	Individual return	Commercial return	Time available
The accountant's time	3	4	240
Computer time	1	2	100

The profit is subject to the following constraints:

$$\begin{cases} x \geq 0 \\ y \geq 0 \\ 3x + 4y \leq 240 \\ x + 2y \leq 100 \end{cases}$$

The inequalities $x \geq 0$ and $y \geq 0$ indicate that the number of individual and commercial returns cannot be negative.

The inequality $3x + 4y \leq 240$ is a constraint on the accountant's time. Each of the x individual hours takes 3 hours of his time, and each of the y commercial returns takes 4 hours. The sum of these two amounts of time must be less than or equal to his available time, which is 240 hours.

The inequality $x + 2y \leq 100$ is a constraint on computer time. Each of the x individual returns takes 1 hour of computer time, and each of the y commercial returns takes 2 hours of computer time. The sum of x and $2y$ must be less than or equal to the available time, which is 100 hours.

We graph each of the constraints to find the feasible region R, as in Figure 7-19. The four corners of region R have coordinates of $(0, 0)$, $(80, 0)$, $(40, 30)$, and $(0, 50)$. We substitute the coordinates of each corner point into the profit function:

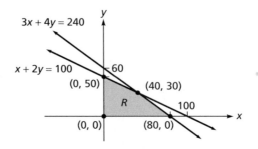

FIGURE 7-19

$$P = 60x + 150y$$

to find the maximum profit P.

Point	$P = 60x + 150y$
(0,0)	$P = 60(0) + 150(0) = 0$
(80,0)	$P = 60(80) + 150(0) = 4800$
(40,30)	$P = 60(40) + 150(30) = 6900$
(0,50)	$P = 60(0) + 150(50) = 7500$

The accountant will maximize his profit at $7500 if he prepares no individual returns and 50 commercial returns.   ■

Applications of Linear Programming

Linear programming has many applications in areas of *optimization*. Subject to constraints on time, energy, money, or other resources, it is often desirable to find an optimal or best solution to a problem. A profit or return is to be maximized; a cost or labor effort is to be minimized. We will look at a diet problem and a production schedule problem in this section.

A Diet Problem

EXAMPLE 4   Vitamix and Megamin are two diet supplements. Each Vitamix tablet contains 3 units of calcium, 20 units of Vitamin C, and 40 units of iron and costs $0.50. Each Megamin tablet contains 4 units of calcium, 40 units of Vitamin C, and 30 units of iron and costs $0.60. At least 24 units of calcium, 200 units of Vitamin C, and 120 units of iron are required for the daily needs of a particular patient. How many tablets of each supplement should be taken daily for a minimum cost? Find the daily minimum cost.

Solution   We can let x represent the number of Vitamix tablets to be taken daily and y represent the number of Megamin tablets to be taken daily. Next we write the objective function. Since each of the x Vitamix tablets costs $0.50, a total of $0.50x$ will be spent on Vitamix. The Megamin tablets cost $0.60y$. Thus, the objective function to be minimized is

$$C = 0.50x + 0.60y$$

We then construct a table displaying the calcium, Vitamin C, iron, and cost information.

	Vitamix	Megamin	Amount required
Calcium	3	4	24
Vitamin C	20	40	200
Iron	40	30	120
Cost	$0.50	$0.60	

Since cost is to be minimized, the cost function is the objective function.

$$C = 0.50x + 0.60y$$

We then write the constraints. Since there are requirements for calcium, Vitamin C, and iron, there is a constraint for each. Note that neither x nor y can be negative.

Calcium	$3x + 4y \geq 24$
Vitamin C	$20x + 40y \geq 200$
Iron	$40x + 30y \geq 120$
Nonnegative constraints	$x \geq 0, y \geq 0$

We graph the inequalities to find the feasible region and the coordinates of its corners, as in Figure 7-20.

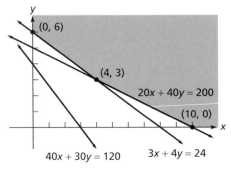

FIGURE 7-20

In this case, the feasible region is not bounded on all sides. The coordinates of the corner points are (0, 6), (4, 3), and (10, 0). To find the minimum cost, we substitute each pair of coordinates into the objective function, as in the following table.

Point	$C = 0.50x + 0.60y$
(0,6)	$C = 0.50(0) + 0.60(6) = 3.60$
(4,3)	$C = 0.50(4) + 0.60(3) = 3.80$
(10,0)	$C = 0.50(10) + 0.60(0) = 5.00$

A minimum cost will occur if no Vitamix and 6 Megamin tablets are taken daily. The minimum daily cost is $3.60. ∎

A Production Scheduling Problem

EXAMPLE 5   A television program director must schedule comedy skits and musical numbers for prime-time variety shows. Each comedy skit requires 2 hours of rehearsal time, costs $3000, and brings in $20,000 from the show's sponsors. Each musical number requires 1 hour of rehearsal time, costs $6000, and generates $12,000. If 250 hours are available for rehearsal and $600,000 is budgeted for comedy and music, how many segments of each type should be produced to maximize income? Find the maximum income.

Solution   We can let x represent the number of comedy skits and y the number of musical numbers to be scheduled. We then construct a table with information about rehearsal time, production cost, and income generated.

	Comedy	Musical	Available
Rehearsal time (hours)	2	1	250
Cost (in \$1000s)	3	6	600
Generated income (in \$1000s)	20	12	

Next we write the objective function. Since each of the x comedy skits generates 20 thousand dollars, the income generated by the comedy skits is $20x$ thousand dollars. The musical numbers produce $12y$ thousand dollars. The objective function to be maximized is

$$I = 20x + 12y$$

Then we write the constraints. Since there are limits on rehearsal time and budget, there is a constraint for each. Note that neither x nor y can be negative.

Constraint on rehearsal time	$2x + y \le 250$
Constraint on cost	$3x + 6y \le 600$
Nonnegative constraints	$x \ge 0, y \ge 0$

We graph the inequalities to find the feasible region shown in Figure 7-21 and find the coordinates of each corner point.

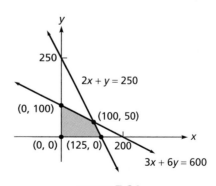

FIGURE 7-21

The coordinates of the corner points of the feasible region are $(0, 0)$, $(0, 100)$, $(100, 50)$, and $(125, 0)$. To find the maximum income, we substitute each pair of coordinates into the objective function, as in the following table.

Corner point	$I = 20x + 12y$
$(0, 0)$	$I = 20(0) + 12(0) = 0$
$(0, 100)$	$I = 20(0) + 12(100) = 1200$
$(100, 50)$	$I = 20(100) + 12(50) = 2600$
$(125, 0)$	$I = 20(125) + 12(0) = 2500$

Maximum income occurs if 100 comedy skits and 50 musical numbers are scheduled. The maximum income will be 2600 thousand dollars, or \$2,600,000. ∎

7.8 EXERCISES

In Exercises 1–8, maximize P subject to the following constraints.

1. $P = 2x + 3y$
$$\begin{cases} x \ge 0 \\ y \ge 0 \\ x + y \le 4 \end{cases}$$

2. $P = 3x + 2y$
$$\begin{cases} x \ge 0 \\ y \ge 0 \\ x + y \le 4 \end{cases}$$

3. $P = y + \dfrac{1}{2}x$
$$\begin{cases} x \ge 0 \\ y \ge 0 \\ 2y - x \le 1 \\ y - 2x \ge -2 \end{cases}$$

4. $P = 4y - x$
$$\begin{cases} x \le 2 \\ y \ge 0 \\ x + y \ge 1 \\ 2y - x \le 1 \end{cases}$$

5. $P = 2x + y$
$$\begin{cases} y \ge 0 \\ y - x \le 2 \\ 2x + 3y \le 6 \\ 3x + y \le 3 \end{cases}$$

6. $P = x - 2y$
$$\begin{cases} x + y \le 5 \\ y \le 3 \\ x \le 2 \\ x \ge 0 \\ y \ge 0 \end{cases}$$

7. $P = 3x - 2y$
$$\begin{cases} x \le 1 \\ x \ge -1 \\ y - x \le 1 \\ x - y \le 1 \end{cases}$$

8. $P = x - y$
$$\begin{cases} 5x + 4y \le 20 \\ y \le 5 \\ x \ge 0 \\ y \ge 0 \end{cases}$$

In Exercises 9–16, minimize P subject to the following constraints.

9. $P = 5x + 12y$
$$\begin{cases} x \ge 0 \\ y \ge 0 \\ x + y \le 4 \end{cases}$$

10. $P = 3x + 6y$
$$\begin{cases} x \ge 0 \\ y \ge 0 \\ x + y \le 4 \end{cases}$$

11. $P = 3y + x$
$$\begin{cases} x \ge 0 \\ y \ge 0 \\ 2y - x \le 1 \\ y - 2x \ge -2 \end{cases}$$

12. $P = 5y + x$
$$\begin{cases} x \le 2 \\ y \ge 0 \\ x + y \ge 1 \\ 2y - x \le 1 \end{cases}$$

13. $P = 6x + 2y$
$$\begin{cases} y \ge 0 \\ x \ge 0 \\ 2x + y \ge 3 \\ x + 2y \ge 3 \end{cases}$$

14. $P = 2y - x$
$$\begin{cases} x \ge 0 \\ y \ge 0 \\ x + y \le 5 \\ x + 2y \ge 2 \end{cases}$$

15. $P = 2x - 2y$
$$\begin{cases} x \le 1 \\ x \ge -1 \\ y - x \le 1 \\ x - y \le 1 \end{cases}$$

16. $P = 2x + 2y$
$$\begin{cases} x \ge 1 \\ y \ge 0 \\ 2x + y \ge 5 \\ x + 2y \ge 4 \end{cases}$$

In Exercises 17–28, write the objective function and the inequalities that describe the constraints in each problem. Graph the feasible region, showing corner points. Then find the maximum or minimum value of the objective function.

17. *Making furniture* Two woodworkers, Tom and Carlos, earn a profit of \$100 for making a table and \$80 for making a chair. On average, Tom must work 3 hours and Carlos 2 hours to make a chair. Tom must work 2 hours and Carlos 6 hours to make a table. If neither wishes to work more than 42 hours per week, how many tables and how many chairs should they make each week to maximize their profits?

18. **Making crafts** Two artists, Nina and Roberta, make winter yard ornaments. They receive a profit of $80 for each wooden snowman they make and $64 for each wooden Santa Claus. On average, Nina must work 4 hours and Roberta 2 hours to make a snowman. Nina must work 3 hours and Roberta 4 hours to make a Santa Claus. If neither wishes to work more than 20 hours per week, how many of each should they make each week to maximize their profits?

19. **Inventories** An electronics store manager stocks from 20 to 30 IBM-compatible computers and from 30 to 50 Macintosh computers. There is room in the store to stock up to 60 computers. The manager receives a commission of $50 on the sale of each IBM-compatible computer and $40 on the sale of each Macintosh computer. If the manager can sell all of the computers, how many should she stock to maximize her commissions?

20. **Food for a trip** Bill packs two foods for his camping trip. One ounce of food x costs 35¢ and provides 150 calories and 21 units of vitamins. One ounce of food y costs 27¢ and provides 60 calories and 42 units of vitamins. Every day Bill needs at least 3000 calories and at least 1260 units of vitamins. However, he does not want to carry more than 60 ounces of food for each day of his trip. How much of each type of food should he carry to minimize cost? Find the minimum cost per day.

21. **Nutrition** The owner of Fast-Food Burger Bar purchases hamburger from two suppliers. Supplier A's hamburger has 2 units of protein and 20 units of carbohydrate per ounce. Supplier B's hamburger has 8 units of protein and 40 units of carbohydrate per ounce. The mix should have at least 360 units of protein and 2000 units of carbohydrate. If A's hamburger is 10% fat and B's hamburger is 30% fat, how many ounces from each supplier should be used in order to minimize fat? Find the minimum fat content.

22. **Diet problem** A diet requires at least 16 units of vitamin C and at least 34 units of vitamin B complex. Two food supplements are available that provide these nutrients in the amounts and costs shown in the table. How much of each should be used to minimize the cost?

Supplement	Vitamin C	Vitamin B	Cost
A	3 units per gram	2 units per gram	3¢ per gram
B	2 units per gram	6 units per gram	4¢ per gram

23. **Production problem** Machines A and B are available 24 hours a day, 7 days a week, to manufacture chewing gum and bubble gum. To make a case of chewing gum, machine A must run 2 hours, and machine B must run 8 hours. To make a case of bubble gum, machine A must run 4 hours, and machine B must run 2 hours. Profit from chewing gum is $200 per case; profit from bubble gum is $160 per case. How many cases of each should be manufactured weekly to maximize profit? Find the maximum profit.

24. **Production problem** Manufacturing VCRs and TVs requires the use of the electronics, assembly, and finishing departments of a factory, according to the following schedule.

	Hours for VCR	Hours for TV	Hours available per week
Electronics	3	4	180
Assembly	2	3	120
Finishing	2	1	60

VCRs have a profit of $40, and TVs have a profit of $32. How many VCRs and TVs should be manufactured weekly to maximize profit? Find the maximum profit.

25. **Financial planning** A stockbroker has $200,000 to invest in stocks and bonds. She wants to invest at least $100,000 in stocks and at least $50,000 in bonds. If stocks have an annual yield of 9% and bonds have an annual yield of 7%, how much should she invest in each to maximize her interest? Find the maximum return.

26. **Production problem** A small country exports soybeans and flowers. Soybeans require 8 workers per acre, flowers require 12 workers per acre, and 100,000 workers are available. Government contracts demand that there be at least 3 times as much soybeans as flowers planted. It costs $250 per acre to plant soybeans and

$300 per acre to plant flowers, and there is a budget of $3 million. If the profit from soybeans is $1600 per acre and the profit from flowers is $2000 per acre, how many acres of each crop should be planted to maximize profit? Find the maximum profit.

27. Band trip A high school band trip will require renting buses and trucks to transport not less than 100 students and 18 or more large instruments. Each bus can accommodate 40 students plus 3 large instruments and costs $350 to rent. Each truck can accommodate 10 students plus 6 large instruments and costs $200 to rent. How many of each type of vehicle should be rented if the cost is minimum? Find the minimum cost.

28. Making ice cream An ice cream store sells two new flavors: Fantasy and Excess. Each barrel of Fantasy requires 4 pounds of nuts and 3 pounds of chocolate and has a profit of $500. Each barrel of Excess requires 4 pounds of nuts and 2 pounds of chocolate and has a profit of $400. There are 18 pounds of nuts and 16 pounds of chocolate in stock, and the owner does not want to buy more for this batch. How many barrels of each should be made for a maximum profit? Find the maximum profit.

---

7 CHAPTER SUMMARY

Key Words

additive identity matrix (7.3)
additive inverse of a matrix (7.3)
augmented matrix (7.2)
back substitution (7.2)
coefficient matrix (7.2)
cofactor of an element of a determinant (7.5)
consistent systems of equations (7.1)
Cramer's rule (7.5)
dependent equations (7.1)
determinant (7.5)

echelon form of a matrix (7.2)
elementary row operations (7.2)
equivalent systems (7.1)
expansion by minors (7.5)
Gaussian elimination (7.2)
Gauss–Jordan elimination (7.2)
identity matrix (7.3)
inconsistent system of equations (7.1)
inverse of a matrix (7.4)
independent equations (7.1)
linear inequalities (7.7)

linear programming (7.8)
m × n matrix (7.3)
minor of an element of a determinant (7.5)
nonsingular matrix (7.4)
partial fractions (7.6)
reduced row echelon form (7.2)
row equivalent matrices (7.2)
scalar (7.3)
singular matrix (7.4)
zero matrix (7.3)

Key Ideas

(7.1) Systems of equations can be solved by graphing, substitution, and addition.

(7.2) The three elementary row operations for matrices:

1. In a type 1 row operation, any two rows of a matrix are interchanged.
2. In a type 2 row operation, the elements of any row of a matrix are multiplied by any nonzero constant.
3. In a type 3 row operation, any row of a matrix is altered by adding a multiple of another row to it.

(7.3) Matrices are equal if and only if they are the same size and have the same corresponding entries.

Two $m \times n$ matrices are added by adding the corresponding elements of those matrices.

Two $m \times n$ matrices are subtracted by the rule $A - B = A + (-B)$.

The product AB of the $m \times n$ matrix A and the $n \times p$ matrix B is the $m \times p$ matrix C. The ith-row, jth-column entry of C is found by keeping a running total of the products of the elements in the ith row of A with the corresponding elements in the jth column of B.

Matrix addition is associative and commutative. Matrix multiplication is associative but not commutative.

(7.4) The inverse (if it exists) of an $n \times n$ matrix A can be found by using elementary row operations to transform the $n \times 2n$ matrix $[A|I]$ into $[I|A^{-1}]$, where I is an identity matrix. If A cannot be transformed into I, then A is not invertible.

If A is invertible, then the solution of $AX = B$ is $X = A^{-1}B$.

(7.5) The determinant of matrix $A = \begin{bmatrix} a & b \\ c & d \end{bmatrix}$ is

$\det(A) = |A| = ad - bc$.

The determinant of an $n \times n$ matrix A is the sum of the products of the elements of any row (or column) and the cofactors of those elements.

Cramer's rule. To solve a system $Ax = B$ of n independent equations in n unknowns, form the quotients of two determinants. The denominator is the determinant of the coefficient matrix, A. The numerator is the determinant of a modification coefficient matrix; when solving for the ith variable, replace the ith column of A with column of constants, B.

If a row (or column) of a matrix is multiplied by a constant k, the value of its determinant is multiplied by k.

If a row (or column) of a matrix is altered by adding a multiple of another row (or column) to it, the value of its determinant is unchanged.

(7.6) The fraction $\frac{P(x)}{Q(x)}$ can be written as the sum of several simpler fractions with denominators determined by the irreducible factors of $Q(x)$.

(7.7) Systems of inequalities in two variables can be solved by graphing. The solution is represented by a plane region with boundaries determined by graphing the inequalities as if they were equations.

(7.8) The maximum and the minimum values of a linear function in two variables, subject to the constraints of a system of linear inequalities, are attained at a vertex or along an entire edge of the region determined by the system of inequalities.

7 CHAPTER REVIEW EXERCISES

In Review Exercises 1–4, solve each system of equations by the method of graphing.

1. $\begin{cases} 2x - y = -1 \\ x + y = 7 \end{cases}$

2. $\begin{cases} 5x + 2y = 1 \\ 2x - y = -5 \end{cases}$

3. $\begin{cases} y = 5x + 7 \\ x = y - 7 \end{cases}$

4. $\begin{cases} x = y + 5 \\ y = -4 + \dfrac{x}{2} \end{cases}$

In Review Exercises 5–8, solve each system of equations by substitution.

5. $\begin{cases} y = 3x + 2 \\ y = 5x \end{cases}$

6. $\begin{cases} 2y + x = 0 \\ x = y + 3 \end{cases}$

7. $\begin{cases} 2x + y = -3 \\ x - y = 3 \end{cases}$

8. $\begin{cases} \dfrac{x + y}{2} + \dfrac{x - y}{3} = 1 \\ y = 3x - 2 \end{cases}$

In Review Exercises 9–12, solve each system of equations by addition.

9. $\begin{cases} x + 5y = 7 \\ 3x + y = -7 \end{cases}$

10. $\begin{cases} 2x + 3y = 11 \\ 3x - 7y = -41 \end{cases}$

11. $\begin{cases} 2(x + y) - x = 0 \\ 3(x + y) + 2y = 1 \end{cases}$

12. $\begin{cases} \dfrac{x + y}{2} + \dfrac{x - y}{3} = \dfrac{7}{2} \\ \dfrac{x + y}{5} + \dfrac{x - y}{2} = \dfrac{5}{2} \end{cases}$

In Review Exercises 13–16, solve each system of equations by any method.

13. $\begin{cases} 3x + 2y - z = 2 \\ x + y - z = 0 \\ 2x + 3y - z = 1 \end{cases}$

14. $\begin{cases} 5x - y + z = 3 \\ 3x + y + 2z = 2 \\ x + y = 2 \end{cases}$

15. $\begin{cases} 2x - y + z = 1 \\ x - y + 2z = 3 \\ x - y + z = 1 \end{cases}$

16. $\begin{cases} x + 2y - z = -6 \\ x + y - z = -4 \\ 3y - 2z = -12 \end{cases}$

17. *Planning inventory* The buyer for a large department store must order 40 coats. He is unsure of the expected sales; he must fill the order with imitation fur coats and leather coats. He can buy 25 fur coats and the rest leather for $9300, or 10 fur coats and the rest leather for $12,600. How much does he pay if he decides to split the order evenly?

18. *Ticket receipts* Adult tickets for the championship game are usually $5, but today is Seniors' Day, and senior citizens pay $4. Children's tickets are $2.50. Sales of 1800 tickets netted $7425, and children and seniors accounted for one-half of the tickets sold. How many of each were sold?

In Review Exercises 19–24, solve each system of equations by matrix methods, if possible.

19. $\begin{cases} 2x + 5y = 7 \\ 3x - y = 2 \end{cases}$

20. $\begin{cases} x + 3y - z = 8 \\ 2x + y - 2z = 11 \\ x - y + 5z = -8 \end{cases}$

21. $\begin{cases} x + 3y + z = 3 \\ 2x - y + z = -11 \\ 3x + 2y + 3z = 2 \end{cases}$

22. $\begin{cases} x + y + z = 4 \\ 3x - 2y - 2z = -3 \\ 4x - y - z = 0 \end{cases}$

23. $\begin{cases} w + x - 3y + z = 2 \\ x - y + 2z = 0 \\ w + y - z = 2 \\ w - x + 2z = -3 \end{cases}$

24. $\begin{cases} w + x + z = 3 \\ x - y + z = 3 \\ w + 2x - y + 2z = 6 \\ y - z = -2 \end{cases}$

In Review Exercises 25–34, perform the indicated matrix arithmetic, where possible.

25. $\begin{bmatrix} 3 & 2 & 1 \\ 3 & 2 & 1 \end{bmatrix} + \begin{bmatrix} -2 & 1 & 3 \\ 1 & -2 & 1 \end{bmatrix}$

26. $\begin{bmatrix} 2 & 3 & 5 \\ 1 & -2 & 4 \\ 2 & 1 & -2 \end{bmatrix} - \begin{bmatrix} 0 & -2 & 1 \\ 3 & 4 & -2 \\ 6 & -4 & 1 \end{bmatrix}$

27. $\begin{bmatrix} 2 & 3 \\ -1 & 2 \end{bmatrix} \begin{bmatrix} 1 & -2 \\ -3 & 1 \end{bmatrix}$

28. $\begin{bmatrix} -2 & 3 & 5 \\ 1 & -2 & -3 \end{bmatrix} \begin{bmatrix} 2 & 1 \\ -1 & 2 \\ -2 & 3 \end{bmatrix}$

29. $\begin{bmatrix} 1 & -3 & 2 \end{bmatrix} \begin{bmatrix} 2 \\ 1 \\ 3 \end{bmatrix}$

30. $\begin{bmatrix} 1 & -1 & -2 \\ 2 & -1 & 1 \end{bmatrix} \begin{bmatrix} 1 & 3 & -1 \\ 2 & -1 & 5 \\ 1 & -5 & 3 \end{bmatrix}$

31. $\begin{bmatrix} 1 \\ 2 \\ 1 \\ 5 \end{bmatrix} \begin{bmatrix} 2 & -1 & 1 & 3 \end{bmatrix}$

32. $\begin{bmatrix} 1 & -5 & 3 \\ 2 & 1 & -1 \end{bmatrix} \begin{bmatrix} 2 \\ -2 \\ 3 \end{bmatrix} \begin{bmatrix} 1 & -1 \\ -1 & 3 \end{bmatrix} \begin{bmatrix} 1 \\ -2 \end{bmatrix}$

33. $\begin{bmatrix} 1 & -3 & 2 \end{bmatrix} \begin{bmatrix} 2 \\ 1 \\ -5 \end{bmatrix} + \begin{bmatrix} 1 & -3 \end{bmatrix} \begin{bmatrix} 2 \\ 5 \end{bmatrix}$

34. $\left(\begin{bmatrix} 1 & -3 \\ 3 & 1 \end{bmatrix} + \begin{bmatrix} -1 & 3 \\ 1 & 1 \end{bmatrix} \right) \begin{bmatrix} 1 \\ -5 \end{bmatrix}$

In Review Exercises 35–42, find the inverse of each matrix, if possible.

35. $\begin{bmatrix} 1 & 3 \\ -3 & 5 \end{bmatrix}$

36. $\begin{bmatrix} 2 & 7 \\ 5 & 17 \end{bmatrix}$

37. $\begin{bmatrix} 1 & 3 & -5 \\ 1 & 4 & 4 \\ 0 & 0 & 1 \end{bmatrix}$

38. $\begin{bmatrix} 1 & 0 & 0 \\ 2 & 0 & -2 \\ 1 & 2 & 2 \end{bmatrix}$

39. $\begin{bmatrix} 1 & 0 & 8 \\ 3 & 7 & 6 \\ 1 & 2 & 3 \end{bmatrix}$ **40.** $\begin{bmatrix} 3 & 2 & 1 \\ -2 & -1 & 0 \\ 1 & 1 & 2 \end{bmatrix}$ **41.** $\begin{bmatrix} -1 & 1 & 0 \\ -2 & 1 & 0 \\ 3 & -1 & -1 \end{bmatrix}$ **42.** $\begin{bmatrix} 4 & 4 & 1 \\ 1 & 1 & 1 \\ -1 & -1 & 0 \end{bmatrix}$

In Review Exercises 43–44, use the inverse of the coefficient matrix to solve each system of equations.

43. $\begin{cases} 4x - y + 2z = 0 \\ x + y + 2z = 1 \\ x \quad\;\; + z = 0 \end{cases}$

44. $\begin{cases} w + 3x + y + 3z = 1 \\ w + 4x + y + 3z = 2 \\ x + y \quad\quad\; = 1 \\ w + 2x - y + 2z = 1 \end{cases}$

In Review Exercises 45–48, evaluate each determinant.

45. $\begin{vmatrix} 3 & -2 \\ 1 & -3 \end{vmatrix}$

46. $\begin{vmatrix} 1 & -2 & 3 \\ 2 & -1 & 3 \\ 1 & -1 & 0 \end{vmatrix}$

47. $\begin{vmatrix} 1 & 3 & -1 \\ 1 & 2 & 1 \\ 1 & 0 & 2 \end{vmatrix}$

48. $\begin{vmatrix} 1 & 2 & 3 & 4 \\ -1 & 3 & -3 & 2 \\ 0 & 0 & 0 & -1 \\ 3 & 3 & 4 & 3 \end{vmatrix}$

in Review Exercises 49–52, use Cramer's rule to solve each system of equations.

49. $\begin{cases} x + 3y = -5 \\ -2x + y = -4 \end{cases}$

50. $\begin{cases} x - y + z = -1 \\ 2x - y + 3z = -4 \\ x - 3y + z = -1 \end{cases}$

51. $\begin{cases} x - 3y + z = 7 \\ x + y - 3z = -9 \\ x + y + z = 3 \end{cases}$

52. $\begin{cases} w + x - y + z = 4 \\ 2w + x \quad\quad + z = 4 \\ x + 2y + z = 0 \\ w \quad\; + y + z = 2 \end{cases}$

In Review Exercises 53–56, decompose each fraction into partial fractions.

53. $\dfrac{7x + 3}{x^2 + x}$

54. $\dfrac{4x^3 + 3x + x^2 + 2}{x^4 + x^2}$

55. $\dfrac{x^2 + 5}{x^3 + x^2 + 5x}$

56. $\dfrac{x^2 + 1}{(x + 1)^3}$

In Review Exercises 57–60, maximize P subject to the given conditions.

57. $P = 2x + y$
$\begin{cases} x \geq 0 \\ y \geq 0 \\ x + y \leq 3 \end{cases}$

58. $P = 2x - 3y$
$\begin{cases} x \geq 0 \\ y \leq 3 \\ x - y \leq 4 \end{cases}$

59. $P = 3x - y$
$\begin{cases} y \geq 1 \\ y \leq 2 \\ y \leq 3x + 1 \\ x \leq 1 \end{cases}$

60. $P = y - 2x$
$\begin{cases} x + x \geq 1 \\ x \leq 1 \\ y \leq \dfrac{x}{2} + 2 \\ x + y \leq 2 \end{cases}$

61. A company manufactures two fertilizers, x and y. Each 50-pound bag of fertilizer requires three ingredients, which are available in the limited quantities shown below.

The profit on each bag of fertilizer x is $6, and on each bag of y, $5. How many bags of each product should be produced to maximize the profit?

Ingredient	Number of pounds in fertilizer x	Number of pounds in fertilizer y	Total number of pounds available
Nitrogen	6	10	20,000
Phosphorus	8	6	16,400
Potash	6	4	12,000

| **7** | **CHAPTER TEST** |

In Questions 1–2, solve each system of equations by the graphing method.

1. $\begin{cases} x - 3y = -5 \\ 2x - y = 0 \end{cases}$

2. $\begin{cases} x = 2y + 5 \\ y = 2x - 4 \end{cases}$

In Questions 3–4, solve each system of equations by the substitution or addition method.

3. $\begin{cases} 3x + y = 0 \\ 3x - 5y = 18 \end{cases}$

4. $\begin{cases} \dfrac{x + y}{2} + x = 7 \\ \dfrac{x - y}{2} - y = -6 \end{cases}$

5. *Mixing solutions* A chemist has two solutions, one 20% concentration and the other 45% concentration. How many liters of each must she mix to obtain 10 liters of 30% concentration?

6. *Wholesale distribution* Ace Electronics, Hi-Fi Stereo, and CD World buy a total of 175 cassette tape decks from the same distributor each month. Because CD World buys 25 more tape decks than the other two stores combined, CDW's cost is only $160 per unit. The decks cost Hi-Fi $165 each and cost Ace $170 each. How many decks does each retailer buy each month if the distributor receives $28,500 each month from the sale of tape decks to the three stores?

In Questions 7–8, write each system of equations as a matrix and solve it by Gaussian elimination.

7. $\begin{cases} 3x - 2y = 4 \\ 2x + 3y = 7 \end{cases}$

8. $\begin{cases} x + 3y - z = 6 \\ 2x - y - 2z = -2 \\ x + 2y + z = 6 \end{cases}$

In Questions 9–10, write each system of equations as a matrix and solve it by Gauss–Jordan elimination. If the system has infinitely many solutions, show how each can be determined.

9. $\begin{cases} x + 2y + z = 0 \\ 3x - 2y - 2z = 7 \\ 4x \quad - z = 7 \end{cases}$

10. $\begin{cases} x + 2y + 3z = -5 \\ 3x + y - 2z = 7 \\ y - z = 2 \end{cases}$

In Questions 11–12, perform the indicated operations.

11. $3\begin{bmatrix} 2 & -3 & 5 \\ 0 & 3 & -1 \end{bmatrix} - 5\begin{bmatrix} -2 & 1 & -1 \\ 0 & 3 & 2 \end{bmatrix}$

12. $[1 \quad 2 \quad 3]\begin{bmatrix} 2 & -2 \\ -2 & 2 \\ 1 & 0 \end{bmatrix}\begin{bmatrix} 3 \\ -2 \end{bmatrix}$

In Questions 13–14, find the inverse of each matrix, if possible.

13. $\begin{bmatrix} 5 & 19 \\ 2 & 7 \end{bmatrix}$

14. $\begin{bmatrix} -1 & 3 & -2 \\ 4 & 1 & 4 \\ 0 & 3 & -1 \end{bmatrix}$

In Questions 15–16, use the inverses found in Questions 13–14 to solve each system.

15. $\begin{cases} 5x + 19y = 3 \\ 2x + 7y = 2 \end{cases}$

16. $\begin{cases} -x + 3y - 2z = 1 \\ 4x + y + 4z = 3 \\ 3y - z = -1 \end{cases}$

In Questions 17–18, evaluate each determinant.

17. $\begin{vmatrix} 3 & -5 \\ -3 & 1 \end{vmatrix}$

18. $\begin{vmatrix} 3 & 5 & -1 \\ -2 & 3 & -2 \\ 1 & 5 & -3 \end{vmatrix}$

In Questions 19–20, use Cramer's rule to solve each system of equations.

19. $\begin{cases} 3x - 5y = 3 \\ -3x + y = 2 \end{cases}$

20. $\begin{cases} 3x + 5y - z = 2 \\ -2x + 3y - 2z = 1 \\ x + 5y - 3z = 0 \end{cases}$

In Questions 21–22, decompose each expression into partial fractions.

21. $\dfrac{5x}{(2x - 3)(x + 1)}$

22. $\dfrac{3x^2 + x + 2}{x^3 + 2x}$

In Questions 23–24, graph the solution set of each system.

23. $\begin{cases} x - 3y \geq 3 \\ x + 3y \leq 3 \end{cases}$

24. $\begin{cases} 3x + 4y \leq 12 \\ 3x + 4y \geq 6 \\ x \geq 0 \\ y \geq 0 \end{cases}$

25. Maximize $P = 3x + 2y$ subject to
$$\begin{cases} y \geq 0 \\ x \geq 0 \\ 2x + y \leq 4 \\ y \leq 2 \end{cases}$$

26. Minimize $P = y - x$ subject to
$$\begin{cases} x \geq 0 \\ y \geq 0 \\ x + y \leq 8 \\ 2x + y \geq 2 \end{cases}$$

CHAPTER 8

CONIC SECTIONS AND QUADRATIC SYSTEMS

The graphs of second-degree equations in x and y are figures that have interested people since the time of the ancient Greeks. However, the equations of these graphs were not carefully studied until the 17th century, when René Descartes (1596–1650) and Blaise Pascal (1623–1662) began investigating them.

General Form of a Second-Degree Equation in x and y	Let A, B, C, D, E, and F be real numbers. If at least one of A, B, and C is not 0, then $$Ax^2 + Bxy + Cy^2 + Dx + Ey + F = 0$$ is called the **general form of a second-degree equation in x and y**.

Descartes found that graphs of second-degree equations fell into one of several categories: a single point, a pair of lines, a circle, a parabola, an ellipse, a hyperbola, or no graph at all. These graphs are called **conic sections** because each is the intersection of a plane and a right-circular cone. (See Figure 8-1.)

Sonya Kovalevskaya (1850–1891)
This talented young Russian woman hoped to study mathematics at the University of Berlin, but strict rules prohibited women from attending lectures. Undaunted, Sonya studied privately with the great mathematician Karl Weierstrauss and published several important papers.

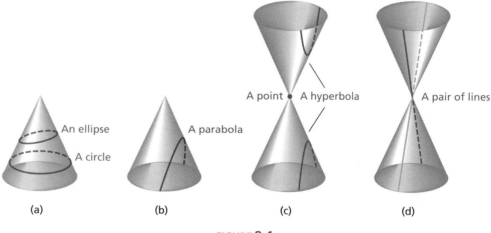

(a) (b) (c) (d)

FIGURE 8-1

8.1 THE CIRCLE AND THE PARABOLA

■ Parabolas   ■ Graphing Equations of Parabolas

In Section 3.4, we saw that the graph of any equation that can be written in the form
$$(x - h)^2 + (y - k)^2 = r^2$$

is a **circle** with **radius** r and **center** at point (h, k). We also saw that the equation $x^2 + y^2 = r^2$ has a graph that is a circle with radius r and center at the origin.

EXAMPLE 1   Find the equation of the circle shown in Figure 8-2.

Solution   To find the radius of the circle, we substitute 1 for x_1, 3 for y_1, -2 for x_2, and -1 for y_2 in the distance formula and simplify:

$$r = \sqrt{(x_2 - x_1)^2 + (y_2 - y_1)^2}$$
$$r = \sqrt{(-2 - 1)^2 + (-1 - 3)^2}$$
$$= \sqrt{(-3)^2 + (-4)^2}$$
$$= \sqrt{9 + 16}$$
$$= \sqrt{25}$$
$$= 5$$

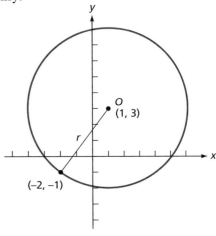

FIGURE 8-2

To find the equation of a circle with radius 5 and center at $(1, 3)$, we substitute 1 for h, 3 for k, and 5 for r in the standard equation of the circle.

$$(x - h)^2 + (y - k)^2 = r^2$$
$$(x - 1)^2 + (y - 3)^2 = 5^2$$

To write the equation in general form, we square the binomials and simplify.

$$x^2 - 2x + 1 + y^2 - 6y + 9 = 25$$
$$x^2 + y^2 - 2x - 6y - 15 = 0 \qquad \text{Subtract 25 from both sides}$$
and simplify.   ■

The final equation in Example 1 can be written as

$$1x^2 + 0xy + 1y^2 - 2x - 6y - 15 = 0$$

which illustrates that the graph of

$$Ax^2 + Bxy + Cy^2 + Dx + Ey + F = 0$$

is a circle whenever $B = 0$ and $A = C$.

EXAMPLE 2   Graph the circle whose equation is $2x^2 + 2y^2 + 4x + 3y = 3$.

Solution   To find the coordinates of the center and the radius, we must complete the square on both x and y. We begin by dividing both sides of the equation by 2 and rearranging terms to get

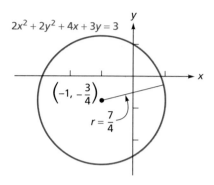

$2x^2 + 2y^2 + 4x + 3y = 3$

$\left(-1, -\dfrac{3}{4}\right)$

$r = \dfrac{7}{4}$

FIGURE 8-3

$$x^2 + 2x + y^2 + \frac{3}{2}y = \frac{3}{2}$$

$$x^2 + 2x + 1 + y^2 + \frac{3}{2}y + \frac{9}{16} = \frac{3}{2} + 1 + \frac{9}{16} \qquad \text{Add 1 and } \frac{9}{16} \text{ to both sides to complete the square.}$$

$$(x + 1)^2 + \left(y + \frac{3}{4}\right)^2 = \frac{49}{16} \qquad \begin{array}{l} \text{Factor } x^2 + 2x + 1 \text{ and} \\ y^2 + \frac{3}{2}y + \frac{9}{16}. \end{array}$$

$$[x - (-1)]^2 + \left[y - \left(-\frac{3}{4}\right)\right]^2 = \left(\frac{7}{4}\right)^2$$

From the equation, we see that the coordinates of the center of the circle are $h = -1$ and $k = -\frac{3}{4}$ and that the radius of the circle is $\frac{7}{4}$. The graph is shown in Figure 8-3. ∎

EXAMPLE 3 The effective broadcast area of a radio station is bounded by the circle $x^2 + y^2 = 2500$, where x and y are measured in miles. Another radio station's broadcast area is bounded by the circle $(x - 100)^2 + (y - 100)^2 = 900$. Is there any location that can receive both stations?

Solution It is possible to receive both stations only if their circular broadcast areas overlap. (See Figure 8-4.) This can happen only when the sum of the radii of the two circles is greater than the distance between their centers.

The center of the circle $x^2 + y^2 = 2500$ is $(0, 0)$ and its radius is 50 miles. The center of the circle $(x - 100)^2 + (y - 100)^2 = 900$ is $(100, 100)$, and its radius is 30 miles.

We can use the distance formula to find the distance between the centers.

$$d = \sqrt{(x_2 - x_1)^2 + (y_2 - y_1)^2}$$
$$d = \sqrt{(100 - 0)^2 + (100 - 0)^2}$$
$$= \sqrt{100^2 + 100^2}$$
$$= \sqrt{100^2 \cdot 2}$$
$$= 100\sqrt{2}$$
$$\approx 141 \text{ miles}$$

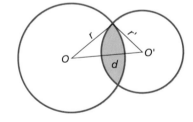

FIGURE 8-4

Since the sum of the radii of the two circles is 80 miles (50 miles + 30 miles), and this is less than the distance between their centers (141 miles), there is no location where both stations can be received. ∎

Parabolas

In Chapter 4, we discussed parabolas that opened upward and downward. These parabolas represent functions. We now discuss parabolas that open to the left and to the right and examine the properties of all parabolas in greater detail.

Parabolas

A **parabola** is the set of all points in a plane equidistant from a line *l*, called the **directrix**, and a fixed point *F*, called the **focus**.

The point on the parabola that is closest to the directrix is called the **vertex**, and the line passing through the vertex and the focus is called the **axis**. In this section, we consider parabolas that have a vertex at point (h, k) and open to the left, to the right, upward, and downward.

The parabola shown in Figure 8-5 opens to the right and passes through its vertex $V(h, k)$ and some point $P(x, y)$. Since each point is equidistant from the focus (point F) and the directrix, we can let $d(DV) = d(VF) = p$, where p is a positive constant.

Because of the geometry of the figure,

$$d(MP) = p - h + x$$

and by the distance formula,

$$d(PF) = \sqrt{[x - (p + h)]^2 + (y - k)^2}$$

By the definition of the parabola, $d(MP) = d(PF)$. Thus,

$$p - h + x = \sqrt{[x - (p + h)]^2 + (y - k)^2}$$
$$(p - h + x)^2 = [x - (p + h)]^2 + (y - k)^2 \qquad \text{Square both sides.}$$

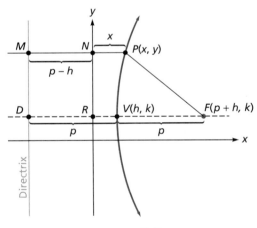

FIGURE 8-5

Finally, we expand the expression on each side of the equation and simplify:

$$p^2 - ph + px - ph + h^2 - hx + px - hx + x^2 = x^2 - 2px - 2hx + p^2 + 2ph + h^2 + (y - k)^2$$
$$-2ph + 2px = -2px + 2ph + (y - k)^2$$
$$4px - 4ph = (y - k)^2$$

1. $$4p(x - h) = (y - k)^2$$

Equation 1 is one of four **standard equations of a parabola**.

The Standard Equation of a Parabola with Vertex at V(h, k)	The **standard equation of a parabola** with vertex at $V(h, k)$ and opening to the right is

$$(y - k)^2 = 4p(x - h)$$

where p is the distance from the vertex to the focus.

If the parabola has its vertex at the origin, both h and k are 0, and we have the following theorem.

The Standard Equation of a Parabola with Vertex at the Origin	The standard equation of a parabola with vertex at the origin and opening to the right is

$$y^2 = 4px$$

where p is the distance from the vertex to the focus.

Equations of parabolas that open to the right, left, upward, and downward are summarized in Table 8-1. If $p > 0$, then

Parabola opening	Vertex at origin	Vertex at $V(h, k)$
Right	$y^2 = 4px$	$(y - k)^2 = 4p(x - h)$
Left	$y^2 = -4px$	$(y - k)^2 = -4p(x - h)$
Upward	$x^2 = 4py$	$(x - h)^2 = 4p(y - k)$
Downward	$x^2 = -4py$	$(x - h)^2 = -4p(y - k)$

TABLE **8-1**

EXAMPLE 4   Find the equation of the parabola with vertex at the origin and focus at (3, 0).

Solution   A sketch of the parabola is shown in Figure 8-6. Because the focus is to the right of the vertex, the parabola opens to the right, and because the vertex is the origin, the standard equation is $y^2 = 4px$. Furthermore, the distance between the focus and the vertex is 3, which is p. We can substitute 3 for p into the standard equation to get

$$y^2 = 4px$$
$$y^2 = 4(\mathbf{3})x$$
$$y^2 = 12x$$

The equation of the parabola is $y^2 = 12x$.   ■

y

F(3, 0)

$y^2 = 12x$

x

FIGURE **8-6**

EXAMPLE 5   Find the equation of the parabola that opens upward, has vertex at the point (4, 5), and passes through the point (0, 7).

Solution   Because the parabola opens upward, we use the equation $(x - h)^2 = 4p(y - k)$. Since $(h, k) = (4, 5)$ and the point (0, 7) is on the curve, we can substitute 4 for h, 5 for k, 0 for x, and 7 for y into the standard equation and solve for p.

$$(x - h)^2 = 4p(y - k)$$
$$(0 - 4)^2 = 4p(7 - 5)$$
$$16 = 8p$$
$$2 = p$$

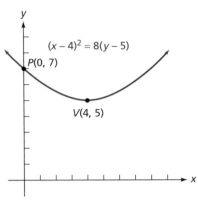

To find the equation of the parabola, we substitute 4 for h, 5 for k, and 2 for p in the standard equation and simplify:

$$(x - h)^2 = 4p(y - k)$$
$$(x - 4)^2 = 4 \cdot 2(y - 5)$$
$$(x - 4)^2 = 8(y - 5)$$

The graph of the equation appears in Figure 8-7.   **FIGURE 8-7**   ■

EXAMPLE 6   Find the equations of two parabolas with a vertex at $(2, 4)$ and passing through the point $(0, 0)$.

Solution   The two parabolas are shown in Figure 8-8.

Part 1. To find the parabola that opens to the left, we use the equation $(y - k)^2 = -4p(x - h)$. Since the curve passes through the point $(x, y) = (0, 0)$ and the vertex $(h, k) = (2, 4)$, we substitute 0 for x, 0 for y, 2 for h, and 4 for k into the standard equation and solve for p:

$$(y - k)^2 = -4p(x - h)$$
$$(0 - 4)^2 = -4p(0 - 2)$$
$$16 = 8p$$
$$2 = p$$

Since $h = 2, k = 4, p = 2$, and the parabola opens to the left, its equation is

$$(y - k)^2 = -4p(x - h)$$
$$(y - 4)^2 = -4(2)(x - 2)$$
$$(y - 4)^2 = -8(x - 2)$$

Part 2. To find the equation of the parabola that opens downward, we use the equation $(x - h)^2 = -4p(y - k)$ and substitute 2 for h, 4 for k, 0 for x, and 0 for y and solve for p:

$$(x - h)^2 = -4p(y - k)$$
$$(0 - 2)^2 = -4p(0 - 4)$$
$$4 = 16p$$
$$p = \frac{1}{4}$$

Norbert Wiener
(1894–1964)
A child prodigy, Norbert Wiener received his Ph.D. from Harvard University at the age of 19. As a professor of mathematics at M.I.T., Wiener analyzed the nature of information and communication, and created a new field called cybernetics. Without this study, modern computers would not exist.

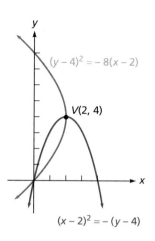

FIGURE **8-8**

Since $h = 2$, $k = 4$, and $p = \frac{1}{4}$, and the parabola opens downward, its equation is

$$(x - h)^2 = -4p(y - k)$$

$$(x - 2)^2 = -4\left(\frac{1}{4}\right)(y - 4) \qquad \text{Substitute 2 for } h, \frac{1}{4} \text{ for } p, \text{ and 4 for } k.$$

$$(x - 2)^2 = -(y - 4)$$ ■

■ **Graphing Equations of Parabolas**

EXAMPLE 7   Find the vertex and y-intercepts of the parabola $y^2 + 8x - 4y = 28$. Then graph the parabola.

Solution   We can complete the square on y to write the equation in standard form:

$$y^2 + 8x - 4y = 28$$
$$y^2 - 4y = -8x + 28 \qquad \text{Subtract } 8x \text{ from both sides.}$$
$$y^2 - 4y + \mathbf{4} = -8x + 28 + \mathbf{4} \qquad \text{Add 4 to both sides.}$$
$$\mathbf{2.} \qquad (y - 2)^2 = -8(x - 4) \qquad \text{Factor both sides.}$$

Equation 2 represents a parabola opening to the left with vertex at $(4, 2)$. To find the y-intercept, we substitute 0 for x in Equation 2 and solve for y.

$$(y - 2)^2 = -8(x - 4)$$
$$(y - 2)^2 = -8(\mathbf{0} - 4) \qquad \text{Substitute 0 for } x.$$
$$y^2 - 4y + 4 = 32 \qquad \text{Remove parentheses.}$$
$$\mathbf{3.} \qquad y^2 - 4y - 28 = 0 \qquad \text{Add } -32 \text{ to both sides.}$$

We can use the quadratic formula to find that the roots of Equation 3 are $y \approx 7.7$ and $y \approx -3.7$. Thus, the points with coordinates of approximately $(0, 7.7)$ and $(0, -3.7)$ lie on the graph of the parabola. We can use this information and the knowledge that the graph opens to the left and has a vertex at $(4, 2)$ to draw the curve as shown in Figure 8-9.   ■

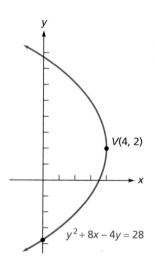

FIGURE **8-9**

EXAMPLE 8   If a stone is thrown straight up, the equation $s = 128t - 16t^2$ expresses its height in feet t seconds after it was thrown. Find the maximum height reached by the stone.

Solution   The graph of $s = 128t - 16t^2$, expressing the height of the stone t seconds after it was thrown, is the parabola shown in Figure 8-10. To find the maximum height

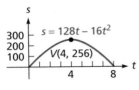

FIGURE 8-10

reached by the stone, we find the y-coordinate k of the vertex of the parabola. To find k, we write the equation of the parabola in standard form by completing the square on t.

$$s = 128t - 16t^2$$

$$16t^2 - 128t = -s \qquad \text{Multiply both sides by } -1.$$

$$t^2 - 8t = \frac{-s}{16} \qquad \text{Divide both sides by } 16.$$

$$t^2 - 8t + 16 = \frac{-s}{16} + 16 \qquad \text{Add 16 to both sides to complete the square.}$$

$$(t - 4)^2 = \frac{-s + 256}{16} \qquad \text{Factor } t^2 - 8t + 16 \text{ and combine terms.}$$

$$(t - 4)^2 = -\frac{1}{16}(s - 256) \qquad \text{Factor out } -\frac{1}{16}.$$

This equation indicates that the maximum height of 256 feet was reached in 4 seconds.

> **Warning!** The parabola shown in Figure 8-10 is not the path of the stone. The stone goes straight up and straight down. ∎

8.1 EXERCISES

In Exercises 1–6, write the equation of each circle.

1.
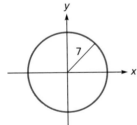

2.
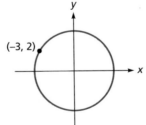

3.
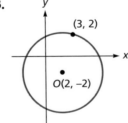

4.
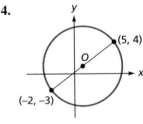

5. Radius of 6; center at the intersection of $3x + y = 1$ and $-2x - 3y = 4$

6. Radius of 8; center at the intersection of $x + 2y = 8$ and $2x - 3y = -5$

In Exercises 7–10, graph each equation.

7. $x^2 + y^2 = 4$

8. $x^2 - 2x + y^2 = 15$

9. $3x^2 + 3y^2 - 12x - 6y = 12$

10. $2x^2 + 2y^2 + 4x - 8y + 2 = 0$

11. Broadcast ranges   A television tower broadcasts a
signal with a circular range, as shown in Illustration 1.
Can a city 50 miles east and 70 miles north of the tower
receive the signal?

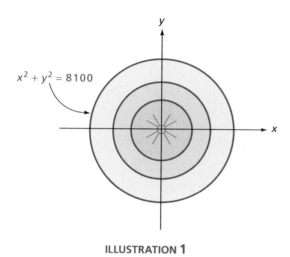

$x^2 + y^2 = 8100$

ILLUSTRATION 1

12. Warning siren   A tornado warning siren can be heard
in the circular range shown in Illustration 2. Can a
person 4 miles west and 5 miles south of the siren hear
its sound?

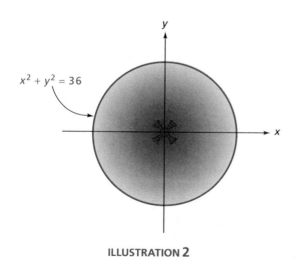

$x^2 + y^2 = 36$

ILLUSTRATION 2

13. Radio translators   Some radio stations extend their
broadcast range by installing a translator—a remote

device that receives a signal and retransmits it. A station
with a broadcast range given by $x^2 + y^2 = 1600$, where
x and y are in miles, installs a translator with a broadcast
area given by $x^2 + y^2 - 70y + 600 = 0$. Find the
greatest distance from the main transmitter that the signal can be received.

14. Ripples in a pond   When a stone is thrown into the
center of a pond, the ripples spread out in a circular
pattern, moving at a rate of 3 feet per second. If the
stone is dropped at the point $(0, 0)$ in Illustration 3,
when will the ripple reach the seagull floating at the
point $(15, 36)$?

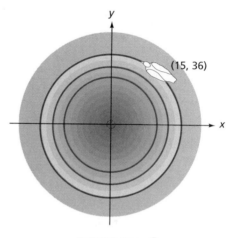

$(15, 36)$

ILLUSTRATION 3

15. Writing equations of circles   Find the equation of the
outer rim of the circular arch shown in Illustration 4.

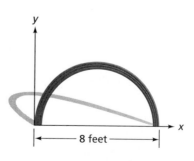

8 feet

ILLUSTRATION 4

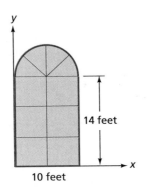

14 feet

10 feet

ILLUSTRATION 5

16. ***Writing equations of circles*** The shape of the window shown in Illustration 5 is a combination of a rectangle and a semicircle. Find the equation of the circle of which the semicircle is a part.

17. Write the equation of the circle passing through $(0, 8)$, $(5, 3)$, and $(4, 6)$.

18. Write the equation of the circle passing through $(-2, 0)$, $(2, 8)$, and $(5, -1)$.

In Exercises 19–34, find the equation of each parabola with the given properties.

19. Vertex at $(0, 0)$; focus at $(0, 3)$

20. Vertex at $(0, 0)$; focus at $(0, -3)$

21. Vertex at $(0, 0)$; focus at $(-3, 0)$

22. Vertex at $(0, 0)$; focus at $(3, 0)$

23. Vertex at $(3, 5)$; focus at $(3, 2)$

24. Vertex at $(3, 5)$; focus at $(-3, 5)$

25. Vertex at $(3, 5)$; focus at $(3, -2)$

26. Vertex at $(3, 5)$; focus at $(6, 5)$

27. Vertex at $(2, 2)$; passes through $(0, 0)$

28. Vertex at $(-2, -2)$; passes through $(0, 0)$

29. Vertex at $(-4, 6)$; passes through $(0, 3)$

30. Vertex at $(-2, 3)$; passes through $(0, -3)$

31. Vertex at $(6, 8)$; passes through $(5, 10)$ and $(5, 6)$

32. Vertex at $(2, 3)$; passes through $\left(1, \dfrac{13}{4}\right)$ and $\left(-1, \dfrac{21}{4}\right)$

33. Vertex at $(3, 1)$; passes through $(4, 3)$ and $(2, 3)$

34. Vertex at $(-4, -2)$; passes through $(-3, 0)$ and $\left(\dfrac{9}{4}, 3\right)$

In Exercises 35–44, change each equation to standard form and graph each parabola.

35. $y = x^2 + 4x + 5$

36. $2x^2 - 12x - 7y = 10$

37. $y^2 + 4x - 6y = -1$

38. $x^2 - 2y - 2x = -7$

39. $y^2 + 2x - 2y = 5$

40. $y^2 - 4y = -8x + 20$

41. $x^2 - 6y + 22 = -4x$

42. $4y^2 - 4y + 16x = 7$

43. $4x^2 - 4x + 32y = 47$

44. $4y^2 - 16x + 17 = 20y$

45. ***Solar furnaces*** A parabolic mirror collects rays of the sun and concentrates them at its focus. In Illustration 6, how far from the vertex of the parabolic mirror will it get the hottest? (Assume that all measurements are in feet.)

46. ***Searchlight reflectors*** A parabolic mirror reflects light in a beam when the light source is placed at its focus. In Illustration 7, how far from the vertex of the

parabolic reflector should the light source be placed? (Assume that all measurements are in feet.)

47. ***Writing equations of parabolas*** Derive the equation of the parabolic arch shown in Illustration 8.

48. ***Design of satellite antennas*** The cross section of the satellite antenna shown in Illustration 9 is a parabola with pickup at its focus. Find the distance d that the pickup is from the center of the dish.

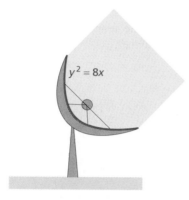

ILLUSTRATION 6

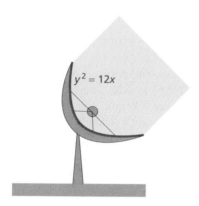

ILLUSTRATION 7

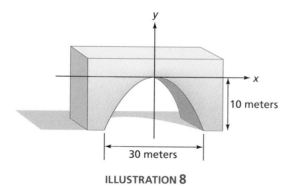

ILLUSTRATION 8

49. **Operating a resort**   A resort owner plans to build and rent n cabins for d dollars per week. The price, d, that she can charge for each cabin depends on the number of cabins she builds, where $d = -45\left(\frac{n}{32} - \frac{1}{2}\right)$. Find the

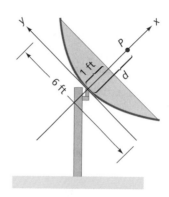

ILLUSTRATION 9

number of cabins she should build to maximize her weekly income.

50. **Toy rockets**   A toy rocket is s feet above the earth at the end of t seconds, where $s = -16t^2 + 80\sqrt{3}t$. Find the maximum height of the rocket.

51. **Design of parabolic reflectors**   Find the outer diameter (the length $\overline{AB}$) of the parabolic reflector shown in Illustration 10.

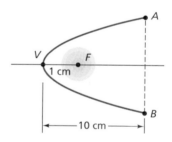

ILLUSTRATION 10

52. **Design of suspension bridges**   The cable between the towers of the suspension bridge shown in Illustration 11

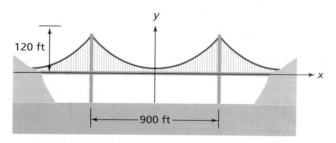

ILLUSTRATION 11

has the shape of a parabola with a vertex 15 feet above the roadway. Find the equation of the parabola.

53. **Gateway Arch** The Gateway Arch in Saint Louis has a shape that approximates a parabola. (See Illustration 12.) Find the width w of the arch 200 feet above the ground.

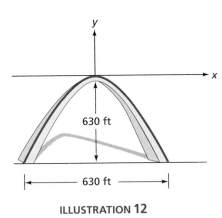

ILLUSTRATION **12**

54. **Building tunnels** A construction firm plans to build a tunnel whose arch is in the shape of a parabola. (See Illustration 13.) The tunnel will span a two-lane highway 8 meters wide. To allow safe passage for most vehi-

cles, the tunnel must be 5 meters high at a distance of 1 meter from the tunnel's edge. Find the maximum height of the tunnel.

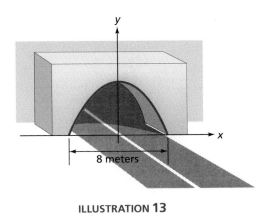

ILLUSTRATION **13**

55. **Ballistics** A stone tossed upward is s feet above the earth after t seconds, where $s = 16t^2 + 128t$. Show that the stone's height x seconds after it is thrown is equal to its height x seconds before it hits the ground.

56. Show that the standard form of the parabola $(y - 2)^2 = 8(x - 1)$ is a special case of the general form of the equation of second degree.

In Exercises 57–58, find the equation of the form $y = ax^2 + bx + c$ that determines a parabola passing through the given points.

57. $(1, 8), (-2, -1)$, and $(2, 15)$

58. $(1, -3), (-2, 12)$, and $(-1, 3)$

8.2 THE ELLIPSE

■ Graphing Equations of Ellipses

A third important conic is the ellipse.

Ellipse	An **ellipse** is the set of all points P in a plane such that the sum of the distances from P to two other fixed points F and F' is a positive constant.

In the ellipse shown in Figure 8-11, the fixed points F and F' are called **foci**, the midpoint of the **chord** FF' is called the **center**, and the chord VV' is called the **major axis**. Each of the endpoints V and V' of the major axis is called a **vertex**. The chord BB', perpendicular to the major axis and passing through the center C of the ellipse, is called the **minor axis**.

To derive the equation of the ellipse shown in Figure 8-12, we note that the origin O is the midpoint of chord FF' and let $d(OF) = d(OF') = c$, where $c > 0$. Then the coordinates of point F are $(c, 0)$, and the coordinates of F' are $(-c, 0)$. We also let $P(x, y)$ be any point on the ellipse.

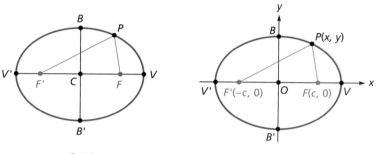

FIGURE **8-11**   FIGURE **8-12**

By the definition of an ellipse, $d(F'P) + d(PF)$ must be a positive constant, which we will call $2a$. Thus,

1. $d(F'P) + d(PF) = 2a$

We can use the distance formula to compute the lengths of $F'P$ and PF:

$$d(F'P) = \sqrt{[x - (-c)]^2 + y^2} \quad \text{and} \quad d(PF) = \sqrt{(x - c)^2 + y^2}$$

and substitute these values into Equation 1 to obtain

$$\sqrt{[x - (-c)]^2 + y^2} + \sqrt{(x - c)^2 + y^2} = 2a$$

or

2. $\sqrt{(x + c)^2 + y^2} = 2a - \sqrt{(x - c)^2 + y^2}$ Subtract $\sqrt{(x - c)^2 + y^2}$ from both sides.

We can square both sides of Equation 2 and simplify to get

$$(x + c)^2 + y^2 = 4a^2 - 4a\sqrt{(x - c)^2 + y^2} + [(x - c)^2 + y^2]$$
$$x^2 + 2cx + c^2 + y^2 = 4a^2 - 4a\sqrt{(x - c)^2 + y^2} + x^2 - 2cx + c^2 + y^2$$
$$4cx = 4a^2 - 4a\sqrt{(x - c)^2 + y^2}$$
$$cx = a^2 - a\sqrt{(x - c)^2 + y^2}$$
$$cx - a^2 = -a\sqrt{(x - c)^2 + y^2}$$

We square both sides again and simplify to get

$$c^2x^2 - 2a^2cx + a^4 = a^2[(x - c)^2 + y^2]$$
$$c^2x^2 - 2a^2cx + a^4 = a^2(x^2 - 2cx + c^2 + y^2)$$

$$c^2x^2 - 2a^2cx + a^4 = a^2x^2 - 2a^2cx + a^2c^2 + a^2y^2$$
$$c^2x^2 + a^4 = a^2x^2 + a^2c^2 + a^2y^2$$
$$a^4 - a^2c^2 = a^2x^2 - c^2x^2 + a^2y^2$$

3. $\qquad a^2(a^2 - c^2) = (a^2 - c^2)x^2 + a^2y^2$

Because the shortest distance between two points is a line segment, $d(F'P) + d(PF) > d(F'F)$. Therefore, $2a > 2c$. Thus, $a > c$, and $a^2 - c^2$ is a positive number, which we will call b^2. Letting $b^2 = a^2 - c^2$ and substituting into Equation 3, we have

4. $\qquad a^2b^2 = b^2x^2 + a^2y^2$

Dividing both sides of Equation 4 by a^2b^2 gives the standard equation for an ellipse with center at the origin and major axis on the x-axis:

$$\frac{x^2}{a^2} + \frac{y^2}{b^2} = 1 \qquad \text{where } a > b > 0$$

To find the coordinates of the vertices V and V', we substitute 0 for y and solve for x:

$$\frac{x^2}{a^2} + \frac{y^2}{b^2} = 1$$

$$\frac{x^2}{a^2} + \frac{0^2}{b^2} = 1$$

$$\frac{x^2}{a^2} = 1$$

$$x^2 = a^2$$

$$x = a \qquad \text{or} \qquad x = -a$$

Since the coordinates of V are $(a, 0)$ and the coordinates of V' are $(-a, 0)$, a is the distance between the center of the ellipse and either of its vertices. Thus, the center of the ellipse is the midpoint of the major axis.

To find the coordinates of B and B', we substitute 0 for x and solve for y:

$$\frac{x^2}{a^2} + \frac{y^2}{b^2} = 1$$

$$\frac{0^2}{a^2} + \frac{y^2}{b^2} = 1$$

$$y^2 = b^2$$

$$y = b \qquad \text{or} \qquad y = -b$$

Since the coordinates of B are $(0, b)$ and the coordinates of B' are $(0, -b)$, the distance between the center of the ellipse and either endpoint of the minor axis is b.

Theorem	The standard equation of an ellipse with center at the origin and major axis on the x-axis is

$$\frac{x^2}{a^2} + \frac{y^2}{b^2} = 1 \qquad \text{where } a > b > 0$$

If the major axis of an ellipse with center at $(0, 0)$ lies on the y-axis, the standard equation of the ellipse is

$$\frac{y^2}{a^2} + \frac{x^2}{b^2} = 1 \qquad \text{where } a > b > 0$$

In either case, the length of the major axis is $2a$, and the length of the minor axis is $2b$.

If we were to develop the equation of the ellipse with center at (h, k), we would obtain the following results.

Theorem	The standard equation of an ellipse with center at (h, k) and major axis parallel to the x-axis is

$$\frac{(x - h)^2}{a^2} + \frac{(y - k)^2}{b^2} = 1 \qquad \text{where } a > b > 0$$

If the major axis of an ellipse with center at (h, k) is parallel to the y-axis, the standard equation of the ellipse is

$$\frac{(y - k)^2}{a^2} + \frac{(x - h)^2}{b^2} = 1 \qquad \text{where } a > b > 0$$

In either case, the length of the major axis is $2a$, and the length of the minor axis is $2b$.

EXAMPLE 1   Find the equation of the ellipse with center at the origin, major axis of length 6 units located on the x-axis, and minor axis of length 4 units.

Solution   Since the center is the origin and the length of the major axis is 6, $a = 3$, and the coordinates of the vertices are $(3, 0)$ and $(-3, 0)$, as shown in Figure 8-13.

   Since the length of the minor axis is 4, $b = 2$, and the coordinates of B and B' are $(0, 2)$ and $(0, -2)$. To find the equation of the ellipse, we substitute 3 for a and 2 for b in the standard equation of an ellipse with center at the origin and major axis on the x-axis. We then simplify the equation.

$$\frac{x^2}{a^2} + \frac{y^2}{b^2} = 1$$

$$\frac{x^2}{3^2} + \frac{y^2}{2^2} = 1$$

$$\frac{x^2}{9} + \frac{y^2}{4} = 1$$

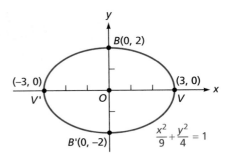

FIGURE 8-13   ∎

EXAMPLE 2   Find the equation of the ellipse with focus at $(0, 3)$ and vertices V and V' at $(3, 3)$ and $(-5, 3)$.

Solution   Since the midpoint of the major axis is the center of the ellipse, the coordinates of the center are $(-1, 3)$, as in Figure 8-14. Because the major axis is parallel to the x-axis, the standard equation to use is

$$\frac{(x - h)^2}{a^2} + \frac{(y - k)^2}{b^2} = 1 \qquad \text{where } a > b > 0$$

From Figure 8-14, we see that the distance between the center of the ellipse and either vertex is $a = 4$. We also see that the distance between the focus and the center is $c = 1$.

Since $b^2 = a^2 - c^2$ is an ellipse, we can substitute 4 for a and 1 for c and find b^2:

$$\begin{aligned} b^2 &= a^2 - c^2 \\ &= 4^2 - 1^2 \\ &= 15 \end{aligned}$$

To find the equation of the ellipse, we substitute -1 for h, 3 for k, 16 for a^2, and 15 for b^2 in the standard equation and simplify:

$$\frac{(x - h)^2}{a^2} + \frac{(y - k)^2}{b^2} = 1$$

$$\frac{[x - (-1)]^2}{16} + \frac{(y - 3)^2}{15} = 1$$

$$\frac{(x + 1)^2}{16} + \frac{(y - 3)^2}{15} = 1$$

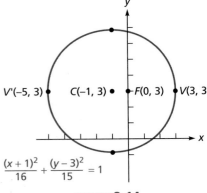

FIGURE 8-14

∎

EXAMPLE 3   The orbit of the earth is approximately an ellipse with the sun at one focus. The ratio of c to a (called the **eccentricity** of the ellipse) is about $\frac{1}{62}$, and the length of the major axis is approximately 186,000,000 miles. How close does the earth get to the sun?

Solution   We will assume that the ellipse has its center at the origin and vertices V' and V at $(-93{,}000{,}000, 0)$ and $(93{,}000{,}000, 0)$, as shown in Figure 8-15. Because the eccentricity, $\frac{c}{a}$, is given to be $\frac{1}{62}$ and $a = 93{,}000{,}000$, we have

$$\frac{c}{a} = \frac{1}{62}$$

$$c = \frac{1}{62}a$$

$$c = \frac{1}{62}(93{,}000{,}000)$$

$$= 1{,}500{,}000$$

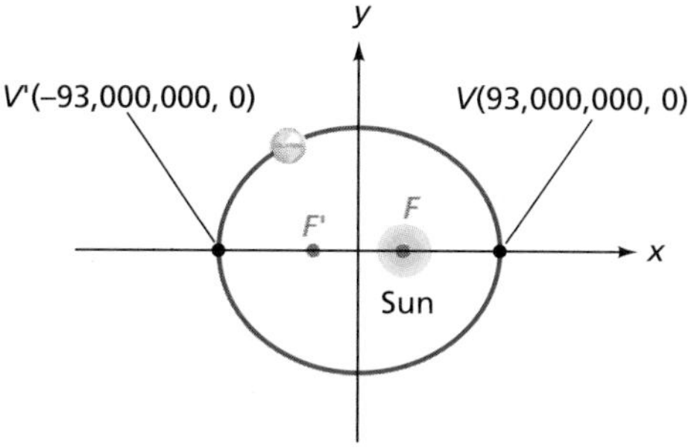

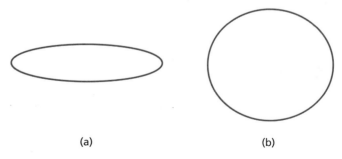

FIGURE **8-15**

The distance $d(FV)$ is the shortest distance between the earth and the sun. (You'll be asked to prove this in the exercises.) Thus,

$$d(FV) = a - c = 93{,}000{,}000 - 1{,}500{,}000 = 91{,}500{,}000 \text{ miles}$$

The earth's point of closest approach to the sun (called **perihelion**) is approximately 91.5 million miles.   ■

We can use the eccentricity of an ellipse to judge its shape. If the eccentricity is close to 1, the ellipse is relatively flat, as in Figure 8-16(a). If the eccentricity is close to 0, the ellipse is more circular, as in Figure 8-16(b). Since the eccentricity of the earth's orbit is $\frac{1}{62}$, the earth's orbit is almost a circle.

(a)   (b)

FIGURE **8-16**

■ Graphing Equations of Ellipses

EXAMPLE 4   Graph the ellipse $\dfrac{(x + 2)^2}{4} + \dfrac{(y - 2)^2}{9} = 1$.

Solution   The center is at $(-2, 2)$, and the major axis is parallel to the y-axis. Because $a = 3$, the vertices are 3 units above and below the center at points $(-2, 5)$ and $(-2, -1)$. Because $b = 2$, the endpoints of the minor axis are 2 units to the right and left of the center at points $(0, 2)$ and $(-4, 2)$. Using these points as guides, we can sketch the ellipse shown in Figure 8-17.

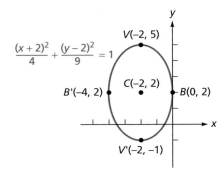

FIGURE 8-17   ∎

EXAMPLE 5   Graph the equation $4x^2 + 9y^2 - 16x - 18y = 11$.

Solution   We write the equation in standard form by completing the square on x and y:

$$4x^2 + 9y^2 - 16x - 18y = 11$$
$$4x^2 - 16x + 9y^2 - 18y = 11$$
$$4(x^2 - 4x) + 9(y^2 - 2y) = 11$$
$$4(x^2 - 4x + 4) + 9(y^2 - 2y + 1) = 11 + 16 + 9$$
$$4(x - 2)^2 + 9(y - 1)^2 = 36$$
$$\frac{(x - 2)^2}{9} + \frac{(y - 1)^2}{4} = 1 \qquad \text{Divide both sides by 36 to make the right-hand side equal to 1.}$$

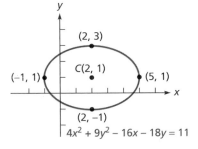

FIGURE 8-18

We can now see that the graph is an ellipse with center at $(2, 1)$ and major axis parallel to the x-axis. Because $a = 3$, the vertices are at $(-1, 1)$ and $(5, 1)$. Because $b = 2$, the endpoints of the minor axis are at $(2, -1)$ and $(2, 3)$. Using these points as guides, we can sketch the ellipse shown in Figure 8-18.   ∎

8.2   EXERCISES

In Exercises 1–6, write the equation of the ellipse that has its center at the origin.

1. Focus at $(3, 0)$; vertex at $(5, 0)$

2. Focus at $(0, 4)$; vertex at $(0, 7)$

3. Focus at $(0, 1)$; $\dfrac{4}{3}$ is one-half of the length of the minor axis.

4. Focus at $(1, 0)$; $\dfrac{4}{3}$ is one-half the length of the minor axis.

5. Focus at $(0, 3)$; major axis equal to 8

6. Focus at $(5, 0)$; major axis equal to 12

In Exercises 7–16, write the equation of each ellipse.

7. Center at $(3, 4)$; $a = 3$, $b = 2$; major axis parallel to the y-axis

8. Center at $(3, 4)$; passes through $(3, 10)$ and $(3, -2)$; $b = 2$

9. Center at $(3, 4)$; $a = 3$, $b = 2$; major axis parallel to the x-axis.

10. Center at $(3, 4)$; passes through $(8, 4)$ and $(-2, 4)$; $b = 2$

11. Foci at $(-2, 4)$ and $(8, 4)$; $b = 4$

12. Foci at $(8, 5)$ and $(4, 5)$; $b = 3$

13. Vertex at $(6, 4)$; foci at $(-4, 4)$ and $(4, 4)$

14. Center at $(-4, 5)$; $\dfrac{c}{a} = \dfrac{1}{3}$; vertex at $(-4, -1)$

15. Foci at $(6, 0)$ and $(-6, 0)$; $\dfrac{c}{a} = \dfrac{3}{5}$

16. Vertices at $(2, 0)$ and $(-2, 0)$; $\dfrac{2b^2}{a} = 2$

In Exercises 17–24, graph each ellipse.

17. $\dfrac{x^2}{25} + \dfrac{y^2}{49} = 1$

18. $4x^2 + y^2 = 4$

19. $\dfrac{x^2}{16} + \dfrac{(y + 2)^2}{36} = 1$

20. $(x - 1)^2 + \dfrac{4y^2}{25} = 4$

21. $x^2 + 4y^2 - 4x + 8y + 4 = 0$

22. $x^2 + 4y^2 - 2x - 16y = -13$

23. $16x^2 + 25y^2 - 160x - 200y + 400 = 0$

24. $3x^2 + 2y^2 + 7x - 6y = -1$

25. *Astronomy*   The moon has an orbit that is an ellipse with the earth at one focus. If the major axis of the orbit is 378,000 miles and the ratio of c to a is approximately $\frac{11}{200}$, how far does the moon get from the earth? (This farthest point in an orbit is called the **apogee**.)

26. *Equation of an arch*   An arch is a semiellipse 10 meters wide and 5 meters high. Write the equation of the ellipse if the ellipse is centered at the origin.

27. *Design of a track*   A track is built in the shape of an ellipse and has a maximum length of 100 meters and a maximum width of 60 meters. Write the equation of the ellipse and find its **focal width**. That is, find the length of a chord that is perpendicular to the major axis and that passes through either focus of the ellipse.

28. *Whispering galleries*   Any sound from one focus of an ellipse reflects off the ellipse directly back to the other focus. This property explains "whispering galleries" such as Statuary Hall in Washington, D.C. The whispering gallery shown in Illustration 1 has the shape of a semiellipse. Find the distance sound travels as it leaves focus F and returns to focus F'.

29. *Finding the width of a mirror*   The oval mirror shown in Illustration 2 is in the shape of an ellipse. Find the width of the mirror 12 inches above its base.

30. *Finding the height of a window*   The window shown in Illustration 3 has the shape of an ellipse. Find the height of the window 20 inches from one end.

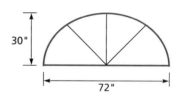

ILLUSTRATION 3

31. If F is a focus of the ellipse shown in Illustration 4 and B is an endpoint of the minor axis, use the distance formula to prove that the length of segment FB is a. (*Hint:* In an ellipse, $a^2 - c^2 = b^2$.)

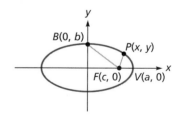

ILLUSTRATION 4

32. If F is a focus of the ellipse shown in Illustration 4 and P is any point on the ellipse, use the distance formula to show that the length of FP is $a - \frac{c}{a}x$. (*Hint:* In an ellipse, $a^2 - c^2 = b^2$.)

33. *Finding the focal width*   In the ellipse shown in Illustration 5, chord AA' passes through the focus F and is perpendicular to major axis. Show that the length of AA' (called the **focal width**) is $\frac{2b^2}{a}$.

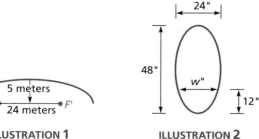

ILLUSTRATION 1

ILLUSTRATION 2

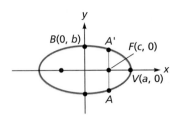

ILLUSTRATION 5

34. Prove that segment *FV* in Example 3 is the shortest distance between the earth and sun. (*Hint:* Refer to Exercise 33.)

35. *Constructing an ellipse* The ends of a piece of string 6 meters long are attached to two thumbtacks that are 2 meters apart. A pencil catches the loop and draws it tight. As the pencil is moved about the thumbtacks (always keeping the tension), an ellipse is produced with the thumbtacks as foci. Write the equation of the ellipse. (*Hint:* You'll have to establish a coordinate system.)

36. Prove that $a > b$ in the development of the standard equation of an ellipse.

37. Show that the expansion of the standard equation of an ellipse is a special case of the general second-degree equation.

38. The distance between point $P(x, y)$ and point $(0, 2)$ is $\frac{1}{3}$ of the distance of point P from the line $y = 18$. Find the equation of the curve on which point P lies.

8.3 THE HYPERBOLA

■ Asymptotes of a Hyperbola   ■ Graphing Equations of Hyperbolas

The definition of a hyperbola is similar to the definition of an ellipse, except that a constant difference of $2a$ is required instead of a constant sum.

Hyperbola	A **hyperbola** is the set of all points P in a plane such that the difference of the distances from point P to two other points in the plane is a positive constant.

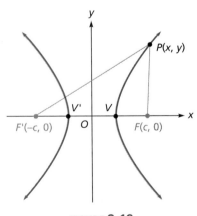

FIGURE 8-19

Points F and F' shown in Figure 8-19 are called the **foci** of the hyperbola, and the midpoint of chord FF' is called the **center**. The points V and V', where the hyperbola intersects FF', are called **vertices**. The segment VV' is called the **transverse axis**.

To develop the equation of the hyperbola centered at the origin, we note that the origin is the midpoint of chord FF' and let $d(F'O) = d(OF) = c$, where $c > 0$. Then F is at $(c, 0)$ and F' is at $(-c, 0)$. The definition of a hyperbola requires that $d(F'P) - d(PF) = 2a$, where $2a$ is a positive constant. Using the distance formula to compute the lengths of $F'P$ and PF gives

$$d(F'P) = \sqrt{[x - (-c)]^2 + y^2}$$
$$d(PF) = \sqrt{(x - c)^2 + y^2}$$

Substituting these values into the equation $d(F'P) - d(PF) = 2a$ gives

$$\sqrt{(x + c)^2 + y^2} - \sqrt{(x - c)^2 + y^2} = 2a$$

or

$$\sqrt{(x + c)^2 + y^2} = 2a + \sqrt{(x - c)^2 + y^2}$$

We square both sides and simplify to obtain

$$(x + c)^2 + y^2 = 4a^2 + 4a\sqrt{(x - c)^2 + y^2} + (x - c)^2 + y^2$$
$$x^2 + 2cx + c^2 + y^2 = 4a^2 + 4a\sqrt{(x - c)^2 + y^2} + x^2 - 2cx + c^2 + y^2$$
$$4cx = 4a^2 + 4a\sqrt{(x - c)^2 + y^2}$$
$$cx - a^2 = a\sqrt{(x - c)^2 + y^2}$$

Squaring both sides again and simplifying gives

$$c^2x^2 - 2a^2cx + a^4 = a^2(x^2 - 2cx + c^2 + y^2)$$
$$c^2x^2 - 2a^2cx + a^4 = a^2x^2 - 2a^2cx + a^2c^2 + a^2y^2$$
$$c^2x^2 + a^4 = a^2x^2 + a^2c^2 + a^2y^2$$

1. $$(c^2 - a^2)x^2 - a^2y^2 = a^2(c^2 - a^2)$$

Because $c > a$ (you will be asked to prove this in the exercises), $c^2 - a^2$ is a positive number. Thus, we can let $b^2 = c^2 - a^2$ and substitute b^2 for $c^2 - a^2$ in Equation 1 to get

$$b^2x^2 - a^2y^2 = a^2b^2$$

Dividing both sides of the previous equation by a^2b^2 gives the standard equation for a hyperbola with center at the origin and foci on the x-axis:

$$\frac{x^2}{a^2} - \frac{y^2}{b^2} = 1$$

If $y = 0$, the previous equation becomes

$$\frac{x^2}{a^2} = 1 \quad \text{or} \quad x^2 = a^2$$

Solving this equation for x gives

$$x = a \quad \text{or} \quad x = -a$$

Because the coordinates of V and V' are $(a, 0)$ and $(-a, 0)$, the distance between the center of the hyperbola and either vertex is a, and the center of the hyperbola is the midpoint of the segment $V'V$ as well as of the segment FF'.

If $x = 0$, the equation becomes

$$\frac{-y^2}{b^2} = 1 \quad \text{or} \quad y^2 = -b^2$$

Since this equation has no real solutions, the hyperbola cannot intersect the y-axis.
    These results suggest the following theorem.

Theorem

The standard equation of a hyperbola with center at the origin and foci on the x-axis is

$$\frac{x^2}{a^2} - \frac{y^2}{b^2} = 1$$

The standard equation of a hyperbola with center at the origin and foci on the y-axis is

$$\frac{y^2}{a^2} - \frac{x^2}{b^2} = 1$$

In either case, the distance between the vertices is $2a$.

The standard equation of the hyperbola with center at (h, k) can be developed with the following results.

Theorem

The standard equation of a hyperbola with center at (h, k) and foci on a line parallel to the x-axis is

$$\frac{(x - h)^2}{a^2} - \frac{(y - k)^2}{b^2} = 1$$

The standard equation of a hyperbola with center at (h, k) and foci on a line parallel to the y-axis is

$$\frac{(y - k)^2}{a^2} - \frac{(x - h)^2}{b^2} = 1$$

EXAMPLE 1 Write the equation of the hyperbola with vertices $(3, -3)$ and $(3, 3)$ and a focus at $(3, 5)$.

Solution First, we plot the vertices and focus, as shown in Figure 8-20. Because the foci lie on a vertical line, the equation to use is

$$\frac{(y - k)^2}{a^2} - \frac{(x - h)^2}{b^2} = 1$$

The center of the hyperbola is midway between the vertices V and V'. Thus, the center is point $(3, 0)$, and $h = 3$ and $k = 0$. The distance between the vertex and the center is $a = 3$, and the distance between the focus and the center is $c = 5$. We can find b^2 by substituting 3 for a and 5 for c in the following equation to get

$$b^2 = c^2 - a^2 \qquad \text{In a hyperbola, } b^2 = c^2 - a^2.$$
$$b^2 = 5^2 - 3^2$$
$$b^2 = 16$$

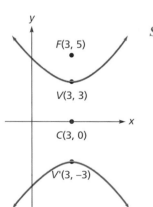

FIGURE **8-20**

Substituting the values for h, k, a^2, and b^2 into the standard equation gives the equation of the hyperbola:

$$\frac{(y-k)^2}{a^2} - \frac{(x-h)^2}{b^2} = 1$$

$$\frac{(y-0)^2}{9} - \frac{(x-3)^2}{16} = 1$$

$$\frac{y^2}{9} - \frac{(x-3)^2}{16} = 1$$

■

Asymptotes of a Hyperbola

The values of a and b are important aids in graphing hyperbolas. To see their value, we consider the hyperbola

$$\frac{x^2}{a^2} - \frac{y^2}{b^2} = 1$$

with center at the origin and vertices at $V(a, 0)$ and $V'(-a, 0)$. We can plot points V, V', $B(0, b)$, and $B'(0, -b)$ and form rectangle $RSQP$, called the **fundamental rectangle**, as shown in Figure 8-21. We can show that the extended diagonals of this rectangle are asymptotes of the hyperbola. In the exercises, you will be asked to show that the equations of these two lines are

$$y = \frac{b}{a}x \qquad \text{and} \qquad y = -\frac{b}{a}x$$

To show that the extended diagonals are asymptotes of the hyperbola, we solve the equation

$$\frac{x^2}{a^2} - \frac{y^2}{b^2} = 1$$

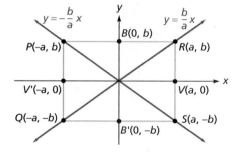

FIGURE 8-21

for y and modify its form as follows:

$$\frac{x^2}{a^2} - \frac{y^2}{b^2} = 1$$

$$b^2x^2 - a^2y^2 = a^2b^2 \qquad \text{Multiply both sides by } a^2b^2.$$

$$y^2 = \frac{b^2x^2 - a^2b^2}{a^2} \qquad \text{Subtract } b^2x^2 \text{ from both sides and divide both sides by } -a^2.$$

$$y^2 = \frac{b^2x^2}{a^2}\left(1 - \frac{a^2}{x^2}\right) \qquad \text{Factor out } b^2x^2 \text{ in the numerator.}$$

2. $\qquad y = \pm\frac{bx}{a}\sqrt{1 - \frac{a^2}{x^2}} \qquad \text{Take the square root of both sides.}$

In Equation 2, if a is constant and $|x| \to \infty$, then $\dfrac{a^2}{x^2} \to 0$, and $\sqrt{1 - \dfrac{a^2}{x^2}}$ approaches 1. Thus, the hyperbola approaches the lines

$$y = \frac{b}{a}x \quad \text{and} \quad y = -\frac{b}{a}x$$

Knowing the asymptotes makes it easy to sketch a hyperbola. We simply convert its equation into standard form, find the coordinates of its vertices, and construct the fundamental rectangle with its extended diagonals. Using the vertices and the asymptotes as guides, we can sketch the hyperbola shown in Figure 8-22. The segment BB' is called the **conjugate axis** of the hyperbola.

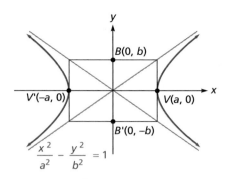

FIGURE 8-22

Graphing Equations of Hyperbolas

EXAMPLE 2 Graph the hyperbola $x^2 - y^2 - 2x + 4y = 12$.

Solution We complete the square on x and y to convert the equation into standard form:

$$x^2 - 2x - y^2 + 4y = 12$$
$$x^2 - 2x - (y^2 - 4y) = 12$$
$$x^2 - 2x + 1 - (y^2 - 4y + 4) = 12 + 1 - 4$$
$$(x - 1)^2 - (y - 2)^2 = 9$$
$$\frac{(x - 1)^2}{9} - \frac{(y - 2)^2}{9} = 1 \qquad \text{Divide both sides by 9.}$$

From the standard equation of a hyperbola, we observe that the center is $(1, 2)$, that $a = 3$ and $b = 3$, and that the vertices are on a line segment parallel to the x-axis, as shown in Figure 8-23. Thus, the vertices V and V' are 3 units to the right and left of the center and have coordinates of $(4, 2)$ and $(-2, 2)$. Points B and B', 3 units above and below the center, have coordinates of $(1, 5)$ and $(1, -1)$. After using points V, V', B, and B' to construct the fundamental rectangle and its extended diagonals, we sketch the graph.

Pierre de Fermat
(1601–1665)
Pierre de Fermat shares the honor of discovering analytic geometry (with Descartes) and of developing the theory of probability (with Pascal). But to Fermat alone goes credit for founding number theory. He is probably most famous for a theorem called Fermat's last theorem. It states that if n represents a number greater than 2, there are no whole numbers a, b, and c that satisfy the equation $a^n + b^n = c^n$. Only recently have mathematicians made significant progress toward proving it.

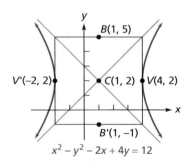

FIGURE 8-23

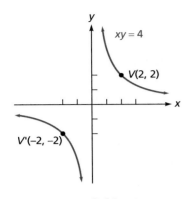

FIGURE **8-24**

We have considered only hyperbolas with a transverse axis that is horizontal or vertical. However, there are hyperbolas with nonhorizontal or nonvertical transverse axes. For example, the graph of the equation $xy = 4$ is a hyperbola with vertices at $(2, 2)$ and $(-2, -2)$, as shown in Figure 8-24.

8.3   EXERCISES

In Exercises 1–12, write the equation of each hyperbola with the given properties.

1. Vertices $(5, 0)$ and $(-5, 0)$; focus $(7, 0)$

2. Focus $(3, 0)$; vertex $(2, 0)$; center $(0, 0)$

3. Center $(2, 4)$; $a = 2, b = 3$; transverse axis is horizontal

4. Center $(-1, 3)$; vertex $(1, 3)$; focus $(2, 3)$

5. Center $(5, 3)$; vertex $(5, 6)$; passes through $(1, 8)$

6. Foci $(0, 10)$ and $(0, -10)$; $\dfrac{c}{a} = \dfrac{5}{4}$

7. Vertices $(0, 3)$ and $(0, -3)$; $\dfrac{c}{a} = \dfrac{5}{3}$

8. Focus $(4, 0)$; vertex $(2, 0)$; center $(0, 0)$

9. Center $(1, -3)$; $a^2 = 4$; $b^2 = 16$

10. Center $(1, 4)$; focus $(7, 4)$; vertex $(3, 4)$

11. Center at the origin; passes through $(4, 2)$ and $(8, -6)$

12. Center $(3, -1)$; y-intercept -1; x-intercept $3 + \dfrac{3\sqrt{5}}{2}$

In Exercises 13–16, find the area of the fundamental rectangle of each hyperbola.

13. $4(x - 1)^2 - 9(y + 2)^2 = 36$

14. $x^2 - y^2 - 4x - 6y = 6$

15. $x^2 + 6x - y^2 + 2y = -11$

16. $9x^2 - 4y^2 = 18x + 24y + 63$

In Exercises 17–20, write the equation of each hyperbola.

17. Center $(-2, -4)$; $a = 2$; area of fundamental rectangle is 36 square units

18. Center $(3, -5)$; $b = 6$; area of fundamental rectangle is 24 square units

19. Vertex $(6, 0)$; one end of conjugate axis at $\left(0, \dfrac{5}{4}\right)$

20. Vertex $(3, 0)$; focus $(-5, 0)$; center $(0, 0)$

In Exercises 21–30, graph each hyperbola.

21. $\dfrac{x^2}{9} - \dfrac{y^2}{4} = 1$

22. $\dfrac{y^2}{4} - \dfrac{x^2}{9} = 1$

23. $4x^2 - 3y^2 = 36$

24. $3x^2 - 4y^2 = 36$

25. $y^2 - x^2 = 1$

26. $9(y + 2)^2 - 4(x - 1)^2 = 36$

27. $4x^2 - 2y^2 + 8x - 8y = 8$

28. $x^2 - y^2 - 4x - 6y = 6$

29. $y^2 - 4x^2 + 6y + 32x = 59$

30. $x^2 + 6x - y^2 + 2y = -11$

In Exercises 31–32, graph each hyperbola by plotting points.

31. $-xy = 6$

32. $xy = 20$

33. *Astronomy* Many comets have a hyperbolic orbit with the sun as one focus. When the comet shown in Illustration 1 is far away from earth, it appears to be approaching earth along the line $y = 2x$. Find the equation of its orbit if the comet comes within 100 million miles of the earth.

35. *Navigation* The **LORAN** system (LOng RAnge Navigation) in Illustration 3 uses two radio transmitters 26 miles apart to send simultaneous signals. The navigator on a ship at $P(x, y)$ receives the closer signal first and determines that the difference of the distances between the ship and each transmitter is 24 miles. That places the ship on a certain curve. Identify the curve and find its equation.

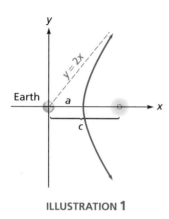

ILLUSTRATION 1

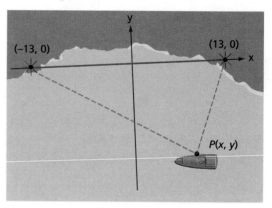

ILLUSTRATION 3

34. *Physics* Parallel beams of similarly charged particles are shot from two atomic accelerators 20 meters apart, as shown in Illustration 2. If the particles were not deflected, the beams would be 2.0×10^{-4} meters apart. However, because the charged particles repel each other, the beams follow the hyperbolic path $y = \frac{k}{x}$, for some k. Find k.

36. *Wave propagation* Stones dropped into a calm pond at points A and B create ripples that propagate in widening circles. In Illustration 4, points A and B are 20 feet apart, and the radii of the circles differ by 12 feet. The point $P(x, y)$ where the circles intersect moves along a curve. Identify the curve and find its equation.

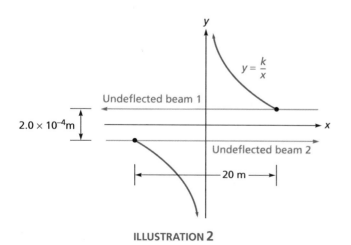

ILLUSTRATION 2

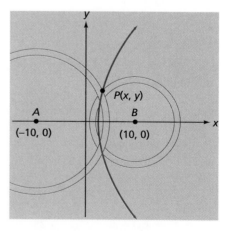

ILLUSTRATION 4

In Exercises 37–40, find the equation of each curve on which point P lies.

37. The difference of the distances between $P(x, y)$ and the points $(-2, 1)$ and $(8, 1)$ is 6.

38. The difference of the distances between $P(x, y)$ and the points $(3, -1)$ and $(3, 5)$ is 5.

39. The distance between point $P(x, y)$ and the point $(0, 3)$ is $\frac{3}{2}$ of the distance between P and the line $y = -2$.

40. The distance between point $P(x, y)$ and the point $(5, 4)$ is $\frac{5}{3}$ of the distance between P and the line $y = -3$.

41. Prove that $c > a$ for a hyperbola with center at $(0, 0)$ and line segment FF' on the x-axis.

42. Show that the extended diagonals of the fundamental rectangle of $\frac{x^2}{a^2} - \frac{y^2}{b^2} = 1$ are $y = \frac{b}{a}x$ and $y = -\frac{b}{a}x$.

43. Show that the expansion of the standard equation of a hyperbola is special case of the general equation of second degree with $B = 0$.

8.4 SOLVING SIMULTANEOUS SECOND-DEGREE EQUATIONS

■ Solving Systems with Graphing Calculators   ■ Solving Systems Algebraically

We now discuss techniques for solving systems of two equations in two variables, where at least one of the equations is of second degree.

EXAMPLE 1   Solve the system $\begin{cases} x^2 + y^2 = 25 \\ 2x + y = 10 \end{cases}$ by graphing.

Solution   The graph of $x^2 + y^2 = 25$ is a circle with center at the origin and radius of 5. The graph of $2x + y = 10$ is straight line. Depending on whether the line is a secant (intersecting the circle at two points) or a tangent (intersecting the circle at one point) or does not intersect the circle at all, there are two, one, or no solutions to the system, respectively.

After graphing the circle and the line, as shown in Figure 8-25, we see that there are two intersection points, $P(3, 4)$ and $P'(5, 0)$. Thus, the solutions to the system are

$$\begin{cases} x = 3 \\ y = 4 \end{cases} \quad \text{and} \quad \begin{cases} x = 5 \\ y = 0 \end{cases}$$

Verify that these are exact solutions.

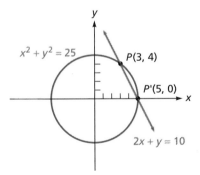

FIGURE 8-25

Solving Systems with Graphing Calculators

To solve the system of equations in Example 1 with a graphing calculator, we must first solve each of the equations for y. From the first equation, $x^2 + y^2 = 25$, we get two equations to graph:

$$y_1 = \sqrt{25 - x^2}$$
$$y_2 = -\sqrt{25 - x^2}$$

From the second equation, $2x + y = 10$, we get a third equation:

$$y_3 = 10 - 2x$$

The graphs of these equations will be similar to those shown in Figure 8-26(a). To find one solution, we use ZOOM and TRACE to read the coordinates of the point of intersection as shown in Figure 8-26(b): $x \approx 3$ and $y \approx 4$. Because the graph is not complete at the other point of intersection shown in Figure 8-26(c), we must use our best judgment. We read the coordinates to be $x \approx 5$ and $y \approx 0$.

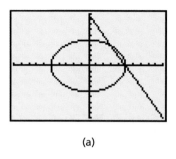

(a)

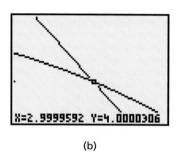

(b)

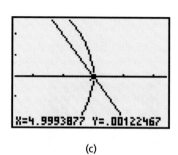

(c)

FIGURE 8-26

■ Solving Systems Algebraically

Algebraic methods can be used to find exact solutions.

EXAMPLE 2   Solve the system $\begin{cases} x^2 + y^2 = 25 \\ 2x \ + y \ = 10 \end{cases}$ algebraically.

Solution   This system contains one equation of second degree and another of first degree. We can solve such systems by substitution. Solving the linear equation for y gives

$$2x + y = 10$$
$$y = -2x + 10$$

We can now substitute $-2x + 10$ for y in the second-degree equation and solve the resulting quadratic equation for x:

$$x^2 + y^2 = 25$$
$$x^2 + (-2x + 10)^2 = 25$$

$x^2 + 4x^2 - 40x + 100 = 25$ Remove parentheses.

$5x^2 - 40x + 75 = 0$ Subtract 25 from both sides and combine terms.

$x^2 - 8x + 15 = 0$ Divide both sides by 5.

$(x - 5)(x - 3) = 0$ Factor $x^2 - 8x + 15$.

$x - 5 = 0$ or $x - 3 = 0$

$x = 5$ | $x = 3$

Because $y = -2x + 10$, if $x = 5$, then $y = 0$, and if $x = 3$, then $y = 4$. The two solutions are

$$\begin{cases} x = 5 \\ y = 0 \end{cases} \quad \text{or} \quad \begin{cases} x = 3 \\ y = 4 \end{cases}$$

■

EXAMPLE 3   Solve the system $\begin{cases} 4x^2 + 9y^2 = 5 \\ y = x^2 \end{cases}$ algebraically.

Solution   We can solve this system by substitution.

$$4x^2 + 9y^2 = 5$$
$$4y + 9y^2 = 5 \qquad \text{Substitute } y \text{ for } x^2.$$
$$9y^2 + 4y - 5 = 0 \qquad \text{Add } -5 \text{ to both sides.}$$
$$(9y - 5)(y + 1) = 0 \qquad \text{Factor } 9y^2 + 4y - 5.$$
$$9y - 5 = 0 \quad \text{or} \quad y + 1 = 0$$
$$y = \frac{5}{9} \qquad\qquad\quad y = -1$$

Because $y = x^2$, we can find the values of x by solving the equations

$$x^2 = \frac{5}{9} \quad \text{and} \quad x^2 = -1$$

The solutions of the equation $x^2 = \frac{5}{9}$ are

$$x = \frac{\sqrt{5}}{3} \quad \text{or} \quad x = -\frac{\sqrt{5}}{3}$$

The equation $x^2 = -1$ has no real solutions. Thus, the solutions of the system are

$$\left(\frac{\sqrt{5}}{3}, \frac{5}{9} \right) \quad \text{and} \quad \left(-\frac{\sqrt{5}}{3}, \frac{5}{9} \right).$$

■

EXAMPLE 4   Solve the system $\begin{cases} 3x^2 + 2y^2 = 36 \\ 4x^2 - y^2 = 4 \end{cases}$ algebraically.

Solution   Here we have two second-degree equations of the form $ax^2 + by^2 = c$. We can solve such systems by eliminating one of the variables by addition. To eliminate the terms involving y^2, we copy the first equation and multiply the second equation by 2 to obtain the following equivalent system of equations

$$\begin{cases} 3x^2 + 2y^2 = 36 \\ 8x^2 - 2y^2 = 8 \end{cases}$$

We can then add the equations and solve the resulting equation for x:

$$11x^2 = 44$$
$$x^2 = 4$$
$$x = 2 \quad \text{or} \quad x = -2$$

To find y, we substitute 2 for x and then -2 for x in the first equation and proceed as follows:

$$
\begin{array}{c|c}
\textbf{For } x = 2 & \textbf{For } x = -2 \\
3x^2 + 2y^2 = 36 & 3x^2 + 2y^2 = 36 \\
3(2)^2 + 2y^2 = 36 & 3(-2)^2 + 2y^2 = 36 \\
12 + 2y^2 = 36 & 12 + 2y^2 = 36 \\
2y^2 = 24 & 2y^2 = 24 \\
y^2 = 12 & y^2 = 12 \\
\end{array}
$$

$$
\begin{array}{c c c | c c c}
y = \sqrt{12} & \text{or} & y = -\sqrt{12} & y = \sqrt{12} & \text{or} & y = -\sqrt{12} \\
y = 2\sqrt{3} & | & y = -2\sqrt{3} & y = 2\sqrt{3} & | & y = -2\sqrt{3}
\end{array}
$$

The four solutions of this system are

$$(2, 2\sqrt{3}), (2, -2\sqrt{3}), (-2, 2\sqrt{3}), \text{ and } (-2, -2\sqrt{3}) \qquad \blacksquare$$

8.4 EXERCISES

In Exercises 1–10, solve each system of equations by graphing.

1. $\begin{cases} 8x^2 + 32y^2 = 256 \\ x = 2y \end{cases}$
2. $\begin{cases} x^2 + y^2 = 2 \\ x + y = 2 \end{cases}$
3. $\begin{cases} x^2 + y^2 = 90 \\ y = x^2 \end{cases}$
4. $\begin{cases} x^2 + y^2 = 5 \\ x + y = 3 \end{cases}$

5. $\begin{cases} x^2 + y^2 = 25 \\ 12x^2 + 64y^2 = 768 \end{cases}$
6. $\begin{cases} x^2 + y^2 = 13 \\ y = x^2 - 1 \end{cases}$
7. $\begin{cases} x^2 - 13 = y^2 \\ y = 2x - 4 \end{cases}$
8. $\begin{cases} x^2 + y^2 = 20 \\ y = x^2 \end{cases}$

9. $\begin{cases} x^2 - 6x - y = -5 \\ x^2 - 6x + y = -5 \end{cases}$
10. $\begin{cases} x^2 - y^2 = -5 \\ 3x^2 + 2y^2 = 30 \end{cases}$

In Exercises 11–14, use a graphing calculator to solve each system of equations.

11. $\begin{cases} y = x + 1 \\ y = x^2 + x \end{cases}$
12. $\begin{cases} y = 6 - x^2 \\ y = x^2 - x \end{cases}$
13. $\begin{cases} 6x^2 + 9y^2 = 10 \\ 3y - 2x = 0 \end{cases}$
14. $\begin{cases} x^2 + y^2 = 68 \\ y^2 - 3x^2 = 4 \end{cases}$

In Exercises 15–40, solve each system of equations algebraically for real values of x and y.

15. $\begin{cases} 25x^2 + 9y^2 = 225 \\ 5x + 3y = 15 \end{cases}$
16. $\begin{cases} x^2 + y^2 = 20 \\ y = x^2 \end{cases}$
17. $\begin{cases} x^2 + y^2 = 2 \\ x + y = 2 \end{cases}$
18. $\begin{cases} x^2 + y^2 = 36 \\ 49x^2 + 36y^2 = 1764 \end{cases}$

19. $\begin{cases} x^2 + y^2 = 5 \\ x + y = 3 \end{cases}$
20. $\begin{cases} x^2 - x - y = 2 \\ 4x - 3y = 0 \end{cases}$
21. $\begin{cases} x^2 + y^2 = 13 \\ y = x^2 - 1 \end{cases}$
22. $\begin{cases} x^2 + y^2 = 25 \\ 2x^2 - 3y^2 = 5 \end{cases}$

23. $\begin{cases} x^2 + y^2 = 30 \\ y = x^2 \end{cases}$
24. $\begin{cases} 9x^2 - 7y^2 = 81 \\ x^2 + y^2 = 9 \end{cases}$
25. $\begin{cases} x^2 + y^2 = 13 \\ x^2 - y^2 = 5 \end{cases}$
26. $\begin{cases} 2x^2 + y^2 = 6 \\ x^2 - y^2 = 3 \end{cases}$

27. $\begin{cases} x^2 + y^2 = 20 \\ x^2 - y^2 = -12 \end{cases}$
28. $\begin{cases} xy = -\dfrac{9}{2} \\ 3x + 2y = 6 \end{cases}$
29. $\begin{cases} y^2 = 40 - x^2 \\ y = x^2 - 10 \end{cases}$
30. $\begin{cases} x^2 - 6x - y = -5 \\ x^2 - 6x + y = -5 \end{cases}$

31. $\begin{cases} y = x^2 - 4 \\ x^2 - y^2 = -16 \end{cases}$
32. $\begin{cases} 6x^2 + 8y^2 = 182 \\ 8x^2 - 3y^2 = 24 \end{cases}$
33. $\begin{cases} x^2 - y^2 = -5 \\ 3x^2 + 2y^2 = 30 \end{cases}$
34. $\begin{cases} \dfrac{1}{x} + \dfrac{1}{y} = 5 \\ \dfrac{1}{x} - \dfrac{1}{y} = -3 \end{cases}$

35. $\begin{cases} \dfrac{1}{x} + \dfrac{2}{y} = 1 \\ \dfrac{2}{x} - \dfrac{1}{y} = \dfrac{1}{3} \end{cases}$ **36.** $\begin{cases} \dfrac{1}{x} + \dfrac{3}{y} = 4 \\ \dfrac{2}{x} - \dfrac{1}{y} = 7 \end{cases}$ **37.** $\begin{cases} 3y^2 = xy \\ 2x^2 + xy - 84 = 0 \end{cases}$ **38.** $\begin{cases} x^2 + y^2 = 10 \\ 2x^2 - 3y^2 = 5 \end{cases}$

39. $\begin{cases} xy = \dfrac{1}{6} \\ y + x = 5xy \end{cases}$ **40.** $\begin{cases} xy = \dfrac{1}{12} \\ y + x = 7xy \end{cases}$

41. *Geometry* The area of a rectangle is 63 square centimeters, and its perimeter is 32 centimeters. Find the dimensions of the rectangle.

42. *Investments* Grant receives $225 annual income from one investment. Jeff invested $500 more than Grant, but at an annual rate of 1% less. Jeff's annual income is $240. Find the amount and rate of Grant's investment.

43. *Investments* Carol receives $67.50 annual income from one investment. John invested $150 more than Carol at an annual rate of $1\frac{1}{2}\%$ more. John's annual income is $94.50. Find the amount and rate of Carol's investment. (*Hint:* There are two answers.)

44. *Finding rates and time* Jim drove 306 miles. Jim's brother made the same trip at a speed 17 miles per hour slower than Jim did and required an extra $1\frac{1}{2}$ hours. Find Jim's rate and time.

45. 🖳 ***Radio reception*** A radio station located 120 miles due east of Collinsville has a listening radius of 100 miles. A straight road joins Collinsville with Harmony, a town 200 miles to the east and 100 miles north. (See Illustration 1.) How far from Collinsville will a driver first pick up the station?

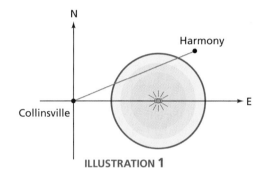

ILLUSTRATION 1

46. 🖳 ***Listening range*** For how many miles will a driver in Exercise 45 continue to receive the signal?

---

8 CHAPTER SUMMARY

Key Words

axis of a parabola (8.1)
center of a circle (8.1)
center of an ellipse (8.2)
center of a hyperbola (8.3)
chord (8.2)
circle (8.1)
conjugate axis of a hyperbola (8.3)
directrix of a parabola (8.1)

eccentricity (8.2)
ellipse (8.2)
foci of an ellipse (8.2)
foci of a hyperbola (8.3)
focus of a parabola (8.1)
fundamental rectangle (8.3)
hyperbola (8.3)
major axis of an ellipse (8.2)

minor axis of an ellipse (8.2)
parabola (8.1)
radius of a circle (8.1)
transverse axis of a hyperbola (8.3)
vertices of an ellipse (8.2)
vertices of a hyperbola (8.3)
vertex of a parabola (8.1)

Key Ideas

(8.1) The graph of any equation that can be written in the form

$$(x - h)^2 + (y - k)^2 = r^2$$

is a circle with center at point $C(h, k)$ and with a radius of r.

The graph of any equation that can be written in the form

$$x^2 + y^2 = r^2$$

is a circle with center at the origin and with radius r.

The standard equations of parabolas that open to the right, left, upward, and downward are as follows (consider $p > 0$):

Parabola opening	Vertex at origin	Vertex at $V(h,k)$
Right	$y^2 = 4px$	$(y-k)^2 = -4p(x-h)$
Left	$y^2 = -4px$	$(y-k)^2 = -4p(x-h)$
Upward	$x^2 = 4py$	$(x-h)^2 = 4p(y-k)$
Downward	$x^2 = -4py$	$(x-h)^2 = -4p(y-k)$

(8.2) The standard equation of an ellipse with center at the origin and major axis on the x-axis is

$$\frac{x^2}{a^2} + \frac{y^2}{b^2} = 1 \qquad \text{where } a > b > 0$$

If the major axis of the ellipse with center at the origin lies on the y-axis, the standard equation of the ellipse is

$$\frac{y^2}{a^2} + \frac{x^2}{b^2} = 1 \qquad \text{where } a > b > 0$$

In either case, the length of the major axis is $2a$, and the length of the minor axis is $2b$.

The standard equation of an ellipse with center at (h, k) and major axis parallel to the x-axis is

$$\frac{(x - h)^2}{a^2} + \frac{(y - k)^2}{b^2} = 1 \qquad \text{where } a > b > 0$$

If the major axis of an ellipse with center (h, k) is parallel to the y-axis, the standard equation of the ellipse is

$$\frac{(y - k)^2}{a^2} + \frac{(x - h)^2}{b^2} = 1 \qquad \text{where } a > b > 0$$

In either case, the length of the major axis is $2a$, and the length of the minor axis is $2b$.

(8.3) The standard equation of a hyperbola with center at the origin and foci on the x-axis is

$$\frac{x^2}{a^2} - \frac{y^2}{b^2} = 1$$

The standard equation of a hyperbola with center at the origin and foci on the y-axis is

$$\frac{y^2}{a^2} - \frac{x^2}{b^2} = 1$$

The standard equation of a hyperbola with center at (h, k) and foci on the line parallel to the x-axis is

$$\frac{(x - h)^2}{a^2} - \frac{(y - k)^2}{b^2} = 1$$

The standard equation of a hyperbola with center at (h, k) and foci on the line parallel to the y-axis is

$$\frac{(y - k)^2}{a^2} - \frac{(x - h)^2}{b^2} = 1$$

The extended diagonals of the fundamental rectangle are asymptotes of the graph of a hyperbola.

(8.4) Good estimates for solutions to systems of simultaneous second-degree equations can be found by graphing.

Exact solutions to systems of simultaneous second-degree equations can often be found by algebraic techniques.

8 CHAPTER REVIEW EXERCISES

In Review Exercises 1–4, write the equation of the circle with the given properties.

1. center $(0, 0)$; passes through $(5, 5)$

2. center $(0, 0)$; passes through $(6, 8)$

3. endpoints of diameter $(-2, 4)$ and $(12, 16)$

4. endpoints of diameter $(-3, -6)$ and $(7, 10)$

5. Write in standard form the equation of a circle $x^2 + y^2 - 6x + 4y = 3$ and graph the circle.

6. Write in standard form the equation of a circle $x^2 + 4x + y^2 - 10y = -13$ and graph the circle.

In Review Exercises 7–8, write the equation of the parabola with the given properties.

7. vertex $(0, 0)$; passes through $(-8, 4)$ and $(-8, -4)$

8. vertex $(0, 0)$; passes through $(-8, 4)$ and $(8, 4)$

9. Write the equation of the parabola $y = ax^2 + bx + c$ in standard form to show that the x-coordinate of the vertex of the parabola is $-\frac{b}{2a}$.

10. Find the equation of the parabola with vertex at $(-2, 3)$, curve passing through point $(-4, -8)$, and opening downward.

In Review Exercises 11–14, graph each equation.

11. $x^2 - 4y - 2x = -9$ **12.** $y^2 - 6y = 4x - 13$

13. $y^2 - 4x - 2y = -13$ **14.** $2x + 5 = x^2 - y$

15. Write the equation of the ellipse with center at the origin, major axis that is horizontal and 12 units long, and minor axis 8 units long.

16. Write the equation of the ellipse with center at point $(-2, 3)$ and curve passing through points $(-2, 0)$ and $(2, 3)$.

In Review Exercises 17–18, graph each equation.

17. $4x^2 + y^2 - 16x + 2y = -13$

18. $x^2 + 9y^2 - 6x - 18y + 9 = 0$

19. Write the equation of the hyperbola with vertices at point $(-3, 3)$ and a focus at point $(5, 3)$.

20. Write the equation of the hyperbola with vertices at point $(3, -3)$ and $(3, 3)$ and a focus at point $(3, 5)$.

In Review Exercises 21–22, graph each equation.

21. $9x^2 - 4y^2 - 16y - 18x = 43$

22. $-2xy = 9$

23. Solve the following system by graphing:

$$\begin{cases} 3x^2 + y^2 = 52 \\ x^2 - y^2 = 12 \end{cases}$$

24. Solve the system in Review Exercise 23 algebraically.

25. Solve the following system algebraically:

$$\begin{cases} x^2 + y^2 = 16 \\ -\sqrt{3}y + 4\sqrt{3} = 3x \end{cases}$$

26. Solve the system in Review Exercise 25 by graphing.

27. Solve the following system by graphing:

$$\begin{cases} \dfrac{x^2}{16} + \dfrac{y^2}{12} = 1 \\ x^2 - \dfrac{y^2}{3} = 1 \end{cases}$$

28. Solve the system in Review Exercise 27 algebraically.

---

8   CHAPTER TEST

In Questions 1–3, write the equation of each circle with the given properties.

1. Center $(2, 3)$; $r = 3$

2. Ends of diameter at $(-2, -2)$ and $(6, 8)$

3. Center $(2, -5)$, passes through $(7, 7)$

4. Change the equation of the circle

$$x^2 + y^2 - 4x + 6y + 4 = 0$$

to standard form and graph it.

In Questions 5–7, find the equation of each parabola with the given properties.

5. Vertex $(3, 2)$; focus at $(3, 6)$

6. Vertex $(4, -6)$; passes through $(3, -8)$ and $(3, -4)$

7. Vertex $(2, -3)$; passes through $(0, 0)$

8. Change the equation of the parabola $x^2 - 6x - 8y = 7$ into standard form and graph it.

In Questions 9–11, find the equation of each ellipse with the given properties.

9. Vertex $(10, 0)$, center at the origin, focus at $(6, 0)$

10. Minor axis 24, center at the origin, focus at $(5, 0)$

11. Center $(2, 3)$; passes through $(2, 9)$ and $(0, 3)$

12. Change the equation of the ellipse

$$9x^2 + 4y^2 - 18x - 16y - 11 = 0$$

into standard form and graph it.

In Questions 13–15, find the equation of each hyperbola with the given properties.

13. Center the origin, focus at $(13, 0)$, vertex at $(5, 0)$

14. Vertices $(6, 0)$ and $(-6, 0)$; $\dfrac{c}{a} = \dfrac{13}{12}$

15. Center $(2, -1)$, major axis horizontal and of length 16, distance of 20 between foci

16. Change the equation of the hyperbola

$$-4x^2 + y^2 - 4y = 4$$

into standard form and graph it.

In Questions 17–18, solve each system of equations algebraically.

17. $\begin{cases} x^2 + y^2 = 23 \\ y = x^2 - 3 \end{cases}$

18. $\begin{cases} 2x^2 - 3y^2 = 9 \\ x^2 + y^2 = 27 \end{cases}$

In Questions 19–20, complete the square to write each equation in standard form, and identify the curve.

19. $y^2 - 4y - 6x - 14 = 0$

20. $2x^2 + 3y^2 - 4x + 12y + 8 = 0$

CHAPTER 9

NATURAL NUMBER FUNCTIONS AND PROBABILITY

We begin this chapter by introducing a way to expand binomials of the form $(x + y)^n$. This method, called the **binomial theorem**, is important when working with probability and statistics.

9.1 THE BINOMIAL THEOREM

■ Factorial Notation ■ Finding a Particular Term of a Binomial Expansion

Consider the following binomial expansions.

$$(a + b)^0 = 1$$
$$(a + b)^1 = a + b$$
$$(a + b)^2 = a^2 + 2ab + b^2$$
$$(a + b)^3 = a^3 + 3a^2b + 3ab^2 + b^3$$
$$(a + b)^4 = a^4 + 4a^3b + 6a^2b^2 + 4ab^3 + b^4$$
$$(a + b)^5 = a^5 + 5a^4b + 10a^3b^2 + 10a^2b^3 + 5ab^4 + b^5$$
$$(a + b)^6 = a^6 + 6a^5b + 15a^4b^2 + 20a^3b^3 + 15a^2b^4 + 6ab^5 + b^6$$

Four patterns are apparent in the above expansions:

- Each expansion has one more term than the power of the binomial.
- The degree of each term in each expansion equals the exponent of the binomial.
- The first term in each expansion is a raised to the power of the binomial.
- The exponents of a decrease by one in each successive term, and the exponents of b, beginning with b^0 in the first term, increase by one in each successive term.

Blaise Pascal
(1623–1662)
Pascal was torn between the fields of religion and mathematics. Each surfaced at times in his life to dominate his interest. In mathematics, Pascal made contributions to the study of conic sections, probability, and differential calculus. At the age of 19, he invented a calculating machine. He is best known for a triangular array of numbers that bears his name.

To see another pattern, we write the coefficients in a triangular array:

$(a + b)^0 =$					1				
$(a + b)^1 =$				1		1			
$(a + b)^2 =$			1		2		1		
$(a + b)^3 =$		1		3		3		1	
$(a + b)^4 =$	1		4		6		4		1
$(a + b)^5 =$	**1**	**5**		10		10		**5**	**1**
$(a + b)^6 =$	1	**6**	15		20		15	**6**	1

In this array, each entry other than the 1's is the sum of the closest pair of numbers in the line above it. For example, the 6's in the bottom row are the sums of the 1's and 5's in the line above. The 10's are the sums of the 6's and 4's in the line above.

This array, called **Pascal's triangle** after Blaise Pascal (1623–1662), continues with the same pattern forever. The next two lines are

$$(a + b)^7 = \quad 1 \quad 7 \quad 21 \quad 35 \quad 35 \quad 21 \quad 7 \quad 1$$
$$(a + b)^8 = \quad 1 \quad 8 \quad 28 \quad 56 \quad 70 \quad 56 \quad 28 \quad 8 \quad 1$$

EXAMPLE 1   Expand the binomial $(x + y)^6$.

Solution   The first term is x^6, and the exponents of x will decrease by one in each successive term. A y will appear in the second term, and the exponents of y will increase in each successive term, concluding when the term y^6 is reached. Thus, the variables in the expansion are

$$x^6 \quad x^5y \quad x^4y^2 \quad x^3y^3 \quad x^2y^4 \quad xy^5 \quad y^6$$

We can use Pascal's triangle to find the coefficients of these variables. Because the binomial is raised to the 6th power, we choose the row where the first entry (other than 1) is 6. The coefficients of the variables are the numbers in that row.

$$1 \quad 6 \quad 15 \quad 20 \quad 15 \quad 6 \quad 1$$

Putting this information together gives the expansion:

$$(x + y)^6 = x^6 + 6x^5y + 15x^4y^2 + 20x^3y^3 + 15x^2y^4 + 6xy^5 + y^6 \qquad \blacksquare$$

EXAMPLE 2   Expand the binomial $(x - y)^6$.

Solution   To expand $(x - y)^6$, we write the binomial in the form $[x + (-y)]^6$ and substitute $-y$ for y in the results of Example 1. The expansion is

$$[x + (-y)]^6 = x^6 + 6x^5(-y) + 15x^4(-y)^2 + 20x^3(-y)^3 + 15x^2(-y)^4 + 6x(-y)^5 + (-y)^6$$
$$= x^6 - 6x^5y + 15x^4y^2 - 20x^3y^3 + 15x^2y^4 - 6xy^5 + y^6$$

In general, the signs in the binomial expansion of $(x - y)^n$ alternate. The sign of the first term is $+$, the sign of the second term is $-$, and so on. $\qquad \blacksquare$

■ Factorial Notation

To expand a binomial with the binomial theorem, we will need **factorial notation**.

Factorial Notation	If n is a natural number, the symbol $n!$ (read either as **"n factorial"** or as **"factorial n"**) is defined as $$n! = n(n - 1)(n - 2)(n - 3) \cdots (3)(2)(1)$$

EXAMPLE 3   Evaluate   **a.** 3!,   **b.** 6!,   and   **c.** 10!.

Solution   **a.** $3! = 3 \cdot 2 \cdot 1 = 6$ **b.** $6! = 6 \cdot 5 \cdot 4 \cdot 3 \cdot 2 \cdot 1 = 720$

c. $10! = 10 \cdot 9 \cdot 8 \cdot 7 \cdot 6 \cdot 5 \cdot 4 \cdot 3 \cdot 2 \cdot 1 = 3{,}628{,}800$ $\qquad \blacksquare$

There are two fundamental properties of factorials.

Property 1 of Factorials	By definition, $0! = 1$.
Property 2 of Factorials	$n(n - 1)! = n!$

EXAMPLE 4   Show that   **a.** $6 \cdot 5! = 6!$ and   **b.** $8 \cdot 7! = 8!$.

Solution   **a.** $6 \cdot 5! = 6(5 \cdot 4 \cdot 3 \cdot 2 \cdot 1) = 6 \cdot 5 \cdot 4 \cdot 3 \cdot 2 \cdot 1 = 6!$

b. $8 \cdot 7! = 8(7 \cdot 6 \cdot 5 \cdot 4 \cdot 3 \cdot 2 \cdot 1) = 8 \cdot 7 \cdot 6 \cdot 5 \cdot 4 \cdot 3 \cdot 2 \cdot 1 = 8!$ ∎

We can now state the binomial theorem.

The Binomial Theorem	If n is any positive number, then

$$(a + b)^n = a^n + \frac{n!}{1!(n - 1)!}a^{n-1}b + \frac{n!}{2!(n - 2)!}a^{n-2}b^2 + \frac{n!}{3!(n - 3)!}a^{n-3}b^3$$

$$+ \cdots + \frac{n!}{r!(n - r)!}a^{n-r}b^r + \cdots + b^n$$

A proof of the binomial theorem appears in Appendix I.

In the binomial theorem, the exponents of the variables in each term on the right-hand side follow the familiar patterns:

- The sum of the exponents of a and b in each term is n.
- The exponents on a decrease by one in each successive term.
- The exponents on b increase by one in each successive term.

Only the method of finding the coefficients is different. Except for the first and last terms, $n!$ is the numerator of each fractional coefficient. When the exponent on b is 2, the factors in the denominator of the fractional coefficient are $2!$ and $(n - 2)!$. When the exponent on b is 3, the factors in the denominator are $3!$ and $(n - 3)!$, and so on. In general, when the exponent on b is r, the factors in the denominator are $r!$ and $(n - r)!$.

EXAMPLE 5   Use the binomial theorem to expand $(a + b)^5$.

Solution   We substitute directly into the binomial theorem to get

$$(a + b)^5 = a^5 + \frac{5!}{1!(5 - 1)!}a^4b + \frac{5!}{2!(5 - 2)!}a^3b^2 + \frac{5!}{3!(5 - 3)!}a^2b^3 + \frac{5!}{4!(5 - 4)!}ab^4 + b^5$$

$$= a^5 + \frac{5 \cdot 4!}{1 \cdot 4!}a^4b + \frac{5 \cdot 4 \cdot 3!}{2 \cdot 1 \cdot 3!}a^3b^2 + \frac{5 \cdot 4 \cdot 3!}{3! \cdot 2 \cdot 1}a^2b^3 + \frac{5 \cdot 4!}{4! \cdot 1}ab^4 + b^5$$

$$= a^5 + 5a^4b + 10a^3b^2 + 10a^2b^3 + 5ab^4 + b^5$$

We note that the coefficients are the same numbers that appear in the sixth row of Pascal's triangle. ∎

EXAMPLE 6   Use the binomial theorem to expand $(2x - 3y)^4$.

Solution   We first find the expansion of $(a + b)^4$.

$$(a + b)^4 = a^4 + \frac{4!}{1!(4 - 1)!}a^3b + \frac{4!}{2!(4 - 2)!}a^2b^2 + \frac{4!}{3!(4 - 3)!}ab^3 + b^4$$

$$= a^4 + \frac{4 \cdot 3!}{1 \cdot 3!}a^3b + \frac{4 \cdot 3 \cdot 2!}{2 \cdot 1 \cdot 2!}a^2b^2 + \frac{4 \cdot 3!}{3! \cdot 1}ab^3 + b^4$$

1. $$= a^4 + 4a^3b + 6a^2b^2 + 4ab^3 + b^4$$

We then substitute $2x$ for a and $-3y$ for b in Equation 1.

$$(a + b)^4 = a^4 + 4a^3b + 6a^2b^2 + 4ab^3 + b^4$$

$$[2x + (-3y)]^4 = (2x)^4 + 4(2x)^3(-3y) + 6(2x)^2(-3y)^2 + 4(2x)(-3y)^3 + (-3y)^4$$

$$(2x - 3y)^4 = 16x^4 - 96x^3y + 216x^2y^2 - 216xy^3 + 81y^4$$ ∎

■ **Finding a Particular Term of a Binomial Expansion**

Suppose that we wish to find the fifth term of the expansion of $(a + b)^{11}$. It would be tedious to raise the binomial to the 11th power and then look at the fifth term. The binomial theorem provides an easier way.

EXAMPLE 7   Find the fifth term of the expansion of $(a + b)^{11}$.

Solution   In the fifth term, the exponent on b is 4, because the exponent on b is always 1 less than the number of the term. Since the exponent on b added to the exponent on a must equal 11, the exponent on a is 7. Thus, the variables of the fifth term are a^7b^4.

The number in the numerator of the fractional coefficient is $n!$, which in this case is 11!. The factors in the denominator are 4! and $(11 - 4)!$. Thus, the complete fifth term is

$$\frac{11!}{4!(11 - 4)!}a^7b^4 = \frac{11!}{4!7!}a^7b^4$$

$$= \frac{11 \cdot 10 \cdot 9 \cdot 8 \cdot 7!}{4 \cdot 3 \cdot 2 \cdot 1 \cdot 7!}a^7b^4$$

$$= 330a^7b^4$$ ∎

EXAMPLE 8   Find the sixth term of the expansion of $(a + b)^9$.

Solution   In the sixth term, the exponent on b is 5, and the exponent on a is $9 - 5$, or 4. The numerator of the fractional coefficient is 9!, and the factors in the denominator are 5! and $(9 - 5)!$. Thus, the sixth term of the expansion is

$$\frac{9!}{5!(9-5)!}a^4b^5 = \frac{9 \cdot 8 \cdot 7 \cdot 6 \cdot 5!}{5! \cdot 4!}a^4b^5$$

$$= \frac{9 \cdot 8 \cdot 7 \cdot 6}{4 \cdot 3 \cdot 2 \cdot 1}a^4b^5 \qquad \text{Divide out } 5!.$$

$$= 126a^4b^5 \qquad\qquad \blacksquare$$

EXAMPLE 9 Find the third term of the expansion of $(3x - 2y)^6$.

Solution We begin by finding the third term in the expansion of $(a + b)^6$.

$$\textbf{2.} \quad \frac{6!}{2!(6-2)!}a^4b^2 = \frac{6 \cdot 5 \cdot 4!}{2 \cdot 1 \cdot 4!}a^4b^2 = 15a^4b^2$$

We can then substitute $3x$ for a and $-2y$ for b in Equation 2 to obtain the third term of the expansion of $(3x - 2y)^6$.

$$15a^4b^2 = 15(3x)^4(-2y)^2$$
$$= 15(3)^4(-2)^2x^4y^2$$
$$= 4860x^4y^2 \qquad\qquad \blacksquare$$

9.1 EXERCISES

In Exercises 1–12, evaluate each expression.

1. $4!$

2. $-5!$

3. $3! \cdot 6!$

4. $0! \cdot 7!$

5. $6! + 6!$

6. $5! - 2!$

7. $\dfrac{9!}{12!}$

8. $\dfrac{8!}{5!}$

9. $\dfrac{5!7!}{9!}$

10. $\dfrac{3!5!7!}{1!8!}$

11. $\dfrac{18!}{6!(18-6)!}$

12. $\dfrac{15!}{9!(15-9)!}$

In Exercises 13–28, use the binomial theorem to expand each binomial.

13. $(a + b)^3$

14. $(a + b)^4$

15. $(a - b)^5$

16. $(x - y)^4$

17. $(2x + y)^3$

18. $(x + 2y)^3$

19. $(x - 2y)^3$

20. $(2x - y)^3$

21. $(2x + 3y)^4$

22. $(2x - 3y)^4$

23. $(x - 2y)^4$

24. $(x + 2y)^4$

25. $(x - 3y)^5$

26. $(3x - y)^5$

27. $\left(\dfrac{x}{2} + y\right)^4$

28. $\left(x + \dfrac{y}{2}\right)^4$

In Exercises 29–48, find the required term in each binomial expansion.

29. $(a + b)^4$; 3rd term

30. $(a - b)^4$; 2nd term

31. $(a + b)^7$; 5th term

32. $(a + b)^5$; 4th term

33. $(a - b)^5$; 6th term

34. $(a - b)^8$; 7th term

35. $(a + b)^{17}$; 5th term

36. $(a - b)^{12}$; 3rd term

37. $(a - \sqrt{2})^4$; 2nd term

38. $(a - \sqrt{3})^8$; 3rd term

39. $(a + \sqrt{3}b)^9$; 5th term

40. $(\sqrt{2}a - b)^7$; 4th term

41. $\left(\dfrac{x}{2} + y\right)^4$; 3rd term

42. $\left(m + \dfrac{n}{2}\right)^8$; 3rd term

43. $\left(\dfrac{r}{2} - \dfrac{s}{2}\right)^{11}$; 10th term

44. $\left(\dfrac{p}{2} - \dfrac{q}{2}\right)^9$; 6th term

45. $(a + b)^n$; 4th term   **46.** $(a - b)^n$; 5th term

47. $(a + b)^n$; rth term   **48.** $(a + b)^n$; $(r + 1)$th term

49. Find the sum of the numbers in each row of the first ten rows of Pascal's triangle. Do you see a pattern?

50. Show that the sum of the coefficients in the binomial expansion of $(x + y)^n$ is 2^n. (*Hint:* Let $x = y = 1$.)

51. Find the constant term in the expansion of $\left(a - \dfrac{1}{a}\right)^{10}$.

52. Find the coefficient of x^5 in the expansion of $\left(x + \dfrac{1}{x}\right)^9$.

53. If we apply the pattern of coefficients to the coefficient of the first term in the binomial theorem, it would be $\dfrac{n!}{0!(n - 0)!}$. Show that this expression equals 1.

54. If we apply the pattern of coefficients to the coefficient of the last term in the binomial theorem, it would be $\dfrac{n!}{n!(n - n)!}$. Show that this expression equals 1.

SEQUENCES, SERIES, AND SUMMATION NOTATION

■ Recursive Definition of a Sequence   ■ Summation Notation

In this section, we will introduce a special function whose domain is the set of natural numbers. This function, called a **sequence**, is a list of numbers in a specific order. The sum of the terms in a sequence is called a **series**.

Sequences

A **sequence** is a function whose domain is the set of natural numbers.

Since a sequence is a function whose domain is the set of natural numbers, we can write its terms as a list of numbers. For example, if n is a natural number, the function defined by $f(n) = 2n - 1$ generates the sequence

$$1, 3, 5, \ldots, 2n - 1, \ldots$$

The number 1 is the first term of the sequence, 3 is the second term, and $2n - 1$ is the **general**, or **nth term**. If n is a natural number, the function $f(n) = 3n^2 + 1$ generates the sequence

$$4, 13, 28, \ldots, 3n^2 + 1, \ldots$$

The number 4 is the first term, 13 is the second term, 28 is the third term, and $3n^2 + 1$ is the general term.

A constant function such as $g(n) = 1$ is a sequence also, because it generates the sequence

$$1, 1, 1, \ldots.$$

We seldom use function notation to denote a sequence, because it is difficult or even impossible to write the general term—the expression that shows how the terms are constructed. In such cases, if there is a pattern that is assumed to be continued, we simply list several terms of the sequence. Some examples of sequences follow:

$$1^2, 2^2, 3^3, \ldots, n^2, \ldots$$

$$3, 9, 19, 33, \ldots, 2n^2 + 1, \ldots$$

$$1, 3, 6, 10, 15, 21, \ldots, \frac{n(n + 1)}{2}, \ldots$$

$$1, 1, 2, 3, 5, 8, 13, 21, \ldots \quad \text{(Fibonacci sequence)}$$

$$2, 3, 5, 7, 11, 13, 17, 19, 23, \ldots \quad \text{(prime numbers)}$$

The **Fibonacci sequence** is named after the twelfth-century mathematician Leonardo of Pisa, also known as Fibonacci. After the two 1's in the Fibonacci sequence, each term is the sum of the two terms that immediately precede it. The Fibonacci sequence occurs in the study of botany (for example) in the growth patterns of plants.

Recursive Definition of a Sequence

Leonardo Fibonacci (late 12th and early 13th centuries) Fibonacci, an Italian mathematician, is also known as Leonardo da Pisa. In his work Liber abaci, *he advocated the adoption of Arabic numerals, the numerals that we use today. He is best known for a sequence of numbers that bears his name.*

A sequence can be defined **recursively** by giving its first term, a_1, and a rule showing how to obtain the $(n + 1)$th term, a_{n+1}, from the nth term, a_n. For example, the information

$$a_1 = 5 \quad \text{and} \quad a_{n+1} = 3a_n - 2$$

defines a sequence recursively. To find the first five terms of this sequence, we proceed as follows:

$$a_2 = 3(a_1) - 2 = 3(5) - 2 = 13$$
$$a_3 = 3(a_2) - 2 = 3(13) - 2 = 37$$
$$a_4 = 3(a_3) - 2 = 3(37) - 2 = 109$$
$$a_5 = 3(a_4) - 2 = 3(109) - 2 = 325$$

To add the terms of a sequence, we replace each comma between its terms with a $+$ sign to form a **series**. Because each sequence is infinite, the number of terms in the series associated with it is infinite also. Two examples of infinite series are

$$1^2 + 2^2 + 3^2 + \cdots + n^2 + \cdots$$

and

$$1 + 2 + 3 + 5 + 8 + 13 + 21 + \cdots$$

If the signs between successive terms of an infinite series alternate, the series is called an **alternating infinite series**. Two examples of alternating infinite series are

$$-3 + 6 - 9 + 12 - \cdots + (-1)^n 3n + \cdots$$

and

$$2 - 4 + 8 - 16 + \cdots + (-1)^{n+1} 2^n + \cdots$$

Summation Notation

Summation notation is a shorthand way to indicate the sum of the first n terms, or the **nth partial sum**, of a sequence. For example, the expression

$$\sum_{n=1}^{3} (2n^2 + 1) \qquad \text{The symbol } \sum \text{ is the capital sigma in the Greek alphabet.}$$

indicates the sum of the three terms obtained if we successively substitute the natural numbers 1, 2, and 3 for n in the expression $2n^2 + 1$. Thus,

$$\sum_{n=1}^{3} (2n^2 + 1) = [2(\mathbf{1})^2 + 1] + [2(\mathbf{2})^2 + 1] + [2(\mathbf{3})^2 + 1]$$
$$= 3 + 9 + 19$$
$$= 31$$

EXAMPLE 1   Evaluate $\sum_{n=1}^{4} (n^2 - 1)$.

Solution   In this example, n is said to run from 1 to 4. Thus, we substitute 1, 2, 3, and 4 for n in the expression $n^2 - 1$ and find the sum of the resulting terms:

$$\sum_{n=1}^{4} (n^2 - 1) = (\mathbf{1}^2 - 1) + (\mathbf{2}^2 - 1) + (\mathbf{3}^2 - 1) + (\mathbf{4}^2 - 1)$$
$$= 0 + 3 + 8 + 15$$
$$= 26$$

■

EXAMPLE 2   Evaluate $\sum_{n=3}^{5} (3n + 2)$.

Solution   Since n runs from 3 to 5, we substitute 3, 4, and 5 for n in the expression $3n + 2$ and find the sum of the resulting terms:

$$\sum_{n=3}^{5} (3n + 2) = [3(\mathbf{3}) + 2] + [3(\mathbf{4}) + 2] + [3(\mathbf{5}) + 2]$$
$$= 11 + 14 + 17$$
$$= 42$$

■

The theorems that follow give three basic properties of summations. The first theorem states that the summation of a constant as k runs from 1 to n is n times the constant.

Theorem

If c is a constant, then $\sum_{k=1}^{n} c = nc$.

Proof   Because c is a constant, each term is c for each value for k as k runs from 1 to n.

$$n \text{ numbers of } c\text{'s}$$

$$\sum_{k=1}^{n} c = \overbrace{c + c + c + c + \cdots + c}^{} = nc$$

□

EXAMPLE 3   Evaluate $\displaystyle\sum_{n=1}^{5} 13$.

Solution $\displaystyle\sum_{n=1}^{5} 13 = 13 + 13 + 13 + 13 + 13$

$$= 5(13)$$
$$= 65 \qquad \blacksquare$$

The second theorem states that a constant factor can be brought outside a summation sign.

Theorem	If c is a constant, then $\displaystyle\sum_{k=1}^{n} cf(k) = c\sum_{k=1}^{n} f(k)$.

Proof $\displaystyle\sum_{k=1}^{n} cf(k) = cf(1) + cf(2) + cf(3) + \cdots + cf(n)$

$$= c[f(1) + f(2) + f(3) + \cdots + f(n)] \qquad \text{Factor out } c.$$

$$= c\sum_{k=1}^{n} f(k) \qquad\qquad\qquad \square$$

EXAMPLE 4   Show that $\displaystyle\sum_{k=1}^{3} 5k^2 = 5\sum_{k=1}^{3} k^2$.

Solution $\displaystyle\sum_{k=1}^{3} 5k^2 = 5(1)^2 + 5(2)^2 + 5(3)^2$ $\displaystyle 5\sum_{k=1}^{3} k^2 = 5[(1)^2 + (2)^2 + (3)^2]$

$\qquad\qquad = 5 + 20 + 45$ $\qquad\qquad = 5[1 + 4 + 9]$

$\qquad\qquad = 70$ $\qquad\qquad = 5(14)$

$\qquad\qquad\qquad\qquad\qquad\qquad\qquad\qquad\qquad = 70$

Thus, the quantities are equal. $\blacksquare$

The third theorem states that the summation of a sum is equal to the sum of the summations.

Theorem	$\displaystyle\sum_{k=1}^{n} [f(k) + g(k)] = \sum_{k=1}^{n} f(k) + \sum_{k=1}^{n} g(k)$

Proof $\displaystyle\sum_{k=1}^{n} [f(k) + g(k)] = [f(1) + g(1)] + [f(2) + g(2)] + [f(3) + g(3)]$

$$+ \cdots + [f(n) + g(n)]$$

$$= [f(1) + f(2) + f(3) + \cdots + f(n)] + [g(1) + g(2)$$
$$+ g(3) + \cdots + g(n)]$$

$$= \sum_{k=1}^{n} f(k) + \sum_{k=1}^{n} g(k)$$

□

EXAMPLE 5   Show that $\sum_{k=1}^{3} (k + k^2) = \sum_{k=1}^{3} k + \sum_{k=1}^{3} k^2.$

Solution $\sum_{k=1}^{3} (k + k^2) = (1 + 1^2) + (2 + 2^2) + (3 + 3^2)$
$$= 2 + 6 + 12$$
$$= 20$$

$\sum_{k=1}^{3} k + \sum_{k=1}^{3} k^2 = (1 + 2 + 3) + (1^2 + 2^2 + 3^2)$
$$= 6 + 14$$
$$= 20$$

■

EXAMPLE 6   Evaluate $\sum_{k=1}^{5} (2k - 1)^2$ directly. Then expand the binomial, apply the previous theorems, and evaluate the expression again.

Solution   ***Part 1.*** $\sum_{k=1}^{5} (2k - 1)^2 = 1 + 9 + 25 + 49 + 81 = 165$

Part 2. $\sum_{k=1}^{5} (2k - 1)^2 = \sum_{k=1}^{5} (4k^2 - 4k + 1)$

$$= \sum_{k=1}^{5} 4k^2 + \sum_{k=1}^{5} (-4k) + \sum_{k=1}^{5} 1$$

The summation of a sum is the sum of the summations.

$$= 4 \sum_{k=1}^{5} k^2 - 4 \sum_{k=1}^{5} k + \sum_{k=1}^{5} 1$$

Bring the constant factors outside the summation signs.

$$= 4 \sum_{k=1}^{5} k^2 - 4 \sum_{k=1}^{5} k + 5$$

The summation of a constant as k runs from 1 to 5 is 5 times that constant.

$$= 4(1 + 4 + 9 + 16 + 25) - 4(1 + 2 + 3 + 4 + 5) + 5$$
$$= 4(55) - 4(15) + 5$$
$$= 220 - 60 + 5$$
$$= 165$$

Either way, the sum is 165.

■

9.2 EXERCISES

In Exercises 1–2, write the first six terms of the sequence defined by each function.

1. $f(n) = 5n(n - 1)$

2. $f(n) = n\left(\dfrac{n - 1}{2}\right)\left(\dfrac{n - 2}{3}\right).$

In Exercises 3–8, find the next term of each sequence.

3. $1, 6, 11, 16 \ldots$

4. $1, 8, 27, 64, \ldots$

5. $a, a + d, a + 2d, a + 3d, \ldots$

6. $a, ar, ar^2, ar^3, \ldots$

7. $1, 3, 6, 10, \ldots$

8. $20, 17, 13, 8, \ldots$

In Exercises 9–16, find the sum of the first five terms of the sequence with the given general term.

9. n

10. $2k$

11. 3

12. $4k^0$

13. $2\left(\dfrac{1}{3}\right)^n$

14. $(-1)^n$

15. $3n - 2$

16. $2k + 1$

In Exercises 17–24, a sequence is defined recursively. Find the first four terms of each sequence.

17. $a_1 = 3$ and $a_{n+1} = 2a_n + 1$

18. $a_1 = -5$ and $a_{n+1} = -a_n - 3$

19. $a_1 = -4$ and $a_{n+1} = \dfrac{a_n}{2}$

20. $a_1 = 0$ and $a_{n+1} = 2a_n^2$

21. $a_1 = k$ and $a_{n+1} = a_n^2$

22. $a_1 = 3$ and $a_{n+1} = ka_n$

23. $a_1 = 8$ and $a_{n+1} = \dfrac{2a_n}{k}$

24. $a_1 = m$ and $a_{n+1} = \dfrac{a_n^2}{m}$

In Exercises 25–28, tell whether each series is an alternating infinite series.

25. $-1 + 2 - 3 + \cdots + (-1)^n n + \cdots$

26. $a + \dfrac{a}{b} + \dfrac{a}{b^2} + \cdots + a\left(\dfrac{1}{b}\right)^{n-1} + \cdots; b = 4$

27. $a + a^2 + a^3 + \cdots + a^n + \cdots; a = 3$

28. $a + a^2 + a^3 + \cdots + a^n + \cdots; a = -2$

In Exercises 29–42, evaluate each sum.

29. $\displaystyle\sum_{k=1}^{5} 2k$

30. $\displaystyle\sum_{k=3}^{6} 3k$

31. $\displaystyle\sum_{k=3}^{4} (-2k^2)$

32. $\displaystyle\sum_{k=1}^{100} 5$

33. $\displaystyle\sum_{k=1}^{5} (3k - 1)$

34. $\displaystyle\sum_{n=2}^{5} (n^2 + 3n)$

35. $\displaystyle\sum_{k=1}^{1000} \dfrac{1}{2}$

36. $\displaystyle\sum_{x=4}^{5} \dfrac{2}{x}$

37. $\displaystyle\sum_{x=3}^{4} \dfrac{1}{x}$

38. $\displaystyle\sum_{x=2}^{6} (3x^2 + 2x) - 3\sum_{x=2}^{6} x^2$

39. $\displaystyle\sum_{x=1}^{4} (4x + 1)^2 - \sum_{x=1}^{4} (4x - 1)^2$

40. $\displaystyle\sum_{x=0}^{10} (2x - 1)^2 + 4\sum_{x=0}^{10} x(1 - x)$

41. $\displaystyle\sum_{x=6}^{8} (5x - 1)^2 + \sum_{x=6}^{8} (10x - 1)$

42. $\displaystyle\sum_{x=2}^{7} (3x + 1)^2 - 3\sum_{x=2}^{7} x(3x + 2)$

43. Find a counterexample to disprove the proposition that the summation of a product is the product of the summations. In other words, prove that

$$\sum_{k=1}^{n} f(k)g(k) \neq \sum_{k=1}^{n} f(k) \sum_{k=1}^{n} g(k)$$

44. Find a counterexample to disprove the proposition that the summation of a quotient is the quotient of the summations. In other words, prove that

$$\frac{\sum_{k=1}^{n} f(k)}{\sum_{k=1}^{n} g(k)} \neq \frac{\sum_{k=1}^{n} f(k)}{\sum_{k=1}^{n} g(k)}$$

9.3 ARITHMETIC AND GEOMETRIC SEQUENCES

- Arithmetic Means
- Geometric Sequences
- Geometric Means
- Infinite Geometric Sequences

The German mathematician Carl Friedrich Gauss (1777–1855) was once a student in the class of a strict teacher. One day, the teacher asked the students to add together all of the natural numbers from 1 through 100. Gauss immediately knew the sum.

His solution was relatively simple. He recognized that in the sum $1 + 2 + 3 + \cdots + 98 + 99 + 100$, the first number (1) added to the last number (100) is 101. Furthermore, the second number (2) added to the second-from-the-last number (99) is also 101, as is the third number (3) added to the third-from-the-last number (98). Gauss knew that this pattern continues, and because there are fifty pairs of numbers, there are fifty sums of 101. He multiplied 101 by 50 to get the correct answer of 5050.

This story illustrates a problem involving the sum of the terms of a special sequence called an **arithmetic sequence**, in which each term (except the first) is found by adding a constant to the preceding term.

Arithmetic Sequence

An **arithmetic sequence** (or **arithmetic progression**) is a sequence of the form

$$a, a + d, a + 2d, a + 3d, \ldots, a + (n - 1)d, \ldots$$

where a is the first term, $a + (n - 1)d$ is the nth term, and d is the common difference.

In this definition, the second term has an addend of d, the third term has an addend of $2d$, the fourth term has an addend of $3d$, and so on. This is why the nth term has an addend of $(n - 1)d$.

EXAMPLE 1 Write the first six terms and the 21st term of an arithmetic sequence with a first term of 7 and a common difference of 5.

Solution Because the first term a is 7 and the common difference d is 5, the first six terms are

$$7, \quad 7 + 5, \quad 7 + 2(5), \quad 7 + 3(5), \quad 7 + 4(5), \quad 7 + 5(5)$$

or

$$7, \quad 12, \quad 17, \quad 22, \quad 27, \quad 32$$

Because we want to find the 21st term, we substitute 21 for n in the formula for the nth term:

$$n\text{th term} = a + (n - 1)d$$
$$\mathbf{21}\text{st term} = \mathbf{7} + (\mathbf{21} - 1)\mathbf{5}$$
$$= 7 + (20)5$$
$$= 107$$

The 21st term is 107. ∎

EXAMPLE 2 Find the 98th term of an arithmetic sequence whose first three terms are 2, 6, and 10.

Solution Here $a = 2$, $n = 98$, and $d = 6 - 2 = 10 - 6 = 4$. Because we want to find the 98th term, we substitute these numbers into the formula for the nth term:

$$n\text{th term} = a + (n - 1)d$$
$$\mathbf{98}\text{th term} = \mathbf{2} + (\mathbf{98} - 1)\mathbf{4}$$
$$= 2 + (97)4$$
$$= 390$$ ∎

■ Arithmetic Means

Numbers inserted between a first and last term to form a segment of an arithmetic sequence are called **arithmetic means**. When finding arithmetic means, we consider the last term, l, to be the nth term:

$$l = a + (n - 1)d$$

EXAMPLE 3 Insert three arithmetic means between -3 and 12.

Solution Because we are inserting three terms between -3 and 12, the total number of terms is five. Thus, $a = -3$, $l = 12$, and $n = 5$. To find the common difference, we substitute -3 for a, 12 for l, and 5 for n in the formula for the last term and solve for d:

$$l = a + (n - 1)d$$
$$\mathbf{12} = \mathbf{-3} + (\mathbf{5} - 1)d$$
$$15 = 4d \qquad \text{Add 3 to both sides and simplify.}$$
$$\frac{15}{4} = d \qquad \text{Divide both sides by 4.}$$

Once d has been found, we can find the second, third, and fourth terms of the arithmetic sequence:

$$a + d = -3 + \frac{15}{4} = \frac{3}{4}$$

$$a + 2d = -3 + 2\left(\frac{15}{4}\right) = -3 + \frac{30}{4} = 4\frac{1}{2}$$

$$a + 3d = -3 + 3\left(\frac{15}{4}\right) = -3 + \frac{45}{4} = 8\frac{1}{4}$$

The three arithmetic means are $\frac{3}{4}$, $4\frac{1}{2}$, and $8\frac{1}{4}$. ■

To find the sum of the first n terms of an arithmetic sequence, we use the following formula.

Sum of the First n Terms of an Arithmetic Sequence	The formula $$S_n = \frac{n(a + l)}{2}$$ gives the sum of the first n terms of an arithmetic sequence. In this formula, a is the first term, l is the last (or nth) term, and n is the number of terms.

Proof   We write the first n terms of an arithmetic sequence (letting S_n represent their sum), rewrite the same sum in reverse order, and add the equations together term by term:

$$S_n = \quad a \quad + \quad (a + d) \quad + \cdots + [a + (n - 2)d] \ + [a + (n - 1)d]$$
$$S_n = [a + (n - 1)d] \ + \ [a + (n - 2)d] + \cdots + \quad (a + d) \quad + \quad a$$
$$\overline{2S_n = [2a + (n - 1)d] + [2a + (n - 1)d] + \cdots + [2a + (n - 1)d] + [2a + (n - 1)d]}$$

Because there are n equal terms on the right-hand side of the previous equation,

$$2S_n = n[2a + (n - 1)d]$$

or

1. $\qquad 2S_n = n\{a + [a + (n - 1)d]\}$

We can substitute l for $a + (n - 1)d$ in the right-hand side of Equation 1 and divide both sides by 2 to get

$$S_n = \frac{n(a + l)}{2}$$

The theorem is proved. □

EXAMPLE 4   Find the sum of the first 30 terms of the arithmetic sequence 5, 8, 11, ...

Solution   Here $a = 5$, $n = 30$, $d = 3$, and $l = 5 + 29(3) = 92$. Substituting these values into the formula for the sum of an arithmetic sequence gives

$$S_{30} = \frac{n(a + l)}{2}$$

$$S_{30} = \frac{30(5 + 92)}{2}$$

$$= 15(97)$$

$$= 1455$$

The sum of the first 30 terms is 1455. ■

■ Geometric Sequences

A **geometric sequence** is a sequence in which each term (except the first) is found by multiplying the preceding term by a constant.

Geometric Sequence	A **geometric sequence** (or **geometric progression**) is a sequence of the form
	$$a, ar, ar^2, ar^3, \ldots, ar^{n-1}, \ldots$$
	where a is the first term, ar^{n-1} is the nth term, and r is the common ratio.

In this definition, the second term of the sequence has a factor of r^1, the third term has a factor of r^2, the fourth term has a factor of r^3, and so on. This explains why the nth term has a factor of r^{n-1}.

EXAMPLE 5 Write the first six terms and the 15th term of the geometric sequence whose first term is 3 and whose common ratio is 2.

Solution We first write the first six terms of the geometric sequence:

$$3, \quad 3(2), \quad 3(2)^2, \quad 3(2)^3, \quad 3(2)^4, \quad 3(2)^5$$

or

$$3, \quad 6, \quad 12, \quad 24, \quad 48, \quad 96$$

To obtain the 15th term, we substitute 15 for n, 3 for a, and 2 for r in the formula for the nth term:

$$n\text{th term} = ar^{n-1}$$
$$\mathbf{15}\text{th term} = \mathbf{3(2)}^{15-1}$$
$$= 3(2)^{14}$$
$$= 3(16{,}384)$$
$$= 49{,}152$$ ■

EXAMPLE 6 Find the eighth term of a geometric sequence whose first three terms are 9, 3, and 1.

Solution Here $a = 9$, $r = \frac{1}{3}$, and $n = 8$. To find the eighth term, we substitute these values into the formula for the nth term.

$$n\text{th term} = ar^{n-1}$$

$$8\text{th term} = 9\left(\frac{1}{3}\right)^{8-1}$$

$$= 9\left(\frac{1}{3}\right)^{7}$$

$$= \frac{1}{243}$$ ∎

Geometric Means

Numbers inserted between a first and last term to form a segment of a geometric sequence are called **geometric means**. When finding geometric means, we consider the last term, l, to be the nth term.

EXAMPLE 7 Insert two geometric means between 4 and 256.

Solution The first term is $a = 4$, and because 256 is the fourth term, $n = 4$ and $l = 256$. To find the common ratio, we substitute these values into the formula for the nth term and solve for r:

$$ar^{n-1} = l$$
$$4r^{4-1} = 256$$
$$r^3 = 64$$
$$r = 4$$

The common ratio is 4. The two geometric means are the second and third terms of the geometric sequence:

$$ar = 4 \cdot 4 = 16$$
$$ar^2 = 4 \cdot 4^2 = 4 \cdot 16 = 64$$

The first four terms of the geometric sequence are 4, 16, 64, and 256. The two geometric means between 4 and 256 are 16 and 64. ∎

To find the sum of the first n terms of a geometric sequence, we use the following formula.

Sum of the First n Terms of a Geometric Sequence	The formula $$S_n = \frac{a - ar^n}{1 - r} \quad (r \neq 1)$$ gives the sum of the first n terms of a geometric sequence. In the formula, S_n is the sum, a is the first term, r is the common ratio, and n is the number of terms.

Proof We write out the sum of the first n terms of the geometric sequence:

2. $S_n = a + ar + ar^2 + \cdots + ar^{n-3} + ar^{n-2} + ar^{n-1}$

Multiplying both sides of this equation by r gives

3. $S_n r = ar + ar^2 + \cdots + ar^{n-2} + ar^{n-1} + ar^n$

We now subtract Equation 3 from Equation 2 and solve for S_n:

$$S_n - S_n r = a - ar^n$$

$$S_n(1 - r) = a - ar^n \qquad \text{Factor out } S_n.$$

$$S_n = \frac{a - ar^n}{1 - r}$$

The theorem is proved. $\qquad\qquad\qquad\qquad\qquad\qquad\qquad\qquad\qquad$ □

EXAMPLE 8 Find the sum of the first six terms of the geometric sequence 8, 4, 2, ...

Solution Here $a = 8$, $n = 6$, and $r = \frac{1}{2}$. Substituting these values into the formula for the sum of the first n terms of a geometric sequence gives

$$S_n = \frac{a - ar^n}{1 - r}$$

$$S_6 = \frac{8 - 8\left(\frac{1}{2}\right)^6}{1 - \frac{1}{2}}$$

$$= 2\left(8 - \frac{1}{8}\right)$$

$$= 2\left(\frac{63}{8}\right)$$

$$= \frac{63}{4}$$

The sum of the first six terms is $\dfrac{63}{4}$. $\qquad\qquad\qquad\qquad\qquad\qquad\qquad$ ■

Infinite Geometric Sequences

Under certain conditions, we can find the sum of all of the terms in an **infinite geometric sequence**. To define this sum, we consider the geometric sequence

$$a, ar, ar^2, \ldots$$

- The first partial sum, S_1, of the sequence is $S_1 = a$.
- The second partial sum, S_2, of the sequence is $S_2 = a + ar$.
- The nth partial sum, S_n, of the sequence is $S_n = a + ar + ar^2 + \cdots + ar^{n-1}$.

If the nth partial sum, S_n, of an infinite geometric sequence approaches some number S as $n \to \infty$, then S is called the **sum of the infinite geometric sequence**. The following symbol denotes the sum, S, of an infinite geometric sequence, provided the sum exists.

$$S = \sum_{n=1}^{\infty} ar^{n-1}$$

To develop a formula for finding the sum of all the terms in an infinite geometric sequence, we consider the formula

4. $S_n = \dfrac{a - ar^n}{1 - r} \qquad (r \neq 1)$

If $|r| < 1$ and a is a constant, then as $n \to \infty$, $ar^n \to 0$ and the term ar^n in Equation 4 can be dropped. This argument leads to the following theorem.

Sum of the Terms of an Infinite Geometric Sequence	If $|r| < 1$, the sum of the terms of an infinite geometric sequence is given by $$S = \dfrac{a}{1 - r}$$ where a is the first term and r is the common ratio.

 Warning!   If $|r| \geq 1$, the terms get larger and larger, and the sum does not approach a limit. In this case, the previous theorem does not apply.

EXAMPLE 9   Change $0.\overline{4}$ to a common fraction.

Solution   We write the decimal as an infinite geometric series and find its sum:

$$S = \frac{4}{10} + \frac{4}{100} + \frac{4}{1000} + \frac{4}{10,000} + \cdots$$

$$S = \frac{4}{10} + \frac{4}{10}\left(\frac{1}{10}\right) + \frac{4}{10}\left(\frac{1}{10}\right)^2 + \frac{4}{10}\left(\frac{1}{10}\right)^3 + \cdots$$

Because the common ratio $r = \frac{1}{10}$ and $\left|\frac{1}{10}\right| < 1$, we can use the formula for the sum of an infinite geometric series:

$$S = \frac{a}{1 - r} = \frac{\dfrac{4}{10}}{1 - \dfrac{1}{10}} = \frac{\dfrac{4}{10}}{\dfrac{9}{10}} = \frac{4}{9}$$

Long division will verify that $\dfrac{4}{9} = 0.\overline{4}$.   ∎

9.3 EXERCISES

In Exercises 1–6, write the first six terms of the arithmetic sequences with the given properties.

1. $a = 1; d = 2$ **2.** $a = -12; d = -5$ **3.** $a = 5$; 3rd term is 2   **4.** $a = 4$; 5th term is 12

5. 7th term is 24; common difference is $\frac{5}{2}$.   **6.** 20th term is -49; common difference is -3.

In Exercises 7–10, find the sum of the first n terms of each arithmetic sequence.

7. $5 + 7 + 9 + \cdots$ (to 15 terms)   **8.** $-3 + (-4) + (-5) + \cdots$ (to 10 terms)

9. $\displaystyle\sum_{n=1}^{20} \left(\frac{3}{2}n + 12\right)$ **10.** $\displaystyle\sum_{n=1}^{10} \left(\frac{2}{3}n + \frac{1}{3}\right)$

11. Find the sum of the first 30 terms of an arithmetic sequence with 25th term of 10 and a common difference of $\frac{1}{2}$.

12. Find the sum of the first 100 terms of an arithmetic sequence with 15th term of 86 and first term of 2.

13. If the fifth term of an arithmetic sequence is 14 and its second term is 5, find the 15th term.

14. Can an arithmetic sequence have a first term of 4, a 25th term of 126, and a common difference of $4\frac{1}{4}$? Explain.

15. Insert three arithmetic means between 10 and 20.

16. Insert five arithmetic means between 5 and 15.

17. Insert four arithmetic means between -7 and $\frac{2}{3}$.

18. Insert three arithmetic means between -11 and -2.

In Exercises 19–26, write the first four terms of each geometric sequence with the given properties.

19. $a = 10; r = 2$ **20.** $a = -3; r = 2$ **21.** $a = -2$ and $r = 3$ **22.** $a = 64; r = \dfrac{1}{2}$

23. $a = 3; r = \sqrt{2}$ **24.** $a = 2; r = \sqrt{3}$ **25.** $a = 2$; 4th term is 54   **26.** 3rd term is 4; $r = \dfrac{1}{2}$

In Exercises 27–32, find the sum of the indicated terms of each geometric sequence.

27. $4, 8, 16, \ldots$ (to 5 terms)   **28.** $9, 27, 81, \ldots$ (to 6 terms)

29. $2, -6, 18, \ldots$ (to 10 terms)   **30.** $\dfrac{1}{8}, \dfrac{1}{4}, \dfrac{1}{2}, \ldots$ (to 12 terms)

31. $\displaystyle\sum_{n=1}^{6} 3\left(\frac{3}{2}\right)^{n-1}$ **32.** $\displaystyle\sum_{n=1}^{6} 12\left(-\frac{1}{2}\right)^{n-1}$

In Exercises 33–36, find the sum of each infinite geometric sequence.

33. $6 + 4 + \dfrac{8}{3} + \cdots$ **34.** $8 + 4 + 2 + 1 + \cdots$ **35.** $\displaystyle\sum_{n=1}^{\infty} 12\left(-\frac{1}{2}\right)^{n-1}$ **36.** $\displaystyle\sum_{n=1}^{\infty} \left(\frac{1}{3}\right)^{n-1}$

37. Insert three positive geometric means between 10 and 20.

38. Insert five geometric means between -5 and 5, if possible.

39. Insert four geometric means between 2 and 2048.

40. Insert three geometric means between 162 and 2. (There are two possibilities.)

In Exercises 41–44, change each decimal to a common fraction.

41. $0.\overline{5}$ **42.** $0.\overline{6}$ **43.** $0.\overline{25}$ **44.** $0.\overline{37}$

9.4 APPLICATIONS OF SEQUENCES

Many of the exponential growth problems discussed in Chapter 6 can be solved using the concepts of geometric sequences.

EXAMPLE 1   A town with a population of 3500 people has a predicted growth rate of 6% over the preceding year for the next 20 years. How many people are expected to live in the town 20 years from now?

Solution   Let p_0 be the initial population of the town. After 1 year, there will be a different population, p_1. The initial population (p_0) plus the growth (the product of p_0 and the rate of growth, r) will equal the new population after 1 year (p_1):

$$p_1 = p_0 + p_0 r = p_0(1 + r)$$

The population of the town at the end of 2 years will be p_2, and

$$p_2 = p_1 + p_1 r$$
$$p_2 = p_1(1 + r) \qquad \text{Factor out } p_1.$$
$$p_2 = p_0(1 + r)(1 + r) \qquad \text{Substitute } p_0(1 + r) \text{ for } p_0.$$
$$p_2 = p_0(1 + r)^2$$

The population at the end of the third year will be $p_3 = p_0(1 + r)^3$. Writing the terms in a sequence gives

$$p_0, \quad p_1(1 + r), \quad p_1(1 + r)^2, \quad p_0(1 + r)^3, \quad p_0(1 + r)^4, \ldots$$

This is a geometric sequence with p_0 as the first term and $1 + r$ as the common ratio. Recall that the nth term is given by the formula $l = ar^{n-1}$. In this example, $p_0 = 3500$, $1 + r = 1.06$, and (because the population after 20 years will be the value of the 21st term of the geometric sequence) $n = 21$. The population after 20 years is $= 3500(1.06)^{20}$. Use a calculator to find that $p \approx 11{,}225$. ∎

EXAMPLE 2   A woman deposits $2500 in a bank at 7% annual interest, compounded daily. If the investment is left untouched for 6 years, how much money will be in the account?

Solution   We let the initial amount in the account be a_0 and r be the rate. At the end of the first day, the account is worth

$$a_1 = a_0 + a_0\left(\frac{r}{365}\right) = a_0\left(1 + \frac{r}{365}\right)$$

After the second day, the account is worth

$$a_2 = a_1 + a_1\left(\frac{r}{365}\right) = a_1\left(1 + \frac{r}{365}\right) = a_0\left(1 + \frac{r}{365}\right)^2$$

The daily amounts form the following geometric sequence

$$a_0, \quad a_0\left(1 + \frac{r}{365}\right), \quad a_0\left(1 + \frac{r}{365}\right)^2, \quad a_0\left(1 + \frac{r}{365}\right)^3, \ldots$$

where a_0 is the initial deposit and r is the annual rate of interest.

Because interest is compounded daily for 6 years (2190 days), the amount at the end of 6 years will be the 2191th term of the sequence.

$$a_{2191} = 2500\left(1 + \frac{0.07}{365}\right)^{2190}$$

We can use a calculator to find that $a_{2191} = \$3804.75$.   ∎

EXAMPLE 3   The equation $s = 16t^2$ represents the distance in feet, s, that an object will fall in t seconds.

- After 1 second, the object has fallen 16 feet.
- After 2 seconds, the object has fallen 64 feet.
- After 3 seconds, the object has fallen 144 feet.

We see that the object fell 16 feet during the first second, 48 feet during the next second, and 80 feet during the third second. Find the distance the object falls during the 12th second.

Solution   The sequence 16, 48, 80, . . . is an arithmetic sequence with $a = 16$ and $d = 32$. To find the 12th term, we substitute these values into the formula for the last term and simplify:

$$l = a + (n - 1)d$$
$$l = 16 + (12 - 1)32$$
$$l = 16 + 11(32)$$
$$l = 368$$

During the 12th second, the object falls 368 feet.   ∎

EXAMPLE 4   A pump can remove 20% of the gas in a container with each stroke. Find the percentage of gas that remains in the container after six strokes.

Solution   We let V represent the volume of the container. Because each stroke of the pump removes 20% of the gas, 80% of the gas remains after each stroke, and we have the geometric sequence

$$V, \quad 0.80\,V, \quad 0.80(0.80\,V), \quad 0.80[0.80(0.80\,V)], \ldots$$

or

$$V, \quad 0.8\,V, \quad (0.8)^2\,V, \quad (0.8)^3\,V, \quad (0.8)^4\,V, \ldots$$

The amount of gas remaining after six strokes is the seventh term, l, of the sequence:

$$l = ar^{n-1}$$
$$l = V(0.8)^{7-1}$$
$$l = V(0.8)^6$$

We can use a calculator to find that approximately 26% of the gas remains after six strokes.   ■

9.4   EXERCISES

Decide whether each of the following exercises involves an arithmetic or geometric sequence, and then solve each problem.

1. **Staffing a mathematics department**   The number of students studying college algebra at State College is 623. The Department Chair expects enrollment to increase 10% each year. How many professors will be needed in 8 years to teach college algebra if one professor can handle 60 students?

2. **Borrowing money**   If Juanita borrows $5500 interest-free from her mother to buy a new car and agrees to pay her mother back at the rate of $105 per month, how much does she still owe after four years?

3. **Bouncing balls**   A Super Ball rebounds to approximately 95% of the height from which it was dropped. If it is dropped from a height of 10 meters, how high will the ball rebound after the 13th bounce?

4. **Investing money**   If a married couple invests $1000 in a 1-year certificate of deposit at $6\frac{3}{4}\%$ annual interest, compounded daily, how much interest will be earned during the year?

5. **Biology**   If a single cell divides into two cells every 30 minutes, how many cells will there be at the end of ten hours?

6. **Depreciation**   A lawn tractor, costing c dollars when new, depreciates 20% of the previous year's value each year. How much is the lawn tractor worth after 5 years?

7. **Financial planning**   Maria can invest $1000 at $7\frac{1}{2}\%$, compounded annually, or at $7\frac{1}{4}\%$, compounded daily. If she invests the money for a year, which is the better investment?

8. **Falling objects**   Find how many feet a brick will travel during the 10th second of its fall.

9. **Population study**   If the population of the earth were to double every 30 years, approximately how many people would there be in the year 3010? (Consider the population in 1990 to be 5 billion and use 1990 as the base year.)

10. **Investing money**   If Linda deposits $1300 in a bank at 7% interest, compounded annually, how much will be in the bank 17 years later? (Assume that there are no other transactions on the account.)

11. **Real estate appreciation**   If a house purchased for $50,000 in 1988 appreciates in value by 6% each year, how much will the house be worth in the year 2010?

12. *Compound interest* Find the value of $1000 left on deposit for 10 years at an annual rate of 7%, compounded annually.

13. *Compound interest* Find the value of $1000 left on deposit for 10 years at an annual rate of 7%, compounded quarterly.

14. *Compound interest* Find the value of $1000 left on deposit for 10 years at an annual rate of 7%, compounded monthly.

15. *Compound interest* Find the value of $1000 left on deposit for 10 years at an annual rate of 7%, compounded daily.

16. *Compound interest* Find the value of $1000 left on deposit for 10 years at an annual rate of 7%, compounded hourly.

17. *Saving for retirement* When John was 20 years old, he opened an individual retirement account by investing $2000 that will earn 11% interest, compounded quarterly. How much will his investment be worth when John is 65 years old?

18. *Biology* One bacterium divides into two bacteria every 5 minutes. If two bacteria multiply enough to completely fill a petri dish in 2 hours, how long will it take one bacterium to fill the dish?

19. *Mathematical myths* A legend tells of a king who offered to grant the inventor of the game of chess any request. The inventor said "Simply place one grain of wheat on the first square of a chessboard, two grains on the second, four on the third, and so on, until the board is full. Then give me the wheat." The king agreed. How many grains did the king need to fill the chessboard?

20. *Mathematical myths* Estimate the size of the wheat pile in Exercise 19. (*Hint:* There are about one-half million grains of wheat in a bushel.)

21. Does $0.999999 = 1$? Explain.

22. Does $0.999 \ldots = 1$? Explain.

MATHEMATICAL INDUCTION

■ The Method of Mathematical Induction

In Section 9.3, we developed the following formula for the sum of the first n terms of an arithmetic sequence:

$$S_n = \frac{n(a + l)}{2}$$

If we apply this formula to the arithmetic series $1 + 2 + 3 + \cdots + n$, we have

$$S_n = 1 + 2 + 3 + \cdots + n = \frac{n(1 + n)}{2}$$

To see that this formula is true, we can check it for some positive numbers n:

For $n = 1$: $1 = \dfrac{1(1 + 1)}{2}$ is a true statement because $1 = 1$.

For $n = 2$: $1 + 2 = \dfrac{2(1 + 2)}{2}$ is a true statement because $3 = 3$.

For $n = 3$: $1 + 2 + 3 = \dfrac{3(1 + 3)}{2}$ is a true statement because $6 = 6$.

For $n = 6$: $1 + 2 + 3 + 4 + 5 + 6 = \dfrac{6(1 + 6)}{2}$ is a true statement because $21 = 21$.

However, since the set of positive numbers is infinite, it is impossible to prove the formula by verifying it for all positive numbers. To verify this sequence formula for all positive numbers n, we must have a method of proof called **mathematical induction**, a method first used extensively by Giuseppe Peano (1858–1932).

The Method of Mathematical Induction

Suppose that we are standing in line for a movie and are worrying about whether we will be admitted. Even if the first person gets in, our worries might still be justified. Perhaps there is only room in the theater for a few people.

It would be good news to hear the theater manager say "If anyone gets in, the next person in line will get in also." However, this promise does not guarantee that anyone will be admitted. Perhaps the theater is already full and no one will get in.

However, when we see the first person in line walk in, we can conclude that everyone will be admitted, because we know two things:

- The first person was admitted.
- Because of the promise, if the first person is admitted, then so is the second, and when the second person is admitted, then so is the third, and so on until everyone gets in.

This situation is similar to a game played with dominoes. Suppose that some dominoes are placed on end as in Figure 9-1. When the first domino is knocked over, it knocks over the second. The second domino, in turn, knocks over the third, which knocks over the fourth, and so on until all of the dominoes fall. Two things must happen to guarantee that all of the dominoes fall:

- The first domino must be knocked over.
- Every domino that falls must knock over the next one.

When both conditions are met, it is certain that all of the dominoes will fall.

FIGURE 9-1

The preceding examples illustrate the principle of mathematical induction.

The Axiom of Mathematical Induction	If a statement involving the natural number n has the following two properties **1.** The statement is true for $n = 1$, and **2.** If the statement is true for $n = k$, then it is true for $n = k + 1$, then the statement is true for all natural numbers.

Mathematical induction provides a way to prove many theorems. Any proof by induction involves two parts. First, we must show that the formula is true for the number 1. Second, we must show that if the formula is true for any natural number k, then it also is true for the natural number $k + 1$. A proof by induction is complete only when both of these properties are established.

EXAMPLE 1   Use mathematical induction to prove that the formula

$$1 + 2 + 3 + \cdots + n = \frac{n(n + 1)}{2}$$

is true for every natural number n.

Solution   **Part 1.** Verify that the formula is true for $n = 1$. When $n = 1$, there is a single term, the number 1, on the left-hand side of the equation. Substituting 1 for n on the right-hand side, we have

$$1 = \frac{n(n + 1)}{2}$$

$$1 = \frac{(1)(1 + 1)}{2}$$

$$1 = 1$$

Thus, the formula is true when $n = 1$. Part 1 of the proof is complete.

Part 2. We assume that the given formula is true when $n = k$. By this assumption, called the **induction hypothesis**, we accept that

1. $$1 + 2 + 3 + \cdots + k = \frac{k(k + 1)}{2}$$

is a true statement. We must show that the induction hypothesis forces the given formula to be true when $n = k + 1$. We can show this by verifying the statement

2. $$1 + 2 + 3 + \cdots + k + (k + 1) = \frac{(k + 1)[(k + 1) + 1]}{2}$$

obtained from the given formula by replacing n with $k + 1$.

Comparing the left-hand sides of Equations 1 and 2 shows that the left-hand side of Equation 2 contains an extra term of $k + 1$. Thus, we add $k + 1$ to both sides of Equation 1 (which was assumed to be true) to obtain the equation

$$1 + 2 + 3 + \cdots + k + (k + 1) = \frac{k(k + 1)}{2} + (k + 1)$$

Because both terms on the right-hand side of this equation have a common factor of $k + 1$, the right-hand side factors, and the equation can be written as follows:

$$1 + 2 + 3 + \cdots + k + (k + 1) = (k + 1)\left(\frac{k}{2} + 1\right)$$

$$= (k + 1)\left(\frac{k + 2}{2}\right)$$

$$= \frac{(k + 1)(k + 2)}{2}$$

$$= \frac{(k + 1)[(k + 1) + 1]}{2}$$

This final result is Equation 2. Because the truth of Equation 1 implies the truth of Equation 2, Part 2 of the proof is complete. Parts 1 and 2 together establish that the formula is true for any natural number n. ∎

EXAMPLE 2   Use mathematical induction to prove the following formula for all natural numbers n.

$$1 + 5 + 9 + \cdots + (4n - 3) = n(2n - 1)$$

Solution   **Part 1.** First, we verify the formula for $n = 1$. When $n = 1$, there is a single term, the number 1, on the left-hand side of the equation. Substituting 1 for n on the right-hand side, we have

$$1 = 1[2(1) - 1]$$
$$1 = 1$$

Thus, the formula is true for $n = 1$. Part 1 of the proof is complete.

Part 2. We assume that the formula is true for $n = k$. Hence,

3. $1 + 5 + 9 + \cdots + (4k - 3) = k(2k - 1)$

is a true statement. To show that the induction hypothesis guarantees the truth of the formula for $k + 1$ terms, we add the $(k + 1)$th term to both sides of Equation 3. Because the terms on the left-hand side increase by 4, the $(k + 1)$th term is $(4k - 3) + 4$, or $4k + 1$. Adding $4k + 1$ to both sides of Equation 3 gives

$$1 + 5 + 9 + \cdots + (4k - 3) + (4k + 1) = k(2k - 1) + (4k + 1)$$

We can simplify the right-hand side and write the previous equation as follows:

$$1 + 5 + 9 + \cdots + (4k - 3) + [4(k + 1) - 3] = 2k^2 + 3k + 1$$
$$= (k + 1)(2k + 1)$$
$$= (k + 1)[2(k + 1) - 1]$$

Since this result has the same form as the given formula, except that $k + 1$ replaces n, the truth of the formula for $n = k$ implies the truth of the formula for $n = k + 1$. Proof 2 of the proof is complete.

Because both of the induction requirements are true, the formula is true for all natural numbers n. ∎

EXAMPLE 3 Prove that $\dfrac{1}{2} + \dfrac{1}{4} + \dfrac{1}{8} + \cdots + \dfrac{1}{2^n} < 1$.

Solution ***Part 1.*** We verify the formula for $n = 1$. When $n = 1$, there is a single term, the fraction $\frac{1}{2}$, on the left-hand side of the equation. Substituting 1 for n on the right-hand side, we have the following true statement:

$$\frac{1}{2} < 1$$

Thus, the formula is true for $n = 1$. Part 1 of the proof is complete.

Part 2. We assume that inequality is true for $n = k$. Thus,

$$\frac{1}{2} + \frac{1}{4} + \frac{1}{8} + \cdots + \frac{1}{2^k} < 1$$

We can multiply both sides of the above inequality by $\dfrac{1}{2}$ to get

$$\frac{1}{2}\left(\frac{1}{2} + \frac{1}{4} + \frac{1}{8} + \cdots + \frac{1}{2^k}\right) < 1\left(\frac{1}{2}\right)$$

or

$$\frac{1}{4} + \frac{1}{8} + \frac{1}{16} + \cdots + \frac{1}{2^{k+1}} < \frac{1}{2}$$

We now add $\dfrac{1}{2}$ to both sides of this inequality to get

$$\frac{1}{2} + \frac{1}{4} + \frac{1}{8} + \frac{1}{16} + \cdots + \frac{1}{2^{k+1}} < \frac{1}{2} + \frac{1}{2}$$

or

$$\frac{1}{2} + \frac{1}{4} + \frac{1}{8} + \frac{1}{16} + \cdots + \frac{1}{2^{k+1}} < 1$$

The resulting inequality is the same as the original inequality, except that $k + 1$ appears in place of n. Thus, the truth of the inequality for $n = k$ implies the truth of the inequality for $n = k + 1$. Part 2 of the proof is complete.

Because both of the induction requirements have been verified, this inequality is true for all natural numbers. ∎

There are statements that are not true when $n = 1$ but are true for all natural numbers equal to or greater than some given natural number, say, q. In these cases, we verify the given statements for $n = q$ in Part 1 of the induction proof. After establishing Part 2 of the induction proof, the given statement is proved for all natural numbers that are greater than or equal to q.

9.5 EXERCISES

In Exercises 1–4, verify each formula for $n = 1, 2, 3,$ and 4.

1. $5 + 10 + 15 + \cdots + 5n = \dfrac{5n(n + 1)}{2}$

2. $1^2 + 2^2 + 3^2 + \cdots + n^2 = \dfrac{n(n + 1)(2n + 1)}{6}$

3. $7 + 10 + 13 + \cdots + (3n + 4) = \dfrac{n(3n + 11)}{2}$

4. $1(3) + 2(4) + 3(5) + \cdots + n(n + 2) = \dfrac{n}{6}(n + 1)(2n + 7)$

In Exercises 5–20, prove each formula by mathematical induction, if possible.

5. $2 + 4 + 6 + \cdots + 2n = n(n + 1)$

6. $1 + 3 + 5 + \cdots + (2n - 1) = n^2$

7. $3 + 7 + 11 + \cdots + (4n - 1) = n(2n + 1)$

8. $4 + 8 + 12 + \cdots + 4n = 2n(n + 1)$

9. $10 + 6 + 2 + \cdots + (14 - 4n) = 12n - 2n^2$

10. $8 + 6 + 4 + \cdots + (10 - 2n) = 9n - n^2$

11. $2 + 5 + 8 + \cdots + (3n - 1) = \dfrac{n(3n + 1)}{2}$

12. $3 + 6 + 9 + \cdots + 3n = \dfrac{3n(n + 1)}{2}$

13. $1^2 + 2^2 + 3^2 + \cdots + n^2 = \dfrac{n(n + 1)(2n + 1)}{6}$

14. $1 + 2 + 3 + \cdots + (n - 1) + n + (n - 1) + \cdots + 3 + 2 + 1 = n^2$

15. $\dfrac{1}{3} + 2 + \dfrac{11}{3} + \cdots + \left(\dfrac{5}{3}n - \dfrac{4}{3}\right) = n\left(\dfrac{5}{6}n - \dfrac{1}{2}\right)$

16. $\dfrac{1}{1 \cdot 2} + \dfrac{1}{2 \cdot 3} + \dfrac{1}{3 \cdot 4} + \cdots + \dfrac{1}{n(n + 1)} = \dfrac{n}{n + 1}$

17. $\dfrac{1}{2} + \dfrac{1}{4} + \dfrac{1}{8} + \cdots + \left(\dfrac{1}{2}\right)^n = 1 - \left(\dfrac{1}{2}\right)^n$

18. $\dfrac{1}{3} + \dfrac{2}{9} + \dfrac{4}{27} + \cdots + \dfrac{1}{3}\left(\dfrac{2}{3}\right)^{n-1} = 1 - \left(\dfrac{2}{3}\right)^n$

19. $2^0 + 2^1 + 2^2 + 2^3 + \cdots + 2^{n-1} = 2^n - 1$

20. $1^3 + 2^3 + 3^3 + \cdots + n^3 = \left[\dfrac{n(n + 1)}{2}\right]^2$

21. Prove that $x - y$ is a factor of $x^n - y^n$. (*Hint:* Consider subtracting and adding xy^k to the binomial $x^{k+1} - y^{k+1}$).

22. Prove that $n < 2^n$.

23. There are $180°$ in the sum of the angles of any triangle. Prove that $(n - 2)180°$ is the sum of the angles of any simple polygon when n is its number of sides. (*Hint:* If a polygon has $k + 1$ sides, it has $k - 2$ sides plus three more sides.)

24. Consider the equation $1 + 3 + 5 + \cdots + 2n - 1 = 3n - 2$.
   a. Is the equation true for $n = 1$?
   b. Is the equation true for $n = 2$?
   c. Is the equation true for all natural numbers n?

25. If $1 + 2 + 3 + \cdots + n = \dfrac{n}{2}(n + 1) + 1$ were true for $n = k$, show that it would be true for $n = k + 1$. Is it true for $n = 1$?

26. Prove that $n + 1 = 1 + n$ for each natural number n.

27. If n is any natural number, prove that $7^n - 1$ is divisible by 6.

28. Prove that $1 + 2n < 3^n$ for $n > 1$.

29. Prove that if r is a real number where $r \neq 1$, then $1 + r + r^2 + \cdots + r^n = \dfrac{1 - r^{n+1}}{1 - r}$.

30. The expression a^m, where m is a natural number, was defined in Section 1.3. An alternative definition of a^m is (Part 1) $a^1 = a$ and (Part 2) $a^{m+1} = a^m \cdot a$. Use induction on n to prove the law of exponents, $a^m a^n = a^{m+n}$.

9.6 PERMUTATIONS AND COMBINATIONS

■ Permutations   ■ Formulas for Permutations   ■ Combinations
■ Formulas for Combinations

Lydia plans to go to dinner and attend a movie. If she has a choice of four restaurants and three movies, in how many ways can she spend her evening? There are four choices of restaurants, and for any one of these choices, there are three choices of movies, as shown in the tree diagram in Figure 9-2.

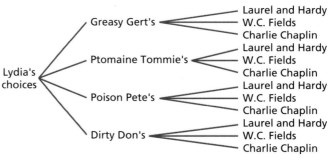

FIGURE 9-2

The diagram shows that Lydia has 12 ways to spend her evening. One possibility is to eat at Ptomaine Tommie's and watch W. C. Fields. Another is to eat at Dirty Don's and watch Laurel and Hardy.

Any situation that has several outcomes is called an **event**. Lydia's first event (choosing a restaurant) can occur in 4 ways. Her second event (choosing a movie) can occur in 3 ways. Thus, she has $4 \cdot 3$, or 12, ways to spend her evening. This example illustrates the **multiplication principle for events**.

The Multiplication Principle for Events	Let E_1 and E_2 be two events. If E_1 can be done in a_1 ways, and if—after E_1 has occurred—E_2 can be done in a_2 ways, then the event "E_1 followed by E_2" can be done in $a_1 \cdot a_2$ ways.

The multiplication principle can be extended to n events.

EXAMPLE 1 If a business executive has four ways to travel from New York to Chicago, three ways to travel from Chicago to Denver, and six ways to travel from Denver to San Francisco, in how many ways can she go from New York to San Francisco?

Solution We can let E_1 be the event "going from New York to Chicago," E_2 be the event "going from Chicago to Denver," and E_3 be the event "going from Denver to San Francisco." Since there are 4 ways to accomplish E_1, 3 ways to accomplish E_2, and 6 ways to accomplish E_3, the number of routes available is

$$4 \cdot 3 \cdot 6 = 72$$

■

■ Permutations

Suppose we want to arrange 7 books on a shelf. We can fill the first space with any of the 7 books, the second space with any of the remaining 6 books, the third space with any of the remaining 5 books, and so on, until there is only one space left to fill with the last book. According to the multiplication principle, the number of ways that we can arrange the books is

$$7 \cdot 6 \cdot 5 \cdot 4 \cdot 3 \cdot 2 \cdot 1 = 5040$$

When finding the number of possible arrangements of books on a shelf, we are finding the number of **permutations**. The number of permutations of 7 books, using all the books, is 5040. The symbol $P(n, r)$ is read as "the number of permutations of n things r at a time." Thus, $P(7, 7) = 5040$.

EXAMPLE 2 Assume that there are 7 signal flags of 7 different colors to hang on a mast. How many different signals can be sent when 3 flags are used?

Solution We are asked to find $P(7, 3)$, the number of permutations of 7 things using 3 of them. Any one of the 7 flags can hang in the top position on the mast. Any one of the 6 remaining flags can hang in the middle position, and any one of the remaining 5 flags can hang in the bottom position. By the multiplication principle, we have

$$P(7, 3) = 7 \cdot 6 \cdot 5 = 210$$

It is possible to send 210 different signals.

■

■ Formulas for Permutations

Although it is proper to write $P(7, 3) = 7 \cdot 6 \cdot 5$, we will change the form of the answer to obtain a convenient formula. To derive this formula, we proceed as follows:

$$P(7, 3) = 7 \cdot 6 \cdot 5 = \frac{7 \cdot 6 \cdot 5 \cdot 4 \cdot 3 \cdot 2 \cdot 1}{4 \cdot 3 \cdot 2 \cdot 1} = \frac{7!}{4!} = \frac{7!}{(7 - 3)!}$$

The generalization of this idea gives the following formula.

The Formula for Finding $P(n, r)$	The number of permutations of n things r at a time is given by $$P(n, r) = \frac{n!}{(n - r)!}$$

EXAMPLE 3 Find **a.** $P(8, 4)$, **b.** $P(n, n)$, and **c.** $P(n, 0)$.

Solution **a.** $P(8, 4) = \dfrac{8!}{(8 - 4)!} = \dfrac{8 \cdot 7 \cdot 6 \cdot 5 \cdot 4!}{4!} = 1680$

b. $P(n, n) = \dfrac{n!}{(n - n)!} = \dfrac{n!}{0!} = n!$

c. $P(n, 0) = \dfrac{n!}{(n - 0)!} = \dfrac{n!}{n!} = 1$ ∎

Parts **b** and **c** of Example 3 establish the following formulas.

The Formulas for Finding $P(n, n)$ and $P(n, 0)$	The number of permutations of n things n at a time and n things 0 at a time are given by the formulas $$P(n, n) = n! \quad \text{and} \quad P(n, 0) = 1$$

EXAMPLE 4 In how many ways can a baseball manager arrange a batting order of 9 players if there are 25 players on the team?

Solution To find the number of permutations of 25 things 9 at a time, we substitute 25 for n and 9 for r in the formula for finding $P(n, r)$.

$$P(n, r) = \frac{n!}{(n - r)!}$$

$$P(25, 9) = \frac{25!}{(25 - 9)!}$$

$$= \frac{25!}{16!}$$

$$= \frac{25 \cdot 24 \cdot 23 \cdot 22 \cdot 21 \cdot 20 \cdot 19 \cdot 18 \cdot 17 \cdot 16!}{16!}$$

$$\approx 741{,}354{,}768{,}000$$

The number of permutations is approximately 741,354,768,000. ∎

EXAMPLE 5 In how many ways can 5 people stand in a line if 2 people refuse to stand next to each other?

Solution The total number of ways that 5 people can stand in line is

$$P(5, 5) = 5! = 5 \cdot 4 \cdot 3 \cdot 2 \cdot 1 = 120$$

To find the number of ways that 5 people can stand in line if 2 people insist on standing together, we consider the two people as one person. Then, there are 4 people to stand in line, and this can be done in $P(4, 4) = 4! = 24$ ways. However, because either could be first, there are two arrangements for the pair who insist on standing together. Thus, there are $2 \cdot 4!$, or 48 ways that 5 people can stand in line if 2 people insist on standing together.

The number of ways that 5 people can stand in line if two people refuse to stand together is $5! = 120$ (the total number of ways to line up 5 people) minus $2 \cdot 4! = 48$ (the number of ways to line up the 5 people if 2 do stand together):

$$120 - 48 = 72$$

There are 72 ways to line up 5 people if 2 people refuse to stand next to each other. ∎

EXAMPLE 6 In how many ways can 5 people be seated at a round table?

Solution If we were to seat 5 people in a row, there would be 5! possible arrangements. However, at a round table, each person has a neighbor to the left and to the right. If each person moves one, two, three, four, or five places to the left, everyone has the same neighbors and the arrangement has not changed. Thus, we must divide 5! by 5 to get rid of these duplications. The number of ways that 5 people can be seated at a round table is

$$\frac{5!}{5} = 4! = 4 \cdot 3 \cdot 2 \cdot 1 = 24$$

∎

The results of Example 6 can be generalized into the following theorem.

Theorem	There are $(n - 1)!$ ways to arrange n things in a circle.

■ Combinations

Suppose that a class of 12 students selects a committee of 3 to plan a party. If a possible committee is made up of John, Maria, and Raul, the order of the 3 is not important. A committee of John, Maria, and Raul is the same as the committee of Maria, Raul, and John. However, if we assume that order is important and find the number of permutations of 12 things 3 at a time, we get

$$P(12, 3) = \frac{12!}{(12 - 3)!} = \frac{12 \cdot 11 \cdot 10 \cdot 9!}{9!} = 1320$$

Thus, there are 1320 ways of arranging 3 people if there are 12 people to choose from. However, with committees we do not care about order. Because there are six ways (3!) of ordering the committee of 3 students, the result of $P(12, 3) = 1320$ is 6 times too big. To get the correct number of committees, we must divide $P(12, 3)$ by 6:

$$\frac{P(12, 3)}{6} = \frac{1320}{6} = 220$$

In cases of selection where order is not important, we are interested in **combinations**, not permutations. The symbols $C(n, r)$ and $\binom{n}{r}$ both mean the number of combinations of n things at a time.

■ Formulas for Combinations

If a committee of r people is chosen from a total of n people, the number of possible committees is $C(n, r)$, and there will be $r!$ arrangements of each committee. If we consider the committee as an ordered grouping, the number of orderings is $P(n, r)$. Thus, we have

1. $r!C(n, r) = P(n, r)$

We can divide both sides of Equation 1 by $r!$ to obtain the formula for finding $C(n, r)$.

$$C(n, r) = \binom{n}{r} = \frac{P(n, r)}{r!} = \frac{n!}{r!(n - r)!}$$

The Formula for Finding $C(n, r)$	The number of combinations of n things r at a time is given by $$C(n, r) = \binom{n}{r} = \frac{n!}{r!(n - r)!}$$

In the exercises, you will be asked to prove the following formulas.

The Formulas for Finding $C(n, n)$ and $C(n, 0)$	If n is a whole number, then $$C(n, n) = 1 \quad \text{and} \quad C(n, 0) = 1$$

EXAMPLE 7   If Carla must read 4 books from a reading list of 10 books, how many choices does she have?

Solution   Because the order in which the books are read is unimportant, we find the number of combinations of 10 things 4 at a time:

$$\begin{aligned}
C(10, 4) &= \frac{10!}{4!(10 - 4)!} = \frac{10 \cdot 9 \cdot 8 \cdot 7 \cdot 6!}{4 \cdot 3 \cdot 2 \cdot 1 \cdot 6!} \\
&= \frac{10 \cdot 9 \cdot 8 \cdot 7}{4 \cdot 3 \cdot 2} \\
&= 210
\end{aligned}$$

Carla has 210 choices.   ■

EXAMPLE 8   A class consists of 15 men and 8 women. In how many ways can a debate team be chosen with 3 men and 3 women?

Solution   There are $C(15, 3)$ ways of choosing three men and $C(8, 3)$ ways of choosing three women. By the multiplication principle, there are $C(15, 3) \cdot C(8, 3)$ ways of choosing members of the debate team:

$$C(15, 3) \cdot C(8, 3) = \frac{15!}{3!(15 - 3)!} \cdot \frac{8!}{3!(8 - 3)!}$$

$$= \frac{15 \cdot 14 \cdot 13}{6} \cdot \frac{8 \cdot 7 \cdot 6}{6}$$

$$= 25{,}480$$

There are 25,480 ways to choose the debate team.   ■

The formula

$$C(n, r) = \frac{n!}{r!(n - r)!}$$

gives the coefficient of the $(r + 1)$th term of the binomial expansion of $(a + b)^n$. This implies that the coefficients of a binomial expansion can be used to solve problems involving combinations. The binomial theorem is restated below using combination notation.

The Binomial Theorem

If n is any positive integer, then

$$(a + b)^n = \binom{n}{0}a^n + \binom{n}{1}a^{n-1}b + \binom{n}{2}a^{n-2}b^2$$

$$+ \cdots + \binom{n}{r}a^{n-r}b^r + \cdots + \binom{n}{n}b^n$$

EXAMPLE 9   Use Pascal's triangle to compute $C(7, 5)$.

Solution   Consider the eighth row of Pascal's triangle and the corresponding combinations:

$$\begin{array}{cccccccc} 1 & 7 & 21 & 35 & 35 & 21 & 7 & 1 \\ \binom{7}{0} & \binom{7}{1} & \binom{7}{2} & \binom{7}{3} & \binom{7}{4} & \binom{7}{5} & \binom{7}{6} & \binom{7}{7} \end{array}$$

$$C(7, 5) = \binom{7}{5} = 21$$ ■

When discussing permutations and combinations, a "word" is a distinguishable arrangement of letters. For example, 6 words can be formed with the letters a, b, and c if all 3 letters are used exactly once. The six words are abc, acb, bac, bca, cab, and cba. If there are n distinct letters and each letter is used once, the number of distinct words that can be formed is $n! = P(n, n)$. It is more complicated to compute the number of distinguishable words that can be formed with n letters if some of the letters appear more than once.

EXAMPLE 10   Find the number of "words" that can be formed if each of the 6 letters of the word *little* is used once.

Solution   For the moment, assume that all letters of the word *little* are distinguishable: "LitTle." The number of words that can be formed using each letter once is $6! = P(6, 6)$. However, in reality we cannot tell the *l*'s or the *t*'s apart. Therefore, we must divide by a number to get rid of these duplications. Because there are $2!$ orderings of the two *l*'s and $2!$ orderings of the two *t*'s, we divide by $2! \cdot 2!$. The number of words that can be formed using each letter of the word *little* is

$$\frac{P(6, 6)}{2!2!} = \frac{6!}{2!2!} = \frac{6 \cdot 5 \cdot 4 \cdot 3 \cdot 2 \cdot 1}{2 \cdot 1 \cdot 2 \cdot 1} = 180 \qquad \blacksquare$$

Example 10 illustrates the following general principle.

Theorem	If a word with n letters has a of one letter, b of another letter, and so on, then the number of distinguishable words that can be formed using each letter of the n-letter word exactly once is $$\frac{n!}{a!b! \cdots}$$

9.6   EXERCISES

1. *Choosing a lunch*   A lunchroom has a machine with eight kinds of sandwiches, a machine with four kinds of soda, a machine with both white and chocolate milk, and a machine with three kinds of ice cream. How many different lunches can be chosen? (Consider a lunch to be one sandwich, one drink, and one ice cream.)

2. *Manufacturing license plates*   How many six-digit license plates can be manufactured if only numbers are used and no license plate number begins with 0?

3. *Available phone numbers*   How many different seven-digit phone numbers can be used in one area code if no phone number begins with 0 or 1?

4. *Arranging letters*   In how many ways can the letters of the word *number* be arranged?

5. *Arranging letters with restrictions*   In how many ways can the letters of the word *number* be arranged if the *e* and *r* must remain next to each other?

6. *Arranging letters with restrictions*   In how many ways can the letters of the word *number* be arranged if the *e* and *r* cannot be side by side?

7. *Arranging letters with repetitions*   In how many ways can five Scrabble tiles bearing the letters, *F, F, F, L,* and *U* be arranged to spell the word *fluff*?

8. *Arranging letters with repetitions*   In how many ways can six Scrabble tiles bearing the letters, *B, E, E, E, F,* and *L* be arranged to spell the word *feeble*?

In Exercises 9–24, evaluate each expression.

9. $P(7, 4)$ 10. $P(8, 3)$ 11. $C(7, 4)$ 12. $C(8, 3)$

13. $P(5, 5)$

14. $P(5, 0)$

15. $\binom{5}{4}$

16. $\binom{8}{4}$

17. $\binom{5}{0}$

18. $\binom{5}{5}$

19. $P(5, 4) \cdot C(5, 3)$

20. $P(3, 2) \cdot C(4, 3)$

21. $\binom{5}{3}\binom{4}{3}\binom{3}{3}$

22. $\binom{5}{5}\binom{6}{6}\binom{7}{7}\binom{8}{8}$

23. $\binom{68}{66}$

24. $\binom{100}{99}$

25. *Placing people in line* In how many arrangements can 8 women be placed in a line?

26. *Placing people in line* In how many arrangements can 5 women and 5 men be placed in a line if the women and men alternate?

27. *Placing people in line* In how many arrangements can 5 women and 5 men be placed in a line if all the men line up first?

28. *Placing people in line* In how many arrangements can 5 women and 5 men be placed in a line if all the women line up first?

29. *Combination locks* How many permutations does a combination lock have if each combination has 3 numbers, no two numbers of the combination are the same, and the lock dial has 30 notches?

30. *Combination locks* How many permutations does a combination lock have if each combination has 3 numbers, no two numbers of the combination are the same, and the lock dial has 100 notches?

31. *Seating at a table* In how many ways can 8 people be seated at a round table?

32. *Seating at a table* In how many ways can 7 people be seated at a round table?

33. *Seating at a table with conditions* In how many ways can 6 people be seated at a round table if 2 of the people insist on sitting together?

34. *Seating arrangements with conditions* In how many ways can 6 people be seated at a round table if 2 of the people refuse to sit together?

35. *Arrangements in a circle* In how many ways can 7 children be arranged in a circle if Sally and John want to sit together and Martha and Peter want to sit together?

36. *Arrangements in a circle* In how many ways can 8 children be arranged in a circle if Laura, Scott, and Paula want to sit together?

37. *Selecting candy bars* In how many ways can 4 candy bars be selected from 10 different candy bars?

38. *Selecting birthday cards* In how many ways can 6 birthday cards be selected from 24 different cards?

39. *Circuit wiring* A wiring harness containing a red, a green, a white, and a black wire must be attached to a control panel. In how many different orders can the wires be attached?

40. *Grading homework* A professor grades homework by randomly checking 7 of the 20 problems assigned. In how many different ways can this be done?

41. *Forming words with distinct letters* How many words can be formed from the letters of the word *plastic* if each letter is to be used once?

42. *Forming words with repeated letters* How many words can be formed from the letters of the word *banana* if each letter is to be used once?

43. *Making license plates* How many license plates can be made using two different letters followed by four different digits if the first digit cannot be 0 and the letter *O* is not used?

44. *Planning class schedules* If there are seven class periods in a school day and a typical student takes 5 classes, how many different time patterns are possible for the student?

45. *Selecting golf balls* From a bucket containing 6 red and 8 white golf balls, in how many ways can we draw 6 golf balls of which 3 are red and 3 are white?

46. *Selecting committees* In how many ways can you select a committee of 3 Republicans and 3 Democrats from a group containing 18 Democrats and 11 Republicans?

47. *Selecting committees* In how many ways can you select a committee of 4 Democrats and 3 Republicans from a group containing 12 Democrats and 10 Republicans?

48. Drawing cards In how many ways can you select a group of 5 red cards and 2 black cards from a deck containing 10 red cards and 8 black cards?

49. Planning dinner In how many ways can a husband and wife choose 2 different dinners from a menu of 17 dinners?

50. Placing people in line In how many ways can 7 people stand in a row if 2 of the people refuse to stand together?

51. Geometry How many lines are determined by 8 points if no 3 points lie on a straight line?

52. Geometry How many lines are determined by 10 points if no 3 points lie on a straight line?

53. Coaching basketball How many different teams can a basketball coach start if the entire squad consists of 10 players? (Assume that a starting team has 5 players and each player can play all positions.)

54. Managing baseball How many different teams can a baseball manager start if the entire squad consists of 25 players? (Assume that a starting team has 9 players and each player can play all positions.)

55. Selecting job applicants There are 30 qualified applicants for 5 openings in the sales department. In how many different ways can the group of 5 be selected?

56. Sales promotions If a customer purchases a new stereo system during the Spring Sale, he may choose any six CDs from 20 classical and 30 jazz selections. In how many ways can the customer choose three of each?

57. Guessing on matching questions Ten words are to be paired with the correct 10 out of 12 possible definitions. How many ways are there of guessing?

58. Guessing on true-false exams How many possible ways are there of guessing on a 10-question true-false exam, if it is known that the instructor usually has 5 true and 5 false responses?

59. Use Pascal's triangle to find $C(8, 5)$.

60. Use Pascal's triangle to find $C(10, 8)$.

61. Prove that $C(n, n) = 1$.

62. Prove that $C(n, 0) = 1$.

63. Prove that $\begin{pmatrix} n \\ r \end{pmatrix} = \begin{pmatrix} n \\ n - r \end{pmatrix}$.

64. Show that the binomial theorem can be expressed in the form

$$(a + b)^n = \sum_{k=0}^{n} \begin{pmatrix} n \\ k \end{pmatrix} a^{n-k} b^k$$

9.7   PROBABILITY

■ Multiplication Property of Probabilities

The probability that a certain event will occur is a measure of the likelihood of that event. A tossed coin, for example, can land in two equally likely ways, either heads or tails. Because one of these two outcomes is heads, we expect that out of several tosses about half will be heads. We say that the probability of obtaining heads in a single toss of the coin is $\frac{1}{2}$.

If records show that out of 100 days with weather conditions like today's, 30 have received rain, the weather service reports, "Today, there is a $\frac{30}{100}$ or 30% probability of rain."

An **experiment** is any process for which the outcome is uncertain. Thus, tossing a coin, rolling a die, drawing a card, and predicting rain are examples of experiments. For any experiment, a list of all possible outcomes is called a **sample space**.

For example, the sample space, S, for the experiment of tossing two coins is the set

$$S = \{(H, H), (H, T), (T, H), (T, T)\}$$

where the ordered pair (H, T), for example, represents the outcome "heads on the first coin and tails on the second." Because there are two possible outcomes for the first coin and two for the second, we know by the multiplication principle for events that there are $2 \cdot 2$, or 4, total possible outcomes. There are 4 elements in the sample space S, and we write

$$n(S) = 4 \qquad \text{Read as "The number of elements in set } S \text{ is 4."}$$

An **event** associated with an experiment is any subset of the sample space of that experiment. For example, if E is the event "getting at least one heads" in the experiment of tossing two coins, then

$$E = \{(H, H), (H, T), (T, H)\}$$

and $n(E) = 3$. Because the outcome of getting at least one heads can occur in 3 out of 4 possible ways, we say that the **probability** of a favorable outcome is $\frac{3}{4}$, and we write

$$P(E) = P(\text{at least one heads}) = \frac{3}{4}$$

We define the probability of an event as follows.

Probability of an Event	If S is the sample space of an experiment with n distinct and equally likely outcomes and E is an event that occurs in s of those ways, then the **probability of E** is $$P(E) = \frac{n(E)}{n(S)} = \frac{s}{n}$$

Because $0 \leq s \leq n$, it follows that $0 \leq \frac{s}{n} \leq 1$. Thus, all probabilities have values from 0 to 1. An event that cannot happen has probability 0. An event that is certain to happen has probability 1.

To say that the probability of tossing heads on one toss of a coin is $\frac{1}{2}$ means that if a fair coin is tossed a large number of times, the ratio of the number of heads to the total number of tosses is nearly $\frac{1}{2}$.

To say that the probability of rolling 5 on one roll of a die is $\frac{1}{6}$ means that as the number of rolls approaches infinity, the ratio of the number of favorable outcomes (rolling a 5) to the total number of outcomes (rolling a 1, 2, 3, 4, 5, or 6) approaches $\frac{1}{6}$.

EXAMPLE 1 Exhibit the sample space of the experiment "rolling two dice a single time."

Solution We can use ordered-pair notation and let the first number be the result on the first die and the second number the result on the second die. The sample space, S, of the experiment is the set with the following elements:

$$
\begin{array}{cccccc}
(1,1) & (1,2) & (1,3) & (1,4) & (1,5) & (1,6) \\
(2,1) & (2,2) & (2,3) & (2,4) & (2,5) & (2,6) \\
(3,1) & (3,2) & (3,3) & (3,4) & (3,5) & (3,6) \\
(4,1) & (4,2) & (4,3) & (4,4) & (4,5) & (4,6) \\
(5,1) & (5,2) & (5,3) & (5,4) & (5,5) & (5,6) \\
(6,1) & (6,2) & (6,3) & (6,4) & (6,5) & (6,6)
\end{array}
$$

Because there are 6 possible outcomes with the first die and 6 possible outcomes with the second die, the multiplication principle for events tells us to expect $6 \cdot 6$, or 36 equally likely possible outcomes. Thus, the sample space contains 36 ordered pairs, and we have $n(S) = 36$. ■

EXAMPLE 2 Find the possibility of the event "tossing a sum of 7 on one toss of two dice."

Solution The sample space for this experiment is listed in Example 1. We let E be the set of favorable outcomes, those that give a sum of 7:

$$E = \{(1,6), (2,5), (3,4), (4,3), (5,2), (6,1)\}$$

Since there are 6 favorable outcomes among the 36 equally likely outcomes, we have that $n(E) = 6$, and

$$P(E) = P(\text{tossing a 7}) = \frac{n(E)}{n(S)} = \frac{6}{36} = \frac{1}{6}$$ ■

A standard playing deck of 52 cards has two red suits, hearts and diamonds, and two black suits, clubs and spades. Each suit has 13 cards, including an ace, king, queen, jack, and cards numbered from 2 to 10. We will refer to a standard deck of cards in many examples and exercises.

EXAMPLE 3 Find the probability of drawing 5 cards, all hearts, from a standard deck of cards.

Solution Because the number of ways to draw 5 hearts from the 13 hearts is $C(13, 5)$, we have $n(E) = C(13, 5)$. Because the number of ways to draw 5 cards from the deck is $C(52, 5)$, we have $n(S) = C(52, 5)$. The probability of drawing 5 hearts is the ratio of the number of favorable outcomes to the number of possible outcomes.

$$P(5 \text{ hearts}) = \frac{C(13, 5)}{C(52, 5)}$$

$$P(5 \text{ hearts}) = \frac{\dfrac{13!}{5!8!}}{\dfrac{52!}{5!47!}}$$

$$= \frac{13!}{5!8!} \cdot \frac{5!47!}{52!}$$

$$= \frac{13 \cdot 12 \cdot 11 \cdot 10 \cdot 9 \cdot 8!}{8!} \cdot \frac{47!}{52 \cdot 51 \cdot 50 \cdot 49 \cdot 48 \cdot 47!}$$

$$= \frac{13 \cdot 12 \cdot 11 \cdot 10 \cdot 9}{52 \cdot 51 \cdot 50 \cdot 49 \cdot 48}$$

$$= \frac{33}{66,640}$$

The probability of drawing 5 hearts is $\frac{33}{66,640}$. ∎

■ Multiplication Property of Probabilities

There is a property for probabilities that is similar to the multiplication principle for events. In the following theorem, we read $P(A \cap B)$ as "the probability of A and B" and $P(B|A)$ as "the probability of B given A." If A and B are events, the set $A \cap B$ contains the outcomes that are in both A and B.

Multiplication Property of Probabilities	If $P(A)$ represents the probability of event A, and $P(B|A)$ represents the probability that event B will occur after event A, then
	$$P(A \cap B) = P(A) \cdot P(B|A)$$

EXAMPLE 4 A box contains 40 cubes of the same size. Of these cubes, 17 are red, 13 are blue, and the rest are yellow. If 2 cubes are drawn at random, without replacement, find the probability that 2 yellow cubes will be drawn.

Solution Of the 40 cubes in the box, 10 are yellow. Thus, the probability of getting a yellow cube on the first draw is

$$P(\text{yellow cube on the first draw}) = \frac{10}{40} = \frac{1}{4}$$

Because there is no replacement after the first drawn, 39 cubes remain in the box, and 9 of these are yellow. Thus, the probability of drawing a yellow cube on the second draw after drawing a yellow cube on the first draw is

$$P(\text{yellow cube on the second draw}) = \frac{9}{39} = \frac{3}{13}$$

The probability of drawing 2 yellow cubes in succession is the product of the probability of drawing a yellow cube on the first draw and the probability of drawing a yellow cube on the second draw.

$$P(\text{drawing two yellow cubes}) = \frac{1}{4} \cdot \frac{3}{13} = \frac{3}{52}$$ ∎

EXAMPLE 5 Repeat Example 3 using the multiplication property of probabilities.

Solution The probability of drawing a heart on the first draw is $\frac{13}{52}$. The probability of drawing a heart on the second draw *given that we got a heart on the first draw* is $\frac{12}{51}$. The probability is $\frac{11}{50}$ on the third draw, $\frac{10}{49}$ on the fourth draw, and $\frac{9}{48}$ on the fifth draw. By the multiplication property of probabilities,

$$P(5 \text{ hearts in a row}) = \frac{13}{52} \cdot \frac{12}{51} \cdot \frac{11}{50} \cdot \frac{10}{49} \cdot \frac{9}{48}$$

$$= \frac{33}{66,640} \qquad \blacksquare$$

EXAMPLE 6   In a school, 30% of the students are gifted in mathematics and 10% are gifted in both art and mathematics. If a student is gifted in mathematics, find the probability that the student is also gifted in art.

Solution   Let $P(M)$ be the probability that a randomly chosen student is gifted in mathematics, and let $P(M \cap A)$ be the probability that the student is gifted in both art and mathematics. We must find $P(A|M)$, the probability that the student is gifted in art, given that he or she is gifted in mathematics. To do so, we substitute the given values

$$P(M) = 0.3 \qquad \text{and} \qquad P(M \cap A) = 0.1$$

in the formula for multiplication of probabilities and solve for $P(A|M)$:

$$P(M \cap A) = P(M) \cdot P(A|M)$$
$$\mathbf{0.1} = \mathbf{(0.3)}P(A|M)$$
$$P(A|M) = \frac{0.1}{0.3}$$
$$= \frac{1}{3}$$

If a student is gifted in mathematics, the probability is $\frac{1}{3}$ that he or she is also gifted in art. $\qquad \blacksquare$

9.7 EXERCISES

In Exercises 1–4, list the sample space of each experiment.

1. Rolling one die and tossing one coin

2. Tossing three coins

3. Selecting a letter of the alphabet

4. Picking a one-digit number

In Exercises 5–8, an ordinary die is tossed. Find the probability of each event.

5. Tossing a 2

6. Tossing a number greater than 4

7. Tossing a number larger than 1 but less than 6

8. Tossing an odd number

In Exercises 9–12, balls numbered from 1 to 42 are placed in a container and stirred. If one is drawn at random, find the probability of each result.

9. The number is less than 20.

10. The number is less than 50.

11. The number is a prime number.

12. The number is less than 10 or greater than 40.

In Exercises 13–16, refer to the spinner in Illustration 1. If the spinner is spun, find the probability of each event. Assume that the spinner never stops on a line.

13. The spinner stops on red.

14. The spinner stops on green.

15. The spinner stops on orange.

16. The spinner stops on yellow.

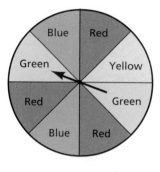

ILLUSTRATION 1

In Exercises 17–34, find the probability of each given event.

17. Rolling a sum of 4 on one roll of two dice

18. Drawing a diamond on one draw from a card deck

19. Drawing two aces in succession from a card deck if the card is replaced and the deck is shuffled after the first draw

20. Drawing two aces from a card deck without replacing the card after the first draw

21. Drawing a red egg from a basket containing 5 red eggs and 7 blue eggs

22. Getting 2 red eggs in a single scoop from a bucket containing 5 red eggs and 7 yellow eggs

23. Drawing a bridge hand of 13 cards, all of one suit

24. Drawing 6 diamonds from a card deck without replacing the cards after each draw

25. Drawing 5 aces from a card deck without replacing the cards after each draw

26. Drawing 5 clubs from the black cards in a card deck

27. Drawing a face card (king, queen, or jack) from a card deck

28. Drawing 6 face cards in a row from a card deck without replacing the cards after each draw

29. Drawing 5 orange cubes from a bowl containing 5 orange cubes and 1 beige cube

30. Rolling a sum of 4 with one roll of three dice

31. Rolling a sum of 11 with one roll of three dice

32. Picking, at random, 5 Republicans from a group containing 8 Republicans and 10 Democrats

33. Tossing 3 heads in 5 tosses of a fair coin

34. Tossing 5 heads in 5 tosses of a fair coin

In Exercises 35–40, assume that the probability that an airplane engine will fail during a torture test is $\frac{1}{2}$ and that the aircraft in question has 4 engines.

35. Construct a sample space for the torture test.

36. Find the probability that all engines will survive the test.

37. Find the probability that exactly 1 engine will survive.

38. Find the probability that exactly 2 engines will survive.

39. Find the probability that exactly 3 engines will survive.

40. Find the probability that no engines will survive.

41. Find the sum of the probabilities in Exercises 36 through 40.

In Exercises 42–44, assume that a survey of 282 people is taken to determine the opinions of doctors, teachers, and lawyers on a proposed piece of legislation, with the following results.

	Number that favor	Number that oppose	Number with no opinion	Total
Doctors	70	32	17	119
Teachers	83	24	10	117
Lawyers	23	15	8	46
Total	176	71	35	282

A person is chosen at random from those surveyed. Refer to the chart to find each probability.

42. The person favors the legislation.

43. A doctor opposes the legislation.

44. A person who opposes the legislation is a lawyer.

45. *Quality control* In a batch of 10 tires, 2 are known to be defective. If 4 tires are chosen at random, find the probability that all 4 tires are good.

46. *Medicine* Out of a group of 9 patients treated with a new drug, 4 suffered a relapse. Find the probability that 3 patients of this group of 9, chosen at random, will remain disease-free.

In Exercises 47–54, use the multiplication property of probabilities.

47. If $P(A) = 0.3$ and $P(B|A) = 0.6$, find $P(A \cap B)$.

48. If $P(A \cap B) = 0.3$ and $P(B|A) = 0.6$, find $P(A)$.

49. *Conditional probability* The probability that a person owns a luxury car is 0.2, and the probability that the owner of such a car also owns a personal computer is 0.7. Find the probability that a person, chosen at random, owns both a luxury car and a personal computer.

50. *Conditional probability* If 40% of the population have completed college, and 85% of the college graduates are registered to vote, what percent of the population are college graduates and registered voters?

51. *Conditional probability* About 25% of the population watches the evening television news coverage as well as the morning soap operas. If 75% of the population watches the news, what percentage of those who watch the news also watch the soaps?

52. *Conditional probability* The probability of rain today is 0.40. If it rains, the probability that Bill will forget his raincoat is 0.70. Find the probability that Bill will get wet.

53. If $P(A \cap B) = 0.7$, is it possible that $P(B|A) = 0.6$? Explain.

54. Is it possible that $P(A \cap B) = P(A)$? Explain.

9.8 COMPUTATION OF COMPOUND PROBABILITIES

Sometimes we must find the probability of one event or another, or the probability of one event and another. Such events are called **compound events**. As we have seen, if A and B are two events, the probability that A and B will both occur is

$P(A \cap B)$. In this section we also discuss $P(A \cup B)$, the probability that either A or B will occur.

Suppose we want to find the probability of drawing a king or a heart from a standard card deck. If K is the event "drawing a king" and H is the event "drawing a heart," then $P(K) = \frac{4}{52}$, and $P(H) = \frac{13}{52}$. However, the probability of drawing a king *or* a heart is not the sum of these two probabilities. Because the king of hearts was counted twice, once as a king and once as a heart, and the probability of drawing the king of hearts is $\frac{1}{52}$, we must subtract $\frac{1}{52}$ from the sum of $\frac{4}{52}$ and $\frac{13}{52}$ to get the correct probability. Thus,

$$P(\text{king } or \text{ heart}) = P(\text{king}) + P(\text{heart}) - P(\text{king of hearts})$$
$$P(K \cup H) = P(K) + P(H) - P(K \cap H)$$
$$= \frac{4}{52} + \frac{13}{52} - \frac{1}{52}$$
$$= \frac{16}{52}$$
$$= \frac{4}{13}$$

In general, we have the following theorem.

Theorem	If A and B are two events, then $$P(A \cup B) = P(A) + P(B) - P(A \cap B)$$

If events A and B have no outcomes in common, then $A \cap B = \emptyset$ and $P(A \cap B) = P(\emptyset) = 0$. Such events are **mutually exclusive** (that is, if one event occurs, the other cannot). The following theorem holds true for mutually exclusive events.

Theorem	If A and B cannot occur simultaneously, then $$P(A \cup B) = P(A) + P(B)$$

The event $\overline{A}$ (read as "not A") contains all outcomes of the sample space that are not elements of event A. Because the events A and $\overline{A}$ are mutually exclusive,

$$P(A \cup \overline{A}) = P(A) + P(\overline{A})$$

Because either event A or event $\overline{A}$ must happen, $P(A \cup \overline{A}) = 1$. Thus,

$$P(A \cup \overline{A}) = 1$$
$$P(A) + P(\overline{A}) = 1$$
$$P(\overline{A}) = 1 - P(A) \qquad \text{Add } -P(A) \text{ to both sides.}$$

This result gives a third theorem about compound probabilities.

Theorem	If A is any event, then $$P(\bar{A}) = 1 - P(A)$$

EXAMPLE 1 A guidance counselor tells a student that his probability of earning a grade of D in algebra is $\frac{1}{5}$ and his probability of earning an F is $\frac{1}{25}$. Find the probability that the student earns a C or better.

Solution Because "earning a D" and "earning an F" are mutually exclusive, the probability of earning a D or F is given by

$$P(D \cup F) = P(D) + P(F) \qquad \text{Note that } P(D \cap F) = 0.$$
$$= \frac{1}{5} + \frac{1}{25}$$
$$= \frac{6}{25}$$

The probability that the student will receive a C or better is

$$P(\text{C or better}) = 1 - P(D \cup F)$$
$$= 1 - \frac{6}{25}$$
$$= \frac{19}{25}$$

The student's probability of earning a C or better is $\frac{19}{25}$. ∎

If two events do not influence each other, they are called **independent events**.

Independent Events	The events A and B are said to be **independent events** if and only if $P(B) = P(B|A)$.

Substituting $P(B)$ for $P(B|A)$ in the multiplication property for probabilities gives a formula for computing probabilities of compound independent events.

Theorem	If A and B are independent events, then $$P(A \cap B) = P(A) \cdot P(B)$$

The event A of "drawing an ace from a standard deck of cards" and the event B of "tossing heads" on one toss of a coin are independent events, because neither event influences the other. Consequently,

$$P(A \cap B) = P(A) \cdot P(B)$$

$$P(\text{drawing an ace and tossing heads}) = P(\text{drawing an ace}) \cdot P(\text{tossing heads})$$

$$P(\text{drawing an ace and tossing heads}) = \frac{4}{52} \cdot \frac{1}{2} = \frac{1}{26}$$

EXAMPLE 2   The probability that a particular baseball player can get a hit is $\frac{1}{3}$. Find the probability that she will get three hits in a row.

Solution   Assume that the three events (each time at bat) are independent; one time at bat does not influence her chances of getting a hit on another turn at bat. Because $P(E_1) = \frac{1}{3}$, $P(E_2) = \frac{1}{3}$, and $P(E_3) = \frac{1}{3}$,

$$P(E_1 \cap E_2 \cap E_3) = \frac{1}{3} \cdot \frac{1}{3} \cdot \frac{1}{3} = \frac{1}{27}$$

The probability that she gets three hits in a row is $\frac{1}{27}$.   ■

EXAMPLE 3   A die is tossed three times. Find the probability that the outcome is six on the first toss, an even number on the second toss, and an odd prime number on the third toss.

Solution   The probability of a six on any toss is $P(6) = \frac{1}{6}$. Because there are three even integers represented on the faces of a die, the probability of tossing an even number is $P(\text{even number}) = P(E) = \frac{3}{6} = \frac{1}{2}$. Since the numbers 3 and 5 are the only odd prime numbers on the faces of a die, the probability of tossing an odd prime is $P(\text{odd prime}) = P(O) = \frac{2}{6} = \frac{1}{3}$.

Because these three events are independent, the probability of the three events happening in succession is the product of the probabilities:

$$\begin{aligned} P(\text{six and even number and odd prime}) &= P(6 \cap E \cap O) \\ &= P(6) \cdot P(E) \cdot P(O) \\ &= \frac{1}{6} \cdot \frac{1}{2} \cdot \frac{1}{3} \\ &= \frac{1}{36} \end{aligned}$$ ■

EXAMPLE 4   The probability that a drug will cure dandruff is $\frac{1}{8}$. However, if the drug is used, the probability that the patient will have side effects is $\frac{1}{6}$. Find the probability that a patient who uses the drug will be cured and will suffer no side effects.

Solution   The probability that the drug will cure dandruff is $P(C) = \frac{1}{8}$. The probability of having side effects is $P(E) = \frac{1}{6}$. The probability that the patient will have no side effects is

$$P(\bar{E}) = 1 - P(E) = 1 - \frac{1}{6} = \frac{5}{6}$$

If we assume that these events are independent, we have

$$P(\text{cure and no side effects}) = P(C \cap \overline{E})$$
$$= P(C) \cdot P(\overline{E})$$
$$= \frac{1}{8} \cdot \frac{5}{6}$$
$$= \frac{5}{48}$$
■

9.8 EXERCISES

In Exercises 1–4, assume that you draw one card from a card deck. Find the probability of each event.

1. Drawing a black card

2. Drawing a jack

3. Drawing a black card or an ace

4. Drawing a red card or a face card

In Exercises 5–8, assume that you draw two cards from a card deck, without replacement. Find the probability of each event.

5. Drawing two aces

6. Drawing three aces

7. Drawing a club and then another black card

8. Drawing a heart and then a spade

In Exercises 9–12, assume that you roll two dice once. Find the probability of each result.

9. Rolling a sum of 7 or 6

10. Rolling a sum of 5 or an even sum

11. Rolling a sum of 10 or an odd sum

12. Rolling a sum of 12 or 1

In Exercises 13–16, assume that you have a bucket containing 7 beige capsules, 3 blue capsules, and 6 green capsules. You make a single draw from the bucket, taking one capsule. Find the probability of each result.

13. Drawing a beige or a blue capsule

14. Drawing a green capsule

15. Not drawing a blue capsule

16. Not drawing either a beige or a blue capsule

In Exercises 17–19, assume that you are using the same bucket of capsules as in Exercises 13–16.

17. On two draws from the bucket, find the probability of drawing a beige capsule followed by a green capsule. (Assume that the capsule is returned to the bucket after the first draw.)

18. On two draws from the bucket, find the probability of drawing one blue capsule and one green capsule. (Assume that the capsule is not returned to the bucket after the first draw.)

19. On three successive draws from the bucket (without replacement), find the probability of failing to draw a beige capsule.

20. Jeff rolls a die and draws one card from a card deck. Find the probability of his rolling a four and drawing a four.

21. *Birthday problem* Three people are in an elevator together. Find the probability that all three were born on the same day of the week.

22. *Birthday problem* Three people are on a bus together. Find the probability that at least one was born on a different day of the week than the others.

23. *Birthday problem* Five people are in a room together. Find the probability that all five were born on a different day of the year.

24. *Birthday problem* Five people are on a bus together. Find the probability that at least two of them were born on the same day of the year.

25. *Sharing homework* If the probability that Rick will solve a problem is $\frac{1}{4}$ and the probability that Dinah will solve it is $\frac{2}{5}$, find the probability that at least one of them will solve it.

26. *Signaling* A bugle is used for communication at camp. The call for dinner is based on four pitches and is five notes long. If a child can play these four pitches on a bugle, find the probability that the first five notes that the child plays will call the camp for food. (Assume that the child is equally likely to play any of the four pitches each time a note is blown.)

In Exercises 27–30, assume that a woman visits her cabin in Canada. The probability that the lawnmower will start is $\frac{1}{2}$, the probability that her gas power saw will start is $\frac{1}{3}$, and the probability that her outboard motor will start is $\frac{3}{4}$. Find each probability.

27. That all three will start

28. That none will start

29. That exactly one will start

30. That exactly two will start

31. *Immigration* Three children will leave Thailand to start a new life in either the United States or France. The probability that May Xao will go to France is $\frac{1}{3}$, that Tou Lia will go to France is $\frac{1}{2}$, and that May Moua will go to France is $\frac{1}{6}$. Find the probability that exactly two of them will end up in the United States.

32. *Preparing for the GED* The administrators of a program to prepare people for the high school equivalency exam have found that 80% of the students require tutoring in math, 60% need help in English, and 45% need work in both math and English. Find the probability that a student selected at random needs help with either math or English.

33. *Insurance losses* Insurance underwriters have determined that in any one year, the probability that George will have a car accident is 0.05 and that if he has an accident, the probability that he will be hospitalized is 0.40. Find the probability that George will have an accident but not be hospitalized.

34. *Grading homework* One instructor grades homework by randomly choosing 3 out of the 15 problems assigned. Bill did only 8 of the problems. Find the probability that he won't get caught.

9.9 ODDS AND MATHEMATICAL EXPECTATION

■ Mathematical Expectation

A concept closely related to probability is **mathematical odds**.

Odds	The **odds for an event** is the probability of a favorable outcome divided by the probability of an unfavorable income.
	The **odds against an event** is the probability of an unfavorable outcome divided by the probability of a favorable outcome.

EXAMPLE 1   The probability that a horse will win a race is $\frac{1}{4}$. Find the odds for and the odds against the horse.

Solution   Because the probability that the horse will win is $\frac{1}{4}$, the probability that the horse will not win is $\frac{3}{4}$. Therefore, the odds for the horse are

$$\frac{\text{probability of a win}}{\text{probability of a loss}} = \frac{\dfrac{1}{4}}{\dfrac{3}{4}} = \frac{1}{3}$$

or 1 to 3. The odds against the horse are

$$\frac{\text{probability of a loss}}{\text{probability of a win}} = \frac{\dfrac{3}{4}}{\dfrac{1}{4}} = 3$$

or 3 to 1. The odds for an event is the reciprocal of the odds against the event.   ■

Mathematical Expectation

Suppose we have a chance to play a simple game with the following rules:

1. Roll a single die once.
2. If a six appears, win $3.
3. If a five appears, win $1.
4. If any other number appears, win 50¢.
5. The cost to play (one roll of the die) is $1.

In this game, the probability of any one of the six outcomes—rolling a 6, 5, 4, 3, 2, or 1—is $\frac{1}{6}$, and the winnings are $3, $1, and 50¢. The expected winnings can be found by using the following equation and simplifying the right-hand side:

$$E = \frac{1}{6}(3) + \frac{1}{6}(1) + \frac{1}{6}(0.50) + \frac{1}{6}(0.50) + \frac{1}{6}(0.50) + \frac{1}{6}(0.50)$$

$$= \frac{1}{6}(3 + 1 + 0.50 + 0.50 + 0.50 + 0.50)$$

$$= \frac{1}{6}(6)$$

$$= 1$$

Over the long run, we could expect to win $1 with every play of the game. Since it costs $1 to play the game, the expected gain or loss is 0. Because the expected winnings are equal to the admission price, the game is fair.

Mathematical Expectation	If a certain event has n different outcomes with probabilities $P_1, P_2, P_3, \ldots, P_n$ and the winnings assigned to each outcome are $x_1, x_2, x_3, \ldots, x_n$, the expected winnings, or **mathematical expectation**, E, is given by $$E = P_1x_1 + P_2x_2 + P_3x_3 + \cdots + P_nx_n$$

EXAMPLE 2   It costs \$1 to play the following game: Roll two dice, collect \$5 if we roll a sum of 7, and collect \$2 if we roll a sum of 11. All other numbers pay nothing. Is it wise to play the game?

Solution   The probability of rolling a 7 on a single roll of two dice is $\frac{6}{36}$, the probability of rolling an 11 is $\frac{2}{36}$, and the probability of rolling something else is $\frac{28}{36}$. The mathematical expectation is

$$E = \frac{6}{36}(5) + \frac{2}{36}(2) + \frac{28}{36}(0) = \frac{17}{18} \approx \$0.944$$

By playing the game over and over for a long period of time, we can expect to get back about 95¢ for every dollar spent. For the fun of playing, the cost is about 5¢ a game. If the game is enjoyable, it might be worth the expected loss. However, the game is slightly unfair.   ■

9.9 EXERCISES

In Exercises 1–6, assume a single roll of a die.

1. Find the probability of rolling a 6.

2. Find the odds in favor of rolling a 6.

3. Find the odds against rolling a 6.

4. Find the probability of rolling an even number.

5. Find the odds in favor of rolling an even number.

6. Find the odds against rolling an even number.

In Exercises 7–12, assume a single roll of two dice.

7. Find the probability of rolling a sum of 6.

8. Find the odds in favor of rolling a sum of 6.

9. Find the odds against rolling a sum of 6.

10. Find the probability of rolling an even sum.

11. Find the odds in favor of rolling an even sum.

12. Find the odds against rolling an even sum.

In Exercises 13–16, assume that you are drawing one card from a standard deck.

13. Find the odds in favor of drawing a queen.

14. Find the odds against drawing a black card.

15. Find the odds in favor of drawing a face card.

16. Find the odds against drawing a diamond.

17. If the odds in favor of victory are 5 to 2, find the probability of victory.

18. If the odds in favor of victory are 5 to 2, find the odds against victory.

19. If the odds against winning are 90 to 1, find the odds in favor of winning.

20. Find the odds in favor of rolling a 7 on a single toss of two dice.

21. Find the odds against tossing four heads in a row with a nickel.

22. Find the odds in favor of a couple having four baby girls in succession. $\left(\text{Assume } P(\text{boy}) = \frac{1}{2}.\right)$

23. The odds against a horse are 8 to 1. Find the probability that the horse will win.

24. The odds against a horse are 1 to 1. Find the probability that the horse will lose.

25. It costs $2 to play the following game:
    a. Draw one card from a card deck.
    b. Collect $5 if an ace is drawn.
    c. Collect $4 if a king is drawn.
    d. Collect nothing for all other cards drawn.
    Is it wise to play this game? Explain.

26. **Lottery tickets** One thousand tickets are sold for a lottery with two grand prizes of $800. Find a fair price for the tickets.

27. Find the odds against a couple having three baby boys in a row. $\left(\text{Assume } P(\text{boy}) = \frac{1}{2}.\right)$

28. Find the odds in favor of tossing at least three heads in five tosses of a coin.

29. Suppose you toss a coin five times and collect $5 if you toss five heads, $4 if you toss four heads, $3 if you toss three heads, and no money for any other combinations. How much should you pay to play the game if the game is to be fair?

30. If you toss two dice one time and collect $10 for double sixes and $1 for double ones, what is a fair price for playing the game?

31. Counting an ace as 1, a face card as 10, and all others at their numerical values, find the expected value if you draw one card from a standard deck.

32. Find the expected sum of one roll of two dice.

33. A multiple-choice test of eight questions gives five possible answers for each question. Only one of the answers for each question is right. Find the probability of getting seven right answers by simple guessing.

34. In the situation described in Exercise 33, find the odds in favor of getting seven answers right.

---

9 CHAPTER SUMMARY

Key Words

alternating infinite series (9.2)
arithmetic means (9.3)
arithmetic sequence (9.3)
binomial theorem (9.1)
combinations (9.6)
compound events (9.8)
factorial notation (9.1)
Fibonacci sequence (9.2)
event (9.6)
experiment (9.7)
general term of a sequence (9.2)
geometric means (9.3)

geometric sequence (9.3)
independent events (9.8)
induction hypothesis (9.5)
infinite geometric sequence (9.3)
mathematical expectation (9.9)
mathematical induction (9.5)
*multiplication principle for
    events* (9.6)
*multiplication property of
    probabilities* (9.7)
odds against an event (9.9)

odds for an event (9.9)
partial sum (9.2)
Pascal's triangle (9.1)
permutations (9.6)
probability (9.7)
sample space (9.7)
sequence (9.2)
series (9.2)
*sum of an infinite geometric
    sequence* (9.3)
summation notation (9.2)

Key Ideas

(9.1) $n! = n(n-1)(n-2) \cdots 3 \cdot 2 \cdot 1$
$0! = 1 \qquad n(n-1)! = n!$

The Binomial Theorem. If n is any positive integer, then

$$(a+b)^n = a^n + \frac{n!}{1!(n-1)!}a^{n-1}b + \frac{n!}{2!(n-2)!}a^{n-2}b^2$$

$$+ \frac{n!}{3!(n-3)!}a^{n-3}b^3 + \cdots + \frac{n!}{r!(n-r)!}a^{n-r}b^r$$

$$+ \cdots + b^n$$

(9.2) If c is a constant, then $\sum\limits_{k=1}^{n} c = nc$.

If c is a constant, then $\sum\limits_{k=1}^{n} cf(k) = c \sum\limits_{k=1}^{n} f(k)$.

$$\sum_{k=1}^{n} [f(k) + g(k)] = \sum_{k=1}^{n} f(k) + \sum_{k=1}^{n} g(k)$$

(9.3) The formula $S_n = \dfrac{n(a + l)}{2}$ gives the sum of the first n terms of an arithmetic sequence, where S_n is the sum, a is the first term, l is the last (or nth) term, and n is the number of terms.

The formula $S_n = \dfrac{a - ar^n}{1 - r}$ $(r \neq 1)$ gives the sum of the first n terms of a geometric sequence, where S_n is the sum, a is the first term, r is the common ratio, and n is the number of terms.

If $|r| < 1$, the formula $S = \dfrac{a}{1 - r}$ gives the sum of the terms of an infinite geometric sequence, where S is the sum, a is the first term, and r is the common ratio.

(9.5) **The Axiom of Mathematical Induction.** If a statement involving the natural number n has the two properties that

1. The statement is true for $n = 1$ and
2. If the statement is true for $n = k$, then it is true for $n = k + 1$, then the statement is true for all natural numbers.

(9.6) **The formulas for computing permutations:**

$$P(n, r) = \frac{n!}{(n - r)!} \qquad P(n, n) = n! \qquad P(n, 0) = 1$$

There are $(n - 1)!$ ways to place n things in a circle.

The formulas for computing combinations:

$$C(n, r) = \binom{n}{r} = \frac{n!}{r!(n - r)!} \qquad C(n, n) = 1 \quad C(n, 0) = 1$$

If an n-letter word has a of one letter, b of another letter, and so on, then the number of distinguishable words that can be formed using each letter exactly once is

$$\frac{n!}{a!b! \cdots}$$

(9.7) An event that cannot happen has a probability of 0. An event that is certain to happen has a probability of 1. All other events have probabilities between 0 and 1.

$$P(A \cap B) = P(A) \cdot P(B|A)$$

(9.8) If a and B are two events, then
$$P(A \cup B) = P(A) + P(B) - P(A \cap B).$$

If A and B cannot occur simultaneously, then
$$P(A \cup B) = P(A) + P(B).$$

If A is any event, then $P(\overline{A}) = 1 - P(A)$.

If A and B are independent events, then
$$P(A \cap B) = P(A) \cdot P(B).$$

(9.9) The concepts of probabilities and odds are related. A game is fair if the cost to play equals the expected winnings.

---

9 CHAPTER REVIEW EXERCISES

In Review Exercises 1–2, find each value.

1. $7! \cdot 0! \cdot 1! \cdot 3!$

2. $\dfrac{5! \cdot 7! \cdot 8!}{6! \cdot 9!}$

In Review Exercises 3–4, use the binomial theorem to find the expansion of each expression.

3. $(2a - b)^3$

4. $(4a - 5b)^4$

In Review Exercises 5–8, find the required term of each expression.

5. $(a + b)^8$; 4th term

6. $(2x - y)^5$; 3rd term

7. $(x - y)^9$; 7th term

8. $(4x + 7)^6$; 4th term

In Review Exercises 9–10, a sequence is defined recursively. Find the first four terms of each sequence.

9. $a_1 = 5$ and $a_{n+1} = 3a_n + 2$

10. $a_1 = -2$ and $a_{n+1} = 2a_n^2$

In Review Exercises 11–14, evaluate each expression.

11. $\displaystyle\sum_{k=1}^{4} 3k^2$

12. $\displaystyle\sum_{k=1}^{10} 6$

13. $\displaystyle\sum_{k=5}^{8} (k^3 + 3k^2)$

14. $\displaystyle\sum_{k=1}^{30} \left(\frac{3}{2}k - 12\right) - \frac{3}{2}\sum_{k=1}^{30} k$

In Review Exercises 15–18, find the required term of each arithmetic sequence.

15. $5, 9, 13, \ldots$; 29th term

16. $8, 15, 22, \ldots$; 40th term

17. $6, -1, -8, \ldots$; 15th term

18. $\dfrac{1}{2}, -\dfrac{3}{2}, -\dfrac{7}{2}, \ldots$; 35th term

In Review Exercises 19–22, find the required term of each geometric sequence.

19. $81, 27, 9, \ldots$; 11th term

20. $2, 6, 18, \ldots$; 9th term

21. $9, \dfrac{9}{2}, \dfrac{9}{4}, \ldots$; 15th term

22. $8, -\dfrac{8}{5}, \dfrac{8}{25}, \ldots$; 7th term

In Review Exercises 23–26, find the sum of the first 40 terms in each sequence.

23. $5, 9, 13, \ldots$

24. $8, 15, 22, \ldots$

25. $6, -1, -8, \ldots$

26. $\dfrac{1}{2}, -\dfrac{3}{2}, -\dfrac{7}{2}, \ldots$

In Review Exercises 27–30, find the sum of the first eight terms in each sequence.

27. $81, 27, 9, \ldots$

28. $2, 6, 18, \ldots$

29. $9, \dfrac{9}{2}, \dfrac{9}{4}, \ldots$

30. $8, -\dfrac{8}{5}, \dfrac{8}{25}, \ldots$

In Review Exercises 31–34, find the sum of each infinite sequence, if possible.

31. $\dfrac{1}{3}, \dfrac{1}{6}, \dfrac{1}{12}, \ldots$

32. $\dfrac{1}{5}, -\dfrac{2}{15}, \dfrac{4}{45}, \ldots$

33. $1, \dfrac{3}{2}, \dfrac{9}{4}, \ldots$

34. $0.5, 0.25, 0.125, \ldots$

In Review Exercises 35–38, use the formula for the sum of the term of an infinite geometric sequence to change each decimal into a common fraction.

35. $0.\overline{3}$

36. $0.\overline{9}$

37. $0.\overline{17}$

38. $0.\overline{45}$

39. Find three arithmetic means between 2 and 8.

40. Find five arithmetic means between 10 and 100.

41. Find three geometric means between 2 and 8.

42. Find four geometric means between -2 and 64.

43. Find the sum of the first eight terms of the sequence $\frac{1}{3}$, 1, 3, ...

44. Find the seventh term of the sequence $2\sqrt{2}$, 4, $4\sqrt{2}$, ...

45. Find the positive geometric mean between 4 and 64.

46. *Investment problem* If Leonard invests \$3000 in a 6-year certificate of deposit at the annual rate of 7.75%, compounded daily, how much money will be in the account when it matures?

47. *College enrollments* The enrollment at Hometown College is growing at a rate of 5% over each previous year's enrollment. If the enrollment is currently 4000 students, what will the enrollment be 10 years from now? What was it 5 years ago?

48. *House trailer depreciation*   A house trailer that originally cost \$10,000 depreciates in value at the rate of 10% per year. How much will the trailer be worth after 10 years?

49. Verify the following formula for $n = 1$, $n = 2$, $n = 3$, and $n = 4$:

$$1^3 + 2^3 + 3^3 + \cdots + n^3 = \frac{n^2(n + 1)^2}{4}$$

50. Prove the formula given in Exercise 49 by mathematical induction.

In Review Exercises 51–62, evaluate each expression.

51. $P(8, 5)$

52. $C(7, 4)$

53. $0! \cdot 1!$

54. $P(10, 2) \cdot C(10, 2)$

55. $P(8, 6) \cdot C(8, 6)$

56. $C(8, 5) \cdot C(6, 2)$

57. $C(7, 5) \cdot P(4, 0)$

58. $C(12, 10) \cdot C(11, 0)$

59. $\dfrac{P(8, 5)}{C(8, 5)}$

60. $\dfrac{C(8, 5)}{C(13, 5)}$

61. $\dfrac{C(6, 3)}{C(10, 3)}$

62. $\dfrac{C(13, 5)}{C(52, 5)}$

63. Make a tree diagram to illustrate the possible results of tossing a coin four times.

64. In how many ways can you draw a five-card poker hand of 3 aces and 2 kings?

65. Find the probability of drawing the hand described in Review Exercise 64.

66. Find the probability of not drawing the hand described in Review Exercise 64.

67. In how many ways can 10 teenagers be seated at a round table if 2 girls wish to sit with their boyfriends?

68. How many distinguishable words can be formed from the letters of the word *casserole* if each letter is used exactly once?

69. Find the probability of having a 13-card bridge hand consisting of 4 aces, 4 kings, 4 queens, and 1 jack.

70. Find the probability of choosing a committee of 3 men and 2 women from a group of 8 men and 6 women.

71. Find the probability of drawing a club or a spade on one draw from a card deck.

72. Find the probability of drawing a black card or a king on one draw from a card deck.

73. Find the probability of getting an ace-high royal flush (ace, king, queen, jack, and ten of hearts) in poker.

74. Find the probability of being dealt 13 cards of one suit in bridge hand.

75. Find the probability of getting 3 heads or fewer on 4 tosses of a fair coin.

76. Find the odds against a horse if the probability that the horse will win is $\frac{7}{8}$.

77. Find the odds in favor of a couple having 4 baby girls in a row.

78. Find the expected earnings if you collect \$1 for every heads you get when you toss a fair coin 4 times.

79. If the probability that Joe will marry is $\frac{5}{6}$ and the probability that John will marry is $\frac{3}{4}$, find the odds against either one becoming a husband.

80. If odds against Priscilla's graduation from college are $\frac{10}{11}$, find the probability that she will graduate.

81. If the probability that a drug cures a certain disease is 0.83 and we give the drug to 800 people with the disease, find the number of people that we can expect to be cured.

82. If the total number of subsets that a set with n elements can have is 2^n, explain why

$$\binom{n}{0} + \binom{n}{1} + \binom{n}{2} + \cdots + \binom{n}{n} = 2^n$$

9 CHAPTER TEST

In Questions 1–2, find each value.

1. $3! \cdot 0! \cdot 4! \cdot 1!$

2. $\dfrac{2! \cdot 4! \cdot 6! \cdot 8!}{3! \cdot 5! \cdot 7!}$

In Questions 3–4, find the required term in each expansion.

3. $(x + 2y)^5$; 2nd term

4. $(2a - b)^8$; 7th term

In Questions 5–6, find each sum.

5. $\displaystyle\sum_{k=1}^{3} (4k + 1)$

6. $\displaystyle\sum_{k=2}^{4} (3k - 21)$

In Questions 7–8, find the sum of the first ten terms of each sequence.

7. $2, 5, 8, \ldots$

8. $5, 1, -3, \ldots$

9. Find three arithmetic means between 4 and 24.

10. Find two geometric means between -2 and -54.

In Questions 11–12, find the sum of the first ten terms of each sequence.

11. $\dfrac{1}{4}, \dfrac{1}{2}, 1, \ldots$

12. $6, 2, \dfrac{2}{3}, \ldots$

13. A car costing \$$c$ when new depreciates 25% of the previous year's value each year. How much is the car worth after 3 years?

14. A house costing \$$c$ when new appreciates 10% of the previous year's value each year. How much will the house be worth after 4 years?

15. Prove by induction:

$$1^3 + 2^3 + 3^3 + \cdots + n^3 = \dfrac{n^2(n + 1)^2}{4}.$$

16. How many six-digit license plates can be made if no plate begins with 0 or 1?

In Questions 17–20, find each value.

17. $P(7, 2)$

18. $P(4, 4)$

19. $C(8, 2)$

20. $C(12, 0)$

21. How many ways can 4 men and 4 women stand in line if all the women are first?

22. How many different ways can 6 people be seated at a round table?

23. How many different words can be formed from the letters of the word *bluff* if each letter is used once?

24. Show the sample space of the experiment: toss a fair coin three times.

In Questions 25–28, find each probability.

25. Rolling a 5 on one roll of a die.

26. Drawing a jack or a queen from a standard card deck.

27. Receiving 5 hearts for a 5-card poker hand.

28. Tossing 2 heads in 5 tosses of a fair coin.

29. If the probability that a person is sick is .1, find the probability that the person is well.

30. If 30% of the population enjoy jazz, 80% enjoy popular music, and 30% don't like either, what percentage of the people like both types of music?

In Questions 31–32, find each probability.

31. Drawing a black card or a face card with one draw from a standard card deck.

32. Rolling a sum of 3 or an even sum with one roll of two dice.

33. If the odds for a horse are 3 to 1, find the probability that the horse will win.

34. Assume the game with the following rules:

Roll a single die once.
If 6 appears, win $4.
If 5 appears, win $2.

Find the fair price to play the game.

CUMULATIVE REVIEW EXERCISES

In Exercises 1–2, solve each system by graphing.

1. $\begin{cases} 2x + y = 8 \\ x - 2y = -1 \end{cases}$

2. $\begin{cases} 3x = -y + 2 \\ y + x - 4 = -2x \end{cases}$

In Exercises 3–4, solve each system.

3. $\begin{cases} 5x = 3y + 12 \\ 2x - 3y = 3 \end{cases}$

4. $\begin{cases} 2x + y - z = 7 \\ x - y + z = 2 \\ x + y - 3z = 2 \end{cases}$

In Exercises 5–6, solve each system with matrices.

5. $\begin{cases} x + y + z = 4 \\ x + 2y - z = 1 \\ 2x - y + z = 3 \end{cases}$

6. $\begin{cases} 2x - 2y + 3z + t = 2 \\ x + y + z + t = 5 \\ -x + 2y - 3z + 2t = 2 \\ x + y + 2z - t = 4 \end{cases}$

In Exercises 7–10, let $A = \begin{bmatrix} 2 & 1 \\ 1 & 4 \end{bmatrix}$, $B = \begin{bmatrix} -1 & 2 \\ 2 & 3 \end{bmatrix}$, and $C = \begin{bmatrix} 2 & 0 & -1 \\ -1 & 2 & 2 \end{bmatrix}$. Find each matrix.

7. $A + B$

8. $B - A$

9. AC

10. $B^2 + 2A$

In Exercises 11–12, find the inverse of each matrix, if possible.

11. $\begin{bmatrix} 2 & 6 \\ 2 & 4 \end{bmatrix}$

12. $\begin{bmatrix} 1 & -1 & 1 \\ 1 & 4 & 0 \\ 2 & 4 & 1 \end{bmatrix}$

In Exercises 13–14, find each determinant.

13. $\begin{vmatrix} -3 & 5 \\ 4 & 7 \end{vmatrix}$

14. $\begin{vmatrix} 2 & -3 & 2 \\ 0 & 1 & -1 \\ 1 & -2 & 1 \end{vmatrix}$

In Exercises 15–16, set up the determinants to find x and y in the system $\begin{cases} 4x + 3y = 11 \\ -2x + 5y = 24 \end{cases}$.

15. $x =$

16. $y =$

In Exercises 17–18, decompose each fraction into partial fractions.

17. $\dfrac{-x + 1}{(x + 1)(x + 2)}$

18. $\dfrac{2x^2 + 3x + 2}{x^3 + x}$

In Exercises 19–20, find each solution by graphing.

19. $y < 2x + 6$

20. $\begin{cases} 2x + 3y \ge 6 \\ 2x - 3y \le 6 \end{cases}$

In Exercises 21–22, write the equation of each circle with the given center O and radius r.

21. $O(0, 0)$; $r = 4$

22. $O(2, -3)$; $r = 11$

In Exercises 23–26, complete the square on x and/or y and graph each equation.

23. $x^2 + y^2 - 4y = 12$

24. $x^2 - 2y - 2x = -7$

25. $x^2 + 4y^2 + 2x = 3$

26. $x^2 - 9y^2 - 4x = 5$

In Exercises 27–28, write the equation of each ellipse.

27. Center $(0, 0)$; major axis of 12; minor axis of 8

28. Center $(2, 3)$; $a = 5$; $c = 2$

In Exercises 29–30, write the equation of each hyperbola.

29. Center $(0, 0)$; focus $(3, 0)$; vertex $(2, 0)$

30. Center $(2, 4)$; area of fundamental rectangle 32 square units; vertex $(4, 4)$

In Exercises 31–32, find the required term of the expansion of $(x + 2y)^8$.

31. 2nd term

32. 6th term

In Exercises 33–34, find each sum.

33. $\displaystyle\sum_{k=1}^{5} 2$

34. $\displaystyle\sum_{k=2}^{6} (3x + 1)$

In Exercises 35–36, find the sum of the first six terms of each sequence.

35. $-2, 1, 4, \ldots$

36. $\dfrac{1}{9}, \dfrac{1}{3}, 1, \ldots$

In Exercises 37–40, find each value.

37. $P(8, 4)$

38. $P(24, 0)$

39. $C(12, 10)$

40. $P(4, 4) \cdot C(6, 6)$

41. In how many ways can 6 men and 4 women be placed in a line if the women line up first?

In Exercises 43–46, find each probability.

43. Rolling 11 on one roll of two dice

45. If the probability that a person is married is 0.6 and the probability that a married person has children is 0.8, find the probability that a randomly chosen person is married with children.

47. Prove the following formula by induction:

$$4 + 7 + 10 + \cdots + (3n + 1) = \frac{n(3n + 5)}{2}$$

42. In how many ways can a committee of 4 people be selected from a group of 12 people?

44. Being dealt an all-red 5-card poker hand from a standard deck

46. The probability that a student earns an A in algebra is 0.3, and the probability that a student earns a B is 0.4. Find the probability that a student earns a C or less.

48. Compute the fair cost to play the following game:

1. Draw a card from a standard card deck.

2. If you draw an ace, you receive $100.

3. If you draw a king, you receive $10.

4. All other cards pay nothing.

A P P E N D I X

I

A PROOF OF THE

BINOMIAL THEOREM

The binomial theorem can be proved for positive integral exponents by using mathematical induction.

The Binomial Theorem	If n is a positive integer, then

$$(a + b)^n = a^n + \frac{n!}{1!(n - 1)!} a^{n-1}b + \frac{n!}{2!(n - 2)!} a^{n-2}b^2 + \cdots$$

$$+ \frac{n!}{r!(n - r)!} a^{n-r}b^r + \cdots + b^n$$

Proof As in all induction proofs, there are two parts.

Part 1. Substituting the number 1 for n on both sides of the equation, we have

$$(a + b)^1 = a^1 + \frac{1!}{1(1 - 1)!} a^{1-1}b^1$$

$$a + b = a + a^0 b$$

$$a + b = a + b$$

and the theorem is true when $n = 1$. Part 1 is complete.

Part 2. We write expressions for two general terms in the statement of the induction hypothesis. We assume that the theorem is true for $n = k$:

$$(a + b)^k = a^k + \frac{k!}{1!(k - 1)!} a^{k-1}b + \frac{k!}{2!(k - 2)!} a^{k-2}b^2 + \cdots$$

$$+ \frac{k!}{(r - 1)!(k - r + 1)!} a^{k-r+1}b^{r-1}$$

$$+ \frac{k!}{r!(k - r)!} a^{k-r}b^r + \cdots + b^k$$

We multiply both sides of this equation by $a + b$ and hope to obtain a similar equation in which the quantity $k + 1$ replaces all of the n values in the binomial theorem:

$$(a + b)^k(a + b)$$

$$= (a + b)\left[a^k + \frac{k!}{1!(k - 1)!} a^{k-1}b + \frac{k!}{2!(k - 2)!} a^{k-2}b^2 + \cdots \right.$$

$$\left. + \frac{k!}{(r - 1)!(k - r + 1)!} a^{k-r+1}b^{r-1} + \frac{k!}{r!(k - r)!} a^{k-r}b^r + \cdots + b^k \right]$$

We distribute the multiplication first by a and then by b:

$$(a + b)^{k+1} = \left[a^{k+1} + \frac{k!}{1!(k - 1)!} a^k b + \frac{k!}{2!(k - 2)!} a^{k-1}b^2 + \cdots \right.$$

$$\left. + \frac{k!}{(r - 1)!(k - r + 1)!} a^{k-r+2}b^{r-1} + \frac{k!}{r!(k - r)!} a^{k-r+1}b^r + \cdots + ab^k \right]$$

$$+ \left[a^k b + \frac{k!}{1!(k - 1)!} a^{k-1}b^2 + \frac{k!}{2!(k - 2)!} a^{k-2}b^3 + \cdots \right.$$

$$\left. + \frac{k!}{(r - 1)!(k - r + 1)!} a^{k-r+1}b^r + \frac{k!}{r!(k - r)!} a^{k-r}b^{r+1} + \cdots + b^{k+1} \right]$$

Combining like terms, we have

$$(a + b)^{k+1} = a^{k+1} + \left[\frac{k!}{1!(k - 1)!} + 1 \right]a^k b$$

$$+ \left[\frac{k!}{2!(k - 2)!} + \frac{k!}{1!(k - 1)!} \right]a^{k-1}b^2 + \cdots$$

$$+ \left[\frac{k!}{r!(k - r)!} + \frac{k!}{(r - 1)!(k - r + 1)!} \right]a^{k-r+1}b^r + \cdots + b^{k+1}$$

These results may be written as

$$(a + b)^{k+1} = a^{k+1} + \frac{(k + 1)!}{1!(k + 1 - 1)!} a^{(k+1)-1}b + \frac{(k + 1)!}{2!(k + 1 - 2)!} a^{(k+1)-2}b^2$$

$$+ \cdots + \frac{(k + 1)!}{r!(k + 1 - r)!} a^{(k+1)-r}b^r + \cdots + b^{k+1}$$

This formula has precisely the same form as the binomial theorem, with the quantity $k + 1$ replacing all of the original n values. Therefore, the truth of the theorem for $n = k$ implies the truth of the theorem for $n = k + 1$. Because both parts of the axiom of mathematical induction are verified, the theorem is proved.   □

A P P E N D I X

TABLES

TABLE A Powers and Roots

n	n^2	$\sqrt{n}$	n^3	$\sqrt[3]{n}$	n	n^2	$\sqrt{n}$	n^3	$\sqrt[3]{n}$
1	1	1.000	1	1.000	51	2,601	7.141	132,651	3.708
2	4	1.414	8	1.260	52	2,704	7.211	140,608	3.733
3	9	1.732	27	1.442	53	2,809	7.280	148,877	3.756
4	16	2.000	64	1.587	54	2,916	7.348	157,464	3.780
5	25	2.236	125	1.710	55	3,025	7.416	166,375	3.803
6	36	2.449	216	1.817	56	3,136	7.483	175,616	3.826
7	49	2.646	343	1.913	57	3,249	7.550	185,193	3.849
8	64	2.828	512	2.000	58	3,364	7.616	195,112	3.871
9	81	3.000	729	2.080	59	3,481	7.681	205,379	3.893
10	100	3.162	1,000	2.154	60	3,600	7.746	216,000	3.915
11	121	3.317	1,331	2.224	61	3,721	7.810	226,981	3.936
12	144	3.464	1,728	2.289	62	3,844	7.874	238,328	3.958
13	169	3.606	2,197	2.351	63	3,969	7.937	250,047	3.979
14	196	3.742	2,744	2.410	64	4,096	8.000	262,144	4.000
15	225	3.873	3,375	2.466	65	4,225	8.062	274,625	4.021
16	256	4.000	4,096	2.520	66	4,356	8.124	287,496	4.041
17	289	4.123	4,913	2.571	67	4,489	8.185	300,763	4.062
18	324	4.243	5,832	2.621	68	4,624	8.246	314,432	4.082
19	361	4.359	6,859	2.668	69	4,761	8.307	328,509	4.102
20	400	4.472	8,000	2.714	70	4,900	8.367	343,000	4.121
21	441	4.583	9,261	2.759	71	5,041	8.426	357,911	4.141
22	484	4.690	10,648	2.802	72	5,184	8.485	373,248	4.160
23	529	4.796	12,167	2.844	73	5,329	8.544	389,017	4.179
24	576	4.899	13,824	2.884	74	5,476	8.602	405,224	4.198
25	625	5.000	15,625	2.924	75	5,625	8.660	421,875	4.217
26	676	5.099	17,576	2.962	76	5,776	8.718	438,976	4.236
27	729	5.196	19,683	3.000	77	5,929	8.775	456,533	4.254
28	784	5.292	21,952	3.037	78	6,084	8.832	474,552	4.273
29	841	5.385	24,389	3.072	79	6,241	8.888	493,039	4.291
30	900	5.477	27,000	3.107	80	6,400	8.944	512,000	4.309
31	961	5.568	29,791	3.141	81	6,561	9.000	531,441	4.327
32	1,024	5.657	32,768	3.175	82	6,724	9.055	551,368	4.344
33	1,089	5.745	35,937	3.208	83	6,889	9.110	571,787	4.362
34	1,156	5.831	39,304	3.240	84	7,056	9.165	592,704	4.380
35	1,225	5.916	42,875	3.271	85	7,225	9.220	614,125	4.397
36	1,296	6.000	46,656	3.302	86	7,396	9.274	636,056	4.414
37	1,369	6.083	50,653	3.332	87	7,569	9.327	658,503	4.431
38	1,444	6.164	54,872	3.362	88	7,744	9.381	681,472	4.448
39	1,521	6.245	59,319	3.391	89	7,921	9.434	704,969	4.465
40	1,600	6.325	64,000	3.420	90	8,100	9.487	729,000	4.481
41	1,681	6.403	68,921	3.448	91	8,281	9.539	753,571	4.498
42	1,764	6.481	74,088	3.476	92	8,464	9.592	778,688	4.514
43	1,849	6.557	79,507	3.503	93	8,649	9.644	804,357	4.531
44	1,936	6.633	85,184	3.530	94	8,836	9.695	830,584	4.547
45	2,025	6.708	91,125	3.557	95	9,025	9.747	857,375	4.563
46	2,116	6.782	97,336	3.583	96	9,216	9.798	884,736	4.579
47	2,209	6.856	103,823	3.609	97	9,409	9.849	912,673	4.595
48	2,304	6.928	110,592	3.634	98	9,604	9.899	941,192	4.610
49	2,401	7.000	117,649	3.659	99	9,801	9.950	970,299	4.626
50	2,500	7.071	125,000	3.684	100	10,000	10.000	1,000,000	4.642

TABLE B   Base-10 Logarithms

N	0	1	2	3	4	5	6	7	8	9
1.0	0000	0043	0086	0128	0170	0212	0253	0294	0334	0374
1.1	0414	0453	0492	0531	0569	0607	0645	0682	0719	0755
1.2	0792	0828	0864	0899	0934	0969	1004	1038	1072	1106
1.3	1139	1173	1206	1239	1271	1303	1335	1367	1399	1430
1.4	1461	1492	1523	1553	1584	1614	1644	1673	1703	1732
1.5	1761	1790	1818	1847	1875	1903	1931	1959	1987	2014
1.6	2041	2068	2095	2122	2148	2175	2201	2227	2253	2279
1.7	2304	2330	2355	2380	2405	2430	2455	2480	2504	2529
1.8	2553	2577	2601	2625	2648	2672	2695	2718	2742	2765
1.9	2788	2810	2833	2856	2878	2900	2923	2945	2967	2989
2.0	3010	3032	3054	3075	3096	3118	3139	3160	3181	3201
2.1	3222	3243	3263	3284	3304	3324	3345	3365	3385	3404
2.2	3424	3444	3464	3483	3502	3522	3541	3560	3579	3598
2.3	3617	3636	3655	3674	3692	3711	3729	3747	3766	3784
2.4	3802	3820	3838	3856	3874	3892	3909	3927	3945	3962
2.5	3979	3997	4014	4031	4048	4065	4082	4099	4116	4133
2.6	4150	4166	4183	4200	4216	4232	4249	4265	4281	4298
2.7	4314	4330	4346	4362	4378	4393	4409	4425	4440	4456
2.8	4472	4487	4502	4518	4533	4548	4564	4579	4594	4609
2.9	4624	4639	4654	4669	4683	4698	4713	4728	4742	4757
3.0	4771	4786	4800	4814	4829	4843	4857	4871	4886	4900
3.1	4914	4928	4942	4955	4969	4983	4997	5011	5024	5038
3.2	5051	5065	5079	5092	5105	5119	5132	5145	5159	5172
3.3	5185	5198	5211	5224	5237	5250	5263	5276	5289	5302
3.4	5315	5328	5340	5353	5366	5378	5391	5403	5416	5428
3.5	5441	5453	5465	5478	5490	5502	5514	5527	5539	5551
3.6	5563	5575	5587	5599	5611	5623	5635	5647	5658	5670
3.7	5682	5694	5705	5717	5729	5740	5752	5763	5775	5786
3.8	5798	5809	5821	5832	5843	5855	5866	5877	5888	5899
3.9	5911	5922	5933	5944	5955	5966	5977	5988	5999	6010
4.0	6021	6031	6042	6053	6064	6075	6085	6096	6107	6117
4.1	6128	6138	6149	6160	6170	6180	6191	6201	6212	6222
4.2	6232	6243	6253	6263	6274	6284	6294	6304	6314	6325
4.3	6335	6345	6355	6365	6375	6385	6395	6405	6415	6425
4.4	6435	6444	6454	6464	6474	6484	6493	6503	6513	6522
4.5	6532	6542	6551	6561	6571	6580	6590	6599	6609	6618
4.6	6628	6637	6646	6656	6665	6675	6684	6693	6702	6712
4.7	6721	6730	6739	6749	6758	6767	6776	6785	6794	6803
4.8	6812	6821	6830	6839	6848	6857	6866	6875	6884	6893
4.9	6902	6911	6920	6928	6937	6946	6955	6964	6972	6981
5.0	6990	6998	7007	7016	7024	7033	7042	7050	7059	7067
5.1	7076	7084	7093	7101	7110	7118	7126	7135	7143	7152
5.2	7160	7168	7177	7185	7193	7202	7210	7218	7226	7235
5.3	7243	7251	7259	7267	7275	7284	7292	7300	7308	7316
5.4	7324	7332	7340	7348	7356	7364	7372	7380	7388	7396

TABLE B   *(continued)*

N	0	1	2	3	4	5	6	7	8	9
5.5	7404	7412	7419	7427	7435	7443	7451	7459	7466	7474
5.6	7482	7490	7497	7505	7513	7520	7528	7536	7543	7551
5.7	7559	7566	7574	7582	7589	7597	7604	7612	7619	7627
5.8	7634	7642	7649	7657	7664	7672	7679	7686	7694	7701
5.9	7709	7716	7723	7731	7738	7745	7752	7760	7767	7774
6.0	7782	7789	7796	7803	7810	7818	7825	7832	7839	7846
6.1	7853	7860	7868	7875	7882	7889	7896	7903	7910	7917
6.2	7924	7931	7938	7945	7952	7959	7966	7973	7980	7987
6.3	7993	8000	8007	8014	8021	8028	8035	8041	8048	8055
6.4	8062	8069	8075	8082	8089	8096	8102	8109	8116	8122
6.5	8129	8136	8142	8149	8156	8162	8169	8176	8182	8189
6.6	8195	8202	8209	8215	8222	8228	8235	8241	8248	8254
6.7	8261	8267	8274	8280	8287	8293	8299	8306	8312	8319
6.8	8325	8331	8338	8344	8351	8357	8363	8370	8376	8382
6.9	8388	8395	8401	8407	8414	8420	8426	8432	8439	8445
7.0	8451	8457	8463	8470	8476	8482	8488	8494	8500	8506
7.1	8513	8519	8525	8531	8537	8543	8549	8555	8561	8567
7.2	8573	8579	8585	8591	8597	8603	8609	8615	8621	8627
7.3	8633	8639	8645	8651	8657	8663	8669	8675	8681	8686
7.4	8692	8698	8704	8710	8716	8722	8727	8733	8739	8745
7.5	8751	8756	8762	8768	8774	8779	8785	8791	8797	8802
7.6	8808	8814	8820	8825	8831	8837	8842	8848	8854	8859
7.7	8865	8871	8876	8882	8887	8893	8899	8904	8910	8915
7.8	8921	8927	8932	8938	8943	8949	8954	8960	8965	8971
7.9	8976	8982	8987	8993	8998	9004	9009	9015	9020	9025
8.0	9031	9036	9042	9047	9053	9058	9063	9069	9074	9079
8.1	9085	9090	9096	9101	9106	9112	9117	9122	9128	9133
8.2	9138	9143	9149	9154	9159	9165	9170	9175	9180	9186
8.3	9191	9196	9201	9206	9212	9217	9222	9227	9232	9238
8.4	9243	9248	9253	9258	9263	9269	9274	9279	9284	9289
8.5	9294	9299	9304	9309	9315	9320	9325	9330	9335	9340
8.6	9345	9350	9355	9360	9365	9370	9375	9380	9385	9390
8.7	9395	9400	9405	9410	9415	9420	9425	9430	9435	9440
8.8	9445	9450	9455	9460	9465	9469	9474	9479	9484	9489
8.9	9494	9499	9504	9509	9513	9518	9523	9528	9533	9538
9.0	9542	9547	9552	9557	9562	9566	9571	9576	9581	9586
9.1	9590	9595	9600	9605	9609	9614	9619	9624	9628	9633
9.2	9638	9643	9647	9652	9657	9661	9666	9671	9675	9680
9.3	9685	9689	9694	9699	9703	9708	9713	9717	9722	9727
9.4	9731	9736	9741	9745	9750	9754	9759	9763	9768	9773
9.5	9777	9782	9786	9791	9795	9800	9805	9809	9814	9818
9.6	9823	9827	9832	9836	9841	9845	9850	9854	9859	9863
9.7	9868	9872	9877	9881	9886	9890	9894	9899	9903	9908
9.8	9912	9917	9921	9926	9930	9934	9939	9943	9948	9952
9.9	9956	9961	9965	9969	9974	9978	9983	9987	9991	9996

TABLE C *Base-e Logarithms*

N	0	1	2	3	4	5	6	7	8	9
1.0	.0000	.0100	.0198	.0296	.0392	.0488	.0583	.0677	.0770	.0862
1.1	.0953	.1044	.1133	.1222	.1310	.1398	.1484	.1570	.1655	.1740
1.2	.1823	.1906	.1989	.2070	.2151	.2231	.2311	.2390	.2469	.2546
1.3	.2624	.2700	.2776	.2852	.2927	.3001	.3075	.3148	.3221	.3293
1.4	.3365	.3436	.3507	.3577	.3646	.3716	.3784	.3853	.3920	.3988
1.5	.4055	.4121	.4187	.4253	.4318	.4383	.4447	.4511	.4574	.4637
1.6	.4700	.4762	.4824	.4886	.4947	.5008	.5068	.5128	.5188	.5247
1.7	.5306	.5365	.5423	.5481	.5539	.5596	.5653	.5710	.5766	.5822
1.8	.5878	.5933	.5988	.6043	.6098	.6152	.6206	.6259	.6313	.6366
1.9	.6419	.6471	.6523	.6575	.6627	.6678	.6729	.6780	.6831	.6881
2.0	.6931	.6981	.7031	.7080	.7129	.7178	.7227	.7275	.7324	.7372
2.1	.7419	.7467	.7514	.7561	.7608	.7655	.7701	.7747	.7793	.7839
2.2	.7885	.7930	.7975	.8020	.8065	.8109	.8154	.8198	.8242	.8286
2.3	.8329	.8372	.8416	.8459	.8502	.8544	.8587	.8629	.8671	.8713
2.4	.8755	.8796	.8838	.8879	.8920	.8961	.9002	.9042	.9083	.9123
2.5	.9163	.9203	.9243	.9282	.9322	.9361	.9400	.9439	.9478	.9517
2.6	.9555	.9594	.9632	.9670	.9708	.9746	.9783	.9821	.9858	.9895
2.7	.9933	.9969	1.0006	.0043	.0080	.0116	.0152	.0188	.0225	.0260
2.8	1.0296	.0332	.0367	.0403	.0438	.0473	.0508	.0543	.0578	.0613
2.9	.0647	.0682	.0716	.0750	.0784	.0818	.0852	.0886	.0919	.0953
3.0	1.0986	.1019	.1053	.1086	.1119	.1151	.1184	.1217	.1249	.1282
3.1	.1314	.1346	.1378	.1410	.1442	.1474	.1506	.1537	.1569	.1600
3.2	.1632	.1663	.1694	.1725	.1756	.1787	.1817	.1848	.1878	.1909
3.3	.1939	.1969	.2000	.2030	.2060	.2090	.2119	.2149	.2179	.2208
3.4	.2238	.2267	.2296	.2326	.2355	.2384	.2413	.2442	.2470	.2499
3.5	1.2528	.2556	.2585	.2613	.2641	.2669	.2698	.2726	.2754	.2782
3.6	.2809	.2837	.2865	.2892	.2920	.2947	.2975	.3002	.3029	.3056
3.7	.3083	.3110	.3137	.3164	.3191	.3218	.3244	.3271	.3297	.3324
3.8	.3350	.3376	.3403	.3429	.3455	.3481	.3507	.3533	.3558	.3584
3.9	.3610	.3635	.3661	.3686	.3712	.3737	.3762	.3788	.3813	.3838
4.0	1.3863	.3888	.3913	.3938	.3962	.3987	.4012	.4036	.4061	.4085
4.1	.4110	.4134	.4159	.4183	.4207	.4231	.4255	.4279	.4303	.4327
4.2	.4351	.4375	.4398	.4422	.4446	.4469	.4493	.4516	.4540	.4563
4.3	.4586	.4609	.4633	.4656	.4679	.4702	.4725	.4748	.4770	.4793
4.4	.4816	.4839	.4861	.4884	.4907	.4929	.4951	.4974	.4996	.5019
4.5	1.5041	.5063	.5085	.5107	.5129	.5151	.5173	.5195	.5217	.5239
4.6	.5261	.5282	.5304	.5326	.5347	.5369	.5390	.5412	.5433	.5454
4.7	.5476	.5497	.5518	.5539	.5560	.5581	.5602	.5623	.5644	.5665
4.8	.5686	.5707	.5728	.5748	.5769	.5790	.5810	.5831	.5851	.5872
4.9	.5892	.5913	.5933	.5953	.5974	.5994	.6014	.6034	.6054	.6074
5.0	1.6094	.6114	.6134	.6154	.6174	.6194	.6214	.6233	.6253	.6273
5.1	.6292	.6312	.6332	.6351	.6371	.6390	.6409	.6429	.6448	.6467
5.2	.6487	.6506	.6525	.6544	.6563	.6582	.6601	.6620	.6639	.6658
5.3	.6677	.6696	.6715	.6734	.6752	.6771	.6790	.6808	.6827	.6845
5.4	.6864	.6882	.6901	.6919	.6938	.6956	.6974	.6993	.7011	.7029

TABLE C *(continued)*

N	0	1	2	3	4	5	6	7	8	9
5.5	1.7047	.7066	.7084	.7102	.7120	.7138	.7156	.7174	.7192	.7210
5.6	.7228	.7246	.7263	.7281	.7299	.7317	.7334	.7352	.7370	.7387
5.7	.7405	.7422	.7440	.7457	.7475	.7492	.7509	.7527	.7544	.7561
5.8	.7579	.7596	.7613	.7630	.7647	.7664	.7681	.7699	.7716	.7733
5.9	.7750	.7766	.7783	.7800	.7817	.7834	.7851	.7867	.7884	.7901
6.0	1.7918	.7934	.7951	.7967	.7984	.8001	.8017	.8034	.8050	.8066
6.1	.8083	.8099	.8116	.8132	.8148	.8165	.8181	.8197	.8213	.8229
6.2	.8245	.8262	.8278	.8294	.8310	.8326	.8342	.8358	.8374	.8390
6.3	.8405	.8421	.8437	.8453	.8469	.8485	.8500	.8516	.8532	.8547
6.4	.8563	.8579	.8594	.8610	.8625	.8641	.8656	.8672	.8687	.8703
6.5	1.8718	.8733	.8749	.8764	.8779	.8795	.8810	.8825	.8840	.8856
6.6	.8871	.8886	.8901	.8916	.8931	.8946	.8961	.8976	.8991	.9006
6.7	.9021	.9036	.9051	.9066	.9081	.9095	.9110	.9125	.9140	.9155
6.8	.9169	.9184	.9199	.9213	.9228	.9242	.9257	.9272	.9286	.9301
6.9	.9315	.9330	.9344	.9359	.9373	.9387	.9402	.9416	.9430	.9445
7.0	1.9459	.9473	.9488	.9502	.9516	.9530	.9544	.9559	.9573	.9587
7.1	.9601	.9615	.9629	.9643	.9657	.9671	.9685	.9699	.9713	.9727
7.2	.9741	.9755	.9769	.9782	.9796	.9810	.9824	.9838	.9851	.9865
7.3	.9879	.9892	.9906	.9920	.9933	.9947	.9961	.9974	.9988	2.0001
7.4	2.0015	.0028	.0042	.0055	.0069	.0082	.0096	.0109	.0122	.0136
7.5	2.0149	.0162	.0176	.0189	.0202	.0215	.0229	.0242	.0255	.0268
7.6	.0281	.0295	.0308	.0321	.0334	.0347	.0360	.0373	.0386	.0399
7.7	.0412	.0425	.0438	.0451	.0464	.0477	.0490	.0503	.0516	.0528
7.8	.0541	.0554	.0567	.0580	.0592	.0605	.0618	.0631	.0643	.0656
7.9	.0669	.0681	.0694	.0707	.0719	.0732	.0744	.0757	.0769	.0782
8.0	2.0794	.0807	.0819	.0832	.0844	.0857	.0869	.0882	.0894	.0906
8.1	.0919	.0931	.0943	.0956	.0968	.0980	.0992	.1005	.1017	.1029
8.2	.1041	.1054	.1066	.1078	.1090	.1102	.1114	.1126	.1138	.1150
8.3	.1163	.1175	.1187	.1199	.1211	.1223	.1235	.1247	.1258	.1270
8.4	.1282	.1294	.1306	.1318	.1330	.1342	.1353	.1365	.1377	.1389
8.5	2.1401	.1412	.1424	.1436	.1448	.1459	.1471	.1483	.1494	.1506
8.6	.1518	.1529	.1541	.1552	.1564	.1576	.1587	.1599	.1610	.1622
8.7	.1633	.1645	.1656	.1668	.1679	.1691	.1702	.1713	.1725	.1736
8.8	.1748	.1759	.1770	.1782	.1793	.1804	.1815	.1827	.1838	.1849
8.9	.1861	.1872	.1883	.1894	.1905	.1917	.1928	.1939	.1950	.1961
9.0	2.1972	.1983	.1994	.2006	.2017	.2028	.2039	.2050	.2061	.2072
9.1	.2083	.2094	.2105	.2116	.2127	.2138	.2148	.2159	.2170	.2181
9.2	.2192	.2203	.2214	.2225	.2235	.2246	.2257	.2268	.2279	.2289
9.3	.2300	.2311	.2322	.2332	.2343	.2354	.2364	.2375	.2386	.2396
9.4	.2407	.2418	.2428	.2439	.2450	.2460	.2471	.2481	.2492	.2502
9.5	2.2513	.2523	.2534	.2544	.2555	.2565	.2576	.2586	.2597	.2607
9.6	.2618	.2628	.2638	.2649	.2659	.2670	.2680	.2690	.2701	.2711
9.7	.2721	.2732	.2742	.2752	.2762	.2773	.2783	.2793	.2803	.2814
9.8	.2824	.2834	.2844	.2854	.2865	.2875	.2885	.2895	.2905	.2915
9.9	.2925	.2935	.2946	.2956	.2966	.2976	.2986	.2996	.3006	.3016

Use the properties of logarithms and ln 10 ≈ 2.3026 to find logarithms of numbers less than 1 or greater than 10.

APPENDIX III

ANSWERS TO SELECTED EXERCISES

Exercise 1.1 (page 11)

1. $\subseteq$ **3.** $\in$ **5.** $\subseteq$ **7.** $\subseteq$ **9.** $\{3, 5, 7\}$ **11.** $\{1, 2, 3, 4, 5, 6, 7, 8, 9, 10\}$ **13.** $\{1, 3, 5, 7, 9\}$
15. $\{2, 3, 4, 5, 6, 7, 8, 10\}$ **17.** $10, 12, 14, 15, 16, 18$ **19.** 2   **21.** infinite   **23.** finite   **25.** finite   **27.** 53
29. false; 27 can be divided by 3 and 9   **31.** true   **33.** false; $0 = \frac{0}{9}$ **35.** true   **37.** false; a prime number must be
greater than 1   **39.** false; the product of two primes is composite: $3 \cdot 5 = 15$ **41.** false; $3^2 = 9$, and 9 is composite
43. false; $4 + 6 = 10$, and 10 is composite   **45.** false; 2 is prime   **47.** true   **49.** true   **51.** false; $3 + 5 + 10 = 18$
53. true   **55.** false; all integers are rational   **57.** false; the set of reals is the union of the rationals and irrationals   **59.** true

61.

63.

65.   **67.** no such numbers   **69.**   **71.**

73.   **75.**   **77.**   **79.**

81.   **83.**   **85.**   **87.**

89.   **91.** 12   **93.** 6   **95.** 36   **97.** -3 **99.** 7   **101.** 1   **103.** $\sqrt{2} - 1 \approx 0.414$

105. $\pi - 1 \approx 2.142$ **107.** 3   **109.** transitive   **111.** reflexive, transitive   **113.** reflexive, symmetric, transitive
115. reflexive, symmetric, transitive   **117.** commutative property of addition   **119.** transitive property
121. reflexive property   **123.** commutative property of addition   **125.** symmetric property
127. associative property of multiplication

Exercise 1.2 (page 20)

1. 169   **3.** -25 **5.** $4xxx$ **7.** $625xxxx$ **9.** $7x^3$ **11.** x^2 **13.** $-27t^3$ **15.** x^3y^2 **17.** x^5 **19.** z^6

21. y^{21} **23.** z^{26} **25.** $27x^3$ **27.** x^6y^3 **29.** $\dfrac{a^6}{b^3}$ **31.** 1   **33.** 1   **35.** $\dfrac{1}{z^4}$ **37.** $\dfrac{1}{y^5}$ **39.** x^2 **41.** x^4

43. a^4 **45.** x **47.** $\dfrac{m^9}{n^6}$ **49.** $\dfrac{1}{a^9}$ **51.** $\dfrac{a^{12}}{b^4}$ **53.** $\dfrac{1}{r^4}$ **55.** $\dfrac{x^{32}}{y^{16}}$ **57.** $\dfrac{9x^{10}}{25}$ **59.** 1   **61.** $\dfrac{4y^{12}}{x^{20}}$ **63.** $-x^{10}y^{5x}$

A-6

65. $-x^{8xy}y^{12xy}$ **67.** x^4 **69.** x^{5n^2} **71.** x^5 **73.** $\dfrac{64z^7}{25y^6}$ **75.** $\dfrac{1}{m^{14}n^{16}p^{12}}$ **77.** 4   **79.** -8 **81.** 216

83. -12 **85.** 20   **87.** 0   **89.** $\frac{3}{64}$ **91.** 0   **93.** 3.72×10^5 **95.** -1.77×10^8 **97.** 7×10^{-3}

99. -6.93×10^{-7} **101.** 1×10^{12} **103.** 1×10^{-12} **105.** 9.97×10^{-3} **107.** 937,000   **109.** 0.0000221

111. 3.2   **113.** -0.0032 **115.** 1.17×10^4 **117.** 7×10^4 **119.** 1.986×10^4 meters per minute

121. 1.67248×10^{-15} g

Exercise 1.3   (page 32)

1. 3   **3.** -2 **5.** $\frac{1}{5}$ **7.** $3r^2$ **9.** $-4p^2$ **11.** $\dfrac{4a^2}{5b}$ **13.** $4|a|$ **15.** $2|a^3|$ **17.** $-2a$ **19.** $-6b^2$ **21.** 8

23. -8 **25.** -100 **27.** $\frac{1}{8}$ **29.** $\frac{1}{512}$ **31.** $-\frac{1}{27}$ **33.** $\frac{32}{243}$ **35.** $\frac{16}{9}$ **37.** $10s^2$ **39.** $\dfrac{1}{2y^2z}$ **41.** x^6y^3

43. $\dfrac{1}{r^6s^{12}}$ **45.** $\dfrac{4a^4}{25b^6}$ **47.** $\dfrac{100s^8}{9r^4}$ **49.** $\dfrac{a^2c^6}{b^5}$ **51.** a **53.** $\dfrac{2s}{a^3}$ **55.** $b^{11x/24}$ **57.** 7   **59.** 5   **61.** -3

63. $-\frac{1}{5}$ **65.** $6x$ **67.** xy^2 **69.** $2y$ **71.** $\dfrac{xy^2}{z^3}$ **73.** $2|x|$ **75.** $-4b^2$ **77.** $3x$ **79.** $|x|y^2$ **81.** $\sqrt{2}$

83. $17x\sqrt{2}$ **85.** $2y^2\sqrt{3y}$ **87.** $12\sqrt[3]{3}$ **89.** $6z\sqrt[4]{3z}$ **91.** $6x\sqrt{2y}$ **93.** 0   **95.** $\frac{x}{2}$ **97.** $\sqrt{3}$ **99.** $\dfrac{2\sqrt{x}}{x}$

101. $\sqrt[3]{4}$ **103.** $\sqrt[3]{5a^2}$ **105.** $\dfrac{2b\sqrt[4]{27a^2}}{3a}$ **107.** $\dfrac{\sqrt{2xy}}{2y}$ **109.** $\dfrac{u\sqrt[3]{6uv^2}}{3v}$ **111.** $\dfrac{2\sqrt{3}}{9}$ **113.** $-\dfrac{\sqrt{2x}}{8}$ **115.** $\dfrac{x\sqrt{y}}{3}$

117. $\dfrac{3y}{x}$ **119.** $\dfrac{x^4y^3\sqrt[5]{x^4y^4}}{2}$ **121.** $\sqrt{3}$ **123.** $\sqrt[5]{4x^3}$ **125.** $\sqrt[6]{32}$ **127.** $\dfrac{\sqrt[4]{12}}{2}$ **129.** $x \geq 0$ **131.** $x \leq 0$

Exercise 1.4   (page 43)

1. 2nd degree; trinomial   **3.** not a polynomial   **5.** 3rd degree; binomial   **7.** 0th degree; monomial   **9.** no degree; monomial   **11.** not a polynomial   **13.** $6x^3 - 3x^2 - 8x$ **15.** $4y^3 + 14$ **17.** $-x^2 + 14$ **19.** $-28t + 96$

21. $-4y^2 + y$ **23.** $-4x^2 + x$ **25.** $2x^2y - 4xy^2$ **27.** $8x^3y^7$ **29.** $\dfrac{m^4n^4}{2}$ **31.** $-4r^3s - 4rs^3$

33. $12a^2b^2c^2 + 18ab^3c^3 - 24a^2b^4c^2$ **35.** $a^2 + 4a + 4$ **37.** $a^2 - 12a + 36$ **39.** $x^2 - 16$ **41.** $a^2 - 4ab + 4b^2$

43. $9m^2 - 16n^2$ **45.** $x^2 + 2x - 15$ **47.** $3u^2 + 4u - 4$ **49.** $6x^2 - x - 15$ **51.** $10x^2 + 13x - 3$

53. $12v^2 - 7v - 12$ **55.** $6y^2 - 16xy + 8x^2$ **57.** $9x^3 - 27xy - yx^2 + 3y^2$ **59.** $5z^3 - 5tz + 2tz^2 - 2t^2$

61. $27x^3 - 27x^2 + 9x - 1$ **63.** $x^3y^3 + 2xy^2 - x^2y - 2$ **65.** $6x^3 + 14x^2 - 5x - 3$ **67.** $6x^3 - 5x^2y + 6xy^2 + 8y^3$

69. $x^4 + 2x^3 + x^2 - 1$ **71.** $6y^{2n} - 4y^{n+2} + 2$ **73.** $-10x^{4n} - 15y^{2n}$ **75.** $x^{2n} - x^n - 12$

77. $6r^{2n} - 25r^n + 14$ **79.** $xy + x^{3/2}y^{1/2}$ **81.** $a - b$ **83.** $\sqrt{3} + 1$ **85.** $x(\sqrt{7} - 2)$ **87.** $2(\sqrt{5} + \sqrt{2})$

89. $\dfrac{x(x + \sqrt{3})}{x^2 - 3}$ **91.** $\dfrac{(y + \sqrt{2})^2}{y^2 - 2}$ **93.** $9 - 4\sqrt{5}$ **95.** $\dfrac{\sqrt{3} + 3 - \sqrt{2} - \sqrt{6}}{2}$ **97.** $\dfrac{x - 2\sqrt{xy} + y}{x - y}$

99. $\dfrac{1}{\sqrt{x + 3} + \sqrt{x}}$ **101.** $\dfrac{2a}{b^3}$ **103.** $-\dfrac{2z^9}{3x^3y^2}$ **105.** $\dfrac{x}{2y} + \dfrac{3xy}{2}$ **107.** $\dfrac{2y^3}{5} - \dfrac{3y}{5x^3} + \dfrac{1}{5x^4y^3}$ **109.** $3x + 2$

111. $x - 7 + \dfrac{2}{2x - 5}$ **113.** $x - 3$ **115.** $x^2 - 2 + \dfrac{-x^2 + 5}{x^3 - 2}$ **117.** $x^4 + 2x^3 + 4x^2 + 8x + 16$

119. $x^2 + 2 + \dfrac{20}{x^2 - 2}$ **121.** $6x^2 + x - 12$ **123.** $3x^2 - x + 2$

Exercise 1.5   (page 51)

1. $3(x - 2)$ **3.** $4x^2(2 + x)$ **5.** $7x^2y^2(1 + 2x)$ **7.** $3abc(a + 2b + 3c)$ **9.** $(x + y)(b - a)$
11. $(4a + b)(1 - 3a)$ **13.** $(x + 1)(3x^2 - 1)$ **15.** $t(y + c)(2x - 3)$ **17.** $(a + b)(x + y + z)$

19. $(2x + y)(3c + d)$ **21.** $(2x + 3)(2x - 3)$ **23.** $(2 + 3r)(2 - 3r)$ **25.** $(x + z + 5)(x + z - 5)$

27. prime   **29.** $(x + y - z)(x - y + z)$ **31.** $-4xy$ **33.** $(x^2 + y^2)(x + y)(x - y)$ **35.** $(x^4 + 8z^2)(x^4 - 8z^2)$

37. $3(x + 2)(x - 2)$ **39.** $2x(3y + 2)(3y - 2)$ **41.** $(x + 4)(x + 4)$ **43.** $(b - 5)(b - 5)$ **45.** $(m + 2n)(m + 2n)$

47. $(x + 7)(x + 3)$ **49.** $(x - 6)(x + 2)$ **51.** prime   **53.** $(4x - 3y)(3x + 2y)$ **55.** $(6y - 5)(4y - 3)$

57. $(6a + 5)(4a - 3)$ **59.** $(3x + 7y)(2x + 5y)$ **61.** $(2x - 5)(3x + 7)$ **63.** $2(6y - 35)(y + 1)$

65. $-x(6x + 7)(x - 5)$ **67.** $x^2(2x - 7)(3x + 5)$ **69.** $-(6x - 5)(x - 7)$ **71.** $(x^2 + 5)(x^2 - 3)$

73. $2(y^2z^2 - 8yz + 40)$ **75.** $(a^n - 3)(a^n + 1)$ **77.** $(3x^n - 2)(2x^n - 1)$ **79.** $(2x^n + 3y^n)(2x^n - 3y^n)$

81. $(5y^n + 2)(2y^n - 3)$ **83.** $(2z - 3)(4z^2 + 6z + 9)$ **85.** $2(x + 10)(x^2 - 10x + 100)$

87. $(x + y - 4)(x^2 + 2xy + y^2 + 4x + 4y + 16)$ **89.** $-x(x^2 + 3x + 3)$

91. $(2a + y)(2a - y)(4a^2 - 2ay + y^2)(4a^2 + 2ay + y^2)$ **93.** $(a - b)(a^2 + ab + b^2 + 1)$

95. $(4x^2 + y^2)(16x^4 - 4x^2y^2 + y^4)$ **97.** $(x - 3 + 12y)(x - 3 - 12y)$ **99.** $(a + b - 5)(a + b + 2)$

101. $(3u + 3v + 4)(2u + 2v + 1)$ **103.** $(x + 2)(x^2 - 2x + 4)(x - 1)(x^2 + x + 1)$ **105.** $(a + b)(c + d + 1)$

107. $(x^2 + x + 2)(x^2 - x + 2)$ **109.** $(x^2 + x + 4)(x^2 - x + 4)$ **111.** $(2a^2 + a + 1)(2a^2 - a + 1)$

113. $2(x^2 + 2x + 2)(x^2 - 2x + 2)$ **115.** $2\left(\frac{3}{2}x + 1\right)$ **117.** $x^{1/2}(x^{1/2} + 1)$ **119.** $ab(b^{1/2} - a^{1/2})$ **121.** $a^2(a^n + a^{n+1})$

Exercise 1.6   (page 58)

1. equal   **3.** not equal   **5.** $\dfrac{8x}{35a}$ **7.** $\dfrac{16}{3}$ **9.** $\dfrac{z}{c}$ **11.** $\dfrac{2x^2y}{a^2b^3}$ **3.** $\dfrac{2}{x + 2}$ **15.** $-\dfrac{x - 5}{x + 5}$ **17.** $\dfrac{3x - 4}{2x - 1}$

19. $\dfrac{x^2 + 2x + 4}{x + a}$ **21.** $\dfrac{x(x - 1)}{x + 1}$ **23.** $\dfrac{x - 1}{x}$ **25.** $\dfrac{x(x + 1)^2}{x + 2}$ **27.** $\dfrac{1}{2}$ **29.** $\dfrac{z - 4}{(z + 2)(z - 2)}$ **31.** 1   **33.** 1

35. $\dfrac{x(x + 3)}{x + 1}$ **37.** $\dfrac{x + 5}{x + 3}$ **39.** 4   **41.** $\dfrac{-1}{x - 5}$ **43.** $\dfrac{5x - 1}{(x + 1)(x - 1)}$ **45.** $\dfrac{2a - 4}{(a + 4)(a - 4)}$ **47.** $\dfrac{2}{(x + 2)(x - 2)}$

49. $\dfrac{2(x^2 - 3x + 1)}{(x + 1)^2(x - 1)}$ **51.** $\dfrac{3y - 2}{y - 1}$ **53.** $\dfrac{1}{x + 2}$ **55.** $\dfrac{2x - 5}{2x(x - 2)}$ **57.** $\dfrac{2x^2 + 19x + 1}{(x - 4)(x + 4)}$ **59.** 0

61. $\dfrac{-x^4 + 3x^3 - 43x^2 - 58x + 697}{(x + 5)(x - 5)(x + 4)(x - 4)}$ **63.** $\dfrac{b}{2c}$ **65.** $81a$ **67.** -1 **69.** $\dfrac{y + x}{x^2y^2}$ **71.** $\dfrac{y + x}{y - x}$ **73.** $\dfrac{a^2(3x - 4ab)}{ax + b}$

75. $\dfrac{x - 2}{x + 2}$ **77.** $\dfrac{3x^2y^2}{xy - 1}$ **79.** $\dfrac{3x^2}{x^2 + 1}$ **81.** $\dfrac{x^2 - 3x - 4}{x^2 + 5x - 3}$ **83.** $\dfrac{x}{x + 1}$ **85.** $\dfrac{5x + 1}{x - 1}$ **87.** $\dfrac{3x}{3 + x}$ **89.** $\dfrac{x + 1}{2x + 1}$

91. $\dfrac{(x - 1)\sqrt{x^2 + 4}}{x^2 + 4}$ **93.** $\dfrac{(z - 2)\sqrt{z^2 - 7}}{z^2 - 7}$

CHAPTER 1 REVIEW EXERCISES   (page 62)

1. $\{2\}$ **2.** $\{1, 2, 3, 4, 5, 7, 11\}$ **3.** $\{2\}$ **4.** $\{2, 3, 5, 7, 11\}$ **5.**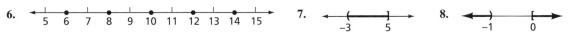

6.    **7.** 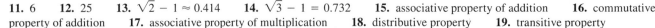   **8.**

9.   **10.**

11. 6   **12.** 25   **13.** $\sqrt{2} - 1 \approx 0.414$ **14.** $\sqrt{3} - 1 = 0.732$ **15.** associative property of addition   **16.** commutative property of addition   **17.** associative property of multiplication   **18.** distributive property   **19.** transitive property

20. reflexive property   **21.** $x^{12}y^8$ **22.** $\dfrac{a^{12}}{b^6}$ **23.** $\dfrac{n^8}{m^6}$ **24.** $\dfrac{q^6}{8p^6}$ **25.** $\dfrac{y^8}{9}$ **26.** $\dfrac{a^8}{b^{10}}$ **27.** $\dfrac{y^4}{9x^4}$ **28.** $-\dfrac{8m^{12}}{n^3}$

29. 2   **30.** 4   **31.** -4 **32.** -32 **33.** $\dfrac{8}{27}$ **34.** $\dfrac{4}{9}$ **35.** $\dfrac{9}{4}$ **36.** $\dfrac{125}{8}$ **37.** 6   **38.** -7 **39.** $\dfrac{3}{5}$

40. $\dfrac{3}{5}$ **41.** x^6y **42.** $\dfrac{y^2}{x^7}$ **43.** $-c$ **44.** a^2 **45.** xy^2 **46.** x **47.** $\dfrac{m^2n}{p^3}$ **48.** $\dfrac{a^3b^2}{c}$ **49.** $x^8y^2|c|$

50. $\left|\dfrac{b^3}{a^7}\right|$ **51.** x^2y^4 **52.** $|a|b^2$ **53.** $\dfrac{2\sqrt{5}}{5}$ **54.** $2\sqrt{2}$ **55.** $\dfrac{\sqrt[3]{4}}{2}$ **56.** $\dfrac{2\sqrt[3]{5}}{5}$ **57.** $\sqrt{3} + 1$

58. $-2(\sqrt{3} + \sqrt{2})$ **59.** $\dfrac{2x(\sqrt{x} + 2)}{x - 4}$ **60.** $\dfrac{x - 2\sqrt{xy} + y}{x - y}$ **61.** $\dfrac{2}{5\sqrt{2}}$ **62.** $\dfrac{1}{\sqrt{5}}$ **63.** $\dfrac{x - 4}{5(\sqrt{x} - 2)}$

64. $\dfrac{1 - a}{a(1 + \sqrt{a})}$ **65.** $7\sqrt{2}$ **66.** 0 **67.** $5 + 2\sqrt{6}$ **68.** -1 **69.** $\sqrt{6} + \sqrt{2} + \sqrt{3} + 1$ **70.** -5

71. third-degree binomial   **72.** second-degree trinomial   **73.** second-degree monomial   **74.** not a polynomial

75. $2x^3 - 7x^2 - 6x$ **76.** $6x^2 - 7xy - 3y^2$ **77.** $8a^2 - 8ab - 6b^2$ **78.** $3z^3 + 10z^2 + 2z - 3$ **79.** $x^2 + 2x + 1$

80. $x^3 + 2x - 3 - \dfrac{6}{x^2 - 1}$ **81.** $3t(t + 1)(t - 1)$ **82.** $5(r - 1)(r^2 + r + 1)$ **83.** $(3x + 8)(2x - 3)$

84. $(3a + x)(a - 1)$ **85.** $(2x - 5)(4x^2 + 10x + 25)$ **86.** $2(3x + 2)(x - 4)$ **87.** $(x + 3 + t)(x + 3 - t)$

88. prime   **89.** $(2z + 7)(4z^2 - 14z + 49)$ **90.** $(7b + 1)^2$ **91.** $(11z - 2)^2$ **92.** $8(2y - 5)(4y^2 + 10y + 25)$

93. $(y - 2z)(2x - w)$ **94.** $(x^2 + 1 + x)(x^2 + 1 - x)(x^4 + 1 - x^2)$ **95.** $(x - 2)(x + 3)$ **96.** $\dfrac{2y - 5}{y - 2}$ **97.** $\dfrac{t + 1}{5}$

98. $\dfrac{p(p + 4)}{(p^2 + 8p + 4)(p - 3)}$ **99.** $\dfrac{(x - 2)(x + 3)(x - 3)}{(x - 1)(x + 2)^2}$ **100.** 1 **101.** $\dfrac{3x^2 - 10x + 10}{(x - 4)(x + 5)}$ **102.** $\dfrac{2(2x^2 + 3x + 6)}{(x + 2)(x - 2)}$

103. $\dfrac{3x^3 - 12x^2 + 11x}{(x - 1)(x - 2)(x - 3)}$ **104.** $\dfrac{-5x - 6}{(x + 1)(x + 2)}$ **105.** $\dfrac{-x^3 + x^2 - 12x - 15}{x^2(x + 1)}$ **106.** $\dfrac{3x}{x + 1}$ **107.** $\dfrac{20}{3x}$ **108.** $\dfrac{y}{2}$

109. $\dfrac{y + x}{xy(x - y)}$ **110.** $\dfrac{y + x}{x - y}$

CHAPTER 1 TEST   (page 64)

1. true   **2.** false   **3.** $\{2, 3, 4, 5, 6\}$ **4.** $\{4\}$ **5.**   **6.**

7. commutative property of addition   **8.** distributive property   **9.** x^{11} **10.** $\dfrac{r}{s}$ **11.** a **12.** x^{24} **13.** 4.5×10^5

14. 3.45×10^{-4} **15.** 3700 **16.** 0.0012 **17.** $5a^2$ **18.** $\dfrac{216}{729}$ **19.** $\dfrac{9s^6}{4t^4}$ **30.** $3a^2$ **21.** $5\sqrt{3}$

22. $4\sqrt[3]{10} - 6\sqrt[3]{3}$ **23.** $\dfrac{x(\sqrt{x} + 2)}{x - 4}$ **24.** $\dfrac{x + 2\sqrt{xy} + y}{x - y}$ **25.** $-a^2 + 7$ **26.** $-6a^6b^6$ **27.** $6x^2 + 13x - 28$

28. $a^{2n} + a^n - 2$ **29.** $x^4 - 16$ **30.** $2x^3 - 5x^2 + 7x - 6$ **31.** $6x + 19 + \dfrac{34}{x - 3}$ **32.** $x^2 + 2x + 1$

33. $3(x + 2y)$ **34.** $(x + 10)(x - 10)$ **35.** $(5t - 2w)(2t - 3w)$ **36.** $3(a - b)(a^2 + 6a + 36)$

37. $(x + 2)(x - 2)(x^2 + 3)$ **38.** $(3x^2 - 2)(2x^2 + 5)$ **39.** 1 **40.** $\dfrac{-2x}{(x + 1)(x - 1)}$ **41.** $\dfrac{(x + 5)^2}{x + 4}$

42. $\dfrac{1}{(x + 1)(x - 2)}$ **43.** $\dfrac{b + a}{a}$ **44.** $\dfrac{y}{y + x}$

Exercise 2.1   (page 74)

1. all real numbers   **3.** all real numbers but 0   **5.** all nonnegative real numbers   **7.** all real numbers but 3 and -2

9. $x = 5$; conditional equation   **11.** no solution   **13.** $x = 7$; conditional equation   **15.** no solution   **17.** identity

19. $b = 6$; conditional equation   **21.** identity   **23.** $x = 1$ **25.** $z = 9$ **27.** $z = 10$ **29.** $x = -3$ **31.** $x = 6$

33. $x = -2$ **35.** $x = \dfrac{5}{2}$ **37.** $x = 1$ **39.** $x = -\dfrac{14}{11}$ **41.** identity   **43.** $y = 2$ **45.** $s = 4$ **47.** $x = -4$

49. no solution   **51.** $a = 17$ **53.** $x = -\dfrac{2}{5}$ **55.** $x = \dfrac{2}{3}$ **57.** $n = 3$ **59.** $y = 5$ **61.** no solution

63. $a = 2$ **65.** $p = \dfrac{h}{2.2}$ **67.** $b_2 = \dfrac{2A}{h} - b_1$ **69.** $r^2 = \dfrac{3V}{\pi h}$ **71.** $s = \dfrac{f(P_n - L)}{i}$ **73.** $w = \dfrac{A - 2hl}{2(l + h)}$

75. $m = \dfrac{r^2 F}{Mg}$ **77.** $y = b\left(1 - \dfrac{x}{a}\right)$ **79.** $x^2 = a^2\left(1 + \dfrac{y^2}{b^2}\right)$ **81.** $r = \dfrac{r_1 r_2}{r_1 + r_2}$ **83.** $n = \dfrac{l - a + d}{d}$

85. $n = \dfrac{360}{180 - a}$ **87.** $a = S - Sr + lr$ **89.** $r_1 = \dfrac{-Rr_2 r_3}{Rr_3 + Rr_2 - r_2 r_3}$

Exercise 2.2   (page 81)

1. 84   **3.** 17 and 37   **5.** 7   **7.** $2\frac{1}{2}$ ft   **9.** 20 ft by 8 ft   **11.** 20 ft   **13.** \$10,000 at 7%, \$12,000 at 6%   **15.** 327
17. \$17,500 at 8%; \$19,500 at $9\frac{1}{2}$%   **19.** \$79.95   **21.** 200 units   **23.** 21   **25.** $\frac{4}{3}$ hr   **27.** $\frac{190}{9}$ hr   **29.** 1 1
31. $\frac{1}{15}$1   **33.** about 4.5 gal   **35.** 4 1   **37.** 39 mph going; 65 mph returning   **39.** $2\frac{1}{2}$ hr   **41.** 50 sec   **43.** 12 mph
45. 8 of each   **47.** 600 lb barley; 1637 lb oats; 163 lb soybean meal   **49.** about 11.2 mm

Exercise 2.3   (page 93)

1. 3, -2 **3.** 12, -12 **5.** 2, $-\frac{5}{2}$ **7.** 2, $\frac{3}{5}$ **9.** $\frac{3}{5}$, $-\frac{5}{3}$ **11.** $\frac{3}{2}$, $\frac{1}{2}$ **13.** 3, -3 **15.** $5\sqrt{2}$, $-5\sqrt{2}$
17. 3, -1 **19.** 2, -4 **21.** $x^2 + 6x + 9$ **23.** $x^2 - 4x + 4$ **25.** $a^2 + 5a + \frac{25}{4}$ **27.** $r^2 - 11r + \frac{121}{4}$
29. $y^2 + \frac{3}{4}y + \frac{9}{64}$ **31.** $q^2 - \frac{1}{5}q + \frac{1}{100}$ **33.** 5, 3   **35.** 2, -3 **37.** 0, 25   **39.** $\frac{2}{3}$, -2 **41.** $\dfrac{-5 \pm \sqrt{5}}{2}$
43. $\dfrac{-2 \pm \sqrt{7}}{3}$ **45.** $\pm 2\sqrt{3}$ **47.** 3, $-\frac{5}{2}$ **49.** 2, $-\frac{1}{5}$ **51.** 1, -2 **53.** $\dfrac{-5 \pm \sqrt{13}}{6}$ **55.** $\dfrac{-1 \pm \sqrt{61}}{10}$
57. rational and equal   **59.** not real numbers   **61.** rational and unequal   **63.** not real   **65.** 2, 10   **67.** yes
69. 3, -4 **71.** $\frac{3}{2}$, $-\frac{1}{4}$ **73.** $\frac{1}{2}$, $-\frac{4}{3}$ **75.** $\frac{5}{6}$, $-\frac{2}{5}$ **77.** $4 \pm 2\sqrt{2}$ **79.** 1, -1 **81.** $-\frac{1}{2}$, 5   **83.** -2
85. 3, $\dfrac{-8}{11}$ **87.** $t = \pm \sqrt{\dfrac{2h}{g}}$ **89.** $t = \dfrac{64 \pm \sqrt{4096 - 64h}}{32} = \dfrac{8 \pm \sqrt{64 - h}}{4}$
91. $y = \pm \sqrt{\dfrac{a^2b^2 - b^2x^2}{a^2}} = \pm \dfrac{b\sqrt{a^2 - x^2}}{a}$ **93.** $a = \pm \sqrt{\dfrac{b^2x^2}{b^2 + y^2}} = \pm \dfrac{bx\sqrt{b^2 + y^2}}{b^2 + y^2}$ **95.** $x = \dfrac{-y \pm y\sqrt{5}}{2}$

Exercise 2.4   (page 98)

1. 6, 8   **3.** 4 ft by 8 ft   **5.** 9 cm   **7.** 2 in.   **9.** 20 mph going and 10 mph returning   **11.** 7 hr   **13.** 25 sec
15. about 9.5 sec   **17.** 10¢   **19.** 1440   **21.** 4 hr   **23.** about 9.5 hr   **25.** no   **27.** Matilda at 8%; Maude at 7%
29. 10   **31.** 20   **33.** 10m and 24m

Exercise 2.5   (page 107)

1. i **3.** -1 **5.** -1 **7.** -1 **9.** $x = 3; y = 5$ **11.** $x = \frac{2}{3}; y = -\frac{2}{9}$ **13.** $5 - 6i$ **15.** $-2 - 10i$
17. $4 + 10i$ **19.** $1 - i$ **21.** $-9 + 19i$ **23.** $-5 + 12i$ **25.** $52 - 56i$ **27.** $-6 + 17i$ **29.** $0 + i$
31. $4 + 0i$ **33.** $\frac{2}{5} - \frac{1}{5}i$ **35.** $\frac{1}{25} + \frac{7}{25}i$ **37.** $\frac{1}{2} + \frac{1}{2}i$ **39.** $-\frac{7}{13} - \frac{22}{13}i$ **41.** $-\frac{12}{17} + \frac{11}{34}i$ **43.** $\dfrac{6 + \sqrt{3}}{10} + \dfrac{3\sqrt{3} - 2}{10}i$
45. 5   **47.** $\sqrt{13}$ **49.** $7\sqrt{2}$ **51.** $\dfrac{\sqrt{2}}{2}$ **53.** 6   **55.** $\sqrt{2}$ **57.** $\dfrac{3\sqrt{5}}{5}$ **59.** 1   **61.** $-1 \pm i$ **63.** $-2 \pm i$
65. $1 \pm 2i$ **67.** $\frac{1}{3} \pm \frac{1}{3}i$ **69.** $(x + 2i)(x - 2i)$ **71.** $(5p + 6qi)(5p - 6qi)$ **73.** $2(y + 2zi)(y - 2zi)$
75. $2(5m + ni)(5m - ni)$

Exercise 2.6   (page 113)

1. 0, -5, -4 **3.** 0, $\frac{4}{3}$, $-\frac{1}{2}$ **5.** 5, -5, 1, -1 **7.** 6, -6, 1, -1 **9.** $3\sqrt{2}$, $-3\sqrt{2}$, $\sqrt{5}$, $-\sqrt{5}$ **11.** 0, 1
13. $\frac{1}{8}$, -8 **15.** 1, 144   **17.** $\frac{1}{64}$ **19.** $\frac{1}{9}$ **21.** 27   **23.** 1   **25.** $-\frac{3}{2}$, $-\frac{5}{2}$ **27.** 9   **29.** 20   **31.** 2   **33.** 2
35. 3, 4   **37.** $\frac{1}{5}$, -1 **39.** 3, 5   **41.** -2, 1   **43.** 2, $-\frac{5}{2}$ **45.** -2 **47.** no solution   **49.** 3   **51.** 1

Exercise 2.7   (page 123)

1. $(-\infty, 1)$ — graph with open bracket at 1

3. $[1, \infty)$ — graph with closed bracket at 1

5. $(-\infty, 1)$ — graph at 1

7. $[1, \infty)$ — graph at 1

9. $(-\infty, 3]$ — graph at 3

11. $(-10/3, \infty)$ — graph at $-10/3$

13. $(5, \infty)$ — graph at 5

15. $[14, \infty)$ — graph at 14

17. $(-\infty, 15/4]$ — graph at $15/4$

19. $(-44/41, \infty)$ — graph at $-44/41$

21. $(6, 9]$ — graph at 6, 9

23. $(8, 22]$ — graph at 8, 22

25. $[-11, 4]$ — graph at -11, 4

27. $[5, 21]$ — graph at 5, 21

29. $(-\infty, 0)$ — graph at 0

31. $(0, \infty)$ — graph at 0

33. $(3, 14/3)$ — graph at 3, $14/3$

35. $(2, \infty)$ — graph at 2

37. $(-4, 5/6)$ — graph at -4, $5/6$

39. $[-2, \infty)$ — graph at -2

41. $(-4, -3)$ — graph at -4, -3

43. $(-\infty, 2] \cup [3, \infty)$ — graph at 2, 3

45. $(-3, -2)$ — graph at -3, -2

47. $(-\infty, -1/2] \cup [-1/3, \infty)$ — graph at $-1/2$, $-1/3$

49. $(1/3, 1/2)$ — graph at $1/3$, $1/2$

51. $(-\infty, -3/2] \cup [1, \infty)$ — graph at $-3/2$, 1

53. $(-3, 2)$ — graph at -3, 2

55. $(-\infty, -1) \cup (-1, 0) \cup (1, \infty)$ — graph at -1, 0, 1

57. $(-\infty, -3) \cup (-3, -2] \cup (2, \infty)$ — graph at -3, -2, 2

59. $(-\infty, -2) \cup (-2, -1/3) \cup (1/2, \infty)$ — graph at -2, $-1/3$, $1/2$

61. $(0, 3/2)$ — graph at 0, $3/2$

63. $(-\infty, 0) \cup (3/2, \infty)$ — graph at 0, $3/2$

65. $(-\infty, 2) \cup [13/5, \infty)$ — graph at 2, $13/5$

67. $(-\infty, -\sqrt{7}) \cup (-1, 1) \cup (\sqrt{7}, \infty)$ — graph at $-\sqrt{7}$, -1, 1, $\sqrt{7}$

69. $5.25 \text{ mi} \le a \text{ mi} \le 6.25 \text{ mi}$ **71.** $16\frac{2}{3} \text{ cm} < s \text{ cm} < 20 \text{ cm}$ **73.** $40 + 2w < P < 60 + 2w$ **75.** 12
77. $\left\{x \mid x \ge \frac{4}{3}\right\}$ **79.** $\left\{x \mid x > -\frac{2}{3}\right\}$ **81.** $\{x \mid x \le -3 \text{ or } x \ge 3\}$ **83.** $\{x \mid x \le -3 \text{ or } -1 \le x < 0 \text{ or } x > 0\}$

Exercise 2.8   (page 130)

1. 7   **3.** 0   **5.** 2   **7.** $\pi - 2 \approx 1.142$ **9.** $x - 5$ **11.** x^3 if $x \ge 0$; $-x^3$ if $x < 0$ **13.** $0, -4$ **15.** $2, -\frac{4}{3}$
17. $\frac{14}{3}, -2$ **19.** $7, -3$ **21.** no solution   **23.** $\frac{2}{7}, 2$ **25.** $x \ge 0$ **27.** $-\frac{3}{2}$ **29.** $0, -6$ **31.** 0   **33.** $\frac{3}{5}, 3$
35. $-\frac{3}{13}, \frac{9}{5}$ **37.** $(-3, 9)$ — graph at -3, 9   **39.** $(-\infty, -9) \cup (3, \infty)$ — graph at -9, 3   **41.** $(-\infty, -7] \cup [3, \infty)$ — graph at -7, 3   **43.** $[-13/3, 1]$ — graph at $-13/3$, 1

45. $(-\infty, -3) \cup (-3, \infty)$ — graph at -3 **47.** $(-1, 1/5)$ — graph at -1, $1/5$ **49.** $(-\infty, -7/9) \cup (13/9, \infty)$ — graph at $-7/9$, $13/9$ **51.** $(-5, 7)$ — graph at -5, 7

53. $(-2, -1/2) \cup (-1/2, 1)$

55. $(-7/3, -2/3) \cup (4/3, 3)$

57. $(-7, -1) \cup (11, 17)$

59. $(-18, -6) \cup (10, 22)$

61. $(-10, -7] \cup [5, 8)$

63. $\left[-\frac{1}{2}, \infty\right)$ **65.** $(-\infty, 0)$ **67.** $\left(-\infty, -\frac{1}{2}\right)$ **69.** $[0, \infty)$

CHAPTER 2 REVIEW EXERCISES (page 132)

1. all real numbers **2.** all real numbers but 0 **3.** all nonnegative real numbers **4.** all real numbers but 2 and 3

5. $\frac{16}{27}$; conditional equation **6.** -14; conditional equation **7.** $\frac{16}{5}$; conditional equation **8.** no solution

9. 7; conditional equation **10.** identity **11.** 7; conditional equation **12.** $\frac{1}{3}$; conditional equation

13. $F = \frac{9}{5}C + 32$ **14.** $f = \dfrac{is}{P_n - l}$ **15.** $f_1 = \dfrac{ff_2}{f_2 - f}$ **16.** $l = \dfrac{a - S + Sr}{r}$ **17.** 1.5 l **18.** about 3.9 hr

19. $5\frac{1}{7}$ hr **20.** $3\frac{1}{3}$ oz **21.** $4500 at 11%; $5500 at 14% **22.** 10 **23.** $2, -\frac{3}{2}$ **24.** $\frac{1}{4}, -\frac{4}{3}$ **25.** $0, \frac{8}{5}$

26. $\frac{2}{3}, \frac{4}{9}$ **27.** either 95 by 110 yd or 55 by 190 yd **28.** 320 mph for prop plane; 440 mph for jet plane **29.** 3, 5

30. $-4, -2$ **31.** $\dfrac{1 \pm \sqrt{21}}{10}$ **32.** $0, \frac{1}{5}$ **33.** $2, -7$ **34.** $9, -\frac{2}{3}$ **35.** $\dfrac{-1 \pm \sqrt{21}}{10}$ **36.** $-1 \pm 2i$ **37.** $\frac{1}{3}$

38. 10, 2 **39.** 1 sec **40.** 2 ft **41.** $0 + i$ **42.** $0 - i$ **43.** $-2 - i$ **44.** $-2 - 5i$ **45.** $5 - 2i$

46. $21 - 9i$ **47.** $0 - 3i$ **48.** $0 - 2i$ **49.** $\frac{3}{2} - \frac{3}{2}i$ **50.** $-\frac{2}{5} + \frac{4}{5}i$ **51.** $\frac{4}{5} + \frac{3}{5}i$ **52.** $\frac{1}{2} - \frac{5}{2}i$ **53.** $\sqrt{10}$

54. 1 **55.** $2, -3$ **56.** $4, -2$ **57.** $1, 1, -1, -1$ **58.** $1, -1, 6, -6$ **59.** 9 **60.** $8, -27$ **61.** 5

62. 0 **63.** $4, -4$ **64.** no solution **65.** $(-\infty, 7)$ **66.** $[-1/5, \infty)$ **67.** $(-\infty, 5/3)$

68. $[-3, 5)$ **69.** $(-\infty, -2) \cup (4, \infty)$ **70.** $(-4, 1)$ **71.** $(-1, 3)$

72. $(-\infty, -3/2) \cup (1, \infty)$ **73.** $(-\infty, -2] \cup (3, \infty)$ **74.** $(-4, 1]$ **75.** $[-2, 1] \cup (3, \infty)$

76. $(-\infty, 0) \cup (5/2, \infty)$ **77.** $5, -7$ **78.** 0 **79.** $(-6, 0)$ **80.** $(-\infty, 2] \cup [8/3, \infty)$

81. $(-5, 1)$ **82.** $(-\infty, -29) \cup (35, \infty)$ **83.** $(-7/2, -2) \cup (-1, 1/2)$

84. $(-1, 4/3) \cup (4/3, 11/3)$

CHAPTER 2 TEST   (page 134)

1. all real numbers but 0 and 1   **2.** all nonnegative real numbers   **3.** $\frac{5}{2}$ **4.** no solution   **5.** 37   **6.** $x = \mu + z\sigma$

7. 87.5   **8.** \$14,000   **9.** $\frac{1}{2}, \frac{3}{2}$ **10.** $\frac{3}{2}, -4$ **11.** $x = \dfrac{-b \pm \sqrt{b^2 - 4ac}}{2a}$ **12.** $\dfrac{5 \pm \sqrt{133}}{6}$ **13.** $5, -3$

14. 8 sec   **15.** i **16.** 1   **17.** $7 - 12i$ **18.** $-23 - 43i$ **19.** $\frac{4}{5} + \frac{2}{5}i$ **20.** $0 + i$ **21.** 13   **22.** $\dfrac{\sqrt{10}}{10}$

23. $2, -2, 3, -3$ **24.** $1, -\frac{1}{32}$ **25.** 139   **26.** -1 **27.**

$(-\infty, 2]$

           2

28. $(-\infty, 5)$

      5

29. $(2, \infty)$

   2

30. $[-2, 1)$

   -2 1

31. $2, \dfrac{-10}{3}$ **32.** 0

33. $(-\infty, 3/2) \cup (7/2, \infty)$

   3/2   7/2

34. $[-9, 6]$

   -9 6

35. $(-3, 2) \cup (2, 7)$

   -3 2 7

Exercise 3.1   (page 145)

1. $A(2, 3)$ **3.** $C(-2, -3)$ **5.** $E(0, 0)$ **7.** $G(-5, -5)$ **9.** QI   **11.** QIII   **13.** QI   **15.** positive x-axis

17. positive y-axis   **19.** negative x-axis   **21.**
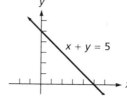
$x + y = 5$

23.
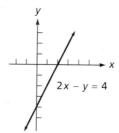
$2x - y = 4$

25.
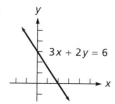
$3x + 2y = 6$

27.
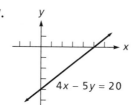
$4x - 5y = 20$

29.
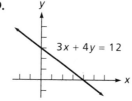
$3x + 4y = 12$

31.
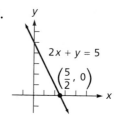
$2x + y = 5$

$\left(\frac{5}{2}, 0\right)$

33.
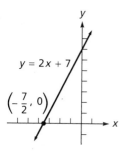
$y = 2x + 7$

$\left(-\frac{7}{2}, 0\right)$

35.
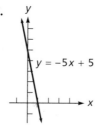
$y = -5x + 5$

37.
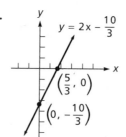
$y = 2x - \dfrac{10}{3}$

$\left(\frac{5}{3}, 0\right)$

$\left(0, -\dfrac{10}{3}\right)$

39.
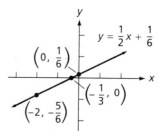
$y = \frac{1}{2}x + \frac{1}{6}$

$\left(0, \frac{1}{6}\right)$

$\left(-\frac{1}{3}, 0\right)$

$\left(-2, -\frac{5}{6}\right)$

41.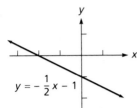

$y = -\dfrac{1}{2}x - 1$

43.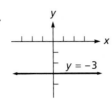

$y = -3$

45.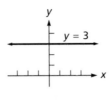

$y = 3$

47.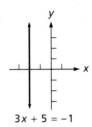

$3x + 5 = -1$

49.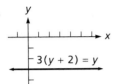

$3(y + 2) = y$

51.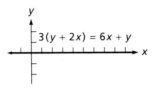

$3(y + 2x) = 6x + y$

53. 5 **55.** $\sqrt{13}$ **57.** $\sqrt{2}$ **59.** 2 **61.** 5 **63.** 13 **65.** 25 **67.** $8\sqrt{2}$ **69.** 2 **71.** 7 **73.** (4, 6)
75. (0, 1) **77.** (0, 0) **79.** $\left(\dfrac{\sqrt{5}}{2}, \dfrac{\sqrt{5}}{2}\right)$ **81.** $\sqrt{a^2 + b^2}$ **83.** $\sqrt{a^2 + b^2}$ **85.** $\left(\dfrac{a}{2}, \dfrac{b}{2}\right)$ **87.** $\left(\dfrac{a + b}{2}, \dfrac{a + b}{2}\right)$
89. (5, 6) **91.** (5, 15) **97.** $(-4, -3)$ **99.** (5, 0), (0, 5), (−5, 0), (0, −5) **101.** $\sqrt{2}$ units
105. approximately 170 mi

Exercise 3.2 (page 154)

1. 5 **3.** $-\dfrac{7}{4}$ **5.** -2 **7.** undefined **9.** $\dfrac{5}{3}$ **11.** -7 **13.** 1 **15.** -1 **17.** 3 **19.** $\dfrac{1}{2}$ **21.** $\dfrac{2}{3}$
23. 0 **25.** perpendicular **27.** parallel **29.** perpendicular **31.** perpendicular **33.** perpendicular
35. parallel **37.** perpendicular **39.** neither **41.** not on same line **43.** on same line **45.** No two are
perpendicular. **47.** PQ and PR are perpendicular. **49.** PQ and PR are perpendicular.

51.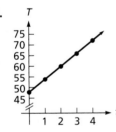

$\dfrac{\Delta T}{\Delta t}$ is the hourly rate
of change of temperature.

53. The slope is the speed of the plane.

D(miles)

$D = 590t$

1200

600 (1, 590)

t(hours)

Exercise 3.3 (page 163)

1. $y - 4 = 2(x - 2)$ **3.** $y - \dfrac{1}{2} = 2\left(x + \dfrac{3}{2}\right)$ **5.** $y + 5 = -5(x - 5)$ **7.** $y = \pi(x - \pi)$ **9.** $y = 3x + 1$
11. $y = -2x + \dfrac{13}{2}$ **13.** $y = -4x + 8$ **15.** $y = \dfrac{1}{2}x$ **17.** $y = 3x - 2$ **19.** $y = 5x - \dfrac{1}{5}$ **21.** $y = ax + \dfrac{1}{a}$
23. $y = ax + a$ **25.** $3x - 2y = 0$ **27.** $3x + y = -4$ **29.** $\sqrt{2}x - y = -\sqrt{2}$ **31.** $sx - ry = 0$
33. $x + y = 5$ **35.** $6x + y = 16$ **37.** $13x + 9y = 0$ **39.** $2x + 3y = 2$ **41.** $-7; 12$ **43.** $\dfrac{7}{2}; -\dfrac{3}{2}$
45. 3; -3 **47.** no slope; no y-intercept **49.** -1 **51.** 3 **53.** $\dfrac{1}{3}$ **55.** 0 **57.** $\dfrac{3}{5}$ **59.** $x - 2y = -5$
61. $8x + 7y = 12$ **63.** $11x + 2y = 0$ **65.** $x - y = -7$ **67.** $2x + 3y = 30$ **69.** $y = -5$ **71.** $3x - y = 5$
73. $x = -4$ **75.** $5x - 3y = 15$ **77.** $y = -\dfrac{3}{2}x$ **79.** $y = \dfrac{5}{3}x + \dfrac{19}{3}$ **81.** $y = \dfrac{s}{r}x$ **83.** $C = \dfrac{5}{9}(F - 32)$
85. 70° **87.** $L = 3.785\,g$ **89.** $y = -9x + 300$ **91.** \$2339 **93.** 301 **95.** 1655 barrels per day
97. $y = 137{,}000 - \dfrac{42{,}500}{3}x$ **105.** $-\dfrac{b}{m}$ **107.** $y = 0$

Exercise 3.4   (page 175)

1. $x = -2, 2; y = -4$ **3.** $x = 0, \frac{1}{2}; y = 0$ **5.** $x = -1, 5; y = -5$ **7.** $x = 1, -2; y = -2$
9. $x = -3, 0, 3; y = 0$ **11.** $x = -1, 1; y = -1$

13. 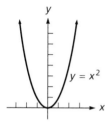   **15.** 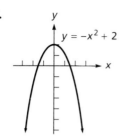   **17.** 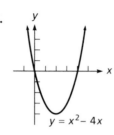   **19.**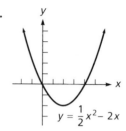

21. symmetric about the y-axis   **23.** symmetric about the x-axis   **25.** symmetric about the x-axis, the y-axis, and the origin
27. symmetric about the y-axis   **29.** not symmetric   **31.** symmetric about the x-axis   **33.** symmetric about the y-axis

35. symmetric about the x-axis   **37.** 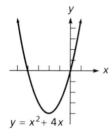   **39.** 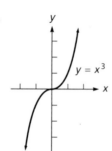   **41.**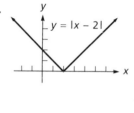

43. 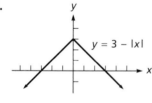   **45.** 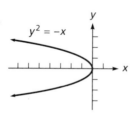   **47.** 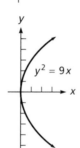   **49.**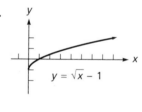

51.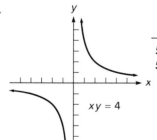

53. $x^2 + y^2 - 1 = 0$ **55.** $x^2 + y^2 - 12x - 16y + 84 = 0$ **57.** $x^2 + y^2 - 6x + 8y + 23 = 0$
59. $x^2 + y^2 - 6x - 6y - 7 = 0$ **61.** $x^2 + y^2 + 6x - 8y = 0$

63. 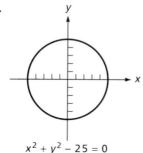   **65.**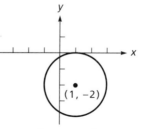

67.

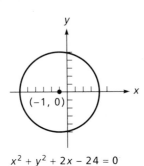

$x^2 + y^2 + 2x - 24 = 0$

69.

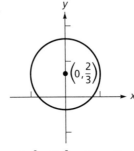

$9x^2 + 9y^2 - 12y = 5$

71.

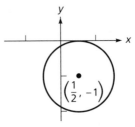

$4x^2 + 4y^2 - 4x + 8y + 1 = 0$

Exercise 3.5   (page 180)

1. $y = x^2 + 1$ **3.** $y = 3 - x^2$ **5.** $y = x^3 - x$ **7.** $y = \dfrac{x^2 - x}{5}$ **9.** $y = x^2 - 10x$ **11.** $y = 9 + x^2$

13. $y = |x|$ **15.** $y = |x| - 3$ **17.** $y^2 = x$ **19.** $x^2 - y^2 = 4$ **21.** $(0.25, 0.88)$ **23.** $(0.50, 7.25)$

25. ± 2.65 **27.** 1.44 **29.** $-3 < x < 2$

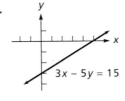 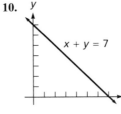 solution

Exercise 3.6   (page 186)

1. 14   **3.** $2, -3$ **5.** 18 women   **7.** $\dfrac{1}{2}$ **9.** 1000   **11.** 1   **13.** $\dfrac{21}{4}$ **15.** -8 **17.** direct variation

19. neither   **21.** $14\dfrac{6}{11}$ cu ft   **23.** 3 sec   **25.** 20 volts   **27.** 432 hertz   **29.** $\dfrac{\sqrt{3}}{4}$

CHAPTER 3 REVIEW EXERCISES   (page 189)

1. $A(2, 0)$ **2.** $B(-2, 1)$ **3.** $C(0, -1)$ **4.** $D(3, -1)$

5.–8.

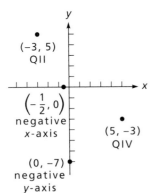

9.

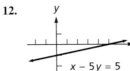

$3x - 5y = 15$

10.

$x + y = 7$

11.

$x + y = -7$

12.

$x - 5y = 5$

13. $10; (0, 3)$ **14.** $13; \left(-6, \frac{15}{2}\right)$ **15.** $2; (\sqrt{3}, 8)$ **16.** $2\sqrt{2}|a|; (0, 0)$ **17.** -6 **18.** 2 **19.** -1 **20.** 1

21. $y = -\frac{7}{5}x$ **22.** $y = -4x - 7$ **23.** $y = -2x + 9$ **24.** $y = \frac{2}{3}x + 3$ **25.** $y = 17$ **26.** $x = -5$

27. $y = -7x + 54$ **28.** $y = \frac{1}{7}x - \frac{22}{7}$ **29.** $3x - 4y = 6$ **30.** $3x - y = 21$

31.

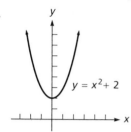

$y = x^2 + 2$

32.

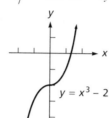

$y = x^3 - 2$

33.

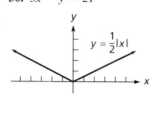

$y = \frac{1}{2}|x|$

34.

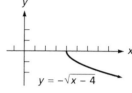

$y = -\sqrt{x - 4}$

35.

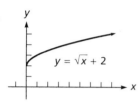

$y = \sqrt{x} + 2$

36.

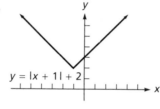

$y = |x + 1| + 2$

37.

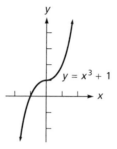

$y = x^3 + 1$

38.

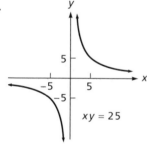

$xy = 25$

39. $(x + 3)^2 + (y - 4)^2 = 144$, or $x^2 + y^2 + 6x - 8y - 119 = 0$

40. $\left(x + \frac{1}{2}\right)^2 + \left(y - \frac{5}{2}\right)^2 = \frac{121}{2}$, or $x^2 + y^2 + x - 5y - 54 = 0$

41.

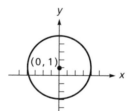

(0, 1)

$x^2 + y^2 - 2y = 15$

42.

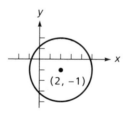

$(2, -1)$

$x^2 + y^2 - 4x + 2y = 4$

43. $y = (x - 2)(x + 5)$ **44.** $y = x + 2|x|$ **45.** $y^2 = x - 3$ **46.** $x^2 - y^2 = 4$

47. $-3.32, 3.32$ **48.** $0, -1, 1$ **49.** $-1.73, 1.73, -1, 1$ **50.** $4.19, -1.19$ **51.** $2, 5$ **52.** $3\sqrt{3}, -3\sqrt{3}$

53. $1, 5$ **54.** $-1, -3$ **55.** $\frac{9}{5}$ lb   **56.** $333\frac{1}{3}$ cc   **57.** $\frac{25}{9}$ **58.** about 117 ohms   **59.** $385

60.

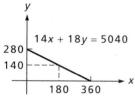

140 hr

$14x + 18y = 5040$

CHAPTER 3 TEST   (page 190)

1. QII   **2.** negative y-axis   **3.**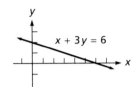

4.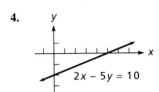

5.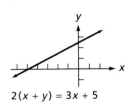

6.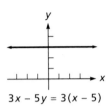

7.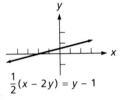

8.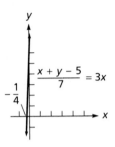

9. $\sqrt{41}$ **10.** approximately 4.44   **11.** $(0, 0)$ **12.** $(\sqrt{2}, 2\sqrt{2})$ **13.** $-\frac{5}{4}$ **14.** $\frac{\sqrt{3}}{3}$ **15.** neither

16. perpendicular   **17.** $y = 2x - 11$ **18.** $y = 3x + \frac{1}{2}$ **19.** $y = 2x + 5$ **20.** $y = -\frac{1}{2}x + 5$ **21.** $y = 2x - \frac{11}{2}$

22. $x = 3$ **23.** $x = -4, 0, 4; y = 0$ **24.** $x = 4; y = 4$ **25.** symmetric about the x-axis

26. symmetric about the y-axis   **27.**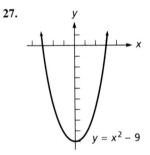

28.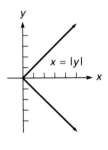

29.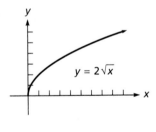

30.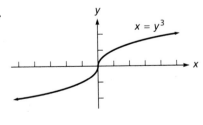

31. $(x - 5)^2 + (y - 7)^2 = 64$ or $x^2 + y^2 - 10x - 14y + 10 = 0$
32. $(x - 2)^2 + (y - 4)^2 = 32$ or $x^2 + y^2 - 4x - 8y - 12 = 0$

33.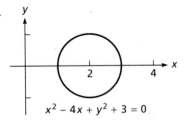

34.

35. $y = kx^2$ **36.** $w = krs^2$ **37.** $P = \frac{35}{2}$ **38.** $x = \frac{27}{32}$ **39.** 2.65   **40.** 5.85

CUMULATIVE REVIEW EXERCISES   (page 192)

1. $\{3\}$ **2.** $\{3, 4, 5, 6\}$ **3.**

$[-4, 7)$

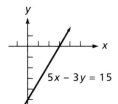

4.

$(-\infty, 0) \cup [2, \infty)$

5. commutative property of addition   **6.** transitive property   **7.** $9a^2$ **8.** $81a^2$ **9.** a^6b^4 **10.** $\dfrac{9y^2}{x^4}$ **11.** $\dfrac{x^4}{16y^2}$

12. $\dfrac{4y^{10}}{9x^6}$ **13.** a^3b^3 **14.** abc^2 **15.** $\sqrt{3}$ **16.** $\dfrac{\sqrt[3]{2x^2}}{x}$ **17.** $\dfrac{3(y + \sqrt{3})}{y^2 - 3}$ **18.** $\dfrac{3x(\sqrt{x} + 1)}{x - 1}$ **19.** $5\sqrt{3} - 3\sqrt{5}$

20. $3\sqrt{2}$ **21.** $5 - 2\sqrt{6}$ **22.** 4   **23.** $-8x + 8$ **24.** $9x^4 - 4x^3$ **25.** $6x^2 + 11x - 35$ **26.** $z^3 + z^2 + 4$

27. $2x^2 - x + 1$ **28.** $3x^2 + 1 - \dfrac{x}{x^2 + 2}$ **29.** $3t(t - 2)$ **30.** $(3x + 2)(x - 4)$

31. $(x^2 + x + 1)(x^2 - x + 1)(x^4 - x^2 + 1)$ **32.** $(x - 1)(x + 1)(x^2 + x + 1)(x^2 - x + 1)$ **33.** $\dfrac{x - 5}{x + 5}$

34. $(2x + 1)(x + 2)$ **35.** $\dfrac{5x^2 + 17x - 6}{(x + 3)(x - 3)}$ **36.** $\dfrac{-x^2 + x + 7}{(x + 2)(x + 3)}$ **37.** $b + a$ **38.** $-\dfrac{1}{xy}$ **39.** $0, 10$ **40.** 34

41. $R = \dfrac{R_1 R_2}{R_1 + R_2}$ **42.** $r = \dfrac{S - a}{S - l}$ **43.** 8 ft by 24 ft or 12 ft by 16 ft   **44.** \$8000   **45.** $\dfrac{3}{5} + \dfrac{4}{5}i$

46. $\dfrac{3}{2} - \dfrac{1}{2}i$ **47.** 5   **48.** $10i$ **49.** 3   **50.** $-2, 2, -3, 3$ **51.** 2, 7   **52.** 729; 64

53.

$(-\infty, 11/5]$

$11/5$

54.

$(-\infty, 3) \cup (5, \infty)$

$3 \quad 5$

55.

$[-3, -1] \cup (2, \infty)$

$-3 \quad -1 \quad 2$

56.

$(-\infty, -3) \cup (0, 3)$

$-3 \quad 0 \quad 3$

57.

$(-\infty, -1] \cup [4, \infty)$

$-1 \quad 4$

58.

$(1/3, 3)$

$1/3 \quad 3$

59.

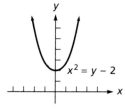

$5x - 3y = 15$

60.

$3x + 2y = 12$

61. $\sqrt{41}; \left(\frac{1}{2}, \frac{3}{2}\right); -\frac{4}{5}$ **62.** $2\sqrt{29}; (-2, 5); \frac{2}{5}$ **63.** $y = -2x - 1$ **64.** $y = \frac{7}{2}x - \frac{11}{4}$ **65.** $y = \frac{3}{5}x + 6$ **66.** $y = -4x$

67.

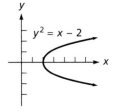

$x^2 = y - 2$

68.

$y^2 = x - 2$

69.

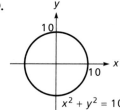

$x^2 + y^2 = 100$

70.

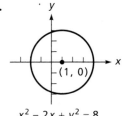

$(1, 0)$

$x^2 - 2x + y^2 = 8$

71. 1, 10   **72.** 2, 8   **73.** \$62.50   **74.** $\frac{25}{4}$

Exercise 4.1   (page 204)

1.

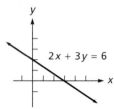

Domain is the set of real numbers.
Range is the set of real numbers.

3.

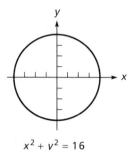

$x^2 + y^2 = 16$

Domain is the set of real numbers from -4 to 4.
Range is the set of real numbers from -4 to 4.

5.

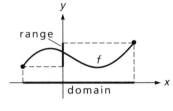

7. a function   **9.** not a function   **11.** a function   **13.** not a function   **15.** not a function   **17.** a function
19. Domain is the set of real numbers; range is the set of real numbers.   **21.** Domain is the set of real numbers; range is the
set of nonnegative real numbers.   **23.** Domain is the set of real numbers except -1; range is the set of real numbers except 0.
25. Domain is the set of nonnegative real numbers; range is the set of nonnegative real numbers.   **27.** a function
29. a function   **31.** not a function; does not pass the vertical line test   **33.** $f(2) = 4; f(-3) = -11; f(k) = 3k - 2;$
$f(k^2 - 1) = 3k^2 - 5$ **35.** $f(2) = 4; f(-3) = \frac{3}{2}; f(k) = \frac{1}{2}k + 3; f(k^2 - 1) = \frac{1}{2}k^2 + \frac{5}{2}$ **37.** $f(2) = 4; f(-3) = 9; f(k) = k^2;$
$f(k^2 - 1) = k^4 - 2k^2 + 1$ **39.** $f(2) = \frac{1}{3}; f(-3) = 2; f(k) = \frac{2}{k + 4}; f(k^2 - 1) = \frac{2}{k^2 + 3}$ **41.** $f(2) = \frac{1}{3}; f(-3) = \frac{1}{8};$
$f(k) = \frac{1}{k^2 - 1}; f(k^2 - 1) = \frac{1}{k^4 - 2k^2}$ **43.** $f(2) = \sqrt{5}; f(-3) = \sqrt{10}; f(k) = \sqrt{k^2 + 1}; f(k^2 - 1) = \sqrt{k^4 - 2k^2 + 2}$

45.

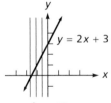

a function

47.

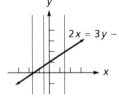

a function

49.

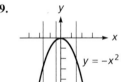

a function

51.

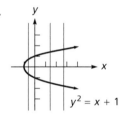

not a function

53.

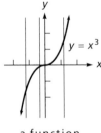

a function

55.

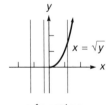

a function

57.

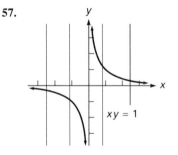

a function

59.

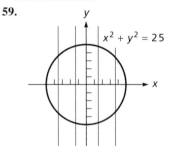

not a function

61. $r = \dfrac{C}{2\pi}$ **63.** $P = 4\sqrt{A}$ **65.** $V = A^{3/2}$ **67.** $A = \dfrac{d^2}{2}$ **69.** $F = \frac{9}{5}C + 32$ **71.** $c = \frac{3}{2500}n + \frac{7}{2}$
73. The domain is the set of real numbers greater than or equal to $\frac{5}{2}$; the range is the set of nonnegative real numbers.
75. The domain is the set of real numbers; the range is the set of real numbers.

Exercise 4.2 (page 210)

1.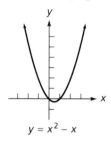
$y = x^2 - x$

3.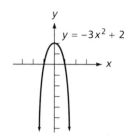
$y = -3x^2 + 2$

5.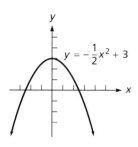
$y = -\frac{1}{2}x^2 + 3$

7.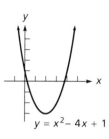
$y = x^2 - 4x + 1$

9. $(0, -1)$ **11.** $(2, 0)$ **13.** $(-3, -12)$ **15.** $(3, 1)$ **17.** 20 ft   **19.** $a = 50\sqrt{2} + 1$ **21.** 25 by 25 ft
23. $w = 12$ in.; $d = 6$ in.   **25.** Both numbers are 3.   **27.** $600   **29.** $95   **31.** $\frac{5}{2}$ sec   **33.** 100 ft
35. 6 by $4\frac{1}{2}$ units   **37.** $(-2.25, 66.13)$ **39.** $(3.3, -68.5)$

Exercise 4.3 (page 217)

1.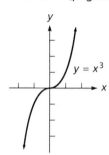
$y = x^3$

3.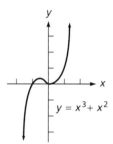
$y = x^3 + x^2$

5.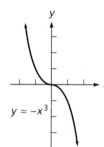
$y = -x^3$

7.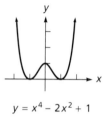
$y = x^4 - 2x^2 + 1$

9. even   **11.** neither   **13.** odd   **15.** odd   **17.** decreasing for $x < 0$; increasing for $x > 0$ **19.** increasing for $x < 0$;
decreasing for $x > 0$ **21.** decreasing for $x < 2$; increasing for $x > 2$

23.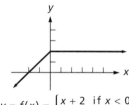
$y = f(x) = \begin{cases} x + 2 & \text{if } x < 0 \\ 2 & \text{if } x \geq 0 \end{cases}$

25.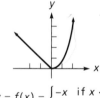
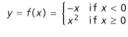
$y = f(x) = \begin{cases} -x & \text{if } x < 0 \\ x^2 & \text{if } x \geq 0 \end{cases}$

27.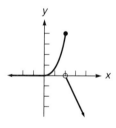
$y = f(x) = \begin{cases} 0 & \text{if } x < 0 \\ x^2 & \text{if } 0 \leq x \leq 2 \\ 4 - 2x & \text{if } x > 2 \end{cases}$

29.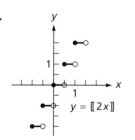
$y = [\![2x]\!]$

31.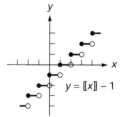
$y = [\![x]\!] - 1$

33.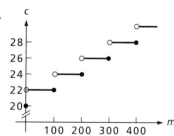
$26

35.

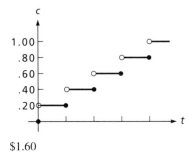

$1.60

37.

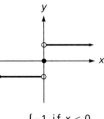

$$y = \begin{cases} -1 & \text{if } x < 0 \\ 0 & \text{if } x = 0 \\ 1 & \text{if } x > 0 \end{cases}$$

Exercise 4.4   (page 223)

1.

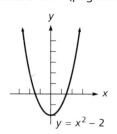

$y = x^2 - 2$

3.

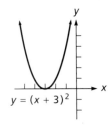

$y = (x + 3)^2$

5.

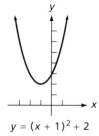

$y = (x + 1)^2 + 2$

7.

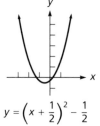

$y = \left(x + \dfrac{1}{2} \right)^2 - \dfrac{1}{2}$

9.

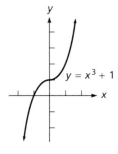

$y = x^3 + 1$

11.

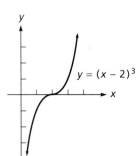

$y = (x - 2)^3$

13.

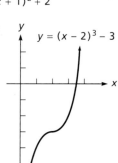

$y = (x - 2)^3 - 3$

15.

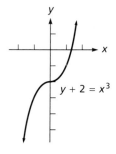

$y + 2 = x^3$

17.

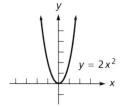

$y = 2x^2$

19.

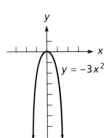

$y = -3x^2$

21.

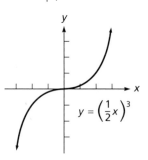

$y = \left(\dfrac{1}{2} x \right)^3$

23.

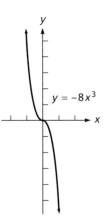

$y = -8x^3$

25.

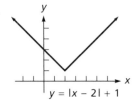

$y = |x - 2| + 1$

27.

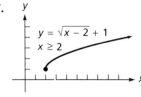

$y = \sqrt{x - 2} + 1$
$x \geq 2$

29.

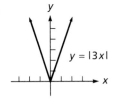

$y = |3x|$

31.

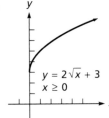

$y = 2\sqrt{x} + 3$
$x \geq 0$

33.

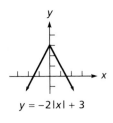

$y = -2|x| + 3$

Exercise 4.5   (page 231)

1. all real numbers except 2   **3.** all real numbers except -5 and 5   **5.** all real numbers except $0, 1, -1$ **7.** all real numbers

9.

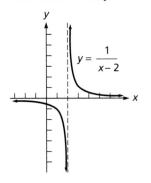

$y = \dfrac{1}{x - 2}$

11.

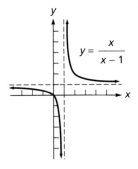

$y = \dfrac{x}{x - 1}$

13.

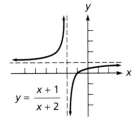

$y = \dfrac{x + 1}{x + 2}$

15.

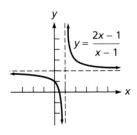

$y = \dfrac{2x - 1}{x - 1}$

17.

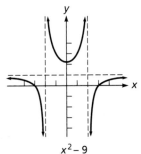

$y = \dfrac{x^2 - 9}{x^2 - 4}$

19.

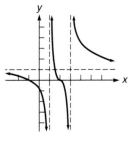

$y = \dfrac{x^2 - x - 2}{x^2 - 4x + 3}$

21.

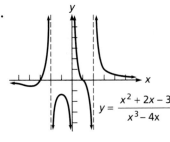

$y = \dfrac{x^2 + 2x - 3}{x^3 - 4x}$

23.

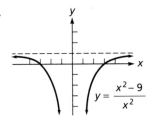

$y = \dfrac{x^2 - 9}{x^2}$

25.

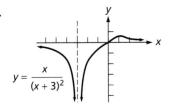

$y = \dfrac{x}{(x + 3)^2}$

27.

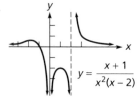

$$y = \frac{x + 1}{x^2(x - 2)}$$

29.

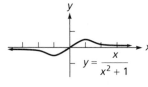

$$y = \frac{x}{x^2 + 1}$$

31.

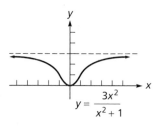

$$y = \frac{3x^2}{x^2 + 1}$$

33.

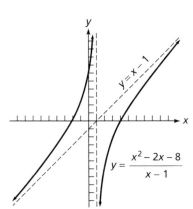

$$y = x - 1$$

$$y = \frac{x^2 - 2x - 8}{x - 1}$$

35.

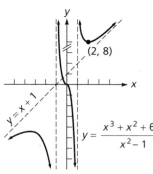

(2, 8)

$$y = x + 1$$

$$y = \frac{x^3 + x^2 + 6x}{x^2 - 1}$$

37.

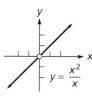

$$y = \frac{x^2}{x}$$

39.

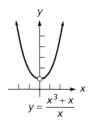

$$y = \frac{x^3 + x}{x}$$

41.

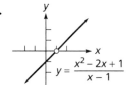

$$y = \frac{x^2 - 2x + 1}{x - 1}$$

43.

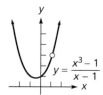

$$y = \frac{x^3 - 1}{x - 1}$$

45. no

Exercise 4.6   (page 239)

1. $(f + g)(x) = 5x - 1$; all real numbers   **3.** $(f \cdot g)(x) = 6x^2 - x - 2$; all real numbers   **5.** $(f - g)(x) = x + 1$; all real

numbers   **7.** $(f/g)(x) = \dfrac{x^2 + x}{x^2 - 1} = \dfrac{x}{x - 1}$; all real numbers except 1 and -1 **9.** 7   **11.** 1   **13.** 12   **15.** no value

17. $f(x) = 3x^2, g(x) = 2x$ **19.** $f(x) = 3x^2, g(x) = x^2 - 1$ **21.** $f(x) = 3x^3, g(x) = -x$ **23.** $f(x) = x + 9,$

$g(x) = x - 2$ **25.** 11   **27.** -17 **29.** 190   **31.** 145   **33.** $(f \circ g)(x) = 3x + 3$; all real numbers

35. $(f \circ f)(x) = 9x$; all real numbers   **37.** $(g \circ f)(x) = 2x^2$; all real numbers   **39.** $(g \circ g)(x) = 4x$; all real numbers

41. $(f \circ g)(x) = \sqrt{x^2 + 1}$; all real numbers   **43.** $(f \circ f)(x) = \sqrt[4]{x}$; all nonnegative real numbers   **45.** $(g \circ f)(x) = x$; all real

numbers greater than or equal to -1 **47.** $(g \circ g)(x) = x^4 - 2x^2$; all real numbers   **49.** $f(x) = x - 2, g(x) = 3x$

51. $f(x) = x - 2, g(x) = x^2$ **53.** $f(x) = x^2, g(x) = x - 2$ **55.** $f(x) = \sqrt{x}, g(x) = x + 2$ **57.** $f(x) = x + 2,$

$g(x) = \sqrt{x}$ **59.** $f(x) = x, g(x) = x$ **61.** $(f + f)(x) = 3x + 3x = 6x; f(x + x) = 3(x + x) = 6x$ **63.** $(f \circ f)(x) = -\dfrac{1}{x}$

Exercise 4.7   (page 246)

1. one-to-one   **3.** not one-to-one   **5.** not one-to-one   **7.** not one-to-one   **9.** not one-to-one   **11.** one-to-one

13. one-to-one   **15.** not one-to-one   **17.** $(f \circ g)(x) = f(g(x)) = 5\left(\frac{1}{5}x\right) = x; (g \circ f)(x) = g(f(x)) = \frac{1}{5}(5x) = x$

19. $(f \circ g)(x) = f(g(x)) = \dfrac{\dfrac{1}{x-1}+1}{\dfrac{1}{x-1}} = x$; $(g \circ f)(x) = g(f(x)) = \dfrac{1}{\dfrac{x+1}{x}-1} = x$ **21.** $f^{-1}(x) = \dfrac{1}{3}x$ **23.** $f^{-1}(x) =$

$\dfrac{x-2}{3}$ **25.** $f^{-1}(x) = \dfrac{1}{x} - 3$ **27.** $f^{-1}(x) = \dfrac{1}{2x}$

29.
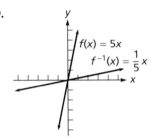

31.
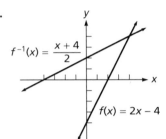

33.
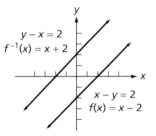

35.
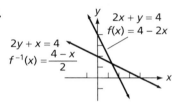

37.
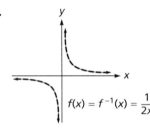

39.
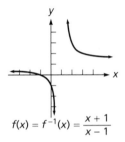

41. $f^{-1}(x) = -\sqrt{x+3}$ $(x \ge -3)$ **43.** $f^{-1}(x) = \sqrt[4]{x+8}$ $(x \ge -8)$ **45.** $f^{-1}(x) = \sqrt{4 - x^2}$ $(0 \le x \le 2)$
47. Domain is all real numbers except 1; range is all real numbers except 1.   **49.** Domain is all real numbers except 0; range is all real numbers except -2.

CHAPTER 4 REVIEW EXERCISES   (page 249)

1. a function   **2.** a function   **3.** not a function   **4.** a function   **5.** Domain is all real numbers; range is all real numbers greater than or equal to $-\dfrac{25}{12}$.   **6.** Domain; all real numbers except 5; range is all real numbers.   **7.** Domain is all real numbers greater than or equal to 1; range is all nonnegative real numbers.   **8.** Domain is all real numbers; range is all real numbers greater than or equal to 1.   **9.** $f(2) = 8, f(-3) = -17, f(0) = -2$ **10.** $f(2) = -2, f(-3) = -\dfrac{3}{4}, f(0) = -\dfrac{6}{5}$
11. $f(2) = 0, f(-3) = 5, f(0) = 2$ **12.** $f(2) = \dfrac{1}{7}, f(-3) = \dfrac{1}{2}, f(0) = -1$

13.
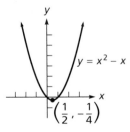

14.
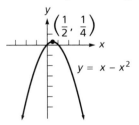

15.
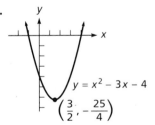

16.
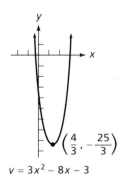

17. 300 units   **18.** Both numbers are $\dfrac{1}{2}$.

19.

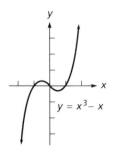

$y = x^3 - x$

an odd function

20.

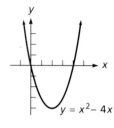

$y = x^2 - 4x$

neither even nor odd

21.

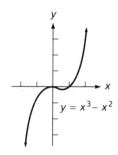

$y = x^3 - x^2$

neither even nor odd

22.

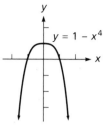

$y = 1 - x^4$

an even function

23.

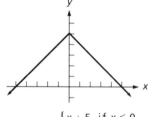

$y = f(x) = \begin{cases} x + 5 & \text{if } x \le 0 \\ 5 - x & \text{if } x > 0 \end{cases}$

increasing for $x < 0$;
decreasing for $x > 0$

24.

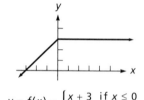

$y = f(x) = \begin{cases} x + 3 & \text{if } x \le 0 \\ 3 & \text{if } x > 0 \end{cases}$

increasing for $x < 0$;
constant for $x > 0$

25.

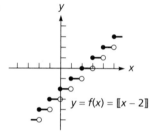

$y = f(x) = [\![x - 2]\!]$

26.

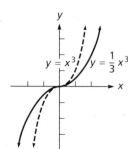

$y = f(x) = [\![x]\!] - 2$

27.

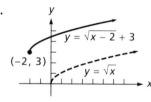

$y = \sqrt{x - 2} + 3$

$(-2, 3)$

$y = \sqrt{x}$

28.

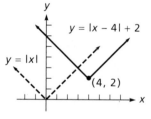

$y = |x - 4| + 2$

$y = |x|$

$(4, 2)$

29.

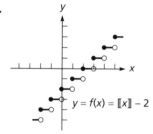

$y = x^3$ $y = \frac{1}{3} x^3$

30.

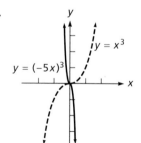

$y = x^3$

$y = (-5x)^3$

31.

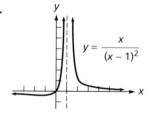

$y = \dfrac{x}{(x - 1)^2}$

32.

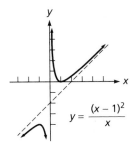

$$y = \frac{(x-1)^2}{x}$$

33.

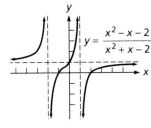

$$y = \frac{x^2 - x - 2}{x^2 + x - 2}$$

34.

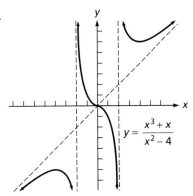

$$y = \frac{x^3 + x}{x^2 - 4}$$

35. $(f + g)(x) = x^2 + 2x$ **36.** $(f \cdot g)(x) = 2x^3 + x^2 - 2x - 1$ **37.** $(f \circ g)(x) = f(g(x)) = 4x^2 + 4x$

38. $(g \circ f)(x) = g(f(x)) = 2x^2 - 1$ **39.** $f^{-1}(x) = \dfrac{x + 1}{7}$ **40.** $f^{-1}(x) = 2 - \dfrac{1}{x}$ **41.** $f^{-1}(x) = \dfrac{x}{x + 1}$

42. $f^{-1}(x) = \sqrt[3]{\dfrac{3}{x}}$

CHAPTER 4 TEST   (page 250)

1. Domain is the set of all real numbers except 5; range is the set of all real numbers except 0.   **2.** Domain is the set of all real numbers greater than or equal to -3; range is the set of all nonnegative real numbers.   **3.** $f(-1) = \frac{1}{2}, f(2) = 2$
4. $f(-1) = \sqrt{6}, f(2) = 3$ **5.** $(7, -3)$ **6.** $(1, -4)$ **7.** $(4, -10)$ **8.** $(-2, 9)$

9.

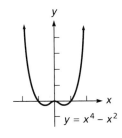

$$y = x^4 - x^2$$

10.

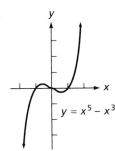

$$y = x^5 - x^3$$

11. $\frac{25}{8}$ sec   **12.** $\frac{625}{4}$ ft   **13.** 10 ft   **14.** 110 ft   **15.** vertical asymptotes: $x = -3$ and $x = 3$; horizontal asymptote; $y = 0$
16. vertical asymptote: $x = 3$; slant asymptote: $y = x - 2$

17.

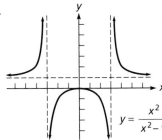

$$y = \frac{x^2}{x^2 - 9}$$

18.

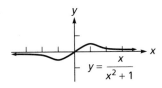

$$y = \frac{x}{x^2 + 1}$$

19.

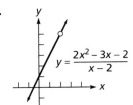

$$y = \frac{2x^2 - 3x - 2}{x - 2}$$

20.

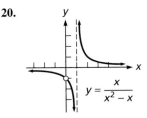

$$y = \frac{x}{x^2 - x}$$

21. $(f + g)(x) = f(x) + g(x) = 3x + x^2 + 2$ **22.** $(g \circ f)(x) = g(f(x)) = 9x^2 + 2$ **23.** $(f/g)(x) = f(x)/g(x) = \dfrac{3x}{x^2 + 2}$

24. $(f \circ g)(x) = f(g(x)) = 3x^2 + 6$ **25.** $f^{-1}(x) = \dfrac{x + 1}{x - 1}$ **26.** $f^{-1}(x) = \sqrt[3]{x + 3}$

27.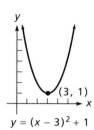

28.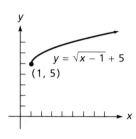

29. Range is all real numbers except -2.   **30.** Range is all real numbers except 3.

Exercise 5.1   (page 258)

1. 4   **3.** 67   **5.** 260   **7.** $\frac{1}{8}$ **9.** 3   **11.** -3 **13.** 70,249   **15.** 128.3085938   **17.** 31   **19.** 142
21. true   **23.** true   **25.** false   **27.** true   **29.** $\{3, -1, -5\}$ **31.** $\{1, 1, \sqrt{3}, -\sqrt{3}\}$ **33.** $\{2, 3, i, -i\}$
35. $x^2 - 9x + 20$ **37.** $x^3 - 3x^2 + 3x - 1$ **39.** $x^3 - 11x^2 + 38x - 40$ **41.** $x^4 - 3x^2 + 2$
43. $x^3 - \sqrt{2}\,x^2 + x - \sqrt{2}$ **45.** $x^3 - 2x^2 + 2x$ **47.** $1, -\dfrac{1}{2} + \dfrac{\sqrt{3}}{2}i, -\dfrac{1}{2} - \dfrac{\sqrt{3}}{2}i$ **49.** $-5, \dfrac{5}{2} + \dfrac{5\sqrt{3}}{2}i, \dfrac{5}{2} - \dfrac{5\sqrt{3}}{2}i$
51. $a_0 = 0$

Exercise 5.2   (page 263)

1. 47   **3.** -569 **5.** 1   **7.** $-1 + 6i$ **9.** 3   **11.** 30   **13.** $\frac{15}{8}$ **15.** 5   **17.** 0   **19.** 384   **21.** 0
23. 9   **25.** 8   **27.** 0   **29.** $16 + 2i$ **31.** $40 - 40i$ **33.** $(x - 1)(3x^2 + x - 5) - 9$
35. $(x - 3)(3x^2 + 7x + 15) + 41$ **37.** $(x + 1)(3x^2 - 5x - 1) - 3$ **39.** $(x + 3)(3x^2 - 11x + 27) - 85$
41. $x^2 + 2x + 3$ **43.** $7x^2 - 10x + 5 + \dfrac{-4}{x + 1}$ **45.** $4x^3 + 9x^2 + 27x + 80 + \dfrac{245}{x - 3}$
47. $3x^4 + 12x^3 + 48x^2 + 192x$ **49.** $x^3 + 6x^2 + 13x + 26 + \dfrac{57}{x - 2}$ **51.** $5x^4 - 10x^3 + 23x^2 - 45x + 90 + \dfrac{-175}{x + 2}$
53. $x[x(3x + 4) + 5] + 12$ **55.** $x\{x[x(4x - 2) - 8] + 3\} + 5$ **57.** $x\{x[x(5x^2 + 3)] + 9\} + 2$ **59.** 4.9643

Exercise 5.3   (page 269)

1. 10   **3.** 4   **5.** 4   **7.** $x^3 - 3x^2 + x - 3 = 0$ **9.** $x^3 - 6x^2 + 13x - 10 = 0$
11. $x^4 - 5x^3 + 7x^2 - 5x + 6 = 0$ **13.** $x^4 - 2x^3 + 3x^2 - 2x + 2 = 0$ **15.** 0 or 2 positive; 1 negative; 0 or 2 nonreal
17. 0 positive; 1 or 3 negative; 0 or 2 nonreal   **19.** 0 positive; 0 negative; 4 nonreal   **21.** 1 positive; 1 negative; 2 nonreal
23. 0 positive; 0 negative; 10 nonreal   **25.** 0 positive; 0 negative; 8 nonreal; yes   **27.** 1 positive; 1 negative; 2 nonreal
29. $-2, 4$ **31.** $-1, 1$ **33.** $-4, 6$ **35.** $-4, 3$ **37.** $-2, 4$

Exercise 5.4   (page 277)

1. $1, -1, 5$ **3.** $3, -3, 2$ **5.** $1, -1, 2$ **7.** $1, 2, 3, 4$ **9.** $2, -5$ **11.** $1, -1, 2, -2, -3$
13. $0, 2, 2, 2, -2, -2, -2$ **15.** $\frac{2}{3}, 2i, -2i$ **17.** $-1, \frac{2}{3}, 2, 3$ **19.** $-3, \frac{1}{3}, \frac{1}{2}, \frac{1}{2}$ **21.** $-3, 2, 2, -\frac{1}{3}, \frac{1}{2}$
23. $-\frac{3}{5}, \frac{2}{3}, \frac{3}{2}$ **25.** $\frac{2}{3}, -\frac{3}{5}, 4$ **27.** $-\frac{1}{2}, \frac{3}{2}, \frac{6}{5}$ **29.** $3, 1 + i, 1 - i$ **31.** $3, -3, 1 + i, 1 - i$ **33.** $-\frac{2}{3}, 3, -1$
35. $\frac{1}{2}, \frac{1}{2}, \frac{1}{2}, 1, 1$ **39.** 10, 20, 60 ohms   **41.** $(3, 7)$ and approx. $(1.54, 13.63)$

Exercise 5.5   (page 282)

1. $P(-2) = 3; P(-1) = -2$ **3.** $P(4) = -40; P(5) = 30$ **5.** $P(1) = 8; P(2) = -1$ **7.** $P(2) = -72; P(3) = 154$
9. $P(0) = 10; P(1) = -60$ **11.** 1.7   **13.** -2.2 **15.** 1.7   **17.** -1.2 **19.** $-2.2, 2.2$ **21.** 1, 2; the bisection method fails to find the root 2   **23.** $(0.83, 0.56), (-0.83, -0.56)$

CHAPTER 5 REVIEW EXERCISES   (page 284)

1. -2 **2.** 66   **3.** 241   **4.** $-\frac{9}{4}$ **5.** false   **6.** false   **7.** true   **8.** true   **9.** $2x^3 - 5x^2 - x + 6$
10. $2x^3 + 3x^2 - 8x + 3$ **11.** $x^4 + 3x^3 - 9x^2 + 3x - 10$ **12.** $x^4 + x^3 - 5x^2 + x - 6$ **13.** $3x^3 + 9x^2 + 29x + 90$
with remainder of 277   **14.** $2x^3 + 4x^2 + 5x + 13$ with remainder of 25   **15.** $5x^4 - 14x^3 + 31x^2 - 64x + 129$ with a
remainder of -259 **16.** $4x^4 - 2x^3 + x^2 + 2x$ with a remainder of 1   **17.** 6   **18.** 6   **19.** 65   **20.** 1984
21. 0 or 2 positive; 0 or 2 negative; 0, 2, or 4 nonreal   **22.** 1 or 3 positive; 1 negative; 0 or 2 nonreal   **23.** 1 positive;
0, 2, or 4 negative; 0, 2, or 4 nonreal   **24.** 1 or 3 positive; 0 or 2 negative; 2, 4, or 6 nonreal   **25.** 0 positive; 0 negative;
4 nonreal   **26.** 0 positive; 1 negative; 6 nonreal   **27.** $-5, -\frac{3}{2}, -2$ **28.** $\frac{1}{3}, -\frac{1}{2} + \frac{\sqrt{3}}{2}i, -\frac{1}{2} - \frac{\sqrt{3}}{2}i$

29. $2, -2, \frac{3}{2}, -\frac{3}{2}$ **30.** $4, 4, -2, -\frac{1}{2}$ **31.** $P(-1) = -9; P(0) = 18$ **32.** $P(1) = -8; P(2) = 21$ **33.** 4.2
34. 2.5   **35.** 1.67   **36.** 0.67   **37.** 10m by 6m; 12m by 5m; 15m by 4m   **38.** 17.27 ft

CHAPTER 5 TEST   (page 286)

1. 1   **2.** -30 **3.** $x^3 - 4x^2 - 5x$ **4.** $x^4 - 2x^2 - 3$ **5.** 5   **6.** -28 **7.** $\frac{11}{3}$ **8.** $6 - 3i$
9. $(x - 2)(2x^2 + x - 2) - 5$ **10.** $(x + 1)(2x^2 - 5x + 1) - 2$ **11.** $2x + 3$ **12.** $3x^2 + x$
13. $x^3 - 2x^2 + x - 2$ **14.** $x^3 - 5x^2 + 9x - 5$ **15.** 1 or 3 positive; 0 or 2 negative; 0, 2, or 4 nonreal
16. 1 positive; 0 or 2 negative; 0 or 2 nonreal   **17.** $-3, 3$ **18.** $-1, 6$ **19.** $2, -3, -\frac{1}{2}$ **20.** 3.3

Exercise 6.1   (page 295)

1. 11.0357   **3.** 451.8079   **5.**   **7.**   **9.**

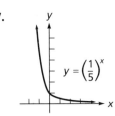

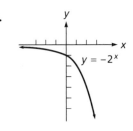

11.   **13.**   **15.**

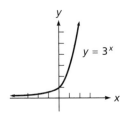

17.   **19.**   **21.**

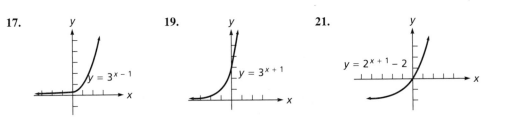

23.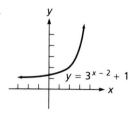
$y = 3^{x-2} + 1$

25.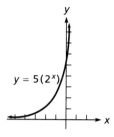
$y = 5(2^x)$

27.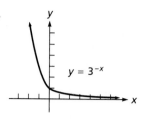
$y = 3^{-x}$

29. $b = 5$ **31.** no value of b **33.** $b = \frac{1}{2}$ **35.** no value of b **37.** 0.1868 gm   **39.** 0.1317 A_0 **41.** 0.8123 g
43. $1104.02   **45.** $16.01   **47.** $2,273,996.13   **49.** $1425.93   **51.** 2 lumens   **53.** location B **55.** 0.3536 unit
57. 0.3150 unit   **59.** 824   **61.** $C_0 = 3 \times 10^{-4}$ **63.** $1847.16   **65.** $4957.69   **67.** $4539.23   **69.** 5.3 lb/in.^{2}
71. about $120   **73.** .75, .50, .27

Exercise 6.2   (page 304)

1.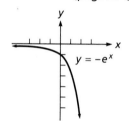
$y = -e^x$

3.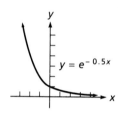
$y = e^{-0.5x}$

5.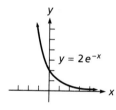
$y = 2e^{-x}$

7.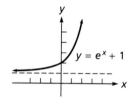
$y = e^x + 1$

9. yes   **11.** no   **13.** no   **15.** no   **17.** 13,375.68   **19.** $6849.16   **21.** $7647.95 from continuous compounding;
$7518.28 from annual compounding   **23.** 315 people   **25.** 51   **27.** about 9.2 billion   **29.** The population will be
multiplied by 2.6.   **31.** about 673,000 people   **33.** 0.076   **35.** about 7350 people   **37.** about 0.07% will remain
39. 49 mps   **41.** e^3 **43.** about 72.2 years   **45.** 2.718253969; 4 decimal places

Exercise 6.3   (page 312)

1. $3^4 = 81$ **3.** $\left(\frac{1}{2}\right)^3 = \frac{1}{8}$ **5.** $4^{-3} = \frac{1}{64}$ **7.** $x^z = y$ **9.** $\log_8 64 = 2$ **11.** $\log_4 \frac{1}{16} = -2$ **13.** $\log_{1/2} 32 = -5$
15. $\log_x z = y$ **17.** 3   **19.** 3   **21.** 3   **23.** $\frac{1}{2}$ **25.** -3 **27.** 64   **29.** 7   **31.** 5   **33.** $\frac{1}{25}$ **35.** $\frac{1}{6}$
37. $-\frac{3}{2}$ **39.** $\frac{2}{3}$ **41.** 5   **43.** $\frac{3}{2}$ **45.** 4   **47.** 8   **49.** 5   **51.** 4   **53.** 3   **55.** 100   **57.** 0.5119
59. -2.3307 **61.** -0.0726 **63.** 3.6348   **65.** -0.2752 **67.** not possible   **69.** 25.25   **71.** 69.41
73. 0.00   **75.** 120.07

77.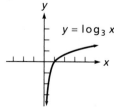
$y = \log_3 x$

79.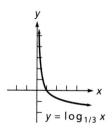
$y = \log_{1/3} x$

81.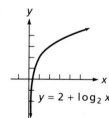
$y = 2 + \log_2 x$

83.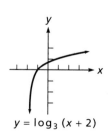
$y = \log_3 (x + 2)$

85.

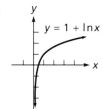

87.

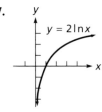

89.

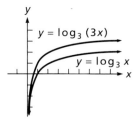

91.

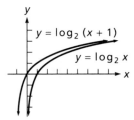

93.

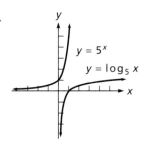

95. 3   **97.** $\frac{1}{2}$ **99.** yes   **101.** no

Exercise 6.4   (page 319)

1. 0   **3.** 7   **5.** 10   **7.** 1   **15.** $\log_b 2xy = \log_b 2 + \log_b x + \log_b y$ **17.** $\log_b \frac{2x}{y} = \log_b 2 + \log_b x - \log_b y$

19. $\log_b x^2 y^3 = 2\log_b x + 3\log_b y$ **21.** $\log_b (xy)^{1/3} = \frac{1}{3}(\log_b x + \log_b y)$ **23.** $\log_b x\sqrt{z} = \log_b x + \frac{1}{2}\log_b z$

25. $\log_b \dfrac{\sqrt[3]{x}}{\sqrt[3]{yz}} = \dfrac{1}{3}\log_b x - \dfrac{1}{3}\log_b y - \dfrac{1}{3}\log_b z$ **27.** $\log_b (x+1) - \log_b x = \log_b \dfrac{x+1}{x}$

29. $2\log_b x + \dfrac{1}{3}\log_b y = \log_b x^2 y^{1/3}$ **31.** $-3\log_b x - 2\log_b y + \dfrac{1}{2}\log_b z = \log_b \dfrac{z^{1/2}}{x^3 y^2}$

33. $\log_b \left(\dfrac{x}{z} + x\right) - \log_b \left(\dfrac{y}{z} + y\right) = \log_b \dfrac{\frac{x}{z} + x}{\frac{y}{z} + y} = \log_b \dfrac{x}{y}$ **35.** true   **37.** false   **39.** true   **41.** false   **43.** true

45. false   **47.** false   **49.** true   **51.** true   **53.** true   **55.** 1.4472   **57.** 0.3521   **59.** 1.1972   **61.** 2.4014
63. 2.0493   **65.** 0.4682   **67.** 1.7712   **69.** 0.9597   **71.** 1.8928   **73.** 2.3219

Exercise 6.5   (page 325)

1. 4.77   **3.** from 0.000501 to 0.00126   **5.** 0.71V   **7.** $10\log\dfrac{P_O}{P_I} = 10\log\dfrac{kE_O^2}{kE_I^2} = 10\log\left(\dfrac{E_O}{E_I}\right)^2 = 20\log\dfrac{E_O}{E_I}$ **9.** 4.4

11. 2500 μm   **13.** 19.9 hr   **15.** $L = L_0 + k\ln 2$, where $L_0 = k\ln I$ **17.** about 5.8 years   **19.** about 3653.5 joules
21. 3 years old   **23.** about 10.8 years   **25.** about 208,000 V   **29.** $\frac{1}{2}$

Exercise 6.6   (page 332)

1. $x = \dfrac{\log 5}{\log 4} \approx 1.1610$ **3.** $x = \dfrac{\log 2}{\log 13} + 1 \approx 1.2702$ **5.** $x = \dfrac{\log 2}{\log 3 - \log 2} \approx 1.7095$ **7.** $x = 0$

9. $x = \pm\sqrt{\dfrac{1}{\log 7}} \approx \pm 1.0878$ **11.** $x = 0$ or $x = \dfrac{\log 9}{\log 8} \approx 1.0566$ **13.** $x = 3$ or $x = -1$ **15.** $x = -2$ or $x = -2$

17. $x = 0$ **19.** $x = \dfrac{\log 2}{2 \log 6 - \log 3} \approx 0.2789$ **21.** $x = 1$ or $x = 3$ **23.** $x = 0$ **25.** $x = 7$ **27.** $x = 4$

29. $x = 10$ or $x = -10$ **31.** $x = 50$ **33.** $x = 20$ **35.** $x = 10$ **37.** $x = 3$ or $x = 4$ **39.** $x = 1$ or $x = 100$
41. no solution   **43.** $x = 6$ **45.** $x = 9$ **47.** $x = 4$ **49.** $y = 7$ or $y = 1$ **51.** about 5.2 years
53. about 42.7 days   **55.** about 4200 years   **57.** about 5.6 years   **59.** about 5.4 years   **61.** because $\ln 2 \approx .70$

63. about 27 m   **65.** $\dfrac{\ln 0.75}{3}$ **67.** $\dfrac{\ln \frac{2}{3}}{5}$ **69.** 20   **71.** about 1.8

CHAPTER 6 REVIEW EXERCISES   (page 334)

1.

2.
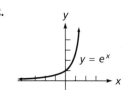

3.
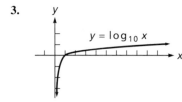

4.
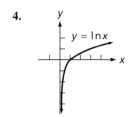

5.
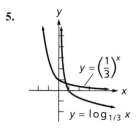

6.
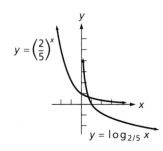

7.
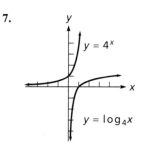

8.
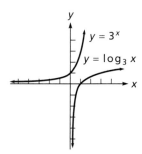

9. 2   **10.** $-\dfrac{1}{2}$ **11.** 0   **12.** -2 **13.** $\dfrac{1}{2}$ **14.** $\dfrac{1}{3}$ **15.** 4   **16.** 0   **17.** 7   **18.** 3   **19.** 4   **20.** 9

21. 8   **22.** $\dfrac{1}{9}$ **23.** 3   **24.** 2   **25.** 1   **26.** $\dfrac{1}{2}$ **27.** $\dfrac{1}{6}$ **28.** 2   **29.** -2 **30.** 0   **31.** 27   **32.** $\dfrac{1}{5}$

33. 32   **34.** 9   **35.** 27   **36.** -1 **37.** $\dfrac{1}{8}$ **38.** 2   **39.** 4   **40.** 2   **41.** 10   **42.** $\dfrac{1}{25}$ **43.** 5   **44.** 3

45. $\log_b \dfrac{x^2 y^3}{z^4} = 2 \log_b x + 3 \log_b y - 4 \log_b z$ **46.** $\log_b \sqrt{\dfrac{x}{yz^2}} = \dfrac{1}{2} \log_b x - \dfrac{1}{2} \log_b y - \log_b z$ **47.** $\log_b \dfrac{x^3 z^7}{y^5}$

48. $\log_b \dfrac{y^3 \sqrt{x}}{z^7}$ **49.** 3.36   **50.** 1.56   **51.** 2.64   **52.** -6.72 **53.** $x = \dfrac{\log 7}{\log 3}$ **54.** $x = 2$

55. $x = \dfrac{\log 3}{\log 3 - \log 2}$ **56.** $x = -3$ or $x = -1$ **57.** $x = 25$ or $x = 4$ **58.** $x = 4$ **59.** $x = 2$

60. $x = 4$ or $x = 3$ **61.** $x = 6$ **62.** $x = 31$ **63.** $x = \dfrac{\ln 9}{\ln 2}$ **64.** no solution   **65.** $x = \dfrac{e}{e - 1}$

66. $x = 1$ **67.** about 3300 years   **68.** about 7.94×10^{-4} gram-ions   **69.** about 34.2 years

70. $\text{pH} = \log_{10} \dfrac{1}{[\text{H}^+]} = \log_{10}[\text{H}^+]^{-1} = -\log_{10}[\text{H}^+]$

CHAPTER 6 TEST   (page 336)

1.

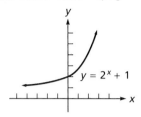
$y = 2^x + 1$

2.

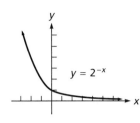
$y = 2^{-x}$

3. $\frac{3}{64}$ gm   **4.** $1060.90

5.

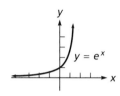
$y = e^x$

6.

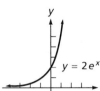
$y = 2e^x$

7. $4451.08   **8.** 3   **9.** -3 **10.** 17   **11.** 2   **12.** -3

13.

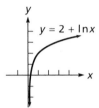
$y = \log(x - 1)$

14.

$y = 2 + \ln x$

15. $\log a^2 bc^3 = 2\log a + \log b + 3\log c$ **16.** $\ln \sqrt{\dfrac{a}{b^2 c}} = \dfrac{1}{2}(\ln a - 2\ln b - \ln c)$ **17.** $\log \dfrac{b\sqrt{a+2}}{c^2}$ **18.** $\log \dfrac{\sqrt[3]{a}}{c\sqrt[3]{b^2}}$

19. 1.3801   **20.** 0.4259   **21.** $\log_7 3 = \dfrac{\log 3}{\log 7}$ **22.** $\log_\pi e = \dfrac{\log e}{\log \pi}$ **23.** true   **24.** false   **25.** false

26. false   **27.** 6.4   **28.** 46   **29.** $x = \dfrac{\log 3}{(\log 3) - 2}$ **30.** $x = 3$ or $x = -1$ **31.** $x = 1$ **32.** $x = 10$

CUMULATIVE REVIEW EXERCISES   (page 337)

1. a function   **2.** a function   **3.** a function   **4.** not a function   **5.** Domain is the set of real numbers; range is the set of real numbers greater than or equal to 5.   **6.** Domain is the set of real numbers except -2; range is the set of real numbers except 0.   **7.** Domain is the set of real numbers greater than or equal to 2; range is the set of nonpositive real numbers.
8. Domain is the set of real numbers greater than or equal to 4; range is the set of nonnegative real numbers.   **9.** $\left(-\frac{5}{2}, -\frac{49}{4}\right)$
10. $\left(\frac{5}{2}, \frac{49}{4}\right)$ **11.**

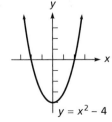
$y = x^2 - 4$

12.

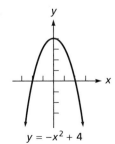
$y = -x^2 + 4$

13.

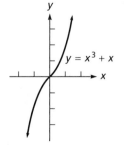
$y = x^3 + x$

14.

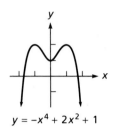

$y = -x^4 + 2x^2 + 1$

15.

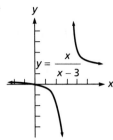

$y = \dfrac{x}{x-3}$

16.

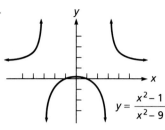

$y = \dfrac{x^2 - 1}{x^2 - 9}$

17. $(f + g)(x) = x^2 + 3x - 3$; domain is the set of all real numbers. **18.** $(f - g)(x) = -x^2 + 3x - 5$; domain is the set of

all real numbers. **19.** $(f \cdot g)(x) = 3x^3 - 4x^2 + 3x - 4$; domain is the set of real numbers. **20.** $(g/f)(x) = \dfrac{x^2 + 1}{3x - 4}$;

domain is the set of all real numbers but $\frac{4}{3}$. **21.** $(f \circ g)(2) = 11$ **22.** $(g \circ f)(2) = 5$ **23.** $(f \circ g)(x) = 3x^2 - 1$

24. $(g \circ f)(x) = 9x^2 - 24x + 17$ **25.** $f^{-1}(x) = \dfrac{x - 2}{3}$ **26.** $f^{-1}(x) = \dfrac{3x + 1}{x}$ **27.** $y = \pm \sqrt{x - 5}$

28. $f^{-1}(x) = \dfrac{x + 1}{3}$ **29.** $y = kwz$ **30.** $y = \dfrac{kx}{t^2}$ **31.** 9 **32.** -36 **33.** 4 **34.** $2 - i$

35. a factor **36.** not a factor **37.** a factor **38.** a factor **39.** 12 **40.** 2000 **41.** 2 positive; 2 negative
42. 1 positive; 3 negative **43.** $3, -3, -1$ **44.** $1, -1, 2$

45.

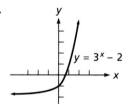

$y = 3^x - 2$

46.

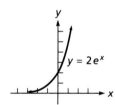

$y = 2e^x$

47.

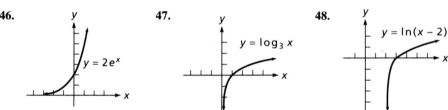

$y = \log_3 x$

48.

$y = \ln(x - 2)$

49. 6 **50.** -3 **51.** 3 **52.** 2 **53.** $\log abc = \log a + \log b + \log c$ **54.** $\log \dfrac{a^2 b}{c} = 2 \log a + \log b - \log c$

55. $\log \sqrt{\dfrac{ab}{c^2}} = \dfrac{1}{2}(\log a + \log b - 2 \log c)$ **56.** $\log \dfrac{\sqrt{ab^2}}{c} = \dfrac{1}{2}(\log a + 2 \log b) - \log c$ **57.** $\log \dfrac{a^3}{b^3}$ **58.** $\log \dfrac{b^3 \sqrt{a}}{\sqrt{c}}$

59. $x = \dfrac{\log 8}{\log 3} - 1$ **60.** no solution **61.** 500 **62.** $x = \sqrt{11}$

Exercise 7.1 (page 348)

1.

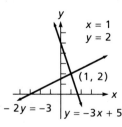

$x = 1$
$y = 2$
$(1, 2)$
$x - 2y = -3$
$y = -3x + 5$

3.

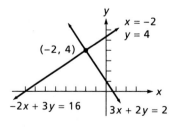

$x = -2$
$y = 4$
$(-2, 4)$
$-2x + 3y = 16$
$3x + 2y = 2$

5. $x = 2.2$; $y = -4.7$ **7.** $x = 1.7, y = 0.3$ **9.** $x = -1, y = -2$ **11.** $x = 3, y = -2$ **13.** $x = \frac{1}{2}, y = \frac{1}{3}$
15. no solution **17.** $x = 3, y = 1$ **19.** $x = 3, y = 2$ **21.** $x = -3, y = 0$ **23.** $x = 1, y = -\frac{1}{2}$

25. $x = \frac{1}{3}, y = \frac{5}{3}$ **27.** no solution; inconsistent system   **29.** $x = 2, y = -3$ **31.** $x = 9, y = -1$

33. $x = 1, y = 2, z = 0$ **35.** $x = 0, y = -\frac{1}{3}, z = -\frac{1}{3}$ **37.** $x = 1, y = 2, z = -1$ **39.** $x = 1, y = 0, z = 5$

41. no solution; inconsistent system   **43.** $x = 0, y = 1, z = 0$ **45.** $x = \frac{2}{3}, y = \frac{1}{4}, z = \frac{1}{2}$ **47.** $x = \frac{1}{2}, y = \frac{1}{2}, z = \frac{1}{2}$

49. $x = 2 - y, y = $ any number, $z = 1$ **51.** 225 acres of corn, 125 acres of soybeans   **53.** \$2500 initiation fee; \$75 per month dues   **55.** 8 kph   **57.** 6 lb and 4 lb   **59.** 1300 mi   **61.** $E(x) = 43.53x + 742.72, R(x) = 89.95x$; 16 pairs per day   **63.** 15 hr cooking hamburgers, 10 hr pumping gas, 5 hr janitorial   **65.** 1.05 million in 0–14 group, 1.56 million in 15–49 group, 0.39 million in 50 and over group   **67.** 30°, 50°, 100°   **69.** Dictionaries are 4.5 in. wide; atlases are 3.5 in. wide; thesauruses are 4 in. wide.

Exercise 7.2   (page 359)

1. $x = \frac{13}{3}, y = \frac{8}{3}$ **3.** $x = 7, y = 6$ **5.** $x = 1, y = 2, z = 3$ **7.** $x = 1, y = 1, z = 3$ **9.** row echelon form **11.** reduced row echelon form   **13.** $x = 2, y = -1$ **15.** $x = -2, y = 0$ **17.** $x = 3, y = 1$ **19.** no solution; inconsistent system   **21.** $x = 1, y = 0, z = 2$ **23.** $x = 2, y = -2, z = 1$ **25.** $x = 1, y = 1, z = 2$ **27.** $x = -1, y = 3, z = 1$ **29.** $x = 13, y = 3$ **31.** $x = -13, y = 7, z = -2$ **33.** $x = 10, y = 3$ **35.** $x = 0, y = 0$ **37.** $x = 3, y = 1, z = -2$ **39.** $x = \frac{1}{4}, y = 1, z = \frac{1}{4}$ **41.** $x = 1, y = 2, z = 1, t = 1$ **43.** $x = 1, y = 2, z = 0, t = 1$ **45.** $x = \frac{9}{4}, y = -3, z = \frac{3}{4}$ **47.** $x = 0, y = \frac{20}{3}, z = -\frac{1}{3}$ **49.** $x = 1, y = -3$ **51.** dependent equations; general solution is $\left(\frac{8}{7} + \frac{1}{7}z, \frac{10}{7} - \frac{4}{7}z, z\right)$ **53.** dependent equations; general solution is $(1 + z, -z, -1 - z, z)$ **55.** no solution; inconsistent system   **57.** $x = \pm 2, y = \pm 1, z = \pm 3$

Exercise 7.3   (page 369)

1. $x = 2, y = 5$ **3.** $x = 1, y = 2$ **5.** $\begin{bmatrix} 15 & -15 \\ 0 & -10 \end{bmatrix}$ **7.** $\begin{bmatrix} 25 & 25 & -10 \\ -10 & -25 & 5 \end{bmatrix}$ **9.** $\begin{bmatrix} -1 & 2 & 1 \\ -6 & 0 & 0 \end{bmatrix}$

11. $\begin{bmatrix} -6 & 5 & 0 \\ 1 & -1 & 0 \end{bmatrix}$ **13.** $\begin{bmatrix} 18 & -1 & -4 \\ -35 & 0 & -1 \end{bmatrix}$ **15.** $\begin{bmatrix} -5 & 2 & -7 \\ 5 & 0 & -3 \\ 2 & -3 & 5 \end{bmatrix}$ **17.** $\begin{bmatrix} 2 & -2 \\ 3 & 10 \end{bmatrix}$

19. $\begin{bmatrix} -22 & -22 \\ -105 & 126 \end{bmatrix}$ **21.** $\begin{bmatrix} 4 & 2 & 10 \\ 5 & -2 & 4 \\ 2 & -2 & 1 \end{bmatrix}$ **23.** $\begin{bmatrix} 32 \\ 2 \end{bmatrix}$ **25.** not possible   **27.** $\begin{bmatrix} -4 & -19 & 13 \\ -11 & 3 & -11 \\ 4 & 6 & -2 \\ 13 & 13 & -1 \end{bmatrix}$

29. $\begin{bmatrix} 4 & 5 \\ -7 & -1 \end{bmatrix}$ **31.** not possible   **33.** $\begin{bmatrix} 24 & 16 \\ 39 & 26 \end{bmatrix}$ **35.** $\begin{bmatrix} 47 \\ 81 \end{bmatrix}$ **37.** $\begin{bmatrix} -36.29 \\ 16.2 \\ -19.26 \end{bmatrix}$

39. $\begin{bmatrix} -16.11 & 4.71 & 33.64 \\ -19.6 & 20.35 & 6.4 \\ -100.6 & 72.82 & 62.71 \end{bmatrix}$ **41.** $QP = \begin{bmatrix} 584.50 \\ 709.25 \\ 1036.75 \end{bmatrix}$; adult males spent \$584.50 on drinks; adult females spent \$709.25 on drinks; children spent \$1036.75 on drinks.   **47.** $\begin{bmatrix} 1 & 1 & 0 \\ 0 & 1 & 1 \\ 1 & 0 & 0 \end{bmatrix}$

49. $A^2 = \begin{bmatrix} 5 & 1 & 2 & 2 \\ 1 & 5 & 2 & 2 \\ 2 & 2 & 6 & 0 \\ 2 & 2 & 0 & 4 \end{bmatrix}$ indicates the number of ways two cities can be linked with exactly one intermediate city to relay messages.

$A + A^2 = \begin{bmatrix} 5 & 3 & 3 & 2 \\ 3 & 5 & 3 & 2 \\ 3 & 3 & 6 & 2 \\ 2 & 2 & 2 & 4 \end{bmatrix}$ indicates the number of ways two cities can be linked with *at most* one intermediate city.

51. No. Let $A = \begin{bmatrix} 1 & 1 \\ 1 & 1 \end{bmatrix}$ and $B = \begin{bmatrix} 1 & 0 \\ 0 & 0 \end{bmatrix}$. Then $(AB)^2 \neq A^2B^2$.   **53.** Let $A = \begin{bmatrix} 1 & 2 \\ 1 & 2 \end{bmatrix}$ and $B = \begin{bmatrix} 2 & 2 \\ -1 & -1 \end{bmatrix}$. Then AB is the zero matrix, and neither A nor B is the zero matrix.

Exercise 7.4   (page 377)

1. $\begin{bmatrix} 3 & 4 \\ 2 & 3 \end{bmatrix}$ **3.** $\begin{bmatrix} 5 & -7 \\ -2 & 3 \end{bmatrix}$ **5.** $\begin{bmatrix} -2 & 3 & -3 \\ -5 & 7 & -6 \\ 1 & -1 & 1 \end{bmatrix}$ **7.** $\begin{bmatrix} 4 & 1 & -3 \\ -5 & -1 & 4 \\ -1 & -1 & 1 \end{bmatrix}$ **9.** no inverse

11. $\begin{bmatrix} 1 & -2 & 1 \\ 0 & 1 & -2 \\ 0 & 0 & 1 \end{bmatrix}$ **13.** no inverse   **15.** $\begin{bmatrix} 1 & -2 & 1 & 0 \\ 0 & 1 & -2 & 1 \\ 0 & 0 & 1 & -2 \\ 0 & 0 & 0 & 1 \end{bmatrix}$ **17.** $\begin{bmatrix} 8 & -2 & -6 \\ -5 & 2 & 4 \\ 2 & 0 & -2 \end{bmatrix}$

19. $\begin{bmatrix} -2.5 & 5 & 3 & 5.5 \\ 5.5 & -8 & -6 & -9.5 \\ -1 & 3 & 1 & 3 \\ -5.5 & 9 & 6 & 10.5 \end{bmatrix}$ **21.** $x = 23, y = 17$ **23.** $x = 0, y = 0$ **25.** $x = 1, y = 2, z = 2$

27. $x = 54, y = -37, z = -49$ **29.** $x = 2, y = 1$ **31.** $x = 1, y = 2, z = 1$ **33.** 2 of model A, 3 of model B

35. $X = \begin{bmatrix} 0 \\ 0 \\ 0 \end{bmatrix}$ **41.** no

Exercise 7.5   (page 389)

1. 8   **3.** 1   **5.** $\begin{vmatrix} -2 & 3 \\ 8 & 9 \end{vmatrix}$ **7.** $\begin{vmatrix} 1 & -2 \\ 4 & 5 \end{vmatrix}$ **9.** $-\begin{vmatrix} -2 & 3 \\ 8 & 9 \end{vmatrix}$ **11.** $\begin{vmatrix} 1 & -2 \\ 4 & 5 \end{vmatrix}$ **13.** -54 **15.** -7 **17.** 86

19. -2 **21.** 2   **23.** 12   **25.** true   **27.** true   **29.** 3   **31.** 3   **33.** $x = 1; y = 2$ **35.** $x = 3, y = 0$
37. $x = 1, y = 0, z = 1$ **39.** $x = 1, y = -1, z = 2$ **41.** $x = 6, y = 6, z = 12$ **43.** $p = \frac{5}{6}, q = \frac{2}{3}, r = \frac{1}{2}, s = \frac{5}{2}$
45. $3x - 2y = 0$ **47.** $6x + 7y = 9$ **49.** 30 sq units   **51.** 73 sq units   **57.** 10   **59.** 24   **61.** $x = 8$
63. $x = -1$ **67.** Domain is the set of $n \times n$ matrices; range is the set of real numbers.   **69.** yes   **71.** $x = \dfrac{5 \pm \sqrt{17}}{2}$
73. 21.468

Exercise 7.6   (page 397)

1. $\dfrac{1}{x} + \dfrac{2}{x-1}$ **3.** $\dfrac{5}{x} - \dfrac{3}{x-3}$ **5.** $\dfrac{1}{x+1} + \dfrac{2}{x-1}$ **7.** $\dfrac{1}{x-3} - \dfrac{3}{x+2}$ **9.** $\dfrac{8}{x+3} - \dfrac{5}{x-1}$

11. $\dfrac{5}{2x-3} + \dfrac{2}{x-5}$ **13.** $\dfrac{3}{x} + \dfrac{2}{x+1} + \dfrac{1}{(x+1)^2}$ **15.** $\dfrac{1}{x} + \dfrac{2}{x^2} - \dfrac{3}{x-1}$ **17.** $\dfrac{2}{x} + \dfrac{3}{x-1} - \dfrac{1}{x+1}$

19. $\dfrac{2}{x} + \dfrac{1}{x-3} + \dfrac{2}{(x-3)^2}$ **21.** $\dfrac{1}{x-1} - \dfrac{4}{(x-1)^3}$ **23.** $\dfrac{1}{x} + \dfrac{1}{x^2+3}$ **25.** $\dfrac{2}{x} + \dfrac{3x+2}{x^2+1}$ **27.** $\dfrac{2}{x} + \dfrac{1}{x^2} + \dfrac{x+2}{x^2+x+1}$

29. $\dfrac{1}{x^2} + \dfrac{1}{x^2+1}$ **31.** $\dfrac{-1}{x+1} - \dfrac{3}{x^2+2}$ **33.** $\dfrac{x+1}{x^2+2} + \dfrac{2}{x^2+x+2}$ **35.** $\dfrac{1}{x} + \dfrac{x}{x^2+2x+5} + \dfrac{x+2}{(x^2+2x+5)^2}$

37. $x - 3 - \dfrac{1}{x+1} + \dfrac{8}{x+2}$ **39.** $1 + \dfrac{2}{3x+1} + \dfrac{1}{x^2+1}$ **41.** $1 + \dfrac{1}{x} + \dfrac{x}{x^2+x+1}$ **43.** $2 + \dfrac{1}{x} + \dfrac{3}{x-1} + \dfrac{2}{x^2+1}$

Exercise 7.7   (page 402)

1.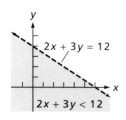
$2x + 3y = 12$
$2x + 3y < 12$

3.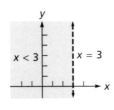
$x < 3$ $x = 3$

5.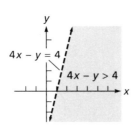
$4x - y = 4$ $4x - y > 4$

7.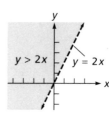
$y > 2x$ $y = 2x$

9.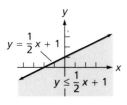
$y = \frac{1}{2}x + 1$ $y \leq \frac{1}{2}x + 1$

11.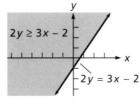
$2y \geq 3x - 2$ $2y = 3x - 2$

13.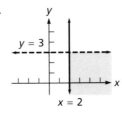
$y = 3$ $x = 2$

15.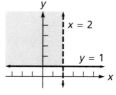
$x = 2$ $y = 1$

17.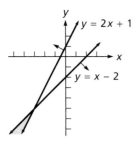
$y = 2x + 1$ $y = x - 2$

19.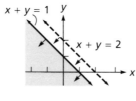
$x + y = 1$ $x + y = 2$

21.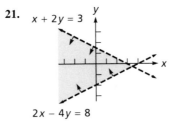
$x + 2y = 3$ $2x - 4y = 8$

23.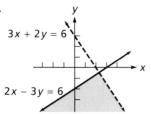
$3x + 2y = 6$ $2x - 3y = 6$

25.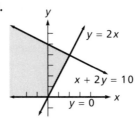
$y = 2x$ $x + 2y = 10$ $y = 0$

27.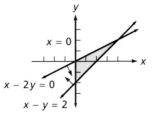
$x = 0$ $x - 2y = 0$ $x - y = 2$

29.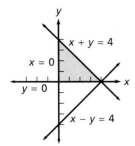
$x + y = 4$ $x = 0$ $y = 0$ $x - y = 4$

31.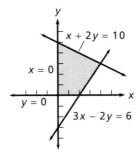
$x + 2y = 10$ $x = 0$ $y = 0$ $3x - 2y = 6$

Exercise 7.8   (page 410)

1. $P = 12$ at $(0, 4)$ **3.** $P = \frac{13}{6}$ at $\left(\frac{5}{3}, \frac{4}{3}\right)$ **5.** $P = \frac{18}{7}$ at $\left(\frac{3}{7}, \frac{12}{7}\right)$ **7.** $P = 3$ at $(1, 0)$ **9.** $P = 0$ at $(0, 0)$

11. $P = 0$ at $(0, 0)$ **13.** $P = 6$ at $(0, 3)$ **15.** $P = -2$ at $(1, 2)$ and $(-1, 0)$ **17.** 3 tables and 12 chairs; $1260
19. 30 IBMs and 30 Macintoshes; $2700   **21.** in batches of 20 oz from A and 40 oz from B; minimum fat per batch is 14 oz
23. 12 cases of chewing gum and 36 cases of bubble gum; $8160   **25.** $150,000 in stocks, $50,000 in bonds; $17,000
27. 2 buses, 2 trucks; $1100

CHAPTER 7 REVIEW EXERCISES   (page 413)

1.
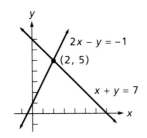

2.
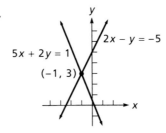

3.
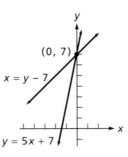

4.
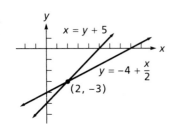

5. $x = 1, y = 5$ **6.** $x = 2, y = -1$ **7.** $x = 0, y = -3$ **8.** $x = 1, y = 1$ **9.** $x = -3, y = 2$ **10.** $x = -2,$ $y = 5$ **11.** $x = 2, y = -1$ **12.** $x = 4, y = 1$ **13.** $x = 1, y = 0, z = 1$ **14.** $x = 1, y = 1, z = -1$
15. $x = 0, y = 1, z = 2$ **16.** $x = 1, y = -2, z = 3$ **17.** $10,400   **18.** 900 adult tickets, 450 senior citizen tickets, 450 children's tickets   **19.** $x = 1, y = 1$ **20.** $x = 3, y = 1, z = -2$ **21.** $x = -10, y = 1, z = 10$ **22.** no solution
23. $w = 1, x = 2, y = 0, z = -1$ **24.** $w = 2 - z, x = 1, y = -2 + z, z =$ any number   **25.** $\begin{bmatrix} 1 & 3 & 4 \\ 4 & 0 & 2 \end{bmatrix}$

26. $\begin{bmatrix} 2 & 5 & 4 \\ -2 & -6 & 6 \\ -4 & 5 & -3 \end{bmatrix}$ **27.** $\begin{bmatrix} -7 & -1 \\ -7 & 4 \end{bmatrix}$ **28.** $\begin{bmatrix} -17 & 19 \\ 10 & -12 \end{bmatrix}$ **29.** $[5]$ **30.** $\begin{bmatrix} -3 & 14 & -12 \\ 1 & 2 & -4 \end{bmatrix}$

31. $\begin{bmatrix} 2 & -1 & 1 & 3 \\ 4 & -2 & 2 & 6 \\ 2 & -1 & 1 & 3 \\ 10 & -5 & 5 & 15 \end{bmatrix}$ **32.** not possible   **33.** $[-24]$ **34.** $\begin{bmatrix} 0 \\ -6 \end{bmatrix}$ **35.** $\begin{bmatrix} \dfrac{5}{14} & -\dfrac{3}{14} \\ \dfrac{3}{14} & \dfrac{1}{14} \end{bmatrix}$ **36.** $\begin{bmatrix} -17 & 7 \\ 5 & -2 \end{bmatrix}$

37. $\begin{bmatrix} 4 & -3 & 32 \\ -1 & 1 & -9 \\ 0 & 0 & 1 \end{bmatrix}$ **38.** $\begin{bmatrix} 1 & 0 & 0 \\ -\dfrac{3}{2} & \dfrac{1}{2} & \dfrac{1}{2} \\ 1 & -\dfrac{1}{2} & 0 \end{bmatrix}$ **39.** $\begin{bmatrix} 9 & 16 & -56 \\ -3 & -5 & 18 \\ -1 & -2 & 7 \end{bmatrix}$ **40.** $\begin{bmatrix} -2 & -3 & 1 \\ 4 & 5 & -2 \\ -1 & -1 & 1 \end{bmatrix}$

41. $\begin{bmatrix} 1 & -1 & 0 \\ 2 & -1 & 0 \\ 1 & -2 & -1 \end{bmatrix}$ **42.** No inverse exists.   **43.** $x = 1, y = 2, z = -1$ **44.** $w = 1, x = 1, y = 0, z = -1$

45. -7 **46.** -6 **47.** 3 **48.** -25 **49.** $x = 1, y = -2$ **50.** $x = 1, y = 0, z = -2$ **51.** $x = 1, y = -1,$ $z = 3$ **52.** $w = 1, x = 0, y = -1, z = 2$ **53.** $\dfrac{3}{x} + \dfrac{4}{x + 1}$ **54.** $\dfrac{3}{x} + \dfrac{2}{x^2} + \dfrac{x - 1}{x^2 + 1}$ **55.** $\dfrac{1}{x} - \dfrac{1}{x^2 + x + 5}$

56. $\dfrac{1}{x + 1} - \dfrac{2}{(x + 1)^2} + \dfrac{2}{(x + 1)^3}$ **57.** $P = 6$ at $(3, 0)$ **58.** $P = 12$ at $(0, -4)$ **59.** $P = 2$ at $(1, 1)$

60. $P = 3$ at $\left(-\dfrac{2}{3}, \dfrac{5}{3}\right)$ **61.** 1000 bags of x, 1400 bags of y

CHAPTER 7 TEST   (page 416)

1. 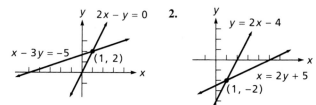   **2.**

3. $x = 1, y = -3$ **4.** $x = 3, y = 5$ **5.** 6 liters of 20% solution, 4 liters of 45% solution   **6.** CD World buys 100 decks; Ace buys 25 decks; Hi Fi buys 50 decks.   **7.** $x = 2, y = 1$ **8.** $x = 1, y = 2, z = 1$ **9.** $x = -\dfrac{2}{5}y + \dfrac{7}{5}, z = -\dfrac{8}{5}y - \dfrac{7}{5},$ $y = $ any number   **10.** $x = 1, y = 0, z = -2$ **11.** $\begin{bmatrix} 16 & -14 & 20 \\ 0 & -6 & -13 \end{bmatrix}$ **12.** $[-1]$ **13.** $\begin{bmatrix} -\dfrac{7}{3} & \dfrac{19}{3} \\ \dfrac{2}{3} & -\dfrac{5}{3} \end{bmatrix}$

14. $\begin{bmatrix} -13 & -3 & 14 \\ 4 & 1 & -4 \\ 12 & 3 & -13 \end{bmatrix}$ **15.** $x = \dfrac{17}{3}, y = -\dfrac{4}{3}$ **16.** $x = -36, y = 11, z = 34$ **17.** -12 **18.** -24

19. $x = -\dfrac{13}{12}, y = -\dfrac{5}{4}$ **29.** $x = -\dfrac{1}{2}, y = 1, z = \dfrac{3}{2}$ **21.** $\dfrac{3}{2x - 3} + \dfrac{1}{x + 1}$ **22.** $\dfrac{1}{x} + \dfrac{2x + 1}{x^2 + 2}$

23. 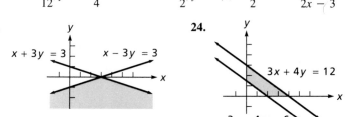   **24.** 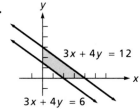   **25.** $P = 7$ at $(1, 2)$ **26.** $P = -8$ at $(8, 0)$

Exercise 8.1   (page 426)

1. $x^2 + y^2 = 49$ **3.** $(x - 2)^2 + (y + 2)^2 = 17$ **5.** $(x - 1)^2 + (y + 2)^2 = 36$

7. 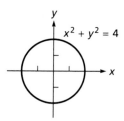   **9.**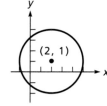

11. yes   **13.** 60 mi   **15.** $(x - 4)^2 + y^2 = 16$ **17.** $x^2 + (y - 3)^2 = 25$ **19.** $x^2 = 12y$ **21.** $y^2 = -12x$
23. $(x - 3)^2 = -12(y - 5)$ **25.** $(x - 3)^2 = -28(y - 5)$ **27.** $(x - 2)^2 = -2(y - 2)$ or $(y - 2)^2 = -2(x - 2)$
29. $(x + 4)^2 = -\frac{16}{3}(y - 6)$ or $(y - 6)^2 = \frac{9}{4}(x + 4)$ **31.** $(y - 8)^2 = -4(x - 6)$ **33.** $(x - 3)^2 = \frac{1}{2}(y - 1)$

35.

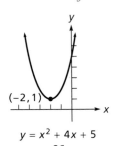

$y = x^2 + 4x + 5$
or
$y - 1 = (x + 2)^2$

37.

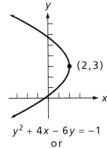

$(2,3)$

$y^2 + 4x - 6y = -1$
or
$(y - 3)^2 = -4(x - 2)$

39.

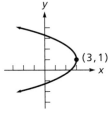

$(3,1)$

$y^2 + 2x - 2y = 5$
or
$(y - 1)^2 = -2(x - 3)$

41.

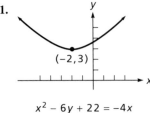

$(-2,3)$

$x^2 - 6y + 22 = -4x$
or
$(x + 2)^2 = 6(y - 3)$

43.

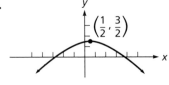

$\left(\frac{1}{2}, \frac{3}{2}\right)$

$4x^2 - 4x + 32y = 47$
or
$\left(x - \frac{1}{2}\right)^2 = -8\left(y - \frac{3}{2}\right)$

45. 2 ft   **47.** $x^2 = -\frac{45}{2}y$ **49.** 8 cabins   **51.** about 12.6 cm   **53.** about 520 ft   **57.** $y = x^2 + 4x + 3$

Exercise 8.2   (page 436)

1. $\dfrac{x^2}{25} + \dfrac{y^2}{16} = 1$ **3.** $\dfrac{9x^2}{16} + \dfrac{9y^2}{25} = 1$ **5.** $\dfrac{x^2}{7} + \dfrac{y^2}{16} = 1$ **7.** $\dfrac{(x - 3)^2}{4} + \dfrac{(y - 4)^2}{9} = 1$ **9.** $\dfrac{(x - 3)^2}{9} + \dfrac{(y - 4)^2}{4} = 1$
11. $\dfrac{(x - 3)^2}{41} + \dfrac{(y - 4)^2}{16} = 1$ **13.** $\dfrac{x^2}{36} + \dfrac{(y - 4)^2}{20} = 1$ **15.** $\dfrac{x^2}{100} + \dfrac{y^2}{64} = 1$

17.

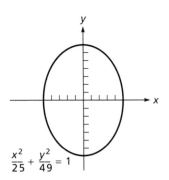

$\dfrac{x^2}{25} + \dfrac{y^2}{49} = 1$

19.

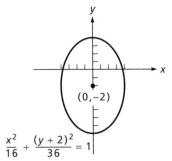

$(0,-2)$

$\dfrac{x^2}{16} + \dfrac{(y + 2)^2}{36} = 1$

21.

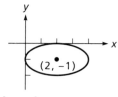

$x^2 + 4y^2 - 4x + 8y + 4 = 0$

23.

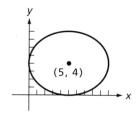

$16x^2 + 25y^2 - 160x - 200y + 400 = 0$

25. 199,395 mi   **27.** $\dfrac{x^2}{2500} + \dfrac{y^2}{900} = 1$; 36 m   **29.** about 20.8 in.   **35.** $\dfrac{x^2}{9} + \dfrac{y^2}{8} = 1$

Exercise 8.3   (page 443)

1. $\dfrac{x^2}{25} - \dfrac{y^2}{24} = 1$ **3.** $\dfrac{(x-2)^2}{4} - \dfrac{(y-4)^2}{9} = 1$ **5.** $\dfrac{(y-3)^2}{9} - \dfrac{(x-5)^2}{9} = 1$ **7.** $\dfrac{y^2}{9} - \dfrac{x^2}{16} = 1$

9. $\dfrac{(x-1)^2}{4} - \dfrac{(y+3)^2}{16} = 1$ or $\dfrac{(y+3)^2}{4} - \dfrac{(x-1)^2}{16} = 1$ **11.** $\dfrac{x^2}{10} - \dfrac{3y^2}{20} = 1$ **13.** 24 sq units   **15.** 12 sq units

17. $\dfrac{(x+2)^2}{4} - \dfrac{4(y+4)^2}{81} = 1$ or $\dfrac{(y+4)^2}{4} - \dfrac{4(x+2)^2}{81} = 1$ **19.** $\dfrac{x^2}{36} - \dfrac{16y^2}{25} = 1$

21.

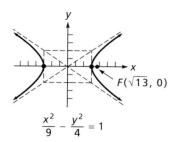

$\dfrac{x^2}{9} - \dfrac{y^2}{4} = 1$

23.

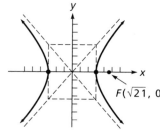

$4x^2 - 3y^2 = 36$

25.

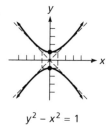

$y^2 - x^2 = 1$

27.

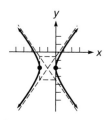

$4x^2 - 2y^2 + 8x - 8y = 8$

29.

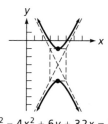

$y^2 - 4x^2 + 6y + 32x = 59$

31.

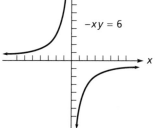

$-xy = 6$

33. $\dfrac{x^2}{100,000,000^2} - \dfrac{y^2}{200,000,000^2} = 1$ **35.** hyperbola; $\dfrac{x^2}{144} - \dfrac{y^2}{25} = 1$ **37.** $\dfrac{(x-3)^2}{9} - \dfrac{(y-1)^2}{16} = 1$

39. $4x^2 - 5y^2 - 60y = 0$

Exercise 8.4 (page 448)

1.
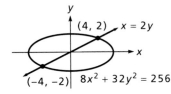

3.
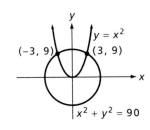

5.
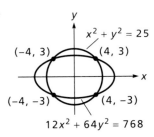

7.
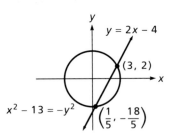

9.
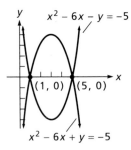

11. $(1, 2), (-1, 0)$ **13.** $(1, 0.67), (-1, -0.67)$ **15.** $(3, 0), (0, 5)$ **17.** $(1, 1)$ **19.** $(1, 2), (2, 1)$ **21.** $(-2, 3)$,
$(2, 3)$ **23.** $(\sqrt{5}, 5), (-\sqrt{5}, 5)$ **25.** $(3, 2), (3, -2), (-3, 2), (-3, -2)$ **27.** $(2, 4), (2, -4), (-2, 4), (-2, -4)$
29. $(-\sqrt{15}, 5), (\sqrt{15}, 5), (-2, -6), (2, -6)$ **31.** $(0, -4), (-3, 5), (3, 5)$ **33.** $(-2, 3), (2, 3), (-2, -3), (2, -3)$
35. $(3, 3)$ **37.** $(6, 2), (-6, -2), (\sqrt{42}, 0), (-\sqrt{42}, 0)$ **39.** $\left(\frac{1}{2}, \frac{1}{3}\right), \left(\frac{1}{3}, \frac{1}{2}\right)$ **41.** 7 by 9 cm
43. either \$750 at 9%, or \$900 at 7.5% **45.** about 23 mi

CHAPTER 8 REVIEW EXERCISES (page 450)

1. $x^2 + y^2 = 50$ **2.** $x^2 + y^2 = 100$ **3.** $(x - 5)^2 + (y - 10)^2 = 85$ **4.** $(x - 2)^2 + (y - 2)^2 = 89$

5.
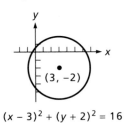

6.
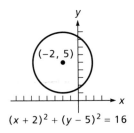

7. $y^2 = -2x$ **8.** $x^2 = 16y$ **9.** $\left(x + \dfrac{b}{2a}\right)^2 = \dfrac{1}{a}\left(y + \dfrac{b^2 - 4ac}{4a}\right)$; vertex at $\left(-\dfrac{b}{2a}, \dfrac{4ac - b^2}{4a}\right)$

10. $(x + 2)^2 = -\dfrac{4}{11}(y - 3)$ **11.**
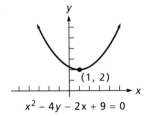

12.
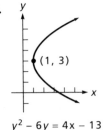

13.

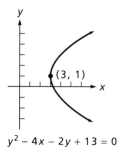

$y^2 - 4x - 2y + 13 = 0$

14.

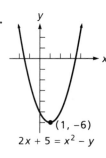

$2x + 5 = x^2 - y$

15. $\dfrac{x^2}{36} + \dfrac{y^2}{16} = 1$ **16.** $\dfrac{(x + 2)^2}{16} + \dfrac{(y - 3)^2}{9} = 1$

17.

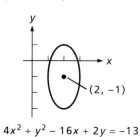

$4x^2 + y^2 - 16x + 2y = -13$

18.

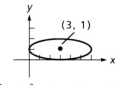

$x^2 + 9y^2 - 6x - 18y + 9 = 0$

19. $\dfrac{x^2}{9} - \dfrac{(y - 3)^2}{16} = 1$ **20.** $\dfrac{y^2}{9} - \dfrac{(x - 3)^2}{16} = 1$

21.

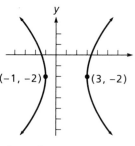

$9x^2 - 4y^2 - 16y - 18x = 43$

22.

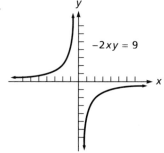

$-2xy = 9$

23.

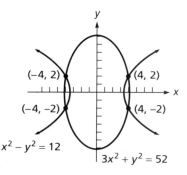

$x^2 - y^2 = 12$ $3x^2 + y^2 = 52$

24. $(4, 2), (4, -2), (-4, 2), (-4, -2)$ **25.** $(0, 4), (2\sqrt{3}, -2)$

26.

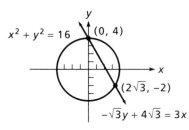

$x^2 + y^2 = 16$ $-\sqrt{3}y + 4\sqrt{3} = 3x$

27.

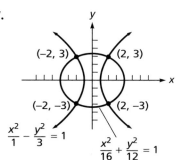

$\dfrac{x^2}{1} - \dfrac{y^2}{3} = 1$ $\dfrac{x^2}{16} + \dfrac{y^2}{12} = 1$

28. $(2, 3), (2, -3), (-2, 3), (-2, -3)$

CHAPTER 8 TEST   (page 451)

1. $(x - 2)^2 + (y - 3)^2 = 9$ **2.** $(x - 2)^2 + (y - 3)^2 = 41$ **3.** $(x - 2)^2 + (y + 5)^2 = 169$

4.

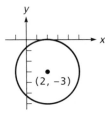

$(x - 2)^2 + (y + 3)^2 = 9$

5. $(x - 3)^2 = 16(y - 2)$ **6.** $(y + 6)^2 = -4(x - 4)$ **7.** $(x - 2)^2 = \frac{4}{3}(y + 3)$ or $(y + 3)^2 = -\frac{9}{2}(x - 2)$

8.

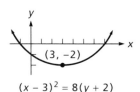

$(x - 3)^2 = 8(y + 2)$

9. $\dfrac{x^2}{100} + \dfrac{y^2}{64} = 1$ **10.** $\dfrac{x^2}{169} + \dfrac{y^2}{144} = 1$ **11.** $\dfrac{x^2}{4} + \dfrac{y^2}{36} = 1$

12.

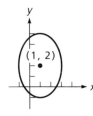

$\dfrac{(x - 1)^2}{4} + \dfrac{(y - 2)^2}{9} = 1$

13. $\dfrac{x^2}{25} - \dfrac{y^2}{144} = 1$ **14.** $\dfrac{x^2}{36} - \dfrac{4y^2}{25} = 1$ **15.** $\dfrac{(x - 2)^2}{64} - \dfrac{(y + 1)^2}{36} = 1$

16.

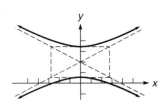

17. $(\sqrt{7}, 4), (-\sqrt{7}, 4)$ **18.** $(3\sqrt{2}, 3), (-3\sqrt{2}, 3), (3\sqrt{2}, -3), (-3\sqrt{2}, -3)$
19. parabola   **20.** ellipse

Exercise 9.1   (page 458)

1. 24   **3.** 4320   **5.** 1440   **7.** $\frac{1}{1320}$ **9.** $\frac{5}{3}$ **11.** 18,564

13. $a^2 + 3a^2b + 3ab^2 + b^3$ **15.** $a^5 - 5a^4b + 10a^3b^2 - 10a^2b^3 + 5ab^4 - b^5$ **17.** $8x^3 + 12x^2y + 6xy^2 + y^3$
19. $x^3 - 6x^2y + 12xy^2 - 8y^3$ **21.** $16x^4 + 96x^3y + 216x^2y^2 + 216xy^3 + 81y^4$
23. $x^4 - 8x^3y + 24x^2y^2 - 32xy^3 + 16y^4$ **25.** $x^5 - 15x^4y + 90x^3y^2 - 270x^2y^3 + 405xy^4 - 243y^5$
27. $\dfrac{x^4}{16} + \dfrac{x^3y}{2} + \dfrac{3x^2y^2}{2} + 2xy^3 + y^4$ **29.** $6a^2b^2$ **31.** $35a^3b^4$ **33.** $-b^5$ **35.** $2380a^{13}b^4$ **37.** $-4\sqrt{2}\,a^3$

39. $1134\,a^5b^4$ **41.** $\dfrac{3x^2y^2}{2}$ **43.** $\dfrac{-55r^2s^9}{2048}$ **45.** $\dfrac{n!}{3!(n - 3)!}a^{n-3}b^3$ **47.** $\dfrac{n!}{(r - 1)!(n - r + 1)!}a^{n-r+1}b^{r-1}$ **51.** -252

Exercise 9.2   (page 464)

1. 0, 10, 30, 60, 100, 150   **3.** 21   **5.** $a + 4d$ **7.** 15   **9.** 15   **11.** 15   **13.** $\dfrac{242}{243}$ **15.** 35   **17.** 3, 7, 15, 31

19. $-4, -2, -1, -\dfrac{1}{2}$ **21.** k, k^2, k^4, k^8 **23.** $8, \dfrac{16}{k}, \dfrac{32}{k^2}, \dfrac{64}{k^3}$ **25.** an alternating series   **27.** not an alternating series

29. 30   **31.** -50 **33.** 40   **35.** 500   **37.** $\dfrac{7}{12}$ **39.** 160   **41.** 3725

Exercise 9.3   (page 472)

1. 1, 3, 5, 7, 9, 11   **3.** $5, \dfrac{7}{2}, 2, \dfrac{1}{2}, -1, -\dfrac{5}{2}$ **5.** $9, \dfrac{23}{2}, 14, \dfrac{33}{2}, 19, \dfrac{43}{2}$ **7.** 285   **9.** 555   **11.** $157\dfrac{1}{2}$ **13.** 44

15. $\dfrac{25}{2}, 15, \dfrac{35}{2}$ **17.** $-\dfrac{82}{15}, -\dfrac{59}{15}, -\dfrac{36}{15}, -\dfrac{13}{15}$ **19.** 10, 20, 40, 80   **21.** $-2, -6, -18, -54$ **23.** $3, 3\sqrt{2}, 6, 6\sqrt{2}$

25. 2, 6, 18, 54   **27.** 124   **29.** $-29, 524$ **31.** $\dfrac{1995}{32}$ **33.** 18   **35.** 8   **37.** $10\sqrt[4]{2}, 10\sqrt{2}, 10\sqrt[4]{8}$

39. 8, 32, 128, 512   **41.** $\dfrac{5}{9}$ **43.** $\dfrac{25}{99}$

Exercise 9.4   (page 475)

1. 23   **3.** 5.13m   **5.** 1,048,576   **7.** She will earn \$0.19 more on the $7\dfrac{1}{4}$% investment.   **9.** about 8.6×10^{19}

11. \$180,176.87   **13.** \$2001.60   **15.** \$2013.62   **17.** \$264,094.58   **19.** 1.8447×10^{19} grains   **21.** no

Exercise 9.5   (page 481)

25. no

Exercise 9.6   (page 488)

1. 144   **3.** 8,000,000   **5.** 240   **7.** 6   **9.** 840   **11.** 35   **13.** 120   **15.** 5   **17.** 1   **19.** 1200   **21.** 40

23. 2278   **25.** 40,320   **27.** 14,400   **29.** 24,360   **31.** 5040   **33.** 48   **35.** 96   **37.** 210   **39.** 24

41. 5040   **43.** 2,721,600   **45.** 1120   **47.** 59,400   **49.** 272   **51.** 28   **53.** 252   **55.** 142,506

57. 66   **59.** 56

Exercise 9.7   (page 494)

1. {(1, H), (2, H), (3, H), (4, H), (5, H), (6, H), (1, T), (2, T), (3, T), (4, T), (5, T), (6, T)}

3. {a, b, c, d, e, f, g, h, i, j, k, l, m, n, o, p, q, r, s, t, u, v, w, x, y, z}   **5.** $\dfrac{1}{6}$ **7.** $\dfrac{2}{3}$ **9.** $\dfrac{19}{42}$ **11.** $\dfrac{13}{42}$ **13.** $\dfrac{3}{8}$ **15.** 0

17. $\dfrac{1}{12}$ **19.** $\dfrac{1}{169}$ **21.** $\dfrac{5}{12}$ **23.** about 6.3×10^{-12} **25.** 0   **27.** $\dfrac{3}{13}$ **29.** $\dfrac{1}{6}$ **31.** $\dfrac{1}{8}$ **33.** $\dfrac{5}{16}$

35. (S = survive; F = fail) SSSS, SSSF, SSFS, SFSS, FSSS, SSFF, SFSF, FSSF, SFFS, FSFS, FFSS, SFFF, FSFF, FFSF,

FFFS, FFFF   **37.** $\dfrac{1}{4}$ **39.** $\dfrac{1}{4}$ **41.** 1   **43.** $\dfrac{32}{119}$ **45.** $\dfrac{1}{3}$ **47.** 0.18   **49.** 0.14   **51.** about 33%   **53.** no

Exercise 9.8   (page 500)

1. $\dfrac{1}{2}$ **3.** $\dfrac{7}{13}$ **5.** $\dfrac{1}{221}$ **7.** $\dfrac{25}{204}$ **9.** $\dfrac{11}{36}$ **11.** $\dfrac{7}{12}$ **13.** $\dfrac{5}{8}$ **15.** $\dfrac{13}{16}$ **17.** $\dfrac{21}{128}$ **19.** $\dfrac{3}{20}$ **21.** $\dfrac{1}{49}$ **23.** 0.973

25. $\dfrac{11}{20}$ **27.** $\dfrac{1}{8}$ **29.** $\dfrac{3}{8}$ **31.** $\dfrac{17}{36}$ **33.** 0.03

Exercise 9.9   (page 503)

1. $\dfrac{1}{6}$ **3.** 5 to 1   **5.** 1 to 1   **7.** $\dfrac{5}{36}$ **9.** 31 to 5   **11.** 1 to 1   **13.** 1 to 12   **15.** 3 to 10   **17.** $\dfrac{5}{7}$ **19.** 1 to 90

21. 15 to 1   **23.** $\dfrac{1}{9}$ **25.** no; the expected winnings are $\$\dfrac{9}{13}$.   **27.** 7 to 1   **29.** \$1.72   **31.** 6.54   **33.** $\dfrac{32}{390,625}$

CHAPTER 9 REVIEW EXERCISES   (page 505)

1. 30,240   **2.** $\frac{280}{3}$ **3.** $8a^3 - 12a^2b + 6ab^2 - b^3$ **4.** $256a^4 - 1280a^3b + 2400a^2b^2 - 2000ab^3 + 625b^4$

5. $56a^5b^3$ **6.** $80x^3y^2$ **7.** $84x^3y^6$ **8.** $439,040x^3$ **9.** 5, 17, 53, 161   **10.** $-2, 8, 128, 32,768$ **11.** 90

12. 60   **13.** 1718   **14.** -360 **15.** 117   **16.** 281   **17.** -92 **18.** $-\frac{135}{2}$ **19.** $\frac{1}{729}$ **20.** 13,122

21. $\frac{9}{16,384}$ **22.** $\frac{8}{15,625}$ **23.** 3320   **24.** 5780   **25.** -5220 **26.** -1540 **27.** $\frac{3280}{27}$ **28.** 6560   **29.** $\frac{2295}{128}$

30. $\frac{520,832}{78,125}$ **31.** $\frac{2}{3}$ **32.** $\frac{3}{25}$ **33.** no sum   **34.** 1   **35.** $\frac{1}{3}$ **36.** 1   **37.** $\frac{17}{99}$ **38.** $\frac{5}{11}$ **39.** $\frac{7}{2}, 5, \frac{13}{2}$

40. 25, 40, 55, 70, 85   **41.** $2\sqrt{2}, 4, 4\sqrt{2}$ **42.** $4, -8, 16, -32$ **43.** $\frac{3280}{3}$ **44.** $16\sqrt{2}$ **45.** 16   **46.** \$4775.81

47. 6516, 3134   **48.** \$3486.78   **51.** 6720   **52.** 35   **53.** 1   **54.** 4050   **55.** 564,480   **56.** 840   **57.** 21

58. 66   **59.** 120   **60.** $\frac{56}{1287}$ **61.** $\frac{1}{6}$ **62.** $\frac{33}{66,640}$ **64.** 24   **65.** $\frac{1}{108,290}$ **66.** $\frac{108,289}{108,290}$ **67.** 20,160

68. 90,720   **69.** about 6.3×10^{-12} **70.** $\frac{420}{1001} = \frac{60}{143}$ **71.** $\frac{1}{2}$ **72.** $\frac{7}{13}$ **73.** $\frac{1}{2,598,960}$ **74.** about $\frac{4}{6.350 \times 10^{11}}$

75. $\frac{15}{16}$ **76.** 1 to 7   **77.** 1 to 15   **78.** \$2.00   **79.** 1 to 23   **80.** $\frac{11}{21}$ **81.** 664

CHAPTER 9 TEST   (page 508)

1. 144   **2.** 384   **3.** $10x^4y$ **4.** $112a^2b^6$ **5.** 27   **6.** -36 **7.** 155   **8.** -130 **9.** 9, 14, 19

10. $-6, -18$ **11.** 255.75   **12.** about 9   **13.** about \$0.42c **14.** about \$1.46c **16.** 800,000   **17.** 42

18. 24   **19.** 28   **20.** 1   **21.** 576   **22.** 120   **23.** 60   **24.** {(H, H, H), (H, H, T), (H, T, H), (H, T, T), (T, H, H),

(T, H, T), (T, T, H), (T, T, T)}   **25.** $\frac{1}{6}$ **26.** $\frac{2}{13}$ **27.** $\frac{429}{866,320}$ **28.** $\frac{5}{16}$ **29.** 0.9   **30.** 40%   **31.** $\frac{8}{13}$

32. $\frac{5}{9}$ **33.** $\frac{3}{4}$ **34.** \$1

CUMULATIVE REVIEW EXERCISES   (page 509)

1.
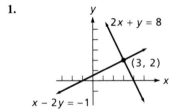

2.
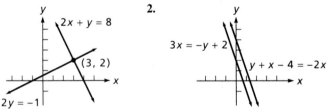

inconsistent system

3. $x = 3, y = 1$ **4.** $x = 3, y = 2, z = 1$ **5.** $x = 1, y = 1, z = 2$ **6.** $x = 1, y = 2, z = 1, t = 1$ **7.** $\begin{bmatrix} 1 & 3 \\ 3 & 7 \end{bmatrix}$

8. $\begin{bmatrix} -3 & 1 \\ 1 & -1 \end{bmatrix}$ **9.** $\begin{bmatrix} 3 & 2 & 0 \\ -2 & 8 & 7 \end{bmatrix}$ **10.** $\begin{bmatrix} 9 & 6 \\ 6 & 21 \end{bmatrix}$ **11.** $\begin{bmatrix} -1 & 1.5 \\ .5 & -.5 \end{bmatrix}$ **12.** $\begin{bmatrix} 4 & 5 & -4 \\ -1 & -1 & 1 \\ -4 & -6 & 5 \end{bmatrix}$

13. -41 **14.** -1 **15.** $x = \dfrac{\begin{vmatrix} 11 & 3 \\ 24 & 5 \end{vmatrix}}{\begin{vmatrix} 4 & 3 \\ -2 & 5 \end{vmatrix}}$ **16.** $y = \dfrac{\begin{vmatrix} 4 & 11 \\ -2 & 24 \end{vmatrix}}{\begin{vmatrix} 4 & 3 \\ -2 & 5 \end{vmatrix}}$ **17.** $\dfrac{2}{x+1} - \dfrac{3}{x+2}$ **18.** $\dfrac{2}{x} + \dfrac{3}{x^2+1}$

19.

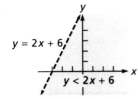

$y = 2x + 6$

$y < 2x + 6$

20.

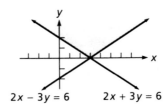

$2x - 3y = 6$ $2x + 3y = 6$

21. $x^2 + y^2 = 16$

22. $(x - 2)^2 + (y + 3)^2 = 121$

23.

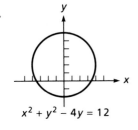

$x^2 + y^2 - 4y = 12$

24.

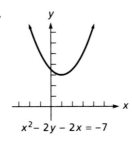

$x^2 - 2y - 2x = -7$

25.

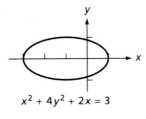

$x^2 + 4y^2 + 2x = 3$

26.

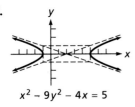

$x^2 - 9y^2 - 4x = 5$

27. $\dfrac{x^2}{36} + \dfrac{y^2}{16} = 1$ or $\dfrac{x^2}{16} + \dfrac{y^2}{36} = 1$ **28.** $\dfrac{(x-2)^2}{25} + \dfrac{(y-3)^2}{21} = 1$ or $\dfrac{(x-2)^2}{21} + \dfrac{(y-3)^2}{25} = 1$ **29.** $\dfrac{x^2}{4} - \dfrac{y^2}{5} = 1$

30. $\dfrac{(x-2)^2}{4} - \dfrac{(y-4)^2}{16} = 1$ **31.** $16x^7y$ **32.** $1792x^3y^5$ **33.** 10   **34.** 65   **35.** 33   **36.** $\dfrac{364}{9}$ **37.** 1680

38. 1   **39.** 66   **40.** 24   **41.** 17,280   **42.** 495   **43.** $\dfrac{1}{18}$ **44.** $\dfrac{506}{19,992} = \dfrac{253}{9996}$ **45.** 0.48   **46.** 0.3

48. about $8.46

INDEX

INDEX OF APPLICATIONS

6.1 EXPONENTIAL FUNCTIONS

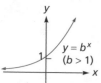
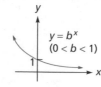

Formula for radioactive decay: $A = A_0 2^{-t/h}$

Formula for compound interest: $A = A_0\left(1 + \dfrac{r}{k}\right)^{kt}$

6.2 BASE-e EXPONENTIAL FUNCTIONS

$e = 2.718281828\ldots$

Formula for continuous compound interest: $A = A_0 e^{rt}$

Formula for population growth: $P = P_0 e^{kt}$

6.3 LOGARITHMIC FUNCTIONS

$y = \log_b x$ is equivalent to $x = b^y$.

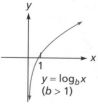
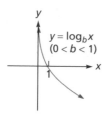

6.4 PROPERTIES OF LOGARITHMS

$\log_b 1 = 0 \qquad \log_b b = 1$

$\log_b b^x = x \qquad b^{\log_b x} = x$

$\log_b MN = \log_b M + \log_b N$

$\log_b \dfrac{M}{N} = \log_b M - \log_b N$

$\log_b M^p = p \log_b M$

If $\log_b x = \log_b y$, then $x = y$.

The change-of-base formula: $\log_b y = \dfrac{\log_a y}{\log_a b}$

7.5 DETERMINANTS

$$\begin{vmatrix} a & b \\ c & d \end{vmatrix} = ad - bc$$

8.1 THE CIRCLE AND THE PARABOLA

$(x - h)^2 + (y - k)^2 = r^2$ Circle with center at (h, k) and radius r

$x^2 + y^2 = r^2$ Circle with center at the origin and radius r

$\left.\begin{aligned}(y - k)^2 &= \pm 4p(x - h) \\ (x - h)^2 &= \pm 4p(y - k)\end{aligned}\right\}$ Parabola with vertex at (h, k)

8.2 THE ELLIPSE

$\left.\begin{aligned}\dfrac{(x - h)^2}{a^2} + \dfrac{(y - k)^2}{b^2} &= 1 \\ \dfrac{(y - k)^2}{a^2} + \dfrac{(x - h)^2}{b^2} &= 1\end{aligned}\right\}$ Ellipse with center at (h, k)

8.3 THE HYPERBOLA

$\left.\begin{aligned}\dfrac{(x - h)^2}{a^2} - \dfrac{(y - k)^2}{b^2} &= 1 \\ \dfrac{(y - k)^2}{a^2} - \dfrac{(x - h)^2}{b^2} &= 1\end{aligned}\right\}$ Hyperbola with center at (h, k)

9.1 THE BINOMIAL THEOREM

$$n! = n(n - 1)(n - 2) \cdot \cdots \cdot 3 \cdot 2 \cdot 1$$

$$0! = 1$$

$$n(n - 1)! = n!$$

$$(a + b)^n = a^n + \frac{n!}{1!(n - 1)!} a^{n-1}b +$$

$$\frac{n!}{2!(n - 2)!} a^{n-2}b^2 + \cdots + b^n$$

9.2 SEQUENCES, SERIES, AND SUMMATION NOTATION

If c is a constant, then

$$\sum_{k=1}^{n} c = nc$$

$$\sum_{k=1}^{n} cf(k) = c \sum_{k=1}^{n} f(k)$$

$$\sum_{k=1}^{n} [f(k) + g(k)] = \sum_{k=1}^{n} f(k) + \sum_{k=1}^{n} g(k)$$

9.3 ARITHMETIC AND GEOMETRIC SEQUENCES

Arithmetic sequence:

$$a, a + d, a + 2d, a + 3d, \ldots, a + (n - 1)d, \ldots$$

$$S_n = \frac{n(a + l)}{2}$$

Geometric sequence:

$$a, ar, ar^2, ar^3, \ldots, ar^{n-1}, \ldots$$

$$S_n = \frac{a - ar^n}{1 - r} \quad (r \neq 1)$$

If $|r| < 1$, then $S = \dfrac{a}{1 - r}$.

9.6 PERMUTATIONS AND COMBINATIONS

$$P(n, r) = \frac{n!}{(n - r)!} \qquad P(n, n) = n!$$

$$C(n, r) = \frac{n!}{r!(n - r)!} \qquad C(n, n) = 1$$

$$C(n, 0) = 1$$

9.7 PROBABILITY

$$P(\text{event}) = \frac{\text{number of favorable outcomes}}{\text{total number of outcomes}}$$

$$P(A \cap B) = P(A) \cdot P(B|A)$$

9.8 COMPUTATION OF COMPOUND PROBABILITIES

$$P(A \cup B) = P(A) + P(B) - P(A \cap B)$$

If A and B cannot occur simultaneously, then

$$P(A \cup B) = P(A) + P(B)$$

$$P(\overline{A}) = 1 - P(A)$$

If $P(B) = P(B|A)$, then A and B are independent events.

If A and B are independent events, then

$$P(A \cap B) = P(A) \cdot P(B)$$

9.9 ODDS AND MATHEMATICAL EXPECTATION

$$\text{Odds for an event} = \frac{P(\text{favorable outcome})}{P(\text{unfavorable outcome})}$$

$$\text{Odds against an event} = \frac{P(\text{unfavorable outcome})}{P(\text{ favorable outcome})}$$

$$= \frac{1}{\text{odds for an event}}$$

If $p_1, p_2, \ldots, p_n$ are the probabilities of n outcomes, and $x_1, x_2, \ldots, x_n$ are the corresponding winnings, the mathematical expectation is

$$E = p_1x_1 + p_2x_2 + \cdots + p_nx_n$$